Intermediate
Algebra
with Early Functions and Graphing

SIXTH EDITION

Margaret L. Lial
American River College

John Hornsby
University of New Orleans

Charles D. Miller

ADDISON-WESLEY

An imprint of Addison Wesley Longman, Inc.

Reading, Massachusetts • Menlo Park, California • New York • Harlow, England
Don Mills, Ontario • Sydney • Mexico City • Madrid • Amsterdam

Publisher: Jason Jordan

Acquisition Editor: Rita Ferrandino

Assistant Editor: K. B. Mello

Developmental Editor: Sandi Goldstein

Managing Editor: Ron Hampton

Production Services: Elm Street Publishing Services, Inc.

Art Development: Meredith Nightingale

Marketing Manager: Liz O'Neil

Prepress Services Buyer: Caroline Fell

Senior Manufacturing Manager: Roy Logan

Manufacturing Manager: Ralph Mattivello

Text and Cover Designer: Susan Carsten

Cover Illustration: © SuperStock

Intermediate Algebra with Early Functions and Graphing, Sixth Edition

Printed in the U.S.A.

Library of Congress Cataloging-in-Publication Data

Lial, Margaret L.
    Intermediate algebra / Margaret L. Lial, John Hornsby, Jr.,
Charles D. Miller. — 8th ed.
        p.   cm.
    Includes index.
    ISBN 0-321-01266-6 (Student Edition)
    ISBN 0-321-40337-1 (Annotated Instructor's Edition)
    1. Algebra.   I. Hornsby, John.   II. Miller, Charles David.
III. Title.
QA152.2.L52   1997
512.9--dc21   96-52939
                                                                CIP

3456789-DOW-01 00 99

Contents

Preface

This new edition of *Intermediate Algebra with Early Functions and Graphing* is designed for college students who have completed a course in introductory algebra or who require review of the basic concepts of algebra before taking additional courses in mathematics, science, business, nursing, or computer science. The primary objectives of this text are to familiarize students with mathematical symbols and operations, and to introduce them early on to the important concept of functions and their graphs. Polynomial, exponential, and logarithmic equations, and both linear and nonlinear systems are studied so that students can solve related applications. In addition, graphing is emphasized throughout the text.

Intermediate Algebra with Early Functions and Graphing retains the successful features that have distinguished previous editions while addressing the evolving needs of today's mathematics students. Building on the proven hallmark qualities of the text's exercises, in this edition we have further refined the exercise sets to thoroughly test conceptual understanding while providing ample opportunity for drill. Other hallmark features include careful exposition, fully developed examples, learning objectives keyed to each section, convenient margin exercises for student practice, the helpful "Cautions" and "Notes" feature, and a well-integrated design that complements the pedagogy, highlighting important definitions, rules, and procedures. Furthermore, in preparing this edition we addressed the concerns of the **National Council of Teachers of Mathematics (NCTM)** and the **American Mathematical Association of Two-Year Colleges (AMATYC)**, by including many new exercises that focus on concepts, writing, and analysis of data obtained from sources in the world around us.

The following pages describe some of the key features of this text.

New and Updated Features

▶ *Numbers in the Real World: A Graphing Calculator Minicourse* Throughout this series of worktexts, we have included a feature titled Numbers in the Real World. In this volume we have chosen to emphasize the influence of technology in our world by including nine lessons on the basics of using graphing calculators. These lessons are listed in the Contents.

▶ *Many new exercises* use data from real-life sources. They are designed to show students how algebra is used to describe and understand data in everyday life.

▶ *Mathematical Connections* exercise groups appear in many sections. These exercises are designed to be worked in sequential order, and they tie together concepts so students can see the connections among different topics in mathematics. For example, an exercise group may show how algebra and geometry are connected, or how a graph of the linear equation in two variables is related to the solution of the corresponding linear equation in one variable.

▶ While graphing calculators are not required for this text, it is likely that students will go on to courses that use them. For this reason, we have included a feature entitled Interpreting Technology in selected exercise sets that illustrates the power of graphing calculators and allows the student to interpret typical results seen on graphing calculator screens.

▶ Since many students using this text may be studying algebra for the first time or the first time in several years, or may have English as a second language, we have provided phonetic spellings for many of the important terms that appear in the book. A separate *Spanish Glossary* is also available to better serve Spanish-speaking students.

▶ Instructors and students will have access to a World Wide Web site (**www.mathnotes.com**), which include additional support material such as exercise sets and an online tutorial.

Continuing Features

We have retained the hallmark features that have distinguished previous editions of the text. Some of these successful features are as follows:

▶ *Ample and varied exercise sets* Students of basic algebra require a large number and variety of exercises. This text contains approximately 5800 exercises, including about 1200 review exercises, 480 conceptual and writing exercises, and 980 margin problems.

▶ *Learning objectives* Each section begins with clearly stated, numbered objectives, and material in the section is keyed to these objectives. In this way, students know exactly what is being covered in each section.

▶ *Margin problems* Margin problems, with answers immediately available, are found in every section. These problems allow the student to practice the material covered in the section in preparation for the exercise set that follows.

▶ *Cautions and Notes* We often give students warnings of common errors and emphasize important ideas in Cautions and Notes that appear throughout the exposition.

▶ *Ample opportunity for review* Each section contains a *chapter summary, chapter review exercises* keyed to individual sections as well as mixed review exercises, and a *chapter test*. Furthermore, following every chapter after Chapter 1, there is a set of cumulative review exercises that covers material going back to the first chapter. Students always have an opportunity to review material that appears earlier in the text, and this provides an excellent way to prepare for the final examination.

▶ *Answers and solutions* Answers to all margin problems are provided at the bottom of the page on which the problem appears. Furthermore, answers to odd-numbered exercises in numbered sections, answers to every exercise in Mathematical Connections groups, answers to all chapter test exercises, review exercises, and cumulative review exercises, and complete, worked-out solutions to every fourth exercise in numbered sections are provided at the back of the text. In this edition, we have also included sample answers to writing exercises.

All-New Supplements Package

Our extensive new supplements package includes an Annotated Instructor's Edition, testing materials, solutions, software, and videotapes.

F O R   T H E   I N S T R U C T O R

Annotated Instructor's Edition

This edition provides instructors with immediate access to the answers to every exercise in the text. Each answer is printed in color next to the corresponding text exercise. Symbols are used to identify the writing (▨) and conceptual (◉) exercises to assist in making homework assignments. Challenging (▲) exercises are also marked for the instructor for this purpose.

Instructor's Resource Guide

The *Instructor's Resource Guide* includes short-answer and multiple-choice versions of a pretest; eight forms of chapter tests for each chapter, including six open-response and two multiple-choice forms; short-answer and multiple-choice forms of a final examination; and an extensive set of additional exercises, providing 10 to 20 exercises for each textbook objective, which instructors can use as an additional source of questions for tests, quizzes, or student review of difficult topics. In addition, a section containing teaching tips is included for the instructor's convenience.

Instructor's Solutions Manual

This book includes solutions to all of the even-numbered section exercises (except the Connections exercises, which are in the *Student's Solutions Manual*). The two solutions manuals plus the solutions given at the back of the textbook provide detailed, worked-out solutions to each exercise and margin problem in the book.

Answer Book

This manual includes answers to all exercises.

TestGen EQ with QuizMaster EQ

This test generation software is available in Windows and Macintosh versions and is fully networkable. TestGen EQ's friendly, graphical interface enables instructors to easily view, edit, and add questions; transfer questions to tests; and print tests in a variety of fonts and forms. Search and sort features let instructors quickly locate questions and arrange them in a preferred order. Six question formats are available, including short-answer, true-false, multiple-choice, essay, matching, and bimodal formats. A built-in question editor gives users power to create graphs, import graphics, insert mathematical symbols and templates, and insert variable numbers or text. Computerized testbanks include algorithmically defined problems organized according to each textbook in the series.

QuizMaster EQ enables instructors to create and save tests using TestGen EQ so students can take them for practice or a grade on a computer network. Instructors can set preferences for how and when tests are administered. QuizMaster EQ automatically grades the exams, stores results on disk, and allows instructors to view or print a variety of reports for individual students, classes, or courses.

InterAct Mathematics Plus—Management System

InterAct Math Plus combines course management and online testing with the features of the basic InterAct Math tutorial software to create an invaluable teaching resource. Consult your Addison Wesley Longman representative for details.

FOR THE STUDENT

Student's Solutions Manual

This book contains solutions to every other odd-numbered section exercise (those not included at the back of the textbook) as well as solutions to all margin problems, Mathematical Connections exercises, chapter review exercises, chapter tests, and cumulative review exercises. (ISBN 0-321-01319-0)

InterAct Mathematics Tutorial Software

InterAct Math tutorial software has been developed and designed by professional software engineers working closely with a team of experienced developmental mathematics educators.

InterAct Math tutorial software includes exercises that are linked with every objective in the textbook and require the same computational and problem-solving skills as their companion exercises in the text. Each exercise has an example and an interactive guided solution that are designed to involve students in the solution process and to help them identify precisely where they are having trouble. In addition, the software recognizes common student errors and provides students with appropriate customized feedback.

With its sophisticated answer-recognition capabilities, InterAct Math tutorial software recognizes appropriate forms of the same answer for any kind of input. It also tracks student activity and scores for each section which can then be printed out. Available for Windows and Macintosh computers, the software is free to qualifying adopters or can be bundled with books for sale to students. (Macintosh: ISBN 0-321-00939-8, Windows: ISBN 0-321-00938-X)

Videotapes

A videotape series has been developed to accompany *Intermediate Algebra with Early Functions and Graphing,* Sixth Edition. In a separate lesson for each section in the book, the series covers all objectives, topics, and problem-solving techniques discussed within the text. (ISBN 0-321-01318-2)

Spanish Glossary

A separate *Spanish Glossary* is now being offered as part of the supplements package for this textbook series. This book contains the key terms from each of the four texts in the series and their Spanish translations. (ISBN 0-321-01647-5)

Acknowledgments

For a textbook to succeed through six editions, it is necessary for the authors to rely on comments and suggestions of users, non-users, instructors, and students. We are grateful for the many responses that we have received over the years. We wish to thank the following individuals who reviewed this edition of the text:

Nancy Ballard, *Mineral Area College*

Richard DeCesare, *Southern Connecticut University*

Joseph Gutel, *Oklahoma State University–Oklahoma City*

Michael Karelius, *American River College*

Linda Laningham, *Rend Lake College*

Keith Lathrop, *Valencia Community College–West Campus*

Janice Rech, *University of Nebraska at Omaha*

Don Reichman, *Mercer County College*

Nancy Ressler, *Oakton Community College*

Steven Tarry, *Ricks College*

Lucy C. Thrower, *Francis Marion University*

No author can write a text of this magnitude without the help of many other individuals. Our sincere thanks go to Rita Ferrandino of Addison Wesley Longman who coordinated the package of texts of which this book is a part. We are appreciative of the support given to us by Greg Tobin, also of Addison Wesley Longman, who helped make the transition from our former publisher go as smoothly as one could ever imagine. We wish to thank Sandi Goldstein for her efforts in coordinating the reviews and working with us in the early stages of preparation for production. Becky Troutman assisted in preparing the solutions that appear in the back of the book. Theresa McGinnis continues to provide help whenever we ask, and assisted in checking for the accuracy of the answers. Mark Coffey and Chris Burditt also assisted in checking the answers. Ron Hampton coordinated the art program. Cathy Wacaser of Elm Street Publishing Services provided her usual excellent production work. She is indeed one of the best in the business. As usual, Paul Van Erden created an accurate, useful index. We are also grateful to Tommy Thompson who made suggestions for the feature "To the Student: Success in Algebra."

Paul Eldersveld of the College of DuPage has coordinated our print supplements for many years. The importance of his job cannot be overestimated, and we want to thank him for his work over the years. He is a wonderful friend and colleague, and we are happy to have him as part of our publishing family.

Margaret L. Lial
John Hornsby

Feature Walk-Through

A new design updates and refreshes the overall flow of the text. New, full-color situational art throughout the text enhances student comprehension of the material.

Section Objectives: Each section begins with stated objectives that guide student learning.

Student Resources: Found at the opening of each section, this feature cross-references relevant material in the student supplements package, providing each student with a rich variety of extra help and resources.

World Wide Web: Instructors and students will have access to a World Wide Web site (www.mathnotes.com), where additional support material is available.

Margin Exercises: Exercises appear in the margin for immediate practice and reinforcement. Answers to the margin exercises are available at the bottom of the page.

Cautions and Notes: Common student errors are anticipated and highlighted with the heading "Caution" or "Note." This helps clarify concepts for students and assists them in identifying common errors.

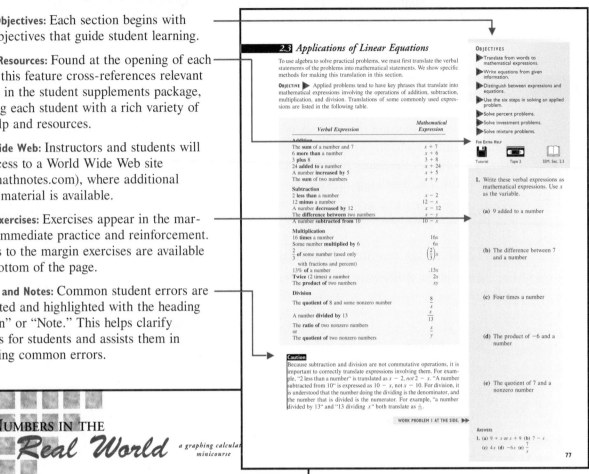

2.3 Applications of Linear Equations

To use algebra to solve practical problems, we must first translate the verbal statements of the problems into mathematical statements. We show specific methods for making this translation in this section.

OBJECTIVE ▶ Applied problems tend to have key phrases that translate into mathematical expressions involving the operations of addition, subtraction, multiplication, and division. Translations of some commonly used expressions are listed in the following table.

Verbal Expression	Mathematical Expression
Addition	
The **sum** of a number and 7	$x + 7$
6 **more than** a number	$x + 6$
3 **plus** 8	$3 + 8$
24 **added to** a number	$x + 24$
A number **increased by** 5	$x + 5$
The **sum** of two numbers	$x + y$
Subtraction	
2 **less than** a number	$x - 2$
12 **minus** a number	$12 - x$
A number **decreased by** 12	$x - 12$
The **difference between** two numbers	$x - y$
A number **subtracted from** 10	$10 - x$
Multiplication	
16 **times** a number	$16x$
Some number **multiplied by** 6	$6x$
$\frac{2}{3}$ **of** some number (used only with fractions and percent)	$\left(\frac{2}{3}\right)x$
13% **of** a number	$.13x$
Twice (2 times) a number	$2x$
The **product of** two numbers	xy
Division	
The **quotient of** 8 and some nonzero number	$\frac{8}{x}$
A number **divided by** 13	$\frac{x}{13}$
The **ratio of** two nonzero numbers or The **quotient of** two nonzero numbers	$\frac{x}{y}$

Caution

Because subtraction and division are not commutative operations, it is important to correctly translate expressions involving them. For example, "2 less than a number" is translated as $x - 2$, *not* $2 - x$. "A number subtracted from 10" is expressed as $10 - x$, not $x - 10$. For division, it is understood that the number doing the dividing is the denominator, and the number that is divided is the numerator. For example, "a number divided by 13" and "13 dividing x" both translate as $\frac{x}{13}$.

WORK PROBLEM 1 AT THE SIDE. ▶▶

OBJECTIVES
- ▶ Translate from words to mathematical expressions.
- ▶ Write equations from given information.
- ▶ Distinguish between expressions and equations.
- ▶ Use the six steps in solving an applied problem.
- ▶ Solve percent problems.
- ▶ Solve investment problems.
- ▶ Solve mixture problems.

FOR EXTRA HELP

Tutorial   Tape 2   SSM, Sec. 2.3

1. Write these verbal expressions as mathematical expressions. Use x as the variable.

   (a) 9 added to a number

   (b) The difference between 7 and a number

   (c) Four times a number

   (d) The product of −6 and a number

   (e) The quotient of 7 and a nonzero number

ANSWERS
1. (a) $9 + x$ or $x + 9$ (b) $7 - x$
   (c) $4x$ (d) $-6x$ (e) $\frac{7}{x}$

77

NUMBERS IN THE Real World
a graphing calculator minicourse

Lesson 2: Solution of Linear Inequalities

In Lesson 1 we observed that the x-intercept of the graph of the line $y = mx + b$ is the solution of the equation $mx + b = 0$. We can extend our observations to consider the solution of the associated inequalities $mx + b > 0$ and $mx + b < 0$. The solution set of $mx + b > 0$ is the set of all x-values for which the graph of $y = mx + b$ is *above* the x-axis. (We consider points above because the symbol is $>$.) On the other hand, the solution set of $mx + b < 0$ is the set of all x-values for which the graph of $y = mx + b$ is *below* the x-axis. (We consider points below because the symbol is $<$.) Therefore, once we know the solution set of the equation and have the graph of the line, we can determine the solution sets of the corresponding inequalities.

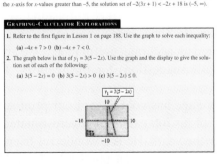

In the first figure (to the left), we see that the x-intercept of $y_1 = 3x - 9$ is 3. Therefore, by the concepts of Lesson 1, the solution set of $3x - 9 = 0$ is $\{3\}$. Because the graph of y_1 lies above the x-axis for x-values greater than 3, the solution set of $3x - 9 > 0$ is $(3, \infty)$. Because the graph lies below the x-axis for x-values less than 3, the solution set of $3x - 9 < 0$ is $(-\infty, 3)$.

Suppose that we wish to solve the equation $-2(3x + 1) = -2x + 18$, and the associated inequalities $-2(3x + 1) > -2x + 18$ and $-2(3x + 1) < -2x + 18$. We begin by considering the equation rewritten so that the right side is equal to 0: $-2(3x + 1) + 2x - 18 = 0$. Graphing $y_1 = -2(3x + 1) + 2x - 18$ yields the x-intercept -5, as shown in the second figure. The first inequality listed is equivalent to $y_1 > 0$. Because the line lies *above* the x-axis for x-values less than -5, the solution set of $-2(3x + 1) > -2x + 18$ is $(-\infty, -5)$. Because the line lies *below* the x-axis for x-values greater than -5, the solution set of $-2(3x + 1) < -2x + 18$ is $(-5, \infty)$.

GRAPHING-CALCULATOR EXPLORATIONS

1. Refer to the first figure in Lesson 1 on page 188. Use the graph to solve each inequality:

   (a) $-4x + 7 > 0$ (b) $-4x + 7 < 0$.

2. The graph below is that of $y_1 = 3(5 - 2x)$. Use the graph and the display to give the solution set of each of the following:

   (a) $3(5 - 2x) = 0$ (b) $3(5 - 2x) > 0$ (c) $3(5 - 2x) \leq 0$.

Numbers in the Real World: A Graphing Calculator Minicourse: To address the increasing use of graphing calculators in the classroom, nine lessons that provide general guidelines on how they can be used effectively in algebra are included. We provide explanations on how solution of equations, inequalities, and systems can be supported, how matrices and determinants can be analyzed and evaluated, and how a table of values of a function can be generated at the touch of a key.

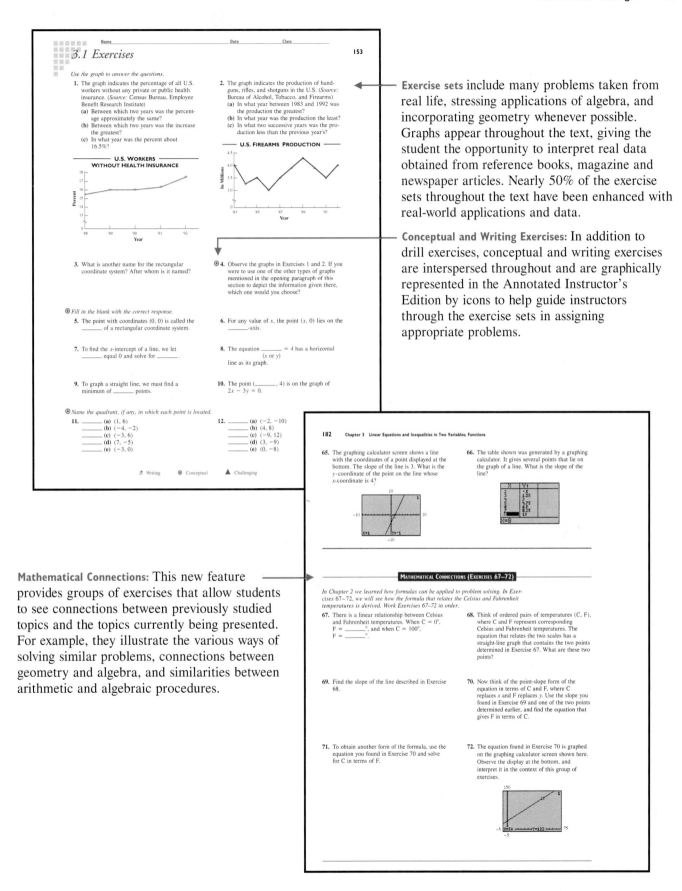

Exercise sets include many problems taken from real life, stressing applications of algebra, and incorporating geometry whenever possible. Graphs appear throughout the text, giving the student the opportunity to interpret real data obtained from reference books, magazine and newspaper articles. Nearly 50% of the exercise sets throughout the text have been enhanced with real-world applications and data.

Conceptual and Writing Exercises: In addition to drill exercises, conceptual and writing exercises are interspersed throughout and are graphically represented in the Annotated Instructor's Edition by icons to help guide instructors through the exercise sets in assigning appropriate problems.

Mathematical Connections: This new feature provides groups of exercises that allow students to see connections between previously studied topics and the topics currently being presented. For example, they illustrate the various ways of solving similar problems, connections between geometry and algebra, and similarities between arithmetic and algebraic procedures.

Art Program: The art program has been refreshed with a new, appealing, pedagogical layout. Extensive additions to the art program keep students interested in the mathematical material.

Interpreting Technology: A new feature, Interpreting Technology demonstrates the power of graphing calculators and allows students to view typical graphing calculator-generated screens and interpret results.

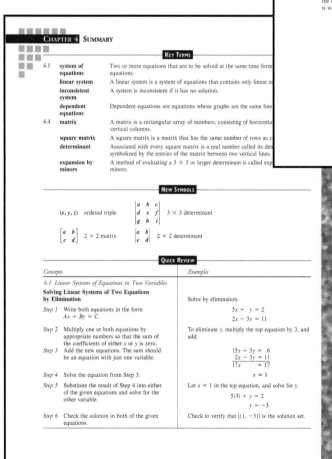

End-of-Chapter Material: In order to reinforce concepts learned and assist students with their retention, the end-of-chapter material includes a summary of Key Terms and New Symbols, a Quick Review of concepts learned, Review Exercises, a Chapter Test, and a Cumulative Review.

Name _____ 3.3 Linear Equations in Two Variables **181**

Assume that the situation described in the figure can be modeled by a straight-line graph, and use the information to find the $y = mx + b$ form of the equation of the line.

61. The number of post offices in the United States has been declining in recent years. Use the information given on the bar graph for the years 1985 and 1993, letting $x = 0$ represent the year 1985 and letting y represent the number of post offices. (*Source:* U.S. Postal Service, Annual Report of the Postmaster General)

U.S. POST OFFICES

62. The number of motor vehicle deaths in the United States declined from 1990 to 1992 as seen in the accompanying bar graph. Use the information given, with $x = 0$ representing 1990 and y representing the number of deaths. (*Source:* National Safety Council)

U.S. MOTOR-VEHICLE DEATHS

INTERPRETING TECHNOLOGY (EXERCISES 63–66)

63. The graphing calculator screen shows two lines. One is the graph of $y_1 = -2x + 3$ and the other is the graph of $y_2 = 3x - 4$. Which is which?

64. The graphing calculator screen shows two lines. One is the graph of $y_1 = 2x - 5$ and the other is the graph of $y_2 = 4x - 5$. Which is which?

CHAPTER 4 SUMMARY

KEY TERMS

4.1	system of equations	Two or more equations that are to be solved at the same time form equations.
	linear system	A linear system is a system of equations that contains only linear e
	inconsistent system	A system is inconsistent if it has no solution.
	dependent equations	Dependent equations are equations whose graphs are the same line
4.4	matrix	A matrix is a rectangular array of numbers, consisting of horizonta vertical columns.
	square matrix	A square matrix is a matrix that has the same number of rows as c
	determinant	Associated with every square matrix is a real number called its det symbolized by the entries of the matrix between two vertical lines.
	expansion by minors	A method of evaluating a 3 × 3 or larger determinant is called exp minors.

NEW SYMBOLS

(x, y, z) ordered triple

$\begin{vmatrix} a & b & c \\ d & e & f \\ g & h & i \end{vmatrix}$ 3 × 3 determinant

$\begin{bmatrix} a & b \\ c & d \end{bmatrix}$ 2 × 2 matrix

$\begin{vmatrix} a & b \\ c & d \end{vmatrix}$ 2 × 2 determinant

QUICK REVIEW

Concepts	Examples
4.1 Linear Systems of Equations in Two Variables	
Solving Linear Systems of Two Equations by Elimination	Solve by elimination.
Step 1 Write both equations in the form $Ax + By = C$.	$5x + y = 2$ $2x - 3y = 11$
Step 2 Multiply one or both equations by appropriate numbers so that the sum of the coefficients of either x or y is zero.	To eliminate y, multiply the top equation by 3, and add.
Step 3 Add the new equations. The sum should be an equation with just one variable.	$\begin{aligned} 15x + 3y &= 6 \\ \underline{2x - 3y} &= \underline{11} \\ 17x \quad\;\; &= 17 \end{aligned}$ $x = 1$
Step 4 Solve the equation from Step 3.	
Step 5 Substitute the result of Step 4 into either of the given equations and solve for the other variable.	Let $x = 1$ in the top equation, and solve for y. $5(1) + y = 2$ $y = -3$
Step 6 Check the solution in both of the given equations.	Check to verify that $\{(1, -3)\}$ is the solution set.

An Introduction to Scientific Calculators

The emphasis placed on paper-and-pencil computation, which has long been a part of the school mathematics curriculum, is not as important as it once was. The search for easier ways to calculate and compute has culminated in the development of hand-held calculators and computers. Professional organizations devoted to the teaching and learning of mathematics recommend that this technology be incorporated throughout the mathematics curriculum. In light of this recommendation, this text assumes that all students have access to calculators, and the authors recommend that students become computer literate to the extent their academic resources and individual finances allow.

In this text we view calculators as a means of allowing students to spend more time on the conceptual nature of mathematics and less time on the drudgery of computation with paper and pencil. Calculators come in a large array of different types, sizes, and prices, making it difficult to decide which machine best fits your needs. For the general population, a calculator that performs the operations of arithmetic and a few other functions is sufficient. These are known as **four-function calculators.** Students who take higher mathematics courses (engineers, for example) usually need the added power of **scientific calculators. Programmable calculators,** which allow short programs to be written and performed, and **graphing calculators,** which actually plot graphs on small screens, are also available. Keep in mind that calculators differ from one manufacturer to the other. For this reason, remember the following.

Always refer to your owner's manual if you need assistance in performing an operation with your calculator.

Because of their relatively inexpensive cost and features that far exceed those of four-function calculators, scientific calculators probably provide students the best value for their money. For this reason, we will give a short synopsis of the major functions of scientific calculators.

Most scientific calculators use *algebraic logic.* (Models sold by Texas Instruments, Sharp, Casio, and Radio Shack, for example, use algebraic logic.) A notable exception is Hewlett Packard, a company whose calculators use *Reverse Polish Notation* (RPN). In this introduction, we explain how to use calculators with algebraic logic.

Arithmetic Operations

To perform an operation of arithmetic, simply enter the first number, touch the operation key ([+],[−],[×], or [÷]), enter the second number, and then touch the [=] key. For example, to add 4 and 3, use the following keystrokes.

(The final answer is displayed in color.)

Change Sign Key

The key marked [±] allows you to change the sign of a display. This is particularly useful when you wish to enter a negative number. For example, to enter -3, use the following keystrokes.

Parentheses Keys

These keys, [(] and [)], are designed to allow you to group numbers in arithmetic operations as you desire. For example, if you wish to evaluate $4 \times (6 + 3)$, use the following keystrokes.

Memory Key

Scientific calculators can hold a number in memory for later use. The label of the memory key varies among models; two versions are [M] and [STO]. [M+] and [M−] allow you to add to or subtract from the value currently in memory. The memory recall key, labeled [MR], [RM], or [RCL], allows you to retrieve the value stored in memory.

Suppose that you wish to store the number 5 in memory. Enter 5, then touch the key for memory. You can then perform other calculations. When you need to retrieve the 5, touch the key for memory recall.

If a calculator has a constant memory feature, the value in memory will be retained even after the power is turned off. Some advanced calculators have more than one memory. It is best to read the owner's manual for your model to see exactly how memory is activated.

Clearing/Clear Entry Keys

These keys allow you to clear the display or clear the last entry entered into the display. They are usually marked [C] and [CE]. In some models, touching the [C] key once will clear the last entry, while touching it twice will clear the entire operation in progress.

Second Function Key

This key is used in conjunction with another key to activate a function that is printed *above* an operation key (and not on the key itself). It is usually marked [2nd]. For example, suppose you wish to find the square of a number, and the squaring function (explained in more detail later) is printed above another key. You would need to touch [2nd] before the desired squaring function can be activated.

Some models of scientific calculators (the TI-35X by Texas Instruments, for example) even provide a third function key, marked [3rd], which is used in a manner similar to the one described for the second function key.

Square Root and Cube Root Keys

Touching the square root key, $\boxed{\sqrt{x}}$, will give the square root (or an approximation of the square root) of the number in the display. For example, to find the square root of 36, use the following keystrokes.

$$\boxed{3}\ \boxed{6}\ \boxed{\sqrt{x}} \qquad \boxed{\qquad\qquad \textbf{6}}$$

The square root of 2 is an example of an irrational number (see Section 7.1). The calculator will give an approximation of its value, since the decimal for $\sqrt{2}$ never terminates and never repeats. The number of digits shown will vary among models. To find an approximation of $\sqrt{2}$, use the following keystrokes.

$$\boxed{2}\ \boxed{\sqrt{x}} \qquad \boxed{\textbf{1.4142136}} \quad \text{An approximation}$$

The cube root key, $\boxed{\sqrt[3]{x}}$, is used in the same manner as the square root key. To find the cube roots of 64 and 93, use the following keystrokes.

$$\boxed{6}\ \boxed{4}\ \boxed{\sqrt[3]{x}} \qquad \boxed{\qquad\qquad \textbf{4}}$$
$$\boxed{9}\ \boxed{3}\ \boxed{\sqrt[3]{x}} \qquad \boxed{\textbf{4.53065490}} \quad \text{An approximation}$$

Note: Calculators differ in the number of digits provided in the display. For example, one calculator gives the approximation of the cube root of 93 as 4.530654896, showing more digits than shown above.

Squaring and Cubing Keys

The squaring key, $\boxed{x^2}$, allows you to square the entry in the display. For example, to square 35.7, use the following keystrokes.

$$\boxed{3}\ \boxed{5}\ \boxed{.}\ \boxed{7}\ \boxed{x^2} \qquad \boxed{\textbf{1274.49}}$$

The squaring key and the square root key are often found on the same key, with one of them being a second function (that is, activated by the second function key, described above).

The cubing key, $\boxed{x^3}$, allows you to cube the entry in the display. To cube 3.5, follow these keystrokes.

$$\boxed{3}\ \boxed{.}\ \boxed{5}\ \boxed{x^3} \qquad \boxed{\textbf{42.875}}$$

Reciprocal Key

The key marked $\boxed{1/x}$ (or $\boxed{x^{-1}}$) is the reciprocal key. (When two numbers have a product of 1, they are called *reciprocals*.) Suppose you wish to find the reciprocal of 5. Use the following keystrokes.

$$\boxed{5}\ \boxed{1/x} \qquad \boxed{\qquad\qquad \textbf{0.2}}$$

Inverse Key

Some calculators have an inverse key, marked $\boxed{\text{INV}}$. Inverse operations are operations that "undo" each other. For example, the operations of squaring and taking the square root are inverse operations. The use of the $\boxed{\text{INV}}$ key varies among different models of calculators, so read your owner's manual carefully.

Exponential Key

This key, marked $\boxed{x^y}$ or $\boxed{y^x}$, allows you to raise a number to a power. For example, if you wish to raise 4 to the fifth power (that is, find 4^5), use the following keystrokes.

$$\boxed{4}\ \boxed{x^y}\ \boxed{5}\ \boxed{=} \qquad \boxed{\qquad\qquad \textbf{1024}}$$

Root Key

Some calculators have this key specifically marked $\boxed{\sqrt[x]{}}$ or $\boxed{\sqrt[y]{}}$; with others, the operation of taking roots is accomplished by using the inverse key in conjunction with the exponential key. Suppose, for example, your calculator is of the latter type, and you wish to find the fifth root of 1024. Use the following keystrokes.

$$\boxed{1}\ \boxed{0}\ \boxed{2}\ \boxed{4}\ \boxed{\text{INV}}\ \boxed{x^y}\ \boxed{5}\ \boxed{=}\ \boxed{4}$$

Notice how this "undoes" the operation explained in the exponential key discussion earlier.

Pi Key

The number π is important in mathematics. It occurs, for example, in the area and circumference formulas for a circle. By touching the $\boxed{\pi}$ key, you can get the display of the first few digits of π. (Because π is irrational, the display shows only an approximation.) One popular model gives the following display when the $\boxed{\pi}$ key is activated: $\boxed{3.1415927}$. As mentioned before, calculators will vary in the number of digits they give for π in their displays.

Other Considerations

When decimal approximations are shown on scientific calculators, they are either *truncated* or *rounded*. To see which of these a particular model is programmed to do, evaluate 1/18 as an example. If the display shows .0555555 (last digit 5), it truncates the display. If it shows .0555556 (last digit 6), it rounds off the display.

When very large or very small numbers are obtained as answers, scientific calculators often express these numbers in scientific notation. For example, if you multiply 6,265,804 by 8,980,591, the display might look like this:

$$\boxed{5.6270623\quad 13}$$

The "13" at the far right means that the number on the left is multiplied by 10^{13}. This means that the decimal point must be moved 13 places to the right if the answer is to be expressed in its usual form. Even then, the value obtained will only be an approximation: 56,270,623,000,000.

Two features of advanced scientific calculators are programmability and graphing capability. A programmable calculator has the capability of running small programs, much like a mini-computer. A graphing calculator can be used to plot graphs of functions on a small screen. One of the issues in mathematics education today deals with how graphing calculators should be incorporated into the curriculum. Their availability in the 1990s parallels the availability of scientific calculators in the 1980s, and they are providing a major influence in mathematics education as we move into the twenty-first century.

To the Student: Success in Algebra

The main reason students have difficulty with mathematics is that they don't know how to study it. Studying mathematics *is* different from studying subjects like English or history. The key to success is regular practice.

This should not be surprising. After all, can you learn to play the piano or to ski well without a lot of regular practice? The same thing is true for learning mathematics. Working problems nearly every day is the key to becoming successful. Here is a list of things that will help you succeed in studying algebra.

1. *Attend class regularly.* Pay attention to what your teacher says and does in class, and take careful notes. In particular, note the problems the teacher works on the board and copy the complete solutions. Keep these notes separate from your homework to avoid confusion when you read them over later.

2. Don't hesitate to ask questions in class. It is not a sign of weakness, but of strength. There are always other students with the same question who are too shy to ask.

3. *Read your text carefully.* Many students read only enough to get by, usually only the examples. Reading the complete section will help you solve the homework problems. Most exercises are keyed to specific examples or objectives that will explain the procedures for working them.

4. Before you start on your homework assignment, rework the problems the teacher worked in class. This will reinforce what you have learned. Many students say, "I understand it perfectly when you do it, but I get stuck when I try to work the problem myself."

5. Do your homework assignment only *after* reading the text and reviewing your notes from class. Check your work against the answers in the back of the book. If you get a problem wrong and are unable to understand why, mark that problem and ask your instructor about it. Then practice working additional problems of the same type to reinforce what you have learned.

6. Work as neatly as you can. Write your symbols clearly, and make sure the problems are clearly separated from each other. Working neatly will help you to think clearly and also make it easier to review the homework before a test.

7. After you complete a homework assignment, look over the text again. Try to identify the main ideas that are in the lesson. Often they are clearly highlighted or boxed in the text.

8. Use the chapter test at the end of each chapter as a practice test. Work through the problems under test conditions, without referring to the text or the answers until you are finished. You may want to time yourself to see how long it takes you. When you finish, check your answers against those in the back of the book, and study the problems you missed. Answers are keyed to the appropriate sections of the text.

9. Keep any quizzes and tests that are returned to you, and use them when you study for future tests and the final exam. These quizzes and tests indicate what your instructor considers most important. Be sure to correct any problems on these tests that you missed, so you will have the corrected work to study.

10. Don't worry if you do not understand a new topic right away. As you read more about it and work through the problems, you will gain understanding. Each time you review a topic you will understand it a little better. No one understands each topic completely right from the start.

The Real Numbers

1.1 Basic Terms

The study of algebra extends and generalizes the rules of arithmetic. In algebra, symbols are used to convey ideas in an economical way. Many of the symbols used throughout the book are introduced in this chapter.

OBJECTIVE 1 A **set** is a well-defined collection of objects. These objects are called the **elements** (ELL-uh-ments) or **members** (MEM-berz) of the set. In algebra, the elements of a set are usually numbers. Sets are written with braces used to enclose the elements. For example, 2 is an element of the set {1, 2, 3}. In our study of algebra, we will find it useful to refer to certain sets of numbers by name. The set

$$N = \{1, 2, 3, 4, 5, 6, \ldots\}$$

is called the **natural numbers** or the **counting numbers.** The three dots show that the list continues in the same pattern indefinitely. A set with no elements is called the **empty set** and is written as ∅.

> **Caution**
> Do not write {∅} instead of ∅, because {∅} represents a *set* that contains one *element,* namely ∅.

WORK PROBLEM 1 AT THE SIDE. ▶▶

Letters called **variables** (VAIR-ee-uh-buls) are often used to represent numbers. Variables can also be used to define sets of numbers. For example,

$$\{x \mid x \text{ is a natural number between 3 and 15}\}$$

(read "the set of all elements x such that x is a natural number between 3 and 15") defines the set

$$\{4, 5, 6, 7, \ldots, 14\}.$$

The notation $\{x \mid x \text{ is a natural number between 3 and 15}\}$ is an example of **set-builder notation,** read as follows.

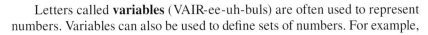 **{*x* | *x* has property *P*}**

The set of all elements *x*   such that   *x* has a given property *P*

The property *P* describes the elements to be included in the set, as illustrated in the next example.

OBJECTIVES

1 ▶ Learn the terminology of sets.
2 ▶ Use number lines.
3 ▶ Know the common sets of numbers.
4 ▶ Find additive inverses.
5 ▶ Use absolute value.

FOR EXTRA HELP

Tutorial   Tape 1   SSM, Sec. 1.1

1. Which elements of the set $\left\{10, \dfrac{3}{10}, 52, 98.6\right\}$ are natural numbers?

ANSWERS

1. 10 and 52

2. (a) List the elements in the set. $\{x \mid x$ is a natural number less than 5$\}$

(b) Use set-builder notation to describe the set $\{13, 14, 15, \ldots\}$.

3. Graph each set.

(a) $\{-4, -2, 0, 2, 4, 6\}$

<hr style="border: 1px solid; width: 60%;" />

(b) $\left\{-1, 0, \dfrac{2}{3}, \dfrac{5}{2}\right\}$

<hr style="border: 1px solid; width: 60%;" />

(c) $\left\{5, \dfrac{16}{3}, 6, \dfrac{13}{2}, 7, \dfrac{29}{4}\right\}$

<hr style="border: 1px solid; width: 60%;" />

ANSWERS

2. (a) $\{1, 2, 3, 4\}$ **(b)** One answer is $\{x \mid x$ is a natural number greater than 12$\}$.

3. (a) ![number line graph] $-4\ -2\ \ 0\ \ 2\ \ 4\ \ 6$

(b) ![number line graph] $-2\ -1\ \ 0\ \ 1\ \ 2\ \ 3$

(c) ![number line graph] $4\ \ 5\ \ 6\ \ 7\ \ 8$

E X A M P L E 1   Using Set-Builder Notation

(a) Write the set $\{x \mid x$ is a natural number less than 4$\}$ by listing its elements.

The natural numbers less than 4 are 1, 2, and 3. The given set may also be written as $\{1, 2, 3\}$.

(b) Use set-builder notation to describe the set $\{2, 4, 6, 8, 10\}$.

There are several ways to describe a set in set-builder notation. One way to describe this set is as $\{y \mid y$ is one of the first five even natural numbers$\}$.

■

◀◀ **WORK PROBLEM 2 AT THE SIDE.**

Objective 2 A good way to get a picture of sets of numbers is to use a diagram called a **number line.** To construct a number line, choose any point on a horizontal line and label it 0. Next, choose a point to the right of 0 and label it 1. The distance from 0 to 1 establishes a scale that can be used to locate more points, with positive numbers to the right of 0 and negative numbers to the left of 0. The number 0 is neither positive nor negative. A number line is shown in Figure 1.

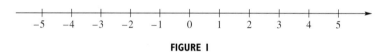

FIGURE 1

Each number is called the **coordinate** (koh-OR-din-et) of the point that it labels, while the point is the **graph** of the number. Figure 2 shows a number line with several selected points graphed on it.

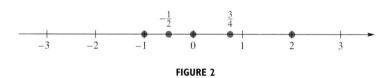

FIGURE 2

Numbers written with a positive or negative sign, such as $+4$, $+8$, -9, -5, and so on, are called **signed numbers.** Positive numbers can be called signed numbers even if the plus sign is left off (as it usually is).

◀◀ **WORK PROBLEM 3 AT THE SIDE.**

Every point on a number line corresponds to a **real number,** and every real number corresponds to a point on the line. Real numbers can be written in decimal form, either as a terminating decimal, such as .6 or .125; a repeating decimal, such as .33333 . . . or .127127 . . . ; or as a decimal that neither repeats nor terminates, such as .12534875 . . . . Repeating decimals are often written with a bar over the repeating digits. Using this notation, .127127 . . . is written $.\overline{127}$. Repeating and terminating decimal numbers are **rational** numbers; they *can* be written as fractions in which the numerator and denominator (di-NAHM-in-ay-ter) are integers. Decimal numbers that do not repeat or terminate are *not* rational, and thus are called **irrational** (ear-RA-shun-ul) numbers. Many square roots are irrational numbers; some examples are $\sqrt{7}$, $\sqrt{11}$, $\sqrt{2}$, and $-\sqrt{5}$. Some square roots *are* rational: $\sqrt{16} = \sqrt{4^2} = 4$, $\sqrt{100} = \sqrt{10^2} = 10$, and so on. A square root is rational only if the number under the radical sign is a square. Another irrational number is π, the ratio of the circumference of a circle to its diameter. All irrational numbers are real numbers.

OBJECTIVE 3▶ The sets of numbers listed below will be used throughout the rest of the book.

Natural numbers or counting numbers	$\{1, 2, 3, 4, 5, 6, 7, 8, \ldots\}$
Whole numbers	$\{0, 1, 2, 3, 4, 5, 6, \ldots\}$
Integers	$\{\ldots, -3, -2, -1, 0, 1, 2, 3, \ldots\}$
Rational numbers	$\left\{\dfrac{p}{q} \mid p \text{ and } q \text{ are integers}, q \neq 0\right\}$ Examples: $\dfrac{4}{1}$, 1.3, $-\dfrac{9}{2}$, $\dfrac{16}{8}$ or 2, $\sqrt{9}$ or 3, $.\overline{6}$
Irrational numbers	$\{x \mid x \text{ is a real number that is not rational}\}$ Examples: $\sqrt{3}$, $-\sqrt{2}$, π
Real numbers	$\{x \mid x \text{ is a coordinate of a point on a number line}\}$*

The relationships among these various sets of numbers are shown in Figure 3; in particular, the figure shows that the set of real numbers includes both the rational and irrational numbers. Every real number is either rational or irrational. Also, notice that the integers are elements of the set of rational numbers, and the whole numbers and the natural numbers are elements of the set of integers.

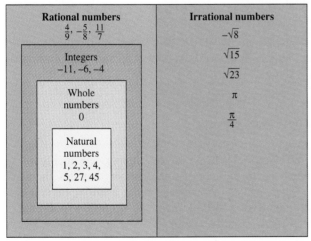

Real numbers

(a)

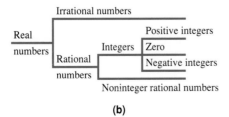

(b)

FIGURE 3 *The Real Numbers*

---

* An example of a number that is not a coordinate of a point on a number line is $\sqrt{-1}$. This number, called an *imaginary* number, is discussed in Chapter 7.

4. Select all the words from the following list that apply to each number.
   Whole number
   Rational number
   Irrational number
   Real number
   Undefined

   (a) −6

   (b) 12

   (c) $.\overline{3}$

   (d) $-\sqrt{15}$

   (e) $-\dfrac{6}{0}$

   (f) π

   (g) $\dfrac{22}{7}$

   (h) 3.14

5. Decide whether the statement is true or false. If false, tell why.

   (a) All whole numbers are integers.

   (b) Some integers are whole numbers.

   (c) Every real number is irrational.

ANSWERS
4. (a) rational, real
   (b) whole, rational, real
   (c) rational, real
   (d) irrational, real
   (e) undefined
   (f) irrational, real
   (g) rational, real
   (h) rational, real
5. (a) true (b) true
   (c) False; some real numbers are irrational, but others are rational numbers.

E X A M P L E   2   Identifying Examples of Number Sets

Which numbers in the list $\{-8, -4/0, -\sqrt{2}, -9/64, 0, .5, 2/3, 1.1212\ldots, \sqrt{3}, 2, 5\}$ are elements of the following sets?

(a) rational numbers
   $-8, -9/64, 0, .5, 2/3, 1.1212\ldots, 2,$ and 5 are rational numbers.

(b) integers
   $-8, 0, 2,$ and 5 are integers.

(c) irrational numbers
   $-\sqrt{2}$ and $\sqrt{3}$ are irrational numbers.

(d) real numbers
   All the numbers listed in (a), (b), and (c) above are real numbers.

(e) undefined
   $-4/0$ is undefined because the definition of a rational number requires the denominator to be nonzero.

◄◄ WORK PROBLEM 4 AT THE SIDE.

E X A M P L E   3   Determining Relationships between Sets of Numbers

Decide whether the statement is true or false.

(a) All irrational numbers are real numbers.
   This is true. As shown in Figure 3, the set of real numbers includes all the irrational numbers.

(b) Every rational number is an integer.
   This statement is false. Some rational numbers are integers (every integer is a rational number). However, other rational numbers such as $2/3$, $-1/4$, and so on are not integers.

◄◄ WORK PROBLEM 5 AT THE SIDE.

OBJECTIVE 4▶ Now look again at the number line in Figure 1. For each positive number, there is a negative number that lies the same distance from zero. These pairs of numbers are called **additive inverses** (ADD-ih-tiv IN-verses), **negatives** (NEG-uh-tivs), or **opposites** (OP-uh-zits) of each other. For example, 5 is the additive inverse of -5, and -5 is the additive inverse of 5.

   For any real number a, the number $-a$ is the additive inverse of a. (Change the sign to get the additive inverse.)

The sum of a number and its additive inverse is always zero.

The symbol "−" can be used to indicate any of the following things:

(a) a negative number, such as −9 or −15;

(b) the additive inverse of a number, as in "−4 is the additive inverse of 4";

(c) subtraction, as in 12 − 3.

In the expression −(−5), the symbol "−" is being used in two ways: the first − indicates the additive inverse of −5, and the second indicates a negative number, −5. The additive inverse of −5 is 5, so −(−5) = 5.

For any real number a, $\quad -(-a) = a$.

EXAMPLE 4 Finding Additive Inverses

The following list shows several signed numbers and the additive inverse of each.

Number	Additive Inverse
6	−6
−4	4
$\frac{2}{3}$	$-\frac{2}{3}$
0	0

WORK PROBLEM 6 AT THE SIDE. ▶▶

OBJECTIVE 5 ▶ Geometrically, the **absolute value** (ab-soh-LOOT VAL-yoo) of a number a, written $|a|$, is the distance on the number line from 0 to a. For example, the absolute value of 5 is the same as the absolute value of −5 because each number lies five units from 0 (see Figure 4). That is, both

$$|5| = 5 \quad \text{and} \quad |-5| = 5.$$

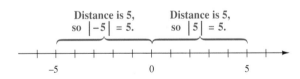

Distance is 5, so $|-5| = 5$. Distance is 5, so $|5| = 5$.

FIGURE 4

Caution
Because absolute value is used to represent distance, and distance is never negative, *the absolute value of a number is never negative.* However, the absolute value of 0 is 0.

The algebraic definition of absolute value is written as follows.

$$|a| = \begin{cases} a & \text{if } a \text{ is positive or zero} \\ -a & \text{if } a \text{ is negative} \end{cases}$$

The second part of this definition, $|a| = -a$ if a is negative, requires careful thought. If a is a *negative* number, then $-a$, the additive inverse or opposite of a, is a positive number, so that $|a| = -a$ is positive. For example, if $a = -3$,

$$|a| = |-3| = -(-3) = 3.$$

6. Give the additive inverse of each number.

(a) 9

(b) −12

(c) $-\dfrac{6}{5}$

(d) 0

7. Find the value of each expression.

(a) $|6|$

(b) $|-3|$

(c) $-|5|$

(d) $-|-2|$

(e) $-(-7)$

(f) $|-6| + |-3|$

(g) $|-9| - |-4|$

(h) $-|9 - 4|$

E X A M P L E 5   Evaluating Absolute Value Expressions

Find the value of each expression.

(a) $|13| = 13$

(b) $|-2| = -(-2) = 2$

(c) $|0| = 0$

(d) $-|8|$

Evaluate the absolute value first. Then find the additive inverse.
$$-|8| = -(8) = -8$$

(e) $-|-8|$

Work as in part (d): $|-8| = 8$, so
$$-|-8| = -(8) = -8.$$

Notice that $-|-8|$ is not the same as $-(-8) = 8$.

(f) $|-2| + |5|$

Evaluate each absolute value first, then add.
$$|-2| + |5| = 2 + 5 = 7$$

(g) $-|5 - 2| = -|3| = -3$

◀◀ WORK PROBLEM 7 AT THE SIDE.

1.1 Exercises

Decide whether each statement concerning real numbers is true or false. If false, tell why.

1. Division of a number by zero is undefined.

2. If zero is divided by a nonzero number, the result is zero.

3. Every number has an additive inverse.

4. The absolute value of any number must be positive.

5. There is a number that has a negative absolute value.

6. The absolute value of a negative number is its additive inverse.

Graph the elements of each set on a number line.

7. $\{-3, -1, 0, 4, 6\}$

8. $\{-4, -2, 0, 3, 5\}$

9. $\left\{-\dfrac{2}{3}, 0, \dfrac{4}{5}, \dfrac{12}{5}, \dfrac{9}{2}, 4.8\right\}$

10. $\left\{-\dfrac{6}{5}, -\dfrac{1}{4}, 0, \dfrac{5}{6}, \dfrac{13}{4}, 5.2, \dfrac{11}{2}\right\}$

11. Explain the difference between the graph of a number and the coordinate of a point.

12. Give two examples of negative irrational numbers.

Write each set by listing its elements. See Example 1.

13. $\{x \mid x \text{ is a natural number less than } 6\}$

14. $\{m \mid m \text{ is a natural number less than } 9\}$

15. $\{z \mid z \text{ is an integer greater than } 4\}$

16. $\{y \mid y \text{ is an integer greater than } 8\}$

17. $\{a \mid a$ is an even integer greater than 8$\}$

18. $\{k \mid k$ is an odd integer less than 1$\}$

19. $\{x \mid x$ is an irrational number that is also rational$\}$

20. $\{r \mid r$ is a number that is both positive and negative$\}$

21. $\{p \mid p$ is a number whose absolute value is 4$\}$

22. $\{w \mid w$ is a number whose absolute value is 7$\}$

23. $\{z \mid z$ is a whole number and a multiple of 3$\}$

24. $\{n \mid n$ is a counting number and a multiple of 7$\}$

25. A student claimed that $\{x \mid x$ is a natural number greater than 3$\}$ and $\{y \mid y$ is a natural number greater than 3$\}$ actually name the same set, even though different variables are used. Was this student correct?

26. Give two different ways of describing the set $\{\ldots, -4, -3, -2\}$ using set-builder notation. (Use different wording, not just different variables.)

Write the set using set-builder notation. (More than one description is possible. We give one possible answer.) See Example 1(b).

27. $\{4, 8, 12, 16, \ldots\}$

28. $\{\ldots, -6, -3, 0, 3, 6, \ldots\}$

29. $\{2, 4, 6, 8\}$

30. $\{11, 12, 13, 14\}$

Which elements of the given set are (a) natural numbers, (b) whole numbers, (c) integers, (d) rational numbers, (e) irrational numbers, (f) real numbers, (g) undefined? See Example 2.

31. $\left\{-8, -\sqrt{5}, -.6, 0, \dfrac{1}{0}, \dfrac{3}{4}, \sqrt{3}, 4, 5, \dfrac{13}{2}, 17, \dfrac{40}{2}\right\}$

32. $\left\{-9, -\sqrt{6}, -.7, 0, \dfrac{2}{0}, \dfrac{6}{7}, \sqrt{7}, 3, 8, \dfrac{21}{2}, 13, \dfrac{75}{5}\right\}$

Give (a) the additive inverse and (b) the absolute value of each number. See Examples 4 and 5.

33. 6

34. 8

35. -12

36. -15

37. $5 - 5$

38. $9 - 9$

39. $\dfrac{6}{5}$

40. .13

Find the value of each expression. See Example 5.

41. $|-8|$ **42.** $|-11|$ **43.** $-|5|$ **44.** $-|17|$

45. $-|-2|$ **46.** $-|-8|$ **47.** $-|4.5|$ **48.** $-|12.6|$

49. $|-2| + |3|$ **50.** $|-16| + |12|$ **51.** $|-9| - |-3|$ **52.** $|-10| - |-5|$

53. $|-9| + |-13|$ **54.** $|-13| + |-21|$

55. $|-1| + |-2| - |-3|$ **56.** $|-6| + |-4| - |-10|$

Decide whether each statement is true or false. If false, tell why. See Example 3.

57. Every rational number is an integer.

58. Every natural number is an integer.

59. Every irrational number is an integer.

60. Every integer is a rational number.

61. Every whole number is a real number.

62. Every natural number is a whole number.

63. Some rational numbers are irrational.

64. Some natural numbers are whole numbers.

65. Some rational numbers are integers.

66. Some real numbers are integers.

67. The absolute value of any number is the same as the absolute value of its additive inverse.

68. The absolute value of any nonzero number is positive.

Sea level refers to the surface of the ocean. The depth of a body of water such as an ocean or sea can be expressed as a negative number, representing average depth in feet below sea level. On the other hand, the altitude of a mountain can be expressed as a positive number, indicating its height in feet above sea level. The following chart gives selected depths and heights.

Bodies of Water	Average Depth in Feet (as a negative number)	Mountain	Altitude in Feet (as a positive number)
Pacific Ocean	−12,925	McKinley	20,320
South China Sea	−4802	Point Success	14,150
Gulf of California	−2375	Matlalcueyetl	14,636
Caribbean Sea	−8448	Ranier	14,410
Indian Ocean	−12,598	Steele	16,644

69. List the bodies of water in order, starting with the deepest and ending with the shallowest.

70. List the mountains in order, starting with the shortest and ending with the tallest.

71. True or false: The absolute value of the depth of the Pacific Ocean is greater than the absolute value of the depth of the Indian Ocean.

72. True or false: The absolute value of the depth of the Gulf of California is greater than the absolute value of the depth of the Caribbean Sea.

Determine the value or values of x for which each statement is true.

73. $|x| = |-x|$ **74.** $x = |x|$ **75.** $|-x| = x$ **76.** $|x| = |2x|$

1.2 Inequality

The statement $4 + 2 = 6$ is an **equation** (ee-KWAY-zhun); it states that two quantities are equal. The statement $4 \neq 6$ (read "4 is not equal to 6") is an **inequality** (in-ee-KWAHL-it-ee), a statement that two expressions are *not* equal.

OBJECTIVE 1 When two numbers are not equal, one must be less than the other. The symbol $<$ means "is less than." For example,

$$8 < 9, \quad -6 < 15, \quad \text{and} \quad 0 < \frac{4}{3}.$$

"Is greater than" is written with the symbol $>$. For example,

$$12 > 5, \quad 9 > -2, \quad \text{and} \quad \frac{6}{5} > 0.$$

Note
Notice that in either case the symbol "points" toward the smaller number.

The number line in Figure 5 shows the numbers 4 and 9, and we know that $4 < 9$. On the graph, 4 is to the left of 9. The smaller of two numbers is always to the left of the other on a number line.

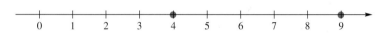

FIGURE 5

The geometric definitions of $<$ and $>$ are as follows.

> $a < b$ if a is to the left of b on the number line.
> $b > a$ if b is to the right of a on the number line.

E X A M P L E I Using a Number Line to Determine Order

(a) As the number line in Figure 6 shows,

$$-6 < 1.$$

Also, $1 > -6$.

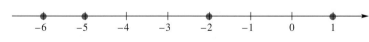

FIGURE 6

(b) From the same number line, $-5 < -2$, or $-2 > -5$.

WORK PROBLEM I AT THE SIDE. ▶▶

The following table summarizes results about positive and negative numbers. The same statement is given both in words and in symbols.

OBJECTIVES

1 Use inequality symbols.

2 Graph sets of real numbers.

FOR EXTRA HELP

Tutorial Tape I SSM, Sec. 1.2

1. Insert $<$ or $>$ in each blank.

(a) 3 _____ 7

(b) 9 _____ 2

(c) -4 _____ -8

(d) -2 _____ -1

(e) 0 _____ -3

2. Answer *true* or *false*.

(a) $-2 \leq -3$

(b) $8 \leq 8$

(c) $-9 \geq -1$

(d) $5 \cdot 8 \leq 7 \cdot 7$

(e) $3(4) > 2(6)$

Words	Symbols
Every negative number is less than 0.	If a is negative, then $a < 0$.
Every positive number is greater than 0.	If a is positive, then $a > 0$.

0 is neither positive nor negative.

In addition to the symbols $<$ and $>$, other inequality symbols are often used. Here is a list of these symbols.

INEQUALITY SYMBOLS

Symbol	Meaning	Example
$\neq$	is not equal to	$3 \neq 7$
$<$	is less than	$-4 < -1$
$>$	is greater than	$3 > -2$
$\leq$	is less than or equal to	$6 \leq 6$
$\geq$	is greater than or equal to	$-8 \geq -10$

E X A M P L E 2   Interpreting Inequality Symbols

The following table shows several statements and the reason that each is true.

Statement	Reason
$6 \leq 8$	$6 < 8$
$-2 \leq -2$	$-2 = -2$
$-9 \geq -12$	$-9 > -12$
$-3 \geq -3$	$-3 = -3$
$6 \cdot 4 \leq 5(5)$	$24 < 25$

In the last line, the dot in $6 \cdot 4$ indicates the product 6×4, or 24. Also, $5(5)$ means 5×5, or 25. The statement $6 \cdot 4 \leq 5(5)$ becomes $24 \leq 25$, which is true.

◀◀ **WORK PROBLEM 2 AT THE SIDE.**

OBJECTIVE 2 Inequality symbols can be used to write sets of real numbers. For example, the set of all real numbers greater than -2 can be written as $\{x \mid x > -2\}$. To show the elements of this set on a number line, draw a line from -2 to the right. Place a parenthesis (per-ENTH-uh-sis) at -2 to show that -2 is not an element of the set. The result, shown in Figure 7, is called the **graph** of the set $\{x \mid x > -2\}$.

FIGURE 7

The set of numbers greater than -2 is an example of an **interval** (IN-ter-vul) on the number line. A simplified notation, called **interval notation,** is used for writing intervals. For example, using this notation, the interval of numbers greater than -2 is written as $(-2, \infty)$. The infinity symbol ∞ does not indicate a number; it is used to show that the interval includes all real numbers greater than -2. The left parenthesis indicates that -2 is not included. A parenthesis is always used next to the infinity symbol in interval notation.

3. Write in interval notation and graph.

(a) $\{x \mid x < -1\}$

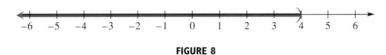

─ **E X A M P L E   3**   **Graphing an Inequality Written in Interval Notation**

Write $\{x \mid x < 4\}$ in interval notation and graph the interval.

In interval notation, $\{x \mid x < 4\}$ is written as $(-\infty, 4)$. The graph is shown in Figure 8. Since the elements of the set are all the numbers *less* than 4, the graph extends to the left.

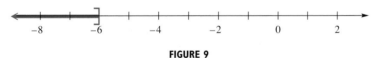

FIGURE 8

(b) $\{x \mid x > 0\}$

> **WORK PROBLEM 3 AT THE SIDE.** ▶▶

4. Write in interval notation and graph.

The elements of the set $\{x \mid x \le -6\}$ are all the real numbers less than or equal to -6. To show that -6 is part of the set, we use a square bracket (BRAK-et) on the graph at -6, as shown in Figure 9. The square bracket is also used in the interval notation $(-\infty, -6]$ to show that -6 is included in the set.

(a) $\{x \mid x \ge -3\}$

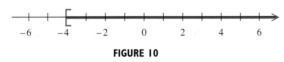

FIGURE 9

─ **E X A M P L E   4**   **Graphing an Inequality Written in Interval Notation**

Write $\{x \mid x \ge -4\}$ in interval notation and graph the interval.

This interval is written as $[-4, \infty)$. The square bracket indicates that -4 is included in the set. The graph is shown in Figure 10. A square bracket is also used on the graph at -4 to show that -4 is part of the set.

(b) $\{x \mid x \le 5\}$

FIGURE 10

> **WORK PROBLEM 4 AT THE SIDE.** ▶▶

Note
In a previous course you may have graphed $\{x \mid x > -2\}$ using an open circle instead of a parenthesis at -2. Also, you may have graphed $\{x \mid x \le -6\}$ using a solid dot instead of a bracket at -6. The interval notation we use in this book is preferred in more advanced courses.

Answers

3. (a) $(-\infty, -1)$

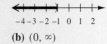

(b) $(0, \infty)$

4. (a) $[-3, \infty)$

(b) $(-\infty, 5]$

5. Write in interval
notation and graph.

(a) $\{x \mid -1 \le x \le 2\}$

The next example shows a set whose elements are *between* two numbers. For example, to indicate that x is between 3 and 6 we write $3 < x < 6$. The three-part inequality $3 < x < 6$ is equivalent (ee-KWIV-uh-lent) to the compound (KAHM-pound) statement

$$3 < x \quad \text{and} \quad x < 6.$$

The solution of a compound statement connected with *and* includes all numbers that satisfy *both* parts of the statement at the same time. In this case, it is all numbers between 3 and 6.

E X A M P L E 5   Graphing a Three-Part Inequality

Write in interval notation and graph $\{x \mid -2 < x < 4\}$.

The inequality $-2 < x < 4$ is read "x is greater than -2 and less than 4," or "x is between -2 and 4." The set $\{x \mid -2 < x < 4\}$ includes all real numbers between but not including -2 and 4. In interval notation, the set is written as $(-2, 4)$. The graph of this set goes from -2 to 4, with parentheses at -2 and 4. See Figure 11.

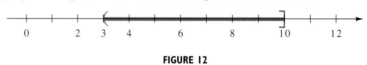

FIGURE 11

(b) $\{x \mid 8 < x < 12\}$

E X A M P L E 6   Graphing a Three-Part Inequality

Write in interval notation and graph $\{x \mid 3 < x \le 10\}$.

In interval notation the set is written as $(3, 10]$. As shown in Figure 12, the graph has a parenthesis at 3 and a square bracket at 10.

FIGURE 12

(c) $\{x \mid -4 \le x < 2\}$

◀◀ **WORK PROBLEM 5 AT THE SIDE.**

ANSWERS

5. (a) $[-1, 2]$

(b) $(8, 12)$

(c) $[-4, 2)$

1.2 Exercises

Use the number line shown to answer true or false to each statement. See Examples 1 and 2.

$$-6 \;-5\; -4\; -3 \;-2 \;-1 \quad 0 \quad 1 \quad 2 \quad 3 \quad 4 \quad 5 \quad 6$$

1. $-6 < -2$

2. $-4 < -3$

3. $-4 > -3$

4. $-2 > -1$

5. $3 > -2$

6. $5 > -3$

7. $-3 \geq -3$

8. $-4 \leq -4$

9. An inequality of the form "$a < b$" may also be written "$b > a$." Write $-3 < 2$ using this alternate form, and explain why both inequalities are true.

10. If $x > 0$ is a false statement, then is $x < 0$ necessarily a true statement? If not, explain why.

11. Describe the difference between the meaning of the symbols $<$ and $\leq$ as used in the expressions $x < 5$ and $x \leq 5$.

12. Describe the difference between the parenthesis and the bracket in interval notation.

Use inequality symbols to write each of the following statements.

13. 7 is greater than y.

14. -4 is less than 12.

15. $3t - 4$ is less than or equal to 10.

16. $5x + 4$ is greater than or equal to 19.

17. 5 is greater than or equal to 5.

18. -3 is less than or equal to -3.

19. t is between -3 and 5.

20. r is between -4 and 12.

21. $3x$ is between -3 and 4, including -3 and excluding 4.

22. $5y$ is between -2 and 6, excluding -2 and including 6.

23. $5x + 3$ is not equal to 0.

24. $6x + 7$ is not equal to -3.

Using your knowledge of arithmetic, first simplify on each side of the inequality. Then tell whether the resulting statement is true or false.

25. $-6 < 7 + 3$

26. $-7 < 4 + 2$

27. $2 \cdot 5 \geq 4 + 6$

28. $8 + 7 \leq 3 \cdot 5$

29. $-|-3| \geq -3$

30. $-|-5| \leq -5$

The slash symbol, /, is used to signify the negation of the meaning of a symbol. We know that if a = b is true, then a ≠ b is false, for example. The slash symbol is also used to negate inequality: "a $\not< b$" is read "a is not less than b" and "a $\not> b$" is read "a is not greater than b." The symbol a $\not< b$ is equivalent to a ≥ b, and a $\not> b$ is equivalent to a ≤ b. In Exercises 31–36, write an equivalent statement based on the explanation above.

31. $3 \not< 2$ **32.** $4 \not> 5$ **33.** $-3 \not> -3$ **34.** $-6 \not< -6$ **35.** $5 \geq 3$ **36.** $6 \leq 7$

Write in interval notation and graph each of the following sets of real numbers. See Examples 3–6.

37. $\{x \mid x > -2\}$

38. $\{x \mid x < 5\}$

39. $\{x \mid x \leq 6\}$

40. $\{x \mid x \geq -3\}$

41. $\{x \mid 0 < x < 3.5\}$

42. $\{x \mid -4 < x < 6.1\}$

43. $\{x \mid 2 \leq x \leq 7\}$

44. $\{x \mid -3 \leq x \leq -2\}$

45. $\{x \mid -4 < x \leq 3\}$

46. $\{x \mid 3 \leq x < 6\}$

The graph shows the number of residential building permits in Oregon and Utah for the period from 1986 through 1995.

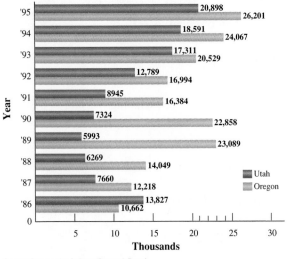

RESIDENTIAL BUILDING PERMITS ISSUED IN UTAH AND OREGON: 1986–1995

'95   20,898   26,201
'94   18,591   24,067
'93   17,311   20,529
'92   12,789   16,994
'91   8945   16,384
'90   7324   22,858
'89   5993   23,089
'88   6269   14,049
'87   7660   12,218
'86   13,827   10,662

■ Utah
■ Oregon

Year / Thousands

Source: Construction Industry Research Board

47. During what years were the number of Oregon permits greater than 20,000?

48. During what years were the number of Utah permits less than 8000?

49. In 1995, were the number of Oregon permits greater than or less than the number of Utah permits?

50. If x represents the number of Utah permits in 1991, and y represents the number in 1986, which one of the following is true: $x > y$ or $x < y$?

1.3 *Operations on Real Numbers*

In this section we review the rules for the four operations with signed numbers: addition, subtraction, multiplication, and division.

OBJECTIVE 1 ▶ To find the **sum** of two real numbers, we add them using the following rules.

> To add two numbers with the *same* sign, first add their absolute values. The sign of the answer (either $+$ or $-$) is the same as the sign of the two numbers.
> To add two numbers with *different* signs, first subtract their absolute values. The sign of the answer is the same as the sign of the number that has the larger absolute value.

E X A M P L E 1   Adding Two Negative Numbers

Find the following sums.

(a) $-12 + (-8)$

First find the absolute values.

$$|-12| = 12 \quad \text{and} \quad |-8| = 8$$

Because these numbers have the same sign, add their absolute values. Both numbers are negative, so the answer is negative.

$$-12 + (-8) = -(12 + 8) = -(20) = -20$$

(b) $-6 + (-3) = -(6 + 3) = -9$

(c) $-1.2 + (-.4) = -(1.2 + .4) = -1.6$

(d) $-\dfrac{5}{6} + \left(-\dfrac{1}{3}\right) = -\left(\dfrac{5}{6} + \dfrac{1}{3}\right) = -\left(\dfrac{5}{6} + \dfrac{2}{6}\right) = -\dfrac{7}{6}$

WORK PROBLEM 1 AT THE SIDE. ▶▶

E X A M P L E 2   Adding Numbers with Different Signs

Add -17 and 11.

To find $-17 + 11$, first find the absolute value of each number:

$$|-17| = 17 \quad \text{and} \quad |11| = 11.$$

Because -17 and 11 have *different* signs, subtract these absolute values:

$$17 - 11 = 6.$$

The number -17 has a larger absolute value than 11 has, so the answer should be negative.

$$-17 + 11 = \underset{\substack{\uparrow \\ \text{Negative because } |-17| > |11|}}{-6}$$

E X A M P L E 3   Adding Numbers with Different Signs

Find the following sums.

(a) $4 + (-1)$

The absolute values are 4 and 1. Because 4 is positive, the sum must be positive: $4 + (-1) = 4 - 1 = 3$

(b) $-9 + 17 = 17 - 9 = 8$

— **CONTINUED ON NEXT PAGE**

CONTINUED ON NEXT PAGE

OBJECTIVES

1 ▶ Add signed numbers.

2 ▶ Subtract signed numbers.

3 ▶ Multiply signed numbers.

4 ▶ Find the reciprocal of a number.

5 ▶ Divide signed numbers.

FOR EXTRA HELP

Tutorial   Tape 1   SSM, Sec. 1.3

1. Add.

(a) $-2 + (-7)$

(b) $-15 + (-6)$

(c) $-1.1 + (-1.2)$

(d) $-\dfrac{3}{4} + \left(-\dfrac{1}{2}\right)$

ANSWERS

1. (a) -9 **(b)** -21 **(c)** -2.3 **(d)** $-\dfrac{5}{4}$

2. Add.

(a) $12 + (-1)$

(b) $3 + (-7)$

(c) $-17 + 5$

(d) $-\dfrac{3}{4} + \dfrac{1}{2}$

(c) $-16 + 12$

The absolute values are 16 and 12. Subtract the absolute values. The negative number has the larger absolute value, so the answer is negative.

$$-16 + 12 = -(16 - 12) = -4$$

(d) $-\dfrac{4}{5} + \dfrac{2}{3}$

Write each number with a common denominator.

$$\frac{4}{5} = \frac{4 \cdot 3}{5 \cdot 3} = \frac{12}{15} \quad \text{and} \quad \frac{2}{3} = \frac{2 \cdot 5}{3 \cdot 5} = \frac{10}{15}$$

$$-\frac{4}{5} + \frac{2}{3} = -\frac{12}{15} + \frac{10}{15}$$

$$= -\left(\frac{12}{15} - \frac{10}{15}\right) \qquad -\tfrac{12}{15} \text{ has the larger absolute value.}$$

$$= -\frac{2}{15} \qquad \text{Subtract.}$$

◀◀ **WORK PROBLEM 2 AT THE SIDE.**

It is important to be able to add numbers accurately and quickly.

OBJECTIVE 2 ▸ The result of subtraction is called the **difference** (DIF-er-ents). Thus, the difference between 7 and 5 is 2. To see how subtraction should be defined, compare the next two statements.

$$7 - 5 = 2$$
$$7 + (-5) = 2$$

Also notice that $9 - 3 = 6$ and $9 + (-3) = 6$ so that $9 - 3 = 9 + (-3)$. These examples suggest the following rule for subtraction of signed numbers.

> For all real numbers a and b,
>
> $$a - b = a + (-b).$$
>
> (Change the sign of the second number and add.)

This method of observing patterns and similarities and generalizing from them is used often in mathematics. Looking at many examples strengthens our confidence in such generalizations; if possible, though, mathematicians prefer to prove the results using previously established facts.

E X A M P L E 4   Subtracting Signed Numbers

Find each difference.

Change to addition.
Change sign of second number.

(a) $6 - 8 = 6 + (-8) = -2$

Changed
Sign changed

(b) $-12 - 4 = -12 + (-4) = -16$

CONTINUED ON NEXT PAGE

ANSWERS

2. (a) 11 **(b)** -4 **(c)** -12 **(d)** $-\dfrac{1}{4}$

Sign changed

(c) $-10 - (-7) = -10 + (7) = -3$

(d) $-2.4 - (-8.1) = -2.4 + (8.1) = 5.7$

(e) $\dfrac{8}{3} - \left(-\dfrac{5}{3}\right) = \dfrac{8}{3} + \dfrac{5}{3} = \dfrac{13}{3}$

WORK PROBLEM 3 AT THE SIDE. ▶▶

When working a problem that involves both addition and subtraction, perform the additions and subtractions in order from left to right, as in the following example. Work inside the brackets or parentheses first.

E X A M P L E 5   Adding and Subtracting Signed Numbers

Perform the indicated operations.

(a) $-8 + 5 - 6 = (-8 + 5) - 6$
$$= -3 - 6$$
$$= -3 + (-6) = -9$$

(b) $15 - (-3) - 5 - 12 = (15 + 3) - 5 - 12$
$$= 18 - 5 - 12$$
$$= 13 - 12 = 1$$

(c) $-4 - (-6) + 7 - 1 = (-4 + 6) + 7 - 1$
$$= 2 + 7 - 1$$
$$= 9 - 1 = 8$$

(d) $-9 - [-8 - (-4)] + 6 = -9 - [-8 + 4] + 6$
$$= -9 - [-4] + 6$$
$$= -9 + 4 + 6$$
$$= -5 + 6 = 1$$

WORK PROBLEM 4 AT THE SIDE. ▶▶

OBJECTIVE 3 ▶ A **product** is the result of multiplying two or more numbers. For example, 24 is the product of 8 and 3. The rules for finding products of signed numbers are given below.

> The product of two numbers with the *same* sign is positive.
> The product of two numbers with *different* signs is negative.

E X A M P L E 6   Finding Products of Signed Numbers

Find each product.

(a) $(-3)(-9) = 27$

(b) $(-.5)(-.4) = .2$

(c) $\left(-\dfrac{3}{4}\right)\left(-\dfrac{5}{3}\right) = \dfrac{5}{4}$

── CONTINUED ON NEXT PAGE

3. Subtract.

(a) $9 - 12$

(b) $-7 - 2$

(c) $-8 - (-2)$

(d) $-6.3 - (-11.5)$

(e) $12 - (-5)$

4. Simplify.

(a) $-6 + 9 - 2$

(b) $12 - (-4) + 8$

(c) $-6 - (-2) - 8 - 1$

(d) $-3 - [(-7) + 15] + 6$

ANSWERS

3. (a) -3 (b) -9 (c) -6 (d) 5.2 (e) 17
4. (a) 1 (b) 24 (c) -13 (d) -5

5. Multiply.

(a) $(-7)(-5)$

(b) $(-.9)(-15)$

(c) $\left(-\dfrac{4}{7}\right)\left(-\dfrac{14}{3}\right)$

(d) $7(-2)$

(e) $(-.8)(.006)$

(f) $\dfrac{5}{8}(-16)$

(g) $\left(-\dfrac{2}{3}\right)(12)$

6. Give the reciprocal of each number.

(a) 15

(b) -7

(c) $\dfrac{8}{9}$

(d) .125

(e) $.0\overline{5}$

(d) $6(-9) = -54$

(e) $(-.05)(.3) = -.015$

(f) $\dfrac{2}{3}(-3) = -2$

(g) $-\dfrac{5}{8}\left(\dfrac{12}{13}\right) = -\dfrac{15}{26}$

◀◀ WORK PROBLEM 5 AT THE SIDE.

OBJECTIVE 4 Earlier, subtraction was defined in terms of addition. Now division is defined in terms of multiplication. This definition depends on the idea of a **multiplicative inverse** (muhl-tih-PLIK-uh-tiv IN-vers) or *reciprocal*; two numbers are *reciprocals* (ree-SIP-ruh-kuls) if they have a product of 1.

> The **reciprocal** of a nonzero number a is $\dfrac{1}{a}$.

Reciprocals (in decimal form) can be found with a calculator that has a key labeled $1/x$ or x^{-1}. For example, a calculator shows that the reciprocal of 25 is .04. (If you do not have a calculator, first write the number as a fraction.)

E X A M P L E   7   Finding the Reciprocal of a Number

The following chart gives several numbers and their reciprocals.

Number	Reciprocal
$-\dfrac{2}{5}$	$-\dfrac{5}{2}$
-6	$-\dfrac{1}{6}$
$\dfrac{7}{11}$	$\dfrac{11}{7}$
.05	20
0	None

There is no reciprocal for 0 because there is no number that can be multiplied by 0 to give a product of 1.

> **Caution**
> A number and its additive inverse have *opposite* signs; however, a number and its reciprocal always have the *same* sign.

◀◀ WORK PROBLEM 6 AT THE SIDE.

OBJECTIVE 5 The result of dividing one number by another is called the **quotient** (KWO-shunt). For example, when 45 is divided by 3, the quotient is 15. To see how to define division of signed numbers, we first write 15 as $\frac{45}{3}$, the quotient of 45 and 3. The same answer will be obtained if 45 and $\frac{1}{3}$ are multiplied, as follows.

$$\frac{45}{3} = 45 \cdot \frac{1}{3} = 15$$

Looking at many other examples of this type suggests the following definition of division of signed numbers.

For all real numbers a and b (where $b \neq 0$),

$$a \div b = \frac{a}{b} = a \cdot \frac{1}{b}.$$

(Multiply the first number by the reciprocal of the second number.)

Note
There is no reciprocal for the number 0, so division by 0 is not defined. (If a problem in this book has a zero denominator, the answer will say "undefined.")

Since division is defined as multiplication by the reciprocal, the rules for quotients resemble those for products.

The quotient of two nonzero real numbers with the same sign is positive.
The quotient of two nonzero real numbers with different signs is negative.

E X A M P L E 8   **Finding the Quotient of Two Signed Numbers**

Find each quotient.

(a) $\dfrac{-12}{4} = -12 \cdot \dfrac{1}{4} = -3$

(b) $\dfrac{6}{-3} = 6\left(-\dfrac{1}{3}\right) = -2$

(c) $\dfrac{-30}{-2} = -30\left(-\dfrac{1}{2}\right) = 15$

(d) $\dfrac{\frac{2}{3}}{\frac{5}{9}} = \dfrac{2}{3} \cdot \dfrac{1}{\frac{5}{9}} = \dfrac{2}{3} \cdot \dfrac{9}{5} = \dfrac{6}{5}$

Recall that the product of a number and its reciprocal is 1. The reciprocal of $\frac{5}{9}$, written

$$\frac{1}{\frac{5}{9}} \quad \text{or} \quad 1 \div \frac{5}{9}$$

is equal to $\frac{9}{5}$ since

$$\frac{5}{9} \cdot \frac{9}{5} = 1.$$

This example shows the origin of the usual rule for dividing by a fraction: "invert the divisor fraction and multiply."

WORK PROBLEM 7 AT THE SIDE. ▶▶

7. Divide.

(a) $\dfrac{-16}{4}$

(b) $\dfrac{8}{-2}$

(c) $\dfrac{-.15}{-.3}$

(d) $\dfrac{\frac{3}{8}}{\frac{11}{16}}$

ANSWERS

7. (a) -4 **(b)** -4 **(c)** $.5$ **(d)** $\dfrac{6}{11}$

The rules for multiplication and division suggest the results given below.

The fractions $\dfrac{-x}{y}$, $-\dfrac{x}{y}$, and $\dfrac{x}{-y}$ are equal.

Also, the fractions $\dfrac{x}{y}$ and $\dfrac{-x}{-y}$ are equal. (Assume $y \neq 0$.)

The forms $\frac{x}{-y}$ and $\frac{-x}{-y}$ are not used very often.

Every fraction has three signs: the sign of the numerator, the sign of the denominator, and the sign of the fraction itself. As shown above, changing any two of these three signs does not change the value of the fraction. (Changing only one sign, or changing all three, does change the value.)

1.3 Exercises

Complete each statement and give an example.

1. The sum of a positive number and a negative number is zero if _____ .

2. The sum of two positive numbers is a _____ number.

3. The sum of two negative numbers is a _____ number.

4. The sum of a negative number and a positive number is negative if _____ .

5. The sum of a positive number and a negative number is positive if _____ .

6. The difference between two positive numbers is negative if _____ .

7. The difference between two negative numbers is negative if _____ .

8. The product of two numbers with like signs is _____ .

9. The product of two numbers with unlike signs is _____ .

10. The quotient formed by any nonzero number divided by zero is _____ , and the quotient formed by zero divided by any nonzero number is _____ .

Add or subtract as indicated. See Examples 1–4.

11. $13 + (-4)$

12. $19 + (-13)$

13. $-6 + (-13)$

14. $-8 + (-15)$

15. $-\dfrac{7}{3} + \dfrac{3}{4}$

16. $-\dfrac{5}{6} + \dfrac{3}{8}$

17. $-2.3 + .45$

18. $-.238 + 4.55$

19. $-6 - 5$ **20.** $-8 - 13$ **21.** $8 - (-13)$ **22.** $13 - (-22)$

23. $-16 - (-3)$ **24.** $-21 - (-8)$ **25.** $-12.31 - (-2.13)$ **26.** $-15.88 - (-9.22)$

27. $\dfrac{9}{10} - \left(-\dfrac{4}{3}\right)$ **28.** $\dfrac{3}{14} - \left(-\dfrac{1}{4}\right)$ **29.** $-2 - |-4|$ **30.** $9 - |-13|$

31. Give an example of a difference between two negative numbers that is equal to 5.

32. Give an example of a sum of a positive number and a negative number that is equal to 4.

33. A common statement is that "Two negatives give a positive." When is this true? When is it false? Make a more precise statement that conveys this message.

34. Explain why the reciprocal of a nonzero number must have the same sign as the number.

Multiply. See Example 6.

35. $5(-7)$ **36.** $6(-6)$ **37.** $-8(-5)$ **38.** $-10(-4)$

39. $-10\left(-\dfrac{1}{5}\right)$ **40.** $-\dfrac{1}{2}(-12)$ **41.** $\dfrac{3}{4}(-16)$ **42.** $\dfrac{4}{5}(-35)$

43. $-\dfrac{5}{2}\left(-\dfrac{12}{25}\right)$ **44.** $-\dfrac{9}{7}\left(-\dfrac{35}{36}\right)$ **45.** $-\dfrac{3}{8}\left(-\dfrac{24}{9}\right)$ **46.** $-\dfrac{2}{11}\left(-\dfrac{99}{4}\right)$

47. $-2.4(-2.45)$ **48.** $-3.45(-2.14)$ **49.** $3.4(-3.14)$ **50.** $5.66(-2.1)$

Give the reciprocal of each number. See Example 7.

51. 6   **52.** 8   **53.** -7 **54.** -11

55. $-\dfrac{2}{3}$ **56.** $-\dfrac{7}{8}$ **57.** $.02$ **58.** $.45$

59. $-.001$ **60.** $-.0003$ **61.** $.08\overline{3}$ **62.** $.41\overline{6}$

Divide where possible. See Example 8.

63. $\dfrac{-14}{2}$ **64.** $\dfrac{-26}{13}$ **65.** $\dfrac{-24}{-4}$ **66.** $\dfrac{-36}{-9}$

67. $\dfrac{100}{-25}$ **68.** $\dfrac{300}{-60}$ **69.** $\dfrac{5}{0}$ **70.** $\dfrac{12}{0}$

71. $-\dfrac{10}{17} \div \left(-\dfrac{12}{5}\right)$ **72.** $-\dfrac{22}{23} \div \left(-\dfrac{33}{4}\right)$ **73.** $\dfrac{\dfrac{12}{13}}{-\dfrac{4}{3}}$ **74.** $\dfrac{\dfrac{5}{6}}{-\dfrac{1}{30}}$

75. $-\dfrac{27.72}{13.2}$ **76.** $\dfrac{-126.7}{36.2}$ **77.** $\dfrac{-100}{-.01}$ **78.** $\dfrac{-50}{-.05}$

Perform the indicated operations. Work inside parentheses or brackets first. Be sure to remember to perform the additions and subtractions in order, from left to right. See Example 5.

79. $-7 + 5 - 9$

80. $-12 + 13 - 19$

81. $6 - (-2) + 8$

82. $7 - (-3) + 12$

83. $-9 - 4 - (-3) + 6$

84. $-10 - 5 - (-12) + 8$

85. $-4 - [(-4 - 6) + 12] - 13$

86. $-10 - [(-2 + 3) - 4] - 17$

87. The highest temperature ever recorded in Juneau, Alaska, was 90° Fahrenheit. The lowest temperature ever recorded there was −22° Fahrenheit. What is the difference between these two temperatures?

88. On August 10, 1936, a temperature of 120° Fahrenheit was recorded in Arkansas. On February 13, 1905, Arkansas recorded a temperature of −29° Fahrenheit. What was the difference between these two temperatures?

The graph shows the percent change in nonagricultural employment from the previous year for the United States and California. As shown by the graph, California lagged behind the U.S. slow down and recovery. Use the graph to find the difference between the percent change for the U.S. and California in each year.

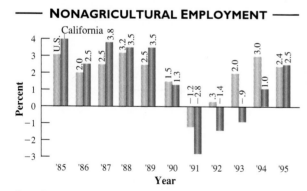

NONAGRICULTURAL EMPLOYMENT

Source: State of California Center for Continuing Study of the California Economy

89. 1986

90. 1991

91. 1993

92. 1994

1.4 Exponents and Roots; Order of Operations

Two or more integers whose product is a third number are **factors** of that third number. For example, 2 and 6 are factors of 12 because $2 \cdot 6 = 12$. Other factors of 12 are 1, 3, 4, 12, -1, -2, -3, -4, -6, and -12.

OBJECTIVE 1▶ In algebra, we use *exponents* as a way of writing products of repeated factors. For example, the product $2 \cdot 2 \cdot 2 \cdot 2 \cdot 2$ is written

$$2 \cdot 2 \cdot 2 \cdot 2 \cdot 2 = 2^5.$$

The number 5 shows that 2 appears as a factor 5 times. The number 5 is the **exponent** (EX-poh-nent), and 2 is the **base.**

$2^5 \longleftarrow$ Exponent
$\qquad \llcorner$ Base

Multiplying out the five 2s gives

$$2^5 = 2 \cdot 2 \cdot 2 \cdot 2 \cdot 2 = 32.$$

If a is a real number and n is a natural number,

$$a^n = \underbrace{a \cdot a \cdot a \cdots a}_{n \text{ factors of } a}$$

where n is the **exponent,** a is the **base,** and a^n is an **exponential** (EX-poh-NEN-shul) or a **power.**

E X A M P L E 1   Evaluating an Exponential

Write each number without exponents.

(a) $5^2 = 5 \cdot 5 = 25$

   Read 5^2 as "5 squared."

(b) $\left(\dfrac{2}{3}\right)^3 = \dfrac{2}{3} \cdot \dfrac{2}{3} \cdot \dfrac{2}{3} = \dfrac{8}{27}$

   Read $\left(\dfrac{2}{3}\right)^3$ as "$\dfrac{2}{3}$ cubed."

(c) $2^6 = 2 \cdot 2 \cdot 2 \cdot 2 \cdot 2 \cdot 2 = 64$

   Read 2^6 as "2 to the sixth."

(d) $(-2)^4 = (-2)(-2)(-2)(-2) = 16$

(e) $(-3)^5 = (-3)(-3)(-3)(-3)(-3) = -243$

Parts (d) and (e) of Example 1 suggest the following generalization.

The product of an *even* number of negative factors is positive; the product of an *odd* number of negative factors is negative.

OBJECTIVES

1▶ Use exponents.

2▶ Identify exponents and bases.

3▶ Find square roots and higher roots.

4▶ Use the order of operations.

5▶ Substitute numbers for variables.

FOR EXTRA HELP

 Tutorial   Tape 1   SSM, Sec. 1.4

1. Write without exponents.

(a) 5^3

(b) 3^4

(c) $(-4)^5$

(d) $(-3)^4$

(e) $(.75)^3$

(f) $\left(\dfrac{2}{5}\right)^4$

2. Identify the exponent and the base in each exponential.

(a) 7^5

(b) m^3

(c) $(-5)^7$

(d) -12^4

(e) $-(.9)^4$

In parts (a) and (b) of Example 1, we used the terms "squared" and "cubed" to refer to powers of 2 and 3, respectively. The term "squared" comes from the figure of a square, which has the same measure for both length and width, as shown in Figure 13(a). Similarly, the term "cubed" comes from the figure of a cube. As shown in Figure 13(b), the length, width, and height of a cube have the same measure.

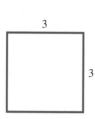

(a) $3 \cdot 3 = 3$ squared, or 3^2

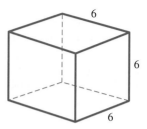

(b) $6 \cdot 6 \cdot 6 = 6$ cubed, or 6^3

FIGURE 13

◀◀ WORK PROBLEM I AT THE SIDE.

OBJECTIVE 2 ▶ The following example shows how to identify exponents and bases.

E X A M P L E 2   Identifying Exponents and Bases

Identify the base and the exponent.

(a) 3^6

The base is 3, and the exponent is 6.

(b) 5^4

The base is 5, and the exponent is 4.

(c) $(-2)^6$

The exponent of 6 refers to the number -2, so the base is -2. Evaluating $(-2)^6$ gives

$$(-2)^6 = (-2)(-2)(-2)(-2)(-2)(-2) = 64.$$

(d) -2^6

Here the lack of parentheses shows that the exponent 6 refers *only* to the number 2, and not to -2, so the base is 2. Evaluate -2^6 as follows:

$$-2^6 = -(2 \cdot 2 \cdot 2 \cdot 2 \cdot 2 \cdot 2) = -64.$$

Caution
As shown by parts (c) and (d) of Example 2, it is important to be careful to distinguish between $-a^n$ and $(-a)^n$.

◀◀ WORK PROBLEM 2 AT THE SIDE.

ANSWERS

1. (a) 125 **(b)** 81 **(c)** -1024
(d) 81 **(e)** .421875 **(f)** $\dfrac{16}{625}$

2. (a) 5; 7 **(b)** 3; m **(c)** 7; -5
(d) 4; 12 **(e)** 4; .9

OBJECTIVE 3 We know that $5^2 = 5 \cdot 5 = 25$, so that 5 squared is 25. The opposite of squaring a number is called taking its **square root.** For example, a square root of 25 is 5. Another square root of 25 is -5 because $(-5)^2 = 25$; thus, 25 has two square roots, 5 and -5.

We write the positive square root of a number with the symbol $\sqrt{}$. For example, the positive square root of 25 is written $\sqrt{25} = 5$. The negative square root of 25 is written $-\sqrt{25} = -5$. Because the square of any nonzero real number is positive, the square root of a negative number, such as $\sqrt{-4}$, is not a real number.

E X A M P L E 3   Finding Square Roots

Find each square root that is a real number.

(a) $\sqrt{36} = 6$ since 6 is positive and $6^2 = 36$.

(b) $\sqrt{144} = 12$ since $12^2 = 144$.

(c) $\sqrt{0} = 0$ since $0^2 = 0$.

(d) $-\sqrt{100} = -10$ since the negative sign is outside the square root radical.

(e) $\sqrt{-16}$ is not a real number because the negative sign is inside the square root radical.

(f) $\sqrt{\dfrac{9}{16}} = \dfrac{3}{4}$

(g) $\sqrt{.16} = .4$

Caution

The symbol $\sqrt{}$ is used only for the *positive* square root, except that $\sqrt{0} = 0$.

WORK PROBLEM 3 AT THE SIDE. ▶▶

Because 6 cubed is $6^3 = 6 \cdot 6 \cdot 6 = 216$, the **cube root** of 216 is 6. We write this as

$$\sqrt[3]{216} = 6.$$

In the same way, the **fourth root** of 81 is 3, written

$$\sqrt[4]{81} = 3.$$

The number -3 is also a fourth root of 81, but the symbol $\sqrt[4]{}$ is reserved for roots that are not negative. We discuss negative roots in Chapter 5.

E X A M P L E 4   Finding Higher Roots

Find each root that is a real number.

(a) $\sqrt[3]{125} = 5$ since $5^3 = 125$.

(b) $\sqrt[3]{\dfrac{8}{27}} = \dfrac{2}{3}$ because $\left(\dfrac{2}{3}\right)^3 = \dfrac{8}{27}$.

(c) $\sqrt[3]{-8} = -2$ because $(-2)^3 = -8$.

(d) $\sqrt[4]{16} = 2$ since $2^4 = 16$.

(e) $\sqrt[5]{32} = 2$ since $2^5 = 32$.

(f) $\sqrt[7]{128} = 2$ since $2^7 = 128$.

(g) $\sqrt[4]{-81}$ is not a real number, since there is no real number whose fourth power is negative.

---

3. Find the following.

(a) $\sqrt{9}$

(b) $\sqrt{49}$

(c) $-\sqrt{81}$

(d) $\sqrt{\dfrac{121}{81}}$

(e) $\sqrt{.25}$

(f) $\sqrt{-9}$

(g) $-\sqrt{-169}$

ANSWERS

3. (a) 3 **(b)** 7 **(c)** -9 **(d)** $\dfrac{11}{9}$
(e) .5 **(f)** not a real number
(g) not a real number

4. Find each root.

(a) $\sqrt[3]{64}$

(b) $\sqrt[3]{-\dfrac{27}{8}}$

(c) $\sqrt[3]{0}$

(d) $\sqrt[4]{625}$

(e) $\sqrt[4]{\dfrac{16}{625}}$

(f) $\sqrt[4]{-16}$

5. Simplify $5 \cdot 9 + 2 \cdot 4$.

6. Simplify
$(4 + 2) - 3^2 - (8 - 3)$.

4. (a) 4 **(b)** $-\dfrac{3}{2}$ **(c)** 0 **(d)** 5 **(e)** $\dfrac{2}{5}$
 (f) not a real number
5. 53
6. -8

◄◄ WORK PROBLEM 4 AT THE SIDE.

We can summarize the conditions for roots to be real numbers as follows.

$\sqrt[n]{a}$, *n* even	$\sqrt[n]{a}$, *n* odd
a must be nonnegative.	If $a > 0$, then $\sqrt[n]{a} > 0$.
	If $a < 0$, then $\sqrt[n]{a} < 0$.

We can use a calculator to find roots. Most calculators have a square root key. Fourth roots also can be found with this key by taking the square root of a square root. To find other roots, some calculators have a $\boxed{\sqrt[x]{\ }}$ key. Any positive integer can be used for *x*. In a later chapter, we will discuss another way to find roots by using exponents.

OBJECTIVE 4 Given a problem such as $5 + 2 \cdot 3$, should 5 and 2 be added first, or should 2 and 3 be multiplied first? When a problem involves more than one operation symbol, use the following **order of operations** (OR-dur of ah-pur-AY-shuns).

Order of Operations

1. Work separately above and below any fraction bar.
2. Use the rules below within each set of parentheses or square brackets. Start with the innermost set and work outward.
3. Evaluate all powers, roots, and absolute values.
4. Do any multiplications or divisions in the order in which they occur, working from left to right.
5. Do any additions or subtractions in the order in which they occur, working from left to right.

E X A M P L E 5   Using Order of Operations

To simplify $5 + 2 \cdot 3$, first multiply and then add.
$$5 + \mathbf{2 \cdot 3} = 5 + \mathbf{6} \quad \text{Multiply.}$$
$$= 11 \quad \text{Add.}$$

◄◄ WORK PROBLEM 5 AT THE SIDE.

E X A M P L E 6   Using Order of Operations

Simplify $4 \cdot 3^2 + 7 - (2 + 8)$.
$$4 \cdot 3^2 + 7 - \mathbf{(2 + 8)}$$
$$= 4 \cdot 3^2 + 7 - \mathbf{10} \quad \text{Work inside parentheses.}$$
$$= 4 \cdot \mathbf{9} + 7 - 10 \quad \text{Evaluate powers.}$$
$$= \mathbf{36} + 7 - 10 \quad \text{Multiply.}$$
$$= \mathbf{43} - 10 \quad \text{Add.}$$
$$= 33 \quad \text{Subtract.}$$

◄◄ WORK PROBLEM 6 AT THE SIDE.

E X A M P L E 7   Using Order of Operations

Simplify $\frac{1}{2} \cdot 4 + (6 \div 3 - 7)$.

Work from left to right inside the parentheses, doing the division before the subtraction.

$$\frac{1}{2} \cdot 4 + (6 \div 3 - 7) = \frac{1}{2} \cdot 4 + (2 - 7)$$ Divide inside parentheses.

$$= \frac{1}{2} \cdot 4 + (-5)$$ Subtract inside parentheses.

$$= 2 + (-5)$$ Multiply.

$$= -3$$ Add.

WORK PROBLEM 7 AT THE SIDE. ▶▶

E X A M P L E 8   Using Order of Operations

Simplify $\dfrac{\frac{2}{3} \cdot 3 + 1}{\frac{1}{2} \cdot 8 - 6}$.

Go through the steps given above. Work above and below the fraction bar separately.

$$\frac{\frac{2}{3} \cdot 3 + 1}{\frac{1}{2} \cdot 8 - 6} = \frac{2 + 1}{4 - 6}$$ Multiply.

$$= \frac{3}{-2}$$ Add and subtract.

$$= -\frac{3}{2}$$

E X A M P L E 9   Using Order of Operations

Simplify $\dfrac{5 + 2^4}{6\sqrt{9} - 9 \cdot 2}$.

$$\frac{5 + 2^4}{6\sqrt{9} - 9 \cdot 2} = \frac{5 + 16}{6 \cdot 3 - 9 \cdot 2}$$ Evaluate powers and roots.

$$= \frac{5 + 16}{18 - 18}$$ Multiply.

$$= \frac{21}{0}$$ Add and subtract.

Because division by zero is undefined, the given expression is undefined.

WORK PROBLEM 8 AT THE SIDE. ▶▶

OBJECTIVE 5 ▶ The final example in this section shows how to evaluate expressions for given values of the variables.

7. Simplify $6 + \dfrac{2}{3}(-9) - \dfrac{5}{8} \cdot 16$.

8. Simplify $\dfrac{\frac{1}{2} \cdot 10 - 6 + \sqrt{9}}{\frac{5}{6} \cdot 12 - 3(2)^2}$.

9. Evaluate if $x = -12$, $y = 64$, and $z = -3$.

(a) $5x - 2 \cdot \sqrt{y}$

(b) $-6(x - \sqrt[3]{y})$

(c) $\dfrac{5x - 3 \cdot \sqrt{y}}{x - 1}$

(d) $x^2 + 2z^3$

E X A M P L E 10   Evaluating Expressions for Given Values of the Variables

Evaluate (ee-VAL-yoo-ate) the following expressions if $m = -4$, $n = 5$, $p = -6$, and $q = \sqrt{25}$.

(a) $5m - 9n$

Replace m with -4 and n with 5.

$$5m - 9n = 5(-4) - 9(5) = -20 - 45 = -65$$

(b) $-2p + 7n = -2(-6) + 7(5) = 47$

(c) $\dfrac{m + 2n}{4p} = \dfrac{-4 + 2(5)}{4(-6)} = \dfrac{-4 + 10}{-24} = \dfrac{6}{-24} = -\dfrac{1}{4}$

(d) $-3m^3 - n^2 q$

$$
\begin{aligned}
-3m^3 - n^2 q &= -3(-4)^3 - (5)^2(\sqrt{25}) & \text{Substitute.} \\
&= -3(-64) - 25(5) & \text{Evaluate powers and roots.} \\
&= 192 - 125 & \text{Multiply.} \\
&= 67 & \text{Subtract.}
\end{aligned}
$$

◀◀ **WORK PROBLEM 9 AT THE SIDE.**

ANSWERS

9. (a) -76 **(b)** 96 **(c)** $\dfrac{84}{13}$ **(d)** 90

1.4 Exercises

Decide whether each statement is true or false. If false, correct the statement so it is true.

1. $-4^6 = (-4)^6$

2. $-4^7 = (-4)^7$

3. $\sqrt{16}$ is a positive number.

4. $3 + 5 \cdot 6 = 3 + (5 \cdot 6)$

5. $(-2)^7$ is a negative number.

6. $(-2)^8$ is a positive number.

7. The product of 8 positive factors and 8 negative factors is positive.

8. The product of 3 positive factors and 3 negative factors is positive.

9. In the exponential -3^5, -3 is the base.

10. $\sqrt[3]{a}$ has the same sign as a for all nonzero real numbers a.

Evaluate. See Example 1.

11. 4^2

12. 2^4

13. $.28^3$

14. $.91^3$

15. $\left(\dfrac{1}{5}\right)^3$

16. $\left(\dfrac{1}{6}\right)^4$

17. $\left(\dfrac{7}{10}\right)^3$

18. $\left(\dfrac{4}{5}\right)^4$

19. $(-5)^3$

20. $(-3)^5$

21. $(-2)^8$

22. $(-3)^6$

23. -3^6

24. -4^6

25. -8^4

26. -10^3

27. Explain the conditions for which $(-a)^n$ is equal to $-a^n$.

28. Explain how the terms "squared" and "cubed" are interpreted geometrically.

Identify the exponent and the base. Do not evaluate. See Example 2.

29. $(-4.1)^7$

30. $(-3.4)^9$

31. -4.1^7

32. -3.4^9

Find the square roots. If the root does not represent a real number, say so. See Example 3.

33. $\sqrt{81}$

34. $\sqrt{64}$

35. $\sqrt{169}$

36. $\sqrt{225}$

37. $-\sqrt{400}$

38. $-\sqrt{900}$

39. $\sqrt{\dfrac{100}{121}}$

40. $\sqrt{\dfrac{225}{169}}$

41. $-\sqrt{.49}$

42. $-\sqrt{.64}$

43. $\sqrt{-25}$

44. $\sqrt{-121}$

45. Why is it incorrect to say that $\sqrt{16}$ is equal to 4 or -4?

46. Explain why $\sqrt[3]{-1000}$ is equal to $-\sqrt[3]{1000}$.

47. If a is a positive number, is $-\sqrt{-a}$ positive, negative, or not a real number?

48. If a is a positive number, is $-\sqrt[3]{-a}$ positive, negative, or not a real number?

Find the roots. See Example 4.

49. $\sqrt[3]{-27}$ **50.** $\sqrt[3]{343}$ **51.** $\sqrt[4]{625}$ **52.** $\sqrt[4]{-10,000}$

53. $\sqrt[5]{243}$ **54.** $\sqrt[5]{-1024}$ **55.** $-\sqrt[6]{729}$ **56.** $-\sqrt[6]{4096}$

Simplify, using the order of operations given in the text. See Examples 5–9.

57. $2[-5 - (-7)]$ **58.** $3[-8 - (-2)]$ **59.** $-12\left(-\dfrac{3}{4}\right) - (-5)$

60. $-7\left(-\dfrac{2}{14}\right) - (-8)$ **61.** $(-3)(5)^2 - (-2)(-8)$ **62.** $(-9)(2)^2 - (-3)(-2)$

63. $(-7)(\sqrt{36}) - (-2)(-3)$ **64.** $(-8)(\sqrt{64}) - (-3)(-7)$

65. $\dfrac{-8 + (-16)}{-3}$ **66.** $\dfrac{-9 + (-11)}{-2}$

67. $\dfrac{(-6 + 3)(-2^2)}{-5 - 1}$ **68.** $\dfrac{(-9 + 4)(-3^2)}{-4 - 1}$

69. $\dfrac{2(-5) + (-3)(-2)}{-6 + 5 + 1}$ **70.** $\dfrac{3(-4) + (-5)(-8)}{8 - 2 - 6}$

71. $\dfrac{\dfrac{1}{4} \cdot 16 + 3}{\dfrac{1}{2} \cdot 12 - 1}$ **72.** $\dfrac{\dfrac{2}{3} \cdot 21 + 7}{\dfrac{1}{2} \cdot 14 - 3}$

73. $-7\left[6 - \dfrac{5}{8}(24) + 3\left(\dfrac{8}{3}\right)\right]$

74. $5\left[1 + \dfrac{3}{4}(-12) - 8 \cdot \dfrac{3}{2}\right]$

Evaluate if a = −3, b = 64, and c = 6. See Example 10.

75. $3a + \sqrt{b}$

76. $-2a - \sqrt{b}$

77. $\sqrt[3]{b} + c - a$

78. $\sqrt[3]{b} - c + a$

79. $4a^3 + 2c$

80. $-3a^4 - 3c$

81. $\dfrac{2c + a^3}{4b + 6a}$

82. $\dfrac{3c + a^2}{2b - 6c}$

INTERPRETING TECHNOLOGY (EXERCISES 83–86)

Graphing calculators are the latest development in the evolution of scientific calculators. Beginning here, we will occasionally include sample screens of graphing calculators for students to interpret.

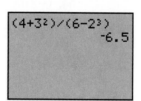

The screens above show how a graphing calculator calculates exponential and algebraic expressions. Note the use of $^\wedge$ for exponentiation. Also, in the second screen, note the careful use of parentheses. Verify the calculations shown. Then give the results of the calculations shown in Exercises 83–86. (In Exercises 85 and 86, give the fraction form.)

83.
```
(-9-5)(-2³+5)
```

84.
```
(-7-3)(-2^4+9)
```

85.
```
(4(-3)-2³)/(5-4(
3))
```

86.
```
(7(-2)-(4²)(-6))
/(12-10(2))
```

1.5 *Properties of Real Numbers*

The study of any object is simplified when we know the properties of the object. A property of water is that it boils when heated to 100°C. Knowing this helps us to predict the behavior of water. The study of numbers is no different. The basic properties of addition and multiplication of real numbers studied in this section will be used in later work in algebra. The properties are results that have been observed to occur consistently in work with numbers, so they have been generalized to apply to expressions with variables as well.

OBJECTIVE 1 The properties we discuss in this section are used in simplifying algebraic expressions. For example,

$$2(3 + 5) = 2 \cdot 8 = 16$$

and

$$2 \cdot 3 + 2 \cdot 5 = 6 + 10 = 16$$

so that

$$2(3 + 5) = 2 \cdot 3 + 2 \cdot 5.$$

This idea is illustrated by the divided rectangle in Figure 14. Similarly,

$$-4[5 + (-3)] = -4(2) = -8$$

and

$$-4(5) + (-4)(-3) = -20 + 12 = -8$$

so

$$-4[5 + (-3)] = -4(5) + (-4)(-3).$$

Area of left part is $2 \cdot 3 = 6$.
Area of right part is $2 \cdot 5 = 10$.
Area of total rectangle is $2(3 + 5) = 16$.

FIGURE 14

These arithmetic examples are generalized to *all* real numbers as the **distributive** (dis-TRIB-yoo-tiv) **property.**

Distributive Property

For any real numbers a, b, and c,

$$a(b + c) = ab + ac \qquad \text{and} \qquad (b + c)a = ba + ca.$$

The distributive property can also be written

$$ab + ac = a(b + c).$$

OBJECTIVES

1 Use the distributive property.

2 Use the inverse properties.

3 Use the identity properties.

4 Use the commutative and associative properties.

5 Use the multiplication property of zero.

FOR EXTRA HELP

Tutorial   Tape 2   SSM, Sec. 1.5

1. Use the distributive property to rewrite each expression.

(a) $8(m + n)$

(b) $-4(p - 5)$

(c) $3k + 6k$

(d) $-6m + 2m$

(e) $2r + 3s$

2. Use the distributive property to calculate each expression.

(a) $14 \cdot 5 + 14 \cdot 85$

(b) $78 \cdot 33 + 22 \cdot 33$

E X A M P L E 1   Using the Distributive Property

Use the distributive property to rewrite each expression.

(a) $3(x + y)$

In the statement of the property, let $a = 3$, $b = x$, and $c = y$. Then
$$3(x + y) = 3x + 3y.$$

(b) $-2(5 + k) = -2(5) + (-2)(k)$
$$= -10 - 2k$$

(c) $4x + 8x$

Use the second form of the property.
$$4x + 8x = (4 + 8)x = 12x$$

(d) $3r - 7r = 3r + (-7)r$ Definition of subtraction
$$= [3 + (-7)]r$$
$$= -4r$$

(e) $5p + 7q$

Because there is no common number or variable here, we cannot use the distributive property to simplify the expression.

As illustrated in Example 1(d), the distributive property can also be used for subtraction, so that
$$a(b - c) = ab - ac.$$

◀◀ **WORK PROBLEM 1 AT THE SIDE.**

E X A M P L E 2   Using the Distributive Property for Calculation

The distributive property is used in arithmetic to mentally perform calculations such as $38 \cdot 17 + 38 \cdot 3$.
$$38 \cdot 17 + 38 \cdot 3 = 38(17 + 3) \quad \text{Distributive property}$$
$$= 38(20)$$
$$= 760$$

◀◀ **WORK PROBLEM 2 AT THE SIDE.**

OBJECTIVE 2 ▶ In Section 1.1 we saw that the additive inverse of a number a is $-a$ and that the sum of a number and its additive inverse is 0. For example, 3 and -3 are additive inverses, as are -8 and 8. The number 0 is its own additive inverse. In Section 1.3, we saw that two numbers with a product of 1 are reciprocals. As mentioned there, another name for a reciprocal is *multiplicative inverse*. This is similar to the idea of an additive inverse. Thus, 4 and $\frac{1}{4}$ are multiplicative inverses, and so are $-\frac{2}{3}$ and $-\frac{3}{2}$. (Note again that a pair of reciprocals has the same sign.) Again, we extend these properties of arithmetic, the **inverse** **properties** of addition and multiplication, to the real numbers of algebra.

ANSWERS

1. (a) $8m + 8n$ (b) $-4p + 20$
   (c) $9k$ (d) $-4m$
   (e) cannot be simplified
2. (a) 1260 (b) 3300

Inverse Properties

For any real number a, there is a single real number $-a$, such that

$$a + (-a) = 0 \quad \text{and} \quad -a + a = 0.$$

The inverse "undoes" addition with the result 0. For any nonzero real number a, there is a single real number $\frac{1}{a}$ such that

$$a \cdot \frac{1}{a} = 1 \quad \text{and} \quad \frac{1}{a} \cdot a = 1.$$

The inverse "undoes" multiplication with result 1.

WORK PROBLEM 3 AT THE SIDE. ▶▶

OBJECTIVE 3 The numbers 0 and 1 each have a special property. Zero is the only number that can be added to any number to get that number. That is, adding 0 leaves the identity of a number unchanged. For this reason, 0 is called the **identity** (eye-DENT-it-ee) **element for addition.** In a similar way, multiplying by 1 leaves the identity of any number unchanged, so 1 is the **identity element for multiplication.** The following **identity properties** summarize this discussion and extend these properties from arithmetic to algebra.

Identity Properties

For any real number a,

$$a + 0 = 0 + a = a.$$

Start with a number a; add 0. The answer is "identical" to a.
Also,

$$a \cdot 1 = 1 \cdot a = a.$$

Start with a number a; multiply by 1. The answer is "identical" to a.

The identity property for 1 is especially useful in simplifying algebraic expressions.

EXAMPLE 3 **Using the Identity Property $1 \cdot a = a$**

Simplify each expression.

(a) $12m + m$

$$
\begin{aligned}
12m + m &= 12m + 1m && \text{Identity property} \\
&= (12 + 1)m && \text{Distributive property} \\
&= 13m && \text{Add in parentheses.}
\end{aligned}
$$

(b)
$$
\begin{aligned}
y + y &= 1y + 1y && \text{Identity property} \\
&= (1 + 1)y && \text{Distributive property} \\
&= 2y && \text{Add in parentheses.}
\end{aligned}
$$

(c)
$$
\begin{aligned}
-(m - 5n) &= -1(m - 5n) && \text{Identity property} \\
&= -1 \cdot m + (-1)(-5n) && \text{Distributive property} \\
&= -m + 5n && \text{Multiply.}
\end{aligned}
$$

WORK PROBLEM 4 AT THE SIDE. ▶▶

3. Complete the following statements.

(a) $4 + \underline{\hspace{1cm}} = 0$

(b) $-7.1 + \underline{\hspace{1cm}} = 0$

(c) $-9 + 9 = \underline{\hspace{1cm}}$

(d) $5 \cdot \underline{\hspace{1cm}} = 1$

(e) $-\dfrac{3}{4} \cdot \underline{\hspace{1cm}} = 1$

(f) $7 \cdot \dfrac{1}{7} = \underline{\hspace{1cm}}$

4. Use the properties to rewrite each expression.

(a) $p - 3p$

(b) $r + r + r$

(c) $-(3 + 4p)$

(d) $-(k - 2)$

ANSWERS

3. (a) -4 (b) 7.1 (c) 0 (d) $\dfrac{1}{5}$

(e) $-\dfrac{4}{3}$ (f) 1

4. (a) $-2p$ (b) $3r$
(c) $-3 - 4p$ (d) $-k + 2$

Expressions such as $12m$ and $5n$ from Example 3 are examples of *terms*. A **term** is a number or the product of a number and one or more variables. Terms with exactly the same variables raised to exactly the same powers are called **like terms.** The number in the product is called the **numerical coefficient** (noo-MAIR-ih-kul koh-uh-FISH-ent) or just the **coefficient.** For example, in the term $5p$, the coefficient is 5.

OBJECTIVE 4 Simplifying expressions as in parts (a) and (b) of Example 3 is called **combining** (cuhm-BY-ning) **like terms.** Only like terms may be combined. To combine like terms in an expression such as

$$-2m + 5m + 3 - 6m + 8,$$

we need two more properties.

Commutative and Associative Properties

For any real numbers a, b, and c,

$$a + b = b + a$$
$$ab = ba$$

Commutative properties

Two terms or factors; reverse the order.

$$a + (b + c) = (a + b) + c$$
$$a(bc) = (ab)c$$

Associative properties

Three terms or factors; same order, parentheses shifted.

The commutative (kuh-MEW-tuh-tiv) properties are used to change the *order* of the terms or factors in an expression. Think of commuting from home to work and then from work to home. The associative (uh-SOH-shuh-tiv) properties are used to *regroup* the terms or factors of an expression. Remember, to *associate* is to be part of a group.

The commutative and associative properties, too, are extensions from our work with arithmetic. We know that

$$3 + 9 = 12 \quad \text{and} \quad 9 + 3 = 12.$$

Also,

$$3 \cdot 9 = 27 \quad \text{and} \quad 9 \cdot 3 = 27.$$

Furthermore,

$$(5 + 7) + (-2) = 12 + (-2) = \mathbf{10}$$

and

$$5 + [7 + (-2)] = 5 + 5 = \mathbf{10}.$$

Also,

$$(5 \cdot 7)(-2) = 35(-2) = \mathbf{-70}$$

and

$$(5)(7 \cdot -2) = 5(-14) = \mathbf{-70}.$$

These arithmetic examples can now be extended to algebraic examples. For example to combine the like terms in the expression $-2m + 5m + 3 - 6m + 8$, we use the properties as shown in the next example.

E X A M P L E   4   Using the Commutative and Associative Properties

Simplify $-2m + 5m + 3 - 6m + 8$.

$$-2m + 5m + 3 - 6m + 8$$
$$= (-2m + 5m) + 3 - 6m + 8 \qquad \text{Order of operations}$$
$$= (-2 + 5)m + 3 - 6m + 8 \qquad \text{Distributive property}$$
$$= 3m + 3 - 6m + 8$$

CONTINUED ON NEXT PAGE

By the order of operations, the next step would be to add $3m$ and 3, but they are unlike terms. To get $3m$ and $-6m$ together, use the associative and commutative properties. Begin by putting in parentheses and brackets according to the order of operations, as shown.

$$[(3m + 3) - 6m] + 8$$

$$= [3m + (3 - 6m)] + 8 \qquad \text{Associative property}$$

$$= [3m + (-6m + 3)] + 8 \qquad \text{Commutative property}$$

$$= [(3m + [-6m]) + 3] + 8 \qquad \text{Associative property}$$

$$= (-3m + 3) + 8 \qquad \text{Combine like terms.}$$

$$= -3m + (3 + 8) \qquad \text{Associative property}$$

$$= -3m + 11 \qquad \text{Add.}$$

In practice, many of the steps are not written down, but you should realize that the commutative and associative properties are used whenever the terms in an expression are rearranged in order to combine like terms.

E X A M P L E 5   Using the Properties of Real Numbers

Simplify each expression.

(a) $5y - 8y - 6y + 11y$

$$5y - 8y - 6y + 11y = (5 - 8 - 6 + 11)y = 2y$$

(b) $-2(m - 3)$

$$-2(m - 3) = -2(m) - (-2)(3) = -2m + 6$$

(c) $3x + 4 - 5(x + 1) - 8$

$$= 3x + 4 - 5x - 5 - 8 \qquad \text{Distributive property}$$

$$= 3x - 5x + 4 - 5 - 8 \qquad \text{Commutative and associative properties}$$

$$= -2x - 9 \qquad \text{Combine like terms.}$$

(d) $8 - (3m + 2)$

$$8 - (3m + 2) = 8 - 1(3m + 2) \qquad \text{Identity property}$$

$$= 8 - 3m - 2 \qquad \text{Distributive property}$$

$$= 6 - 3m \qquad \text{Combine like terms.}$$

(e) $(3x)(5)(y) = [(3x)(5)]y \qquad \text{Order of operations}$

$$= [3(x \cdot 5)]y \qquad \text{Associative property}$$

$$= [3(5x)]y \qquad \text{Commutative property}$$

$$= [(3 \cdot 5)x]y \qquad \text{Associative property}$$

$$= (15x)y \qquad \text{Multiply.}$$

$$= 15(xy) \qquad \text{Associative property}$$

$$= 15xy$$

As mentioned above, many of these steps usually are not written out.

WORK PROBLEM 5 AT THE SIDE. ▶▶

Caution
Be careful. Notice that the distributive property does not apply in Example 5(e), because there is no addition involved.

$$(3x)(5)(y) \neq (3x)(5) \cdot (3x)(y)$$

---

5. Simplify each expression.

(a) $12b - 9b + 4b - 7b + b$

(b) $-3w + 7 - 8w - 2$

(c) $-3(6 + 2t)$

(d) $9 - 2(a - 3) + 4 - a$

(e) $(4m)(2n)$

6. Complete the following.

(a) $197 \cdot 0 =$ _____

OBJECTIVE 5 The additive identity property gives a special property of zero, namely that $a + 0 = a$ for any real number a. The **multiplication property of zero** gives a special property of zero that involves multiplication: The product of any real number and zero is zero.

Multiplication Property of Zero

For all real numbers a,

$$a \cdot 0 = 0 \quad \text{and} \quad 0 \cdot a = 0.$$

This property just extends to all real numbers what is true for positive numbers multiplied by 0.

◀◀ WORK PROBLEM 6 AT THE SIDE.

(b) $(0)\left(-\dfrac{8}{9}\right) =$ _____

(c) $0 \cdot$ _____ $= 0$

1.5 Exercises

Use the distributive property to calculate the following values mentally. See Example 2.

1. $96 \cdot 19 + 4 \cdot 19$

2. $27 \cdot 60 + 27 \cdot 40$

3. $58 \cdot \dfrac{3}{2} - 8 \cdot \dfrac{3}{2}$

4. $\dfrac{8}{5} \cdot 17 + \dfrac{8}{5} \cdot 13$

Answer each question in Exercises 5–10 in a complete sentence.

5. What is the identity element for addition?

6. What is the identity element for multiplication?

7. What is meant by *like terms*?

8. What is the coefficient in the term $-6x^2z$?

9. What is the distinction between the commutative and associative properties?

10. What is the multiplication property of zero?

Use the properties of real numbers to simplify each expression. See Example 1.

11. $5k + 3k$

12. $6a + 5a$

13. $-9r + 7r$

14. $-4n + 6n$

15. $-8z + 4w$

16. $-12k + 3r$

17. $-a + 7a$

18. $-s + 9s$

19. $2(m + p)$

20. $3(a + b)$

21. $-5(2d - f)$

22. $-2(3m - n)$

Simplify each of the following expressions by removing parentheses and combining terms. See Examples 1 and 3–5.

23. $-12y + 4y + 3 + 2y$

24. $-5r - 9r + 8r - 5$

25. $-6p + 11p - 4p + 6 + 5$

26. $-8x - 5x + 3x - 12 + 9$

27. $3(k + 2) - 5k + 6 + 3$

28. $5(r - 3) + 6r - 2r + 4$

29. $-2(m + 1) + 3(m - 4)$

30. $6(a - 5) - 4(a + 6)$

31. $.25(8 + 4p) - .5(6 + 2p)$

32. $.4(10 - 5x) - .8(5 + 10x)$

33. $-(2p + 5) + 3(2p + 4) - 2p$

34. $-(7m - 12) - 2(4m + 7) - 8m$

35. $2 + 3(2z - 5) - 3(4z + 6) - 8$

36. $-4 + 4(4k - 3) - 6(2k + 8) + 7$

Each of the following shows half of a statement illustrating the indicated property. Complete each statement and simplify the answer, if possible.

37. $5x + 8x = $ _____
(distributive property)

38. $9y - 6y = $ _____
(distributive property)

39. $5(9r) = $ _____
(associative property)

40. $-4 + (12 + 8) = $ _____
(associative property)

41. $5x + 9y = $ _____
(commutative property)

42. $-5 \cdot 7 = $ _____
(commutative property)

43. $1 \cdot 7 = $ _____
(identity property)

44. $-12x + 0 = $ _____
(identity property)

45. $8(-4 + x) = $ _____
(distributive property)

46. $3(x - y + z) = $ _____
(distributive property)

47. Give an "everyday" example of a commutative operation.

48. Give an "everyday" example of inverse operations.

49. Are there *any* two different numbers a and b for which $a/b = b/a$? Give an example if your answer is yes.

50. Do *any* different numbers satisfy the statement $a - b = b - a$? Give an example if your answer is yes.

MATHEMATICAL CONNECTIONS (EXERCISES 51–56)

While it may seem that simplifying the expression $3x + 4 + 2x + 7$ to $5x + 11$ is fairly easy, there are several important steps that require mathematical justification. These steps are usually done mentally. For now, provide the property that justifies the statement in the simplification. (These steps could be done in other orders.)

51. $3x + 4 + 2x + 7 = (3x + 4) + (2x + 7)$ _____

52. $\qquad\qquad = 3x + (4 + 2x) + 7$ _____

53. $\qquad\qquad = 3x + (2x + 4) + 7$ _____

54. $\qquad\qquad = (3x + 2x) + (4 + 7)$ _____

55. $\qquad\qquad = (3 + 2)x + (4 + 7)$ _____

56. $\qquad\qquad = 5x + 11$ _____

─────────────────────────── **KEY TERMS** ───────────────────────────

1.1	**set**	A set is a collection of objects.
	empty set	The set with no elements is called the empty set.
	variable	A variable is a letter used to represent a number or a set of numbers.
	set-builder notation	Set-builder notation is used to describe a set of numbers without listing them.
	number line	A number line is a line with a scale to indicate the set of real numbers.
	coordinate	The number that corresponds to a point on the number line is its coordinate.
	graph	The point on the number line that corresponds to a number is its graph.
	signed numbers	Positive and negative numbers are signed numbers.
	additive inverse	The additive inverse (**negative, opposite**) of a number a is $-a$.
	absolute value	The absolute value of a number is its distance from 0 or its magnitude without regard to sign.
1.2	**inequality**	An inequality is a mathematical statement that two quantities are not equal.
	interval	An interval is a portion of a number line.
	interval notation	Interval notation uses symbols to describe an interval on the number line.
1.3	**sum**	The result of addition is called the sum.
	difference	The result of subtraction is called the difference.
	product	The result of multiplication is called the product.
	reciprocals	Two numbers whose product is 1 are reciprocals (**multiplicative inverses**).
	quotient	The result of division is called the quotient.
1.4	**factors**	Two numbers whose product is a third number are factors of that third number.
	exponent	An exponent is a number that shows how many times a factor is repeated in a product.
	base	The base is a number that is a repeated factor in a product.
	exponential	A base with an exponent is called an exponential or a **power.**
	square root	A square root of a number r is a number that can be squared to get r.
	cube root	The cube root of a number r is the number that can be cubed to get r.
1.5	**term**	A term is a number or the product of a number and one or more variables.
	like terms	Like terms are terms with the same variables raised to the same powers.
	coefficient	A coefficient (**numerical coefficient**) is the numerical factor of a term.
	combining like terms	Combining like terms is a method of adding or subtracting like terms by using the properties of real numbers.

NEW SYMBOLS

$\{a, b\}$	set containing the elements a and b
$\emptyset$	the empty set
$\{x \mid x$ has property $P\}$	set-builder notation
$\|x\|$	the absolute value of x
$\neq$	is not equal to
$<$	is less than
$>$	is greater than
(a, ∞)	the interval $\{x \mid x > a\}$
$(-\infty, a)$	the interval $\{x \mid x < a\}$
$(a, b]$	the interval $\{x \mid a < x \le b\}$
a^m	m factors of a
$\sqrt{a}$	the positive square root of a
$\sqrt[n]{a}$	the nth root of a

QUICK REVIEW

Concepts	Examples
1.1 Basic Terms	
Sets of Numbers	
Natural Numbers $\{1, 2, 3, 4, \ldots\}$	10, 25, 143
Whole Numbers $\{0, 1, 2, 3, 4, \ldots\}$	0, 8, 47
Integers $\{\ldots, -2, -1, 0, 1, 2, \ldots\}$	$-22, -7, 0, 4, 9$
Rational Numbers $\left\{\dfrac{p}{q} \mid p \text{ and } q \text{ are integers, } q \neq 0\right\}$ (all terminating or repeating decimals)	$-\dfrac{2}{3}, -.14, 0, 6, \dfrac{5}{8}, .33333\ldots$
Irrational Numbers $\{x \mid x$ is a real number that is not rational$\}$ (all nonterminating, nonrepeating decimals)	$\pi, .125469\ldots, \sqrt{3}, -\sqrt{22}$
Real Numbers $\{x \mid x$ is a coordinate of a point on a number line$\}$	$-3, .7, \pi, -\dfrac{2}{3}$

Concepts	Examples

1.2 Inequality

Inequality Symbols

Symbol	Meaning	
≠	is not equal to	$-3 \neq 3$
<	is less than	$-4 < -1$
>	is greater than	$3 > -2$
≤	is less than or equal to	$6 \leq 6$
≥	is greater than or equal to	$-8 \geq -10$

1.3 Operations on Real Numbers

Addition

Same sign: Add the absolute values. The answer has the same sign as the numbers.

$$-2 + (-7) = -(2 + 7)$$
$$= -9$$

Different signs: Subtract the absolute values. The answer has the sign of the number with the larger absolute value.

$$-5 + 8 = 8 - 5 = 3$$
$$-12 + 4 = -(12 - 4) = -8$$

Subtraction

Change the sign of the second number and add.

$$-5 - (-3) = -5 + 3 = -2$$

Multiplication

Same sign: The answer is positive.

$$(-3)(-8) = 24$$

Different signs: The answer is negative.

$$(-7)(5) = -35$$

Division

Same sign: The answer is positive.

$$\frac{-15}{-5} = 3$$

Different signs: The answer is negative.

$$\frac{-24}{12} = -2$$

Concepts	Examples
1.4 Exponents and Roots; Order of Operations	
The product of an even number of negative factors is positive.	$(-5)^6$ is positive.
The product of an odd number of negative factors is negative.	$(-5)^7$ is negative.
Order of Operations	
1. Simplify above and below fraction bars.	$$\frac{12 + 3}{5 \cdot 2} = \frac{15}{10} = \frac{3}{2}$$
2. Simplify within parentheses.	$(-6)[2^2 - (3 + 4)] + 3$
	$= (-6)[2^2 - 7] + 3$
3. Calculate exponents, roots, and absolute values.	$= (-6)[4 - 7] + 3$
	$= (-6)[-3] + 3$
4. Multiply or divide from left to right.	$= 18 + 3$
5. Add or subtract from left to right.	$= 21$
1.5 Properties of Real Numbers	
Distributive Property	
$a(\overset{\frown}{b + c}) = ab + ac$ (Remove parentheses.)	$12(4 + 2) = 12 \cdot 4 + 12 \cdot 2$
Inverse Properties	
The additive inverse "undoes" addition to give 0.	$5 + (-5) = 0$
The multiplicative inverse "undoes" multiplication to give 1.	$-\dfrac{1}{3} \cdot -3 = 1$
Identity Properties	
Start with a number *a*, add 0; the answer is identical to *a*.	$-32 + 0 = -32$
Start with a number *a*, multiply by 1; the answer is identical to *a*.	$17.5 \cdot 1 = 17.5$
Associative Properties	
Three terms or factors: same order, parentheses shifted.	$7 + (5 + 3) = (7 + 5) + 3$
	$-4(6 \cdot 3) = (-4 \cdot 6)3$
Commutative Properties	
Two terms or factors: reverse the order.	$9 + (-3) = -3 + 9$
	$6(-4) = (-4)6$
Multiplication Property of Zero	
Multiplying any number by 0 gives 0.	$4 \cdot 0 = 0$
	$0 \cdot (-3) = 0$

CHAPTER 1 REVIEW EXERCISES

[1.1] *Graph each set on the number line.**

1. $\left\{-4, -1, 2, \frac{9}{4}, 4\right\}$

2. $\left\{-5, -\frac{11}{4}, -.5, 0, 3, \frac{13}{3}\right\}$

Find the value of each expression.

3. $|-16|$ **4.** $|23|$ **5.** $-|-4|$ **6.** $-|-8| + |-3|$

Let set $S = \left\{-9, -\frac{4}{3}, -\sqrt{4}, -.25, 0, .\overline{35}, \frac{5}{3}, \sqrt{7}, \sqrt{-9}, \frac{12}{3}\right\}$. *Simplify the elements of S as necessary and then list the elements that belong to each set listed below.*

7. Whole numbers   **8.** Integers

9. Rational numbers   **10.** Real numbers

Write each set by listing its elements.

11. $\{x \mid x$ is a natural number between 3 and 9$\}$ **12.** $\{y \mid y$ is a whole number less than 4$\}$

[1.2] *Write* true *or* false *for each inequality.*

13. $4 \cdot 2 \le |12 - 4|$ **14.** $2 + |-2| > 4$ **15.** $4(3 + 7) > -|40|$

Write in interval notation and graph.

16. $\{x \mid x < -5\}$ **17.** $\{x \mid -2 < x \le 3\}$

[1.3] *Add or subtract, as indicated.*

18. $-\frac{5}{8} - \left(-\frac{7}{3}\right)$ **19.** $-\frac{4}{5} - \left(-\frac{3}{10}\right)$ **20.** $-5 + (-11) + 20 - 7$

21. $-9.42 + 1.83 - 7.6 - 1.9$ **22.** $-15 + (-13) + (-11)$ **23.** $-1 - 3 - (-10) + (-7)$

---

* For help with any of these exercises, look in the section given in brackets.

24. $\dfrac{3}{4} - \left(\dfrac{1}{2} - \dfrac{9}{10}\right)$

25. $-\dfrac{2}{3} - \left(\dfrac{1}{6} - \dfrac{5}{9}\right)$

26. $-|-12| - |-9| + (-4) - |10|$

27. State in your own words how to determine the sign of the sum of two numbers.

28. How is subtraction related to addition?

Find each product.

29. $2(-5)(-3)(-3)$

30. $-\dfrac{3}{7}\left(-\dfrac{14}{9}\right)$

31. $-4.6(2.48)$

Find each quotient.

32. $\dfrac{-38}{-19}$

33. $\dfrac{75}{-5}$

34. $\dfrac{\dfrac{2}{3}}{-\dfrac{1}{6}}$

35. $\dfrac{-2.3754}{-.74}$

36. Which one of the following is undefined: $\dfrac{5}{7-7}$ or $\dfrac{7-7}{5}$?

[1.4] *Evaluate.*

37. 10^4

38. $\left(\dfrac{3}{7}\right)^3$

39. $(-5)^3$

40. -5^3

41. $(1.7)^2$

Find each root. If it is not a real number, say so.

42. $\sqrt{400}$

43. $\sqrt[3]{27}$

44. $\sqrt[3]{-343}$

45. $\sqrt[4]{81}$

46. $\sqrt[6]{-64}$

Use the order of operations to simplify.

47. $-14\left(\dfrac{3}{7}\right) + 6 \div 3$

48. $-\dfrac{2}{3}[5(-2) + 8 - 4^3]$

49. $\dfrac{-4(\sqrt{25}) - (-3)(-5)}{3 + (-6)(\sqrt{9})}$

50. $\dfrac{-5(3^2) + 9(\sqrt{4}) - 5}{6 - 5(\sqrt[3]{-8})}$

Evaluate. Assume that $k = -4$, $m = 2$, and $n = 16$.

51. $4k - 7m$

52. $-3(\sqrt{n}) + m + 5k$

53. $-2(3k^2 + 5m)$

54. $\dfrac{4m^3 - 3n}{7k^2 - 10}$

55. In order to evaluate $(3 + 2)^2$, should you work within the parentheses first, or should you square 3 and square 2 and then add?

56. By replacing a with 4 and b with 6, show that $(a + b)^2 \neq a^2 + b^2$.

[1.5] *Use the properties of real numbers to simplify each expression.*

57. $2q + 19q$

58. $13z - 17z$

59. $-m + 6m$

60. $5p - p$

61. $-2(k + 3)$

62. $6(r + 3)$

63. $9(2m + 3n)$

64. $-(3k - 4h)$

65. $-(-p + 6q) - (2p - 3q)$

66. $-2x + 5 - 4x + 1$

67. $-3y + 6 - 5 + 4y$

68. $2a + 3 - a - 1 - a - 2$

69. $-2(k - 1) + 3k - k$

70. $-3(4m - 2) + 2(3m - 1) - 4(3m + 1)$

Each of the following exercises shows half of a statement illustrating the indicated property. Complete each statement. Simplify all answers.

71. $2x + 3x =$ _____
(distributive property)

72. $-4 \cdot 1 =$ _____
(identity property)

73. $2(4x) =$ _____
(associative property)

74. $-3 + 13 =$ _____
(commutative property)

75. $-3 + 3 =$ _____
(inverse property)

76. $5(x + z) =$ _____
(distributive property)

77. $0 + 7 =$ _____
(identity property)

78. $8 \cdot \dfrac{1}{8} =$ _____
(inverse property)

79. $3a + 5a + 6a =$ _____
(distributive property)

80. $\dfrac{9}{28} \cdot 0 =$ _____
(multiplication property of 0)

MIXED REVIEW EXERCISES*

Perform the indicated operations.

81. $\left(-\dfrac{4}{5}\right)^4$

82. $-\dfrac{5}{8}(-40)$

83. $-25\left(-\dfrac{4}{5}\right) + 3^3 - 32 \div \sqrt{4}$

84. $-8 + |-14| + |-3|$

85. $\dfrac{6 \cdot \sqrt{4} - 3 \cdot \sqrt{16}}{-2 \cdot 5 + 7(-3) - 10}$

86. $-\sqrt[5]{32}$

87. $-\dfrac{10}{21} \div \left(-\dfrac{5}{14}\right)$

88. $.8 - 4.9 - 3.2 + 1.14$

89. -3^2

---

*The order of exercises in this final group does not correspond to the order in which topics occur in the chapter. This random ordering should help you prepare for the chapter test in yet another way.

1. Graph $\{-3, .75, \frac{5}{3}, 5, 6.3\}$ on the number line.

Let $A = \{-\sqrt{6}, -1, -.5, 0, 3, 7.5, \sqrt{25}, \frac{24}{2}, \sqrt{-4}\}$. First simplify each element as needed, and then list the elements from A that belong to each set.

2. Whole numbers

3. Integers

4. Rational numbers

5. Real numbers

Write each set in interval notation, and graph it on the number line provided.

6. $\{x \mid x < -3\}$

7. $\{y \mid -4 < y \le 2\}$

Perform the indicated operations.

8. $-6 + 14 + (-11) - (-3)$

9. $10 - 4 \cdot 3 + 6(-4)$

10. $7 - 4^2 + 2(6) + (-4)^2$

11. $\dfrac{10 - 24 + (-6)}{\sqrt{16}(-5)}$

12. $\dfrac{-2[3 - (-1 - 2) + 2]}{\sqrt{9}(-3) - (-2)}$

13. $\dfrac{8 \cdot 4 - 3^2 \cdot 5 - 2(-1)}{-3 \cdot 2^3 + 1}$

Find each of the following roots. If the number is not real, say so.

14. $\sqrt{196}$

15. $-\sqrt{225}$

16. $\sqrt[3]{-27}$

17. $\sqrt[4]{-16}$

---

1. ┼┼┼┼┼┼┼┼┼┼┼┼►

2. _____

3. _____

4. _____

5. _____

6. ──────────────►

7. ──────────────►

8. _____

9. _____

10. _____

11. _____

12. _____

13. _____

14. _____

15. _____

16. _____

17. _____

18. _____

18. Under what conditions will $\sqrt[n]{a}$ represent a real number if a is negative?

Evaluate each expression if $k = -3$, $m = -3$, and $r = 25$.

19. _____

19. $\sqrt{r} + 2k - m$

20. _____

20. $\dfrac{8k + 2m^2}{r - 2}$

Use the properties of real numbers to simplify the following.

21. _____

21. $-3(2k - 4) + 4(3k - 5) - 2 + 4k$.

22. _____

22. How does the subtraction sign affect the terms $-4r$ and 6 when simplifying $(3r + 8) - (-4r + 6)$? What is the simplified form?

Match each statement with the appropriate property. Answers may be used more than once.

23. _____

23. $6 + (-6) = 0$ A. Distributive Property

24. _____

24. $4 + 5 = 5 + 4$ B. Inverse Property

25. _____

25. $-2 + (3 + 6) = (-2 + 3) + 6$ C. Identity Property

26. _____

26. $5x + 15x = (5 + 15)x$ D. Associative Property

27. _____

27. $13 \cdot 0 = 0$ E. Commutative Property

28. _____

28. $-9 + 0 = -9$ F. Multiplication Property of Zero

29. _____

29. $4 \cdot 1 = 4$

30. _____

30. $(a + b) + c = (b + a) + c$

Linear Equations and Inequalities

2.1 Linear Equations in One Variable

In the previous chapter we began to use **algebraic expressions** (al-juh-BRAY-ik eks-PRESH-uns), such as

$$8x + 9, \quad y - 4, \quad \text{and} \quad \frac{x^3 y^8}{z}.$$

A balance scale allows us to compare the weights of two quantities. Equations and inequalities compare algebraic expressions.

Many applications of mathematics lead to **equations** (ee-KWAY-zhuns), statements that two algebraic expressions are equal. A *linear equation in one variable* involves only real numbers and one variable. Examples include the following.

$$x + 1 = -2 \qquad y - 3 = 5 \qquad 2k + 5 = 10$$

It is very important to be able to distinguish between algebraic expressions and equations. An equation always includes an equals sign, while an expression does not.

> **Linear Equation**
>
> An equation in the variable x is **linear** (LIN-ee-er) if it can be written in the form
>
> $$ax + b = c$$
>
> where a, b, and c are real numbers, with $a \neq 0$.

A linear equation is also called a **first-degree** equation, since the highest power on the variable is one.

OBJECTIVE 1 ▶ If the variable in an equation can be replaced by a real number that makes the statement true, then that number is a **solution** (suh-LOO-shun) of the equation. For example, 8 is a solution of the equation $y - 3 = 5$, since replacing y with 8 gives a true statement. An equation is **solved** by finding its **solution set,** the set of all solutions. The solution set of the equation $y - 3 = 5$ is $\{8\}$.

  WORK PROBLEM I AT THE SIDE. ▶▶

Equivalent (ee-KWIV-uh-lent) **equations** are equations that have the same solution set. To solve an equation we usually start with the given equation and replace it with a series of simpler equivalent equations. For example, $5x + 2 = 17$, $5x = 15$, and $x = 3$ are all equivalent, since each has the solution set $\{3\}$.

www.mathnotes.com

OBJECTIVES

1 ▶ Decide whether a number is a solution of a linear equation.

2 ▶ Solve linear equations using the addition and multiplication properties of equality.

3 ▶ Solve linear equations using the distributive property.

4 ▶ Solve linear equations with fractions and decimals.

5 ▶ Identify conditional equations, contradictions, and identities.

FOR EXTRA HELP

Tutorial   Tape 2   SSM, Sec. 2.1

1. Are the given numbers solutions for the given equations?

   (a) $3k = 15$; 5

   (b) $r + 5 = 4$; 1

   (c) $-8m = 12; \dfrac{3}{2}$

ANSWERS

1. (a) yes **(b)** no **(c)** no

2. Solve and check.

(a) $3p + 2p + 1 = -24$

(b) $3p = 2p + 4p + 5$

(c) $4a + 8a = 17a - 9 - 1$

(d) $-7 + 3y - 9y = 12y - 5$

OBJECTIVE 2 Two important properties that are used in producing equivalent equations are the **addition and multiplication properties of equality.**

Addition and Multiplication Properties of Equality

Addition Property of Equality
For all real numbers, a, b, and c, the equations

$$a = b \quad \text{and} \quad a + c = b + c$$

are equivalent. (The same number may be added to both sides of an equation without changing the solution set.)

Multiplication Property of Equality
For all real numbers a, b, and $c \neq 0$, the equations

$$a = b \quad \text{and} \quad ac = bc$$

are equivalent. (Both sides of an equation may be multiplied by the same nonzero number without changing the solution set.)

Because subtraction and division are defined in terms of addition and multiplication, respectively, these properties can be extended: The same number may be subtracted from both sides of an equation, and both sides may be divided by the same nonzero number.

EXAMPLE 1 Solving a Linear Equation

Solve $4y - 2y - 5 = 4 + 6y + 3$.

The goal is to use the addition and multiplication properties to get y alone on one side of the equation. First, combine terms on both sides of the equation to get

$$2y - 5 = 7 + 6y.$$

Next, use the addition property to get the terms with y on the same side of the equation and the remaining terms (the numbers) on the other side. One way to do this is first to add 5 to both sides.

$$2y - 5 + 5 = 7 + 6y + 5 \qquad \text{Add 5.}$$
$$2y = 12 + 6y$$
$$2y - 6y = 12 + 6y - 6y \qquad \text{Subtract } 6y.$$
$$-4y = 12$$
$$\frac{-4y}{-4} = \frac{12}{-4} \qquad \text{Divide by } -4.$$
$$y = -3$$

To be sure that -3 is the solution, check by substituting back into the *original* equation.

$$4y - 2y - 5 = 4 + 6y + 3 \qquad \text{Given equation}$$
$$4(-3) - 2(-3) - 5 = 4 + 6(-3) + 3 \qquad ? \text{ Let } y = -3.$$
$$-12 + 6 - 5 = 4 - 18 + 3 \qquad ? \text{ Multiply.}$$
$$-11 = -11 \qquad \text{True}$$

A true sentence indicates that $\{-3\}$ is the solution set.

(Note that in the above example, the equality symbols are lined up in a column. Do not use more than one equality symbol in a horizontal line of work.)

ANSWERS

2. (a) $\{-5\}$ **(b)** $\left\{-\frac{5}{3}\right\}$ **(c)** $\{2\}$ **(d)** $\left\{-\frac{1}{9}\right\}$

◀◀ WORK PROBLEM 2 AT THE SIDE.

The steps needed to solve a linear equation in one variable are as follows. (Some equations may not require all of these steps.)

Solving a Linear Equation in One Variable

Step 1 Eliminate any fractions by multiplying both sides of the equation by a common denominator.

Step 2 Simplify each side of the equation as much as possible by using the distributive property to clear parentheses and to combine like terms.

Step 3 Use the addition property of equality to get all terms with the variable on one side of the equation and all constants on the other side. Combine terms.

Step 4 Use the multiplication property of equality to make the coefficient of the variable 1.

Step 5 Check by substituting back in the *original* equation.

The distributive property is used in two different ways to simplify equations. We use the form $a(b + c) = ab + ac$ to *clear parentheses*. The form $ab + ac = a(b + c)$ is used to combine terms or to **factor,** to write a sum as a product. We will have much more to say about factoring in a later chapter.

Objective 3 In Example 1 we did not use the Steps 1 and 2. Solving many other equations will require one or both of these steps, however, as shown in the next examples.

┌───
E X A M P L E   2   Using the Distributive Property to Solve a Linear Equation

Solve $2(k - 5) + 3k = k + 6$.

Step 1 does not apply here because there are no fractions. We use the distributive property to simplify the left side, and then combine like terms.

Step 2 $\quad 2(k - 5) + 3k = k + 6$

$\quad\quad\quad 2k - 10 + 3k = k + 6$ \qquad Clear parentheses.

$\quad\quad\quad\quad\quad 5k - 10 = k + 6$ \qquad Combine terms.

Step 3 $\quad\quad 5k - k = k + 16 - k$ \quad Subtract k; add 10.

$\quad\quad\quad\quad\quad\quad 4k = 16$

Step 4 $\quad\quad\quad\quad \dfrac{4k}{4} = \dfrac{16}{4}$ \qquad Divide by 4.

$\quad\quad\quad\quad\quad\quad k = 4$

Step 5 Check: $\quad 2(4 - 5) + 3(4) = 4 + 6$ \quad ? Substitute 4 for k.

$\quad\quad\quad\quad\quad 2(-1) + 12 = 10$ \quad\quad ?

$\quad\quad\quad\quad\quad\quad\quad 10 = 10$ \quad\quad\quad True

Because the check leads to a true statement, the solution set is $\{4\}$.
└───

Note

Because of space limitations, we will not always show the check step when solving an equation. To be sure that your solution is correct, you should always check your work.

WORK PROBLEM 3 AT THE SIDE. ▶▶

---

3. Solve and check.

(a) $5p + 4(3 - 2p)$
$= 2 + p - 10$

(b) $3(z - 2) + 5z = 2$

(c) $-2 + 3(y + 4) = 8y$

(d) $6 - (4 + m)$
$= 8m - 2(3m + 5)$

Answers

3. (a) $\{5\}$ **(b)** $\{1\}$ **(c)** $\{2\}$ **(d)** $\{4\}$

4. Solve and check.

(a) $\dfrac{2p}{7} - \dfrac{p}{2} = -3$

(b) $\dfrac{k+1}{2} + \dfrac{k+3}{4} = \dfrac{1}{2}$

OBJECTIVE 4 When fractions (or decimals) appear in an equation, we can simplify our work by multiplying both sides of the equation by the least common denominator of all the fractions.

E X A M P L E 3 Solving a Linear Equation with Fractions

Solve $\dfrac{x+7}{6} + \dfrac{2x-8}{2} = -4$.

Start by eliminating the fractions (Step 1). The least common denominator is 6.

$$6\left[\dfrac{x+7}{6} + \dfrac{2x-8}{2}\right] = 6 \cdot (-4) \qquad \text{Multiply both sides by 6.}$$

$$6\left(\dfrac{x+7}{6}\right) + 6\left(\dfrac{2x-8}{2}\right) = 6(-4) \qquad \text{Clear parentheses.}$$

$$x + 7 + 3(2x - 8) = -24 \qquad \text{Simplify.}$$

$$x + 7 + 6x - 24 = -24 \qquad \text{Clear parentheses.}$$

$$7x - 17 = -24 \qquad \text{Combine terms.}$$

$$7x - 17 + 17 = -24 + 17 \qquad \text{Add 17.}$$

$$7x = -7$$

$$\dfrac{7x}{7} = \dfrac{-7}{7} \qquad \text{Divide by 7.}$$

$$x = -1$$

Check to see that $\{-1\}$ is the solution set.

◀◀ **WORK PROBLEM 4 AT THE SIDE.**

E X A M P L E 4 Solving a Linear Equation with Decimals

Solve $.06x + .09(15 - x) = .07(15)$.

Because each decimal number is given in hundredths, multiply both sides of the equation by 100. (This is done by moving the decimal points two places to the right.)

$$.06x + .09(15 - x) = .07(15)$$

$$6x + 9(15 - x) = 7(15) \qquad \text{Multiply by 100.}$$

$$6x + 9(15) - 9x = 105 \qquad \text{Clear parentheses.}$$

$$-3x + 135 = 105 \qquad \text{Combine like terms.}$$

$$-3x + 135 - 135 = 105 - 135 \qquad \text{Subtract 135.}$$

$$-3x = -30$$

$$\dfrac{-3x}{-3} = \dfrac{-30}{-3} \qquad \text{Divide by } -3.$$

$$x = 10$$

Check to verify that the solution set is $\{10\}$.

ANSWERS

4. (a) $\{14\}$ (b) $\{-1\}$

Caution
When multiplying the term $.09(15 - x)$ by 100 in Example 4, do not multiply both $.09$ and $15 - x$ by 100. This step is not an application of the distributive property, but of the associative property. The correct procedure is

$$100[.09(15 - x)] = [100(.09)](15 - x) \quad \text{Associative property}$$
$$= 9(15 - x). \quad \text{Multiply.}$$

WORK PROBLEM 5 AT THE SIDE. ▶▶

OBJECTIVE 5 ▶ All the equations above had a solution set containing one element; for example, $2(k - 5) + 3k = k + 6$ has solution set $\{4\}$. Some linear equations, however, have no solutions, while others have an infinite number of solutions. The chart below gives the names of these types of equations.

Type of Equation	Number of Solutions
Conditional	One
Contradiction	None, solution set $\emptyset$
Identity	Infinite, solution set $\{$all real numbers$\}$

The next example shows how to recognize these types of equations.

E X A M P L E 5   Recognizing Conditional Equations, Identities, and Contradictions

Solve each equation. Decide whether it is a conditional equation, an identity, or a contradiction.

(a) $5x - 9 = 4(x - 3)$

$$5x - 9 = 4(x - 3)$$
$$5x - 9 = 4x - 12 \quad \text{Clear parentheses.}$$
$$5x - 9 - 4x = 4x - 12 - 4x \quad \text{Subtract } 4x.$$
$$x - 9 = -12 \quad \text{Combine terms.}$$
$$x - 9 + 9 = -12 + 9 \quad \text{Add 9.}$$
$$x = -3$$

The solution set, $\{-3\}$, has only one element, so $5x - 9 = 4(x - 3)$ is a *conditional* (kun-DISH-un-ul) *equation*.

(b) $5x - 15 = 5(x - 3)$
Clear parentheses on the right side.
$$5x - 15 = 5x - 15$$
Both sides of the equation are *exactly the same,* so any real number would make the equation true. For this reason, the solution set is $\{$all real numbers$\}$, and the equation $5x - 15 = 5(x - 3)$ is an *identity*.

— **CONTINUED ON NEXT PAGE**

5. Solve and check.

(a) $.04x + .06(20 - x)$
$\quad = .05(50)$

(b) $.10(x - 6) + .05x$
$\quad = .06(50)$

ANSWERS

5. (a) $\{-65\}$ **(b)** $\{24\}$

6. Solve each equation. Decide whether it is a conditional equation, an identity, or a contradiction. Give the solution set.

(a) $5(x + 2) - 2(x + 1)$
$= 3x + 1$

(c) $5x - 15 = 5(x - 4)$

$$5x - 15 = 5x - 20 \qquad \text{Clear parentheses.}$$
$$5x - 15 - 5x = 5x - 20 - 5x \qquad \text{Subtract } 5x \text{ from each side.}$$
$$-15 = -20 \qquad \text{False}$$

Since the result, $-15 = -20$, is *false*, the equation has no solution. The solution set is $\emptyset$, so the equation $5x - 15 = 5(x - 4)$ is a *contradiction* (kahn-truh-DIK-shun).

◀◀ WORK PROBLEM 6 AT THE SIDE.

(b) $\dfrac{x + 1}{3} + \dfrac{2x}{3} = x + \dfrac{1}{3}$

(c) $5(3x + 1) = x + 5$

ANSWERS

6. (a) contradiction; $\emptyset$
  (b) identity; {all real numbers}
  (c) conditional; {0}

2.1 Exercises

1. Show that 6 is a solution of $3(x + 4) = 5x$ by substituting 6 for x leading to a true statement. (This is the procedure used to check the solution of an equation.)

2. Determine whether -2 is a solution of $5(x + 4) - 3(x + 6) = 9(x + 1)$ by substitution. If it is not, explain why.

3. The equation $4[x + (2 - 3x)] = 2(4 - 4x)$ is an identity. Let x represent the number of letters in your last name. Is this number a solution of this equation? Verify your answer.

4. Explain in your own words the steps used to solve a linear equation.

Solve and check each of the following equations. See Examples 1 and 2.

5. $7k + 8 = 1$

6. $5m - 4 = 21$

7. $8 - 8x = -16$

8. $9 - 2r = 15$

9. $7y - 5y + 15 = y + 8$

10. $2x + 4 - x = 4x - 5$

11. $12w + 15w - 9 + 5 = -3w + 5 - 9$

12. $-4t + 5t - 8 + 4 = 6t - 4$

13. $2(x + 3) = -4(x + 1)$

14. $4(y - 9) = 8(y + 3)$

15. $3(2w + 1) - 2(w - 2) = 5$

16. $4(x - 2) + 2(x + 3) = 6$

17. $2x + 3(x - 4) = 2(x - 3)$

18. $6y - 3(5y + 2) = 4(1 - y)$

19. $6p - 4(3 - 2p) = 5(p - 4) - 10$

20. $-2k - 3(4 - 2k) = 2(k - 3) + 2$

21. $-[2z - (5z + 2)] = 2 + (2z + 7)$

22. $-[6x - (4x + 8)] = 9 + (6x + 3)$

23. $-(9 - 3a) - (4 + 2a) - 3 = -(2 - 5a) + (-a) + 1$

24. $-(-2 + 4x) - (3 - 4x) + 5 = -(-3 + 6x) + x + 1$

25. In order to solve the linear equation $\frac{8y}{3} - \frac{2y}{4} = -13$, we may multiply both sides by the least common denominator of all the fractions in the equation. What is this least common denominator?

26. In order to solve the linear equation $.05y + .12(y + 5000) = 940$, we may multiply both sides by a power of 10 so that all coefficients are integers. What is the smallest power of 10 that will accomplish this goal?

27. Suppose that in solving the equation

$$\frac{1}{3}y + \frac{1}{2}y = \frac{1}{6}y,$$

you begin by multiplying both sides by 12, rather than the *least* common denominator, 6. Would you get the correct solution anyway? Explain.

28. Show all steps in the check for the solution of the equation in Example 4.

Solve and check each of the following equations. See Examples 3 and 4.

29. $\dfrac{3x}{4} + \dfrac{5x}{2} = 13$

30. $\dfrac{8y}{3} - \dfrac{2y}{4} = -13$

31. $\dfrac{x - 8}{5} + \dfrac{8}{5} = -\dfrac{x}{3}$

32. $\dfrac{2r - 3}{7} + \dfrac{3}{7} = -\dfrac{r}{3}$

33. $\dfrac{4t + 1}{3} = \dfrac{t + 5}{6} + \dfrac{t - 3}{6}$

34. $\dfrac{2x + 5}{5} = \dfrac{3x + 1}{2} + \dfrac{-x + 7}{2}$

35. $.05y + .12(y + 5000) = 940$

36. $.09k + .13(k + 300) = 61$

37. $.02(50) + .08r = .04(50 + r)$

38. $.20(14,000) + .14t = .18(14,000 + t)$

39. $.05x + .10(200 - x) = .45x$

40. $.08x + .12(260 - x) = .48x$

41. Explain the distinction between conditional equation, identity, and contradiction.

42. A student tried to solve the equation $8x = 7x$ by dividing both sides by x, obtaining $8 = 7$. He gave the solution set to be $\emptyset$. Why is this incorrect?

Decide whether each equation is conditional, an identity, or a contradiction. Give the solution set. See Example 5.

43. $-2p + 5p - 9 = 3(p - 4) - 5$

44. $-6k + 2k - 11 = -2(2k - 3) + 4$

45. $-11m + 4(m - 3) + 6m = 4m - 12$

46. $3p - 5(p + 4) + 9 = -11 + 15p$

47. $7[2 - (3 + 4r)] - 2r = -9 + 2(1 - 15r)$

48. $4[6 - (1 + 2m)] + 10m = 2(10 - 3m) + 8m$

MATHEMATICAL CONNECTIONS (EXERCISES 49–54)

Work Exercises 49–54 in order.

49. Use the methods of this section to solve the equation $2[3x + (x - 2)] = 9x + 4$.

50. If an equation is *equivalent* to the one in Exercise 49, what must its solution set be?

51. Let us now consider the following linear equation in x:

$$-4(x + 2) - 3(x + 5) = k.$$

Assume that k is some real number constant. Evaluate the left side of the equation for the value of x you found in Exercise 49. What is your result?

52. What must be the value of k in the equation in Exercise 51 for that equation to be equivalent to the one in Exercise 49?

53. Solve the equation in Exercise 51 with the value of k you found in Exercise 52.

54. Use the concepts presented in Exercises 49–53 to find the value of k that will make the equations $3x + k = 11$ and $5x - 8 = 22$ equivalent.

55. If two equations are equivalent, they have the same _____ _____ .

56. Which one of the following equations is equivalent to $3x = 9$?
    (a) $x = -3$ **(b)** $x^2 = 9$ **(c)** $4x = -1$ **(d)** $5x = 15$

In Exercises 57–60, the given pair of equations are not equivalent. Tell why this is so.

57. $y = -8$

$$\frac{y}{y + 8} = \frac{-8}{y + 8}$$

58. $m = 1$

$$\frac{m + 1}{m - 1} = \frac{2}{m - 1}$$

59. $k = 4$
$k^2 = 16$

60. $x = -6$
$x^2 = 36$

Two methods of representing data are equations and graphs. Exercises 61–64 deal with these two methods.

61. The linear equation $y = 420x + 720$ models the worldwide credit card fraud losses between the years 1989 and 1993, where $x = 0$ corresponds to 1989, $x = 1$ corresponds to 1990, and so on, and y is in millions of dollars. Based on this model, what would be the approximate amount of credit card fraud losses in 1994? In what year did losses reach 3660 million dollars (that is, \$3,660,000,000).

62. According to research done by the Beverage Marketing Corporation, ready-to-drink iced tea sales have boomed during the past few years. The model $y = 310x + 260$ approximates the revenue generated, where $x = 0$ corresponds to 1991, and y is in millions of dollars. Based on this model, what would be the revenue generated in 1992? In what year would revenue reach 2430 million dollars?

63. The accompanying bar graph gives a pictorial representation of bank credit card charges in each year from 1989 to 1992. Use the graph to estimate the charges in each of the following years:
(a) 1989 (b) 1990
(c) 1991 (d) 1992.

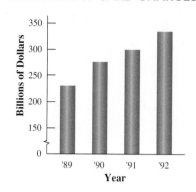

BANK CREDIT CARD CHARGES

64. Based on information from RAM Research, the amount of bank credit card charges in each year between 1989 and 1992 can be approximated by the linear model $y = 37x + 230$, where $x = 0$ corresponds to 1989, and y is in billions of dollars.
(a) According to this model, what would be the charges in 1993?
(b) In 1993, the actual charges were approximately 422 billion dollars. How might you explain the discrepancy between the predicted amount from part (a) and the actual amount charged?

2.2 *Formulas*

Solving a problem in algebra often requires a formula. A **formula** is an equation that tells how to find the value of some quantity. Examples of formulas are

$$d = rt, \quad I = prt, \quad \text{and} \quad P = 2L + 2W.$$

A list of the formulas used in this book is given in the inside covers. Metric units are used in some of the exercises for this section, but no knowledge of the metric system is needed to work the problems.

OBJECTIVE 1 We say a formula is *solved for a specified variable* when that variable is isolated on one side of the equation. In some applications, a particular formula may be useful, but the formula, as given, does not isolate the desired variable. We can isolate the variable we want by "undoing" operations, using inverses. For example, if the variable is multiplied by 2, we undo that by dividing by 2. Addition is undone by subtraction, and so on.

The following examples show how to solve a formula for any one of its variables. The steps used in these examples are the same as those we used in solving a linear equation.

E X A M P L E   I   Solving for a Specified Variable

Solve the formula $I = prt$ for t.

This formula gives the amount of simple interest, I, in terms of the principal (the amount of money deposited), p, the yearly rate of interest, r, and time in years, t. To solve this formula for t, we assume that I, p, and r are constants (having a fixed value) and that t is the only variable. Then we use the properties of equality to get t alone on one side of the equation.

$$I = prt \qquad \text{We want to solve for } t.$$

$$I = (pr)t \qquad \text{Associative property}$$

$$\frac{I}{pr} = \frac{(pr)t}{pr} \qquad \text{Divide by } pr.$$

$$\frac{I}{pr} = t$$

This result is a formula for t, time in years.

WORK PROBLEM I AT THE SIDE. ▶▶

The process of solving for a specified variable uses the steps given in Section 2.1 for solving a linear equation. Some formulas, however, may require the following alternative steps.

Solving for a Specified Variable

Step 1 Use the addition property to get all terms with the specified variable on one side of the equation.

Step 2 Use the addition property to get all terms without the specified variable on the other side of the equation.

Step 3 Use the distributive property to factor the terms with the specified variable.*

Step 4 Complete the solution using the steps listed in Section 2.1.

---

* Writing $ab + ac$ as $a(b + c)$ is called *factoring*.

OBJECTIVES

1 ▶ Solve a formula for a specified variable.

2 ▶ Solve applied problems using formulas.

3 ▶ Solve percent problems.

FOR EXTRA HELP

Tutorial   Tape 2   SSM, Sec. 2.2

1. Solve $I = prt$ for each of the following.

   (a) p

   (b) r

ANSWERS

1. **(a)** $p = \dfrac{I}{rt}$ **(b)** $r = \dfrac{I}{pt}$

2. Solve the formula

$$m = 2k + 3b$$

for each of the following.

(a) k

(b) b

3. Solve the formula

$$P = a + b + c$$

for each of the following.

(a) b

(b) c

4. Solve each equation for x.

(a) $2x + ky = p - qx$

(b) $m = \dfrac{2x - z}{x}$

E X A M P L E 2 Solving for a Specified Variable

Solve the formula $P = 2L + 2W$ for W.

This formula gives the relationship between the perimeter (P) of a rectangle, and the length (L) and width (W) of the rectangle. See Figure 1.

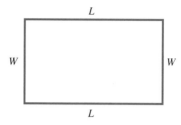

FIGURE 1

To solve the formula for W, we want W alone on one side of the equals sign. Start by subtracting $2L$ from both sides.

$$P = 2L + 2W$$
$$P - 2L = 2L + 2W - 2L \qquad \text{Subtract } 2L.$$
$$P - 2L = 2W$$
$$\frac{P - 2L}{2} = \frac{2W}{2} \qquad \text{Divide by 2.}$$
$$\frac{P - 2L}{2} = W$$

Caution

In Example 2, you cannot reduce the fraction by dividing 2 into the term $2L$. The subtraction in the numerator must be done before the division.

$$\frac{P - 2L}{2} \neq P - L$$

◀◀ WORK PROBLEMS 2 AND 3 AT THE SIDE.

E X A M P L E 3 Solving for a Variable Requiring the Distributive Property

Solve $ax + 4y = 12 - by$ for y.

Use the steps given above.

$$ax + 4y = 12 - by$$

Step 1 $\qquad ax + 4y + by = 12 \qquad$ Add by.

Step 2 $\qquad\qquad 4y + by = 12 - ax \qquad$ Subtract ax.

Step 3 $\qquad\qquad (4 + b)y = 12 - ax \qquad$ Distributive property

$$y = \frac{12 - ax}{4 + b} \qquad \text{Divide by } 4 + b.$$

◀◀ WORK PROBLEM 4 AT THE SIDE.

ANSWERS

2. (a) $k = \dfrac{m - 3b}{2}$ (b) $b = \dfrac{m - 2k}{3}$

3. (a) $b = P - a - c$
    (b) $c = P - a - b$

4. (a) $x = \dfrac{p - ky}{2 + q}$ (b) $x = \dfrac{-z}{m - 2}$

OBJECTIVE 2 The next examples show how we use formulas to solve applied problems.

EXAMPLE 4 Finding Average Speed

Juan Ortega found that on the average it took $\frac{3}{4}$ hour each day to drive a distance of 15 miles to work. What was his average speed?

This problem requires the formula for distance, $d = rt$, where d represents distance traveled; r, rate or speed; and t, time elapsed. We can find the rate or speed, r, by first solving the formula for r.

$$d = rt$$

$$\frac{d}{t} = \frac{rt}{t} \qquad \text{Divide by } t.$$

$$\frac{d}{t} = r$$

Now find r by substituting the given values of d and t into this formula.

$$r = \frac{15}{\frac{3}{4}} = \frac{15}{1} \cdot \frac{4}{3} = 20$$

Ortega's average speed was 20 miles per hour.

WORK PROBLEM 5 AT THE SIDE. ▶▶

OBJECTIVE 3 An important everyday use of mathematics involves the concept of percent. Percent is written with the symbol %. The word **percent** means "per one hundred." One percent means "one per one hundred" or "one one-hundredth."

$$1\% = .01 \quad \text{or} \quad 1\% = \frac{1}{100}$$

A **percentage** is part of a whole. For example, because

$$25\% = 25\left(\frac{1}{100}\right) = \frac{25}{100} = \frac{1}{4}$$

of a whole, 25% of 400 is $\frac{1}{4}$ of 400, or 100.

Solving Percent Problems

Let a represent the (partial) amount, b represent the base (whole amount), and P represent the percentage.

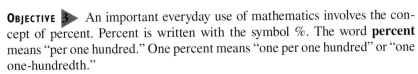

$$\frac{\text{amount}}{\text{base}} = \frac{\text{percentage}}{100}$$

$$\frac{a}{b} = \frac{P}{100} = \text{percent (per 100)}$$

EXAMPLE 5 Solving Percent Problems

(a) Fifty liters of a mixture of acid and water contains 10 liters of acid. What is the percent of acid in the mixture?

The given amount of the mixture is 50 liters and the part that is acid (percentage) is 10 liters. Let x represent the percent of acid. Then, the

── CONTINUED ON NEXT PAGE

5. (a) A triangle has an area of 36 square inches and a base of 12 inches. Find its height.

(b) In 1995 Jacques Villeneuve won the Indianapolis 500 (mile) with a speed of 153.6 miles per hour. Find his time.

6. (a) A mixture of gasoline and oil contains 20 ounces, 1 ounce of which is oil. What percent of the mixture is oil?

(b) An automobile salesman earns an 8% commission on every car he sells. How much does he earn on a car that sells for $12,000?

7. Refer to Figure 2. In a group of 4000 users, how many would we expect to use their card more than 5 times?

percent of acid in the mixture is

$$\frac{10}{50} = x$$

$$x = .20 = \frac{20}{100} = 20\%.$$

(b) If a savings account balance of $3550 earns 8% interest in one year, how much interest is earned?

Here we must find percentage; that is, what part of the whole is earned as interest? One way to do this is to multiply the percent interest by the account balance. Since 8% = .08,

$$\frac{x}{3550} = 8\% = .08$$

$$x = .08(3550)$$

$$x = 284.00.$$

The amount of interest is $284.00.

◀◀ **WORK PROBLEM 6 AT THE SIDE.**

Graphs are sometimes used to represent the percents of a population satisfying certain conditions, as shown in the final example.

E X A M P L E 6   Interpreting Percents from a Graph

According to information provided by Research Partnership, 60% of consumers in the United States have automatic teller cards. (These are known also as ATM cards.) The graph shown in Figure 2 indicates how many times they are used per month. In a typical group of 3500 automatic teller card users, how many would we expect to use their cards 6–9 times?

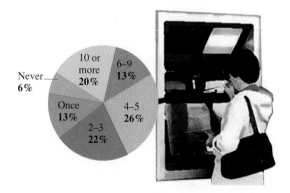

FIGURE 2

According to the graph, 13% of the population use their cards 6–9 times. We must find 13% of 3500. Multiplying .13 by 3500 gives

$$.13(3500) = 455.$$

Therefore, we would expect about 455 people in the group of 3500 to use their cards 6–9 times.

◀◀ **WORK PROBLEM 7 AT THE SIDE.**

———————————————————— **MATHEMATICAL CONNECTIONS (EXERCISES 1–6)** ————————————————————

Consider the following equations:

First Equation

$$x = \frac{5x + 8}{3}$$

Second Equation

$$t = \frac{bt + k}{c} \quad (c \neq 0).$$

Solving the second equation for t follows the same logic as solving the first equation for x. When solving for t, we treat all other variables as though they were constants.

The following group of exercises illustrates the "parallel logic" of solving for x and solving for t. All must be worked sequentially.

1. **(a)** Clear the first equation of fractions by multiplying through by 3.
   (b) Clear the second equation of fractions by multiplying through by c.

2. **(a)** Get the terms involving x on the left side of the first equation by subtracting $5x$ from both sides.
   (b) Get the terms involving t on the left side of the second equation by subtracting bt from both sides.

3. **(a)** Combine like terms on the left side of the first equation. What property allows us to write $3x - 5x$ as $(3 - 5)x = -2x$?
   (b) Write the expression on the left side of the second equation so that t is a factor. What property allows us to do this?

4. **(a)** Divide both sides of the first equation by the coefficient of x.
   (b) Divide both sides of the second equation by the coefficient of t.

5. Look at your answer for the second equation. What restriction must be placed on the variables? Why is this necessary?

6. Write a short paragraph summarizing what you have learned in this group of exercises.

7. One source for geometric formulas gives the formula for the perimeter of a rectangle as $P = 2L + 2W$, while another gives it as $P = 2(L + W)$. Are these equivalent? If so, what property justifies their equivalence?

8. When a formula is solved for a particular variable, several different equivalent forms may be possible. If we solve $A = (1/2)bh$ for h, one possible correct answer is

$$h = \frac{2A}{b}.$$

Which one of the following is *not* equivalent to this?

(a) $h = 2\left(\dfrac{A}{b}\right)$ **(b)** $h = 2A\left(\dfrac{1}{b}\right)$ **(c)** $h = \dfrac{A}{\dfrac{1}{2}b}$ **(d)** $h = \dfrac{\dfrac{1}{2}A}{b}$

Solve each formula for the specified variable. See Examples 1 and 2.

9. $I = prt$ for r (simple interest)

10. $d = rt$ for r (distance)

11. $P = 2L + 2W$ for L
(perimeter of a rectangle)

12. $A = bh$ for b
(area of a parallelogram)

13. $V = LWH$ for W
(volume of a rectangular solid)

14. $P = a + b + c$ for a
(perimeter of a triangle)

15. $C = 2\pi r$ for r
(circumference of a circle)

16. $A = \dfrac{1}{2}bh$ for h (area of a triangle)

17. $A = \dfrac{1}{2}h(B + b)$ for B

(area of a trapezoid)

18. $S = 2\pi rh + 2\pi r^2$ for h

(surface area of a right circular cylinder)

19. $F = \dfrac{9}{5}C + 32$ for C

(Celsius to Fahrenheit)

20. $C = \dfrac{5}{9}(F - 32)$ for F

(Fahrenheit to Celsius)

Solve each of the following for the specified variable. In each case, use the distributive property to factor as necessary. See Example 3.

21. $2k + ar = r - 3y$ for r

22. $4s + 7p = tp - 7$ for p

23. $w = \dfrac{3y - x}{y}$ for y

24. $c = \dfrac{-2t + 4}{t}$ for t

25. Refer to margin problem 4(b) in this section. A student obtained the answer

$$x = \dfrac{z}{2 - m}$$

but was concerned that this answer does not match the one given. Was her answer correct? Why or why not?

26. Suppose the formula $A = 2HW + 2LW + 2LH$ is "solved for L" as follows.

$$A = 2HW + 2LW + 2LH$$
$$A - 2LW - 2HW = 2LH$$
$$\dfrac{A - 2LW - 2HW}{2H} = L$$

While there are no algebraic errors here, what is wrong with the final line, if we are interested in solving for L?

Solve each of the following problems. See Example 4.

27. In 1994 Al Unser, Jr., won the Indianapolis 500 (mile) with a speed of 160.9 miles per hour. Find his time.

28. In 1975, rain shortened the Indianapolis 500 to 435 miles. It was won by Bobby Unser who averaged 149.2 miles per hour. What was his time?

29. The lowest temperature ever recorded in Arizona was −40 degrees Celsius on January 7, 1971. Find the corresponding Fahrenheit temperature.

30. The melting point of brass is 900 degrees Celsius. Find the corresponding Fahrenheit temperature.

31. The base of the Great Pyramid of Cheops is a square whose perimeter is 920 meters. What is the length of each side of this square?

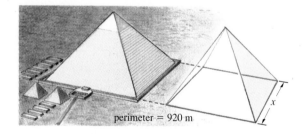

perimeter = 920 m

32. The Peachtree Plaza Hotel in Atlanta is in the shape of a cylinder with radius 46 meters and height 220 meters. Find its volume to the nearest tenth.

33. The circumference of a circle is 480π inches. What is its radius? What is its diameter?

34. The radius of a circle is 2.5 inches. What is its diameter? What is its circumference?

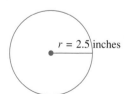

$r = 2.5$ inches

35. A cord of wood contains 128 cubic feet of wood. If a stack of wood is 4 feet wide and 4 feet high, how long must it be if it contains exactly 1 cord?

36. Give one set of possible dimensions for a stack of wood that contains 1.5 cords. (See Exercise 35.)

Exercises 37–40 involve percent. Use your own method to solve each problem. See Examples 5 and 6.

37. The pie chart shows the approximate percents for different sizes of Health Maintenance Organizations (HMOs) for the year 1991. During that year there were approximately 560 such organizations. Find the approximate number for each size group.

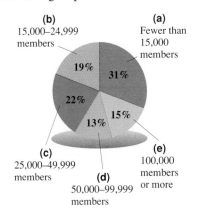

(b)
15,000–24,999
members

(a)
Fewer than
15,000
members

19%   31%

22%

13%   15%

(c)
25,000–49,999
members

(d)
50,000–99,999
members

(e)
100,000
members
or more

38. The pie chart shows the breakdown, by approximate percents, of age groups buying aerobic shoes in 1990. According to figures of the National Sporting Goods Association, $582,000,000 was spent on aerobic shoes that year. How much was spent by each age group?
(a) younger than 14   **(b)** 18 to 24
(c) 45 to 64

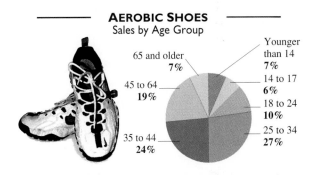

AEROBIC SHOES
Sales by Age Group

65 and older
7%

45 to 64
19%

35 to 44
24%

Younger
than 14
7%

14 to 17
6%

18 to 24
10%

25 to 34
27%

39. The pie chart shows the breakdown for the year 1990 regarding the sales and purchases of exercise equipment. If $2,295,000,000 was spent that year, determine the amount spent by each age group.
(a) 24 to 34   **(b)** 35 to 44
(c) 65 and older

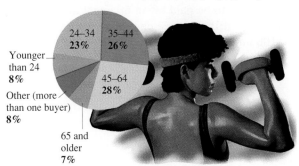

PUMPING SALES
Sales of exercise equipment by age group

24–34
23%

35–44
26%

Younger
than 24
8%

Other (more
than one buyer)
8%

45–64
28%

65 and
older
7%

40. At the start of play on September 18, 1995, the standings of the Central Division of the American League were as shown. "Winning percentage" is commonly expressed as a decimal rounded to the nearest thousandth. To find the winning percentage of a team, divide the number of wins by the total number of games played. Find the winning percentage of each of the following teams.
(a) Cleveland   **(b)** Chicago
(c) Milwaukee

	Won	Lost
Cleveland	91	41
Kansas City	67	63
Milwaukee	62	69
Chicago	60	70
Minnesota	48	81

41. A mixture of alcohol and water contains a total of 36 ounces of liquid. There are 9 ounces of pure alcohol in the mixture. What percent of the mixture is water? What percent is alcohol?

42. A mixture of acid and water is 35% acid. If the mixture contains a total of 40 liters, how many liters of pure acid are in the mixture? How many liters of pure water are in the mixture?

43. In a survey of 263,000 consumers aged 14 years or older, 24% did not read the Sunday paper. How many of these consumers did not read the Sunday paper? (*Source:* Impact Resources)

44. A study showed that 17% of 18–24-year-olds snore. In a group of 2500 people in this age bracket, how many could we expect to snore? (*Source:* The Better Sleep Council)

45. A real estate agent earned $6300 commission on a property sale of $210,000. What is her rate of commission?

46. A certificate of deposit for one year pays $85 simple interest on a principal of $3400. What is the interest rate being paid on this deposit?

The formula

$$A = \frac{24f}{b(p + 1)}$$

gives the approximate annual interest rate for a consumer loan paid off with equal monthly payments. Here f is the finance charge on the loan, p is the number of payments, and b is the original amount of the loan. Solve the following problems using this formula.

47. Find the approximate annual interest rate on a loan that is to be paid off in 48 monthly payments (all equal), if the finance charge on the loan is $400 and the original loan balance is $3840.

48. Find the approximate annual interest rate on a loan that is to be paid off in 60 equal monthly payments, if the finance charge on the loan is $1100 and the amount financed is $6200.

2.3 *Applications of Linear Equations*

To use algebra to solve practical problems, we must first translate the verbal statements of the problems into mathematical statements. We show specific methods for making this translation in this section.

OBJECTIVE 1 Applied problems tend to have key phrases that translate into mathematical expressions involving the operations of addition, subtraction, multiplication, and division. Translations of some commonly used expressions are listed in the following table.

Verbal Expression	Mathematical Expression
Addition	
The **sum** of a number and 7	$x + 7$
6 **more than** a number	$x + 6$
3 **plus** 8	$3 + 8$
24 **added to** a number	$x + 24$
A number **increased by** 5	$x + 5$
The **sum** of two numbers	$x + y$
Subtraction	
2 **less than** a number	$x - 2$
12 **minus** a number	$12 - x$
A number **decreased by** 12	$x - 12$
The **difference between** two numbers	$x - y$
A number **subtracted from** 10	$10 - x$
Multiplication	
16 **times** a number	$16x$
Some number **multiplied by** 6	$6x$
$\frac{2}{3}$ **of** some number (used only with fractions and percent)	$\left(\frac{2}{3}\right)x$
13% **of** a number	$.13x$
Twice (2 times) a number	$2x$
The **product of** two numbers	xy
Division	
The **quotient of** 8 and some nonzero number	$\dfrac{8}{x}$
A number **divided by** 13	$\dfrac{x}{13}$
The **ratio of** two nonzero numbers or The **quotient of** two nonzero numbers	$\dfrac{x}{y}$

Caution
Because subtraction and division are not commutative operations, it is important to correctly translate expressions involving them. For example, "2 less than a number" is translated as $x - 2$, *not* $2 - x$. "A number subtracted from 10" is expressed as $10 - x$, not $x - 10$. For division, it is understood that the number doing the dividing is the denominator, and the number that is divided is the numerator. For example, "a number divided by 13" and "13 dividing x" both translate as $\frac{x}{13}$.

WORK PROBLEM I AT THE SIDE. ▶▶

OBJECTIVES

1 ▶ Translate from words to mathematical expressions.

2 ▶ Write equations from given information.

3 ▶ Distinguish between expressions and equations.

4 ▶ Use the six steps in solving an applied problem.

5 ▶ Solve percent problems.

6 ▶ Solve investment problems.

7 ▶ Solve mixture problems.

FOR EXTRA HELP

Tutorial   Tape 2   SSM, Sec. 2.3

1. Write these verbal expressions as mathematical expressions. Use x as the variable.

   (a) 9 added to a number

   (b) The difference between 7 and a number

   (c) Four times a number

   (d) The product of -6 and a number

   (e) The quotient of 7 and a nonzero number

ANSWERS

1. **(a)** $9 + x$ or $x + 9$ **(b)** $7 - x$
   (c) $4x$ **(d)** $-6x$ **(e)** $\dfrac{7}{x}$

2. Translate each verbal sentence into an equation.

(a) The sum of a number and 6 is 28.

(b) If twice a number is decreased by 3, the result is 17.

(c) The product of a number and 7 is twice the number plus 12.

(d) The quotient of a number and 6, added to twice the number, is 7.

(e) The quotient of a number and 3 is 8.

(f) The sum of a number and twice the number is 3.

3. For each of the following, decide whether it is an equation or an expression.

(a) $5x - 3(x + 2) = 7$

(b) $5x - 3(x + 2)$

ANSWERS

2. **(a)** $x + 6 = 28$ **(b)** $2x - 3 = 17$

   (c) $7x = 2x + 12$ **(d)** $\dfrac{x}{6} + 2x = 7$

   (e) $\dfrac{x}{3} = 8$ **(f)** $x + 2x = 3$

3. **(a)** equation
   (b) expression

OBJECTIVE 2 The symbol for equality, $=$, is often indicated by the word "is." In fact, because equal mathematical expressions represent names for the same number, any words that indicate the idea of "sameness" indicate translation to $=$.

E X A M P L E   1   Translating Words into Equations

Translate the following verbal sentences into equations.

Verbal Sentence	*Equation*
Twice a number, **decreased by 3**, **is** 42.	$2x - 3 = 42.$
If the **product of a number and 12** is decreased by 7, the result **is** 105.	$12x - 7 = 105$
The **quotient of a number and the number plus 4 is** 28.	$\dfrac{x}{x + 4} = 28$
The **quotient of a number and 4**, plus the number, **is** 10.	$\dfrac{x}{4} + x = 10$

◄◄ **WORK PROBLEM 2 AT THE SIDE.**

OBJECTIVE 3 It is important to be able to distinguish between algebraic expressions and equations. An expression translates as a phrase; an equation includes the $=$ symbol and translates as a sentence.

E X A M P L E   2   Distinguishing between Equations and Expressions

For each of the following, decide whether it is an equation or an expression.

(a) $2(3 + x) - 4x + 7$

There is no equals sign, so this is an expression.

(b) $2(3 + x) - 4x + 7 = -1$

Because of the equals sign, this is an equation. Verify that the expression in part (a) simplifies to $-2x + 13$, and the equation in part (b) has solution set $\{7\}$.

◄◄ **WORK PROBLEM 3 AT THE SIDE.**

The remainder of this section consists of examples of some common types of applied problems. The following steps are helpful in solving applied problems, and are used in Examples 3–6.

Solving an Applied Problem

Step 1 Read the problem carefully. Decide what is given and what must be found. Use sketches as necessary.

Step 2 Choose a variable and write down exactly what it represents. (Do not make the mistake of omitting this step.)

Step 3 Write an equation using the information given in the problem and the variable from Step 2.

Step 4 Solve the equation.

Step 5 Make sure you have answered the question of the problem.

Step 6 Check your answer *in the words of the original problem*, not in the equation you obtained from the words.

OBJECTIVE 4 We now see how these steps are applied. The first example comes from geometry.

EXAMPLE 3 Solving a Geometry Problem

The length of a rectangle is 1 centimeter more than twice the width. The perimeter of the rectangle is 110 centimeters. Find the length and the width of the rectangle.

Step 1 What must be found? The length and width of the rectangle. What is given? The length is 1 more than twice the width; the perimeter is 110.
Make a sketch, as in Figure 3.

W

$1 + 2W$

FIGURE 3

Step 2 Choose a variable: let W = the width; then $1 + 2W$ = the length.

Step 3 Write an equation. The perimeter of a rectangle is given by the formula $P = 2L + 2W$.

$$P = 2L + 2W$$
$$110 = 2(1 + 2W) + 2W. \qquad \text{Let } L = 1 + 2W \text{ and } P = 110.$$

Step 4 Solve the equation obtained in Step 3.

$$110 = 2(1 + 2W) + 2W$$
$$110 = 2 + 4W + 2W \qquad \text{Distributive property}$$
$$110 = 2 + 6W \qquad \text{Combine terms.}$$
$$110 - 2 = 2 + 6W - 2 \qquad \text{Subtract 2.}$$
$$108 = 6W$$
$$18 = W \qquad \text{Divide by 6.}$$

Step 5 Answer the question. The width of the rectangle is 18 centimeters and the length is $1 + 2(18) = 37$ centimeters.

Step 6 Check the answer by substituting these dimensions into the words of the original problem.

The next example shows how the six steps are used in a problem that requires finding two unknown quantities.

EXAMPLE 4 Finding Unknown Numerical Quantities

Two outstanding major league pitchers of the past decade are Roger Clemens and Greg Maddux. Between 1984 and 1996, they pitched a total of 719 games. Clemens pitched 47 more games than Maddux. How many did each player pitch?

Step 1 We are asked to find the number of games each player pitched. We must choose a variable to represent the number of games of one of the men.

Let m = the number of games for Maddux.

CONTINUED ON NEXT PAGE

4. (a) The length of a rectangle is 5 centimeters more than its width. The perimeter is five times the width. What are the dimensions of the rectangle?

Step 2 We must also find the number of games for Clemens. Since he pitched 47 games more than Maddux,

$$m + 47 = \text{Clemens' number of games.}$$

Step 3 The sum of the numbers of games is 719, so we can now write an equation.

$$\underbrace{\text{Maddux's games}}_{m} + \underbrace{\text{Clemens' games}}_{(m + 47)} = \underbrace{\text{total}}_{719}$$

Step 4 Solve the equation.

$$m + (m + 47) = 719$$

$2m + 47 = 719$	Combine like terms.
$2m = 672$	Subtract 47.
$m = 336$	Divide by 2.

Step 5 Since m represents the number of Maddux's games, $m + 47 = \mathbf{336} + 47 = 383$ is the number of games pitched by Clemens.

Step 6 383 is 47 more than 336, and the sum of 336 and 383 is 719.

The words of the problem are satisfied, and our solution checks.

(b) Cindy is 5 years older than three times the age of her son, Cody. The sum of their ages is 45. Find their ages.

> **Caution**
> A common error in solving applied problems is forgetting to answer all the questions asked in the problem. In Example 4, we were asked for the number of games for *each* player, so there was an extra step at the end in order to find Clemens' number.

◀◀ **WORK PROBLEM 4 AT THE SIDE.**

OBJECTIVE 5 ▶ The next example involves percent. Recall from the previous section that percent means "per one hundred," so 5% means .05, 14% means .14, and so on. (We do not number the steps in the remaining examples.)

E X A M P L E 5   Solving a Percent Problem

In 1993 there were 154 long distance area codes in the United States. This was an increase of 79% over the number when the area code plan originated in 1947. How many area codes were there in 1947?

We must find the number of area codes in 1947, and we are given that the number in 1947, increased by 79%, was 154. Let x represent the number in 1947, and thus $.79x$ represents the increase. From this, we get the equation

$$\text{the number in 1947} + \text{the increase} = 154$$
$$x + .79x = 154.$$

Solve the equation.

$1x + .79x = 154$	Identity property
$1.79x = 154$	Combine terms.
$x = 86$	Divide by 1.79. (Use a calculator if you wish.)

There were 86 area codes in 1947. Check that the increase, $154 - 86 = 68$, is 79% of 86.

ANSWERS
4. (a) 10 centimeters by 15 centimeters
   (b) Cindy is 35 years old; Cody is 10 years old.

Caution
Watch for two common errors that occur in solving problems like the one in Example 5. First, do not try to find 79% of 154 and subtract that from 154. The 79% should be applied to the *amount in 1947 and not the amount in 1993*. Second, avoid writing the equation as

$$x + .79 = 154.$$

A percent must always be multiplied times some amount, $.79x$ in the example.

WORK PROBLEM 5 AT THE SIDE. ▶▶

OBJECTIVE ▶6▶ The next example shows how we can use linear equations to solve certain types of investment problems.

EXAMPLE 6   Solving an Investment Problem

After winning the state lottery, Michael Chin has $40,000 to invest. He will put part of the money in an account paying 4% interest and the remainder into stocks paying 6% interest. His accountant tells him that the total annual income from these investments should be $2040. How much should he invest at each rate?

Let x = the amount to invest at 4%;

$40,000 - x$ = the amount to invest at 6%.

The formula for interest is $I = prt$. Here the time, t, is 1 year. Make a chart to organize the given information.

% as a Decimal	Amount Invested	Interest in 1 Year
.04	x	$.04x$
.06	$40,000 - x$	$.06(40,000 - x)$
		2040

The last column gives the equation.

interest at 4%   +   interest at 6%   =   total interest

$.04x$ + $.06(40,000 - x)$ =   2040

Solve the equation.

$.04x + .06(40,000) - .06x = 2040$	Clear parentheses.
$-.02x + 2400 = 2040$	Combine terms.
$-.02x = -360$	Subtract 2400.
$x = 18,000$	Divide by $-.02$.

Chin should invest $18,000 at 4% and $40,000 - $18,000 = $22,000 at 6%. Check by finding the annual interest at each rate; they should total $2040.

WORK PROBLEM 6 AT THE SIDE. ▶▶

5. (a) A number increased by 15% is 287.5. Find the number.

(b) Michelle Raymond was paid $162 for a week's work at her part-time job after 10% deductions for taxes. How much did she earn before the deductions were made?

6. (a) A woman invests $72,000 in two ways—some at 5% and some at 3%. Her total annual interest income is $3160. Find the amount invested at each rate.

(b) A man has $34,000 to invest. He invests some at 5% and the balance at 4%. His total annual interest income is $1545. Find the amount invested at each rate.

ANSWERS

5. (a) 250 **(b)** $180
6. (a) $50,000 at 5%; $22,000 at 3%
   (b) $18,500 at 5%; $15,500 at 4%

7. (a) How many liters of a 10% solution should be mixed with 60 liters of a 25% solution to get a 15% solution?

OBJECTIVE 7 ▶ Mixture problems involving rates of concentration may be solved with linear equations.

E X A M P L E 7   Solving a Mixture Problem

A chemist must mix 8 liters of a 40% solution of acid with some 70% solution to get a 50% solution. How much of the 70% solution should be used?

The information in the problem is illustrated in Figure 4.

FIGURE 4

Let x = the number of liters of 70% solution to be used.

Use the given information to complete the following table.

Strength	Liters of Solution	Liters of Pure Acid
40%	8	.40(8) = 3.2
70%	x	.70x
50%	8 + x	.50(8 + x)

Sum must equal

The numbers in the right-hand column were found by multiplying the strengths and the numbers of liters. The number of liters of pure acid in the 40% solution plus the number of liters in the 70% solution must equal the number of liters in the 50% solution.

$$3.2 + .70x = .50(8 + x)$$
$$3.2 + .70x = 4 + .50x \qquad \text{Clear parentheses.}$$
$$.20x = .8 \qquad \text{Subtract 3.2 and .50}x.$$
$$x = 4 \qquad \text{Divide by .20.}$$

Check that the chemist should use 4 liters of the 70% solution.

(b) How many pounds of candy worth $8 per pound should be mixed with 100 pounds of candy worth $4 per pound to get a mixture that can be sold for $7 per pound?

◀◀ **WORK PROBLEM 7 AT THE SIDE.**

2.3 Exercises

1. In your own words, list the six steps suggested for solving applied problems.

Step 1 _____

Step 2 _____

Step 3 _____

Step 4 _____

Step 5 _____

Step 6 _____

2. If x represents the number, express the following using algebraic symbols:
(a) 9 less than a number   **(b)** 9 is less than a number
Which one of these is an *expression*?

Translate the following verbal phrases into mathematical expressions. Use x to represent the unknown number.

3. A number decreased by 13

4. A number decreased by 12

5. 7 increased by a number

6. 12 more than a number

7. The product of 8 and 12 more than a number

8. The product of 9 less than a number and 6 more than the number

9. The quotient of a number and 6

10. The quotient of 6 and a nonzero number

For each of the following, select a variable for the unknown, and write an equation representing the verbal sentence. Then solve the problem. See Example 1.

11. The sum of a number and 6 is -31. Find the number.

12. The sum of a number and -4 is 12. Find the number.

13. If the product of a number and -4 is subtracted from the number, the result is 9 more than the number. Find the number.

14. If the quotient of a number and 6 is added to twice the number, the result is 8 less than the number. Find the number.

15. When $\frac{2}{3}$ of a number is subtracted from 12, the result is 10. Find the number.

16. When 75% of a number is added to 6, the result is 3 more than the number. Find the number.

17. Which one of the following is *not* a valid translation of "20% of a number"?

(a) $.20x$ (b) $.2x$ (c) $\dfrac{x}{5}$ (d) $20x$

18. Explain why $13 - x$ is *not* a correct translation of "13 less than a number."

Decide whether each of the following is an equation or an expression. See Example 2.

19. $5(x + 3) - 8(2x - 6)$

20. $-7(y + 4) + 13(y - 6)$

21. $5(x + 3) - 8(2x - 6) = 12$

22. $-7(y + 4) + 13(y - 6) = 18$

23. $\dfrac{r}{2} - \dfrac{r + 9}{6} - 8$

24. $\dfrac{r}{2} - \dfrac{r + 9}{6} = 8$

Use the six-step method of solving applied problems to solve each of the following. See Examples 3 and 4.

25. The John Hancock Center in Chicago has a rectangular base. The length of the base measures 65 feet less than twice the width. The perimeter of this base is 860 feet. What are the dimensions of the base?

26. The John Hancock Center (see Exercise 25) tapers as it rises. The top floor is rectangular and has perimeter 520 feet. The width of the top floor measures 20 feet more than one-half its length. What are the dimensions of the top floor?

27. The Bermuda Triangle supposedly causes trouble for aircraft pilots. It has perimeter 3075 miles. The shortest side measures 75 miles less than the middle side, and the longest side measures 375 miles more than the middle side. Find the lengths of the three sides.

28. The Vietnam Memorial in Washington, D.C., is in the shape of an unenclosed isosceles triangle. If the two walls of equal length were joined by a straight line of 438 feet, the perimeter of the resulting triangle would be 931.5 feet. Find the lengths of the two walls.

29. According to figures provided by the Air Transport Association of America, the Boeing B747–400 and the McDonnell Douglas L1011–100/200 are among the air carriers with maximum passenger seating. The Boeing seats 110 more passengers than the McDonnell Douglas, and together the two models seat 696 passengers. What is the seating capacity of each model?

30. Two of the highest paid business executives in a recent year were Mike Eisner, chairman of Disney, and Ed Horrigan, vice chairman of RJR Nabisco. Together their salaries totaled 61.8 million dollars. Eisner earned 18.4 million dollars more than Horrigan. What was the salary for each executive?

31. In a recent year, the two U.S. industrial corporations with the highest sales were General Motors and Ford Motor. Their sales together totaled 213.5 billion dollars. Ford Motor sales were 28.7 billion dollars less than General Motors. What were the sales for each corporation?

32. Babe Ruth and Rogers Hornsby were two great hitters. Together they got 5803 base hits in their careers. Hornsby got 57 more hits than Ruth. How many base hits did each get?

Solve the following problems involving percent. See Example 5.

33. From 1970 to 1980, composite scores on the ACT exam dropped from 19.9 to 18.5. What percent decrease was this? (*Source:* The American College Testing Program)

34. In 1986, the number of participants in the ACT exam was 730,000. In 1990, a total of 817,000 people took the exam. What percent increase was this? (*Source:* The American College Testing Program)

35. In 1980, the average tuition for public four-year universities in the United States was $840 for full-time students. By 1990, it had risen approximately 139%. What was the approximate cost in 1990? (*Source:* National Center for Education Statistics, U.S. Dept. of Education)

36. The consumer price index (CPI) in June, 1991, was 136.0. This represented a 4.7% increase from June, 1990. What was the CPI in June, 1990?

37. At the end of a day, Jeff Hornsby found that the total cash register receipts at the motel where he works amounted to $1650.78. This included the 8% sales tax charged. Find the amount of the tax.

38. Dongming Wei sold his house for $159,000. He got this amount knowing that he would have to pay a 6% commission to his agent. What amount did he have after the agent was paid?

Solve the following investment problems. See Example 6.

39. George Duda earned $12,000 last year by giving tennis lessons. He invested part at 3% simple interest and the rest at 4%. He earned a total of $440 in interest. How much did he invest at each rate?

40. Rita Ferrandino won $60,000 on a slot machine in Las Vegas. She invested part at 2% simple interest and the rest at 3%. She earned a total of $1600 in interest. How much was invested at each rate?

41. Greg Tobin invested some money at 4.5% simple interest and $1000 less than twice this amount at 3%. His total annual income from the interest was $1020. How much was invested at each rate?

42. Brenda Bravener invested some money at 3.5% simple interest, and $5000 more than 3 times this amount at 4%. She earned $1440 in interest. How much did she invest at each rate?

43. Ed Moura has $29,000 invested in stocks paying 5%. How much additional money should he invest in certificates of deposit paying 2% so that the average return on the two investments is 3%?

44. Karen Guardino placed $15,000 in an account paying 6%. How much additional money should she deposit at 4% so that the average return on the two investments is 5.5%?

Solve the following mixture problems involving rates of concentration and mixtures. See Example 7.

45. Ten liters of a 4% acid solution must be mixed with a 10% solution to get a 6% solution. How many liters of the 10% solution are needed?

46. How many liters of a 14% alcohol solution must be mixed with 20 liters of a 50% solution to get a 30% solution?

47. In a chemistry class, 12 liters of a 12% alcohol solution must be mixed with a 20% solution to get a 14% solution. How many liters of the 20% solution are needed?

48. How many liters of a 10% alcohol solution must be mixed with 40 liters of a 50% solution to get a 40% solution?

49. How much pure dye must be added to 4 gallons of a 25% dye solution to increase the solution to 40%? (*Hint:* Pure dye is 100% dye.)

50. How much water must be added to 6 gallons of a 4% insecticide solution to reduce the concentration to 3%? (*Hint:* Water is 0% insecticide.)

51. Mr. Meyer wants to mix 50 pounds of nuts worth $2 per pound with some nuts worth $6 per pound to make a mixture worth $5 per pound. How many pounds of $6-nuts must he use?

52. Lucy Thrower wants to mix tea worth 2¢ per ounce with 100 ounces of tea worth 5¢ per ounce to make a mixture worth 3¢ per ounce. How much 2¢-tea should be used?

53. Why is it impossible to add two mixtures of candy worth $4 per pound and $5 per pound to obtain a final mixture worth $6 per pound?

54. Write an equation based on the following problem, solve the equation, and explain why the problem has no solution.

How much 30% acid should be mixed with 15 liters of 50% acid to obtain a mixture that is 60% acid?

MATHEMATICAL CONNECTIONS (EXERCISES 55–59)

Consider the following problems.

Problem A
Jack has $800 invested in two accounts. One pays 5% interest per year and the other pays 10% interest per year. The amount of yearly interest is the same as he would get if the entire $800 was invested at 8.75%. How much does he have invested at each rate?

Problem B
Jill has 800 liters of acid solution. She obtained it by mixing some 5% acid with some 10% acid. Her final mixture of 800 liters is 8.75% acid. How much of each of the 5% and 10% mixtures did she use to get her final mixture?

In Problem A, let x represent the amount invested at 5% interest, and in Problem B, let y represent the amount of 5% acid used. Work the following problems in sequence.

55. (a) Write an expression in x that represents Jack's amount of money invested at 10% in Problem A.
  (b) Write an expression in y that represents Jill's amount of 10% acid mixture used in Problem B.

56. (a) Write expressions that represent the amount of interest Jack earns per year at 5% and at 10%.
  (b) Write expressions that represent the amount of pure acid in Jill's 5% and 10% acid mixtures.

57. (a) The sum of the two expressions in part (a) of Exercise 56 must equal the total amount of interest earned in one year. Write an equation representing this fact.
  (b) The sum of the two expressions in part (b) of Exercise 56 must equal the amount of pure acid in the final mixture. Write an equation representing this fact.

58. (a) Solve Problem A.
  (b) Solve Problem B.

59. Explain the similarities between the solution processes used in solving Problems A and B.

2.4 *More Applications of Linear Equations*

There are three common applications of linear equations that we did not discuss in Section 2.3: money problems, uniform motion problems, and problems involving angles of a triangle. Money problems are very similar to the investment problems seen in Section 2.3.

OBJECTIVE 1▶ In problems involving money, we use the basic fact that

$$\begin{bmatrix} \text{Number of} \\ \text{monetary units} \\ \text{of the same kind} \end{bmatrix} \times [\text{Denomination}] = \begin{bmatrix} \text{Total monetary} \\ \text{value.} \end{bmatrix}$$

For example, 30 dimes have a monetary value of $30(.10) = 3.00$ dollars. Fifteen five-dollar bills have a value of $15(5) = 75$ dollars.

┌─ **E X A M P L E I   Solving a Money Problem**

For a bill totaling $5.65, a cashier received 25 coins consisting of nickels and quarters. How many of each type of coin did the cashier receive?

Let x represent the number of nickels;

then $25 - x$ represents the number of quarters.

We can organize the information in a chart as we did with investment problems.

Denomination	Number of Coins	Value
$.05	x	$.05x$
$.25	$25 - x$	$.25(25 - x)$
		5.65

We get the equation from the last column.

$$.05x + .25(25 - x) = 5.65$$
$$5x + 25(25 - x) = 565 \qquad \text{Multiply by 100.}$$
$$5x + 625 - 25x = 565 \qquad \text{Clear parentheses.}$$
$$-20x = -60 \qquad \text{Subtract 625; combine terms.}$$
$$x = 3 \qquad \text{Divide by } -20.$$

The cashier has 3 nickels and $25 - 3 = 22$ quarters. Check to see that the total value of these coins is $5.65. ■

Caution
Be sure that your answer is reasonable when working problems like Example 1. Because you are dealing with a number of coins, the correct answer can neither be negative nor a fraction.

WORK PROBLEM I AT THE SIDE. ▶▶

OBJECTIVE 2▶ Uniform motion problems use the formula Distance = Rate × Time. In this formula, when rate (or speed) is given in miles per hour, time must be given in hours. When solving such problems, draw a sketch to illustrate what is happening in the problem, and make a chart to summarize the information of the problem.

OBJECTIVES

1▶ Solve problems involving money.

2▶ Solve problems involving uniform motion.

3▶ Solve problems involving the angles of a triangle.

FOR EXTRA HELP

Tutorial   Tape 3   SSM, Sec. 2.4

1. At the end of a day, a cashier had 26 dimes and half-dollars. The total value of these coins was $8.60. How many of each type did he have?

ANSWERS

1. 11 dimes, 15 half-dollars

2. Two cars leave the same location at the same time. One travels north at 60 miles per hour and the other south at 45 miles per hour. In how many hours will they be 420 miles apart?

EXAMPLE 2   Solving a Motion Problem

Two cars leave the same place at the same time, one going east and the other west. The eastbound car averages 40 miles per hour, while the westbound car averages 50 miles per hour. In how many hours will they be 300 miles apart?

A sketch shows what is happening in the problem: The cars are going in *opposite* directions. See Figure 5. Let x represent the time traveled by each car. Again we can summarize the information of the problem in a chart. We can find the distances traveled by the cars, $40x$ and $50x$, by using the formula $d = rt$. When the expressions for rate and time are entered in the chart, *fill in the distance expression by multiplying rate by time.*

	Rate	*Time*	*Distance*
Eastbound car	40	x	$40x$
Westbound car	50	x	$50x$

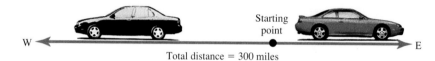

50 mph   40 mph

Starting point

W ⟵ ——————————————————————— ⟶ E

Total distance = 300 miles

FIGURE 5

From the sketch in Figure 5, the sum of the two distances is 300.

$$40x + 50x = 300$$
$$90x = 300 \qquad \text{Combine terms.}$$
$$x = \frac{300}{90} \qquad \text{Divide by 90.}$$
$$x = \frac{10}{3}$$

The cars travel $\frac{10}{3} = 3\frac{1}{3}$ hours, or 3 hours and 20 minutes.

Caution
It is a common error to write 300 as the distance for each car in Example 2. Three hundred miles is the *total* distance traveled.

◀◀ **WORK PROBLEM 2 AT THE SIDE.**

Example 2 involved motion in opposite directions. The next example deals with motion in the same direction.

EXAMPLE 3   Solving a Motion Problem

Jeff Bezzone can bike to work in $\frac{3}{4}$ hour. When he takes the bus, the trip takes $\frac{1}{4}$ hour. If the bus travels 20 miles per hour faster than Jeff rides his bike, how far is it to his workplace?

Although the problem asks for a distance, it is easier here to let x be Jeff's speed when he rides his bike to work. Then the speed of the bus is $x + 20$. Summarize the information of the problem in a chart, and use $d = rt$ to get the distances.

ANSWERS

2. 4 hours

CONTINUED ON NEXT PAGE

	Rate	Time	Distance	
Bike	x	$\frac{3}{4}$	$\frac{3}{4}x$	⎤
Bus	$x + 20$	$\frac{1}{4}$	$\frac{1}{4}(x + 20)$	⎦ Same

The key to setting up the correct equation is to understand that the distance in each case is the same. See Figure 6.

Home

Workplace

FIGURE 6

Since the distance is the same in both cases,

$$\frac{3}{4}x = \frac{1}{4}(x + 20).$$

$$4\left(\frac{3}{4}x\right) = 4\left(\frac{1}{4}\right)(x + 20) \qquad \text{Multiply by 4.}$$

$$3x = x + 20 \qquad \text{Simplify.}$$

$$2x = 20 \qquad \text{Subtract } x.$$

$$x = \mathbf{10} \qquad \text{Divide by 2.}$$

Since x represents Jeff's speed on his bike, he rides his bike at 10 miles per hour.

Now answer the question in the problem. The required distance is given by

$$d = \frac{3}{4}x = \frac{3}{4}(\mathbf{10}) = \frac{30}{4} = \mathbf{7.5}$$

Check by finding the distance using

$$d = \frac{1}{4}(x + 20) = \frac{1}{4}(10 + 20) = \frac{30}{4} = \mathbf{7.5},$$

the same result. Jeff travels 7.5 miles to work.

> **Note**
> As mentioned in Example 3, it was easier to let the variable represent a quantity other than the one that we were asked to find. This is the case in some problems. It takes practice to learn when this approach is the best, and practice means working lots of problems!

WORK PROBLEM 3 AT THE SIDE. ▶▶

OBJECTIVE 3 ▶ One of the important results of Euclidean geometry (the geometry of the Greek mathematician Euclid) is that the sum of the angle measures of any triangle is 180°. This property is used in the next example.

---

3. Elayn begins jogging at 5:00 A.M., averaging 3 miles per hour. Clay leaves at 5:30 A.M., following her, averaging 5 miles per hour. How long will it take him to catch up to her? (*Hint:* 30 minutes = $\frac{1}{2}$ hour.)

4. One angle in a triangle is 15° larger than a second angle. The third angle is 25° larger than twice the second angle. Find the measure of each angle.

E X A M P L E 4   **Finding Angle Measures**

Find the value of x, and determine the measure of each angle in Figure 7.

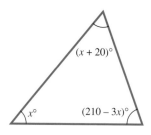

FIGURE 7

Since the three marked angles are angles of a triangle, their sum must be 180°. Write the equation indicating this, and then solve.

$$x + (x + 20) + (210 - 3x) = 180$$
$$-x + 230 = 180 \qquad \text{Combine like terms.}$$
$$-x = 180 - 230 \qquad \text{Subtract 230.}$$
$$-x = -50$$
$$x = 50 \qquad \text{Divide by } -1.$$

One angle measures 50°, another measures $x + 20 = 50 + 20 = 70°$, and the third measures $210 - 3x = 210 - 3(50) = 60°$. Since $50° + 70° + 60° = 180°$, the answers are correct.

■

◀◀ **WORK PROBLEM 4 AT THE SIDE.**

2.4 Exercises

Provide a short answer to each of the following.

_____ **1.** What amount of money is found in a coin purse containing 46 nickels and 18 dimes?

_____ **2.** The distance from Melbourne to London is 10,500 miles. If a jet averages 500 miles per hour between the two cities, what is its travel time in hours?

_____ **3.** Anh Nguyen traveled from Louisville to Kansas City, a distance of 520 miles, in 10 hours. What was his rate in miles per hour?

_____ **4.** A square has perimeter 20 inches. What would be the perimeter of an equilateral triangle whose sides each measure the same length as the side of the square?

Write a short explanation in Exercises 5 and 6.

5. Read over Example 3 in this section. The solution of the equation is 10. Why is *10 miles per hour* not the answer to the problem?

6. Suppose that you know that two angles of a triangle have equal measures, and the third angle measures 26°. Explain in a few words the strategy you would use to find the measures of the equal angles without actually writing an equation.

Solve each of the following problems. See Example 1.

7. Lisa Wunderle's piggy bank has 36 coins. Some are quarters, and the rest are half-dollars. If the total value of the coins is $14.75, how many of each denomination does she have?

8. Jim Camp has a jar in his office that contains 47 coins. Some are pennies, and the rest are dimes. If the total value of the coins is $2.18, how many of each denomination does he have?

9. Ricardo Gutierrez has a box of coins that he uses when playing poker with his friends. The box currently contains 44 coins, consisting of pennies, dimes, and quarters. The number of pennies is equal to the number of dimes, and the total value is $4.37. How many of each denomination of coin does he have in the box?

10. Carlotta Valdes found some coins while looking under her sofa pillows. There were equal numbers of nickels and quarters, and twice as many half-dollars as quarters. If she found $2.60 in all, how many of each denomination of coin did she find?

11. In the nineteenth century, the United States minted two-cent and three-cent pieces. Frances Steib has three times as many three-cent pieces as two-cent pieces, and the face value of these coins is $1.21. How many of each denomination does she have?

12. Pat Kelley collects U.S. gold coins. He has a collection of 80 coins. Some are $10 coins, and the rest are $20 coins. If the face value of the coins is $1060, how many of each denomination does he have?

13. The school production of *Our Town* was a big success. For opening night, 410 tickets were sold. Students paid $1.50 each, while nonstudents paid $3.50 each. If a total of $825 was collected, how many students and how many nonstudents attended?

14. A total of 550 people attended a Boston Pops concert. Floor tickets cost $20 each, while balcony tickets cost $14 each. If a total of $10,400 was collected, how many of each type of ticket were sold?

In Exercises 15–18, find the rate based on the information provided. Use a calculator and round your answer to the nearest hundredth.

Event and Year	Participant	Distance	Time
15. Atlanta Olympics, 200-meter dash, Men, 1996	Michael Johnson	200 meters	19.32 seconds
16. Los Angeles Olympics, 100-meter dash, Women, 1988	Florence Griffith Joyner	100 meters	10.62 seconds
17. Atlanta Olympics, 400-meter dash, Women, 1996	Marie-Jose Perec	400 meters	48.25 seconds
18. Atlanta Olympics, 50-meter freestyle, Women, 1996	Amy Van Dyken	50 meters	24.87 seconds

Solve each of the following problems. See Examples 2 and 3.

19. A train leaves Little Rock, Arkansas, and travels north at 85 kilometers per hour. Another train leaves at the same time and travels south at 95 kilometers per hour. How long will it take before they are 315 kilometers apart?

r	t	d

20. Two steamers leave a port on a river at the same time, traveling in opposite directions. Each is traveling 22 miles per hour. How long will it take for them to be 110 miles apart?

r	t	d

21. Lois and Clark are covering separate stories and have to travel in opposite directions. Lois leaves the *Daily Planet* at 8:00 A.M. and travels 35 miles per hour. Clark leaves at 8:15 A.M. and travels at 40 miles per hour. At what time will they be 140 miles apart?

	r	t	d

22. Agents Mulder and Scully are driving to Georgia to investigate "Big Blue," a giant aquatic reptile reported to inhabit one of the local lakes. Mulder leaves Washington at 8:30 A.M. and averages 65 miles per hour. His partner, Scully, leaves at 9:00 A.M., following the same path and averaging 68 miles per hour. At what time will Scully catch up with Mulder?

	r	t	d

23. When Tri drives his car to work, the trip takes 30 minutes. When he rides the bus, it takes 45 minutes. The average speed of the bus is 12 miles per hour less than his speed when driving. Find the distance he travels to work.

	r	t	d

24. Lakeisha can get to school in 15 minutes if she rides her bike. It takes her 45 minutes if she walks. Her speed when walking is 10 miles per hour slower than her speed when riding. What is her speed when she rides?

	r	t	d

25. On an automobile trip, Angel Vice maintained a steady speed for the first two hours. Rush hour traffic slowed her speed by 25 miles per hour for the last part of the trip. The entire trip, a distance of 125 miles, took $2\frac{1}{2}$ hours. What was her speed during the first part of the trip?

	r	t	d

26. Steve leaves Nashville to visit his cousin David in Napa, 80 miles away. He travels at an average speed of 50 miles per hour. One-half hour later David leaves to visit Steve, traveling at an average speed of 60 miles per hour. How long after David leaves will it be before they meet?

	r	t	d

Find the measure of each angle in the triangle shown. (Be sure to substitute your value of x into each angle expression.) See Example 4.

27.

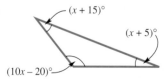

$(x + 15)°$

$(x + 5)°$

$(10x - 20)°$

28.

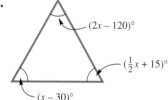

$(2x - 120)°$

$(\frac{1}{2}x + 15)°$

$(x - 30)°$

MATHEMATICAL CONNECTIONS (EXERCISES 29–32)

Consider the following two figures. Work Exercises 29–32 in order.

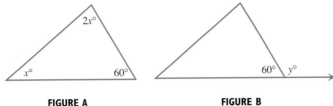

FIGURE A   **FIGURE B**

29. Solve for the measures of the unknown angles in Figure A.

30. Solve for the measure of the unknown angle marked $y°$ in Figure B.

31. Add the measures of the two angles you found in Exercise 29. How does the sum compare to the measure of the angle you found in Exercise 30?

32. From Exercises 29–31, make a conjecture (an educated guess) about the relationship among the angles marked ①, ②, and ③ in the figure shown here.

---

*In Exercises 33 and 34, the angles marked with variable expressions are called **vertical angles**. It is shown in geometry that vertical angles have equal measures. Find the measure of each angle.*

33.

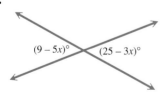

$(9 - 5x)°$ $(25 - 3x)°$

34.

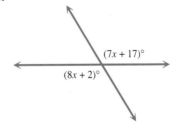

$(7x + 17)°$

$(8x + 2)°$

35. Two angles that form a straight angle (an angle measuring 180°) are called **supplementary angles.** Find the measures of the supplementary angles shown in the figure.

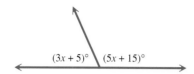

36. Two angles that form a right angle (an angle measuring 90°) are called **complementary angles.** Find the measures of the complementary angles shown in the figure.

Another type of application often studied in introductory and intermediate algebra courses involves consecutive integers. Consecutive integers are integers that follow each other in counting order, such as 8, 9, and 10. Suppose we wish to solve the following problem:

    Find three consecutive integers such that the sum of the first and third, increased by 3, is 50 more than the second.

    Let x represent the first of the unknown integers. Then x + 1 will be the second, and x + 2 will be the third. The equation we need can be found by going back to the words of the original problem.

$$\underset{\substack{\text{Sum of the}\\\text{first and third}\\\downarrow\\x + (x + 2)}}{} \quad \underset{\substack{\text{increased}\\\text{by 3}\\\downarrow\\+\ 3}}{} \quad \underset{\substack{\text{is}\\\downarrow\\=}}{} \quad \underset{\substack{\text{50 more than}\\\text{the second.}\\\downarrow\\(x + 1) + 50}}{}$$

    The solution of this equation is 46, meaning that the first integer is x = 46, the second is x + 1 = 47, and the third is x + 2 = 48. The three integers are 46, 47, and 48. Check by substituting these numbers back into the words of the original problem.

Solve each of the following problems involving consecutive integers.

37. Find four consecutive integers such that the sum of the first three is 54 more than the fourth.

38. Find three consecutive integers such that the sum of the first and twice the second is 17 more than twice the third.

39. Two pages facing each other in this book have 153 as the sum of their page numbers. What are the two page numbers?

40. If I add my current age to the age I will be next year on this date, the sum will be 87 years. How old will I be ten years from today?

2.5 *Linear Inequalities in One Variable*

Now we can extend our work with linear equations to *inequalities*. An **inequality** (in-ee-KWAHL-it-ee) says that two expressions are *not* equal. A **linear inequality in one variable** is an inequality such as

$$x + 5 < 2, \qquad y - 3 \geq 5, \qquad \text{or} \qquad 2k + 5 > 10.$$

(Throughout this section the definitions and rules are given only for $<$, but they are also valid for $>$, $\leq$, and $\geq$.)

OBJECTIVE We solve an inequality by finding all the numbers that make the inequality true. Usually, an inequality has an infinite number of solutions. These solutions are found, as are solutions of equations, by producing a series of simpler equivalent inequalities. **Equivalent inequalities** are inequalities with the same solution set. The inequalities in this chain of equivalent inequalities can be found with the **addition and multiplication properties of inequality.**

Addition Property of Inequality

For all real numbers a, b, and c, the inequalities

$$a < b \quad \text{and} \quad a + c < b + c$$

are equivalent. (The same number may be added to both sides of an inequality.)

As with equations, the addition property can be used to *subtract* the same number from both sides of an inequality.

EXAMPLE 1   Solving a Linear Inequality

Solve each inequality, and graph the solution set.

(a) $x - 7 \leq -12$

$$x - 7 + 7 \leq -12 + 7 \qquad \text{Add 7.}$$
$$x \leq -5$$

The solution set, $(-\infty, -5]$, is graphed in Figure 8.

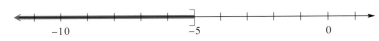

FIGURE 8

(b) $3m > 2m + 14$

$$3m - 2m > 2m + 14 - 2m \qquad \text{Subtract } 2m.$$
$$m > 14$$

The solution set, $(14, \infty)$, is shown in Figure 9.

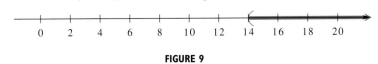

FIGURE 9

WORK PROBLEM 1 AT THE SIDE. ▶▶

---

1 ▶ Solve linear inequalities using the addition property.

2 ▶ Solve linear inequalities using the multiplication property.

3 ▶ Solve linear inequalities with three parts.

4 ▶ Solve applied problems with inequalities.

FOR EXTRA HELP

Tutorial   Tape 3   SSM, Sec. 2.5

---

1. Find the solution set of each inequality, and graph the solution set.

(a) $p + 6 \leq 8$

(b) $k - 5 \geq 1$

(c) $8y < 7y - 6$

ANSWERS

1. (a) $(-\infty, 2]$

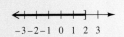

−3 −2 −1 0 1 2 3

(b) $[6, \infty)$

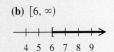

4 5 6 7 8 9

(c) $(-\infty, -6)$

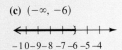

−10 −9 −8 −7 −6 −5 −4

2. Multiply both sides of each of the following by -5. Then insert the correct symbol, either $<$ or $>$, in the middle blank.

(a) $7 < 8$

-35 _____ -40

OBJECTIVE ▶ Solving an inequality such as $3x \leq 15$ requires dividing both sides by 3. This is done with the multiplication property of inequality, which is a little more involved than the corresponding property for equality. To see how this property works, start with the true statement

$$-2 < 5.$$

Multiply both sides by, say, 8.

$$-2(8) < 5(8) \qquad \text{Multiply by 8.}$$
$$-16 < 40 \qquad \text{True}$$

This gives a true statement. Start again with $-2 < 5$, and this time multiply both sides by -8.

$$-2(-8) < 5(-8) \qquad \text{Multiply by } -8.$$
$$16 < -40 \qquad \text{False}$$

The result, $16 < -40$, is false. To make it true, change the direction of the inequality symbol to get

$$16 > -40.$$

◀◀ WORK PROBLEM 2 AT THE SIDE.

As these examples suggest, *multiplying* both sides of an inequality by a negative number requires that we reverse the direction of the inequality symbol. Because division is defined in terms of multiplication, we must also reverse the direction of the inequality symbol when *dividing* by a negative number.

Caution

Avoid the common error of forgetting to reverse the direction of the inequality sign when multiplying or dividing by a negative number!

(b) $-1 > -4$

5 _____ _____

The results of multiplying inequalities by positive and negative numbers are summarized here.

Multiplication Property of Inequality

For all real numbers a, b, and c, with $c \neq 0$,
the inequalities

$$a < b \quad \text{and} \quad ac < bc$$

are equivalent if $c > 0$;
and the inequalities

$$a < b \quad \text{and} \quad ac > bc$$

are equivalent if $c < 0$.

(Both sides of an inequality may be multiplied by a *positive* number without changing the direction of the inequality symbol. Multiplying [or dividing] by a *negative* number requires that we reverse the inequality symbol.)

E X A M P L E   2   Solving a Linear Inequality

Solve each inequality, and graph the solution set.

(a) $5m \leq -30$

Use the multiplication property to divide both sides by 5. Since $5 > 0$, do *not* reverse the inequality symbol.

$$5m \leq -30$$

$$\frac{5m}{5} \leq \frac{-30}{5} \qquad \text{Divide by 5.}$$

$$m \leq -6$$

The solution set, $(-\infty, -6]$, is graphed in Figure 10.

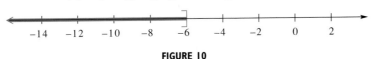

FIGURE 10

(b) $-4k \geq 32$

Divide both sides by -4. Since $-4 < 0$, the inequality symbol must be reversed.

$$-4k \geq 32$$

$$\frac{-4k}{-4} \leq \frac{32}{-4} \qquad \text{Change } \geq \text{ to } \leq.$$

$$k \leq -8$$

Figure 11 shows the graph of the solution set, $(-\infty, -8]$.

FIGURE 11

WORK PROBLEM 3 AT THE SIDE. ▶▶

The steps required for solving a linear inequality are given below.

Solving a Linear Inequality

Step 1 Simplify each side of the inequality separately by combining like terms and clearing parentheses as needed.

Step 2 Use the addition property of inequality to write the inequality so that the variable is on one side.

Step 3 Use the multiplication property to write the inequality in the form $x < k$ or $x > k$.

Remember to change the direction of the inequality symbol **only** when multiplying or dividing both sides of an inequality by a **negative** number, and never otherwise.

3. Graph the solution set of each inequality.

(a) $2y < -10$

_____→

(b) $-7k \geq 8$

_____→

(c) $-9m < -81$

_____→

4. Graph the solution set of each inequality.

(a) $-y \leq 2$

(b) $-z > -11$

(c) $4(x + 2) \geq 6x - 8$

(d) $5 - 3(m - 1)$
$\leq 2(m + 3) + 1$

4. (a)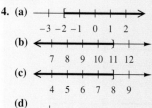
$-3 -2 -1 \quad 0 \quad 1 \quad 2$
(b)
$7 \quad 8 \quad 9 \quad 10 \quad 11 \quad 12$
(c)
$4 \quad 5 \quad 6 \quad 7 \quad 8 \quad 9$
(d) $\frac{1}{5}$
$0 \quad 1 \quad 2 \quad 3 \quad 4 \quad 5$

E X A M P L E   3   Solving a Linear Inequality

Solve $-3(x + 4) + 2 \geq 8 - x$. Graph the solution set.

$$-3x - 12 + 2 \geq 8 - x \qquad \text{Clear parentheses.}$$
$$-3x - 10 \geq 8 - x \qquad \text{Combine terms.}$$
$$-3x - 10 + x \geq 8 - x + x \qquad \text{Add } x.$$
$$-2x - 10 \geq 8$$
$$-2x - 10 + 10 \geq 8 + 10 \qquad \text{Add 10.}$$
$$-2x \geq 18$$
$$\frac{-2x}{-2} \leq \frac{18}{-2} \qquad \text{Divide by } -2 \text{ and change the}$$
$$\text{direction of the inequality symbol.}$$
$$x \leq -9$$

Notice that the inequality symbol was changed from $\geq$ to $\leq$ in the next to last step. Figure 12 shows the graph of the solution set, $(-\infty, -9]$.

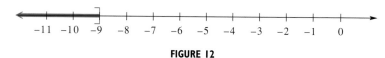

FIGURE 12

E X A M P L E   4   Solving a Linear Inequality

Solve $2 - 4(r - 3) < 3(5 - r) + 5$. Graph the solution set.

$$2 - 4(r - 3) < 3(5 - r) + 5$$
$$2 - 4r + 12 < 15 - 3r + 5 \qquad \text{Clear parentheses.}$$
$$14 - 4r < 20 - 3r \qquad \text{Combine terms.}$$
$$14 - r < 20 \qquad \text{Add } 3r.$$
$$-r < 6 \qquad \text{Subtract 14.}$$

We want r, not $-r$. To get r, multiply both sides of the inequality by -1. Since -1 is negative, change the direction of the inequality symbol.

$$-r < 6$$
$$(-1)(-r) > (-1)(6) \qquad \text{Multiply by } -1 \text{ and change } < \text{ to } >.$$
$$r > -6$$

The solution set, $(-6, \infty)$, is graphed in Figure 13.

FIGURE 13

Note
In this text we will use the interval notation $(-\infty, \infty)$ if the solution set contains all real numbers.

◀◀ WORK PROBLEM 4 AT THE SIDE.

OBJECTIVE 3 In further work in mathematics, it is sometimes necessary to work with an inequality such as $3 < x + 2 < 8$ where $x + 2$ is *between* 3 and 8. To solve a three-part inequality, we add or multiply all three parts, using the properties, as shown in the next example.

E X A M P L E 5   **Solving a Three-Part Inequality**

Solve the inequality $-2 \leq 3k - 1 \leq 5$, and graph the solution set.

Begin by adding 1 to each of the three parts to isolate the variable term in the middle.

$$-2 + 1 \leq 3k - 1 + 1 \leq 5 + 1 \qquad \text{Add 1.}$$

$$-1 \leq 3k \leq 6$$

$$\frac{-1}{3} \leq \frac{3k}{3} \leq \frac{6}{3} \qquad \text{Divide by 3.}$$

$$-\frac{1}{3} \leq k \leq 2$$

A graph of the solution, $[-\frac{1}{3}, 2]$, is shown in Figure 14.

FIGURE 14

WORK PROBLEM 5 AT THE SIDE. ▶▶

The three-part inequality $-2 \leq 3k - 1 \leq 5$ can be written as the *compound* inequality

$$-2 \leq 3k - 1 \quad \text{and} \quad 3k - 1 \leq 5.$$

Compound inequalities are discussed further in the next section.

The types of solutions to be expected from solving linear equations or linear inequalities are shown below.

Solutions of Linear Equations and Inequalities

Equation or Inequality	Typical Solution Set	Graph of Solution Set
Linear equation $ax + b = c$	$\{p\}$	●——→ p
Linear inequality $ax + b < c$	either $(-\infty, p)$ or (p, ∞)	←———) p (———→ p
Three-part inequality $c < ax + b < d$	(p, q)	(———) p q

OBJECTIVE 4▶ There are several phrases that denote inequality. Some of them were discussed in Chapter 1. In addition to the familiar "is less than" and "is greater than" (which are examples of **strict** inequalities), the expressions "is no more than," "is at least," and others also denote inequalities. (These are called **nonstrict.**) Expressions like these sometimes appear in applied problems that we solve using inequalities. The following chart shows how these expressions are interpreted.

5. Graph the solution set of each inequality.

(a) $-3 \leq x - 1 \leq 7$

———————————→

(b) $5 < 3x - 4 < 9$

———————————→

ANSWERS

5. (a) ├─[─┼─┼─┼─┼─┼─]─┼─→
          -4 -2 0 2 4 6 8 10

(b)
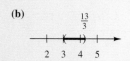

6. John can rent a car from Ames for $48 per day plus 10¢ per mile, or from Hughes at $40 per day plus 15¢ per mile. He plans to use the car for 3 days. What number of miles would make Hughes cost at most as much as Ames?

Word Expression	Interpretation	Word Expression	Interpretation
a **is at least** *b*	$a \geq b$	*a* **is at most** *b*	$a \leq b$
a **is no less than** *b*	$a \geq b$	*a* **is no more than** *b*	$a \leq b$

In Examples 6 and 7, we solve applied problems with inequalities.

E X A M P L E 6 Solving a Rental Problem

A rental company charges $15.00 to rent a chain saw, plus $2.00 per hour. Al Ghandi can spend no more than $35.00 to clear some logs from his yard. What is the *maximum* amount of time he can use the rented saw?

Let h = the number of hours he can rent the saw. He must pay $15.00, plus $2.00h$, to rent the saw for h hours, and this amount must be *no more than* $35.00.

$$\underset{\text{renting}}{\underset{\text{Cost of}}{15 + 2h}} \quad \underset{\text{more than}}{\underset{\text{is no}}{\leq}} \quad \underset{\text{35 dollars}}{35}$$

$15 + 2h - 15$	$\leq$	$35 - 15$	Subtract 15.
$2h$	$\leq$	20	
h	$\leq$	10	Divide by 2.

He can use the saw for a maximum of 10 hours. (Of course, he may use it for less time, as indicated by the inequality $h \leq 10$.)

◀◀ **WORK PROBLEM 6 AT THE SIDE.**

7. A student has grades of 92, 90, and 84 on his first three tests. What grade must the student make on his fourth test in order to keep an average of 90 or greater?

E X A M P L E 7 Solving a Grade-Averaging Problem

A student has grades of 88, 86, and 90 on her first three algebra tests. An average of 90 or above will earn an A in the class. What grade must the student make on her fourth test in order to have an A average?

Let x represent the score on the fourth test. Her average must be at least 90. To find the average of four numbers, add them and divide by 4.

$$\frac{88 + 86 + 90 + x}{4} \geq 90$$

$$\frac{264 + x}{4} \geq 90 \qquad \text{Add the scores.}$$

$$264 + x \geq 360 \qquad \text{Multiply by 4.}$$

$$x \geq 96 \qquad \text{Subtract 264.}$$

She must score *at least* 96 on her fourth test to keep an A average.

◀◀ **WORK PROBLEM 7 AT THE SIDE.**

Answers

6. 480 miles or less

7. at least 94

2.5 Exercises

Match each inequality with the correct graph or interval notation.

_____ **1.** $x \leq 3$

A.

_____ **2.** $x > 3$

B.

_____ **3.** $x < 3$

C. $(3, \infty)$

_____ **4.** $x \geq 3$

D. $(-\infty, 3]$

_____ **5.** $-3 \leq x \leq 3$

E. $(-3, 3)$

_____ **6.** $-3 < x < 3$

F. $[-3, 3]$

7. Explain how you will determine whether to use parentheses or brackets when graphing the solution set of an inequality.

8. When is it necessary to reverse the direction of the inequality sign when solving an inequality?

Solve each inequality, giving its solution set in both interval and graph forms. See Examples 1–4.

9. $5r \leq -15$

10. $12m \leq -36$

11. $4x + 1 \geq 21$

12. $5t + 2 \geq 52$

13. $\dfrac{3k - 1}{4} > 5$

14. $\dfrac{5z - 6}{8} < 8$

15. $-4x < 16$

16. $-2m > 10$

17. $-\dfrac{3}{4}r \geq 30$

18. $-\dfrac{2}{3}y \leq 12$

19. $-\dfrac{3}{2}y \leq -\dfrac{9}{2}$

20. $-\dfrac{2}{5}x \geq -\dfrac{4}{25}$

21. $-1.3m \geq -5.2$

22. $-2.5y \leq -1.25$

23. $\dfrac{2k - 5}{-4} > 5$

24. $\dfrac{3z - 2}{-5} < 6$

25. $y + 4(2y - 1) \geq y$

26. $m - 2(m - 4) \leq 3m$

27. $-(4 + r) + 2 - 3r < -14$

28. $-(9 + k) - 5 + 4k \geq 4$

29. $-3(z - 6) > 2z - 2$

30. $-2(y + 4) \leq 6y + 16$

MATHEMATICAL CONNECTIONS (EXERCISES 31–36)

Work Exercises 31–36 in order.

31. Solve the linear equation
$5(x + 3) - 2(x - 4) = 2(x + 7)$, and graph
the solution on a number line.

32. Solve the linear inequality
$5(x + 3) - 2(x - 4) > 2(x + 7)$, and graph
the solutions on a number line.

33. Solve the linear inequality
$5(x + 3) - 2(x - 4) < 2(x + 7)$, and graph
the solutions on a number line.

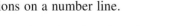

34. Graph all the solution sets of the equation and
inequalities in Exercises 31–33 on the same
number line. What set do you obtain?

35. Based on the results of Exercises 31–33, complete the following using a conjecture (educated guess): The solution set of $-3(x + 2) = 3x + 12$ is $\{-3\}$, and the solution set of $-3(x + 2) < 3x + 12$ is $(-3, \infty)$. Therefore the solution set of $-3(x + 2) > 3x + 12$ is _____.

36. Comment on the following statement: Equality is the boundary between less than and greater than.

Solve each inequality, giving its solution set in both interval and graph forms. See Example 5.

37. $-4 < x - 5 < 6$

38. $-1 < x + 1 < 8$

39. $-9 \leq k + 5 \leq 15$

40. $-4 \leq m + 3 \leq 10$

41. $-6 \leq 2z + 4 \leq 16$

42. $-15 < 3p + 6 < -12$

43. $-19 \leq 3x - 5 \leq 1$

44. $-16 < 3t + 2 < -10$

45. $-1 \leq \dfrac{2x - 5}{6} \leq 5$

46. $-3 \leq \dfrac{3m + 1}{4} \leq 3$

47. $4 \leq 5 - 9x < 8$

48. $4 \leq 3 - 2x < 8$

Everyday words and phrases like "exceed," "at least," and "fewer than" are closely related to mathematical inequality concepts. Based on the given graph, answer the questions in Exercises 49–52.

49. In which months did the percent of tornadoes exceed 7.7%?

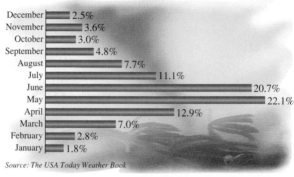

─────── **WHEN TORNADOES STRIKE** ───────

December — 2.5%
November — 3.6%
October — 3.0%
September — 4.8%
August — 7.7%
July — 11.1%
June — 20.7%
May — 22.1%
April — 12.9%
March — 7.0%
February — 2.8%
January — 1.8%

Source: The USA Today Weather Book

50. In which months was the percent of tornadoes at least 12.9%?

51. The data used to determine the graph was based on the number of tornadoes sighted in the United States during the last twenty years. A total of 17,252 tornadoes were reported. In which months were fewer than 1500 reported?

52. How many more tornadoes occurred during March than October? (Use the total given in Exercise 51.)

Solve each problem. See Examples 6 and 7.

53. Margaret Westmoreland earned scores of 90 and 82 on her first two tests in English Literature. What score must she make on her third test to keep an average of 84 or greater?

54. Shannon d'Hemecourt scored 92 and 96 on her first two tests in Methods in Teaching Mathematics. What score must she make on her third test to keep an average of 90 or greater?

55. A couple wishes to rent a car for one day while on vacation. Ford Automobile Rental wants $15.00 per day and 14¢ per mile, while Chevrolet-For-A-Day wants $14.00 per day and 16¢ per mile. After how many miles would the price to rent the Chevrolet exceed the price to rent a Ford?

56. Jane and Terry Brandsma went to Mobile for a week. They needed to rent a car, so they checked out two rental firms. Avis wanted $28 per day, with no mileage fee. Downtown Toyota wanted $108 per week and 14¢ per mile. How many miles would they have to drive before the Avis price is less than the Toyota price?

A product will produce a profit only when the revenue (R) from selling the product exceeds the cost (C) of producing it. Find the smallest whole number of units x that must be sold for the business to show a profit for the item described.

57. Peripheral Visions, Inc. finds that the cost to produce x studio-quality videotapes is $C = 20x + 100$, while the revenue produced from them is $R = 24x$ (C and R in dollars).

58. Speedy Delivery finds that the cost to make x deliveries is $C = 3x + 2300$, while the revenue produced from them is $R = 5.50x$ (C and R in dollars).

2.6 Set Operations and Compound Inequalities

The words *and* and *or* are very important in interpreting certain kinds of equations and inequalities in algebra. They also occur in work with sets. In this section we study the use of these two words as they relate to sets and inequalities.

OBJECTIVE 1 We start by looking at the use of the word "and" with sets. The intersection of sets is defined below.

> For any two sets *A* and *B*, the **intersection** (IN-tur-sek-shun) of *A* and *B*, symbolized $A \cap B$, is defined as follows:
>
> $A \cap B = \{x \mid x$ is an element of *A* **and** *x* is an element of *B*$\}$.

EXAMPLE 1 Finding the Intersection of Two Sets

Let $A = \{1, 2, 3, 4\}$ and $B = \{2, 4, 6\}$. Find $A \cap B$.

The set $A \cap B$ contains those elements that belong to both *A* *and B:* the numbers 2 and 4. Therefore,

$$A \cap B = \{1, 2, 3, 4\} \cap \{2, 4, 6\}$$
$$= \{2, 4\}.$$

WORK PROBLEM 1 AT THE SIDE. ▶▶

OBJECTIVE 2 A **compound inequality** consists of two inequalities linked by a connective word such as *and* or *or*. Examples of compound inequalities are

$$x + 1 \leq 9 \quad \textbf{and} \quad x - 2 \geq 3$$

and

$$2x > 4 \quad \textbf{or} \quad 3x - 6 < 5.$$

To solve a compound inequality with the word *and*, we use the following steps.

> **Solving a Compound Inequality with *and***
>
> *Step 1* Solve each inequality in the compound inequality individually.
>
> *Step 2* Since the inequalities are joined with *and*, the solution will include all numbers that satisfy both solutions in Step 1 (the intersection of the solutions).

The next example shows how a compound inequality with *and* is solved.

EXAMPLE 2 Solving a Compound Inequality with *and*

Solve the compound inequality

$$x + 1 \leq 9 \quad \text{and} \quad x - 2 \geq 3.$$

Step 1 directs that we solve each inequality in the compound inequality individually.

$$
\begin{array}{ccc}
x + 1 \leq 9 & \text{and} & x - 2 \geq 3 \\
x + 1 - 1 \leq 9 - 1 & \text{and} & x - 2 + 2 \geq 3 + 2 \\
x \leq 8 & \text{and} & x \geq 5
\end{array}
$$

— **CONTINUED ON NEXT PAGE**

— **CONTINUED ON NEXT PAGE**

---

1. Let $A = \{3, 4, 5, 6\}$ and $B = \{5, 6, 7\}$. Find $A \cap B$.

ANSWERS

1. $\{5, 6\}$

2. Graph the solution set of each compound inequality.

(a) $x < 10$ and $x > 2$

(b) $x + 3 < 1$ and $x - 4 > -12$

Now we apply Step 2. Because the inequalities are joined with the word *and*, the solution will include all numbers that satisfy both solutions in Step 1. Thus, the compound inequality is true whenever $x \le 8$ and $x \ge 5$ are both true. The top graph in Figure 15 shows $x \le 8$, and the middle graph shows $x \ge 5$. The bottom graph shows the numbers common to the first two graphs. As shown by this third graph, the solution consists of all numbers between 5 and 8, including both 5 and 8. This is the intersection of the two graphs, and we write it as $[5, 8]$.

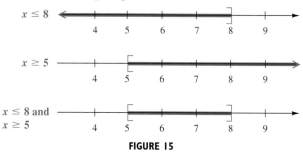

FIGURE 15

◀◀ WORK PROBLEM 2 AT THE SIDE.

E X A M P L E   3   Solving a Compound Inequality with *and*

Solve the compound inequality

$$-3x - 2 > 4 \quad \text{and} \quad 5x - 1 \le -21.$$

Begin by solving each part separately.

$$-3x - 2 > 4 \quad \text{and} \quad 5x - 1 \le -21$$
$$-3x > 6 \quad \text{and} \quad 5x \le -20$$
$$x < -2 \quad \text{and} \quad x \le -4$$

Now find all values of x that satisfy both conditions; that is, the real numbers that are less than -2 and also less than or equal to -4. As shown by the graphs in Figure 16, the solution set is $(-\infty, -4]$.

3. Solve and graph.
$2x \ge x - 1$ and $3x \ge 3 + 2x$

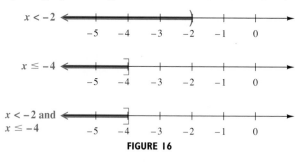

FIGURE 16

◀◀ WORK PROBLEM 3 AT THE SIDE.

E X A M P L E   4   Solving a Compound Inequality with *and*

Solve $x + 2 < 5$ and $x - 10 > 2$.
First solve each inequality separately.

$$x + 2 < 5 \quad \text{and} \quad x - 10 > 2$$
$$x < 3 \quad \text{and} \quad x > 12$$

ANSWERS

2. (a)

0 2 4 6 8 10 12

(b)

−10 −8 −6 −4 −2 0

3.

−1 0 1 2 3 4 5

CONTINUED ON NEXT PAGE

There is no number that is both less than 3 and greater than 12, so the given compound sentence has no solution (solution set is ∅). See Figure 17.

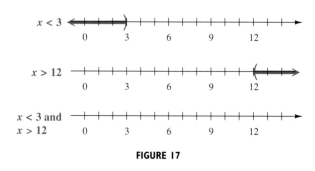

FIGURE 17

WORK PROBLEM 4 AT THE SIDE. ▶▶

OBJECTIVE 3▶ We now discuss the union of two sets, which involves the use of the word "or."

> For any two sets A and B, the **union** (YOON-yun) of A and B, symbolized $A \cup B$, is defined as follows:
>
> $A \cup B = \{x \mid x$ is an element of A **or** x is an element of $B\}$.

E X A M P L E 5   Finding the Union of Two Sets

Find the union of the sets $A = \{1, 2, 3, 4\}$ and $B = \{2, 4, 6\}$.

Begin by listing all the elements of set A: 1, 2, 3, 4. Then list any additional elements from set B. In this case the elements 2 and 4 are already listed, so the only additional element is 6. Therefore,

$$A \cup B = \{1, 2, 3, 4\} \cup \{2, 4, 6\} = \{1, 2, 3, 4, 6\}.$$

The union consists of all elements in either A *or* B (or both).

Notice in Example 5, that even though the elements 2 and 4 appeared in both sets A and B, they are only written once in $A \cup B$. It is not necessary to write them more than once in the union.

WORK PROBLEM 5 AT THE SIDE. ▶▶

OBJECTIVE 4▶ To solve compound inequalities with the word *or,* we use the following steps.

Solving a Compound Inequality with *or*

Step 1 Solve each inequality in the compound inequality individually.

Step 2 Since the inequalities are joined with *or,* the solution will include all numbers that satisfy either one of the solutions in Step 1 (the union of the solutions).

The next examples show how to solve a compound inequality with *or.*

4. Solve $x + 2 > 3$ and $2x + 1 < -3$.

5. Let $A = \{3, 4, 5, 6\}$ and $B = \{5, 6, 7\}$. Find $A \cup B$.

6. Graph each solution set.

(a) $x + 2 > 3$ or
$2x + 1 < -3$

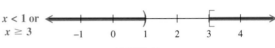

(b) $y - 1 > 2$ or
$3y + 5 < 2y + 6$

7. Solve
$3x - 2 \le 13$ or
$x + 5 \ge 7$.

 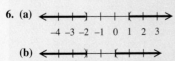
E X A M P L E 6 **Solving a Compound Inequality with *or***

Solve $6x - 4 < 2x$ or $-3x \le -9$.

Solve each inequality separately:

$$6x - 4 < 2x \quad \text{or} \quad -3x \le -9$$
$$4x < 4$$
$$x < 1 \quad \text{or} \quad x \ge 3.$$

The graphs of these results are shown in Figure 18. The third graph gives the combination of these two solutions. This final solution set is written

$$(-\infty, 1) \cup [3, \infty).$$

FIGURE 18

> **Caution**
> When interval notation is used to write the solution of Example 6, it *must* be written as
> $$(-\infty, 1) \cup [3, \infty).$$
> There is no short-cut way to write this solution.

◄◄ **WORK PROBLEM 6 AT THE SIDE.**

E X A M P L E 7 **Solving a Compound Inequality with *or***

Solve $-4x + 1 \ge 9$ or $5x + 3 \ge -12$.

Solve each inequality separately.

$$-4x + 1 \ge 9 \quad \text{or} \quad 5x + 3 \ge -12$$
$$-4x \ge 8 \quad \text{or} \quad 5x \ge -15$$
$$x \le -2 \quad \text{or} \quad x \ge -3$$

The graphs of these solutions are shown in Figure 19. As shown in the figure, every real number is a solution of the compound sentence because every real number satisfies at least one of the two inequalities.

The solution set is $(-\infty, \infty)$.

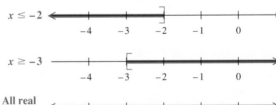

FIGURE 19

◄◄ **WORK PROBLEM 7 AT THE SIDE.**

2.6 Exercises

Decide whether the statement is true or false. If it is false, explain why.

1. The union of the set of rational numbers and the set of irrational numbers is the set of real numbers.

2. The intersection of the set of rational numbers and the set of irrational numbers is the set $\{0\}$.

3. The union of the solution sets of $2x + 1 = 3$, $2x + 1 > 3$, and $2x + 1 < 3$ is the set of real numbers.

4. The intersection of the sets $\{x \mid x \geq 5\}$ and $\{x \mid x \leq 5\}$ is $\{5\}$.

5. The union of the sets $(-\infty, 6)$ and $(6, \infty)$ is $(-\infty, \infty)$.

6. The intersection of the sets $[6, \infty)$ and $(-\infty, 6]$ is $\emptyset$.

Let $A = \{1, 2, 3, 4, 5, 6\}$, $B = \{1, 3, 5\}$, $C = \{1, 6\}$, and $D = \{4\}$. Specify each of the following sets. See Examples 1 and 5.

7. $A \cap D$ 8. $B \cap C$ 9. $B \cap \emptyset$ 10. $A \cap \emptyset$

11. $A \cup B$ 12. $B \cup D$ 13. $B \cup C$ 14. $C \cup B$

15. $C \cup D$ 16. $D \cup C$

17. Use the sets A, B, and C for Exercises 7–16 to show that $A \cap (B \cap C)$ is equal to $(A \cap B) \cap C$. This is true for any choices of sets. What property does this illustrate? (*Hint:* See Section 1.5.)

18. Repeat Exercise 17, showing that $A \cup (B \cup C)$ is equal to $(A \cup B) \cup C$. What property does this illustrate?

19. How can intersection be applied to a real-life situation?

20. A compound inequality uses one of the words *and* or *or*. Explain how you will determine whether to use *intersection* or *union* when graphing the solution set.

For each compound inequality, give the solution set in both interval and graph form. See Examples 2, 3, and 4.

21. $x < 2$ and $x > -3$

22. $x < 5$ and $x > 0$

23. $x \leq 2$ and $x \leq 5$

24. $x \geq 3$ and $x \geq 6$

25. $x \leq 3$ and $x \geq 6$

26. $x \leq -1$ and $x \geq 3$

27. $x \geq -1$ and $x \geq 4$

28. $x \leq -2$ and $x \leq 2$

29. $x \geq -1$ and $x \leq 3$

30. $x \geq 4$ and $x \leq 5$

31. $x - 3 \leq 6$ and $x + 2 \geq 7$

32. $x + 5 \leq 11$ and $x - 3 \geq -1$

33. $3x - 4 \leq 8$ and $4x - 1 \leq 15$

34. $7x + 6 \leq 48$ and $-4x \geq -24$

For each compound inequality, give the solution set in both interval and graph form. See Examples 6 and 7.

35. $x \leq 2$ or $x \geq 4$

36. $x \leq -5$ or $x \geq 6$

37. $x \leq 1$ or $x \leq 8$

38. $x \geq 1$ or $x \geq 8$

39. $x \leq 1$ or $x \geq 10$

40. $x \leq 2$ or $x \geq 9$

41. $x \geq -2$ or $x \geq 5$

42. $x \leq -2$ or $x \leq 6$

43. $x \geq -2$ or $x \leq 4$

44. $x \geq 5$ or $x \leq 7$

45. $x + 2 > 7$ or $x - 1 < -6$

46. $x + 1 > 3$ or $x + 4 < 2$

47. $4x - 8 > 0$ or $4x - 1 < 7$

48. $3x < x + 12$ or $3x - 8 > 10$

For each compound inequality, decide whether intersection or union should be used. Then give the solution set in both interval and graph form. See Examples 2, 3, 4, 6, and 7.

49. $x < -1$ and $x > -5$

50. $x > -1$ and $x < 7$

51. $x < 4$ or $x < -2$

52. $x < 5$ or $x < -3$

53. $x + 1 \geq 5$ and $x - 2 \leq 10$

54. $2x - 6 \leq -18$ and $2x \geq -18$

55. $-3x \leq -6$ or $-3x \geq 0$

56. $-8x \leq -24$ or $-5x \geq 15$

MATHEMATICAL CONNECTIONS (EXERCISES 57–60)

The figures represent the backyards of neighbors Luigi, Mario, Than, and Joe. Find the area and the perimeter of each yard. Suppose that each resident has 150 feet of fencing, and enough sod to cover 1400 square feet of lawn. Give the name or names of the residents whose yards satisfy the following descriptions.

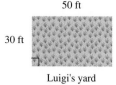
50 ft
30 ft
Luigi's yard

40 ft
35 ft
Mario's yard

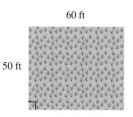
60 ft
50 ft
Than's yard

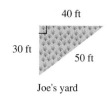
40 ft
30 ft
50 ft
Joe's yard

57. the yard can be fenced *and* the yard can be sodded

58. the yard can be fenced *and* the yard cannot be sodded

59. the yard cannot be fenced *and* the yard can be sodded

60. the yard cannot be fenced *and* the yard cannot be sodded

2.7 *Absolute Value Equations and Inequalities*

OBJECTIVE ▶ In Chapter 1 it was shown that the absolute value of a number x, written $|x|$, represents the distance from x to 0 on the number line. For example, the solutions of $|x| = 4$ are 4 and -4, as shown in Figure 20.

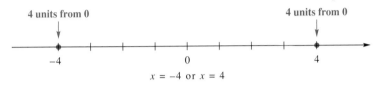

$$x = -4 \text{ or } x = 4$$

FIGURE 20

Because absolute value represents distance from 0, it is reasonable to interpret the solutions of $|x| > 4$ to be all numbers that are *more* than 4 units from 0. The set $(-\infty, -4) \cup (4, \infty)$ fits this description. Figure 21 shows the graph of the solution set of $|x| > 4$.

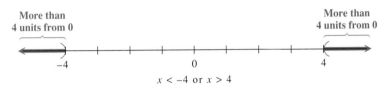

$$x < -4 \text{ or } x > 4$$

FIGURE 21

The solution set of $|x| < 4$ consists of all numbers that are *less* than 4 units from 0 on the number line. Another way of thinking of this is to think of all numbers *between* -4 and 4. This set of numbers is given by $(-4, 4)$, as shown in Figure 22.

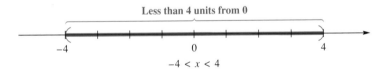

$$-4 < x < 4$$

FIGURE 22

> **WORK PROBLEM 1 AT THE SIDE.** ▶▶

The equation and inequalities just described are examples of **absolute value equations and inequalities.** These are equations and inequalities that involve the absolute value of a variable expression. They generally take the form

$$|ax + b| = k, \qquad |ax + b| > k, \qquad \text{or} \qquad |ax + b| < k$$

where k is a positive number. We may solve them by rewriting them as compound equations or inequalities, as shown in the following summary.

1. Graph the solution set of each equation or inequality.

(a) $|x| = 3$

_____→

(b) $|x| > 3$

_____→

(c) $|x| < 3$

_____→

ANSWERS

1. (a)

117

2. Solve each equation, and graph the solution set.

(a) $|x + 2| = 3$

(b) $|3x - 4| = 11$

Solving Absolute Value Equations and Inequalities

Let k be a positive number, and p and q be two numbers.

1. To solve $|ax + b| = k$, solve the compound equation

$$ax + b = k \quad \text{or} \quad ax + b = -k.$$

The solution set is usually of the form $\{p, q\}$, with two numbers.

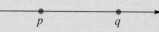

2. To solve $|ax + b| > k$, solve the compound inequality

$$ax + b > k \quad \text{or} \quad ax + b < -k.$$

The solution set is of the form $(-\infty, p) \cup (q, \infty)$, which consists of two separate intervals.

3. To solve $|ax + b| < k$, solve the compound inequality

$$-k < ax + b < k.$$

The solution set is of the form (p, q), a single interval.

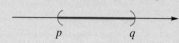

Note

Some people prefer to write the compound statements in parts 1 and 2 of the summary as

$$ax + b = k \quad \text{or} \quad -(ax + b) = k$$

and

$$ax + b > k \quad \text{or} \quad -(ax + b) > k.$$

These forms are equivalent to those we give in the summary and produce the same results.

OBJECTIVE 2 The next example shows how we use a compound equation to solve a typical absolute value equation. Remember that because absolute value refers to distance from the origin, each absolute value equation will have two parts.

EXAMPLE 1 **Solving an Absolute Value Equation**

Solve $|2x + 1| = 7$.

For $|2x + 1|$ to equal 7, $2x + 1$ must be 7 units from 0 on the number line. This can happen only when $2x + 1 = 7$ or $2x + 1 = -7$. Solve this compound equation as follows.

$$2x + 1 = 7 \quad \text{or} \quad 2x + 1 = -7$$
$$2x = 6 \quad \text{or} \qquad 2x = -8$$
$$x = 3 \quad \text{or} \qquad x = -4$$

The solution set is $\{-4, 3\}$. Its graph is shown in Figure 23.

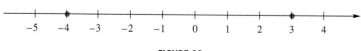

FIGURE 23

ANSWERS

2. (a) $\{-5, 1\}$

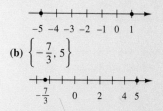

(b) $\left\{-\dfrac{7}{3}, 5\right\}$

◀◀ WORK PROBLEM 2 AT THE SIDE.

OBJECTIVE ▶3▶ We now discuss how to solve absolute value inequalities.

┌─ **E X A M P L E   2**   **Solving an Absolute Value Inequality with >**

Solve $|2x + 1| > 7$.

As shown in the summary, this absolute value inequality should be rewritten as

$$2x + 1 > 7 \quad \text{or} \quad 2x + 1 < -7,$$

because $2x + 1$ must represent a number that is *more* than 7 units from 0 on the number line. Now solve the compound inequality.

$$
\begin{array}{llll}
2x + 1 > 7 & \text{or} & 2x + 1 < -7 \\
2x > 6 & \text{or} & 2x < -8 \\
x > 3 & \text{or} & x < -4
\end{array}
$$

The solution set, $(-\infty, -4) \cup (3, \infty)$, is graphed in Figure 24. Notice that the graph consists of two intervals.

FIGURE 24

WORK PROBLEM 3 AT THE SIDE. ▶▶

┌─ **E X A M P L E   3**   **Solving an Absolute Value Inequality with <**

Solve $|2x + 1| < 7$.

The expression $2x + 1$ must represent a number that is less than 7 units from 0 on the number line. Another way of thinking of this is to realize that $2x + 1$ must be between -7 and 7. As the summary shows, this is written as the three-part inequality

$$-7 < 2x + 1 < 7.$$

We solved such inequalities in Section 2.5 by working with all three parts at the same time.

$$
\begin{array}{ll}
-7 < 2x + 1 < 7 & \\
-8 < 2x < 6 & \text{Subtract 1 from each part.} \\
-4 < x < 3 & \text{Divide each part by 2.}
\end{array}
$$

The solution set is $(-4, 3)$, and the graph consists of a single interval as shown in Figure 25.

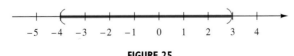

FIGURE 25

WORK PROBLEM 4 AT THE SIDE. ▶▶

3. Solve each inequality, and graph the solution set.

(a) $|x + 2| > 3$

_____→

(b) $|3x - 4| \geq 11$

_____→

4. Solve each inequality, and graph the solution set.

(a) $|x + 2| < 3$

_____→

(b) $|3x - 4| \leq 11$

_____→

ANSWERS

3. (a) $(-\infty, -5) \cup (1, \infty)$

←─┤ + + + + + ├─→
−5 −4 −3 −2 −1 0 1

(b) $\left(-\infty, -\dfrac{7}{3}\right] \cup [5, \infty)$

$-\dfrac{7}{3}$

←─┤│ + + + + + ├─→
  −2   0   2   4 5

4. (a) $(-5, 1)$

←─(+ + + + +)──→
−5 −4 −3 −2 −1 0 1

(b) $\left[-\dfrac{7}{3}, 5\right]$

──┤[+ + + + +]├──
  $-\dfrac{7}{3}$ 0   2   4 5

5. (a) Solve $|5a + 2| - 9 = -7$.

(b) Solve, and graph the solution set.

$$|m + 2| - 3 > 2$$

⟵————————————➤

(c) Solve, and graph the solution set.

$$|3a + 2| + 4 \leq 15$$

⟵————————————➤

Look back at Figures 23, 24, and 25. These are the graphs of $|2x + 1| = 7, |2x + 1| > 7,$ and $|2x + 1| < 7$. If we find the union of the three sets, we get the set of all real numbers. This is because for any value of x, $|2x + 1|$ will satisfy one and only one of the following: it is equal to 7, greater than 7, or less than 7.

> **Caution**
> When solving absolute value equations and inequalities of the types in Examples 1, 2, and 3, be sure to remember the following.
>
> 1. The methods described apply when the constant is alone on one side of the equation or inequality and is *positive*.
> 2. Absolute value equations and absolute value inequalities in the form $|ax + b| > k$ translate into "or" compound statements.
> 3. Absolute value inequalities in the form $|ax + b| < k$ translate into "and" compound statements, which may be written as three-part inequalities.
> 4. An "or" statement *cannot* be written in three parts. It would be incorrect to use
>
> $$-7 > 2x + 1 > 7$$
>
> in Example 2, because this would imply that $-7 > 7$, which is *false*.

OBJECTIVE 4▶ Sometimes an absolute value equation or inequality is given in a form that requires some rewriting before it can be set up as a compound statement. The next example illustrates this process for an absolute value equation.

EXAMPLE 4 Solving an Absolute Value Equation Requiring Rewriting

Solve the equation $|x + 3| + 5 = 12$.

First get the absolute value alone on one side of the equals sign. Do this by subtracting 5 on each side.

$$|x + 3| + 5 - 5 = 12 - 5 \quad \text{Subtract 5.}$$
$$|x + 3| = 7$$

Then use the method shown in Example 1.

$$x + 3 = 7 \quad \text{or} \quad x + 3 = -7$$
$$x = 4 \quad \text{or} \quad x = -10$$

Check that the solution set is $\{4, -10\}$ by substituting in the original equation.

Solving an absolute value *inequality* requiring rewriting is done in a similar manner.

> **Caution**
> A common error in solving an absolute value equation such as the one in Example 4 is to forget about the absolute value symbols and solve $x + 3 + 5 = 12$. Do not make this error.

◀◀ **WORK PROBLEM 5 AT THE SIDE.**

OBJECTIVE 5▶ The next example shows how to solve an equation with two absolute value expressions. For two expressions to have the same absolute value, they must either be equal or be negatives of each other.

ANSWERS

5. (a) $\left\{-\dfrac{4}{5}, 0\right\}$
(b) $(-\infty, -7) \cup (3, \infty)$

⟵++++++++++⟶
-7 -4 -2 0   3

(c) $\left[-\dfrac{13}{3}, 3\right]$

+[++++++++]+⟶
$-\frac{13}{3}$ -2 0 2 3

To solve an absolute value equation of the form

$$|ax + b| = |cx + d|,$$

solve the compound equation

$$ax + b = cx + d \quad \textbf{or} \quad ax + b = -(cx + d).$$

EXAMPLE 5   Solving an Equation with Two Absolute Values

Solve the equation $|z + 6| = |2z - 3|$.

This equation is satisfied either if $z + 6$ and $2z - 3$ are equal to each other, or if $z + 6$ and $2z - 3$ are negatives of each other. Thus, we have

$$z + 6 = 2z - 3 \quad \text{or} \quad z + 6 = -(2z - 3).$$

Solve each equation.

$$z + 6 = 2z - 3 \quad \text{or} \quad z + 6 = -2z + 3$$
$$9 = z \qquad\qquad 3z = -3$$
$$z = -1$$

The solution set is $\{9, -1\}$.

WORK PROBLEM 6 AT THE SIDE. ▶▶

OBJECTIVE ▶ 6 When an absolute value equation or inequality involves a *negative* constant or *zero* alone on one side, simply use the properties of absolute value to solve. Keep in mind the following.

1. The absolute value of an expression can never be negative: $|a| \geq 0$ for all real numbers a.
2. The absolute value of an expression equals 0 only when the expression is equal to 0.

The next two examples illustrate these special cases.

EXAMPLE 6   Solving Special Cases of Absolute Value Equations

Solve each equation.

(a) $|5r - 3| = -4$

Since the absolute value of an expression can never be negative, there are no solutions for this equation. The solution set is $\emptyset$.

(b) $|7x - 3| = 0$

The expression $7x - 3$ will equal 0 *only* for the solution of the equation

$$7x - 3 = 0.$$

The solution of this equation is $\frac{3}{7}$. The solution set is $\left\{\frac{3}{7}\right\}$. It consists of only one element.

WORK PROBLEM 7 AT THE SIDE. ▶▶

---

6. Solve each equation.

(a) $|k - 1| = |5k + 7|$

(b) $|4r - 1| = |3r + 5|$

7. Solve.

(a) $|6x + 7| = -5$

(b) $\left|\dfrac{1}{4}x - 3\right| = 0$

ANSWERS

6. (a) $\{-1, -2\}$

(b) $\left\{-\dfrac{4}{7}, 6\right\}$

7. (a) $\emptyset$ **(b)** $\{12\}$

8. Solve.

(a) $|x| > -1$

(b) $|y| < -5$

(c) $|k + 2| \leq 0$

E X A M P L E 7 **Solving Special Cases of Absolute Value Inequalities**

Solve each of the following inequalities.

(a) $|x| \geq -4$

The absolute value of a number is always nonnegative. For this reason, $|x| \geq -4$ is true for *all* real numbers. The solution set is $(-\infty, \infty)$.

(b) $|k + 6| < -2$

There is no number whose absolute value is less than -2, so this inequality has no solution. The solution set is $\emptyset$.

(c) $|m - 7| \leq 0$

The value of $|m - 7|$ will never be less than 0. However, $|m - 7|$ will equal 0 when $m = 7$. Therefore, the solution set is $\{7\}$.

◀◀ **WORK PROBLEM 8 AT THE SIDE.**

ANSWERS

8. **(a)** $(-\infty, \infty)$ **(b)** $\emptyset$ **(c)** $\{-2\}$

2.7 Exercises

*Keeping in mind that the absolute value of a number can be interpreted as the distance
between the graph of the number and 0 on the number line, match the absolute value equation
or inequality with the graph of its solution set.*

Choices   Choices

1. $|x| = 5$ _____   A.    **2.** $|x| = 9$ _____   A.

$|x| < 5$ _____   B. $|x| > 9$ _____   B.

$|x| > 5$ _____   C. $|x| \geq 9$ _____   C.

$|x| \leq 5$ _____   D. $|x| < 9$ _____   D.

$|x| \geq 5$ _____   E. 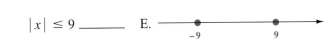 $|x| \leq 9$ _____   E.

3. Explain when to use *and* and when to use *or* if
you are solving an absolute value equation or
inequality of the form $|ax + b| = k$,
$|ax + b| < k$, or $|ax + b| > k$, where k is a
positive number.

4. How many solutions will $|ax + b| = k$ have
if **(a)** $k = 0$;   **(b)** $k > 0$;   **(c)** $k < 0$?

Solve each equation. See Example 1.

5. $|x| = 12$ **6.** $|k| = 14$ **7.** $|4x| = 20$ **8.** $|5x| = 30$

9. $|y - 3| = 9$ **10.** $|p - 5| = 13$ **11.** $|2x + 1| = 7$ **12.** $|2y + 3| = 19$

13. $|4r - 5| = 17$ **14.** $|5t - 1| = 21$ **15.** $|2y + 5| = 14$ **16.** $|2x - 9| = 18$

17. $\left|\dfrac{1}{2}x + 3\right| = 2$ **18.** $\left|\dfrac{2}{3}q - 1\right| = 5$ **19.** $\left|1 - \dfrac{3}{4}k\right| = 7$ **20.** $\left|2 - \dfrac{5}{2}m\right| = 14$

Solve each inequality, and graph the solution set. See Example 2.

21. $|x| > 3$

22. $|y| > 5$

23. $|k| \geq 4$

24. $|r| \geq 6$

25. $|t + 2| > 10$

26. $|r + 5| > 20$

27. $|3x - 1| \geq 8$

28. $|4x + 1| \geq 21$

29. $|3 - x| > 5$

30. $|5 - x| > 3$

31. The graph of the solution set of $|2x + 1| = 9$ is given below.

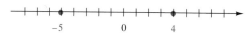

Without actually doing the algebraic work, graph the solution set of each inequality, referring to the graph above.

(a) $|2x + 1| < 9$

(b) $|2x + 1| > 9$

32. The graph of the solution set of $|3y - 4| < 5$ is given below.

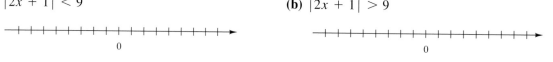

Without actually doing the algebraic work, graph the solution set of each inequality, referring to the graph above.

(a) $|3y - 4| = 5$

(b) $|3y - 4| > 5$

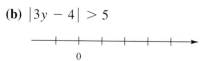

Solve each inequality and graph its solution set. See Example 3. (Hint: Compare your
answers to those in Exercises 21–30.)

33. $|x| \leq 3$

34. $|y| \leq 5$

35. $|k| < 4$

36. $|r| < 6$

37. $|t + 2| \leq 10$

38. $|r + 5| \leq 20$

39. $|3x - 1| < 8$

40. $|4x + 1| < 21$

41. $|3 - x| \leq 5$

42. $|5 - x| \leq 3$

Exercises 43–50 represent a sampling of the various types of absolute value equations and
inequalities covered in Exercises 1–42. Decide which method of solution applies, find the
solution set, and graph. See Examples 1, 2, and 3.

43. $|-4 + k| > 9$

44. $|-3 + t| > 8$

45. $|7 + 2z| = 5$

46. $|9 - 3p| = 3$

47. $|3r - 1| \leq 11$

48. $|2s - 6| \leq 6$

49. $|-6x - 6| \leq 1$

50. $|-2x - 6| \leq 5$

Solve each equation or inequality. Give the solution set in set notation for equations and in interval notation for inequalities. See Example 4.

51. $|x| - 1 = 4$

52. $|y| + 3 = 10$

53. $|x + 4| + 1 = 2$

54. $|y + 5| - 2 = 12$

55. $|2x + 1| + 3 > 8$

56. $|6x - 1| - 2 > 6$

57. $|x + 5| - 6 \leq -1$

58. $|r - 2| - 3 \leq 4$

Solve each equation. See Example 5.

59. $|3x + 1| = |2x + 4|$

60. $|7x + 12| = |x - 8|$

61. $\left| m - \dfrac{1}{2} \right| = \left| \dfrac{1}{2}m - 2 \right|$

62. $\left| \dfrac{2}{3}r - 2 \right| = \left| \dfrac{1}{3}r + 3 \right|$

63. $|6x| = |9x + 1|$

64. $|13y| = |2y + 1|$

65. $|2p - 6| = |2p + 11|$

66. $|3x - 1| = |3x + 9|$

Solve each equation or inequality. See Examples 6 and 7.

67. $|12t - 3| = -8$

68. $|13w + 1| = -3$

69. $|4x + 1| = 0$

70. $|6r - 2| = 0$

71. $|2q - 1| < -6$

72. $|8n + 4| < -4$

73. $|x + 5| > -9$

74. $|x + 9| > -3$

75. $|7x + 3| \leq 0$

76. $|4x - 1| \leq 0$

77. $|5x - 2| \geq 0$

78. $|4 + 7x| \geq 0$

79. $|10z + 7| > 0$

80. $|4x + 1| > 0$

MATHEMATICAL CONNECTIONS (EXERCISES 81–84)

The ten tallest buildings in Kansas City, Missouri, are listed along with their heights.

Building	Height (in feet)
One Kansas City Place	626
AT&T Town Pavilion	590
Hyatt Regency	504
Kansas City Power and Light	476
City Hall	443
Federal Office Building	413
Commerce Tower	402
City Center Square	402
Southwest Bell Telephone	394
Pershing Road Associates	352

Use this information to work through Exercises 81–84 in order.

81. To find the average of a group of numbers, we add the numbers and then divide by the number of items added. Use a calculator to find the average of the heights.

82. Let k represent the average height of these buildings. If a height x satisfies the inequality

$$|x - k| < t,$$

then the height is said to be within t feet of the average. Using your result from Exercise 81, list the buildings that are within 50 feet of the average.

83. Repeat Exercise 82, but find the buildings that are within 75 feet of the average.

84. (a) Write an absolute value inequality that describes the height of a building that is *not* within 75 feet of the average.
   (b) Solve the inequality you wrote in part (a).
 ◉ **(c)** Use the result of part (b) to find the buildings that are not within 75 feet of the average.
   (d) Confirm that your answer to part (c) makes sense by comparing it with your answer to Exercise 83.

SUMMARY SOLVING LINEAR AND ABSOLUTE VALUE EQUATIONS AND INEQUALITIES

Students often have difficulty distinguishing between the various types of equations and inequalities introduced in this chapter. This section of miscellaneous equations and inequalities provides practice in solving all such types. You might wish to refer to the boxes in this chapter that summarize the various methods of solution. Solve each equation or inequality.

1. $4z + 1 = 49$

2. $|m - 1| = 6$

3. $6q - 9 = 12 + 3q$

4. $3p + 7 = 9 + 8p$

5. $|a + 3| = -4$

6. $2m + 1 \le m$

7. $8r + 2 \ge 5r$

8. $4(a - 11) + 3a = 20a - 31$

9. $2q - 1 = -7$

10. $|3q - 7| - 4 = 0$

11. $6z - 5 \le 3z + 10$

12. $|5z - 8| + 9 \ge 7$

13. $9y - 3(y + 1) = 8y - 7$

14. $|y| \ge 8$

15. $9y - 5 \ge 9y + 3$

16. $13p - 5 > 13p - 8$

17. $|q| < 5.5$

18. $4z - 1 = 12 + z$

19. $\frac{2}{3}y + 8 = \frac{1}{4}y$

20. $-\frac{5}{8}y \ge -20$

21. $\frac{1}{4}p < -6$

22. $7z - 3 + 2z = 9z - 8z$

23. $\frac{3}{5}q - \frac{1}{10} = 2$

24. $|r - 1| < 7$

25. $r + 9 + 7r = 4(3 + 2r) - 3$ **26.** $6 - 3(2 - p) < 2(1 + p) + 3$ **27.** $|2p - 3| > 11$

28. $\dfrac{x}{4} - \dfrac{2x}{3} = -10$ **29.** $|5a + 1| \leq 0$ **30.** $5z - (3 + z) \geq 2(3z + 1)$

31. $-2 \leq 3x - 1 \leq 8$ **32.** $-1 \leq 6 - x \leq 5$ **33.** $|7z - 1| = |5z + 3|$

34. $|p + 2| = |p + 4|$ **35.** $|1 - 3x| \geq 4$ **36.** $\dfrac{1}{2} \leq \dfrac{2}{3}r \leq \dfrac{5}{4}$

37. $-(m + 4) + 2 = 3m + 8$ **38.** $\dfrac{p}{6} - \dfrac{3p}{5} = p - 86$ **39.** $-6 \leq \dfrac{3}{2} - x \leq 6$

40. $|5 - y| < 4$ **41.** $|y - 1| \geq -6$ **42.** $|2r - 5| = |r + 4|$

43. $8q - (1 - q) = 3(1 + 3q) - 4$ **44.** $8y - (y + 3) = -(2y + 1) - 12$

45. $|r - 5| = |r + 9|$ **46.** $|r + 2| < -3$

47. $2x + 1 > 5$ or $3x + 4 < 1$ **48.** $1 - 2x \geq 5$ and $7 + 3x \geq -2$

CHAPTER 2 SUMMARY

2.1	**algebraic expression**	An algebraic expression is an expression indicating any combination of the following operations: addition, subtraction, multiplication, division (except by 0), and taking roots on any collection of variables and numbers.
	equation	An equation is a statement that two algebraic expressions are equal.
	linear equation or first-degree equation in one variable	An equation is linear or first-degree in the variable x if it can be written in the form $ax + b = c$, where a, b, and c are real numbers, with $a \neq 0$.
	solution	A solution of an equation is a number that makes the equation true when substituted for the variable.
	solution set	The solution set of an equation is the set of all its solutions.
	equivalent equations	Equivalent equations are equations that have the same solution set.
	addition and multiplication properties of equality	These properties state that the same number may be added to (or subtracted from) both sides of an equation to obtain an equivalent equation; and the same nonzero number may be multiplied by or divided into both sides of an equation to obtain an equivalent equation.
	factor	To factor an expression is to write a sum as a product.
	conditional equation	An equation that has a finite (but nonzero) number of elements in its solution set is called a conditional equation.
	contradiction	An equation that has no solutions (that is, its solution set is $\emptyset$) is called a contradiction.
	identity	An equation that is satisfied by every number for which both sides are defined is called an identity.
2.2	**formula**	A formula is an equation that tells how to find the value of some quantity.
	percent	One percent means "one per hundred."
2.5	**linear inequality in one variable**	An inequality is linear in the variable x if it can be written in the form $ax + b < c$, $ax + b \leq c$, $ax + b > c$, or $ax + b \geq c$, where a, b, and c are real numbers, with $a \neq 0$.
	equivalent inequalities	Equivalent inequalities are inequalities with the same solution set.
	addition and multiplication properties of inequality	The same number may be added to (or subtracted from) both sides of an inequality to obtain an equivalent inequality. Both sides of an inequality may be multiplied or divided by the same positive number. If both sides are multiplied by or divided by a negative number, the inequality symbol must be reversed.
	strict inequality	An inequality that involves $>$ or $<$ is called a strict inequality.
	nonstrict inequality	An inequality that involves $\geq$ or $\leq$ is called a nonstrict inequality.
2.6	**intersection**	The intersection of two sets A and B is the set of elements that belong to both A and B.
	union	The union of two sets A and B is the set of elements that belong to either A or B (or both).
	compound inequality	A compound inequality is formed by joining two inequalities with a connective word, such as *and* or *or*.
2.7	**absolute value equation; absolute value inequality**	Absolute value equations and inequalities are equations and inequalities that involve the absolute value of a variable expression.

NEW SYMBOLS

- ∩   set intersection
- ∪   set union

QUICK REVIEW

Concepts	Examples

2.1 Linear Equations in One Variable

Solving a Linear Equation

If necessary, eliminate fractions by multiplying both sides by the LCD. Simplify each side, and then use the addition property of equality to get the variables on one side and the numbers on the other. Combine terms if possible, and then use the multiplication property of equality to make the coefficient of the variable equal to 1. Check by substituting into the original equation.

Solve the equation $4(8 - 3t) = 32 - 8(t + 2)$.

$$32 - 12t = 32 - 8t - 16$$
$$32 - 12t = 16 - 8t$$
$$32 - 12t + 12t = 16 - 8t + 12t$$
$$32 = 16 + 4t$$
$$32 - 16 = 16 + 4t - 16$$
$$16 = 4t$$
$$\frac{16}{4} = \frac{4t}{4}$$
$$4 = t \quad \text{or} \quad t = 4$$

Check:

$$4(8 - 3t) = 32 - 8(t + 2)$$
$$4(8 - 3 \cdot 4) = 32 - 8(4 + 2)$$
$$4(8 - 12) = 32 - 8(6)$$
$$4(-4) = 32 - 48$$
$$-16 = -16$$

The solution set is {4}.

2.2 Formulas

Solving for a Specified Variable

Use the addition or multiplication properties as necessary to get all terms with the specified variable on one side of the equals sign, and all other terms on the other side. If necessary, use the distributive property to write the terms with the specified variable as the product of that variable and a sum of terms. Complete the solution.

Solve for h: $A = \frac{1}{2}bh$.

$$A = \frac{1}{2}bh$$
$$2A = 2\left(\frac{1}{2}bh\right) \qquad \text{Multiply by 2.}$$
$$2A = bh$$
$$\frac{2A}{b} = h \qquad \text{Divide by } b.$$

Concepts	Examples

2.3 Applications of Linear Equations
2.4 More Applications of Linear Equations

Solving an Applied Problem

Step 1 Read the problem carefully. Decide what is given and what must be found. Use sketches as necessary.

Step 2 Choose a variable and write down exactly what it represents. (Do not make the mistake of omitting this step.)

Step 3 Write an equation using the information given in the problem.

Step 4 Solve the equation.

Step 5 Make sure you have answered the question of the problem.

Step 6 Check your answer in the words of the original problem, not in the equation you obtained from the words.

Two cars start from towns 400 miles apart and travel toward each other. They meet after 4 hours. Find the speed of each car if one travels 20 miles per hour faster than the other.

Let x = speed of the slower car in miles per hour;

$x + 20$ = speed of the faster car.

Use the information in the problem, and $d = rt$ to complete the chart.

	r	t	d
Slower Car	x	4	$4x$
Faster Car	$x + 20$	4	$4(x + 20)$

A sketch shows that the sum of the distances, $4x$ and $4(x + 20)$, must be 400.

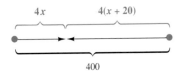

The equation is

$$4x + 4(x + 20) = 400.$$

Solving this equation gives $x = 40$. The slower car travels 40 miles per hour and the faster car travels $40 + 20 = 60$ miles per hour.

2.5 Linear Inequalities in One Variable

Solving a Linear Inequality

Simplify each side separately, combining like terms and removing parentheses. Use the addition property of inequality to get the variables on one side of the inequality sign and the numbers on the other. Combine like terms, and then use the multiplication property to change the inequality to the form $x < k$ or $x > k$.

If an inequality is multiplied or divided by a *negative* number, the inequality symbol *must be reversed.*

Solve $3(x + 2) - 5x \le 12$.

$$3x + 6 - 5x \le 12$$
$$-2x + 6 \le 12$$
$$-2x \le 6$$
$$\frac{-2x}{-2} \ge \frac{6}{-2}$$
$$x \ge -3$$

The solution set is $[-3, \infty)$ and is graphed below.

Concepts	Examples
2.6 Set Operations and Compound Inequalities	

Solving a Compound Inequality
Solve each inequality in the compound inequality individually. If the inequalities are joined with *and,* the solution is the intersection of the two individual solutions. If the inequalities are joined with *or,* the solution is the union of the two individual solutions.

Solve $x + 1 > 2$ and $2x < 6$.

$$x + 1 > 2 \quad \text{and} \quad 2x < 6$$
$$x > 1 \quad \text{and} \quad x < 3$$

The solution set is $(1, 3)$.

Solve $x \geq 4$ or $x \leq 0$.

The solution set is $(-\infty, 0] \cup [4, \infty)$.

2.7 Absolute Value Equations and Inequalities
Let k be a positive number.
To solve $|ax + b| = k$, solve the compound equation

$$ax + b = k \quad \text{or} \quad ax + b = -k.$$

Solve $|x - 7| = 3$.

$$x - 7 = 3 \quad \text{or} \quad x - 7 = -3$$
$$x = 10 \qquad\qquad x = 4$$

The solution set is $\{4, 10\}$.

To solve $|ax + b| > k$, solve the compound inequality

$$ax + b > k \quad \text{or} \quad ax + b < -k.$$

Solve $|x - 7| > 3$.

$$x - 7 > 3 \quad \text{or} \quad x - 7 < -3$$
$$x > 10 \quad \text{or} \quad x < 4$$

The solution set is $(-\infty, 4) \cup (10, \infty)$.

To solve $|ax + b| < k$, solve the compound inequality

$$-k < ax + b < k.$$

Solve $|x - 7| < 3$.

$$-3 < x - 7 < 3$$
$$4 < x < 10$$

The solution set is $(4, 10)$.

To solve an absolute value equation of the form

$$|ax + b| = |cx + d|,$$

solve the compound equation

$$ax + b = cx + d \quad \text{or}$$
$$ax + b = -(cx + d).$$

Solve $|x + 2| = |2x - 6|$.

$$x + 2 = 2x - 6 \quad \text{or} \quad x + 2 = -(2x - 6)$$
$$x = 8 \qquad\qquad x + 2 = -2x + 6$$
$$3x = 4$$
$$x = \frac{4}{3}$$

The solution set is $\left\{\frac{4}{3}, 8\right\}$.

[2.1] *Solve each equation.*

1. $-(8 + 3y) + 5 = 2y + 6$

2. $-(r + 5) - (2 + 7r) + 8r = 3r - 8$

3. $\dfrac{m - 2}{4} + \dfrac{m + 2}{2} = 8$

4. $\dfrac{2q + 1}{3} - \dfrac{q - 1}{4} = 0$

5. $5(2x - 3) = 6(x - 1) + 4x$

6. $-3x + 2(4x + 5) = 10$

7. $-\dfrac{3}{4}x = -12$

8. $.05x + .03(1200 - x) = 42$

9. Which one of the following equations has $\{0\}$ as its solution set?
    (a) $x - 5 = 5$ **(b)** $4x = 5x$ **(c)** $x + 3 = -3$ **(d)** $6x - 6 = 6$

10. Give the steps you would use to solve the equation $-2x + 5 = 7$.

Decide whether the given equation is conditional, an identity, or a contradiction. Give the solution set.

11. $7r - 3(2r - 5) + 5 + 3r = 4r + 20$

12. $8p - 4p - (p - 7) + 9p + 6 = 12p - 7$

13. $-2r + 6(r - 1) + 3r - (4 - r) = -(r + 5) - 5$

[2.2] *Solve the formula for the indicated variable.*

14. $V = LWH$ for H

15. $A = \dfrac{1}{2}h(B + b)$ for h

16. $C = \pi d$ for d

Solve each of the following problems.

17. An incinerator has a volume of 180 cubic feet. Its length is 6 feet and its width is 5 feet. Find its height.

18. Research scientists project that segments of the U.S. weight loss and diet control market will expand dramatically in the next few years. Between 1996 and 2000, the health club market is expected to rise from $16.18 billion to $17.76 billion. What percent increase does this represent?

19. Find the simple interest rate that Francesco Castellucio is getting, if a principal of $30,000 earns $7800 interest in 4 years.

20. If the Fahrenheit temperature is 68°, what is the corresponding Celsius temperature?

21. The economic reform plan proposed by President Clinton in 1993 was projected to save $108 billion by the end of 1999. The accompanying chart shows the sources of savings and how much would be saved. What percent of the total savings is represented by staff cuts?

22. The Sioux drum that Logan purchased has a circumference of 200π millimeters. Find the measure of its radius.

— **SOURCES OF SAVINGS** —
(in billions)

Source: New Orleans Times Picayune

[2.3] *Write the following as mathematical expressions, using x as the variable.*

23. One-third of a number, subtracted from 9

24. The product of 4 and a number, divided by 9 more than the number

Solve each of the following problems.

25. The length of a rectangle is 3 meters less than twice the width. The perimeter of the rectangle is 42 meters. Find the length and width of the rectangle.

26. In a triangle with two sides of equal length, the third side measures 15 inches less than the sum of the two equal sides. The perimeter of the triangle is 53 inches. Find the lengths of the three sides.

27. A candy clerk has three times as many kilograms of chocolate creams as peanut clusters. The clerk has 48 kilograms of the two candies altogether. How many kilograms of peanut clusters does the clerk have?

28. How many liters of a 20% solution of a chemical should be mixed with 15 liters of a 50% solution to get a 30% mixture?

29. Kevin Connors invested some money at 6% and $4000 less than this amount at 4%. Find the amount invested at each rate if his total annual interest income is $840.

30. In 1975 there were 161,927 federally licensed firearm dealers in the United States. In 1993 there were 269,712 such dealers. What percent increase does this represent? (*Source:* Bureau of Alcohol, Tobacco, and Firearms)

[2.4]

31. Which of the following choices is the best *estimate* for the average speed of a trip of 405 miles that lasted 8.2 hours?
  (a) 50 miles per hour
  (b) 30 miles per hour
  (c) 60 miles per hour
  (d) 40 miles per hour

32. (a) A driver averaged 53 miles per hour and took 10 hours to travel from Memphis to Chicago. What is the distance between Memphis and Chicago?
  (b) A small plane traveled from Warsaw to Rome, averaging 164 miles per hour. The trip took two hours. What is the distance from Warsaw to Rome?

33. A passenger train and a freight train leave a town at the same time and go in opposite directions. They travel at 60 miles per hour and 75 miles per hour, respectively. How long will it take for them to be 297 miles apart?

34. Two cars leave towns 230 kilometers apart at the same time, traveling directly toward one another. One car travels 15 kilometers per hour slower than the other. They pass one another 2 hours later. What are their speeds?

35. An automobile averaged 45 miles per hour for the first part of a trip and 50 miles per hour for the second part. If the entire trip took 4 hours and covered 195 miles, for how long was the rate of speed 45 miles per hour?

36. An 85-mile trip to the beach took the Rodriguez family 2 hours. During the second hour, a rainstorm caused them to average 7 miles per hour less than they traveled during the first hour. Find their average speed for the first hour.

37. A total of 1096 people attended a Randy Travis concert. Reserved seats cost $15 each, and general admission seats cost $12 each. If $15,702 was collected, how many of each type of seat were sold?

38. There were 311 tickets sold for a soccer game, some for students and some for nonstudents. Student tickets cost 25¢ each, and nonstudent tickets cost 75¢ each. The total receipts were $108.75. How many of each type of ticket were sold?

39. Find the measure of each marked angle.

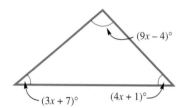

40. What is the best estimate of the number of new cars sold for personal use that were leased in 1993, if a *total* of 4 million cars were sold?
   (a) 2 million   **(b)** 2.5 million
   (c) 1 million   **(d)** 1.5 million

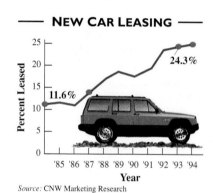

Source: CNW Marketing Research

41. The angles marked in this figure are called **vertical angles,** and their measures are equal. Find the measure of each marked angle.

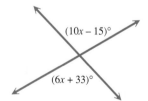

42. The sum of the smallest and largest of three consecutive integers is 47 more than the middle integer. What are the integers?

[2.5] *Solve each inequality. Give the solution set in both interval and graph forms.*

43. $-\dfrac{2}{3}k < 6$

→

44. $-5x - 4 \geq 11$

→

45. $\dfrac{6a + 3}{-4} < -3$

→

46. $\dfrac{9y + 5}{-3} > 3$

→

47. $5 - (6 - 4k) \geq 2k - 7$

→

48. $-6 \leq 2k \leq 24$

→

49. $8 \leq 3y - 1 < 14$

→

50. $-4 < 3 - 2k < 9$

→

51. To pass algebra, a student must have an average of at least 70% on five tests. On the first four tests, a student has grades of 75%, 79%, 64%, and 71%. What possible grades on the fifth test would guarantee a passing grade in the class?

52. While solving the inequality
$10x + 2(x - 4) < 12x - 13$, a student did all the work correctly and obtained the statement $-8 < -13$. The student did not know what to do at this point, because the variable "disappeared." How would you explain to the student the interpretation of this result?

[2.6] *In Exercises 53 and 54, let $A = \{1, 3, 5, 7, 9\}$ and let $B = \{3, 6, 9, 12\}$.*

53. Find $A \cap B$.

54. Find $A \cup B$.

Solve each compound inequality. Give the solution set in both interval and graph forms.

55. $x > 6$ and $x < 9$

56. $x + 4 > 12$ and $x - 2 < 12$

57. $x > 5$ or $x \leq -3$

58. $x \geq -2$ or $x < 2$

59. $x - 4 > 6$ and $x + 3 \leq 10$

60. $-5x + 1 \geq 11$ or $3x + 5 \geq 26$

Use the graphs to answer the questions in Exercises 61 and 62.

U.S. AIDS CASES REPORTED EACH YEAR

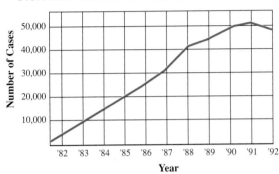

Source: Surgeon General's Report to the American
Public on HIV Infection and AIDS

**NEW AIDS CASES REPORTED AMONG
CHILDREN UNDER 13 YEARS OF AGE**

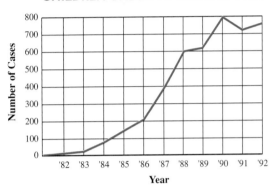

61. In which years did the number of U.S. AIDS cases exceed 30,000 *and* the new AIDS cases among children under thirteen years of age exceed 400?

62. In which years was the number of U.S. AIDS cases greater than 40,000 *or* the new AIDS cases among children under thirteen years of age less than 200?

[2.7] *Solve each absolute value equation.*

63. $|x| = 7$

64. $|y + 2| = 9$

65. $|3k - 7| = 8$

66. $|z - 4| = -12$

67. $|2k - 7| + 4 = 11$

68. $|4a + 2| - 7 = -3$

69. $|3p + 1| = |p + 2|$

70. $|2m - 1| = |2m + 3|$

Solve each absolute value inequality. Give the solution set in both interval and graph forms.

71. $|p| < 14$

72. $|-y + 6| \le 7$

73. $|2p + 5| \leq 1$

⎯⎯⎯⎯⎯⎯⎯⎯⎯⟶

74. $|x + 1| \geq -3$

⎯⎯⎯⎯⎯⎯⎯⎯⎯⟶

75. $|5r - 1| > 9$

⎯⎯⎯⎯⎯⎯⎯⎯⟶

76. $|3k + 6| \geq 0$

⎯⎯⎯⎯⎯⎯⎯⎯⟶

MIXED REVIEW EXERCISES*

Solve.

77. $(7 - 2k) + 3(5 - 3k) \geq k + 8$

78. $x < 5$ and $x \geq -4$

79. $-5(6p + 4) - 2p = -32p + 14$

80. The perimeter of a triangle is 34 inches. The middle side is twice as long as the shortest side. The longest side is 2 inches less than three times the shortest side. Find the lengths of the three sides.

81. $-5r \geq -10$

82. $|7x - 2| > 9$

83. $|2x - 10| = 20$

84. $|m + 3| \leq 13$

85. A square is such that if each side were increased by 4 inches, the perimeter would be 8 inches less than twice the perimeter of the original square. Find the length of a side of the original square.

86. In an election, one candidate received 151 more votes than the other. The total number of votes cast in the election was 1215. Find the number of votes received by each candidate.

⎯⎯⎯⎯⎯⎯

*The order of exercises in this final group does not correspond to the order in which topics occur in the chapter. This random ordering should help you prepare for the chapter test in yet another way.

Solve each equation.

  1. $3(2y - 2) - 4(y + 6) = 3y + 8 + y$

1. _____

  2. $.08x + .06(x + 9) = 1.24$

2. _____

  3. $\dfrac{x + 6}{10} + \dfrac{x - 4}{15} = \dfrac{x + 2}{6}$

3. _____

  4. Solve for L: $P = 2L + 2W$

4. _____

Solve each problem.

  5. The Daytona 500 (mile) race was shortened to 450 miles in 1974 due to the energy crisis. In that year it was won by Richard Petty, who averaged 140.9 miles per hour. What was Petty's time?

5. _____

  6. A certificate of deposit pays \$862.50 in simple interest for one year on a principal of \$23,000. What is the rate of interest?

6. _____

  7. In a certain South Dakota county, 6118 residents live in poverty. This represents 63.1% of the population of the county. What is the population of the county?

7. _____

  8. Charles Dawkins invested some money at 3% simple interest and some at 5% simple interest. The total amount of his investments was \$28,000, and the interest he earned during the first year was \$1240. How much did he invest at each rate?

8. _____

  9. Two cars leave from the same point at the same time, traveling in opposite directions. One travels 15 miles per hour slower than the other. After 3 hours, they are 315 miles apart. Find the rate of each car.

9. _____

  10. Find the measure of each angle.

10. _____

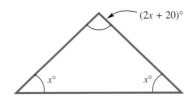

$(2x + 20)°$

$x°$ $x°$

11. _____

11. What is the special rule that must be remembered when multiplying or dividing both sides of an inequality by a negative number?

Solve each inequality. Give the solution set in both interval and graph forms.

12. _____→

12. $4 - 6(x + 3) \leq -2 - 3(x + 6) + 3x$

13. _____→

13. $-\dfrac{4}{7}x > -16$

14. _____→

14. $-6 \leq \dfrac{4}{3}x - 2 \leq 2$

15. _____

15. A student must have an average grade of 90 or greater on the three tests in Intermediate Algebra to earn a grade of A. A student had scores of 84 and 92 on her first two tests. What possible scores on her third test would assure her an A?

Solve each compound or absolute value inequality. Give the solution set in both interval and graph forms.

16. _____→

16. $3k \geq 6$ and $k - 4 < 5$

17. _____→

17. $|4x + 3| \leq 7$

18. _____→

18. $|5 - 6x| > 12$

Solve each absolute value equation.

19. _____

19. $|3k - 2| + 1 = 8$

20. _____

20. $|3 - 5x| = |2x + 8|$

Let $A = \{-8, -\frac{2}{3}, -\sqrt{6}, 0, \frac{4}{5}, 9, \sqrt{36}\}$. *Simplify the elements of A as necessary, and then list the elements that belong to each set listed below.*

1. Natural numbers

2. Whole numbers

3. Integers

4. Rational numbers

5. Irrational numbers

6. Real numbers

Add or subtract, as indicated.

7. $-\dfrac{4}{3} - \left(-\dfrac{2}{7}\right)$

8. $|-4| - |2| + |-6|$

9. $(-2)^4 + (-2)^3$

10. $\sqrt{25} - \sqrt[3]{125}$

Evaluate each of the following.

11. $(-3)^5$

12. $\left(\dfrac{6}{7}\right)^3$

13. $(x^2 + 1)^0$

14. -4^6

15. Which one of the following is not a real number: $-\sqrt{36}$ or $\sqrt{-36}$?

16. Which one of the following is undefined: $\dfrac{4-4}{4+4}$ or $\dfrac{4+4}{4-4}$?

Evaluate if $a = 2$, $b = -3$, and $c = 4$.

17. $-3a + 2b - c$

18. $-2b^2 - 4c$

19. $-8(a^2 + b^3)$

20. $\dfrac{3a^3 - b}{4 + 3c}$

Use the properties of real numbers to simplify each expression.

21. $-7r + 5 - 13r + 12$

22. $-(3k + 8) - 2(4k - 7) + 3(8k + 12)$

Identify the property of real numbers illustrated in each of the following.

23. $(a + b) + 4 = 4 + (a + b)$

24. $4x + 12x = (4 + 12)x$

25. $-9 + 9 = 0$

26. What is the reciprocal, or multiplicative inverse, of $-\frac{2}{3}$?

Solve each equation.

27. $-4x + 7(2x + 3) = 7x + 36$

28. $-\dfrac{3}{5}x + \dfrac{2}{3}x = 2$

29. $.06x + .03(100 + x) = 4.35$

30. $P = a + b + c$ for b

Solve each inequality. Give the solution set in both interval and graph forms.

31. $3 - 2(x + 7) \le -x + 3$

32. $-4 < 5 - 3x \le 0$

33. $2x + 1 > 5$ or $2 - x > 2$

34. $|-7k + 3| \ge 4$

According to figures provided by the Equal Employment Opportunity Commission, Bureau of Labor Statistics, the following are the median weekly earnings of full-time workers by occupation for men and women.

Occupation	Men	Women
Managerial and professional specialty	$753	$527
Mathematical and computer scientists	$923	$707
Waiters and waitresses	$281	$205
Bus drivers	$411	$321

Give the occupation that satisfies the description.

35. The median earnings for men are less than $900 *and* for women are greater than $500.

36. The median earnings for men are greater than $900 *or* for women are greater than $600.

Solve each problem.

37. How much pure alcohol should be added to 7 liters of 10% alcohol to increase the concentration to 30% alcohol?

38. A coin collection contains 29 coins. It consists of cents, nickels, and quarters. The number of quarters is 4 less than the number of nickels, and the face value of the collection is $2.69. How many of each denomination are there in the collection?

Linear Equations and Inequalities in Two Variables; Functions

3.1 The Rectangular Coordinate System

Graphs are widely used in the media. Newspapers and magazines, television, reports to stockholders, and newsletters all present information in the form of a graph. Figure 1(a) shows a **line graph** representing the federal debt (that is, the amount the government owes because it has borrowed to finance its purchasing) for selected years since 1980. The **bar graph** in Figure 1(b) gives the revenue of home-improvement retailers during the years 1990 through 1996. And the **circle graph,** or **pie chart,** in Figure 1(c) shows a breakdown of the sources of credit card fraud. Graphs are used so widely because they show a lot of information in a form that makes it easy to understand. As the saying goes, "A picture is worth a thousand words." In this section, we show how to graph equations of lines.

www.mathnotes.com

OBJECTIVES

1. Plot ordered pairs.
2. Find ordered pairs that satisfy a given equation.
3. Graph lines.
4. Find *x*- and *y*-intercepts.
5. Recognize equations of vertical or horizontal lines.

FOR EXTRA HELP

Tutorial   Tape 3   SSM, Sec. 3.1

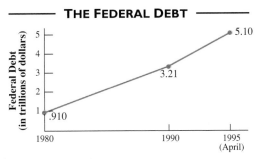

THE FEDERAL DEBT

Source: U.S. Dept. of the Treasury

(a)

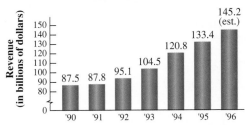

REVENUE OF HOME-IMPROVEMENT RETAILERS

Sources: National Retail Hardware Association; Home Center Institute

(b)

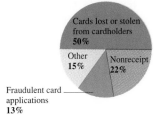

CREDIT CARD FRAUD

Lost or stolen cards account for half of the $15 million in fraud losses reported by banks each year.

Source: Nilsson Report

(c)

FIGURE 1

147

1. Plot the following points.

(a) $(-4, 2)$ **(b)** $(3, -2)$

(c) $(-5, -6)$ **(d)** $(4, 6)$

(e) $(-3, 0)$ **(f)** $(0, -5)$

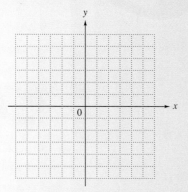

An **ordered pair** (OR-durd PAIR) is a pair of numbers written within parentheses in which the order of the numbers matters. By this definition, the ordered pairs $(2, 5)$ and $(5, 2)$ are different. The two numbers are called **components** of the ordered pair. It is customary for x to represent the first component and y the second component. We graph an ordered pair by using two perpendicular number lines that intersect at the zero points, as shown in Figure 2. The common zero point is called the **origin** (OR-ih-gin). The horizontal line, the **x-axis,** represents the first number in an ordered pair, and the vertical line, the **y-axis,** represents the second. The x-axis and the y-axis make up a **rectangular coordinate system.** It is also called the **Cartesian system,** named after the French mathematician René Descartes (1596–1650).

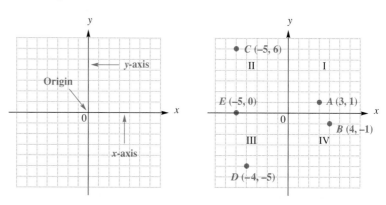

FIGURE 2   **FIGURE 3**

OBJECTIVE 1 To locate, or **plot,** the point on the graph that corresponds to the ordered pair $(3, 1)$, go three units from zero to the right along the x-axis, and then go one unit up parallel to the y-axis. The point corresponding to the ordered pair $(3, 1)$ is labeled A in Figure 3. The point $(4, -1)$ is labeled B, $(-5, 6)$ is labeled C, and $(-4, -5)$ is labeled D. Point E corresponds to $(-5, 0)$. The phrase "the point corresponding to the ordered pair $(2, 1)$" is often abbreviated as "the point $(2, 1)$." The numbers in the ordered pairs are called the **coordinates** (koh-OR-din-ets) of the corresponding point.

The four regions of the graph, shown in Figure 3, are called **quadrants** (KWAD-runts) **I, II, III,** and **IV,** reading counterclockwise from the upper right quadrant. The points on the x-axis and y-axis do not belong to any quadrant. For example, point E in Figure 3 belongs to no quadrant.

◀◀ **WORK PROBLEM I AT THE SIDE.**

OBJECTIVE 2 Each solution to an equation with two variables will include two numbers, one for each variable. To keep track of which number goes with which variable, we write the solutions as ordered pairs. For example, we can show that $(6, -2)$ is a solution of $2x + 3y = 6$ by substitution.

$$2x + 3y = 6$$
$$2(6) + 3(-2) = 6 \quad ?$$
$$12 - 6 = 6 \quad ?$$
$$6 = 6 \quad \text{True}$$

Because the pair of numbers $(6, -2)$ makes the equation true, it is a solution. On the other hand, because

$$2(5) + 3(1) = 10 + 3 = 13 \neq 6,$$

$(5, 1)$ is not a solution of the equation.

ANSWERS

1.

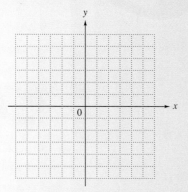

To find ordered pairs that satisfy an equation, we select any number for one of the variables, substitute it into the equation for that variable, and then solve for the other variable. For example, suppose we choose $x = 0$ in the equation $2x + 3y = 6$. Then, by substitution,

$$2x + 3y = 6$$

becomes
$$2(0) + 3y = 6 \qquad \text{Let } x = 0.$$
$$0 + 3y = 6$$
$$3y = 6$$
$$y = 2,$$

giving the ordered pair $(0, 2)$. Some other ordered pairs satisfying $2x + 3y = 6$ are $(6, -2)$, as shown above, and $(3, 0)$. Because every real number could be selected for one variable and would lead to a real number for the other variable, linear equations with two variables have an infinite number of solutions.

EXAMPLE 1 Completing Ordered Pairs

Complete the following ordered pairs for $2x + 3y = 6$.

(a) $(-3, \quad)$

We are given $x = -3$. Substitute into the equation.

$$2(-3) + 3y = 6 \qquad \text{Given } x = -3$$
$$-6 + 3y = 6$$
$$3y = 12$$
$$y = 4$$

The ordered pair is $(-3, 4)$.

(b) $(\quad, -4)$

Replace y with -4.

$$2x + 3y = 6$$
$$2x + 3(-4) = 6 \qquad \text{Given } y = -4$$
$$2x - 12 = 6$$
$$2x = 18$$
$$x = 9$$

The ordered pair is $(9, -4)$.

WORK PROBLEM 2 AT THE SIDE. ▶▶

OBJECTIVE 3 The **graph** of an equation is the set of points that correspond to all the ordered pairs that satisfy the equation. It gives a "picture" of the equation. Most equations with two variables have an infinite set of ordered pairs, so their graphs include an infinite number of points. To graph an equation, we plot a number of ordered pairs that satisfy the equation until we have enough points to suggest the shape of the graph. For example, to graph

2. (a) Complete the following ordered pairs for $3x - 4y = 12$: $(0, \quad), (\quad, 0), (\quad, -2), (-4, \quad)$

(b) Find one other ordered pair that satisfies the equation.

ANSWERS

2. (a) $(0, -3), (4, 0), \left(\frac{4}{3}, -2\right),$
$(-4, -6)$
(b) Many answers are possible; for example, $\left(-6, -\frac{15}{2}\right)$.

3. Graph $3x - 4y = 12$. Use the points from Problem 2.

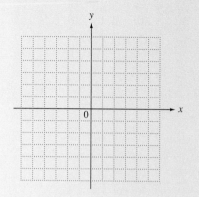

$2x + 3y = 6$, first graph all the ordered pairs found in Example 1 and Objective 2. These points are shown in Figure 4(a). The points appear to lie on a straight line. If all the ordered pairs that satisfy the equation $2x + 3y = 6$ were graphed, they would form a straight line. In fact, the graph of any first-degree equation in two variables is always a straight line. The graph of $2x + 3y = 6$ is the line shown in Figure 4(b).

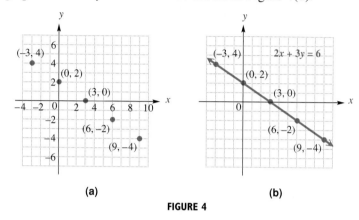

(a)   (b)

FIGURE 4

◀◀ WORK PROBLEM 3 AT THE SIDE.

OBJECTIVE ▶4▶ Because first-degree equations with two variables have straight-line graphs, they are called **linear equations in two variables.** (We discussed linear equations in one variable in Chapter 2.)

> **Standard Form of a Linear Equation**
>
> An equation that can be written in the **standard form**
>
> $$Ax + By = C$$
>
> is a linear equation.

A straight line is determined if any two different points on the line are known; finding two different points is enough to graph the line, but it is wise to find a third point as a check. Two points that are useful for graphing are the x- and y-intercepts. The **x-intercept** (IN-ter-sept) is the point (if any) where the line crosses the x-axis; likewise, the **y-intercept** is the point (if any) where the line crosses the y-axis. In Figure 4(b), the y-value of the point where the line crosses the x-axis is 0. Similarly, the x-value of the point where the line crosses the y-axis is 0. This suggests a method for finding the x- and y-intercepts.

> In the equation of a line, choose $y = 0$ and solve for x to find the x-intercept; choose $x = 0$ and solve for y to find the y-intercept.

E X A M P L E 2   Finding Intercepts

Find the x- and y-intercepts of $4x - y = -3$, and graph the equation.
    To find the x-intercept, let $y = 0$.

$$4x - 0 = -3 \qquad \text{Let } y = 0.$$
$$4x = -3$$
$$x = -\frac{3}{4} \qquad x\text{-intercept is } \left(-\frac{3}{4}, 0\right)$$

CONTINUED ON NEXT PAGE

ANSWERS

3.

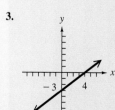

For the *y*-intercept, let $x = 0$.

$$4(0) - y = -3 \qquad \text{Let } x = 0.$$
$$-y = -3$$
$$y = 3 \qquad \text{y-intercept is } (0, 3)$$

We show the two intercepts in the form of a table, called a *table of values,* next to Figure 5. Plot the intercepts and draw a line through them to get the graph in Figure 5.

x	y
$-\frac{3}{4}$	0
0	3

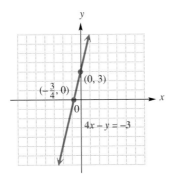

FIGURE 5

Note

While two points, such as the two intercepts in Figure 5, are sufficient to graph a straight line, it is a good idea to use a third point to guard against errors. Verify that $(-1, -1)$ also lies on the graph of $4x - y = -3$.

WORK PROBLEM 4 AT THE SIDE. ▶▶

OBJECTIVE 5 ▶ The next example shows that a graph may not have an *x*-intercept.

— **E X A M P L E 3** **Graphing a Horizontal Line**

Graph $y = 2$.

Writing $y = 2$ in standard form as $0x + 1y = 2$ shows that any value of *x*, including $x = 0$, gives $y = 2$, making the *y*-intercept $(0, 2)$. Every ordered pair that satisfies this equation has a *y*-coordinate of 2. Because *y* is always 2, there is no value of *x* corresponding to $y = 0$, and the graph has no *x*-intercept. The graph, shown in Figure 6, is a horizontal line.

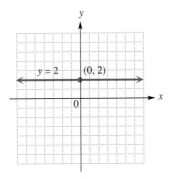

FIGURE 6

4. Find the intercepts, and graph $2x - y = 4$.

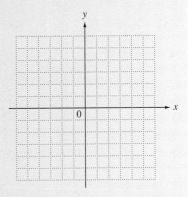

ANSWERS

4. *x*-intercept is $(2, 0)$; *y*-intercept is $(0, -4)$.

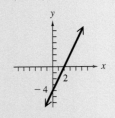

5. Find the intercepts and graph each line.

(a) $y + 4 = 0$

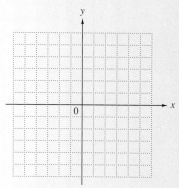

(b) $x = 2$

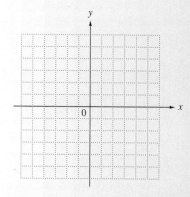

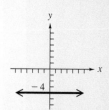
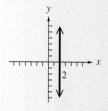
It is also possible for a graph to have no y-intercept, as in the next example.

E X A M P L E 4   Graphing a Vertical Line

Graph $x + 1 = 0$.

The standard form $1x + 0y = -1$ shows that *every* value of y leads to $x = -1$, so no value of y makes x equal to 0. The only way a straight line can have no y-intercept is to be vertical, as in Figure 7. Notice that every point on the line has $x = -1$, while all real numbers are used for the y-values.

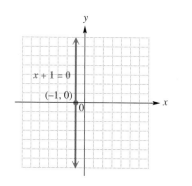

FIGURE 7

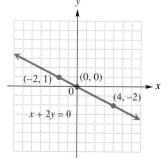

FIGURE 8

◀◀ WORK PROBLEM 5 AT THE SIDE.

The line graphed in the next example has both x-intercept and y-intercept at the origin.

E X A M P L E 5   Graphing a Line That Passes through the Origin

Graph $x + 2y = 0$.

Find the x-intercept by letting $y = 0$.

$$x + 2y = 0$$
$$x + 2(0) = 0 \quad \text{Let } y = 0.$$
$$x + 0 = 0$$
$$x = 0 \quad x\text{-intercept is } (0, 0)$$

To find the y-intercept, let $x = 0$.

$$x + 2y = 0$$
$$0 + 2y = 0 \quad \text{Let } x = 0.$$
$$y = 0 \quad y\text{-intercept is } (0, 0)$$

Both intercepts are the same ordered pair, $(0, 0)$. Another point is needed to graph the line. Choose any number for x, say $x = 4$, and solve for y. (You could also choose any number for y to solve for x.)

$$x + 2y = 0$$
$$4 + 2y = 0 \quad \text{Let } x = 4.$$
$$2y = -4$$
$$y = -2$$

This gives the ordered pair $(4, -2)$. These two points lead to the graph shown in Figure 8. As a check, verify that $(-2, 1)$ also lies on the line.

3.1 *Exercises*

Use the graph to answer the questions.

1. The graph indicates the percentage of all U.S. workers without any private or public health insurance. (*Source:* Census Bureau, Employee Benefit Research Institute)
   (a) Between which two years was the percentage approximately the same?
   (b) Between which two years was the increase the greatest?
   (c) In what year was the percent about 16.5%?

2. The graph indicates the production of handguns, rifles, and shotguns in the U.S. (*Source:* Bureau of Alcohol, Tobacco, and Firearms)
   (a) In what year between 1983 and 1992 was the production the greatest?
   (b) In what year was the production the least?
   (c) In what two successive years was the production less than the previous year's?

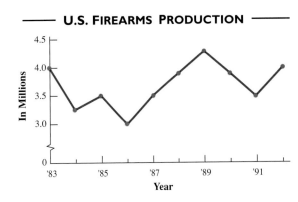

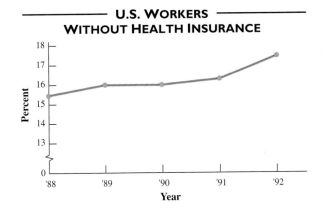

3. What is another name for the rectangular coordinate system? After whom is it named?

4. Observe the graphs in Exercises 1 and 2. If you were to use one of the other types of graphs mentioned in the opening paragraph of this section to depict the information given there, which one would you choose?

Fill in the blank with the correct response.

5. The point with coordinates $(0, 0)$ is called the _____ of a rectangular coordinate system.

6. For any value of x, the point $(x, 0)$ lies on the _____-axis.

7. To find the x-intercept of a line, we let _____ equal 0 and solve for _____ .

8. The equation _____ = 4 has a horizontal
   $(x \text{ or } y)$
   line as its graph.

9. To graph a straight line, we must find a minimum of _____ points.

10. The point (_____, 4) is on the graph of $2x - 3y = 0$.

Name the quadrant, if any, in which each point is located.

11. _____ (a) $(1, 6)$
    _____ (b) $(-4, -2)$
    _____ (c) $(-3, 6)$
    _____ (d) $(7, -5)$
    _____ (e) $(-3, 0)$

12. _____ (a) $(-2, -10)$
    _____ (b) $(4, 8)$
    _____ (c) $(-9, 12)$
    _____ (d) $(3, -9)$
    _____ (e) $(0, -8)$

13. Use the given information to determine the possible quadrants in which the point (x, y) must lie.

    (a) $xy > 0$ **(b)** $xy < 0$ **(c)** $\dfrac{x}{y} < 0$ **(d)** $\dfrac{x}{y} > 0$

14. What must be true about the coordinates of any point that lies on an axis?

Locate the following points on the rectangular coordinate system.

15. $(2, 3)$ **16.** $(-1, 2)$ **17.** $(-3, -2)$ **18.** $(1, -4)$

19. $(0, 5)$ **20.** $(-2, -4)$ **21.** $(-2, 4)$ **22.** $(3, 0)$

23. $(-2, 0)$ **24.** $(3, -3)$

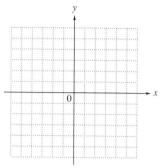

In each exercise, complete the given ordered pairs for the equation, and then graph the equation. See Example 1.

25. $x - y = 3$
$(0, \underline{\ \ }), (\underline{\ \ }, 0)$
$(5, \underline{\ \ }), (2, \underline{\ \ })$

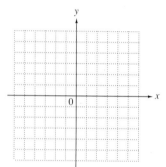

26. $x - y = 5$
$(0, \underline{\ \ }), (\underline{\ \ }, 0)$
$(1, \underline{\ \ }), (3, \underline{\ \ })$

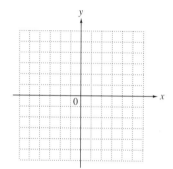

27. $x + 2y = 5$
$(0, \underline{\ \ }), (\underline{\ \ }, 0)$
$(2, \underline{\ \ }), (\underline{\ \ }, 2)$

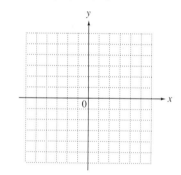

28. $x + 3y = -5$
$(0, \underline{\ \ }), (\underline{\ \ }, 0)$
$(1, \underline{\ \ }), (\underline{\ \ }, -1)$

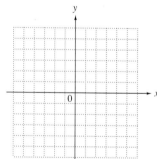

29. $4x - 5y = 20$
$(0, \underline{\ \ }), (\underline{\ \ }, 0)$
$(2, \underline{\ \ }), (\underline{\ \ }, -3)$

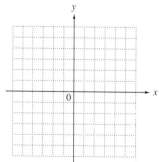

30. $6x - 5y = 30$
$(0, \underline{\ \ }), (\underline{\ \ }, 0)$
$(3, \underline{\ \ }), (\underline{\ \ }, -2)$

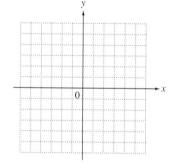

31. What is the equation of a line that coincides with the y-axis?

32. What is the equation of a line that coincides with the x-axis?

For each equation, find the x-intercept and the y-intercept, and graph. See Examples 2–5.

33. $2x + 3y = 12$

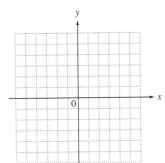

34. $5x + 2y = 10$

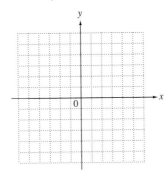

35. $x - 3y = 6$

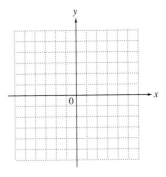

36. $x - 2y = -4$

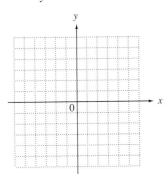

37. $3x - 7y = 9$

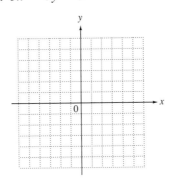

38. $5x + 6y = -10$

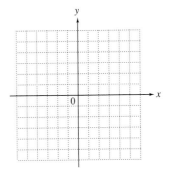

39. $y = 5$

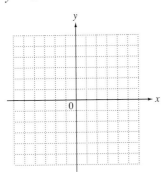

40. $y = -3$

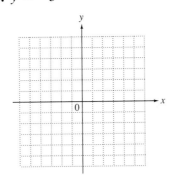

41. $x = 2$

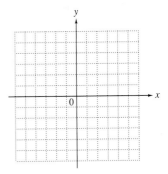

42. $x = -3$

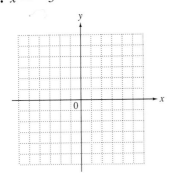

43. $x + 5y = 0$

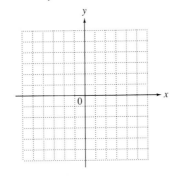

44. $x - 3y = 0$

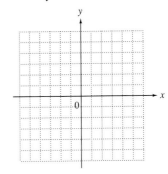

A linear equation can be used as a model to describe data in some cases. Exercises 45 and 46 are based on this idea.

45. Between the years 1980 and 1992, the linear model $y = 2.503x + 198.729$ approximated the winning speed for the Indianapolis 500 race. In the model, $x = 0$ corresponds to 1980, $x = 1$ to 1981, and so on, and y is in miles per hour. Use this model to approximate the speed of the 1988 winner, Rick Mears.

46. According to information provided by Families USA Foundation, the national average family health care cost in dollars between 1980 and 2000 (projected) can be approximated by the linear model $y = 382.75x + 1742$, where $x = 0$ corresponds to 1980 and $x = 20$ corresponds to 2000. Based on this model, what would be the expected national average health care cost in 1998?

INTERPRETING TECHNOLOGY (EXERCISES 47–50)

47. The screen shows the graph of one of the equations below, along with the coordinates of a point on the graph. Which one of the equations is it?

(a) $x + 2y = 4$ (b) $-3x + 5y = 15$
(c) $y = 4x - 2$ (d) $y = -2$

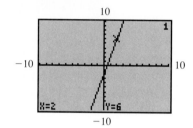

48. The screen shows the graph of one of the equations below. Two views of the graph are given, along with the intercepts. Which one of the equations is it?

(a) $3x + 2y = 6$ (b) $-3x + 2y = 6$ (c) $-3x - 2y = 6$ (d) $3x - 2y = 6$

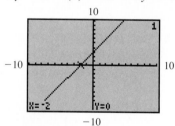 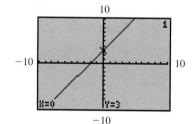

49. The table of points shown was generated by a graphing calculator with a *table* feature. Which one of the equations below corresponds to this table of points?

(a) $y = 2x - 3$ (b) $y = -2x - 3$
(c) $y = 2x + 3$ (d) $y = -2x + 3$

X	Y1
0	3
.5	2
1	1
1.5	0
2	-1
2.5	-2
3	-3

X=0

50. Refer to the model equation in Exercise 46. A portion of its graph is shown on the accompanying screen, along with the coordinates of a point on the line displayed at the bottom. How is this point interpreted in the context of the model?

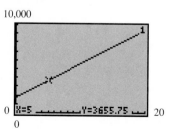

3.2 The Slope of a Line

OBJECTIVES

1. Find the slope of a line, given two points on the line.
2. Find the slope of a line, given an equation of the line.
3. Graph a line, using its slope and a point on the line.
4. Use slopes to determine whether two lines are parallel, perpendicular, or neither.
5. Solve problems involving average rate of change.

FOR EXTRA HELP

Tutorial   Tape 4   SSM, Sec. 3.2

Slope is used in many ways in our everyday world. The slope of a hill (sometimes called the *grade*) is often given as a percent. For example, a 10% (or $\frac{10}{100} = \frac{1}{10}$) slope means the hill rises 1 unit for every 10 horizontal units. Stairs and roofs have slopes, too, as shown in Figure 9.

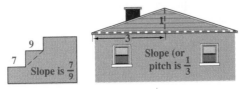

FIGURE 9

OBJECTIVE 1 To formulate a definition of the slope of a line, suppose (x_1, y_1) and (x_2, y_2) are two different points on a line as shown in Figure 10. (The notation x_1 (read "x-sub-one"), x_2, y_1, and y_2 represents specific x-values or y-values.) Then, as we move along the line from (x_1, y_1) to (x_2, y_2), the y-value changes from y_1 to y_2, an amount equal to $y_2 - y_1$. This vertical change is called the *rise*. As y changes from y_1 to y_2, the value of x changes from x_1 to x_2 by the amount $x_2 - x_1$. This horizontal change is called the *run*.

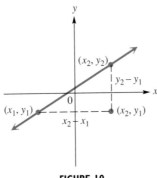

FIGURE 10

The ratio of the change in y to the change in x is called the **slope** of the line, with the letter m used for slope.

Slope

If $x_1 \neq x_2$, the slope of the line through the distinct points (x_1, y_1) and (x_2, y_2) is

$$m = \frac{\text{rise}}{\text{run}} = \frac{\text{change in } y}{\text{change in } x} = \frac{y_2 - y_1}{x_2 - x_1}.$$

The larger the absolute value of m, the steeper the corresponding line will be, because a larger value of m indicates a greater change in y as compared to the change in x.

EXAMPLE 1 Using the Definition of Slope

Find the slope of the line through the points $(2, -1)$ and $(-5, 3)$. Let $(2, -1) = (x_1, y_1)$ and $(-5, 3) = (x_2, y_2)$; then

$$m = \frac{y_2 - y_1}{x_2 - x_1} = \frac{3 - (-1)}{-5 - 2} = \frac{4}{-7} = -\frac{4}{7}.$$

On the other hand, if the pairs are reversed, so that $(2, -1) = (x_2, y_2)$ and $(-5, 3) = (x_1, y_1)$, the slope is

$$m = \frac{-1 - 3}{2 - (-5)} = \frac{-4}{7} = -\frac{4}{7},$$

the same answer.

1. Find the slope of the line through each pair of points.

(a) $(-2, 7), (4, -3)$

(b) $(1, 2), (8, 5)$

(c) $(8, -2), (3, -2)$

Example 1 suggests that the slope is the same no matter which point we consider first. Also, using similar triangles from geometry, we can show that the slope is the same no matter which two different points on the line we choose.

> **Caution**
>
> In calculating the slope, be careful to subtract the y-values and the x-values in the *same* order.
>
> | Correct | Incorrect |
> |---------|-----------|
> | $\dfrac{y_2 - y_1}{x_2 - x_1}$ or $\dfrac{y_1 - y_2}{x_1 - x_2}$ | $\dfrac{y_2 - y_1}{x_1 - x_2}$ or $\dfrac{y_1 - y_2}{x_2 - x_1}$ |

◀◀ **WORK PROBLEM 1 AT THE SIDE.**

OBJECTIVE ▶ When an equation of a line is given, we can find the slope using the definition of slope by first finding two different points on the line.

E X A M P L E 2   Finding the Slope of a Line

Find the slope of the line $4x - y = 8$.

The intercepts can be used as the two different points needed to find the slope. Replace y with 0 to find that the x-intercept is $(2, 0)$; replace x with 0 to find that the y-intercept is $(0, -8)$. The slope is

$$m = \frac{-8 - 0}{0 - 2} = \frac{-8}{-2} = 4.$$

E X A M P L E 3   Finding the Slope of a Line

Find the slope of the following lines.

(a) $x = -3$

By inspection, $(-3, 5)$ and $(-3, -4)$ are two points that satisfy the equation $x = -3$. Use these two points to find the slope.

$$m = \frac{-4 - 5}{-3 - (-3)} = \frac{-9}{0}$$

Division by zero is undefined; therefore the slope is undefined. As shown in the previous section, the graph of an equation such as $x = -3$ is a vertical line (parallel to the y-axis). See Figure 11(a).

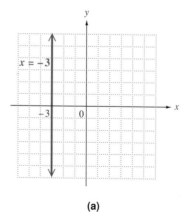

(a)   (b)

FIGURE 11

ANSWERS

1. (a) $-\dfrac{5}{3}$ **(b)** $\dfrac{3}{7}$ **(c)** 0

(b) $y = 5$

Find the slope by selecting two different points on the line, such as $(3, 5)$ and $(-1, 5)$, and using the definition of slope.

$$m = \frac{5 - 5}{3 - (-1)} = \frac{0}{4} = 0$$

The graph of $y = 5$ is the horizontal line in Figure 11(b).

Example 3 suggests the following generalization.

> The slope of a vertical line is undefined; the slope of a horizontal line is 0.

WORK PROBLEM 2 AT THE SIDE. ▶▶

OBJECTIVE 3 ▶ The following example shows how to graph a straight line by using the slope and one point on the line.

E X A M P L E   4   Using the Slope and a Point to Graph a Line

Graph the line that has slope $\frac{2}{3}$ and goes through the point $(-4, 2)$.

First locate the point $(-4, 2)$ on a graph (see Figure 12). Then, from the definition of slope,

$$m = \frac{\text{change in } y}{\text{change in } x} = \frac{2}{3},$$

move 2 units *up* in the y direction and then 3 units to the *right* in the x direction to locate another point on the graph, P. The line through $(-4, 2)$ and P is the required graph. An additional point Q, can be found by moving 2 units *down* and 3 units to the *left* because,

$$\frac{2}{3} = \frac{-2}{-3}.$$

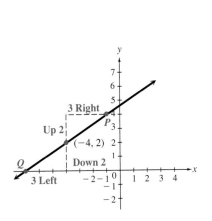

FIGURE 12

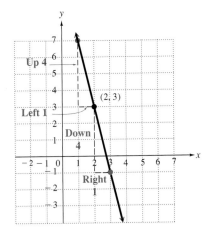

FIGURE 13

2. Find the slope of each line.

(a) $2x + y = 6$

(b) $3x - 4y = 12$

(c) $x = -6$

(d) $y + 5 = 0$

3. Graph the following lines.

(a) Through $(1, -3)$; $m = -\dfrac{3}{4}$

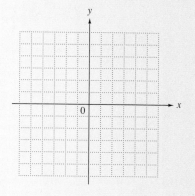

(b) Through $(-1, -4)$; $m = 2$

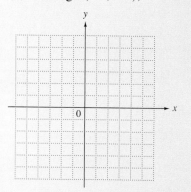

3. (a)

(b)

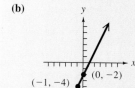

EXAMPLE 5 Using the Slope and a Point to Graph a Line

Graph the line through $(2, 3)$ with slope -4.

Start by locating the point $(2, 3)$ on the graph. Use the definition of slope to find a second point on the line, writing -4 as $\frac{-4}{1}$.

$$\text{Slope} = \frac{\text{change in } y}{\text{change in } x} = \frac{-4}{1}$$

Move *down* 4 units from $(2, 3)$ and then 1 unit to the *right*. Draw a line through this second point and $(2, 3)$, as in Figure 13. We could also write the slope as $\frac{4}{-1}$ and move 4 units *up* and 1 unit *left* to get another point.

■

◀◀ **WORK PROBLEM 3 AT THE SIDE.**

In Problem 3 at the side, the slope for part (b) is *positive*. As shown in the answer, the graph for this line goes up from left to right. The line in part (a) has a *negative* slope, and the graph goes down from left to right. This suggests the following conclusion.

> A positive slope indicates that the line goes up from left to right; a negative slope indicates that the line goes down from left to right.

Figure 14 shows lines of positive, zero, negative, and undefined slopes.

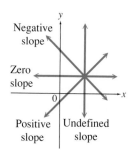

FIGURE 14

OBJECTIVE 4 The slopes of a pair of parallel or perpendicular lines are related in a special way. The slope of a line measures the steepness of a line. Because parallel lines have equal steepness, their slopes must be equal; also, lines with the same slope are parallel.

Slopes of Parallel Lines

Two nonvertical lines with the same slope are parallel; two nonvertical parallel lines have the same slope.

EXAMPLE 6 Determining Parallel Lines

Are the two lines L_1, through $(-2, 1)$ and $(4, 5)$, and L_2, through $(3, 0)$ and $(0, -2)$, parallel?

The slope of L_1 is

$$m_1 = \frac{5 - 1}{4 - (-2)} = \frac{4}{6} = \frac{2}{3}.$$

CONTINUED ON NEXT PAGE

The slope of L_2 is

$$m_2 = \frac{-2 - 0}{0 - 3} = \frac{-2}{-3} = \frac{2}{3}.$$

Because the slopes are equal, the two lines are parallel.

To see how the slopes of perpendicular lines are related, consider a nonvertical line with slope $\frac{a}{b}$. If this line is rotated 90°, the rise and run are exchanged, and the slope is $-\frac{b}{a}$ because the run is now negative. See Figure 15. Thus, the slopes of perpendicular lines have a product of -1 and are negative reciprocals of each other. For example, if the slopes of two lines are $\frac{3}{4}$ and $-\frac{4}{3}$, then the lines are perpendicular because $(\frac{3}{4})(-\frac{4}{3}) = -1$.

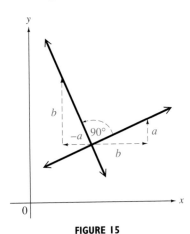

FIGURE 15

Slopes of Perpendicular Lines

If neither line is vertical, the slopes of perpendicular lines are negative reciprocals, so that their product is -1. Also, lines with slopes that are negative reciprocals are perpendicular.

E X A M P L E 7   Determining Perpendicular Lines

Are the lines with equations $2y = 3x - 6$ and $2x + 3y = -6$ perpendicular?

Find the slope of each line by first finding two points on the line. The points $(0, -3)$ and $(2, 0)$ are on the first line. The slope is

$$m_1 = \frac{0 - (-3)}{2 - 0} = \frac{3}{2}.$$

The second line goes through $(-3, 0)$ and $(0, -2)$ and has slope

$$m_2 = \frac{-2 - 0}{0 - (-3)} = -\frac{2}{3}.$$

Because these slopes are negative reciprocals, the product of the slopes is $(\frac{3}{2})(-\frac{2}{3}) = -1$, so the lines are perpendicular.

WORK PROBLEM 4 AT THE SIDE. ▶▶

---

4. Write *parallel*, *perpendicular,* or *neither* for each pair of two distinct lines.

a) The line through $(-1, 2)$ and $(3, 5)$ and the line through $(4, 7)$ and $(8, 10)$

(b) The line through $(5, -9)$ and $(3, 7)$ and the line through $(0, 2)$ and $(8, 3)$

(c) $2x - y = 4$ and $2x + y = 6$

(d) $3x + 5y = 6$ and $5x - 3y = 2$

Note

When deciding whether or not a pair of lines is parallel, in addition to checking for the same slope, be sure they are not the *same line*. The slopes of $4x + 2y = 6$ and $2x + y = 3$ are both -2; however, these equations describe the same line with y-intercept $(0, 3)$.

OBJECTIVE 5 We have seen how the slope of a line is the ratio of the change in y (vertical change) to the change in x (horizontal change). This idea can be extended to real-life situations as follows: the slope gives the average rate of change in y per unit of change in x, where the value of y depends on the value of x. The next example illustrates this idea of average rate of change. We assume a linear relationship between x and y.

EXAMPLE 8   **Interpreting Slope as Average Rate of Change**

The bar graph in Figure 16* shows the number of multimedia personal computers (PCs), in millions, in U.S. homes. Find the average rate of change in the number of multimedia PCs per year.

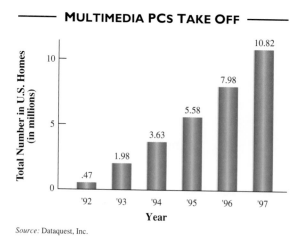

MULTIMEDIA PCS TAKE OFF

FIGURE 16

Since connecting the tops of the bars would closely approach a straight line, we can use the slope formula. We need two pairs of data. If we let 1992 represent 0, then 1993 represents 1, 1994 represents 2, and so on. Then the ordered pair for 1993 is $(1, 1.98)$ and the pair for 1996 is $(4, 7.98)$. The average rate of change in the number of multimedia PCs is found by using the slope formula.

$$\text{Average rate of change} = \frac{y_2 - y_1}{x_2 - x_1} = \frac{7.98 - 1.98}{4 - 1} = 2$$

The result, 2, indicates that the number of multimedia PCs increases by 2 million each year.

---

* Graph for Figure 16, "Multimedia PC's Take Off," from *The Wall Street Journal*, March 21, 1994. (The figures for 1995 to 1997 are estimates.) Reprinted by permission of The Wall Street Journal, Copyright © 1994 Dow Jones & Company, Inc. All Rights Reserved Worldwide.

3.2 *Exercises*

Use the given figure to determine the slope of the line segment described, by counting the number of units of "rise," the number of units of "run," and then finding the quotient.

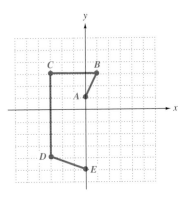

1. *AB*   **2.** *BC*   **3.** *CD*   **4.** *DE*

5. If a walkway rises 2 feet for every 10 feet of horizontal distance, which of the following expresses its slope (or grade)? (There may be more than one correct answer.)

(a) .2   **(b)** $\dfrac{2}{10}$ **(c)** $\dfrac{1}{5}$ **(d)** 20%   **(e)** 5   **(f)** $\dfrac{20}{100}$ **(g)** 500%   **(h)** $\dfrac{10}{2}$

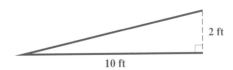

6. If the pitch of a roof is $\frac{1}{4}$, how many feet in the horizontal direction corresponds to a rise of 3 feet?

Find the slope of the line through each pair of points using the slope formula. See Example 1.

7. $(-2, -3)$ and $(-1, 5)$ **8.** $(-4, 3)$ and $(-3, -4)$ **9.** $(-4, 1)$ and $(2, 6)$

10. $(-3, -3)$ and $(5, 6)$ **11.** $(2, 4)$ and $(-4, 4)$ **12.** $(-6, 3)$ and $(2, 3)$

Graph the line with the given equation, and then find its slope based on the graph you have sketched. See Example 3.

13. $x = 5$

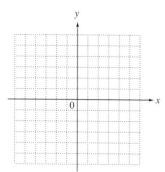

14. $x = -2$

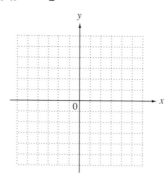

15. $y = 2$

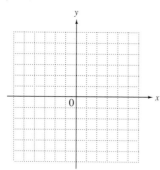

16. $y = 6$

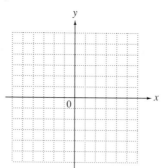

Based on the figure shown here, determine which line satisfies the given description.

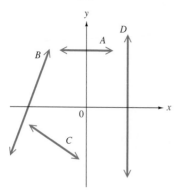

17. The line has positive slope.

18. The line has negative slope.

19. The line has slope 0.

20. The line has undefined slope.

Find the slope of each of the following lines, and sketch the graph. See Examples 1, 2, 4, and 5.

21. $x + 2y = 4$

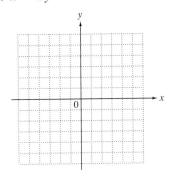

22. $x + 3y = -6$

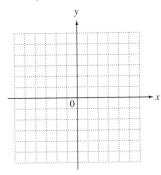

23. $-x + y = 4$

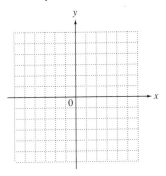

24. $-x + y = 6$

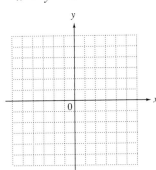

25. $6x + 5y = 30$

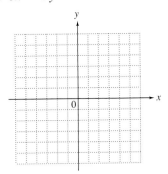

26. $3x + 4y = 12$

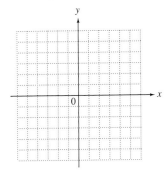

27. $5x - 2y = 10$

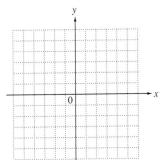

28. $4x - y = 4$

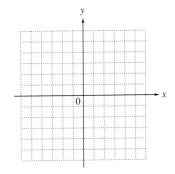

29. $y = 4x$

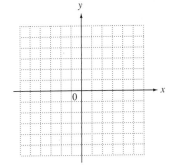

30. $y = -3x$

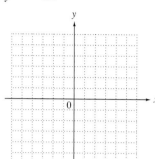

31. $y - 3 = 0$

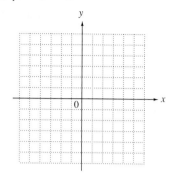

32. $y + 5 = 0$

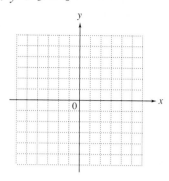

Use the methods shown in Examples 4 and 5 to graph each of the following lines.

33. Through $(-4, 2)$; $m = \dfrac{1}{2}$

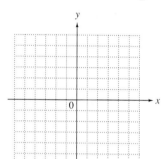

34. Through $(-2, -3)$; $m = \dfrac{5}{4}$

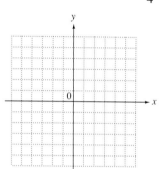

35. Through $(0, -2)$; $m = -\dfrac{2}{3}$

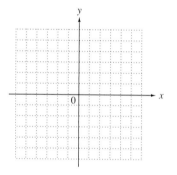

36. Through $(0, -4)$; $m = -\dfrac{3}{2}$

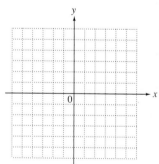

37. Through $(-1, -2)$; $m = 3$

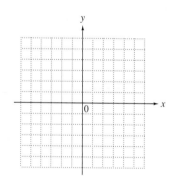

38. Through $(-2, -4)$; $m = 4$

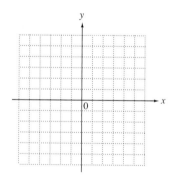

Decide whether the pair of lines is parallel, perpendicular, *or* neither parallel nor perpendicular. *See Examples 6 and 7.*

39. $2x + 5y = -7$ and $5x - 2y = 1$

40. $x + 4y = 7$ and $4x - y = 3$

41. The line through $(4, 6)$ and $(-8, 7)$ and the line through $(-5, 5)$ and $(7, 4)$

42. The line through $(15, 9)$ and $(12, -7)$ and the line through $(8, -4)$ and $(5, -20)$

43. $2x + y = 6$ and $x - y = 4$

44. $4x - 3y = 6$ and $3x - 4y = 2$

Use the concept of slope to solve the problem.

45. The upper deck at the new Comiskey Park in Chicago has produced, among other complaints, displeasure with its steepness. It's been compared to a ski jump. It is 160 feet from home plate to the front of the upper deck and 250 feet from home plate to the back. The top of the upper deck is 63 feet above the bottom. What is its slope?

46. When designing the new arena in Boston to replace the old Boston Garden, architects were careful to design the ramps leading up to the entrances so that circus elephants would be able to march up the ramps. The maximum grade (or slope) that an elephant will walk on is 13%. Suppose that such a ramp was constructed with a horizontal run of 150 feet. What would be the maximum vertical rise the architects could use?

Use the idea of average rate of change to solve the problem. See Example 8.

47. The graph shows how average monthly rates for cable television increased from 1980 to 1992. The graph can be approximated by a straight line.
 (a) Use the information provided for 1980 and 1992 to determine the average rate of change in price per year.
 (b) In your own words, explain how a *positive* rate of change affects the consumer in a situation such as the one illustrated by this graph.

48. Assuming a linear relationship, what is the average rate of change for cable industry revenues over the period from 1990 to 1992?

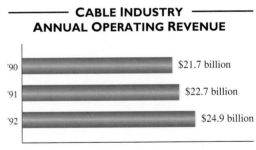

───── **CABLE INDUSTRY** ─────
ANNUAL OPERATING REVENUE

'90 — $21.7 billion
'91 — $22.7 billion
'92 — $24.9 billion

Sources: Census Bureau; Paul Kagan Associates

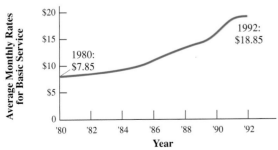

───── **CABLE TV RATES** ─────

1980: $7.85
1992: $18.85

Sources: Census Bureau, Paul Kagan Associates

49. The 1993 Annual Report of AT&T cited the following figures concerning the global growth of international telephone calls: From 47.5 billion minutes in 1993, traffic was expected to rise to 60 billion minutes in 1995. Assuming a linear relationship, what was the average rate of change for this time period? (*Source:* TeleGeography, 1993, Washington, D.C.)

50. The market for international phone calls during the ten-year period from 1986 to 1995 is depicted in the accompanying bar graph. The tops of the bars approximate a straight line. Assuming that the traffic volume at the beginning of this period was 18 billion minutes and at the end was 60 billion minutes, what was the average rate of change for the ten-year period?

— **INTERNATIONAL CALLS: 1986–1995** —

MATHEMATICAL CONNECTIONS (EXERCISES 51–56)

---

*In these exercises we examine a method of determining whether three points lie on the same straight line. (Such points are said to be **collinear**.) The points we consider are A(3, 1), B(6, 2), and C(9, 3). Work these exercises in order.*

51. Find the slope of segment *AB*.

52. Find the slope of segment *BC*.

53. Find the slope of segment *AC*.

54. If slope of *AB* = slope of *BC* = slope of *AC*, then *A*, *B*, and *C* are collinear. Use the results of Exercises 51–53 to show that this statement is satisfied.

55. Use the slope formula to determine whether the points $(1, -2)$, $(3, -1)$, and $(5, 0)$ are collinear.

56. Repeat Exercise 55 for the points $(0, 6)$, $(4, -5)$, and $(-2, 12)$.

3.3 Linear Equations in Two Variables

OBJECTIVES

1. Write the equation of a line, given its slope and a point on the line.

2. Write the equation of a line, given two points on the line.

3. Write the equation of a line, given its slope and *y*-intercept.

4. Find the slope and *y*-intercept of a line, given its equation.

5. Write the equation of a line parallel or perpendicular to a given line through a given point.

6. Apply concepts of linear equations to realistic examples.

Many real-world situations can be described by straight-line graphs. In this section we see how to write a linear equation in such situations.

OBJECTIVE 1 We saw earlier that a straight line is a set of points in the plane such that the slope between any two points is the same. In Figure 17, point P is on the line through P_1 and P_2 if the slope of the line through points P_1 and P equals the slope of the line through points P and P_2. If these slopes are equal to m, then

$$\frac{y - y_1}{x - x_1} = \frac{y - y_2}{x - x_2} = m.$$

$$\frac{y - y_1}{x - x_1} = m$$

$$y - y_1 = m(x - x_1) \qquad \text{Multiply both sides by } x - x_1.$$

This last equation gives the *point-slope form* of the equation of the line, which shows the coordinates of a point (x_1, y_1) on the line and the slope of the line.

FOR EXTRA HELP

Tutorial   Tape 4   SSM, Sec. 3.3

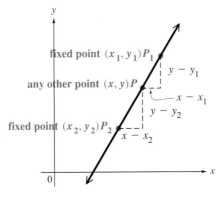

FIGURE 17

Point-slope Form

The **point-slope form** of the equation of a line is

$$y - y_1 = m(x - x_1).$$

Slope ↓   ↑ Given point ↑

Given a point on a line and the slope of the line, we can use the point-slope form to write the equation of the line.

EXAMPLE 1 Using the Point-slope Form to Find the Equation of a Line

Find an equation of the line with slope $\frac{1}{3}$ that goes through the point $(-2, 5)$.

Use the point-slope form of the equation of a line, with $(x_1, y_1) = (-2, 5)$ and $m = \frac{1}{3}$.

— CONTINUED ON NEXT PAGE

1. Write equations of the following lines in standard form.

(a) Through $(-2, 7)$; $m = 3$

$$y - y_1 = m(x - x_1) \qquad \text{Point-slope form}$$

$$y - 5 = \frac{1}{3}[x - (-2)] \qquad y_1 = 5, \ m = \tfrac{1}{3}, \ x_1 = -2$$

$$y - 5 = \frac{1}{3}(x + 2)$$

$$3(y - 5) = 3 \cdot \frac{1}{3}(x + 2) \qquad \text{Multiply by 3.}$$

$$3y - 15 = x + 2$$

$$-x + 3y = 17 \qquad \text{Subtract } x; \text{ add 15.}$$

$$x - 3y = -17 \qquad \text{Standard form}$$

(b) Through $(1, 3)$; $m = -\dfrac{5}{4}$

◀◀ **WORK PROBLEM 1 AT THE SIDE.**

Notice that the point-slope form does not apply to a vertical line because the slope of a vertical line is undefined. A vertical line through the point (k, y), where k is a constant, has equation $x = k$.

A horizontal line has slope 0. From the point-slope form, the equation of a horizontal line through the point (x, k), where k is a constant, is

$$y - y_1 = m(x - x_1)$$

$$y - k = 0(x - x) \qquad \text{Let } m = 0, \ x_1 = x, \text{ and } y_1 = k.$$

$$y - k = 0$$

$$y = k.$$

In summary, horizontal and vertical lines have special equations as follows.

2. Write equations of the following lines.

(a) Through $(8, -2)$; $m = 0$

Horizontal and Vertical Lines

If k is a constant, the vertical line through (k, y) has equation $x = k$, and the horizontal line through (x, k) has equation $y = k$.

◀◀ **WORK PROBLEM 2 AT THE SIDE.**

(b) The vertical line through $(3, 5)$

OBJECTIVE 2 ▶ When two points on a line are known, we can use the definition of slope to find the slope of the line. Then the point-slope form can be used to write an equation for the line.

E X A M P L E 2   Finding the Equation of a Line Given Two Points

Find an equation of the line through the points $(-4, 3)$ and $(5, -7)$.

First find the slope by using the definition.

$$m = \frac{-7 - 3}{5 - (-4)} = -\frac{10}{9}$$

Either $(-4, 3)$ or $(5, -7)$ can be used as (x_1, y_1) in the point-slope form of the equation of a line. If we choose $(-4, 3)$, then $-4 = x_1$ and $3 = y_1$.

CONTINUED ON NEXT PAGE

ANSWERS

1. (a) $3x - y = -13$ **(b)** $5x + 4y = 17$

2. (a) $y = -2$ **(b)** $x = 3$

$$y - y_1 = m(x - x_1) \qquad \text{Point-slope form}$$

$$y - 3 = -\frac{10}{9}[x - (-4)] \qquad y_1 = 3,\ m = -\tfrac{10}{9},\ x_1 = -4$$

$$9(y - 3) = 9\left(-\frac{10}{9}\right)(x + 4) \qquad \text{Multiply by 9.}$$

$$9y - 27 = -10x - 40 \qquad \text{Distributive property}$$

$$10x + 9y = -13 \qquad \text{Standard form}$$

On the other hand, if $(5, -7)$ were used, the equation would be

$$y - (-7) = -\frac{10}{9}(x - 5) \qquad y_1 = -7,\ m = -\tfrac{10}{9},\ x_1 = 5$$

$$9y + 63 = -10x + 50 \qquad \text{Multiply by 9.}$$

$$10x + 9y = -13, \qquad \text{Standard form}$$

the same equation. Either way, the line through $(-4, 3)$ and $(5, -7)$ has equation $10x + 9y = -13$.

WORK PROBLEM 3 AT THE SIDE. ▶▶

OBJECTIVE 3 Suppose a line has slope m, and we know that the y-intercept is $(0, b)$. Using the slope-intercept form gives

$$y - y_1 = m(x - x_1)$$

$$y - b = m(x - 0)$$

$$y = mx + b.$$

When we solve the equation for y, the coefficient of x is the slope, m, and the constant is the y-value of the y-intercept, b. Because this form of the equation shows the slope and the y-intercept, it is called the *slope-intercept form.*

Slope-intercept Form

The equation of a line with slope m and y-intercept $(0, b)$ is written in **slope-intercept form** as

$$y = mx + b.$$
$$\quad\ \uparrow \qquad \uparrow$$
Slope y-intercept is $(0, b)$.

EXAMPLE 3 Using the Slope-intercept Form to Find the Equation of a Line

Find an equation of the line with slope $-\frac{4}{5}$ and y-intercept $(0, -2)$.

Here $m = -\frac{4}{5}$ and $b = -2$. Substitute these values into the slope-intercept form.

$$y = mx + b \qquad \text{Slope-intercept form}$$

$$y = -\frac{4}{5}x - 2 \qquad m = -\tfrac{4}{5},\ b = -2$$

$$5y = -4x - 10 \qquad \text{Multiply by 5.}$$

$$4x + 5y = -10 \qquad \text{Standard form}$$

3. Write equations in standard form of the following lines.

(a) Through $(-1, 2)$ and $(5, 7)$

(b) Through $(-2, 6)$ and $(1, 4)$

ANSWERS

3. (a) $5x - 6y = -17$ **(b)** $2x + 3y = 14$

4. Write an equation in standard form for each line with the given slope and *y*-intercept.

(a) Slope 2; *y*-intercept $(0, -3)$

(b) Slope $-\frac{2}{3}$; *y*-intercept $(0, 0)$

(c) Slope 0; *y*-intercept $(0, 3)$

| Note |

The importance of the slope-intercept form of a linear equation cannot be overemphasized. First, every linear equation (of a nonvertical line) has a *unique* (one and only one) slope-intercept form. Second, at the end of this chapter, we will study linear *functions*. The slope-intercept form is necessary in specifying such functions.

◄◄ WORK PROBLEM 4 AT THE SIDE.

It is convenient to agree on the form for writing a linear equation. In Section 3.1, we defined *standard* form for a linear equation as

$$Ax + By = C.$$

In addition, from now on, let us agree that *A*, *B*, and *C* will be integers, with $A \geq 0$. For example, the final equation found in Example 3, $4x + 5y = -10$, is written in standard form.

| Caution |

The definition of "standard form" varies from one text to another. Any linear equation can be written in many different (all equally correct) forms. For example, the equation $2x + 3y = 8$ can be written as $2x = 8 - 3y, 3y = 8 - 2x, x + \frac{3}{2}y = 4, 4x + 6y = 16$, and so on. In addition to writing it in the form $Ax + By = C$ (with $A \geq 0$), let us agree that the form $2x + 3y = 8$ is preferred over any multiples of both sides, such as $4x + 6y = 16$.

OBJECTIVE 4 We can also use the slope-intercept form to determine the slope and *y*-intercept of a line from its equation by writing the equation in slope-intercept form.

5. Find the slope and the *y*-intercept of each line.

(a) $x + y = 2$

EXAMPLE 4 Writing a Linear Equation in Slope-intercept Form

Write $3y + 2x = 9$ in slope-intercept form; then find the slope and *y*-intercept.

We solve for *y* to put the equation in slope-intercept form.

$$3y + 2x = 9$$
$$3y = -2x + 9 \qquad \text{Subtract } 2x.$$
$$y = -\frac{2}{3}x + 3 \qquad \text{Divide by 3.}$$

Slope ——↑ └— *y*-intercept is $(0, 3)$.

The slope-intercept form gives $-\frac{2}{3}$ for the slope, with *y*-intercept $(0, 3)$.

(b) $2x - 5y = 1$

◄◄ WORK PROBLEM 5 AT THE SIDE.

OBJECTIVE 5 The previous section showed that parallel lines have the same slope, and perpendicular lines have slopes that are negative reciprocals. These results are used in the next two examples.

ANSWERS

4. (a) $2x - y = 3$ **(b)** $2x + 3y = 0$
  (c) $y = 3$

5. (a) $-1, (0, 2)$ **(b)** $\frac{2}{5}, \left(0, -\frac{1}{5}\right)$

─ E X A M P L E 5   **Finding the Equation of a Line Parallel to a Given Line**

Find an equation of the line going through the point $(-2, -3)$ and parallel to the line $2x + 3y = 6$.

The slope of a line parallel to the line $2x + 3y = 6$ can be found by solving for y.

$$2x + 3y = 6$$

$$3y = -2x + 6 \qquad \text{Subtract } 2x \text{ on both sides.}$$

$$y = -\frac{2}{3}x + 2 \qquad \text{Divide both sides by 3.}$$

The slope is given by the coefficient of x, so $m = -\frac{2}{3}$. This means that the line through $(-2, -3)$ and parallel to $2x + 3y = 6$ has slope $-\frac{2}{3}$. Use the point-slope form, with $(x_1, y_1) = (-2, -3)$ and $m = -\frac{2}{3}$ to write the equation.

$$y - y_1 = m(x - x_1) \qquad \text{Point-slope form}$$

$$y - (-3) = -\frac{2}{3}[x - (-2)] \qquad y_1 = -3, m = -\frac{2}{3}, x_1 = -2$$

$$y + 3 = -\frac{2}{3}(x + 2)$$

$$3(y + 3) = -2(x + 2) \qquad \text{Multiply by 3.}$$

$$3y + 9 = -2x - 4 \qquad \text{Distributive property}$$

$$2x + 3y = -13 \qquad \text{Standard form}$$

6. Write an equation in standard form for the line through $(5, 7)$ parallel to $2x - 5y = 15$.

> **WORK PROBLEM 6 AT THE SIDE.** ▶▶

─ E X A M P L E 6   **Finding the Equation of a Line Perpendicular to a Given Line**

Find an equation of the line perpendicular to $2x + 5y = 8$ and going through $(2, 3)$.

First, we find the slope of the line $2x + 5y = 8$ by solving for y.

$$5y = -2x + 8$$

$$y = -\frac{2}{5}x + \frac{8}{5}$$

The slope of this line is $-\frac{2}{5}$. Because the negative reciprocal of $-\frac{2}{5}$ is $\frac{5}{2}$, a line perpendicular to $2x + 5y = 8$ has slope $\frac{5}{2}$. We want an equation of a line with slope $\frac{5}{2}$ that goes through $(2, 3)$. Use the point-slope form.

$$y - 3 = \frac{5}{2}(x - 2) \qquad y_1 = 3, m = \frac{5}{2}, x_1 = 2$$

$$2(y - 3) = 5(x - 2) \qquad \text{Multiply by 2.}$$

$$2y - 6 = 5x - 10 \qquad \text{Distributive property}$$

$$-5x + 2y = -4 \qquad \text{Subtract } 5x; \text{ add 6.}$$

$$5x - 2y = 4 \qquad \text{Standard form}$$

7. Write equations in standard form of lines satisfying the given conditions.

(a) Through $(1, 6)$; perpendicular to $x + y = 9$

(b) Through $(-8, 3)$; perpendicular to $2x - 3y = 10$

> **WORK PROBLEM 7 AT THE SIDE.** ▶▶

ANSWERS
6. $2x - 5y = -25$
7. (a) $x - y = -5$ **(b)** $3x + 2y = -18$

A summary of the forms of linear equations follows.

$Ax + By = C$	**Standard form**
(A, B, and C integers, neither A nor B equal to 0)	Slope is $-\dfrac{A}{B}$.
	x-intercept is $\left(\dfrac{C}{A}, 0\right)$.
	y-intercept is $\left(0, \dfrac{C}{B}\right)$.
$x = k$	**Vertical line** Slope is undefined. x-intercept is $(k, 0)$.
$y = k$	**Horizontal line** Slope is 0. y-intercept is $(0, k)$.
$y = mx + b$	**Slope-intercept form** Slope is m. y-intercept is $(0, b)$.
$y - y_1 = m(x - x_1)$	**Point-slope form** Slope is m. Line passes through (x_1, y_1).

OBJECTIVE 6 Suppose that it is time to fill up your car with gasoline. You drive into your local station and notice that the 89-octane gas that you will use is selling for $1.20 per gallon. Experience has taught you that the final price you will pay can be determined by the number of gallons you buy multiplied by the price per gallon (in this case, $1.20). As you pump the gas you observe two sets of numbers spinning by: one is the number of gallons you have pumped, and the other is the price you will pay for that number of gallons.

The table below shows how ordered pairs can be used to illustrate this situation.

Number of Gallons Pumped	*Price You Will Pay for this Number of Gallons*
0	$0.00 = 0($0.00)
1	$1.20 = 1($1.20)
2	$2.40 = 2($1.20)
3	$3.60 = 3($1.20)
4	$4.80 = 4($1.20)

If we let x denote the number of gallons pumped, then the price y that we will pay can be found by the linear equation $y = 1.20x$, where y is in dollars. This is a simple realistic application of linear equations. Theoretically, there are infinitely many ordered pairs (x, y) that satisfy this equation, but in this application we are limited to nonnegative values for x, since we cannot have a negative number of gallons. The ordered pairs corresponding to the table above are $(0, 0.00)$, $(1, 1.20)$, $(2, 2.40)$, $(3, 3.60)$, and $(4, 4.80)$. There is also a practical maximum value for x in this situation, that will vary from one car to another. What do you think determines this maximum value?

E X A M P L E 7 **Interpreting Ordered Pairs in an Application**

Name other ordered pairs that satisfy the equation $y = 1.20x$, choosing values for x that are not whole numbers. Interpret the ordered pair in the context of the gas-buying situation described earlier.

Let us arbitrarily choose x values of 1.5, 2.4, and 8.25. Substituting these values into $y = 1.20x$ gives the following:

When $x = 1.5$, $y = 1.20(1.5) = 1.80$.   Ordered pair: (1.5, 1.80)

When $x = 2.4$, $y = 1.20(2.4) = 2.88$.   Ordered pair: (2.4, 2.88)

When $x = 8.25$, $y = 1.20(8.25) = 9.90$.   Ordered pair: (8.25, 9.90)

These ordered pairs are interpreted as follows: When 1.5 gallons have been pumped, the meter shows that we owe $1.80. When 2.4 gallons have been pumped, we owe $2.88. And when 8.25 gallons have been pumped, the price is $9.90.

WORK PROBLEM 8 AT THE SIDE. ▶▶

Figure 18 shows the graphs of the eight ordered pairs found in Example 7 and the discussion preceding it. Notice that the points lie on a straight line. If we draw the line we obtain the graph of $y = 1.20x$. See Figure 19. Notice that we have used only nonnegative values for x in the graphs.

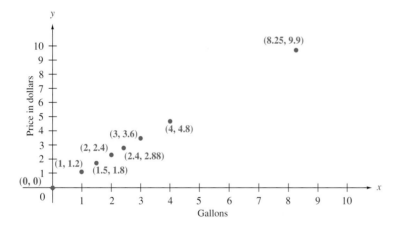

FIGURE 18

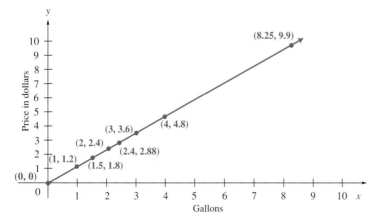

FIGURE 19

8. Repeat Example 7 for the x-value 5.5.

9. Suppose that a rental firm charges a flat rate of $15 plus $10 per day to rent a lawnmower. What equation defines the price y in dollars you will pay if you rent the lawnmower for x days?

In Example 7, the ordered pair $(0, 0)$ lies on the graph, meaning that the graph passes through the origin. If a realistic situation involves an initial charge plus a charge per unit of item purchased, the graph will not go through the origin. The next example illustrates this.

EXAMPLE 8 Determining an Equation and Graphing Ordered Pairs Satisfying It

Suppose that you can get a car wash at the gas station in Example 7 if you pay an additional $3.00.

(a) What is the equation that defines the price you will pay?

Since an additional $3.00 will be charged, you will pay $1.20x + 3.00$ dollars for x gallons of gasoline. Thus, if y represents the price, the equation is $y = 1.2x + 3$. (We deleted the unnecessary zeros here.)

(b) Graph the equation for x values between 0 and 10.

To graph this equation, we need only find two points, and possibly a third point as a check. Let's choose $x = 0$, $x = 2$, and $x = 10$:

When $x = 0$, $y = 1.2(0) + 3 = 3$. Ordered pair: $(0, 3)$

When $x = 2$, $y = 1.2(2) + 3 = 5.4$. Ordered pair: $(2, 5.4)$

When $x = 10$, $y = 1.2(10) + 3 = 15$. Ordered pair: $(10, 15)$

The three ordered pairs are plotted and joined in Figure 20.

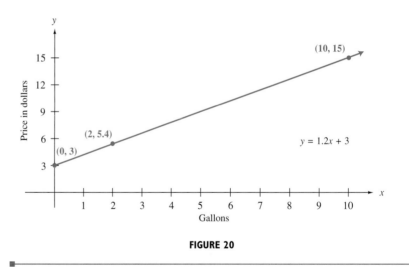

FIGURE 20

◀◀ WORK PROBLEM 9 AT THE SIDE.

3.3 Exercises

Match each equation with the graph that it most closely resembles. (Hint: Determining the signs of m and b will help you make your decision.)

1. $y = 2x + 3$

A.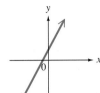

B.

2. $y = -2x + 3$

3. $y = -2x - 3$

C.

D.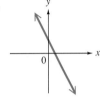

4. $y = 2x - 3$

5. $y = 2x$

E.

F.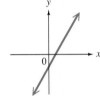

6. $y = -2x$

7. $y = 3$

G.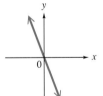

H.

8. $y = -3$

Write equations in standard form of lines satisfying the following conditions. See Example 1.

9. Through $(-2, 4)$; $m = -\dfrac{3}{4}$

10. Through $(-1, 6)$; $m = -\dfrac{5}{6}$

11. Through $(5, 8)$; $m = -2$

12. Through $(12, 10)$; $m = 1$

13. Through $(-5, 4)$; $m = \dfrac{1}{2}$

14. Through $(7, -2)$; $m = \dfrac{1}{4}$

15. Through $(-4, 12)$; horizontal

16. Through $(1, 5)$; horizontal

Write equations in standard form for the following lines. (Hint: What kind of line has undefined slope?)

17. Through $(9, 10)$; undefined slope

18. Through $(-2, 8)$; undefined slope

19. Through $(.5, .2)$; vertical

20. Through $\left(\frac{5}{8}, \frac{2}{9}\right)$; vertical

Write equations in standard form of the lines passing through the following pairs of points. See Example 2.

21. $(3, 4)$ and $(5, 8)$

22. $(5, -2)$ and $(-3, 14)$

23. $(6, 1)$ and $(-2, 5)$

24. $(-2, 5)$ and $(-8, 1)$

25. $\left(-\frac{2}{5}, \frac{2}{5}\right)$ and $\left(\frac{4}{3}, \frac{2}{3}\right)$

26. $\left(\frac{3}{4}, \frac{8}{3}\right)$ and $\left(\frac{2}{5}, \frac{2}{3}\right)$

27. $(2, 5)$ and $(1, 5)$

28. $(-2, 2)$ and $(4, 2)$

29. $(7, 6)$ and $(7, -8)$

30. $(13, 5)$ and $(13, -1)$

Write each equation in slope-intercept form; then give the slope of the line and the y-intercept. See Example 4.

	Equation	*Slope*	*y-intercept*
31. $5x + 2y = 20$	_____	_____	_____
32. $6x + 5y = 40$	_____	_____	_____

	Equation	*Slope*	*y-intercept*

33. $2x - 3y = 10$ _____ _____ _____

34. $4x - 3y = 7$ _____ _____ _____

Write equations in slope-intercept form of lines satisfying the following conditions. See Example 3.

35. $m = 5; b = 15$

36. $m = -2; b = 12$

37. $m = -\dfrac{2}{3}; b = \dfrac{4}{5}$

38. $m = -\dfrac{5}{8}; b = -\dfrac{1}{3}$

39. Slope $\dfrac{2}{5}$; y-intercept $(0, 5)$

40. Slope $-\dfrac{3}{4}$; y-intercept $(0, 7)$

Write equations in standard form of the lines satisfying the following conditions. See Examples 5 and 6.

41. Through $(7, 2)$; parallel to $3x - y = 8$

42. Through $(4, 1)$; parallel to $2x + 5y = 10$

43. Through $(-2, -2)$; parallel to $-x + 2y = 10$

44. Through $(-1, 3)$; parallel to $-x + 3y = 12$

45. Through $(8, 5)$; perpendicular to $2x - y = 7$

46. Through $(2, -7)$; perpendicular to $5x + 2y = 18$

47. Through $(-2, 7)$; perpendicular to $x = 9$

48. Through $(8, 4)$; perpendicular to $x = -3$

Write an equation in the form y = ax for each of the following situations. Then give the three ordered pairs associated with the equation for x-values of 0, 5, and 10. For example, if a car travels 60 miles per hour, then after x hours it will have traveled y = 60x miles. The three ordered pairs are (0, 0), (5, 300), and (10, 600). See Example 7.

49. *x* represents the number of miles traveling at 45 miles per hour, and *y* represents the distance traveled (in miles).

50. *x* represents the number of compact discs sold at $16 each, and *y* represents the total cost of the discs (in dollars).

51. *x* represents the number of gallons of gas sold at $1.30 per gallon, and *y* represents the total cost of the gasoline (in dollars).

52. *x* represents the number of days a videocassette is rented at $1.50 per day, and *y* represents the total charge for the rental (in dollars).

Write an equation in the form y = ax + b for each of the following situations. Then give three ordered pairs associated with the equation for the x-values of 0, 5, and 10. For example, if it costs a flat rate of $.20 to make a certain long-distance call and the charge is $.10 per minute, then after x minutes the cost y in dollars is given by y = .10x + .20. The three ordered pairs are (0, .20), (5, .70), and (10, 1.20). See Example 8.

53. It costs a $15 flat fee to rent a chain saw, plus $3 per day starting with the first day. Let *x* represent the number of days rented, so that *y* represents the charge to the user (in dollars).

54. It costs a borrower $.05 per day for an overdue book, plus a flat $.50 charge for all books borrowed, to be contributed to a fund for building a new library. Let *x* represent the number of days the book is overdue, so that *y* represents the total fine to the tardy user.

55. A rental car costs $25.00 plus $.10 per mile. Let *x* represent the number of miles driven, so that *y* represents the total charge to the user.

56. It costs a flat fee of $450 to drill a well, plus $10 per foot. Let *x* represent the depth of the well, in feet, so that *y* represents the total cost of the well.

Write a linear equation and solve it in order to solve the problem.

57. Refer to Exercise 53. Suppose that the total charge is $69.00. For how many days was the saw rented?

58. Refer to Exercise 54. Suppose that the tardy user paid a $1.30 fine. How many days did this user keep the book past the due date?

59. Refer to Exercise 55. The renter paid $42.30. How many miles was the car driven?

60. Refer to Exercise 56. A certain well costs $6950. How deep is the well?

Assume that the situation described in the figure can be modeled by a straight-line graph, and use the information to find the $y = mx + b$ form of the equation of the line.

61. The number of post offices in the United States has been declining in recent years. Use the information given on the bar graph for the years 1985 and 1993, letting $x = 0$ represent the year 1985 and letting y represent the number of post offices. (*Source:* U.S. Postal Service, Annual Report of the Postmaster General)

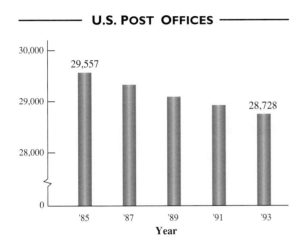

U.S. POST OFFICES

62. The number of motor vehicle deaths in the United States declined from 1990 to 1992 as seen in the accompanying bar graph. Use the information given, with $x = 0$ representing 1990 and y representing the number of deaths. (*Source:* National Safety Council)

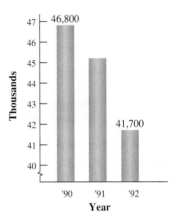

U.S. MOTOR-VEHICLE DEATHS

INTERPRETING TECHNOLOGY (EXERCISES 63–66)

63. The graphing calculator screen shows two lines. One is the graph of $y_1 = -2x + 3$ and the other is the graph of $y_2 = 3x - 4$. Which is which?

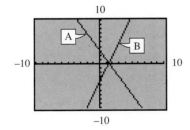

64. The graphing calculator screen shows two lines. One is the graph of $y_1 = 2x - 5$ and the other is the graph of $y_2 = 4x - 5$. Which is which?

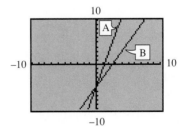

65. The graphing calculator screen shows a line with the coordinates of a point displayed at the bottom. The slope of the line is 3. What is the y-coordinate of the point on the line whose x-coordinate is 4?

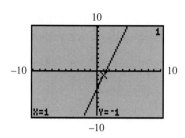

66. The table shown was generated by a graphing calculator. It gives several points that lie on the graph of a line. What is the slope of the line?

X	Y₁
2	-.5
3	1.25
4	3
5	4.75
6	6.5
7	8.25
8	10

X=8

MATHEMATICAL CONNECTIONS (EXERCISES 67–72)

In Chapter 2 we learned how formulas can be applied to problem solving. In Exercises 67–72, we will see how the formula that relates the Celsius and Fahrenheit temperatures is derived. Work Exercises 67–72 in order.

67. There is a linear relationship between Celsius and Fahrenheit temperatures. When C = 0°, F = _____°, and when C = 100°, F = _____°.

68. Think of ordered pairs of temperatures (C, F), where C and F represent corresponding Celsius and Fahrenheit temperatures. The equation that relates the two scales has a straight-line graph that contains the two points determined in Exercise 67. What are these two points?

69. Find the slope of the line described in Exercise 68.

70. Now think of the point-slope form of the equation in terms of C and F, where C replaces x and F replaces y. Use the slope you found in Exercise 69 and one of the two points determined earlier, and find the equation that gives F in terms of C.

71. To obtain another form of the formula, use the equation you found in Exercise 70 and solve for C in terms of F.

72. The equation found in Exercise 70 is graphed on the graphing calculator screen shown here. Observe the display at the bottom, and interpret it in the context of this group of exercises.

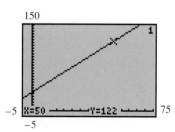

3.4 *Linear Inequalities in Two Variables*

OBJECTIVES

1 ▶ Graph linear inequalities.

2 ▶ Graph the intersection of two linear inequalities.

3 ▶ Graph the union of two linear inequalities.

4 ▶ Graph first-degree absolute value inequalities.

FOR EXTRA HELP

Tutorial   Tape 4   SSM, Sec. 3.4

In Chapter 2 we graphed linear inequalities with one variable on the number line. In this section we will graph linear inequalities in two variables on a rectangular coordinate system.

Linear Inequality

An inequality that can be written as

$$Ax + By < C \quad \text{or} \quad Ax + By > C,$$

where A, B, and C are real numbers and A and B are not both 0, is a **linear inequality in two variables.**

Also, $\leq$ and $\geq$ may replace $<$ and $>$ in the definition.

OBJECTIVE 1 ▶ A line divides the plane into three regions: the line itself and the two half-planes on either side of the line. Recall that the graphs of linear inequalities in one variable are intervals on the number line that sometimes include an endpoint. The graphs of linear inequalities in two variables are *regions* in the real number plane and may include a *boundary line*. The **boundary** (BOUN-dery) **line** for the inequality $Ax + By < C$ or $Ax + By > C$ is the graph of the *equation* $Ax + By = C$. To graph a linear inequality, we go through the following steps.

Graphing a Linear Inequality

Step 1 Draw the graph of the straight line that is the boundary. Make the line solid if the inequality involves $\leq$ or $\geq$; make the line dashed if the inequality involves $<$ or $>$.

Step 2 Choose any point not on the line as a test point.

Step 3 Shade the region that includes the test point if the test point satisfies the original inequality; otherwise, shade the region on the other side of the boundary line.

E X A M P L E I Graphing a Linear Inequality

Graph the solutions of $3x + 2y \geq 6$.

Graph the linear inequality $3x + 2y \geq 6$ by first graphing the straight line $3x + 2y = 6$. The graph of this line, the boundary of the graph of the inequality, is shown in Figure 21.

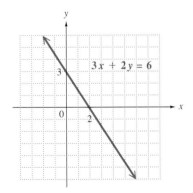

FIGURE 21

CONTINUED ON NEXT PAGE

1. Graph the solutions of each inequality.

(a) $x + y \le 4$

(b) $3x + y \ge 6$

The graph of $3x + 2y \ge 6$ includes the points of the line $3x + 2y = 6$, and either the points *above* the line $3x + 2y = 6$ or the points *below* that line. To decide which side belongs to the graph, first select any point not on the line $3x + 2y = 6$. The origin, $(0, 0)$, is often a good choice. Substitute the values from the test point $(0, 0)$ for x and y in the inequality $3x + 2y > 6$:

$$3(0) + 2(0) > 6$$
$$0 > 6. \quad \text{False}$$

Because the result is false, $(0, 0)$ does not satisfy the inequality, and the solutions include all points on the other side of the line. This region is shaded in Figure 22.

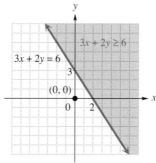

FIGURE 22

◀◀ **WORK PROBLEM I AT THE SIDE.**

   If the inequality is written in the form $y > mx + b$ or $y < mx + b$, the inequality symbol indicates which half-plane to shade.

If $y > mx + b$, shade **above** the boundary line;

if $y < mx + b$, shade **below** the boundary line.

E X A M P L E   2   Graphing an Inequality

Graph the solutions of $x - 3y < 4$.

   First graph the boundary line, shown in Figure 23. The points of the boundary line do not belong to the inequality $x - 3y < 4$ (because the inequality symbol is $<$, not $\le$). For this reason, the line is dashed. Now solve the inequality for y.

$$x - 3y < 4$$
$$-3y < -x + 4$$
$$y > \frac{x}{3} - \frac{4}{3} \qquad \text{Multiply by } -\tfrac{1}{3}; \text{ change the inequality.}$$

Because of the *is greater than* symbol, we should shade *above* the line. As a check, we can choose any point not on the line, say $(1, 2)$, and substitute for x and y in the original inequality.

$$1 - 3(2) < 4$$
$$-5 < 4 \quad \text{True}$$

This result agrees with our decision to shade above the line. The solutions, graphed in Figure 23, include only those points in the shaded half-plane (not those on the line).

ANSWERS

1. (a)

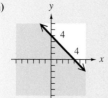

(b)

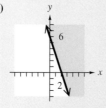

CONTINUED ON NEXT PAGE

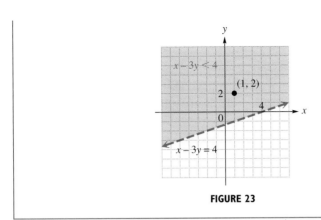

FIGURE 23

WORK PROBLEM 2 AT THE SIDE. ▶▶

OBJECTIVE 2 In Section 2.6 we discussed how the words "and" and "or" are used with compound inequalities. In that section, the inequalities were in a single variable. Those ideas can be extended to include inequalities in two variables. If a pair of inequalities is joined with the word "and," it is interpreted as the intersection of the solutions of the inequalities. The graph of the intersection of two or more inequalities is the region of the plane where all points satisfy all of the inequalities at the same time.

EXAMPLE 3 Graphing the Intersection of Two Inequalities

Graph the intersection of the solutions of $2x + 4y \geq 5$ and $x \geq 1$.

To begin, graph each of the two inequalities $2x + 4y \geq 5$ and $x \geq 1$ separately. The graph of $2x + 4y \geq 5$ is shown in Figure 24(a), and the graph of $x \geq 1$ is shown in Figure 24(b). The graph of the intersection is the region common to both graphs, as shown in Figure 24(c).

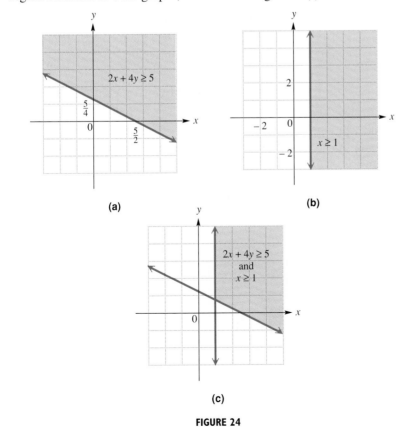

FIGURE 24

2. Graph the solutions of each inequality.

(a) $x - y > 2$

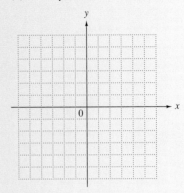

(b) $3x + 4y < 12$

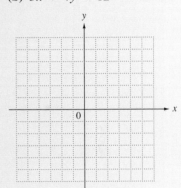

ANSWERS

2. (a)

(b)

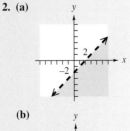

3. Graph the intersection of the solutions of the inequalities: $x - y \leq 4$ and $x \geq -2$.

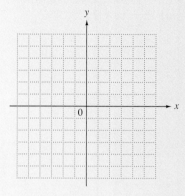

4. Graph the union of the solutions of the inequalities: $7x - 3y < 21$ or $x > 2$.

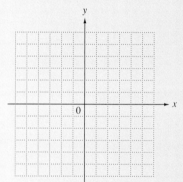

3.

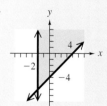

4.

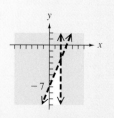

◄◄ WORK PROBLEM 3 AT THE SIDE.

OBJECTIVE 3 When two inequalities are joined by the word "or," we must find the union of the graphs of the inequalities. The graph of the union of two inequalities includes all of the points that satisfy either inequality.

E X A M P L E   4   Graphing the Union of Two Inequalities

Graph the union of the solutions of $2x + 4y \geq 5$ or $x \geq 1$.

The graphs of the two inequalities are shown in Figures 24(a) and 24(b). The graph of the union is shown in Figure 25.

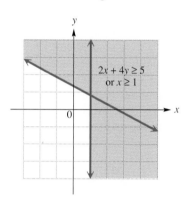

FIGURE 25

◄◄ WORK PROBLEM 4 AT THE SIDE.

OBJECTIVE 4 An absolute value inequality should first be written without absolute value bars as shown in Chapter 2. Then the methods given in the previous examples can be used to graph the inequality.

E X A M P L E   5   Graphing an Absolute Value Inequality

Graph the solutions of $|x| \leq 4$.

Rewrite $|x| \leq 4$ as $-4 \leq x \leq 4$. The topic of this section is linear inequalities in *two* variables, so the boundary lines are the vertical lines $x = -4$ and $x = 4$. Points from the region between these lines satisfy the inequality, so that region is shaded. See Figure 26.

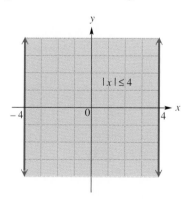

FIGURE 26

EXAMPLE 6 Graphing an Absolute Value Inequality

Graph the solutions of $|y + 2| > 3$.

As shown in Section 2.7, the equation of the boundary, $|y + 2| = 3$, can be rewritten as

$$y + 2 = 3 \quad \text{or} \quad y + 2 = -3$$
$$y = 1 \quad \text{or} \quad y = -5.$$

This shows that $y = 1$ and $y = -5$ are boundary lines. Checking points from each of the three regions determined by the horizontal boundary lines gives the graph shown in Figure 27.

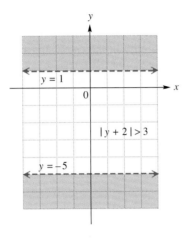

FIGURE 27

WORK PROBLEM 5 AT THE SIDE. ▶▶

5. Graph the solutions of each absolute value inequality.

(a) $|x| \geq 4$

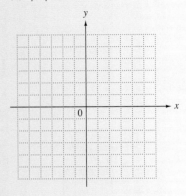

(b) $|x - 2| < 1$

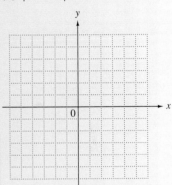

ANSWERS

5. (a)

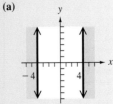

(b)

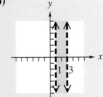

Numbers in the

Real World *a graphing calculator minicourse*

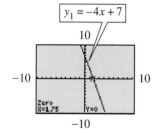

$y_1 = -4x + 7$

Lesson 1: Solution of Linear Equations

A graphing calculator can be used to solve a linear equation of the form $y = 0$. First, we observe that the graph of $y = mx + b$ is a straight line, and if $m \neq 0$, the x-intercept of the graph is the solution of the equation $mx + b = 0$. For a simple example, suppose we want to solve $-4x + 7 = 0$ graphically. We begin by graphing $y_1 = -4x + 7$ in a viewing window that shows the x-intercept. Then we use the capability of the calculator to find the x-intercept. As seen in the first figure to the left, the x-intercept of the graph, which is also the solution or root of the equation, is 1.75. (For convenience, we will refer to the number as the x-intercept rather than the point.) Therefore, the solution set of $-4x + 7 = 0$ is $\{1.75\}$. This can easily be verified using strictly algebraic methods.

If we want to solve a more complicated linear equation, such as

$$-2x - 4(2 - x) = 3x + 4,$$

we begin by writing it as an equivalent equation with 0 on one side. If we subtract $3x$ and subtract 4 from both sides, we get

$$-2x - 4(2 - x) - 3x - 4 = 0.$$

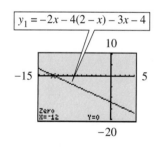

$y_1 = -2x - 4(2 - x) - 3x - 4$

Then we graph $y_1 = -2x - 4(2 - x) - 3x - 4$, and find the x-intercept. Notice that the viewing window must be altered from the one shown in the first figure, because the x-intercept does not lie in the interval $[-10, 10]$. As seen in the second figure, the x-intercept of the line, and thus the solution or root of the equation, is -12. The solution set is $\{-12\}$.

GRAPHING CALCULATOR EXPLORATIONS

1. Solve the equation $-2x - 4(2 - x) = 3x + 4$ using traditional algebraic methods, and show that the solution set is $\{-12\}$, supporting the graphical approach used in the second figure.

2. Use traditional algebraic methods to solve $2(x - 5) + 3x = 0$. Then use a graphing calculator to support your result by showing that the x-intercept of $y_1 = 2(x - 5) + 3x$ corresponds to your solution.

3. Use traditional algebraic methods to solve $6x - 4(3 - 2x) = 5(x - 4) - 10$. Then use a graphing calculator to support your result by showing that the x-intercept of $y_1 = 6x - 4(3 - 2x) - 5(x - 4) + 10$ corresponds to your solution.

4. In Example 3 of Section 2.1, we show that the solution of $\dfrac{x + 7}{6} + \dfrac{2x - 8}{2} = -4$ is -1.

   Graph $y_1 = \dfrac{x + 7}{6} + \dfrac{2x - 8}{2} + 4$, being careful to use parentheses around the numerators as you make the keystrokes. Show that the x-intercept of the graph is -1.

3.4 Exercises

In each statement, fill in the first blank with either solid *or* dashed. *Fill in the second blank with* above *or* below.

1. The boundary of the graph of $y \le -x + 2$ will be a _____ line, and the shading will be _____ the line.

2. The boundary of the graph of $y < -x + 2$ will be a _____ line, and the shading will be _____ the line.

3. The boundary of the graph of $y > -x + 2$ will be a _____ line, and the shading will be _____ the line.

4. The boundary of the graph of $y \ge -x + 2$ will be a _____ line, and the shading will be _____ the line.

Graph each linear inequality in two variables. See Examples 1 and 2.

5. $x + y \le 2$

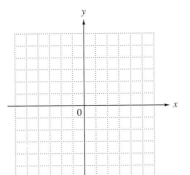

6. $x + y \le -3$

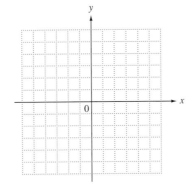

7. $4x - y < 4$

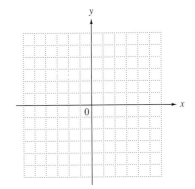

8. $3x - y < 3$

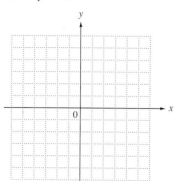

9. $x + 3y \geq -2$

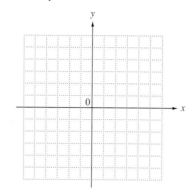

10. $x + 4y \geq -3$

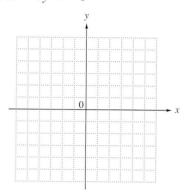

11. $x + y > 0$

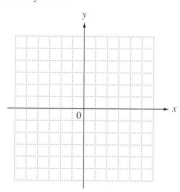

12. $x + 2y > 0$

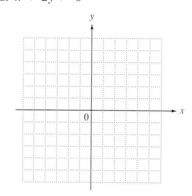

13. $x - 3y \leq 0$

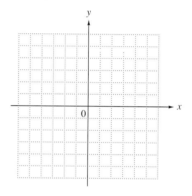

14. $x - 5y \leq 0$

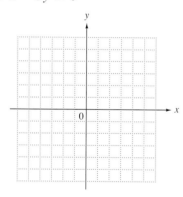

15. $y < x$

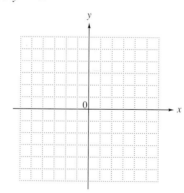

16. $y \leq 4x$

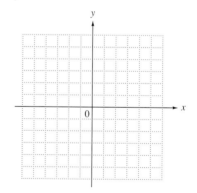

17. Explain how you will know whether to use intersection or union in graphing inequalities like those found in Examples 3 and 4 in this section.

18. Explain how you will know whether to use "and" or "or" in graphing absolute value inequalities like those found in Examples 5 and 6 in this section.

Graph the intersection or union, as appropriate, of the solutions of the following pairs of linear inequalities. See Examples 3 and 4.

19. $x + y \le 1$ and $x \ge 1$

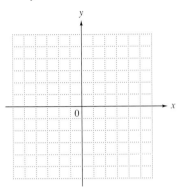

20. $x - y \ge 2$ and $x \ge 3$

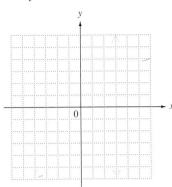

21. $2x - y \ge 2$ and $y < 4$

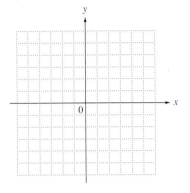

22. $3x - y \ge 3$ and $y < 3$

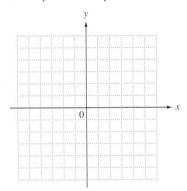

23. $x + y > -5$ and $y < -2$

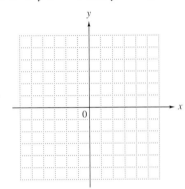

24. $6x - 4y < 10$ and $y > 2$

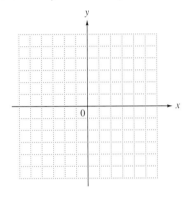

25. $x - y \ge 1$ or $y \ge 2$

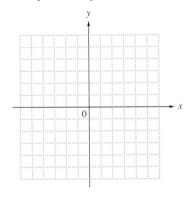

26. $x + y \le 2$ or $y \ge 3$

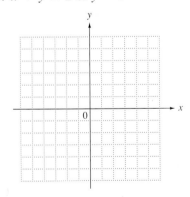

27. $x - 2 > y$ or $x < 1$

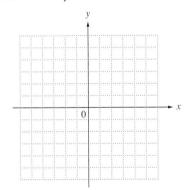

28. $x + 3 < y$ or $x > 3$

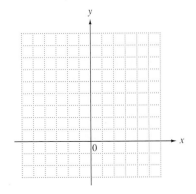

29. $3x + 2y < 6$ or $x - 2y > 2$

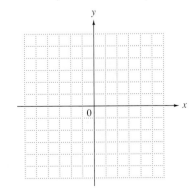

30. $x - y \ge 1$ or $x + y \le 4$

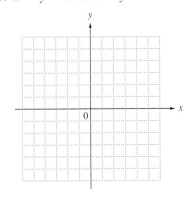

Graph each of the following inequalities involving absolute value. See Examples 5 and 6.

31. $|x| \geq 3$

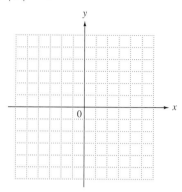

32. $|y| < 5$

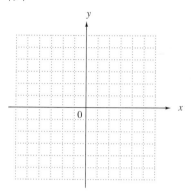

33. $|y + 1| < 2$

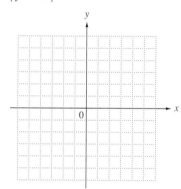

34. $|x - 2| \geq 1$

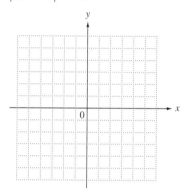

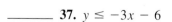

INTERPRETING TECHNOLOGY (EXERCISES 35–38)

Match each inequality with its calculator-generated graph. (Hint: Use the slope, y-intercept, and inequality symbol in making your choice.)

_____ **35.** $y \leq 3x - 6$ **A.**

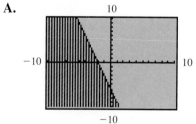

B.

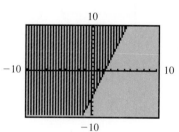

_____ **36.** $y \geq 3x - 6$

_____ **37.** $y \leq -3x - 6$ **C.**

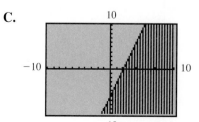

D.

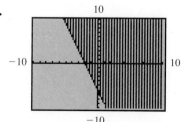

_____ **38.** $y \geq -3x - 6$

3.5 *An Introduction to Functions*

One of the most important concepts in algebra is that of a *function* (FUNK-shun). In this section we will examine functions, and see how they can be expressed in several forms. The table below indicates the number of people on Earth for various years from 1800 through 2050. These figures come from The Population Reference Bureau, Inc., and figures after 1987 are projections.

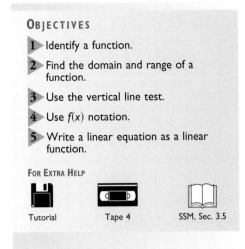

OBJECTIVES

1 Identify a function.
2 Find the domain and range of a function.
3 Use the vertical line test.
4 Use $f(x)$ notation.
5 Write a linear equation as a linear function.

FOR EXTRA HELP

Tutorial   Tape 4   SSM, Sec. 3.5

Year	Population in Billions
1800	1
1927	2
1960	3
1975	4
1987	5
1998	6
2008	7
2019	8
2032	9
2050	10

The information in the table can be expressed as a set of ordered pairs, where the first entry in the ordered pair is the year, and the second entry is the population in billions. For example, three such ordered pairs are (1800, 1), (1960, 3), and (1998, 6).

It has been shown that under certain conditions, the Fahrenheit temperature determines the number of times a cricket will chirp per minute. If x represents the temperature and y represents the number of chirps per minute, the equation $y = .25x + 40$ gives the relationship between them. For example, when $x = 40$, $y = 50$, meaning that at 40°F the cricket chirps 50 times per minute. This can be represented by the ordered pair (40, 50). Some other such ordered pairs are (20, 45) and (60, 55).

A third way of representing data that leads to ordered pairs is a graph. The graph in Figure 28 shows the public debt of the United States for the fiscal years 1990 through 1994.

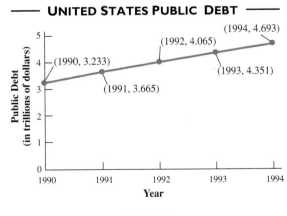

FIGURE 28

If x represents the year and y the public debt in trillions of dollars, one ordered pair describing their relationship is (1990, 3.233).

1. (a) Name two other ordered pairs in the world population example given in the table.

In each example we have given, the second entry in the ordered pair is dependent on the first. That is, the population is dependent on the year, the number of chirps is dependent on the temperature, and the public debt is dependent on the year. In general, when the value of y depends on the value of x, y is called the **dependent variable** (DEE-pen-dent VAIR-ee-uh-bul) and x is called the **independent variable** (IN-dih-pen-dent VAIR-ee-uh-bul).

◀◀ WORK PROBLEM 1 AT THE SIDE.

OBJECTIVE 1▶ We are now ready to introduce the function concept. First, we will examine a more general concept, the relation. Since related quantities can be represented by ordered pairs (as we have just seen), we define a *relation* (ree-LAY-shun) as follows.

(b) Name two other ordered pairs in the cricket example given by the equation.

> **Relation**
>
> A **relation** is a set of ordered pairs of real numbers.

In each of the examples of relations just discussed, for every value of x, there was one and only one value of y. A relation of this type has an important role in algebra, and it is called a *function*.

> **Function**
>
> A **function** is a relation in which, for each value of the first component of the ordered pairs, there is exactly one value of the second component.

(c) Name two other ordered pairs in the public debt example given in Figure 28.

EXAMPLE 1   **Identifying a Function Expressed as a Set of Ordered Pairs**

Determine whether each of the following relations is a function.

$$F = \{(1, 2), (-2, 5), (3, -1)\}$$
$$G = \{(-2, 1), (-1, 0), (0, 1), (1, 2), (2, 2)\}$$
$$H = \{(-4, 1), (-2, 1), (-2, 0)\}$$

Relations F and G are functions, because for each x-value, there is only one y-value. Notice that in G, the last two ordered pairs have the same y-value. This does not violate the definition of function because each first component (x-value) has only one second component (y-value).

Relation H is not a function because the last two ordered pairs have the same x-value, but different y-values.

In addition to sets of ordered pairs, tables, equations, and graphs, a function can be represented by a mapping, as described in Example 2.

EXAMPLE 2   **Recognizing a Function Expressed as a Mapping**

A function can also be expressed as a correspondence or *mapping* from one set to another. The mapping in Figure 29 is a function that assigns to a state its population (in millions) expected by the year 2000.

CONTINUED ON NEXT PAGE

ANSWERS

(Answers will vary.)

1. (a) (1927, 2), (1975, 4)
   (b) (44, 51), (48, 52)
   (c) (1991, 3.665), (1992, 4.065)

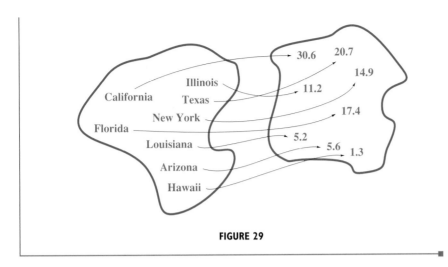

FIGURE 29

WORK PROBLEM 2 AT THE SIDE. ▶▶

OBJECTIVE 2 ▶ The set of all first components (*x*-values) of the ordered pairs of a relation is called the **domain** (doh-MAIN) of the relation, and the set of all second components (*y*-values) is called the **range.** Another way of interpreting this is that the domain is the set of inputs, and the range is the set of outputs. The domain of function F in Example 1 is $\{1, -2, 3\}$; the range is $\{2, 5, -1\}$. Also, the domain of function G is $\{-2, -1, 0, 1, 2\}$, and the range is $\{0, 1, 2\}$. Domains and ranges can also be defined in terms of independent and dependent variables.

> **Domain and Range**
>
> In a relation, the set of all values of the independent variable (x) is the **domain;** the set of all values of the dependent variable (y) is the **range.**

E X A M P L E 3   Finding Domains and Ranges

Give the domain and range of each function.

(a) $\{(3, -1), (4, 2), (0, 5)\}$

The domain, the set of *x*-values, is $\{3, 4, 0\}$; the range is the set of *y*-values, $\{-1, 2, 5\}$.

(b) The function in Figure 29

The domain is {Illinois, Texas, California, New York, Florida, Louisiana, Arizona, Hawaii} and the range is $\{1.3, 5.2, 5.6, 11.2, 14.9, 17.4, 20.7, 30.6\}$.

WORK PROBLEM 3 AT THE SIDE. ▶▶

2. Which are functions?

(a) $\{(1, 2), (2, 4), (3, 3), (4, 2)\}$

(b) $\{(0, 3), (-1, 2), (-1, 3)\}$

(c)

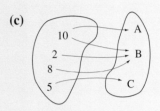

(d)

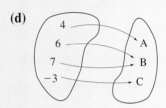

3. Give the domain and range of each function.

(a) $\{(1, 2), (2, 4), (3, 3), (4, 2)\}$

(b)

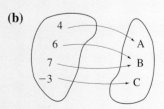

ANSWERS

2. (a) and **(d)** are functions.
3. (a) domain: $\{1, 2, 3, 4\}$
       range: $\{2, 3, 4\}$
   (b) domain: $\{-3, 4, 6, 7\}$
       range: $\{A, B, C\}$

4. Give the domain and range of each relation from its graph.

(a)

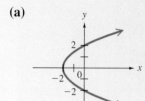

E X A M P L E   4   Finding Domains and Ranges from Graphs

Three relations are graphed in Figure 30. Give the domain and range of each.

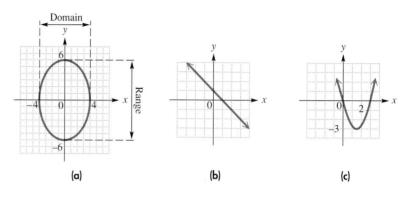

FIGURE 30

(a) In Figure 30(a), the x-values of the points on the graph include all numbers between -4 and 4, inclusive. The y-values include all numbers between -6 and 6, inclusive. Using interval notation,

the domain is $[-4, 4]$; the range is $[-6, 6]$.

(b) In Figure 30(b), the arrowheads indicate that the line extends indefinitely left and right, as well as up and down. Therefore, both the domain and the range are the set of all real numbers, written $(-\infty, \infty)$.

(c) In Figure 30(c), the arrowheads indicate that the graph extends indefinitely left and right, as well as upward. The domain is $(-\infty, \infty)$. Because there is a least y-value, -3, the range includes all numbers greater than or equal to -3, written $[-3, \infty)$.

◀◀ WORK PROBLEM 4 AT THE SIDE.

Relations are often defined by equations such as $y = 2x + 3$ and $y^2 = x$. Sometimes we need to determine the domain of a relation from its equation. The domain of a relation is assumed to be all real numbers that produce real numbers for the dependent variable when substituted for the independent variable. For example, because any real number can be used as a replacement for x in $y = 2x + 3$, the domain of this function is the set of real numbers. As another example, the function defined by $y = \frac{1}{x}$ has all real numbers except 0 as a domain, because y is undefined if $x = 0$. In general, the domain of a function defined by an algebraic expression is all real numbers except those numbers that lead to division by zero or an even root of a negative number.

(b)

E X A M P L E   5   Identifying a Function from an Equation or Inequality

For each of the following decide whether it defines a function, and give the domain.

(a) $y = \sqrt{2x - 1}$

For any choice of x in the domain, there is exactly one corresponding value for y, so this equation defines a function. We saw earlier that the square root of a negative number is not a real number, so we must have

CONTINUED ON NEXT PAGE

$$2x - 1 \geq 0$$
$$2x \geq 1$$
$$x \geq \frac{1}{2}.$$

The domain is $\left[\frac{1}{2}, \infty\right)$.

(b) $y^2 = x$

The ordered pairs $(16, 4)$ and $(16, -4)$ both satisfy this equation. Because one value of x, 16, corresponds to two values of y, 4 and -4, this equation does not define a function. If $y^2 = x$, then $y = \sqrt{x}$ or $y = -\sqrt{x}$, which shows that two values of y correspond to each positive value of x. Because x is equal to the square of y, the values of x must always be nonnegative. The domain is $[0, \infty)$.

(c) $y \leq x - 1$

By definition, y is a function of x if a value of x leads to exactly one value of y. In this example, a particular value of x, say 1, corresponds to many values of y. The ordered pairs $(1, 0)$, $(1, -1)$, $(1, -2)$, $(1, -3)$, and so on, all satisfy the inequality. For this reason, this inequality does not define a function. Any number can be used for x, so the domain is the set of real numbers, $(-\infty, \infty)$.

(d) $y = \dfrac{5}{x - 1}$

Given any value of x, we find y by subtracting 1, then dividing the result into 5. This process produces exactly one value of y for each x-value, so this equation defines a function. The domain includes all real numbers except those that make the denominator zero. We find these numbers by setting the denominator equal to zero and solving for x.

$$x - 1 = 0$$
$$x = 1$$

Thus, the domain includes all real numbers except 1. In interval notation this is written as

$$(-\infty, 1) \cup (1, \infty).$$

Caution
The parentheses used to represent an ordered pair are also used to represent an open interval (introduced in Chapter 1.) In general, there is no confusion between these symbols because the context of the discussion tells us whether we are discussing ordered pairs or open intervals.

WORK PROBLEM 5 AT THE SIDE. ▶▶

OBJECTIVE 3 In a function each value of x leads to only one value of y, so any vertical line drawn through the graph of a function would intersect the graph in at most one point. This is the **vertical line test for a function.**

Vertical Line Test

If a vertical line intersects the graph of a relation in more than one point, then the relation is not a function.

---

5. Decide whether or not each equation or inequality defines a function, and give the domain.

(a) $y = 6x + 12$

(b) $y \leq 4x$

(c) $y = -\sqrt{3x - 2}$

(d) $y^2 = 25x$

ANSWERS

5. **(a)** yes; $(-\infty, \infty)$ **(b)** no; $(-\infty, \infty)$
(c) yes; $\left[\dfrac{2}{3}, \infty\right)$
(d) no; $[0, \infty)$

6. Which of the following graphs represent functions?

(a)

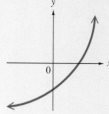

(b)

(c)

For example, the graph shown in Figure 31(a) is not the graph of a function, because a vertical line intersects the graph in more than one point. The graph of Figure 31(b) does represent a function. Any vertical line will cross the graph at most once.

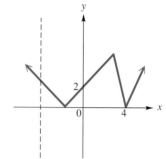

(a)   (b)

FIGURE 31

◀◀ WORK PROBLEM 6 AT THE SIDE.

OBJECTIVE 4 To say that y is a function of x means that for each value of x from the domain of the function, there is exactly one value of y. To emphasize that y *is a function of* x, or that y depends on x, it is common to write

$$y = f(x),$$

with $f(x)$ read "f of x." (In this special notation, the parentheses do not indicate multiplication.) The letter f stands for *function.* (Other letters, such as g and h are often used as well.) For example, if $y = 9x - 5$, we emphasize that y is a function of x by writing $y = 9x - 5$ as

$$f(x) = 9x - 5.$$

We can use this **function notation** to simplify certain statements. For example, if $y = 9x - 5$, then replacing x with 2 gives

$$y = 9 \cdot 2 - 5$$
$$= 18 - 5$$
$$= 13.$$

The statement "if $x = 2$, then $y = 13$" is abbreviated with function notation as

$$f(2) = 13.$$

We read this as "f of 2 equals 13." Also, $f(0) = 9 \cdot 0 - 5 = -5$, and $f(-3) = -32$.

These ideas and the symbols used to represent them can be explained as follows.

Name of the function   Defining expression

$$y = \overbrace{f(x)} = \overbrace{9x - 5}$$

Value of the function   Name of the independent variable

Caution
The symbol $f(x)$ *does not* indicate "f times x," but represents the y-value for the indicated x-value. As shown above, $f(2)$ is the y-value that corresponds to the x-value 2.

ANSWERS

6. (a) and (c) are the graphs of functions.

┌─ **E X A M P L E 6   Using Function Notation**

Let $f(x) = -x^2 + 5x - 3$. Find the following.

(a) $f(2)$

Replace x with 2.

$$f(2) = -2^2 + 5 \cdot 2 - 3 = -4 + 10 - 3 = 3$$

(b) $f(-1)$

$$f(-1) = -(-1)^2 + 5(-1) - 3 = -1 - 5 - 3 = -9$$

(c) $f(q)$

Replace x with q.

$$f(q) = -q^2 + 5q - 3$$

The replacement of one variable with another is important in later courses.

WORK PROBLEM 7 AT THE SIDE. ▶▶

OBJECTIVE 5▶ By the vertical line test, linear equations (except for the type where $x = k$) define functions because their graphs are non-vertical straight lines.

Linear Function

A function that can be written in the form

$$f(x) = mx + b$$

is a **linear function.**

As mentioned earlier, this form of the equation is the slope-intercept form, where m is the slope, $(0, b)$ is the y-intercept, and $f(x)$ is just another name for y.

┌─ **E X A M P L E 7   Graphing a Linear Function**

Graph $f(x) = -2x + 4$. Give the domain and the range.

To graph a linear function $f(x) = mx + b$, we simply replace $f(x)$ with y and then graph the linear equation $y = mx + b$. The graph of $f(x) = -2x + 4$ is a line with slope -2 and y-intercept $(0, 4)$, as shown in Figure 32. The domain and the range are both $(-\infty, \infty)$.

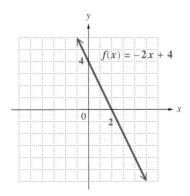

FIGURE 32

WORK PROBLEM 8 AT THE SIDE. ▶▶

7. Find $f(-3)$ and $f(p)$.

(a) $f(x) = 6x - 2$

(b) $f(x) = \dfrac{-3x + 5}{2}$

(c) $f(x) = \dfrac{1}{6}x^2 - 1$

8. Graph the linear function $f(x) = 3x + 1$. Give the domain and the range.

ANSWERS

7. **(a)** -20; $6p - 2$

(b) 7; $\dfrac{-3p + 5}{2}$

(c) $\dfrac{1}{2}$; $\dfrac{1}{6}p^2 - 1$

8.

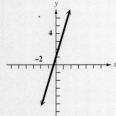

The domain and the range are both $(-\infty, \infty)$.

NUMBERS IN THE

Real World

a graphing calculator minicourse

Lesson 2: Solution of Linear Inequalities

In Lesson 1 we observed that the *x*-intercept of the graph of the line $y = mx + b$ is the solution of the equation $mx + b = 0$. We can extend our observations to consider the solution of the associated inequalities $mx + b > 0$ and $mx + b < 0$. The solution set of $mx + b > 0$ is the set of all *x*-values for which the graph of $y = mx + b$ is *above* the *x*-axis. (We consider points above because the symbol is >.) On the other hand, the solution set of $mx + b < 0$ is the set of all *x*-values for which the graph of $y = mx + b$ is *below* the *x*-axis. (We consider points below because the symbol is <.) Therefore, once we know the solution set of the equation and have the graph of the line, we can determine the solution sets of the corresponding inequalities.

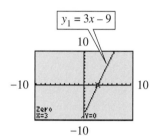

In the first figure to the left, we see that the *x*-intercept of $y_1 = 3x - 9$ is 3. Therefore, by the concepts of Lesson 1, the solution set of $3x - 9 = 0$ is {3}. Because the graph of y_1 lies above the *x*-axis for *x*-values greater than 3, the solution set of $3x - 9 > 0$ is (3, ∞). Because the graph lies below the *x*-axis for *x*-values less than 3, the solution set of $3x - 9 < 0$ is (−∞, 3).

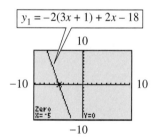

Suppose that we wish to solve the equation $-2(3x + 1) = -2x + 18$, and the associated inequalities $-2(3x + 1) > -2x + 18$ and $-2(3x + 1) < -2x + 18$. We begin by considering the equation rewritten so that the right side is equal to 0: $-2(3x + 1) + 2x - 18 = 0$. Graphing $y_1 = -2(3x + 1) + 2x - 18$ yields the *x*-intercept −5, as shown in the second figure. The first inequality listed is equivalent to $y_1 > 0$. Because the line lies *above* the *x*-axis for *x*-values less than −5, the solution set of $-2(3x + 1) > -2x + 18$ is (−∞, −5). Because the line lies *below* the *x*-axis for *x*-values greater than −5, the solution set of $-2(3x + 1) < -2x + 18$ is (−5, ∞).

GRAPHING CALCULATOR EXPLORATIONS

1. Refer to the first figure in Lesson 1 on page 188. Use the graph to solve each inequality:

   (a) $-4x + 7 > 0$ (b) $-4x + 7 < 0$.

2. The graph below is that of $y_1 = 3(5 - 2x)$. Use the graph and the display to give the solution set of each of the following:

   (a) $3(5 - 2x) = 0$ (b) $3(5 - 2x) > 0$ (c) $3(5 - 2x) \leq 0$.

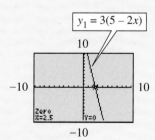

200

3.5 *Exercises*

For the various methods of representing a function shown in Exercises 1–3, give the three ordered pairs with the first entries 1990, 1992, and 1993.

1. Commissioned Officers in the U.S. Army

Year	Number
1990	746,220
1992	661,391
1993	590,324
1994	553,627
1995	521,036

Sources: Department of the Army, U.S. Dept. of Defense

2. $y = 18x - 35{,}753$ gives an approximation for Medicare costs, in billions of dollars, where x is the year and y is the cost.

3. PERSONAL CONSUMPTION EXPENDITURES IN THE UNITED STATES

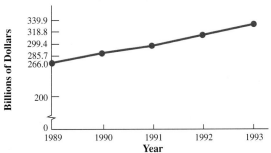

Sources: Bureau of Economic Analysis; U.S. Dept. of Commerce

4. Explain what is meant by each of the following terms.
  (a) relation   **(c)** range of a relation
  (b) domain of a relation **(d)** function

5. Describe the use of the vertical line test.

6. Give an example of a relation that is not a function, having domain $\{-3, 2, 6\}$ and range $\{4, 6\}$. (There are many possible correct answers.)

Decide whether the relation is a function, and give the domain and the range of the relation.
Use the vertical line test in Exercises 13–16. See Examples 1–4.

7. {(5, 1), (3, 2), (4, 9), (7, 3)}

8. {(8, 0), (5, 4), (9, 3), (3, 9)}

9. {(2, 4), (0, 2), (2, 6)}

10. {(9, −2), (−3, 5), (9, 1)}

11.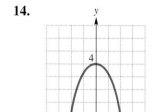

12.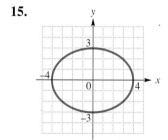

13.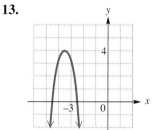

14.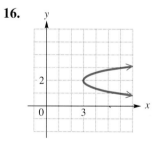

15.

16.

Decide whether the given equation defines y as a function of x. Give the domain. Identify any linear functions. See Example 5.

17. $y = x^2$ **18.** $y = x^3$ **19.** $x = y^6$ **20.** $x = y^4$

21. $x + y < 4$ **22.** $x - y < 3$ **23.** $y = \sqrt{x}$ **24.** $y = -\sqrt{x}$

25. $xy = 1$ **26.** $xy = -3$ **27.** $y = 2x - 6$ **28.** $y = -6x + 8$

29. $y = \sqrt{4x + 2}$ **30.** $y = \sqrt{9 - 2x}$ **31.** $y = \dfrac{2}{x - 9}$ **32.** $y = \dfrac{-7}{x - 16}$

Let f(x) = -3x + 4 and g(x) = -x² + 4x + 1. Find each of the following. See Example 6.

33. $f(0)$ **34.** $f(-3)$ **35.** $g(-2)$ **36.** $g(10)$

37. $f(p)$

38. $g(k)$

39. $f(-x)$

40. $g(-x)$

41. $f(x + 2)$

42. $g(2p)$

43. $f(g(1))$

44. $g(f(1))$

45. Compare the answers to Exercises 43 and 44. Do you think that $f(g(x))$ is, in general, equal to $g(f(x))$?

46. Make up two linear functions f and g such that $f(g(2)) = 4$. (There are many ways to do this.)

47. Fill in the blanks with the correct responses. The equation $2x + y = 4$ has a straight _____ as its graph. One point that lies on the graph is (3, _____). If we solve the equation for y and use function notation, we have a _____ function $f(x) =$ _____. For this function, $f(3) =$ _____, meaning that the point (_____, _____) lies on the graph of the function.

48. Which one of the following defines a linear function?

   (a) $y = \dfrac{x - 5}{4}$ **(b)** $y = \sqrt[3]{x}$

   (c) $y = x^2$ **(d)** $y = \sqrt{x}$

Sketch the graph of the linear function. Give the domain and range. See Example 7.

49. $f(x) = -2x + 5$

50. $g(x) = 4x - 1$

51. $h(x) = \dfrac{1}{2}x + 2$

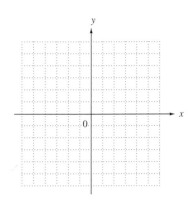

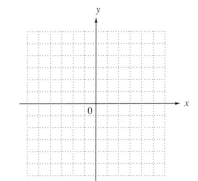

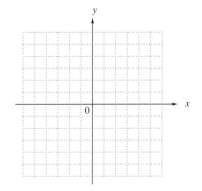

52. $F(x) = -\frac{1}{4}x + 1$ **53.** $G(x) = 2$ **54.** $H(x) = -3x$

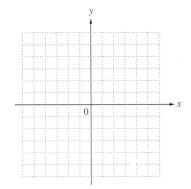   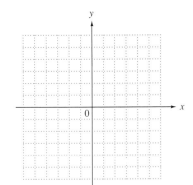   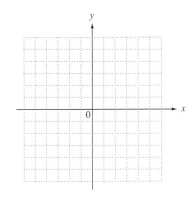

55. The linear function $f(x) = -123x + 29,685$ provides a model for the number of post offices in the United States from 1984 to 1990, where $x = 0$ corresponds to 1984, $x = 1$ corresponds to 1985, and so on. Use this model to give the approximate number of post offices during the given year. (*Source:* U.S. Postal Service, Annual Report of the Postmaster General)
(a) 1985   **(b)** 1987   **(c)** 1990

56. The linear function $f(x) = 1650x + 3817$ provides a model for the United States defense budget for the decade of the 1980s, where $x = 0$ corresponds to 1980, $x = 1$ corresponds to 1981, and so on, with $f(x)$ representing the budget in millions of dollars. Use this model to approximate the defense budget during the given year. (*Source:* U.S. Office of Management and Budget)
(a) 1983   **(b)** 1985   **(c)** 1988

INTERPRETING TECHNOLOGY (EXERCISES 57–62)

---

57. Refer to the linear function in Exercise 55. The graphing calculator screen shows a portion of the graph of $y = f(x)$ with the coordinates of a point on the graph displayed at the bottom of the screen. Interpret the meaning of the display in the context of Exercise 55.

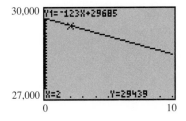

58. Refer to the linear function in Exercise 56. The graphing calculator screen shows a portion of the graph of $y = f(x)$ with the coordinates of a point on the graph displayed at the bottom of the screen. Interpret the meaning of the display in the context of Exercise 56.

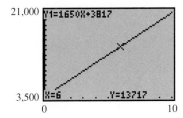

59. The graphing calculator screen shows the graph of a linear function $y = f(x)$, along with the display of coordinates of a point on the graph. Use function notation to write what the display indicates.

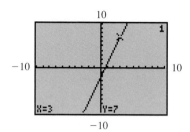

60. The table was generated by a graphing calculator for a linear function $y = f(x)$. Use the table to answer the following questions.
 (a) What is $f(2)$?
 (b) If $f(x) = -3.7$, what is the value of x?
 (c) What is the slope of the line?
 (d) What is the y-intercept of the line?
 (e) Find the expression for $f(x)$.

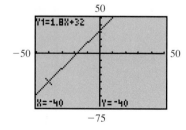

61. The two screens show the graph of the same linear function $y = f(x)$. Find the expression for $f(x)$.

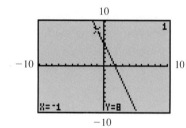 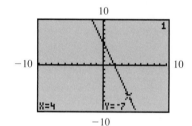

62. The formula for converting Celsius to Fahrenheit is $F = 1.8C + 32$. If we graph $y = f(x) = 1.8x + 32$ on a graphing calculator screen, we obtain the accompanying picture. (We used the interval $[-50, 50]$ on the x-axis and $[-75, 50]$ on the y-axis.) The point $(-40, -40)$ lies on the graph, as indicated by the display. Interpret the meaning of this in the context of this exercise.

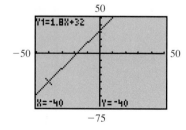

3.6 An Application of Functions; Variation

Certain types of functions are very common, especially in the physical sciences. These are functions where y depends on a multiple of x, or y depends on a number divided by x. In such situations, y is said to *vary directly* as x (in the first case) or *vary inversely* as x (in the second case). For example, the period of a pendulum varies directly as the square root of the length of the pendulum and inversely as the square root of the acceleration due to gravity. In this section we discuss several types of variation.

OBJECTIVE 1 The circumference of a circle is a function of the radius, defined by the formula $C = 2\pi r$, where r is the radius of the circle. As the formula shows, the circumference is always a constant multiple of the radius (C is always found by multiplying r by the constant 2π). Because of this, the circumference is said to *vary directly* as the radius.

> **Direct Variation**
>
> **y varies directly as x** if there exists some constant k such that
>
> $$y = kx.$$

Also, y is said to be **proportional to** x. The number k is called the **constant of variation.** In direct variation, for $k > 0$, as the value of x increases, the value of y also increases. Similarly, as x decreases, y decreases.

The direct variation equation defines a linear function. In applications, functions are often defined by variation equations. For example, if Tom earns $8 per hour, his wages vary directly as, or are proportional to, the number of hours he works. If y represents his total wages and x the number of hours he has worked, then

$$y = 8x.$$

Here k, the constant of variation, is 8. We can see that the constant of variation is the slope of the line $y = 8x$. The graph of a function that is an example of a direct variation will always be a line through the origin.

OBJECTIVE 2 The following examples show how to find the value of the constant k.

EXAMPLE 1   Finding the Constant of Variation

Steven is paid an hourly wage. One week he worked 43 hours and was paid $795.50. Find the constant of variation and the direct variation equation.

Let h represent the number of hours he works and T represent his corresponding pay. Then, T varies directly as h, and

$$T = kh.$$

Here k represents Steven's hourly wage. Since $T = 795.50$ when $h = 43$,

$$795.50 = 43k$$

$$k = 18.50. \quad \text{Use a calculator.}$$

His hourly wage is $18.50. Thus, T and h are related by

$$T = 18.50h.$$

OBJECTIVES

1. Write an equation expressing direct variation.
2. Find the constant of variation, and solve direct variation problems.
3. Solve inverse variation problems.
4. Solve joint variation problems.
5. Solve combined variation problems.

FOR EXTRA HELP

Tutorial   Tape 5   SSM, Sec. 3.6

1. Vicki is paid a daily wage. One month she worked 17 days and earned $1334.50. Find the constant of variation and write a direct variation equation.

E X A M P L E   2   Finding the Constant of Variation

Suppose y varies directly as z, and $y = 50$ when $z = 100$. Find k and the equation relating y and z.

Because y varies directly as z,

$$y = kz,$$

for some constant k. We know that $y = 50$ when $z = 100$. Substituting these values into the equation $y = kz$ gives

$$y = kz$$
$$50 = k \cdot 100. \qquad \text{Let } y = 50 \text{ and } z = 100.$$

Now solve for k.

$$k = \frac{50}{100} = \frac{1}{2}$$

The variables y and z are related by the equation

$$y = \frac{1}{2}z.$$

2. In Example 2, find y for each of the following values of z.

(a) $z = 80$

◀◀ **WORK PROBLEMS 1 AND 2 AT THE SIDE.**

(b) $z = 6$

E X A M P L E   3   Solving a Direct Variation Problem

Power consumption is measured in kilowatt-hours (kwh). The charge to customers varies directly as the number of hours of consumption. If it costs $76.50 to use 850 kwh, how much will 1000 kwh cost?

If c represents the cost and h is the number of kilowatt-hours, then $c = kh$ for some constant k. Because 850 kwh cost $76.50, let $c = 76.50$ and $h = 850$ in the equation $c = kh$ to find k.

$$c = kh$$
$$76.50 = k(850)$$
$$k = \frac{76.50}{850}$$
$$k = .09$$

For 1000 kwh,

$$c = (.09)(1000)$$
$$c = 90.$$

It will cost $90 to use 1000 kilowatts of power.

3. It costs $52 to use 800 kwh of electricity. How much would the following kilowatt-hours cost?

(a) 1000

(b) 650

◀◀ **WORK PROBLEM 3 AT THE SIDE.**

The direct variation equation $y = kx$ defines a linear function. However, other kinds of variation are defined by nonlinear functions. Often, one variable is directly proportional to a *power* of another variable.

> **Direct Variation as a Power**
>
> **y varies directly as the nth power of x** if there exists a real number k such that
>
> $$y = kx^n.$$

An example of direct variation as a power is the formula for the area of a circle, $A = \pi r^2$. Here, π is the constant of variation, and the area varies directly as the square of the radius.

E X A M P L E 4   Solving a Direct Variation Problem

The distance a body falls from rest varies directly as the square of the time it falls (disregarding air resistance). If an object falls 64 feet in 2 seconds, how far will it fall in 8 seconds?

If d represents the distance the object falls, and t the time it takes to fall, then d is a function of t, and

$$d = kt^2$$

for some constant k. To find the value of k, use the fact that the object falls 64 feet in 2 seconds.

$$d = kt^2 \qquad \text{Formula}$$
$$64 = k(2)^2 \qquad \text{Let } d = 64 \text{ and } t = 2.$$
$$k = 16 \qquad \text{Find } k.$$

Using 16 for k, the variation equation becomes

$$d = 16t^2.$$

Now let $t = 8$ to find the number of feet the object will fall in 8 seconds.

$$d = 16(8)^2 \qquad \text{Let } t = 8.$$
$$= 1024$$

The object will fall 1024 feet in 8 seconds.

WORK PROBLEM 4 AT THE SIDE. ▶▶

OBJECTIVE 3▶ Another type of variation is *inverse variation*. If $k > 0$, with inverse variation, as one variable increases, the other decreases.

Inverse Variation

y varies inversely as x if there exists a real number k such that

$$y = \frac{k}{x}.$$

Also, **y varies inversely as the nth power of x** if there exists a real number k such that

$$y = \frac{k}{x^n}.$$

Notice that the inverse variation equations also define functions. Because x is in the denominator, these functions are called *rational functions*.

---

4. The area of a circle varies directly as the square of its radius. A circle with a radius of 3 inches has an area of 28.278 square inches.

(a) Find k and write a variation equation.

(b) What is the area of a circle with a radius of 4.1 inches?

ANSWERS

4. (a) $A = 3.142r^2$
  (b) 52.817 square inches (to the nearest thousandth)

5. If the temperature is constant, the volume of a gas varies inversely as the pressure. For a certain gas, the volume is 10 cubic centimeters when the pressure is 6 kilograms per square centimeter.

(a) Find the variation equation.

(b) Find the volume when the pressure is 12 kilograms per square centimeter.

EXAMPLE 5   Solving an Inverse Variation Problem

The weight of an object above the earth varies inversely as the square of its distance from the center of the earth. A space vehicle in an elliptical orbit has a maximum distance from the center of the earth (apogee) of 6700 miles. Its minimum distance from the center of the earth (perigee) is 4090 miles. See Figure 33. If an astronaut in the vehicle weighs 57 pounds at its apogee, what does the astronaut weigh at its perigee?

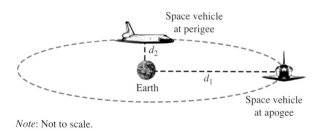

Note: Not to scale.

FIGURE 33

If w is the weight and d is the distance from the center of the earth, then

$$w = \frac{k}{d^2}$$

for some constant k. At the apogee the astronaut weighs 57 pounds, and the distance from the center of the earth is 6700 miles. Use these values to find k.

$$57 = \frac{k}{(6700)^2} \qquad \text{Let } w = 57 \text{ and } d = 6700.$$

$$k = 57(6700)^2$$

Then the weight at the perigee with $d = 4090$ miles is

$$w = \frac{57(6700)^2}{(4090)^2} \approx 153 \text{ pounds.}$$

Note
The approximate answer in Example 5, 153 pounds, was obtained by using a calculator. A calculator will often be helpful in performing operations required in variation problems.

◀◀ WORK PROBLEM 5 AT THE SIDE.

OBJECTIVE 4▶ It is common for the value of one variable to depend on the values of several others. For example, if the value of one variable varies directly as the product of the values of several other variables (perhaps raised to powers), the first variable is said to **vary jointly** as the others. The next example illustrates joint variation.

EXAMPLE 6   Solving a Joint Variation Problem

The strength of a rectangular beam varies jointly as its width and the square of its depth. If the strength of a beam 2 inches wide and 10 inches deep is 1000 pounds per square inch, what is the strength of a beam 4 inches wide and 8 inches deep?

ANSWERS

5. (a) $V = \dfrac{60}{P}$ **(b)** 5 cubic centimeters

CONTINUED ON NEXT PAGE

If S represents the strength, w the width, and d the depth, then,

$$S = kwd^2$$

for some constant k. Because $S = 1000$ if $w = 2$ and $d = 10$,

$$1000 = k(2)(10)^2. \qquad S = 1000, w = 2, \text{ and } d = 10.$$

Solving this equation for k gives

$$1000 = k \cdot 2 \cdot 100$$
$$1000 = 200k,$$

or

$$k = 5,$$

so that

$$S = 5wd^2.$$

Now find S when $w = 4$ and $d = 8$.

$$S = 5(4)(8)^2 \qquad w = 4 \text{ and } d = 8.$$
$$= 1280$$

The strength of the beam is 1280 pounds per square inch.

OBJECTIVE 5 There are many combinations of direct and inverse variation. The final example shows a typical **combined variation** problem.

EXAMPLE 7 Solving a Combined Variation Problem

The body-mass index, or BMI, is used by physicians to assess a person's level of fatness.* The BMI varies directly as an individual's weight in pounds and inversely as the square of the individual's height in inches. A person who weighs 118 pounds and is 64 inches tall has a BMI of 20. (The BMI is rounded to the nearest whole number.) Find the BMI of a person who weighs 165 pounds with a height of 70 inches.

Let B represent the BMI, w the weight, and h the height. Then

$$B = \frac{kw}{h^2}. \quad \begin{array}{l} \leftarrow \text{BMI varies directly as the weight.} \\ \leftarrow \text{BMI varies inversely as the square of the height.} \end{array}$$

To find k, let $B = 20$, $w = 118$, and $h = 64$.

$$20 = \frac{k(118)}{64^2}$$

$$k = \frac{20(64^2)}{118} \qquad \text{Multiply by } 64^2; \text{ divide by 118.}$$

$$= 694 \qquad \text{Use a calculator.}$$

Now find B when $k = 694$, $w = 165$, and $h = 70$.

$$B = \frac{694(165)}{70^2} = 23 \qquad \text{(rounded)}$$

The required BMI is 23. A BMI from 20 through 26 is considered desirable.

WORK PROBLEM 6 AT THE SIDE. ▶▶

---

* From *Reader's Digest*, October 1993.

6. The maximum load that a cylindrical column with a circular cross section can hold varies directly as the fourth power of the diameter of the cross section and inversely as the square of the height. A 9-meter column 1 meter in diameter will support 8 metric tons. How many metric tons can be supported by a column 12 meters high and $\frac{2}{3}$ meter in diameter?

Numbers in the Real World

a graphing calculator minicourse

Lesson 3: Tables

In addition to the obvious capability of graphing functions, graphing calculators are also capable of generating tables of values. For example, suppose that we are interested in finding a table of x- and y-values for the linear equation $2x + y = 6$. We begin by solving the equation for y to get $y = -2x + 6$. This expression for y is then used to define y_1. We can set up the table by directing the calculator to start (TblStart) at 0 and have an increment (ΔTbl) of 1. (See the figure at the left below.) Then the table can be viewed, as seen in the middle figure. We can scroll up or down to find other values automatically. Another option is to have the calculator ask for specific x-values; it will then return the corresponding y-value. (See the figure at the right.) The x-values -100, -45.7, π, 3.33, $\sqrt{23}$, 100, and 10^7 have been entered, and the y-values returned.

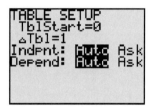

Some graphing calculator models allow the user to see the graph of a function and a table of values on the same screen. Two such screens are seen below for the function $y_1 = -2x + 6$.

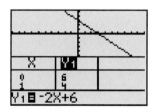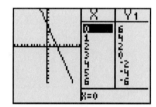

GRAPHING CALCULATOR EXPLORATIONS

1. Use a graphing calculator to generate a table of values for $y = -3x + 7$, starting at $x = 0$ and having increment 1.

2. Use a graphing calculator to generate a table of values for $4x - y = 6$, starting at $x = -5$ and having increment 2. (*Hint:* You will have to solve for y first.)

3. Use the table function of a graphing calculator to find the value of y when $x = 2$ for the equation $3x - 6y = 12$.

3.6 Exercises

Solve each of the following. See Examples 2–6.

1. If x varies directly as y, and $x = 9$ when $y = 3$, find x when $y = 12$.

2. If x varies directly as y, and $x = 10$ when $y = 7$, find y when $x = 50$.

3. If z varies inversely as w, and $z = 10$ when $w = .5$, find z when $w = 8$.

4. If t varies inversely as s, and $t = 3$ when $s = 5$, find s when $t = 5$.

5. Assume p varies jointly as q and r^2, and $p = 200$ when $q = 2$ and $r = 3$. Find p when $q = 5$ and $r = 2$.

6. Assume f varies jointly as g^2 and h, and $f = 50$ when $g = 4$ and $h = 2$. Find f when $g = 3$ and $h = 6$.

7. For $k > 0$, if y varies directly as x, when x increases, y _____ , and when x decreases, y _____ .

8. For $k > 0$, if y varies inversely as x, when x increases, y _____ , and when x decreases, y _____ .

Solve each problem involving variation. See Examples 1–7.

9. Todd bought 8 gallons of gasoline and paid $8.79. To the nearest tenth of a cent, what is the price of gasoline per gallon?

10. Melissa gives horseback rides at Shadow Mountain Ranch. A 2.5-hour ride costs $50.00. What is the price per hour?

11. The weight of an object on Earth is directly proportional to the weight of that same object on the moon. A 200-pound astronaut would weigh 32 pounds on the moon. How much would a 50-pound dog weigh on the moon?

12. In the study of electricity, the resistance of a conductor of uniform cross-sectional area is directly proportional to its length. Suppose that the resistance of a certain type of copper wire is .640 ohm per 1000 feet. What is the resistance of 2500 feet of the wire?

13. The frequency of a vibrating string varies inversely as its length. That is, a longer string vibrates fewer times in a second than a shorter string. Suppose a piano string 2 feet long vibrates 250 cycles per second. What frequency would a string 5 feet long have?

14. The current in a simple electrical circuit is inversely proportional to the resistance. If the current is 20 amperes (an *ampere* is a unit for measuring current) when the resistance is 5 ohms, find the current when the resistance is 7.5 ohms.

15. The illumination produced by a light source varies inversely as the square of the distance from the source. If the illumination produced 4 meters from a light source is 48 footcandles, find the illumination produced 16 meters from the same source.

16. The force with which the earth attracts an object above the earth's surface varies inversely with the square of the object's distance from the center of the earth. If an object 4000 miles from the center of the earth is attracted with a force of 160 pounds, find the force of attraction on an object 6000 miles from the center of the earth.

17. For a given interest rate, simple interest varies jointly as principal and time. If $2000 left in an account for 4 years earned interest of $280, how much interest would be earned in 6 years?

18. The collision impact of an automobile varies jointly as its mass and the square of its speed. Suppose a 2000-pound car traveling at 55 miles per hour has a collision impact of 6.1. What is the collision impact of the same car at 65 miles per hour?

19. The amount of water emptied by a pipe varies directly as the square of the diameter of the pipe. For a certain constant water flow, a pipe emptying into a canal will allow 200 gallons of water to escape in an hour. The diameter of the pipe is 6 inches. How much water would a 12-inch pipe empty into the canal in an hour, assuming the same water flow?

20. The number of long distance phone calls between two cities in a certain time period varies jointly as the populations of the cities, p_1 and p_2, and inversely as the distance between them. If 80,000 calls are made between two cities 400 miles apart, with populations of 70,000 and 100,000, how many calls are made between cities with populations of 50,000 and 75,000 that are 250 miles apart?

21. Ken Griffey, Jr. weighs 205 pounds and is 6 feet, 3 inches tall. Use the information given in Example 7 to find his body-mass index.

22. A body-mass index from 27 through 29 carries a slight risk of weight-related health problems, while one of 30 or more indicates a great increase in risk. Use your own height and weight and the information in Example 7 to determine whether you are at risk.

INTERPRETING TECHNOLOGY (EXERCISES 23–24)

23. The graphing calculator screen shows a portion of the graph of a function $y = f(x)$ that satisfies the conditions for direct variation. What is $f(36)$?

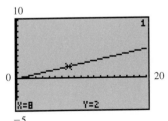

24. The accompanying table of points was generated by a graphing calculator. The points lie on the graph of a function $y_1 = f(x)$ that satisfies the conditions for direct variation. What is $f(36)$?

X	Y1	
0	0	
1	1.5	
2	3	
3	4.5	
4	6	
5	7.5	
6	9	
X=0		

MATHEMATICAL CONNECTIONS (EXERCISES 25–30)

A routine activity such as pumping gasoline can be related to many of the concepts studied in this chapter. Suppose that premium unleaded costs $1.25 per gallon. Work Exercises 25–30 in order.

25. 0 gallons of gasoline cost $0.00, while 1 gallon costs $1.25. Represent these two pieces of information as ordered pairs of the form (gallons, price).

26. Use the information from Exercise 25 to find the slope of the line on which the two points lie.

27. Write the slope-intercept form of the equation of the line on which the two points lie.

28. Using function notation, if $f(x) = ax + b$ represents the line from Exercise 27, what are the values of a and b?

29. How does the value of a from Exercise 28 relate to gasoline in this situation? With relationship to the line, what do we call this number?

30. Why does the equation from Exercise 28 satisfy the conditions for direct variation? In the context of variation, what do we call the value of a?

Chapter 3 Summary

Key Terms

3.1	**line graph, bar graph, circle graph, pie chart**	These are various methods of depicting data using pictorial representations.
	ordered pair	An ordered pair is a pair of numbers written in parentheses in which the order of the numbers matters.
	origin	When two number lines intersect at a right angle, the origin is the common zero point.
	rectangular (Cartesian) coordinate system	Two number lines that intersect at a right angle at their zero points form a rectangular coordinate system, also called the Cartesian system.
	x-axis	The horizontal number line in a rectangular coordinate system is called the x-axis.
	y-axis	The vertical number line in a rectangular coordinate system is called the y-axis.
	plot	To plot an ordered pair is to locate it on a rectangular coordinate system.
	coordinates	The numbers in an ordered pair are called the coordinates of the corresponding point.
	quadrant	A quadrant is one of the four regions in the plane determined by a rectangular coordinate system.
	linear equation in two variables	A first-degree equation with two variables is a linear equation in two variables.
	standard form	A linear equation is in standard form when written as $Ax + By = C$, with $A \geq 0$, and A, B, and C integers.
	x-intercept	The point where a line crosses the x-axis is the x-intercept.
	y-intercept	The point where a line crosses the y-axis is the y-intercept.
3.2	**slope**	The ratio of the change in y compared to the change in x along a line is the slope of the line.
3.4	**linear inequality in two variables**	A first-degree inequality with two variables is a linear inequality in two variables.
	boundary line	In the graph of a linear inequality, the boundary line separates the region that satisfies the inequality from the region that does not satisfy the inequality.
3.5	**dependent variable**	If the quantity y depends on x, then y is called the dependent variable in an equation relating x and y.
	independent variable	If y depends on x, then x is the independent variable in an equation relating x and y.
	relation	A relation is a set of ordered pairs of real numbers.
	function	A function is a set of ordered pairs in which each value of the first component, x, corresponds to exactly one value of the second component, y.
	domain	The domain of a relation is the set of first components (x-values) of the ordered pairs of the relation.
	range	The range of a relation is the set of second components (y-values) of the ordered pairs of the relation.

graph of a relation	The graph of a relation is the graph of the ordered pairs of the relation.
vertical line test	The vertical line test says that if a vertical line cuts the graph of a relation in more than one point, then the relation is not a function.
linear function	A function that can be written in the form $f(x) = mx + b$ is a linear function.

NEW SYMBOLS

(a, b)	ordered pair
x_1	a specific value of the variable x (read "x sub one")
m	slope
$f(x)$	function of x (read "f of x")

QUICK REVIEW

Concepts	*Examples*
3.1 The Rectangular Coordinate System	
Finding Intercepts To find the x-intercept, let $y = 0$. To find the y-intercept, let $x = 0$.	The graph of $2x + 3y = 12$ has x-intercept $(6, 0)$ and y-intercept $(0, 4)$.
3.2 The Slope of a Line	
Slope $$m = \frac{\text{rise}}{\text{run}} = \frac{\text{change in } y}{\text{change in } x} = \frac{y_2 - y_1}{x_2 - x_1}$$	For $2x + 3y = 12$, $$m = \frac{4 - 0}{0 - 6} = -\frac{2}{3}.$$
A vertical line has undefined slope.	$x = 3$ has undefined slope.
A horizontal line has 0 slope.	$y = -5$ has $m = 0$.
Distinct parallel lines have equal slopes.	$\begin{array}{cc} y = 2x + 5 & 4x - 2y = 6 \\ m = 2 & m = 2 \end{array}$ Lines are **parallel**.
The slopes of perpendicular lines are negative reciprocals with a product of -1.	$\begin{array}{cc} y = 3x - 1 & x + 3y = 4 \\ m = 3 & m = -\dfrac{1}{3} \end{array}$ Lines are **perpendicular**.

Concepts	Examples
3.3 Linear Equations in Two Variables	
Standard form $Ax + By = C$	$2x - 5y = 8$
Vertical line $x = k$	$x = -1$
Horizontal line $y = k$	$y = 4$
Slope-intercept form $$y = mx + b$$	$y = 2x + 3$ $m = 2$, y-intercept is $(0, 3)$.
Point-slope form $$y - y_1 = m(x - x_1)$$	$y - 3 = 4(x - 5)$ $(5, 3)$ is on the line, $m = 4$.

3.4 Linear Inequalities in Two Variables	
Graphing a Linear Inequality	
Step 1 Draw the graph of the line that is the boundary. Make the line solid if the inequality involves $\leq$ or $\geq$; make the line dashed if the inequality involves $<$ or $>$.	Graph $2x - 3y \leq 6$. Draw the graph of $2x - 3y = 6$. Use a solid line because the symbol $\leq$ is used.
Step 2 Choose any point not on the line as a test point.	Choose $(1, 2)$. $$2(1) - 3(2) = 2 - 6 \leq 6 \quad \text{True}$$
Step 3 Shade the region that includes the test point if the test point satisfies the original inequality; otherwise, shade the region on the other side of the boundary line.	Shade the side of the line that includes $(1, 2)$. 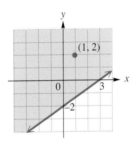

3.5 An Introduction to Functions	
To evaluate a function using function notation (that is, $f(x)$ notation) for a given value of x, substitute the value wherever x appears.	If $f(x) = x^2 - 7x + 12$, then $$f(1) = 1^2 - 7(1) + 12 = 6.$$
To graph the linear function $f(x) = mx + b$, replace $f(x)$ with y and graph $y = mx + b$, as in Section 3.2.	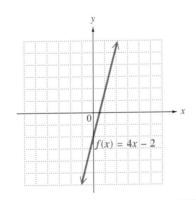

Concepts	Examples
3.6 An Application of Functions; Variation	Area of a circle **varies directly as** the square of the radius.
If there is a constant of variation k such that:	
$y = kx^n$, then y varies directly as, or is proportional to, x^n;	$$A = kr^2$$
$y = \dfrac{k}{x^n}$, then y varies inversely as x^n.	Pressure **varies inversely as** volume.
	$$P = \dfrac{k}{V}$$

CHAPTER 3 REVIEW EXERCISES

[3.1] *Complete the given ordered pairs for each equation, and then graph the equation.*

1. $3x + 2y = 6$
$(0, \underline{}), (\underline{}, 0), (2, \underline{}), (\underline{}, -2)$

2. $x - y = 6$
$(2, \underline{}), (\underline{}, -3), (1, \underline{}), (\underline{}, -2)$

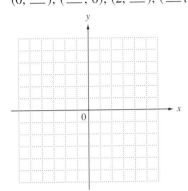

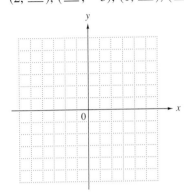

Find the x- and y-intercepts, and graph each of the following equations.

3. $4x + 3y = 12$

4. $5x + 7y = 15$

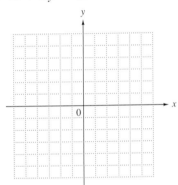

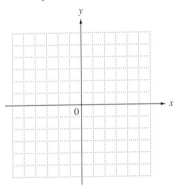

[3.2] *Find the slope for each line in Exercises 5–10.*

5. Through $(-1, 2)$ and $(4, -6)$

6. $y = 2x + 3$

7. $-3x + 4y = 5$

8. $y = 4$

9. A line parallel to
$3y = -2x + 5$

10. A line perpendicular to
$3x - y = 6$

Tell whether the line has positive, negative, zero, or undefined slope.

11.

12.

13.

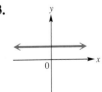

14.

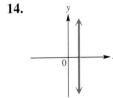

[3.3] *Write an equation for each line. In Exercises 21–24, write the equations in standard form.*

15. Slope $\frac{3}{5}$; y-intercept $(0, -8)$ **16.** Slope $-\frac{1}{3}$; y-intercept $(0, 5)$ **17.** Slope 0; y-intercept $(0, 12)$

18. Undefined slope; through $(2, 7)$ **19.** Horizontal; through $(-1, 4)$ **20.** Vertical; through $(.3, .6)$

21. Through $(2, -5)$ and $(1, 4)$ **22.** Through $(-3, -1)$ and $(2, 6)$

23. Parallel to $4x - y = 3$ and through $(6, -2)$ **24.** Perpendicular to $2x - 5y = 7$ and through $(0, 1)$

Solve each problem.

25. The national average for family health care cost in dollars between 1980 and 2000 (projected) can be approximated by the linear equation

$$y = 382.75x + 1742$$

where $x = 0$ corresponds to 1980 and $x = 20$ corresponds to the year 2000.
(a) What would be the national average for family health care cost in 1999 according to this model?
(b) In what year was the cost $3273?
(c) Graph this linear equation model.

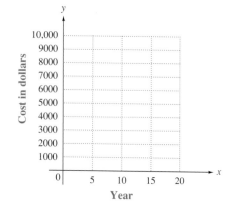

26. For the years 1980 through 1992, the average winning speed in miles per hour in the Indianapolis 500 can be approximated by the linear equation

$$y = 2.503x + 198.729,$$

where $x = 0$ corresponds to 1980, $x = 12$ corresponds to 1992, and y is in miles per hour.
(a) Based on this model, what would have been A1 Unser's winning speed in 1987?
(b) Unser's actual winning speed in 1987 was 215.390 miles per hour. By how much does your answer in part (a) differ from this? Why do you think there is a discrepancy?

[3.4] *Graph the solution of each inequality.*

27. $3x - 2y \leq 12$

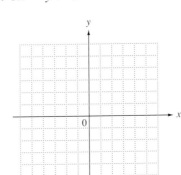

28. $5x - y > 6$

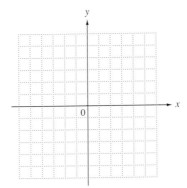

29. $x \geq 2$

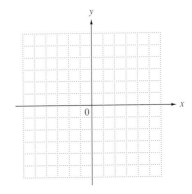

30. $2x + y \leq 1$ and $x \geq 2y$

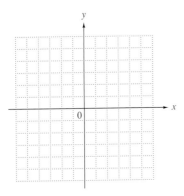

31. $x - 2y < 4$ or $x + y < 3$

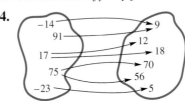

32. $|x - 1| < 4$

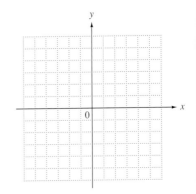

[3.5] *Give the domain and range of each relation. Identify any functions.*

33. $\{(-4, 2), (-4, -2),$
      $(1, 5), (1, -5)\}$

34.

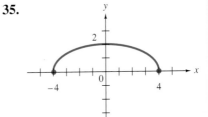

35.

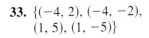

Given $f(x) = -2x^2 + 3x - 6$, *find each of the following.*

36. $f(0)$

37. $f(3)$

38. $f[f(0)]$

39. $f(2p)$

Determine whether the equation defines y as a function of x. Identify any linear functions. Give the domain in each case.

40. $y = 3x - 3$

41. $y < x + 2$

42. $y = |x - 4|$

43. $y = \sqrt{4x + 7}$

44. $x = y^2$

45. $y = \dfrac{7}{x - 36}$

46. Graph the linear function $f(x) = -\dfrac{3}{2}x + \dfrac{7}{2}$.

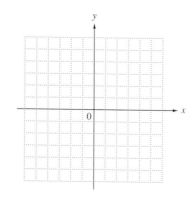

[3.6] *Solve each problem involving variation.*

47. m varies directly as p^2 and inversely as q, and $m = 32$ when $p = 8$ and $q = 10$. Find q when $p = 12$ and $m = 48$.

48. x varies jointly as y and z and inversely as $\sqrt{w}$. If $x = 12$ when $y = 3$, $z = 8$, and $w = 36$, find y when $x = 12$, $z = 4$, and $w = 25$.

49. The resistance in ohms of a platinum wire temperature sensor varies directly as the temperature in *degrees Kelvin* (°K). If the resistance is .646 ohm at a temperature of 190°K, find the resistance at a temperature of 250°K.

50. For the subject in a photograph to appear in the same perspective in the photograph as in real life, the viewing distance must be properly related to the amount of enlargement. For a particular camera, the viewing distance varies directly as the amount of enlargement. A picture taken with this camera that is enlarged 5 times should be viewed from a distance of 250 millimeters. Suppose a print 8.6 times the size of the negative is made. From what distance should it be viewed?

51. A meteorite approaching the earth has velocity inversely proportional to the square root of its distance from the center of the earth. If the velocity is 5 kilometers per second when the distance is 8100 kilometers from the center of the earth, find the velocity at a distance of 6400 kilometers.

52. The period of a pendulum varies directly as the square root of the length of the pendulum and inversely as the square root of the acceleration due to gravity. Find the period when the length is 4 feet and the acceleration due to gravity is 32 feet per second per second, if the period is 1.06π seconds when the length is 9 feet and the acceleration due to gravity is 32 feet per second per second.

CHAPTER 3 TEST

1. Find the slope of the line through $(6, 4)$ and $(-4, -1)$.

1. _____

For each line, find the slope and the x- and y-intercepts.

2. $3x - 2y = 13$

2. _____

3. $y = 5$

3. _____

4. Describe the graph of a line with undefined slope in a rectangular coordinate system.

4. _____

Write the equation of each line in standard form.

5. Through $(-3, 14)$; horizontal

5. _____

6. Through $(4, -1)$; $m = -5$

6. _____

7. Through $(-7, 2)$;
   (a) parallel to $3x + 5y = 6$;
   (b) perpendicular to $y = 2x$

7. (a) _____

   (b) _____

Graph each of the following.

8. $4x - 3y = -12$

8.

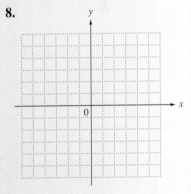

9. $y - 2 = 0$

9.

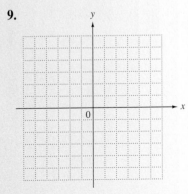

10.

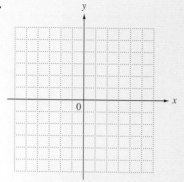

11.

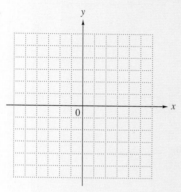

12. _____

13. (a) _____

(b) _____

14. _____

15. _____

16. _____

10. $f(x) = -2x$

11. $3x - 2y > 6$

12. Which one of the following is the graph of a function?

A.

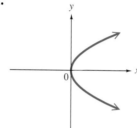

B.

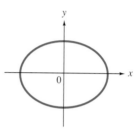

C.

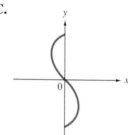

D.

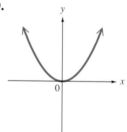

13. For the function $f(x) = \sqrt{x - 3}$,
(a) find $f(7)$, (b) give the domain.

14. The linear function $f(x) = 825x + 8689$ provides a model for the number of cases served by the Child Support Enforcement program from 1985 to 1990, where $x = 0$ corresponds to 1985, $x = 1$ corresponds to 1986, and so on. Use this model to approximate the number of cases served during 1989. (*Source*: Office of Child Support Enforcement)

15. For a body falling freely from rest (disregarding air resistance), the distance the body falls varies directly as the square of the time. If an object is dropped from the top of a tower 576 feet high and hits the ground in 6 seconds, how far did it fall in the first 4 seconds?

16. The force of the wind blowing on a vertical surface varies jointly as the area of the surface and the square of the velocity. If a wind blowing at 40 miles per hour exerts a force of 50 pounds on a surface of $\frac{1}{2}$ square foot, how much force will a wind of 80 miles per hour place on a surface of 2 square feet?

Decide which of the following are true.

1. $5 \cdot 6 \geq |32 - 20|$

2. $5 - |-4| \leq 9$

3. $-4(4 - 8) \geq |-20|$

Perform each operation.

4. $-|-2| - 4 + |-3| + 7$

5. $(-.8)^2$

6. $\sqrt[3]{-64}$

7. $-\dfrac{2}{3}\left(-\dfrac{12}{5}\right)$

Use the properties of real numbers to simplify.

8. $-2(m - 3)$

9. $-(-4m + 3)$

10. $3x^2 - 4x + 4 + 9x - x^2$

Write in interval notation.

11. $\{x \mid x > 2\}$

12. $\{x \mid x \leq 1\}$

13. $\{x \mid -3 < x \leq 5\}$

14. Is $\sqrt{\dfrac{-2 + 4}{-5}}$ a real number?

Evaluate if $p = -4$, $q = -2$, and $r = 5$.

15. $-3(2q - 3p)$

16. $8r^2 + q^2$

17. $|p|^3 - |q^3|$

18. $\dfrac{\sqrt{r}}{-p + 2q}$

19. $\dfrac{5p + 6r^2}{p^2 + q - 1}$

Solve.

20. $2z - 5 + 3z = 4 - (z + 2)$

21. $\dfrac{3a - 1}{5} + \dfrac{a + 2}{2} = -\dfrac{3}{10}$

22. $-\dfrac{4}{3}d \geq -5$

23. $3 - 2(m + 3) < 4m$

24. $2k + 4 < 10$ and $3k - 1 > 5$

25. $2k + 4 > 10$ or $3k - 1 < 5$

26. $|5x + 3| = 13$

27. $|x + 2| < 9$

28. $|2y - 5| \geq 9$

29. Two planes leave the Dallas–Fort Worth airport at the same time. One travels east at 550 miles per hour, and the other travels west at 500 miles per hour. Assuming no wind, how long will it take for the planes to be 2100 miles apart?

30. Ms. Bell must take at least 30 units of a certain medication each day. She can get the medication from white pills or yellow pills, each of which contains 3 units of the drug. To provide other benefits, she needs to take twice as many of the yellow pills as white pills. Find the smallest number of white pills that will satisfy these requirements.

31. Complete the ordered pairs $(0, \ )$, $(\ , 0)$, and $(2, \ )$ for the equation $3x - 4y = 12$.

32. Graph $-4x + 2y = 8$ and give the intercepts.

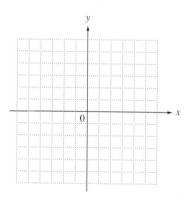

Find the slope of each line described.

33. Through $(-5, 8)$ and $(-1, 2)$

34. Perpendicular to $4x - 3y = 12$

In Exercises 35–37, write an equation in standard form for each line.

35. Slope $-\dfrac{3}{4}$; y-intercept $(0, -1)$

36. Horizontal; through $(2, -2)$

37. Through $(4, -3)$ and $(1, 1)$

38. For the function $f(x) = -4x + 10$,
 (a) what is the domain?
 (b) what is $f(-3)$?

Use the graph to answer the questions in Exercises 39 and 40.

39. What is the slope of the line segment joining the points for 1988 and 1991?

40. Which one of the two line segments shown has a greater slope?

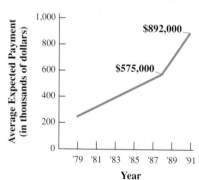

PAYMENT OF CATASTROPHIC CLAIMS

Source: Insurance Research Council

Systems of Linear Equations

4.1 Linear Systems of Equations in Two Variables

Many of the applied problems studied earlier can be solved more easily by writing two or more equations and finding a solution common to these equations. Methods shown in this chapter help to identify numbers that make two or more equations true at the same time. Such a set of equations is called a **system** (SISS-tem) **of equations.**

OBJECTIVE 1 Recall from the previous chapter that the graph of a first-degree equation of the form $Ax + By = C$ is a straight line. For this reason, such an equation is called a linear equation. Two or more linear equations form a **linear system.** An example of a linear system is

$$x + y = 6$$
$$4x - y = 14.$$

The solution set of a linear system of two equations in two variables contains all ordered pairs that satisfy all the equations of the system at the same time.

www.mathnotes.com

OBJECTIVES

1. Decide whether an ordered pair is a solution of a linear system.

2. Solve linear systems by graphing.

3. Solve linear systems with two equations and two unknowns using the elimination method.

4. Solve linear systems with two equations and two unknowns using the substitution method.

FOR EXTRA HELP

Tutorial   Tape 5   SSM, Sec. 4.1

E X A M P L E   I   **Determining Whether an Ordered Pair Is a Solution**

Decide whether the given ordered pair is a solution of the system.

(a) $x + y = 6$ $(4, 2)$
$4x - y = 14$

To make the determination, replace x with 4 and y with 2 in each equation of the system.

$$x + y = 6 \qquad\qquad 4x - y = 14$$
$$4 + 2 = 6 \quad \text{True} \qquad 4(4) - 2 = 14 \quad \text{True}$$

Since $(4, 2)$ makes both equations true, $(4, 2)$ is a solution of the system.

(b) $3x + 2y = 11$ $(-1, 7)$
  $x + 5y = 36$

$$3x + 2y = 11 \qquad\qquad x + 5y = 36$$
$$3(-1) + 2(7) = 11 \quad ? \qquad -1 + 5(7) = 36 \quad ?$$
$$-3 + 14 = 11 \quad \text{True} \qquad -1 + 35 = 36 \quad \text{False}$$

The ordered pair $(-1, 7)$ is not a solution of the system, since it does not make *both* equations true.

1. Are the ordered pairs solutions of the given systems?

  (a) $2x + y = -6$ $(-4, 2)$
     $x + 3y = 2$

  (b) $9x - y = -4$ $(-1, 5)$
     $4x + 3y = 11$

2. Solve each system by graphing.

  (a) $x - y = 3$
     $2x - y = 4$

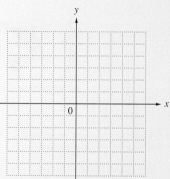

  (b) $2x + y = -5$
     $-x + 3y = 6$

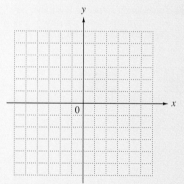

◀◀ **WORK PROBLEM 1 AT THE SIDE.**

OBJECTIVE ▶2 The solution set of a linear system of equations sometimes can be estimated by graphing the equations of the system on the same axes and then estimating the coordinates of any point of intersection.

E X A M P L E 2   Solving a System by Graphing

Solve the following system by graphing.

$$x + y = 5$$
$$2x - y = 4$$

The graphs of these linear equations are shown in Figure 1. The graph suggests that the point of intersection is the ordered pair $(3, 2)$. Check this by substituting these values for x and y in both of the equations. As the check shows, the solution set of the system is $\{(3, 2)\}$.

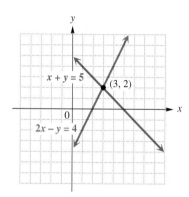

FIGURE 1

◀◀ **WORK PROBLEM 2 AT THE SIDE.**

Since the graph of a linear equation is a straight line, there are three possibilities for the solution of a system of two linear equations, as shown in Figure 2 and described in the box that follows.

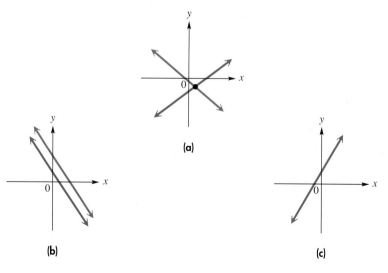

FIGURE 2

ANSWERS
1. (a) yes (b) no
2. (a) $\{(1, -2)\}$

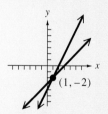

  (b) $\{(-3, 1)\}$

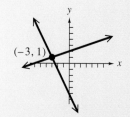

Graphs of a Linear System

1. The two graphs intersect in a single point. The coordinates of this point give the only solution of the system. This is the most common case encountered in basic algebra courses. See Figure 2(a).
2. The graphs are parallel lines. In this case the system is **inconsistent** (IN-kuhn-SISS-tent); that is, there is no solution common to both equations of the system and the solution set is ∅. See Figure 2(b).
3. The graphs are the same line. In this case the equations are **dependent** (DEE-pen-dent), since any solution of one equation of the system is also a solution of the other. The solution set is an infinite set of ordered pairs. See Figure 2(c).

OBJECTIVE 3 ▶ It is possible to find the solution of a system of equations by graphing. However, since it can be hard to read exact coordinates from a graph, an algebraic method is usually used to solve a system. One such algebraic method, called the **elimination method** (or the **addition method**), is explained in the following examples.

┌ **E X A M P L E 3   Solving a System by Elimination**

Solve the system

$$2x + 3y = -6 \tag{1}$$
$$x - 3y = 6. \tag{2}$$

The elimination method involves combining the two equations so that one variable is eliminated. This is done using the following fact.

If $a = b$ and $c = d$, then $a + c = b + d$.

Adding corresponding sides of equations (1) and (2) gives

$$\begin{array}{r} 2x + 3y = -6 \\ x - 3y = 6 \\ \hline 3x = 0, \end{array}$$

and dividing both sides of the equation $3x = 0$ by 3 gives

$$x = 0.$$

To find y, replace x with 0 in either equation (1) or equation (2). Choosing equation (1) gives

$$2x + 3y = -6$$
$$2(0) + 3y = -6 \quad \text{Let } x = 0.$$
$$0 + 3y = -6$$
$$3y = -6$$
$$y = -2.$$

The solution of the system is $x = 0$ and $y = -2$, written as the ordered pair $(0, -2)$. Check this solution by substituting 0 for x and -2 for y in both equations of the original system. The solution set is $\{(0, -2)\}$.

WORK PROBLEM 3 AT THE SIDE. ▶▶

3. Solve each system by elimination.

(a) $3x - y = -7$
    $2x + y = -3$

(b) $2x - 3y = 12$
    $-2x + y = -4$

4. Solve each system.

(a) $x + 3y = 8$
$2x - 5y = -17$

(b) $4x - y = 14$
$3x + 4y = 20$

(c) $2x + 3y = 19$
$3x - 7y = -6$

By adding the equations in Example 3, we eliminated the variable y because the coefficients of y were opposites. In many cases the coefficients will *not* be opposites. In these cases it is necessary to transform one or both equations so that the coefficients of one of the variables are opposites. The general method of solving a system by the elimination method is summarized as follows.

> **Solving Linear Systems of Two Equations by Elimination**
>
> *Step 1* Write both equations in the form $Ax + By = C$.
>
> *Step 2* Multiply one or both equations by appropriate numbers so that the sum of the coefficients of either x or y is zero.
>
> *Step 3* Add the new equations. The sum should be an equation with just one variable.
>
> *Step 4* Solve the equation from Step 3.
>
> *Step 5* Substitute the result of Step 4 into either of the given equations and solve for the other variable.
>
> *Step 6* Check the solution in both of the given equations.

EXAMPLE 4 Solving a System by Elimination

Solve the system

$$5x - 2y = 4 \qquad (3)$$
$$2x + 3y = 13. \qquad (4)$$

Both equations are in the form $Ax + By = C$. Suppose that we wish to eliminate the variable x. One way to do this is to multiply equation (3) by 2 and equation (4) by -5.

$$10x - 4y = 8 \qquad \text{2 times each side of equation (3)}$$
$$-10x - 15y = -65 \qquad \text{-5 times each side of equation (4)}$$

Now add.

$$\begin{array}{r} 10x - 4y = 8 \\ -10x - 15y = -65 \\ \hline -19y = -57 \\ y = 3 \end{array}$$

To find x, substitute 3 for y in either equation (3) or (4). Substituting in equation (4) gives

$$2x + 3y = 13$$
$$2x + 3(3) = 13 \qquad \text{Let } y = 3.$$
$$2x + 9 = 13$$
$$2x = 4 \qquad \text{Subtract 9.}$$
$$x = 2. \qquad \text{Divide by 2.}$$

The solution set of the system is $\{(2, 3)\}$. Check this solution in both equations of the given system.

◀◀ **WORK PROBLEM 4 AT THE SIDE.**

ANSWERS
4. (a) $\{(-1, 3)\}$ **(b)** $\{(4, 2)\}$ **(c)** $\{(5, 3)\}$

─ E X A M P L E 5   **Solving a System
                       of Dependent Equations**

Solve the system

$$2x - y = 3 \qquad \textbf{(5)}$$

$$6x - 3y = 9. \qquad \textbf{(6)}$$

Multiply each side of equation (5) by -3, and then add the result to equation (6).

$$
\begin{array}{ll}
-6x + 3y = -9 & \text{Each side of equation (5) multiplied by } -3 \\
\underline{6x - 3y = 9} & \\
0 = 0 & \text{True}
\end{array}
$$

Adding these equations gave the true statement $0 = 0$. In the original system, we could get equation (6) from equation (5) by multiplying both sides of equation (5) by 3. Because of this, equations (5) and (6) are equivalent and have the same line as the graph shown in Figure 3. The equations are dependent. The solution set is the set of all points on the line with equation $2x - y = 3$. We will write this kind of solution set as $\{(x, y) \mid 2x - y = 3\}$, read "the set of all ordered pairs x, y, such that $2x - y = 3$."

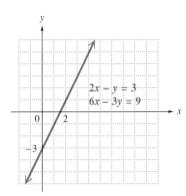

FIGURE 3

WORK PROBLEM 5 AT THE SIDE. ▶▶

─ E X A M P L E 6   **Solving an Inconsistent System**

Solve the system

$$x + 3y = 4 \qquad \textbf{(7)}$$

$$-2x - 6y = 3. \qquad \textbf{(8)}$$

Multiply both sides of equation (7) by 2, and then add the result to equation (8).

$$
\begin{array}{ll}
2x + 6y = 8 & \text{Equation (7) multiplied by 2} \\
\underline{-2x - 6y = 3} & \\
0 = 11 & \text{False}
\end{array}
$$

─ CONTINUED ON NEXT PAGE

5. Solve the system below. Graph both equations.

$$2x + y = 6$$

$$-8x - 4y = -24$$

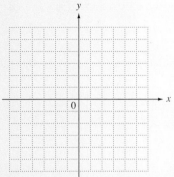

ANSWERS

5. $\{(x, y) \mid 2x + y = 6\}$

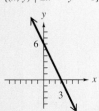

6. Solve the system below. Graph both equations.

$$4x - 3y = 8$$
$$8x - 6y = 14$$

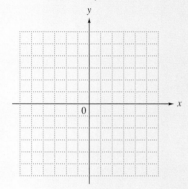

The result of the addition step here is a false statement, which shows that the system is inconsistent. As shown in Figure 4, the graphs of the equations of the system are parallel lines. There are no ordered pairs that satisfy both equations, so there is no solution for the system and the solution set is ∅.

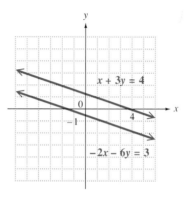

FIGURE 4

The results of Examples 5 and 6 are generalized as follows.

7. Write the equations of Example 5 in slope-intercept form. Use function notation.

Criteria for Special Cases of Linear Systems

If both variables are eliminated when a system of linear equations is solved,

1. there is no solution if the resulting statement is *false*;
2. there are infinitely many solutions if the resulting statement is *true*.

◀◀ WORK PROBLEM 6 AT THE SIDE.

8. Write the equations of Example 6 in slope-intercept form. Use function notation.

Slopes and y-intercepts can be used to decide if the graphs of a system of equations are parallel lines or if they coincide. For the system of Example 5, writing each equation in slope-intercept form shows that both lines have a slope of 2 and a y-intercept of $(0, -3)$, so the graphs are the same line. Such systems have infinite solution sets.

◀◀ WORK PROBLEM 7 AT THE SIDE.

In Example 6, both equations have a slope of $-\frac{1}{3}$, but the y-intercepts are $\left(0, \frac{4}{3}\right)$ and $\left(0, -\frac{1}{2}\right)$, showing that the graphs are two distinct parallel lines. This type of system has ∅ as its solution set.

ANSWERS

6. ∅

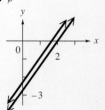

7. Both equations are $f(x) = 2x - 3$.

8. $f(x) = -\frac{1}{3}x + \frac{4}{3}; f(x) = -\frac{1}{3}x - \frac{1}{2}$

◀◀ WORK PROBLEM 8 AT THE SIDE.

OBJECTIVE **4** Linear systems can also be solved using the **substitution method.** The substitution method is most useful in solving linear systems in which one variable has a coefficient of 1. However, as shown in a later chapter, the substitution method is the best choice for solving many *nonlinear* systems.

The method of solving a system by substitution is summarized as follows.

Solving Linear Systems by Substitution

Step 1 Solve one of the equations for either variable. (If one of the variables has coefficient 1 or −1, choose it, since the substitution method is usually easier this way.)

Step 2 Substitute for that variable in the other equation. The result should be an equation with just one variable.

Step 3 Solve the equation from Step 2.

Step 4 Substitute the result from Step 3 into the equation from Step 1 to find the value of the other variable.

Step 5 Check the solution in both of the given equations.

The next two examples illustrate this method.

E X A M P L E   7   Solving a System by Substitution

Solve the system

$$3x + 2y = 13 \tag{9}$$
$$4x - y = -1. \tag{10}$$

To use the substitution method, first solve one of the equations for either x or y. Since the coefficient of y in equation (10) is −1, it is easiest to solve for y in equation (10).

$$-y = -1 - 4x \qquad \text{Subtract } 4x.$$
$$y = 1 + 4x \qquad \text{Multiply by } -1.$$

Substitute $1 + 4x$ for y in equation (9) and solve for x.

$$3x + 2(\mathbf{1 + 4x}) = 13 \qquad \text{Let } y = 1 + 4x.$$
$$3x + 2 + 8x = 13 \qquad \text{Distributive property}$$
$$11x = 11 \qquad \text{Combine terms; subtract 2.}$$
$$x = 1 \qquad \text{Divide by 11.}$$

Since $y = 1 + 4x$,

$$y = 1 + 4(\mathbf{1}) = 5. \qquad \text{Let } x = 1.$$

Check that the solution set is $\{(1, 5)\}$.

WORK PROBLEM 9 AT THE SIDE. ▶▶

If one or more of the equations in the system contains fractions as coefficients, we begin the solution by multiplying both sides by their least common denominator. For example, if the first equation in the system of Example 7 read

$$\frac{x}{2} + \frac{y}{3} = \frac{13}{6},$$

we would multiply both sides by 6.

$$\mathbf{6} \cdot \left(\frac{x}{2} + \frac{y}{3} \right) = \mathbf{6} \cdot \frac{13}{6} \qquad \text{Multiply by the common denominator 6.}$$

$$\mathbf{6} \cdot \frac{x}{2} + \mathbf{6} \cdot \frac{y}{3} = \mathbf{6} \cdot \frac{13}{6} \qquad \text{Distributive property}$$

$$3x + 2y = 13$$

This result is the same as equation (9).

9. Solve by substitution.

$$3x - y = 10$$
$$2x + 5y = 1$$

ANSWERS

9. $\{(3, -1)\}$

10. Solve by substitution.

$$\frac{x}{5} + \frac{2y}{3} = -\frac{8}{5}$$

$$3x - y = 9$$

◀◀ WORK PROBLEM 10 AT THE SIDE.

E X A M P L E 8   Solving a System by Substitution

Solve the system

$$4x - 3y = 7 \tag{11}$$

$$3x - 2y = 6. \tag{12}$$

If the substitution method is to be used, one equation must be solved for one of the two variables. Let us solve equation (12) for x.

$$3x = 2y + 6$$

$$x = \frac{2y + 6}{3}$$

Now substitute $\frac{2y + 6}{3}$ for x in equation (11).

$$4x - 3y = 7 \tag{11}$$

$$4\left(\frac{2y + 6}{3}\right) - 3y = 7$$

Multiply both sides of the equation by 3 to eliminate the fraction.

$4(2y + 6) - 9y = 21$	Multiply by the common denominator, 3.
$8y + 24 - 9y = 21$	Distributive property
$24 - y = 21$	Combine terms.
$-y = -3$	Add -24.
$y = 3$	Divide by -1.

Since $x = \frac{2y + 6}{3}$ and $y = 3$,

$$x = \frac{2(3) + 6}{3} = \frac{6 + 6}{3} = 4.$$

The solution set is $\{(4, 3)\}$.

11. Solve by substitution.

(a) $7x - 2y = -2$

$y = 3x$

(b) $2x - 3y = 1$

$3x = 2y + 9$

◀◀ WORK PROBLEM 11 AT THE SIDE.

The substitution method is not usually the best choice for a system like the one in Example 8. However, it is sometimes necessary when solving a system of *nonlinear* equations to proceed as shown in this example.

ANSWERS

10. $\{(2, -3)\}$

11. (a) $\{(-2, -6)\}$ (b) $\{(5, 3)\}$

4.1 Exercises

Fill in the blanks with the correct responses.

1. If $(3, -6)$ is a solution of a linear system in two variables, then substituting _____ for x and substituting _____ for y leads to a true statement in *both* equations.

2. A solution of a system of linear equations in two variables is a(n) _____ _____ .

3. If the solution process leads to a false statement such as $0 = 5$ when solving a system, the solution set is _____ .

4. If the solution process leads to a true statement such as $0 = 0$ when solving a system, the system has _____ equations.

5. If the two lines forming a system have the same slope and different y-intercepts, the system has _____ solution(s).
   (how many?)

6. If the two lines forming a system have different slopes, the system has _____ solution(s).
   (how many?)

Decide whether the ordered pair is a solution of the given system. See Example 1.

7. $x + y = 6$
   $x - y = 4$ $(5, 1)$

8. $x - y = 17$
   $x + y = -1$ $(8, -9)$

9. $2x - y = 8$
   $3x + 2y = 20$ $(5, 2)$

10. $3x - 5y = -12$
    $x - y = 1$ $(-1, 2)$

Solve the following systems by graphing. See Example 2.

11. $x + y = 4$
     $2x - y = 2$

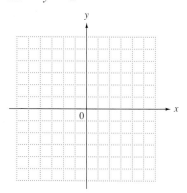

12. $x + y = -5$
     $-2x + y = 1$

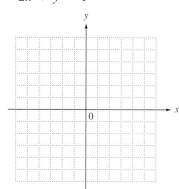

Solve each of the following systems by elimination. See Examples 3–6.

13. $2x - 5y = 11$
     $3x + y = 8$

14. $-2x + 3y = 1$
     $-4x + y = -3$

15. $3x + 4y = -6$
     $5x + 3y = 1$

16. $4x + 3y = 1$
     $3x + 2y = 2$

17. $3x + 3y = 0$
     $4x + 2y = 3$

18. $8x + 4y = 0$
     $4x - 2y = 2$

19. $7x + 2y = 6$
     $-14x - 4y = -12$

20. $x - 4y = 2$
     $4x - 16y = 8$

21. $\dfrac{x}{2} + \dfrac{y}{3} = -\dfrac{1}{3}$

     $\dfrac{x}{2} + 2y = -7$

22. $\dfrac{x}{5} + y = \dfrac{6}{5}$

     $\dfrac{x}{10} + \dfrac{y}{3} = \dfrac{5}{6}$

23. $5x - 5y = 3$
     $x - y = 12$

24. $2x - 3y = 7$
     $-4x + 6y = 14$

Write each equation in slope-intercept form and then tell how many solutions the system has.
Do not actually solve the system.

25. $3x + 7y = 4$
$6x + 14y = 3$

26. $-x + 2y = 8$
$4x - 8y = 1$

27. $2x = -3y + 1$
$6x = -9y + 3$

28. $5x = -2y + 1$
$10x = -4y + 2$

Solve each of the following systems by substitution. See Examples 7 and 8.

29. $4x + y = 6$
$y = 2x$

30. $2x - y = 6$
$y = 5x$

31. $3x - 4y = -22$
$-3x + y = 0$

32. $-3x + y = -5$
$x + 2y = 0$

33. $5x - 4y = 9$
$3 - 2y = -x$

34. $6x - y = -9$
$4 + 7x = -y$

35. $x = 3y + 5$
$x = \dfrac{3}{2}y$

36. $x = 6y - 2$
$x = \dfrac{3}{4}y$

37. $\dfrac{1}{2}x + \dfrac{1}{3}y = 3$
$y = 3x$

38. $\dfrac{1}{4}x - \dfrac{1}{5}y = 9$
$y = 5x$

MATHEMATICAL CONNECTIONS (EXERCISES 39–42)

Work Exercises 39–42 in order to see the connections between systems of linear equations and the graphs of linear functions.

39. Solve the system

$$3x + \ y = 6$$
$$-2x + 3y = 7.$$

Use elimination or substitution.

40. For the first equation in the system of Exercise 39, solve for y and rename it $f(x)$. What special kind of function is f?

41. For the second equation in the system of Exercise 39, solve for y and rename it $g(x)$. What special kind of function is g?

42. Use the result of Exercise 39 to fill in the blanks with the appropriate responses: Because the graphs of f and g are straight lines that are neither parallel nor coincide, they intersect in exactly _____ point. The coordinates of the point are (_____ , _____). Using function notation, this is given by $f(\text{_____}) = \text{_____}$ and $g(\text{_____}) = \text{_____}$.

Answer the questions in Exercises 43 and 44 by observing the graphs provided.

43. Eboni Perkins compared the monthly payments she would incur for two types of mortgages: fixed-rate and variable-rate. Her observations led to the following graphs.

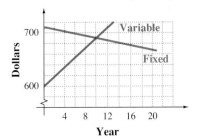

(a) For which years would the monthly payment be more for the fixed-rate mortgage than for the variable-rate mortgage?

(b) In what year would the payments be the same, and what would those payments be?

44. The following figure shows graphs that represent the supply and demand for a certain brand of low-fat frozen yogurt at various prices per half-gallon (in dollars).

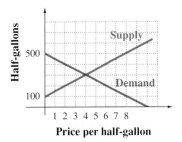

(a) At what price does supply equal demand?

(b) For how many half-gallons does supply equal demand?

(c) What are the supply and the demand at a price of $2 per half-gallon?

45. The accompanying graph shows the trends during the years 1966–1990 relating to bachelor's degrees awarded in the United States. (*Source:* National Science Foundation)

(a) Between what years shown on the horizontal axis did the number of degrees in all fields for men and women reach equal numbers?

(b) When the number of degrees for men and women reached equal numbers, what was that number (approximately)?

46. The accompanying graph shows how the production of vinyl LPs, audiocassettes, and compact discs (CDs) changed over the years from 1983 to 1993. (*Source:* Recording Industry of America)

(a) In what year did cassette production and CD production reach equal levels? What was that level?

(b) Express as an ordered pair of the form (year, production level) the point of intersection of the graphs of LP production and CD production.

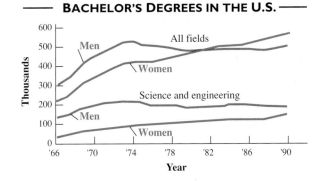

BACHELOR'S DEGREES IN THE U.S.

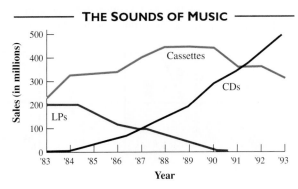

THE SOUNDS OF MUSIC

INTERPRETING TECHNOLOGY (EXERCISES 47–50)

---

47. Which one of the ordered pairs listed could be the only possibility for the solution of the system whose graphs are shown in the standard viewing window of a graphing calculator?

   (a) $(15, -15)$ **(b)** $(15, 15)$

   (c) $(-15, 15)$ **(d)** $(-15, -15)$

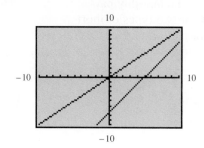

48. The table shown was generated by a graphing calculator. The functions defined by y_1 and y_2 are linear. Based on the table, what are the coordinates of the point of intersection of the graphs?

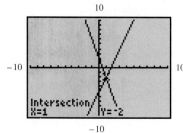

49. The table shown was generated by a graphing calculator. The functions defined by y_1 and y_2 are linear.

   (a) Use the methods of Chapter 3 to find the equation for y_1.

   (b) Use the methods of Chapter 3 to find the equation for y_2.

   (c) Solve the system of equations formed by y_1 and y_2.

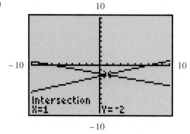

50. The solution set of the system

$$y_1 = 3x - 5$$
$$y_2 = -4x + 2$$

is $\{(1, -2)\}$. Using slopes and y-intercepts, determine which one of the two calculator-generated graphs is the appropriate one for this system.

   (a)   **(b)**

4.2 Linear Systems of Equations in Three Variables

A solution of an equation in three variables, such as $2x + 3y - z = 4$, is called an **ordered triple** and is written (x, y, z). For example, the ordered triple $(1, 1, 1)$ is a solution of the equation, because

$$2(1) + 3(1) - (1) = 2 + 3 - 1 = 4.$$

Verify that another solution of this equation is $(10, -3, 7)$.

In the rest of this chapter, the term *linear equation* is extended to equations of the form $Ax + By + Cz + \cdots + Dw = K$, where not all the coefficients $A, B, C, \ldots, D$ equal zero. For example, $2x + 3y - 5z = 7$ and $x - 2y - z + 3u - 2w = 8$ are linear equations, the first with three variables and the second with five variables.

In this section we discuss the solution of a system of linear equations in three variables such as

$$\begin{aligned} 4x + 8y + z &= 2 \\ x + 7y - 3z &= -14 \\ 2x - 3y + 2z &= 3. \end{aligned}$$

Theoretically, a system of this type can be solved by graphing. However, the graph of a linear equation with three variables is a *plane* and not a line. Since the graph of each equation of the system is a plane, which requires three-dimensional graphing, this method is not practical. However, it does illustrate the number of solutions possible for such systems, as Figure 5 shows.

OBJECTIVES

1 ▶ Solve linear systems with three equations and three unknowns by the elimination method.

2 ▶ Solve linear systems with three equations and three unknowns where some of the equations have missing terms.

3 ▶ Solve linear systems with three equations and three unknowns that are inconsistent or that include dependent equations.

FOR EXTRA HELP

Tutorial   Tape 5   SSM, Sec. 4.2

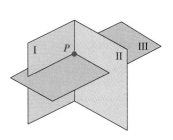

(a) A single solution

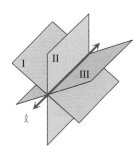

(b) Points of a line in common

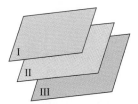

(c) No points in common

(d) All points in common

FIGURE 5

243

Figure 5 illustrates the following cases.

Graphs of Linear Systems in Three Variables

1. The three planes may meet at a single, common point that is the solution of the system. See Figure 5(a).
2. The three planes may have the points of a line in common so that the infinite set of points that satisfy the equation of the line is the solution of the system. See Figure 5(b).
3. The planes may have no points common to all three, in which case there is no solution for the system. See Figure 5(c).
4. The three planes may coincide, in which case the solution of the system is the set of all points on a plane. See Figure 5(d).

OBJECTIVE 1 ▶ Because a graphic solution of a system of three equations in three variables is impractical, these systems can be solved with an extension of the elimination method, summarized as follows.

Solving Linear Systems in Three Variables

Step 1 Use the elimination method to eliminate any variable from any two of the given equations. The result is an equation in two variables.

Step 2 Eliminate the *same* variable from any *other* two equations. The result is an equation in the same two variables as in Step 1.

Step 3 Use the elimination or the substitution method to eliminate a variable from the two equations in two variables that result from Steps 1 and 2. The result is an equation in one variable that gives the value of that variable.

Step 4 Substitute the value of the variable found in Step 3 into either of the equations in two variables to find the value of the second variable.

Step 5 Use the values of the two variables from Steps 3 and 4 to find the value of the third variable by substituting into any of the original equations.

E X A M P L E   1   Solving a System in Three Variables

Solve the system

$$4x + 8y + z = 2 \qquad (1)$$
$$x + 7y - 3z = -14 \qquad (2)$$
$$2x - 3y + 2z = 3. \qquad (3)$$

As before, the elimination method involves eliminating a variable from the sum of two equations. To begin, choose equations (1) and (2) and eliminate z by multiplying both sides of equation (1) by 3 and then adding the result to equation (2).

$$
\begin{array}{rl}
12x + 24y + 3z = 6 & \quad \text{3 times both sides of equation (1)} \\
\underline{x + 7y - 3z = -14} & \quad\quad (2)\\
13x + 31y \phantom{{}- 3z{}} = -8 & \quad\quad (4)
\end{array}
$$

CONTINUED ON NEXT PAGE

Equation (4) has only two variables, x and y. To get another equation without z, multiply both sides of equation (1) by -2 and add the result to equation (3). It is important at this point to eliminate the *same variable, z*.

$$
\begin{array}{ll}
-8x - 16y - 2z = -4 & \quad \text{-2 times both sides of equation (1)} \\
\underline{2x - 3y + 2z = 3} & \quad \text{(3)} \\
-6x - 19y = -1 & \quad \text{(5)}
\end{array}
$$

Now solve the system of equations (4) and (5) for x and y.

WORK PROBLEM 1 AT THE SIDE. ▶▶

As shown in Problem 1 at the side, the solution of the system of equations (4) and (5) is $x = -3$ and $y = 1$. To find z, substitute -3 for x and 1 for y in equation (1). (Any of the three given equations could be used.)

$$
\begin{array}{ll}
4x + 8y + z = 2 & \quad \text{(1)} \\
4(-3) + 8(1) + z = 2 & \\
z = 6 &
\end{array}
$$

The ordered triple $(-3, 1, 6)$ is the only solution of the system. Check that the solution satisfies all three equations of the system so the solution set is $\{(-3, 1, 6)\}$.

WORK PROBLEM 2 AT THE SIDE. ▶▶

OBJECTIVE 2 When one or more of the equations of a system has a missing term, one elimination step can be omitted.

EXAMPLE 2 Solving a System of Equations with Missing Terms

Solve the system

$$
\begin{array}{ll}
6x - 12y = -5 & \quad \text{(6)} \\
8y + z = 0 & \quad \text{(7)} \\
9x - z = 12. & \quad \text{(8)}
\end{array}
$$

Since equation (8) is missing the variable y, one way to begin the solution is to eliminate y again with equations (6) and (7). Multiply both sides of equation (6) by 2 and both sides of equation (7) by 3, and then add.

$$
\begin{array}{ll}
12x - 24y = -10 & \quad \text{2 times both sides of equation (6)} \\
\underline{ 24y + 3z = 0} & \quad \text{3 times both sides of equation (7)} \\
12x + 3z = -10 & \quad \text{(9)}
\end{array}
$$

Use this result, together with equation (8), to eliminate z. Multiply both sides of equation (8) by 3. This gives

$$
\begin{array}{ll}
27x - 3z = 36 & \quad \text{3 times both sides of equation (8)} \\
\underline{12x + 3z = -10} & \quad \text{(9)} \\
39x = 26 &
\end{array}
$$

$$
x = \frac{26}{39} = \frac{2}{3}.
$$

— CONTINUED ON NEXT PAGE

1. Solve the system of equations for x and y.

$$
\begin{array}{l}
13x + 31y = -8 \\
-6x - 19y = -1
\end{array}
$$

2. Solve the system.

$$
\begin{array}{l}
x + y + z = 2 \\
x - y + 2z = 2 \\
-x + 2y - z = 1
\end{array}
$$

ANSWERS
1. $\{(-3, 1)\}$
2. $\{(-1, 1, 2)\}$

3. Solve the system.

$$x - y = 6$$
$$2y + 5z = 1$$
$$3x - 4z = 8$$

4. Solve each system.

(a) $3x - 5y + 2z = 1$
$5x + 8y - z = 4$
$-6x + 10y - 4z = 5$

(b) $7x - 9y + 2z = 0$
$y + z = 0$
$8x - z = 0$

Substitution into equation (8) gives

$$9x - z = 12 \tag{8}$$
$$9\left(\frac{2}{3}\right) - z = 12 \qquad x = \tfrac{2}{3}$$
$$6 - z = 12$$
$$z = -6.$$

Substitution of -6 for z in equation (7) gives

$$8y + z = 0 \tag{7}$$
$$8y - 6 = 0 \qquad z = -6$$
$$8y = 6$$
$$y = \frac{3}{4}.$$

Check in each of the original equations of the system to verify that the solution set of the system is $\{(\tfrac{2}{3}, \tfrac{3}{4}, -6)\}$.

◀◀ **WORK PROBLEM 3 AT THE SIDE.**

OBJECTIVE 3 Linear systems with three variables may be inconsistent or may include dependent equations. The next two examples illustrate these cases.

EXAMPLE 3 **Solving an Inconsistent System with Three Variables**

Solve the following system.

$$2x - 4y + 6z = 5 \tag{10}$$
$$-x + 3y - 2z = -1 \tag{11}$$
$$x - 2y + 3z = 1 \tag{12}$$

Eliminate x by adding equations (11) and (12) to get the equation

$$y + z = 0.$$

Now to eliminate x again, multiply both sides of equation (12) by -2 and add the result to equation (10).

$$-2x + 4y - 6z = -2$$
$$\underline{2x - 4y + 6z = 5}$$
$$ 0 = 3 \qquad \text{False}$$

The resulting false statement indicates that equations (10) and (12) have no common solution; the system is inconsistent and the solution set is $\emptyset$. The graph of the equations of the system would show at least two of the planes parallel to one another.

Note
If you get a false statement from the addition step, as in Example 3, you do not need to go any further with the solution. Since two of the three planes are parallel, it is not possible for the three planes to have any common points.

◀◀ **WORK PROBLEM 4 AT THE SIDE.**

E X A M P L E   4   **Solving a System of Dependent Equations with Three Variables**

Solve the system.

$$2x - 3y + 4z = 8 \tag{13}$$

$$-x + \frac{3}{2}y - 2z = -4 \tag{14}$$

$$6x - 9y + 12z = 24 \tag{15}$$

Multiplying both sides of equation (13) by 3 gives equation (15). Multiplying both sides of equation (14) by -6 also results in equation (15). Because of this, the three equations are dependent. All three equations have the same graph. The solution set is written

$$\{(x, y, z) \mid 2x - 3y + 4z = 8\}.$$

Although we could use any one of the three equations to write the solution set, we prefer to use the equation with coefficients that are integers with no common factor (except 1). This is similar to our choice of a standard form for a linear equation earlier.

WORK PROBLEM 5 AT THE SIDE. ▶▶

We can extend the method discussed in this section to solve larger systems. For example, to solve a system of four equations in four variables, eliminate the same variable from three pairs of equations to get a system of three equations in three unknowns. Then proceed as shown above.

5. Solve the system.

$$x - y + z = 4$$
$$-3x + 3y - 3z = -12$$
$$2x - 2y + 2z = 8$$

NUMBERS IN THE *Real World* a graphing calculator minicourse

Lesson 4: Solving a Linear System of Equations

In Section 4.1 we learned algebraic methods of solving linear systems. Example 4 in that section illustrated how the system

$$5x - 2y = 4$$
$$2x + 3y = 13$$

can be solved by elimination. As shown there, the solution set of the system is $\{(2, 3)\}$, meaning that the graphs of the two lines intersect at the point $(2, 3)$.

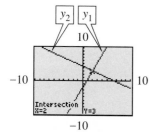

This can be supported with a graphing calculator by graphing the two lines and using the capabilities of the calculator to find the point of intersection. However, we must first solve each equation for y so that the equations can be entered, because this is a requirement for most graphing calculators. If we solve $5x - 2y = 4$ for y, we get $y = \frac{5}{2}x - 2$, and solving $2x + 3y = 13$ for y yields $y = -\frac{2}{3}x + \frac{13}{3}$. If we enter these equations as y_1 and y_2, respectively, we get the graphics shown in the figure at the left. The display at the bottom of the figure indicates that the point of intersection is $(2, 3)$, as determined algebraically.

GRAPHING CALCULATOR EXPLORATIONS

1. The system $4x - 3y = 7$, $3x - 2y = 6$ was solved by substitution in Example 8 of Section 4.1.

   (a) Solve it by elimination, and verify that your solution is the same as shown in Section 4.1.

   (b) Solve $4x - 3y = 7$ for y, and graph it as y_1.

   (c) Solve $3x - 2y = 6$ for y, and graph it as y_2.

   (d) Use your calculator to find the point of intersection of the two lines. It should be the same as the one you found algebraically in part (a).

2. The system $2x - y = 3$, $6x - 3y = 9$ was considered in Example 5 of Section 4.1.

   (a) Solve $2x - y = 3$ for y, and graph it as y_1.

   (b) Solve $6x - 3y = 9$ for y, and graph it as y_2.

   (c) Compare the display on your screen to Figure 3 of Section 4.1, and discuss how the display supports the algebraic conclusion that the system has infinitely many solutions.

3. The system $x + 3y = 4$, $-2x - 6y = 3$ was considered in Example 6 of Section 4.1.

   (a) Solve $x + 3y = 4$ for y, and graph it as y_1.

   (b) Solve $-2x - 6y = 3$ for y, and graph it as y_2.

   (c) Compare the display on your screen to Figure 4 of Section 4.1, and discuss how the display supports the algebraic conclusion that the system has no solutions.

4.2 Exercises

1. Explain what the following statement means: The solution set of the system
$$2x + y + z = 3$$
$$3x - y + z = -2 \quad \text{is } \{(-1, 2, 3)\}.$$
$$4x - y + 2z = 0$$

2. Write a system of three linear equations in three variables that has the solution set $\{(3, 1, 2)\}$. Then solve the system. (*Hint:* Start with the solution and make up three equations that are satisfied by the solution. There are many ways to do this.)

Solve each of the following systems of equations. See Example 1.

3. $3x + 2y + z = 8$
   $2x - 3y + 2z = -16$
   $x + 4y - z = 20$

4. $-3x + y - z = -10$
   $-4x + 2y + 3z = -1$
   $2x + 3y - 2z = -5$

5. $2x + 5y + 2z = 0$
   $4x - 7y - 3z = 1$
   $3x - 8y - 2z = -6$

6. $5x - 2y + 3z = -9$
   $4x + 3y + 5z = 4$
   $2x + 4y - 2z = 14$

7. $x + y - z = -2$
   $2x - y + z = -5$
   $-x + 2y - 3z = -4$

8. $x + 2y + 3z = 1$
   $-x - y + 3z = 2$
   $-6x + y + z = -2$

Solve each system of equations. See Example 2.

9. $2x - 3y + 2z = -1$
$x + 2y + z = 17$
$2y - z = 7$

10. $2x - y + 3z = 6$
$x + 2y - z = 8$
$2y + z = 1$

11. $4x + 2y - 3z = 6$
$x - 4y + z = -4$
$-x + 2z = 2$

12. $2x + 3y - 4z = 4$
$x - 6y + z = -16$
$-x + 3z = 8$

13. $2x + y = 6$
$3y - 2z = -4$
$3x - 5z = -7$

14. $4x - 8y = -7$
$4y + z = 7$
$-8x + z = -4$

15. Using your immediate surroundings, give an example of three planes that
(a) intersect in a single point;
(b) do not intersect;
(c) intersect in infinitely many points.

16. Suppose that a system has infinitely many ordered pair solutions of the form

(x, y, z) such that $x + y + 2z = 1$.

Give three specific ordered pairs that are solutions of the system.

Solve each of the following systems of equations. See Examples 1, 3, and 4.

17. $2x + 2y - 6z = 5$
$-3x + y - z = -2$
$-x - y + 3z = 4$

18. $-2x + 5y + z = -3$
$5x + 14y - z = -11$
$7x + 9y - 2z = -5$

19. $-5x + 5y - 20z = -40$
$\quad\quad x - y + 4z = 8$
$\quad\; 3x - 3y + 12z = 24$

20. $\quad\; x + 4y - z = 3$
$\quad -2x - 8y + 2z = -6$
$\quad\; 3x + 12y - 3z = 9$

21. $2x + y - z = 6$
$\quad 4x + 2y - 2z = 12$
$\quad -x - \dfrac{1}{2}y + \dfrac{1}{2}z = -3$

22. $2x - 8y + 2z = -10$
$\quad -x + 4y - z = 5$
$\quad \dfrac{1}{8}x - \dfrac{1}{2}y + \dfrac{1}{8}z = -\dfrac{5}{8}$

23. $x + y - 2z = 0$
$\quad 3x - y + z = 0$
$\quad 4x + 2y - z = 0$

24. $2x + 3y - z = 0$
$\quad x - 4y + 2z = 0$
$\quad 3x - 5y - z = 0$

─────────────── **MATHEMATICAL CONNECTIONS (EXERCISES 25–30)** ───────────────

Suppose that on a distant planet a function of the form

$$f(x) = ax^2 + bx + c \quad (a \neq 0)$$

describes the height in feet of a projectile x seconds after it has been projected upward. Work through Exercises 25–30 in order to see how this can be related to a system of three equations in three variables a, b, and c.

25. After 1 second, the height of a certain projectile is 128 feet. Thus, $f(1) = 128$. Use this information to find one equation in the variables a, b, and c. (*Hint:* Substitute 1 for x and 128 for $f(x)$.)

26. After 1.5 seconds, the height is 140 feet. Find a second equation in a, b, and c.

27. After 3 seconds, the height is 80 feet. Find a third equation in a, b, and c.

28. Write a system of three equations in a, b, and c, based on your answers in Exercises 25–27. Solve the system.

29. What is the function f for this particular projectile?

30. What was the initial height of the projectile? (*Hint:* Find $f(0)$.)

Systems involving more than three equations can be solved by extending the methods shown in this section. Solve each of the following systems.

31.
$$\begin{aligned}
x + y + z - w &= 5 \\
2x + y - z + w &= 3 \\
x - 2y + 3z + w &= 18 \\
x + y - z - 2w &= -8
\end{aligned}$$

32.
$$\begin{aligned}
3x + y - z + 2w &= 9 \\
x + y + 2z - w &= 10 \\
x - y - z + 3w &= -2 \\
x - y + z - w &= 6
\end{aligned}$$

4.3 Applications of Linear Systems

Many applied problems involve more than one unknown quantity. Although some problems with two unknowns can be solved using just one variable, many times it is easier to use two variables. To solve a problem with two unknowns, we write two equations that relate the unknown quantities. We can then solve the system formed by the pair of equations using the methods given at the beginning of this chapter.

OBJECTIVES

1. Solve problems requiring values of unknown quantities using two variables.

2. Solve money problems using two variables.

3. Solve mixture problems using two variables.

4. Solve problems about distance, rate, and time using two variables.

5. Solve problems with three unknowns using a system of three equations.

FOR EXTRA HELP

Tutorial   Tape 5   SSM, Sec. 4.3

Solving Problems with More than One Variable

Step 1 **Determine what you are to find.** Assign a variable for each unknown and *write down* what it represents.

Step 2 **Write down other information.** If appropriate, draw a figure or a diagram and label it using the variables from Step 1. Use a chart to summarize the information.

Step 3 **Write a system of equations.** Write as many equations as there are unknowns.

Step 4 **Solve the system.**

Step 5 **Answer the question(s).** Be sure you have answered all questions posed.

Step 6 **Check.** Check your solution(s) in the original problem. Be sure your answer makes sense.

OBJECTIVE 1 The next example shows how to write a system of equations to solve a problem that requires finding two unknown values.

EXAMPLE 1   **Solving a Problem to Find Unknown Values**

The length of the foundation of a rectangular house is to be 6 meters more than its width. Find the length and width of the house if the perimeter must be 48 meters.

Begin by sketching a rectangle to represent the foundation of the house. Let

$$x = \text{the length}$$

and

$$y = \text{the width}.$$

See Figure 6.

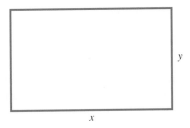

FIGURE 6

The length, x, is 6 meters more than the width, y. Therefore,

$$x = 6 + y.$$

CONTINUED ON NEXT PAGE

1. Solve the system shown in Example 1 to find the width and the length.

The formula for the perimeter of a rectangle is $P = 2L + 2W$. Here $P = 48$, $L = x$, and $W = y$, so

$$48 = 2x + 2y.$$

The length and width can now be found by solving the system

$$x = 6 + y$$
$$48 = 2x + 2y.$$

2. Write a system of equations to solve the following problem. Do not solve it.

The perimeter of a rectangle is 76 inches. If the width were doubled, it would be 13 inches more than the length. Find the width and length.

◀◀ **WORK PROBLEMS 1 AND 2 AT THE SIDE.**

OBJECTIVE ▶ Another type of problem that often leads to a system of equations is one about different amounts of money.

E X A M P L E   2   Solving a Problem about Money

For an art project Kay bought 8 pieces of poster board and 3 marker pens for $6.50. She later needed 2 pieces of poster board and 2 pens. These items cost $3.00. Find the cost of 1 marker pen and 1 sheet of poster board.

Let x represent the cost of a piece of poster board and y represent the cost of a pen. For the first purchase, $8x$ represents the cost of the pieces of poster board and $3y$ the cost of the pens. The total cost was $6.50, so

$$8x + 3y = 6.50.$$

For the second purchase,

$$2x + 2y = 3.00.$$

To solve the system, multiply both sides of the second equation by -4 and add the result to the first equation.

$$
\begin{array}{rcr}
8x + 3y &=& 6.50 \\
-8x - 8y &=& -12.00 \\
\hline
-5y &=& -5.50 \\
y &=& 1.10
\end{array}
$$

3. (a) Jamilla bought 4 pounds of peaches and 2 pounds of apricots, paying $5. Later, she bought 7 pounds of peaches and 3 pounds of apricots for $8.25. Find the cost per pound for each fruit.

By substituting 1.10 for y in either of the equations, verify that $x = .40$. Kay paid $.40 for a piece of poster board and $1.10 for a pen.

> **Note**
> In Example 2, x and y represented costs in *dollars* because the right side of each equation was in dollars, so the left side had to agree. Therefore, $x = .40$ represents $.40, not $.40¢.

(b) A cashier has $1260 in tens and twenties, with a total of 98 bills. How many of each type are there?

◀◀ **WORK PROBLEM 3 AT THE SIDE.**

OBJECTIVE ▶ We solved mixture problems earlier using one variable. For many mixture problems it seems more natural to use more than one variable and a system of equations.

E X A M P L E   3   Solving a Mixture Problem

How many ounces each of 5% hydrochloric acid and 20% hydrochloric acid must be combined to get 10 ounces of solution that is 12.5% hydrochloric acid?

CONTINUED ON NEXT PAGE

Let x represent the number of ounces of 5% solution and y represent the number of ounces of 20% solution. A table summarizes the information from the problem.

Kind of Solution	Ounces of Solution	Ounces of Acid
5%	x	$.05x$
20%	y	$.20y$
12.5%	10	$(.125)10$

When the x ounces of 5% solution and the y ounces of 20% solution are combined, the total number of ounces is 10, so that

$$x + y = 10. \qquad \textbf{(1)}$$

The ounces of acid in the 5% solution $(.05x)$ plus the ounces of acid in the 20% solution $(.20y)$ should equal the total ounces of acid in the mixture, which is $(.125)10$. That is,

$$.05x + .20y = (.125)10. \qquad \textbf{(2)}$$

Eliminate x by first multiplying both sides of equation (2) by 100 to clear it of decimals and then multiplying both sides of equation (1) by -5.

$$
\begin{array}{ll}
5x + 20y = 125 & \text{100 times both sides of equation (2)} \\
\underline{-5x - 5y = -50} & \text{-5 times both sides of equation (1)} \\
 15y = 75 &
\end{array}
$$

$$y = 5$$

Because $y = 5$ and $x + y = 10$, x is also 5, so the desired mixture will require 5 ounces of the 5% solution and 5 ounces of the 20% solution.

■

WORK PROBLEM 4 AT THE SIDE. ▶▶

OBJECTIVE ▶4 Constant rate applications require the distance formula, $d = rt$, where d is distance, r is rate (or speed), and t is time. These applications often lead naturally to a system of equations, as in the next example.

EXAMPLE 4 Solving a Motion Problem

A car travels 250 kilometers in the same time that a truck travels 225 kilometers. If the speed of the car is 8 kilometers per hour faster than the speed of the truck, find both speeds.

As we saw earlier, a table is useful to organize the information in problems about distance, rate, and time. Fill in the given information for each vehicle (in this case, distance) and use variables for the unknown speeds (rates) as follows.

	d	r	t
Car	250	x	
Truck	225	y	

The table shows nothing about time. Get an expression for time by solving the distance formula, $d = rt$, for t.

$$\frac{d}{r} = t$$

The two times can be written as $\frac{250}{x}$ and $\frac{225}{y}$.

CONTINUED ON NEXT PAGE

4. (a) A grocer has some \$4 per pound coffee and some \$8 per pound coffee, which he will mix to make 50 pounds of \$5.60 per pound coffee. How many pounds of each should be used?

(b) Some 40% ethyl alcohol solution is to be mixed with some 80% solution to get 200 liters of a 50% mixture. How many liters of each should be used?

ANSWERS

4. (a) 30 pounds of \$4; 20 pounds of \$8
  (b) 150 liters of 40%; 50 liters of 80%

5. A train travels 600 miles in the same time that a truck travels 520 miles. Find the speed of each vehicle if the train's average speed is 8 miles per hour faster than the truck's.

The problem states that the car travels 8 kilometers per hour faster than the truck. Since the two speeds are x and y,

$$x = y + 8.$$

Both vehicles travel for the same time, so

$$\frac{250}{x} = \frac{225}{y}.$$

This is not a linear equation. However, multiplying both sides by xy gives

$$250y = 225x,$$

which is linear. Now solve the system.

$$x = y + 8 \tag{3}$$
$$250y = 225x \tag{4}$$

The substitution method can be used. Replace x with $y + 8$ in equation (4).

$$250y = 225(y + 8) \qquad \text{Let } x = y + 8.$$
$$250y = 225y + 1800 \qquad \text{Distributive property}$$
$$25y = 1800 \qquad \text{Subtract } 225y.$$
$$y = 72 \qquad \text{Divide by 25.}$$

Because $x = y + 8$, the value of x is $72 + 8 = 80$. It is important to check the solution in the original problem since one of the equations had variable denominators. Checking verifies that the speeds are 80 kilometers per hour for the car and 72 kilometers per hour for the truck.

◄◄ WORK PROBLEM 5 AT THE SIDE.

OBJECTIVE 5 Some applications involve three unknowns. When three variables are used, three equations are necessary to solve the problem. We can then use the methods of the previous section to solve the system. The next two examples illustrate this.

EXAMPLE 5   Solving a Problem about Food Prices

Joe Schwartz bought apples, hamburger, and milk at the grocery store. Apples cost \$.70 a pound, hamburger was \$1.50 a pound, and milk was \$.80 a quart. He bought twice as many pounds of apples as hamburger. The number of quarts of milk was one more than the number of pounds of hamburger. If his total bill was \$8.20, how much of each item did he buy?

First choose variables to represent the three unknowns.

Let x = the number of pounds of apples;
   y = the number of pounds of hamburger;
   z = the number of quarts of milk.

Next, use the information in the problem to write three equations. Since Joe bought twice as many pounds of apples as hamburger,

$$x = 2y$$

or $x - 2y = 0.$

CONTINUED ON NEXT PAGE

ANSWERS

5. The train travels at 60 miles per hour and the truck at 52 miles per hour.

The number of quarts of milk amounted to one more than the number of pounds of hamburger, so

$$z = 1 + y$$

or

$$-y + z = 1.$$

Multiplying the cost of each item by the amount of that item and adding gives the total bill.

$$.70x + 1.50y + .80z = 8.20$$

Multiply both sides of this equation by 10 to clear it of decimals.

$$7x + 15y + 8z = 82$$

Use the method shown in the previous section to solve the system

$$x - 2y \qquad = 0$$
$$-y + z = 1$$
$$7x + 15y + 8z = 82.$$

Verify that the solution is (4, 2, 3). Now go back to the statements defining the variables to decide what the numbers of the solution represent. Doing this shows that Joe bought 4 pounds of apples, 2 pounds of hamburger, and 3 quarts of milk.

WORK PROBLEM 6 AT THE SIDE. ▶▶

Business problems involving production sometimes require the solution of a system of equations. The final example shows how to set up such a system.

E X A M P L E 6   Solving a Business Production Problem

A company produces three color television sets, models X, Y, and Z. Each model X set requires 2 hours of electronics work, 2 hours of assembly time, and 1 hour of finishing time. Each model Y requires 1, 3, and 1 hours of electronics, assembly, and finishing time, respectively. Each model Z requires 3, 2, and 2 hours of the same work, respectively. There are 100 hours available for electronics, 100 hours available for assembly, and 65 hours available for finishing per week. How many of each model should be produced each week if all available time must be used?

Let x = the number of model X produced per week;
    y = the number of model Y produced per week;
    z = the number of model Z produced per week.

A chart is useful for organizing the information in a problem of this type.

	Each Model X	*Each Model Y*	*Each Model Z*	*Totals*
Hours of electronics work	2	1	3	100
Hours of assembly time	2	3	2	100
Hours of finishing time	1	1	2	65

CONTINUED ON NEXT PAGE

6. A department store has three kinds of perfume: cheap, better, and best. It has 10 more bottles of cheap than better, and 3 fewer bottles of the best than better. Each bottle of cheap costs $8, better costs $15, and best costs $32. The total value of all the perfume is $589. How many bottles of each are there?

7. A paper mill makes newsprint, bond, and copy machine paper. Each ton of newsprint requires 3 tons of recycled paper and 1 ton of wood pulp. Each ton of bond requires 2 tons of recycled paper, 4 tons of wood pulp, and 3 tons of rags. A ton of copy machine paper requires 2 tons of recycled paper, 3 tons of wood pulp, and 2 tons of rags. The mill has 4200 tons of recycled paper, 5800 tons of wood pulp, and 3900 tons of rags. How much of each kind of paper can be made from these supplies?

The x model X sets require $2x$ hours of electronics, the y model Y sets require $1y$ (or y) hours of electronics, and the z model Z sets require $3z$ hours of electronics. Since 100 hours are available for electronics,

$$2x + y + 3z = 100.$$

Similarly, from the fact that 100 hours are available for assembly,

$$2x + 3y + 2z = 100,$$

and the fact that 65 hours are available for finishing leads to the equation

$$x + y + 2z = 65.$$

The system

$$2x + y + 3z = 100$$
$$2x + 3y + 2z = 100$$
$$x + y + 2z = 65$$

may be solved to find $x = 15$, $y = 10$, and $z = 20$. The company should produce 15 model X, 10 model Y, and 20 model Z sets per week.

Notice the advantage of setting up the chart in Example 6. By reading across, we can easily determine the coefficients and the constants in the system.

◀◀ **WORK PROBLEM 7 AT THE SIDE.**

4.3 Exercises

For each application in this exercise set, select variables to represent the unknown quantities, write equations using the variables, and solve the resulting systems. The applications in Exercises 1–34 require solving systems with two variables. Exercises 35–42 require solving systems with three variables.

Solve each problem. See Example 1.

1. In a recent year, the number of daily newspapers in Texas was 52 more than the number of daily newspapers in Florida. Together the two states had a total of 134 dailies. How many did each state have? (*Source:* Editor & Publisher International Yearbook)

2. In the United States, the number of nuclear power plants in the South exceeds that of the Midwest by 11. Together these two areas of the country have a total of 73 nuclear power plants. How many plants does each region have? (*Source:* U.S. Energy Information Administration)

3. Find the measures of the angles marked x and y.

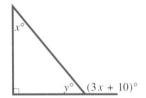

4. Find the measures of the angles marked x and y.

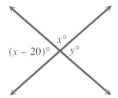

5. Pete and Monica measured the perimeter of a tennis court and found that it was 42 feet longer than it was wide and had a perimeter of 228 feet. What were the length and the width of the tennis court?

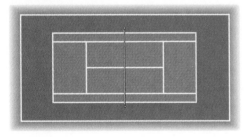

6. Michael and Penny found that the width of their basketball court was 44 feet less than the length. If the perimeter was 288 feet, what were the length and the width of their court?

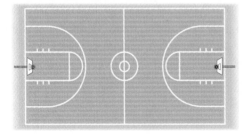

Solve each problem. See Examples 2 and 6.

7. Valencia Community College has decided to supply its mathematics labs with color monitors. A trip to the local electronics outlet provides the following information: 4 CGA monitors and 6 VGA monitors can be purchased for $4600, while 6 CGA monitors and 4 VGA monitors will cost $4400. What are the prices of a single CGA monitor and a single VGA monitor?

8. The Texas Rangers Gift Shop will sell 5 New Era baseball caps and 3 Diamond Collection jerseys for $430, or 3 New Era caps and 4 Diamond Collection jerseys for $500. What are the prices of a single cap and a single jersey?

9. A factory makes use of two basic machines, A and B, which turn out two different products, yarn and thread. Each unit of yarn requires 1 hour on machine A and 2 hours on machine B, while each unit of thread requires 1 hour on A and 1 hour on B. Machine A runs 8 hours per day, while machine B runs 14 hours per day. How many units each of yarn and thread should the factory make to keep its machines running at capacity?

10. The Doors Company makes personal computers. It has found that each standard model requires 4 hours to manufacture electronics and 2 hours for the case. The top-of-the-line model requires 5 hours for the electronics and 1.5 hours for the case. On a particular production run, the company has available 200 hours in the electronics department and 76 hours in the cabinet department. How many of each model can be made?

11. Theodis bought 2 kilograms of dark clay and 3 kilograms of light clay, paying $22 for the clay. He later needed 1 kilogram of dark clay and 2 kilograms of light clay, costing $13 altogether. How much did he pay for each type of clay?

12. Laronda wants to grow two types of algae, green and brown. She has 15 kilograms of nutrient X and 26 kilograms of nutrient Y. A vat of green algae needs 2 kilograms of nutrient X and 3 kilograms of nutrient Y, while a vat of brown algae needs 1 kilogram of nutrient X and 2 kilograms of nutrient Y. How many vats of each type of algae should she grow in order to use all the nutrients?

The formulas $p = br$ (percentage = base $\times$ rate) and $I = prt$ (simple interest = principal $\times$ rate $\times$ time) are used in the applications found in Exercises 17–28. To prepare for the use of these formulas, answer the questions in Exercises 13 and 14.

13. If a container of liquid contains 32 ounces of solution, what is the number of ounces of pure acid if the given solution contains the following acid concentrations?
(a) 10%   (b) 25%   (c) 40%   (d) 50%

14. If $4000 is invested in an account paying simple annual interest, how much interest will be earned during the first year at the following rates?
(a) 2%   (b) 3%   (c) 4%   (d) 3.5%

15. If a pound of oranges costs $.89, how much will x pounds cost?

16. If a ticket to a movie costs $5.50, and y tickets are sold, how much money is collected from the sale?

Solve each problem. See Examples 2 and 3.

17. How many gallons each of 25% alcohol and 35% alcohol should be mixed to get 20 gallons of 32% alcohol?

Kind	Amount	Pure Alcohol
25%	*x*	
35%	*y*	
32%	20	

18. How many liters each of 15% acid and 33% acid should be mixed to get 120 liters of 21% acid?

Kind	Amount	Pure Acid
15%	*x*	
33%	*y*	
21%	120	

19. Pure acid is to be added to a 10% acid solution to obtain 54 liters of a 20% acid solution. What amounts of each should be used?

20. A truck radiator holds 36 liters of fluid. How much pure antifreeze must be added to a mixture that is 4% antifreeze in order to fill the radiator with a mixture that is 20% antifreeze?

21. A party mix is made by adding nuts that sell for $2.50 a kilogram to a cereal mixture that sells for $1 a kilogram. How much of each should be added to get 30 kilograms of a mix that will sell for $1.70 a kilogram?

Ingredient	Amount	Value of the Ingredients
Nuts	*x*	2.50*x*
Cereal	*y*	1.00*y*
Mixed	30	1.70(30)

22. A popular fruit drink is made by mixing fruit juices. Such a mixture with 50% juice is to be mixed with another mixture that is 30% juice to get 200 liters of a mixture that is 45% juice. How much of each should be used?

Ingredient	Amount	Amount of Pure Juice
50% juice	*x*	.50*x*
30% juice	*y*	.30*y*
Mixed	200	.45(200)

23. Tickets to a production of *King Lear* at Delgado Community College cost $5.00 for general admission or $4.00 with a student ID. If 184 people paid to see a performance and $812 was collected, how many of each type of admission were sold?

24. Carol Britz plans to mix pecan clusters that sell for $3.60 per pound with chocolate truffles that sell for $7.20 per pound to get a mixture that she can sell in Valentine boxes for $4.95 per pound. How much of the $3.60 clusters and the $7.20 truffles should she use to create 80 pounds of the mix?

25. Cliff Morris has been saving dimes and quarters. He has 94 coins in all. If the total value is $19.30, how many dimes and how many quarters does he have?

26. A teller at the Bank of New Roads received a checking account deposit in twenty-dollar bills and fifty-dollar bills. She received a total of 70 bills, and the amount of the deposit was $3200. How many of each denomination were deposited?

27. A total of $3000 is invested, part at 2% simple interest and part at 4%. If the total annual return from the two investments is $100, how much is invested at each rate?

Rate	Amount	Interest
2%	x	
4%	y	
Total	$3000	$100

28. An investor must invest a total of $15,000 in two accounts, one paying 4% annual simple interest, and the other 3%. If he wants to earn $550 annual interest, how much should he invest at each rate?

Rate	Amount	Interest
4%	x	
3%	y	
Total	$15,000	$550

The formula d = rt (distance = rate × time) is used in the applications found in Exercises 31–34. To prepare for the use of this formula, answer the questions in Exercises 29 and 30.

29. If the speed of a killer whale is 25 miles per hour, and the whale swims for y hours, how many miles does the whale travel?

30. If the speed of a boat in still water is 20 miles per hour, and the speed of the current of a river is x miles per hour, what is the speed of the boat
(a) going upstream (against the current) and
(b) going downstream (with the current)?

Solve each problem. See Example 4.

31. A freight train and an express train leave towns 390 kilometers apart, traveling toward one another. The freight train travels 30 kilometers per hour slower than the express train. They pass one another 3 hours later. What are their speeds?

32. A train travels 150 kilometers in the same time that a plane covers 400 kilometers. If the speed of the plane is 20 kilometers per hour less than 3 times the speed of the train, find both speeds.

33. Braving blizzard conditions on the planet Hoth, Luke Skywalker sets out at top speed in his snow speeder for a rebel base 3600 miles away. He travels into a steady headwind, and makes the trip in 2 hours. Returning, he finds that the trip back, still at top speed but now with a tailwind, takes only 1.5 hours. Find the top speed of Luke's snow speeder and the speed of the wind.

34. Traveling for three hours into a steady headwind, a plane flies 1650 miles. The pilot determines that flying *with* the same wind for two hours, he could make a trip of 1300 miles. What is the speed of the plane and the speed of the wind?

Solve each problem involving three unknowns. See Examples 5 and 6. (In Exercises 35–38, remember that the sum of the measures of the angles of a triangle is 180°.)

35. In the figure shown, $z = x + 10$ and $x + y = 100$. Determine a third equation involving x, y, and z, and then find the measures of the three angles.

36. In the figure shown, x is 10 less than y and 20 less than z. Write a system of equations, and find the measures of the three angles.

37. In a certain triangle, the measure of the second angle is 10° more than three times the first. The third angle measure is equal to the sum of the measures of the other two. Find the measures of the three angles.

38. The measure of the largest angle of a triangle is 12° less than the sum of the measures of the other two. The smallest angle measures 58° less than the largest. Find the measures of the angles.

39. The perimeter of a triangle is 70 centimeters. The longest side is 4 centimeters less than the sum of the other two sides. Twice the shortest side is 9 centimeters less than the longest side. Find the length of each side of the triangle.

40. The perimeter of a triangle is 56 inches. The longest side measures 4 inches less than the sum of the other two sides. Three times the shortest side is 4 inches more than the longest side. Find the lengths of the three sides.

41. A Mardi Gras trinket manufacturer supplies three wholesalers, A, B, and C. The output from a day's production is 320 cases of trinkets. She must send wholesaler A three times as many cases as she sends B, and she must send wholesaler C 160 cases less than she provides A and B together. How many cases should she send to each wholesaler to distribute the entire day's production to them?

42. A motorcycle manufacturer produces three different models: the Avalon, the Durango, and the Roadripper. Production restrictions require them to make, on a monthly basis, 10 more Roadrippers than the total of the other models, and twice as many Durangos as Avalons. The shop must produce a total of 490 cycles per month. How many cycles of each type should be made per month?

MATHEMATICAL CONNECTIONS (EXERCISES 43–47)

Thus far in this text we have studied only linear *equations. In later chapters we will study the graphs of other kinds of equations. One such graph is a* circle. *We will see that an equation of the form*

$$x^2 + y^2 + ax + by + c = 0$$

may have a circle as its graph. It is a fact from geometry that given three noncollinear points (that is, points that do not all lie on the same straight line), there will be a circle that contains them. For example, the points $(4, 2)$, $(-5, -2)$, *and* $(0, 3)$ *lie on the circle whose equation is*

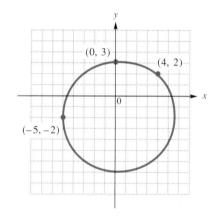

$$x^2 + y^2 - \frac{7}{5}x + \frac{27}{5}y - \frac{126}{5} = 0.$$

The circle is shown in the figure.

Work Exercises 43–47 in order, so that the equation of the circle passing through the points $(2, 1)$, $(-1, 0)$, *and* $(3, 3)$ *can be found.*

43. Let $x = 2$ and $y = 1$ in the equation $x^2 + y^2 + ax + by + c = 0$ to find an equation in a, b, and c.

44. Let $x = -1$ and $y = 0$ to find a second equation in a, b, and c.

45. Let $x = 3$ and $y = 3$ to find a third equation in a, b, and c.

46. Solve the system of equations formed by your answers in Exercises 43–45 to find the values of a, b, and c. What is the equation of the circle?

47. Explain why the relation whose graph is a circle is not a function.

4.4 Determinants

The topics of *matrices* and *determinants* are studied in many branches of mathematics. While matrices are important mathematical entities in their own right, we will not study their applications in this text. However, we will examine the uses of determinants.

An ordered array of numbers within square brackets, such as

$$\text{Rows} \left[\begin{array}{ccc} 2 & 3 & 5 \\ 7 & 1 & 2 \end{array} \right] \quad \text{(Columns)}$$

is called a **matrix** (MAY-triks). Matrices (the plural of matrix) are named according to the number of horizontal rows and vertical columns they contain. The number of rows is given first and the number of columns second, so that the matrix above is "two by three," written 2×3. A matrix with the same number of rows as columns, such as

$$\left[\begin{array}{cc} -1 & 0 \\ 1 & -2 \end{array} \right] \quad \text{and} \quad \left[\begin{array}{ccc} 8 & -1 & -3 \\ 2 & 1 & 6 \\ 0 & 5 & -3 \end{array} \right],$$

is called a **square matrix.**

Every square matrix is associated with a real number called its **determinant** (dee-TERM-in-ent). A determinant is symbolized by the entries of the matrix placed between two vertical lines, such as

$$\left| \begin{array}{cc} 2 & 3 \\ 7 & 1 \end{array} \right| \quad \text{or} \quad \left| \begin{array}{ccc} 7 & 4 & 3 \\ 0 & 1 & 5 \\ -6 & 0 & -1 \end{array} \right|.$$

Determinants are also named according to the number of rows and columns they contain. For example, the first determinant above is a 2×2 (read "two by two") determinant and the second is a 3×3 determinant.

OBJECTIVE 1 The value of the 2×2 determinant

$$\left| \begin{array}{cc} a & b \\ c & d \end{array} \right|$$

is defined as follows.

Value of a 2 × 2 Determinant

$$\left| \begin{array}{cc} a & b \\ c & d \end{array} \right| = ad - bc.$$

EXAMPLE 1 Evaluating a 2 × 2 Determinant

Evaluate the determinant.

$$\left| \begin{array}{cc} -1 & -3 \\ 4 & -2 \end{array} \right|$$

Here $a = -1$, $b = -3$, $c = 4$, and $d = -2$. Using these values, the determinant equals

$$\left| \begin{array}{cc} -1 & -3 \\ 4 & -2 \end{array} \right| = (-1)(-2) - (-3)(4)$$

$$= 2 + 12 = 14.$$

1. Evaluate each determinant.

(a) $\begin{vmatrix} -4 & 6 \\ 2 & 3 \end{vmatrix}$

◀◀ **WORK PROBLEM I AT THE SIDE.**

A 3 × 3 determinant can be evaluated in a similar way.

Value of a 3 × 3 Determinant

$$\begin{vmatrix} a_1 & b_1 & c_1 \\ a_2 & b_2 & c_2 \\ a_3 & b_3 & c_3 \end{vmatrix} = (a_1 b_2 c_3 + b_1 c_2 a_3 + c_1 a_2 b_3) \\ - (a_3 b_2 c_1 + b_3 c_2 a_1 + c_3 a_2 b_1)$$

This rule for evaluating a 3 × 3 determinant is difficult to remember. A method for calculating a 3 × 3 determinant that is easier to use is based on the rule above. Rearranging terms and using the distributive property gives

$$\begin{vmatrix} a_1 & b_1 & c_1 \\ a_2 & b_2 & c_2 \\ a_3 & b_3 & c_3 \end{vmatrix} = a_1(b_2 c_3 - b_3 c_2) - a_2(b_1 c_3 - b_3 c_1) \\ + a_3(b_1 c_2 - b_2 c_1). \quad \textbf{(1)}$$

Each of the quantities in parentheses represents a 2 × 2 determinant that is the part of the 3 × 3 determinant remaining when the row and column of the multiplier are eliminated, as shown below.

(b) $\begin{vmatrix} 3 & -1 \\ 0 & 2 \end{vmatrix}$

$$a_1(b_2 c_3 - b_3 c_2) \qquad \begin{vmatrix} a_1 & b_1 & c_1 \\ a_2 & b_2 & c_2 \\ a_3 & b_3 & c_3 \end{vmatrix}$$

$$a_2(b_1 c_3 - b_3 c_1) \qquad \begin{vmatrix} a_1 & b_1 & c_1 \\ a_2 & b_2 & c_2 \\ a_3 & b_3 & c_3 \end{vmatrix}$$

$$a_3(b_1 c_2 - b_2 c_1) \qquad \begin{vmatrix} a_1 & b_1 & c_1 \\ a_2 & b_2 & c_2 \\ a_3 & b_3 & c_3 \end{vmatrix}$$

These 2 × 2 determinants are called **minors** of the elements in the 3 × 3 determinant. In the determinant above, the minors of a_1, a_2, and a_3 are, respectively,

$$\begin{vmatrix} b_2 & c_2 \\ b_3 & c_3 \end{vmatrix}, \quad \begin{vmatrix} b_1 & c_1 \\ b_3 & c_3 \end{vmatrix}, \quad \begin{vmatrix} b_1 & c_1 \\ b_2 & c_2 \end{vmatrix}.$$

(c) $\begin{vmatrix} -2 & 5 \\ 1 & 5 \end{vmatrix}$

OBJECTIVE 2 ▶ A 3 × 3 determinant can be evaluated by multiplying each element in the first column by its minor and combining the products as indicated in equation (1). This is called the **expansion of the determinant by minors** (ex-PAN-shun of the dee-TERM-in-ent by MY-ners) about the first column.

E X A M P L E   2   Evaluating a 3 × 3 Determinant

Evaluate the determinant using expansion by minors about the first column.

$$\begin{vmatrix} 1 & 3 & -2 \\ -1 & -2 & -3 \\ 1 & 1 & 2 \end{vmatrix}$$

In this determinant, $a_1 = 1$, $a_2 = -1$, and $a_3 = 1$. Multiply each of these numbers by its minor, and combine the three terms using the definition. Notice that the second term in the definition is *subtracted*.

ANSWERS

1. (a) -24 **(b)** 6 **(c)** -15

CONTINUED ON NEXT PAGE

$$\begin{vmatrix} 1 & 3 & -2 \\ -1 & -2 & -3 \\ 1 & 1 & 2 \end{vmatrix} = 1 \begin{vmatrix} -2 & -3 \\ 1 & 2 \end{vmatrix} - (-1) \begin{vmatrix} 3 & -2 \\ 1 & 2 \end{vmatrix} + 1 \begin{vmatrix} 3 & -2 \\ -2 & -3 \end{vmatrix}$$

$$= 1[(-2)(2) - (-3)(1)] + 1[(3)(2) - (-2)(1)]$$
$$+ 1[(3)(-3) - (-2)(-2)]$$
$$= 1(-1) + 1(8) + 1(-13)$$
$$= -1 + 8 - 13$$
$$= -6$$

WORK PROBLEM 2 AT THE SIDE. ▶▶

OBJECTIVE **3** ▶ To get equation (1) we could have rearranged terms in the definition of the determinant and used the distributive property to factor out the three elements of the second or third columns or of any of the three rows. Therefore, expanding by minors about any row or any column results in the same value for a 3 × 3 determinant. To determine the correct signs for the terms of other expansions, the following **array of signs** is helpful.

Array of Signs for a 3 × 3 Determinant

$$\begin{matrix} + & - & + \\ - & + & - \\ + & - & + \end{matrix}$$

The signs alternate for each row and column beginning with a + in the first row, first column position. For example, if the expansion is to be about the second column, the first term would have a minus sign associated with it, the second term a plus sign, and the third term a minus sign.

E X A M P L E   3   Evaluating a 3 × 3 Determinant

Evaluate the determinant of Example 2 using expansion by minors about the second column.

$$\begin{vmatrix} 1 & 3 & -2 \\ -1 & -2 & -3 \\ 1 & 1 & 2 \end{vmatrix}$$

$$= -3 \begin{vmatrix} -1 & -3 \\ 1 & 2 \end{vmatrix} + (-2) \begin{vmatrix} 1 & -2 \\ 1 & 2 \end{vmatrix} - 1 \begin{vmatrix} 1 & -2 \\ -1 & -3 \end{vmatrix}$$

$$= -3(1) - 2(4) - 1(-5)$$
$$= -3 - 8 + 5$$
$$= -6$$

As expected, the result is the same as in Example 2.

WORK PROBLEM 3 AT THE SIDE. ▶▶

2. Expand by minors about the first column.

(a) $\begin{vmatrix} 0 & -1 & 0 \\ 2 & 4 & 2 \\ 3 & 1 & 5 \end{vmatrix}$

(b) $\begin{vmatrix} 2 & 1 & 4 \\ -3 & 0 & 2 \\ -2 & 1 & 5 \end{vmatrix}$

3. Evaluate each determinant using expansion by minors about the second column.

(a) $\begin{vmatrix} 2 & 1 & 3 \\ -1 & 0 & 4 \\ 2 & 4 & 3 \end{vmatrix}$

(b) $\begin{vmatrix} 5 & -1 & 2 \\ 0 & 4 & 3 \\ -1 & 2 & 0 \end{vmatrix}$

ANSWERS

2. (a) 4 (b) −5
3. (a) −33 (b) −19

4. Evaluate.

$$\begin{vmatrix} 1 & 0 & 2 & 0 \\ 3 & 0 & 0 & 4 \\ 0 & -1 & 1 & 0 \\ 2 & 0 & -1 & 0 \end{vmatrix}$$

OBJECTIVE 4 ▶ The method of expansion by minors can be extended to evaluate larger determinants, such as 4 × 4 or 5 × 5. For a larger determinant, the sign array also is extended. For example, the signs for a 4 × 4 determinant are arranged as follows.

Array of Signs for a 4 × 4 Determinant

$$\begin{array}{cccc} + & - & + & - \\ - & + & - & + \\ + & - & + & - \\ - & + & - & + \end{array}$$

E X A M P L E   4   Evaluating a 4 × 4 Determinant

Evaluate the determinant below.

$$\begin{vmatrix} -1 & -2 & 3 & 2 \\ 0 & 1 & 4 & -2 \\ 3 & -1 & 4 & 0 \\ 2 & 1 & 0 & 3 \end{vmatrix}$$

The work can be reduced by choosing a row or column with zeros, say the fourth row. Expand by minors about the fourth row using the elements of the fourth row and the signs from the fourth row of the sign array, as shown below. The minors are 3 × 3 determinants.

$$\begin{vmatrix} -1 & -2 & 3 & 2 \\ 0 & 1 & 4 & -2 \\ 3 & -1 & 4 & 0 \\ 2 & 1 & 0 & 3 \end{vmatrix} = -2\begin{vmatrix} -2 & 3 & 2 \\ 1 & 4 & -2 \\ -1 & 4 & 0 \end{vmatrix} + 1\begin{vmatrix} -1 & 3 & 2 \\ 0 & 4 & -2 \\ 3 & 4 & 0 \end{vmatrix}$$

$$- 0\begin{vmatrix} -1 & -2 & 2 \\ 0 & 1 & -2 \\ 3 & -1 & 0 \end{vmatrix} + 3\begin{vmatrix} -1 & -2 & 3 \\ 0 & 1 & 4 \\ 3 & -1 & 4 \end{vmatrix}$$

Now evaluate each 3 × 3 determinant.

$$= -2(6) + 1(-50) - 0 + 3(-41)$$
$$= -185$$

◀◀ **WORK PROBLEM 4 AT THE SIDE.**

Each of the four 3 × 3 determinants in Example 4 is evaluated by expansion of three 2 × 2 minors. Thus, a great deal of work is needed to evaluate a 4 × 4 or larger determinant. However, such large determinants can be evaluated quickly with the aid of a computer. Many graphing calculators also have this capability.

ANSWERS

4. 20

4.4 Exercises

Decide whether the statement is true or false.

1. A matrix is an array of numbers, while a determinant is just a number.

2. A square matrix has the same number of rows as columns.

3. The determinant $\begin{vmatrix} a & b \\ c & d \end{vmatrix}$ is equal to $ad - bc$.

4. The value of $\begin{vmatrix} 0 & 0 \\ x & y \end{vmatrix}$ is zero for any replacements for x and y.

5. The value of $\begin{vmatrix} a & a \\ a & a \end{vmatrix}$ is a for any replacement for a.

6. $\begin{vmatrix} a & b \\ c & d \end{vmatrix} = \begin{vmatrix} b & a \\ c & d \end{vmatrix}$ for any replacements for the variables.

Evaluate the following determinants. See Example 1.

7. $\begin{vmatrix} -2 & 5 \\ -1 & 4 \end{vmatrix}$

8. $\begin{vmatrix} 3 & -6 \\ 2 & -2 \end{vmatrix}$

9. $\begin{vmatrix} 1 & -2 \\ 7 & 0 \end{vmatrix}$

10. $\begin{vmatrix} -5 & -1 \\ 1 & 0 \end{vmatrix}$

11. $\begin{vmatrix} 0 & 4 \\ 0 & 4 \end{vmatrix}$

12. $\begin{vmatrix} 8 & -3 \\ 0 & 0 \end{vmatrix}$

Evaluate the following determinants using expansion by minors about the first column. See Example 2.

13. $\begin{vmatrix} -1 & 2 & 4 \\ -3 & -2 & -3 \\ 2 & -1 & 5 \end{vmatrix}$

14. $\begin{vmatrix} 2 & -3 & -5 \\ 1 & 2 & 2 \\ 5 & 3 & -1 \end{vmatrix}$

15. $\begin{vmatrix} 1 & 0 & -2 \\ 0 & 2 & 3 \\ 1 & 0 & 5 \end{vmatrix}$

16. $\begin{vmatrix} 2 & -1 & 0 \\ 0 & -1 & 1 \\ 1 & 2 & 0 \end{vmatrix}$

17. $\begin{vmatrix} 1 & 0 & 0 \\ 0 & 1 & 0 \\ 0 & 0 & 1 \end{vmatrix}$

18. $\begin{vmatrix} 0 & 0 & 1 \\ 0 & 1 & 0 \\ 1 & 0 & 0 \end{vmatrix}$

Evaluate the following determinants by expansion about any row or column. (Hint: If possible, choose a row or column with zeros.) See Example 3.

19. $\begin{vmatrix} 4 & 4 & 2 \\ 1 & -1 & -2 \\ 1 & 0 & 2 \end{vmatrix}$

20. $\begin{vmatrix} 3 & -1 & 2 \\ 1 & 5 & -2 \\ 0 & 2 & 0 \end{vmatrix}$

21. $\begin{vmatrix} 2 & 0 & 1 \\ -1 & 0 & 2 \\ 5 & 0 & 4 \end{vmatrix}$

22. $\begin{vmatrix} 2 & -4 & 0 \\ 3 & -5 & 0 \\ 6 & -7 & 0 \end{vmatrix}$

23. $\begin{vmatrix} -6 & 3 & 5 \\ -3 & 2 & 2 \\ 0 & 0 & 0 \end{vmatrix}$

24. $\begin{vmatrix} 0 & 0 & 0 \\ 4 & 0 & -2 \\ 2 & -1 & 3 \end{vmatrix}$

25. $\begin{vmatrix} 3 & 5 & -2 \\ 1 & -4 & 1 \\ 3 & 1 & -2 \end{vmatrix}$

26. $\begin{vmatrix} 1 & 3 & 2 \\ 3 & -1 & -2 \\ 1 & 10 & 20 \end{vmatrix}$

27. $\begin{vmatrix} 1 & 3 & -2 \\ -1 & 4 & 5 \\ 2 & 6 & -4 \end{vmatrix}$

Evaluate the following. Expand by minors about the second row. See Example 4.

28. $\begin{vmatrix} 1 & 4 & 2 & 0 \\ 0 & 2 & 0 & -1 \\ 3 & -1 & 2 & 0 \\ 1 & 4 & -1 & 3 \end{vmatrix}$

29. $\begin{vmatrix} 4 & 1 & 0 & 2 \\ 1 & 0 & 0 & -2 \\ 3 & 4 & 1 & -3 \\ -2 & 1 & 1 & -1 \end{vmatrix}$

30.
$$\begin{vmatrix} 3 & -5 & 1 & 9 \\ 0 & 5 & 2 & 0 \\ 2 & -1 & -1 & -1 \\ -4 & 2 & 2 & 2 \end{vmatrix}$$

31.
$$\begin{vmatrix} 4 & 2 & 2 & 2 \\ 1 & -1 & 0 & 0 \\ 2 & 1 & 1 & 1 \\ 0 & 0 & -3 & -2 \end{vmatrix}$$

32. Consider the following statement: For every square matrix, there is one and only one determinant associated with the matrix.

Explain how this statement illustrates the concept of function.

--- **MATHEMATICAL CONNECTIONS (EXERCISES 33–36)** ---

Recall the formula for slope and the point-slope form of the equation of a line, as found in Chapter 3. Use these formulas in working Exercises 33–36 in order, so that you can see how a determinant can be used in writing the equation of a line.

33. Write the expression for the slope of a line passing through the points (x_1, y_1) and (x_2, y_2).

34. Using the expression from Exercise 33 as m, and the point (x_1, y_1), write the point-slope form of the equation of the line.

35. Using the equation obtained in Exercise 34, multiply both sides by $x_2 - x_1$, and write the equation so that 0 is on the right side.

36. Consider the *determinant equation*

$$\begin{vmatrix} x & y & 1 \\ x_1 & y_1 & 1 \\ x_2 & y_2 & 1 \end{vmatrix} = 0.$$

Expand by minors on the left and show that this determinant equation yields the same result that you obtained in Exercise 35.

INTERPRETING TECHNOLOGY (EXERCISES 37–40)

In Example 2 we showed how to evaluate the determinant

$$\begin{vmatrix} 1 & 3 & -2 \\ -1 & -2 & -3 \\ 1 & 1 & 2 \end{vmatrix}$$

by expanding by minors. Modern graphing calculators have the capability of finding determinants at the stroke of a key. The display on the left is a graphing calculator-generated depiction of the matrix with the same entries as shown in Example 2, and the display on the right shows that its determinant is indeed −6.

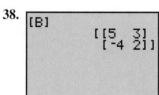

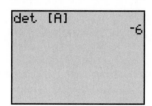

Predict the display that the calculator would give if the determinant of the matrix shown were evaluated.

37.

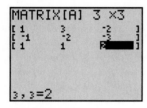

38.

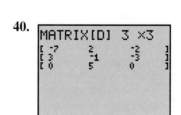

39.

40.

4.5 Solution of Linear Systems of Equations by Determinants—Cramer's Rule

In the previous section we discussed how to evaluate determinants. In this section we will show how determinants can be used to solve systems of equations. They will be important in the development of Cramer's rule, stated later in this section.

OBJECTIVE 1 ▶ We begin by using the elimination method to solve the general system of two equations in two variables,

$$a_1x + b_1y = c_1 \qquad (1)$$
$$a_2x + b_2y = c_2. \qquad (2)$$

The result will be a formula that can be used directly for any system of two equations with two unknowns. To get this general solution, eliminate y and solve for x by first multiplying both sides of equation (1) by b_2 and both sides of equation (2) by $-b_1$. Add these results and solve for x.

$$
\begin{array}{ll}
a_1b_2x + b_1b_2y = c_1b_2 & b_2 \text{ times both sides of equation (1)} \\
\underline{-a_2b_1x - b_1b_2y = -c_2b_1} & -b_1 \text{ times both sides of equation (2)} \\
(a_1b_2 - a_2b_1)x = c_1b_2 - c_2b_1 &
\end{array}
$$

$$x = \frac{c_1b_2 - c_2b_1}{a_1b_2 - a_2b_1}, \quad a_1b_2 - a_2b_1 \neq 0$$

(What happens if $a_1b_2 - a_2b_1 = 0$? We will have more to say about this later in this section.)

To solve for y, multiply both sides of equation (1) by $-a_2$ and both sides of equation (2) by a_1 and add.

$$
\begin{array}{ll}
-a_1a_2x - a_2b_1y = -a_2c_1 & -a_2 \text{ times both sides of equation (1)} \\
\underline{a_1a_2x + a_1b_2y = a_1c_2} & a_1 \text{ times both sides of equation (2)} \\
(a_1b_2 - a_2b_1)y = a_1c_2 - a_2c_1 &
\end{array}
$$

$$y = \frac{a_1c_2 - a_2c_1}{a_1b_2 - a_2b_1}, \quad a_1b_2 - a_2b_1 \neq 0$$

Both numerators and the common denominator of these values for x and y can be written as determinants because

$$a_1c_2 - a_2c_1 = \begin{vmatrix} a_1 & c_1 \\ a_2 & c_2 \end{vmatrix},$$

$$c_1b_2 - c_2b_1 = \begin{vmatrix} c_1 & b_1 \\ c_2 & b_2 \end{vmatrix},$$

and $\qquad a_1b_2 - a_2b_1 = \begin{vmatrix} a_1 & b_1 \\ a_2 & b_2 \end{vmatrix}.$

Using these results, the solutions for x and y become

$$x = \frac{\begin{vmatrix} c_1 & b_1 \\ c_2 & b_2 \end{vmatrix}}{\begin{vmatrix} a_1 & b_1 \\ a_2 & b_2 \end{vmatrix}} \quad \text{and} \quad y = \frac{\begin{vmatrix} a_1 & c_1 \\ a_2 & c_2 \end{vmatrix}}{\begin{vmatrix} a_1 & b_1 \\ a_2 & b_2 \end{vmatrix}}, \quad \begin{vmatrix} a_1 & b_1 \\ a_2 & b_2 \end{vmatrix} \neq 0.$$

For convenience, denote the three determinants in the solution as

$$\begin{vmatrix} a_1 & b_1 \\ a_2 & b_2 \end{vmatrix} = D, \quad \begin{vmatrix} c_1 & b_1 \\ c_2 & b_2 \end{vmatrix} = D_x, \quad \begin{vmatrix} a_1 & c_1 \\ a_2 & c_2 \end{vmatrix} = D_y.$$

Notice that the elements of D are the four coefficients of the variables in the given system; the elements of D_x are obtained by replacing the

1. Solve by Cramer's rule.

(a) $x + y = 5$
$\quad\ x - y = 1$

(b) $2x - 3y = -26$
$\quad\ 3x + 4y = 12$

(c) $4x - 5y = -8$
$\quad\ 3x + 7y = -6$

coefficients of x by the respective constants; the elements of D_y are obtained by replacing the coefficients of y by the respective constants.

These results are summarized as **Cramer's rule.**

Cramer's Rule for 2 × 2 Systems

Given the system

$$a_1 x + b_1 y = c_1$$
$$a_2 x + b_2 y = c_2 \quad \text{with} \quad a_1 b_2 - a_2 b_1 = D \neq 0,$$

then

$$x = \frac{\begin{vmatrix} c_1 & b_1 \\ c_2 & b_2 \end{vmatrix}}{\begin{vmatrix} a_1 & b_1 \\ a_2 & b_2 \end{vmatrix}} = \frac{D_x}{D} \quad \text{and} \quad y = \frac{\begin{vmatrix} a_1 & c_1 \\ a_2 & c_2 \end{vmatrix}}{\begin{vmatrix} a_1 & b_1 \\ a_2 & b_2 \end{vmatrix}} = \frac{D_y}{D}.$$

OBJECTIVE 2 To use Cramer's rule to solve a system of equations, find the three determinants, D, D_x, and D_y and then write the necessary quotients for x and y.

Caution
As indicated above, Cramer's rule does not apply if $D = a_1 b_2 - a_2 b_1$ is 0. When $D = 0$, the system is inconsistent or has dependent equations. For this reason, it is a good idea to evaluate D first.

E X A M P L E 1   Using Cramer's Rule for a 2 × 2 System

Use Cramer's rule to solve the system

$$5x + 7y = -1$$
$$6x + 8y = 1.$$

By Cramer's rule, $x = \dfrac{D_x}{D}$ and $y = \dfrac{D_y}{D}$. As mentioned above, it is a good idea to find D first, since if $D = 0$, Cramer's rule does not apply. If $D \neq 0$, then find D_x and D_y.

$$D = \begin{vmatrix} 5 & 7 \\ 6 & 8 \end{vmatrix} = 5(8) - 6(7) = -2;$$

$$D_x = \begin{vmatrix} -1 & 7 \\ 1 & 8 \end{vmatrix} = (-1)8 - 7(1) = -15;$$

$$D_y = \begin{vmatrix} 5 & -1 \\ 6 & 1 \end{vmatrix} = 5(1) - (-1)6 = 11.$$

From Cramer's rule,

$$x = \frac{D_x}{D} = \frac{-15}{-2} = \frac{15}{2},$$

and

$$y = \frac{D_y}{D} = \frac{11}{-2} = -\frac{11}{2}.$$

The solution set is $\{(\frac{15}{2}, -\frac{11}{2})\}$, as can be verified by checking in the given system.

OBJECTIVE ▶ In a similar manner, Cramer's rule can be applied to systems of three equations with three variables.

2. Find D_y and D_z.

Cramer's Rule for 3 × 3 Systems

Given the system

$$a_1 x + b_1 y + c_1 z = d_1$$
$$a_2 x + b_2 y + c_2 z = d_2$$
$$a_3 x + b_3 y + c_3 z = d_3$$

with

$$D_x = \begin{vmatrix} d_1 & b_1 & c_1 \\ d_2 & b_2 & c_2 \\ d_3 & b_3 & c_3 \end{vmatrix}, \quad D_y = \begin{vmatrix} a_1 & d_1 & c_1 \\ a_2 & d_2 & c_2 \\ a_3 & d_3 & c_3 \end{vmatrix},$$

$$D_z = \begin{vmatrix} a_1 & b_1 & d_1 \\ a_2 & b_2 & d_2 \\ a_3 & b_3 & d_3 \end{vmatrix}, \quad D = \begin{vmatrix} a_1 & b_1 & c_1 \\ a_2 & b_2 & c_2 \\ a_3 & b_3 & c_3 \end{vmatrix} \neq 0,$$

then

$$x = \frac{D_x}{D}, \quad y = \frac{D_y}{D}, \quad z = \frac{D_z}{D}.$$

E X A M P L E   2   **Using Cramer's Rule for a 3 × 3 System**

Use Cramer's rule to solve the system

$$x + y - z + 2 = 0$$
$$2x - y + z + 5 = 0$$
$$x - 2y + 3z - 4 = 0.$$

To use Cramer's rule, we must first rewrite the system in the form

$$x + y - z = -2$$
$$2x - y + z = -5$$
$$x - 2y + 3z = 4.$$

Expand by minors about row 1 to find D.

$$D = \begin{vmatrix} 1 & 1 & -1 \\ 2 & -1 & 1 \\ 1 & -2 & 3 \end{vmatrix} = 1 \begin{vmatrix} -1 & 1 \\ -2 & 3 \end{vmatrix} - 1 \begin{vmatrix} 2 & 1 \\ 1 & 3 \end{vmatrix} + (-1) \begin{vmatrix} 2 & -1 \\ 1 & -2 \end{vmatrix}$$

$$= 1(-1) - 1(5) - 1(-3) = -3$$

Expanding D_x by minors about row 1 gives

$$D_x = \begin{vmatrix} -2 & 1 & -1 \\ -5 & -1 & 1 \\ 4 & -2 & 3 \end{vmatrix}$$

$$= -2 \begin{vmatrix} -1 & 1 \\ -2 & 3 \end{vmatrix} - 1 \begin{vmatrix} -5 & 1 \\ 4 & 3 \end{vmatrix} + (-1) \begin{vmatrix} -5 & -1 \\ 4 & -2 \end{vmatrix}$$

$$= -2(-1) - 1(-19) - 1(14) = 7.$$

CONTINUED ON NEXT PAGE

WORK PROBLEM 2 AT THE SIDE. ▶▶

ANSWERS

2. $D_y = -22$ and $D_z = -21$

3. Solve by Cramer's rule.

(a)
$$x + y + z = 2$$
$$2x \quad - z = -3$$
$$y + 2z = 4$$

(b)
$$3x - 2y + 4z = 5$$
$$4x + y + z = 14$$
$$x - y - z = 1$$

4. Solve by Cramer's rule if applicable.

$$x - y + z = 6$$
$$3x + 2y + z = 4$$
$$2x - 2y + 2z = 14$$

Using the results for D and D_x and the results from Problem 2 at the side, apply Cramer's rule to get

$$x = \frac{D_x}{D} = \frac{7}{-3} = -\frac{7}{3}, \quad y = \frac{D_y}{D} = \frac{-22}{-3} = \frac{22}{3},$$

$$z = \frac{D_z}{D} = \frac{-21}{-3} = 7.$$

The solution set is $\{(-\frac{7}{3}, \frac{22}{3}, 7)\}$.

◀◀ **WORK PROBLEM 3 AT THE SIDE.**

OBJECTIVE ▶ 4 As mentioned earlier, Cramer's rule does not apply when $D = 0$. The next example illustrates this case.

EXAMPLE 3 **Determining When Cramer's Rule Does Not Apply**

If possible, use Cramer's rule to solve the following system.

$$2x - 3y + 4z = 8$$
$$6x - 9y + 12z = 24$$
$$x + 2y - 3z = 5$$

Find D, D_x, D_y, and D_z. Here

$$D = \begin{vmatrix} 2 & -3 & 4 \\ 6 & -9 & 12 \\ 1 & 2 & -3 \end{vmatrix} = 2 \begin{vmatrix} -9 & 12 \\ 2 & -3 \end{vmatrix} - 6 \begin{vmatrix} -3 & 4 \\ 2 & -3 \end{vmatrix} + 1 \begin{vmatrix} -3 & 4 \\ -9 & 12 \end{vmatrix}$$

$$= 2(3) - 6(1) + 1(0)$$

$$= 0.$$

Since $D = 0$ here, Cramer's rule does not apply and we must use another method to solve the system. Multiplying the first equation on both sides by 3 shows that the first two equations have the same solutions, so this system has dependent equations and an infinite solution set.

◀◀ **WORK PROBLEM 4 AT THE SIDE.**

Cramer's rule can be extended to 4×4 or larger systems. See a standard college algebra text for details.

ANSWERS
3. (a) $\{(-1, 2, 1)\}$ **(b)** $\{(3, 2, 0)\}$
4. Cramer's rule does not apply.

4.5 Exercises

1. For the system

$$8x - 4y = 8$$
$$x + 3y = 22,$$

$D_x = 112$, $D_y = 168$, and $D = 28$. What is the solution set of the system?

2. For the system

$$x + 3y - 6z = 7$$
$$2x - y + z = 1$$
$$x + 2y + 2z = -1,$$

the solution set is $\{(1, 0, -1)\}$ and $D = -43$. Find the values of D_x, D_y, and D_z.

Use Cramer's rule to solve each of the following linear systems in two variables. See Example 1.

3. $3x + 5y = -5$
    $-2x + 3y = 16$

4. $5x + 2y = -3$
    $4x - 3y = -30$

5. $3x + 2y = 3$
    $2x - 4y = 2$

6. $7x - 2y = 6$
    $4x - 5y = 15$

7. $8x + 3y = 1$
    $6x - 5y = 2$

8. $3x - y = 9$
    $2x + 5y = 8$

Use Cramer's rule where applicable to solve the following linear systems. If Cramer's rule does not apply, say so. See Examples 2 and 3.

9. $2x + 3y + 2z = 15$
    $x - y + 2z = 5$
    $x + 2y - 6z = -26$

10. $x - y + 6z = 19$
    $3x + 3y - z = 1$
    $x + 9y + 2z = -19$

11. $2x + 2y + z = 10$
    $4x - y + z = 20$
    $-x + y - 2z = -5$

12. $x + 3y - 4z = -12$
    $3x + y - z = -5$
    $5x - y + z = -3$

13. $2x - 3y + 4z = 8$
    $6x - 9y + 12z = 24$
    $-4x + 6y - 8z = -16$

14. $7x + y - z = 4$
    $2x - 3y + z = 2$
    $-6x + 9y - 3z = -6$

15. $3x + 5z = 0$
    $2x + 3y = 1$
    $-y + 2z = -11$

16. $-x + 2y = 4$
    $3x + y = -5$
    $2x + z = -1$

17. $x - 3y = 13$
    $2y + z = 5$
    $-x + z = -7$

18. $-5x - y \qquad = -10$
    $3x + 2y + z = -3$
    $\qquad -y - 2z = -13$

19. $3x + 2y \qquad - w = 0$
    $2x \qquad + z + 2w = 5$
    $x + 2y - z \qquad = -2$
    $2x - y + z + w = 2$

20. $x + 2y - z + w = 8$
    $2x - y \qquad - w = 12$
    $\qquad y + 3z \qquad = 11$
    $x \qquad - z - w = 4$

21. Make up a system of two equations in two variables having consecutive integers as coefficients and constants. For example, one such system would be

$$2x + 3y = 4$$
$$5x + 6y = 7.$$

Now solve the system using Cramer's rule. Repeat, using six different consecutive integers. Compare the two solution sets. What do you notice?

MATHEMATICAL CONNECTIONS (EXERCISES 22–24)

In this section we have seen how determinants can be used to solve systems of equations. In the Mathematical Connections in the exercises for Section 4.4, we saw how a determinant can be used to write the equation of a line given two points on a line. Here, we show how a determinant can be used to find the area of a triangle if we know the coordinates of its vertices.

Suppose that $A(x_1, y_1)$, $B(x_2, y_2)$, and $C(x_3, y_3)$ are the coordinates of the vertices of triangle ABC in the coordinate plane. Then it can be shown that the area of the triangle is given by the absolute value of

$$\frac{1}{2}\begin{vmatrix} x_1 & y_1 & 1 \\ x_2 & y_2 & 1 \\ x_3 & y_3 & 1 \end{vmatrix}.$$

Work Exercises 22–24 in order.

22. Sketch triangle ABC in the coordinate plane, given that the coordinates of A are (0, 0), of B are $(-3, -4)$, and of C are $(2, -2)$.

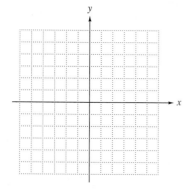

23. Write the determinant expression as described above that gives the area of triangle ABC described in Exercise 22.

24. Evaluate the absolute value of the determinant expression in Exercise 23 to find the area.

4.1 **system of equations**   Two or more equations that are to be solved at the same time form a system of equations.

linear system   A linear system is a system of equations that contains only linear equations.

inconsistent system   A system is inconsistent if it has no solution.

dependent equations   Dependent equations are equations whose graphs are the same line.

4.4 **matrix**   A matrix is a rectangular array of numbers, consisting of horizontal rows and vertical columns.

square matrix   A square matrix is a matrix that has the same number of rows as columns.

determinant   Associated with every square matrix is a real number called its determinant, symbolized by the entries of the matrix between two vertical lines.

expansion by minors   A method of evaluating a 3 × 3 or larger determinant is called expansion by minors.

NEW SYMBOLS

(x, y, z) ordered triple $\begin{vmatrix} a & b & c \\ d & e & f \\ g & h & i \end{vmatrix}$ 3 × 3 determinant

$\begin{bmatrix} a & b \\ c & d \end{bmatrix}$ 2 × 2 matrix $\begin{vmatrix} a & b \\ c & d \end{vmatrix}$ 2 × 2 determinant

---

QUICK REVIEW

Concepts	Examples

4.1 Linear Systems of Equations in Two Variables

Solving Linear Systems of Two Equations by Elimination

Solve by elimination.

$$5x + y = 2$$
$$2x - 3y = 11$$

Step 1   Write both equations in the form $Ax + By = C$.

Step 2   Multiply one or both equations by appropriate numbers so that the sum of the coefficients of either x or y is zero.

To eliminate y, multiply the top equation by 3, and add.

Step 3   Add the new equations. The sum should be an equation with just one variable.

$$\begin{array}{r} 15x + 3y = 6 \\ 2x - 3y = 11 \\ \hline 17x = 17 \end{array}$$

$$x = 1$$

Step 4   Solve the equation from Step 3.

Step 5   Substitute the result of Step 4 into either of the given equations and solve for the other variable.

Let $x = 1$ in the top equation, and solve for y.

$$5(1) + y = 2$$
$$y = -3$$

Step 6   Check the solution in both of the given equations.

Check to verify that $\{(1, -3)\}$ is the solution set.

Concepts	*Examples*
4.1 Linear Systems of Equations in Two Variables (continued)	
Solving Linear Systems of Two Equations by Substitution	Solve by substitution. $$4x - y = 7$$ $$3x + 2y = 30$$
Step 1 Solve one of the equations for either variable.	Solve for y in the top equation. $$y = 4x - 7$$
Step 2 Substitute for that variable in the other equation. The result should be an equation with just one variable.	Substitute $4x - 7$ for y in the bottom equation, and solve for x.
Step 3 Solve the equation from Step 2.	$$3x + 2(4x - 7) = 30$$ $$3x + 8x - 14 = 30$$ $$11x - 14 = 30$$ $$11x = 44$$ $$x = 4$$
Step 4 Substitute the result from Step 3 into the equation from Step 1 to find the value of the other variable.	Substitute 4 for x in the equation $y = 4x - 7$ to find that $y = 9$.
Step 5 Check the solution in both of the given equations.	Check to see that $\{(4, 9)\}$ is the solution set.
4.2 Linear Systems of Equations in Three Variables	
Solving Linear Systems in Three Variables	Solve the system $$x + 2y - z = 6$$ $$x + y + z = 6$$ $$2x + y - z = 7.$$
Step 1 Use the elimination method to eliminate any variable from any two of the given equations. The result is an equation in two variables.	Add the first and second equations; z is eliminated and the result is $2x + 3y = 12$.
Step 2 Eliminate the *same* variable from any *other* two equations. The result is an equation in the same two variables as in Step 1.	Eliminate z again by adding the second and third equations to get $3x + 2y = 13$. Now solve the system $$2x + 3y = 12 \qquad (*)$$ $$3x + 2y = 13.$$
Step 3 Use the elimination method to eliminate a second variable from the two equations in two variables that result from Steps 1 and 2. The result is an equation in one variable that gives the value of that variable.	To eliminate x, multiply the top equation by -3 and the bottom equation by 2. $$-6x - 9y = -36$$ $$\underline{6x + 4y = 26}$$ $$-5y = -10$$ $$y = 2$$

Concepts	Examples

4.2 Linear Systems of Equations in Three Variables (continued)

Step 4   Substitute the value of the variable found in Step 3 into either of the equations in two variables to find the value of the second variable.

Let $y = 2$ in equation (*).

$$2x + 3(2) = 12$$
$$2x + 6 = 12$$
$$2x = 6$$
$$x = 3$$

Step 5   Use the values of the two variables from Steps 3 and 4 to find the value of the third variable by substituting into any of the original equations.

Let $y = 2$ and $x = 3$ in any of the original equations to find $z = 1$. The solution set is $\{(3, 2, 1)\}$.

4.3 Applications of Linear Systems

To solve an applied problem with two (three) unknowns, write two (three) equations that relate the unknowns. Then solve the system.

The perimeter of a rectangle is 18 feet. The length is 3 feet more than twice the width. Find the dimensions of the rectangle.

Let x represent the length and y represent the width. From the perimeter formula, one equation is $2x + 2y = 18$. From the problem, another equation is $x = 3 + 2y$. Now solve the system

$$2x + 2y = 18$$
$$x = 3 + 2y.$$

The solution of the system is $(7, 2)$. Therefore, the length is 7 feet and the width is 2 feet.

4.4 Determinants

Value of a 2 × 2 Determinant

$$\begin{vmatrix} a & b \\ c & d \end{vmatrix} = ad - bc.$$

Determinants larger than 2 × 2 are evaluated by expansion by minors about a column or row.

Array of Signs for a 3 × 3 Determinant

$$\begin{matrix} + & - & + \\ - & + & - \\ + & - & + \end{matrix}$$

$$\begin{vmatrix} 3 & 4 \\ -2 & 6 \end{vmatrix} = (3)(6) - (4)(-2) = 26$$

Evaluate the following determinant by expanding about the second column.

$$\begin{vmatrix} 2 & -3 & -2 \\ -1 & -4 & -3 \\ -1 & 0 & 2 \end{vmatrix} = 3(-5) + (-4)(2) - (0)(-8)$$
$$= -15 - 8 + 0$$
$$= -23$$

Concepts	Examples
4.5 Solution of Linear Systems of Equations by Determinants—Cramer's Rule	Solve using Cramer's rule.

Cramer's Rule for 2 × 2 Systems

Given the system

$$a_1 x + b_1 y = c_1$$
$$a_2 x + b_2 y = c_2$$

with $a_1 b_2 - a_2 b_1 = D \neq 0$, then

$$x = \frac{\begin{vmatrix} c_1 & b_1 \\ c_2 & b_2 \end{vmatrix}}{\begin{vmatrix} a_1 & b_1 \\ a_2 & b_2 \end{vmatrix}} = \frac{D_x}{D}$$

and

$$y = \frac{\begin{vmatrix} a_1 & c_1 \\ a_2 & c_2 \end{vmatrix}}{\begin{vmatrix} a_1 & b_1 \\ a_2 & b_2 \end{vmatrix}} = \frac{D_y}{D}.$$

For a 3 × 3 system, D is the determinant of the matrix of coefficients; D_x is found by replacing the coefficients of x in D with the constants; D_y is found by replacing the coefficients of y in D with the constants; D_z is found by replacing the coefficients of z in D with the constants.

Examples

Solve using Cramer's rule.

$$x - 2y = -1$$
$$2x + 5y = 16$$

$$x = \frac{\begin{vmatrix} -1 & -2 \\ 16 & 5 \end{vmatrix}}{\begin{vmatrix} 1 & -2 \\ 2 & 5 \end{vmatrix}} = \frac{-5 + 32}{5 + 4} = \frac{27}{9} = 3$$

$$y = \frac{\begin{vmatrix} 1 & -1 \\ 2 & 16 \end{vmatrix}}{\begin{vmatrix} 1 & -2 \\ 2 & 5 \end{vmatrix}} = \frac{16 + 2}{5 + 4} = \frac{18}{9} = 2$$

The solution set is $\{(3, 2)\}$.

Solve using Cramer's rule.

$$3x + 2y + z = -5$$
$$x - y + 3z = -5$$
$$2x + 3y + z = 0$$

Using expansion by minors, it can be shown that $D_x = 45$, $D_y = -30$, $D_z = 0$, and $D = -15$. Therefore,

$$x = \frac{D_x}{D} = \frac{45}{-15} = -3,$$

$$y = \frac{D_y}{D} = \frac{-30}{-15} = 2,$$

$$z = \frac{D_z}{D} = \frac{0}{-15} = 0.$$

The solution set is $\{(-3, 2, 0)\}$.

[4.1] *Solve the following systems of equations using the elimination method. In Exercises 1 and 2, also graph the system.*

1. $x + 3y = 8$
$2x - y = 2$

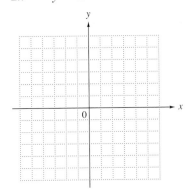

2. $x - 4y = -4$
$3x + y = 1$

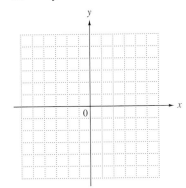

3. $6x + 5y = 4$
$-4x + 2y = 8$

4. $\dfrac{x}{6} + \dfrac{y}{6} = -\dfrac{1}{2}$
$x - y = -9$

5. $4x + 5y = 9$
$3x + 7y = -1$

6. $9x - y = -4$
$y = x + 4$

7. $-3x + y = 6$
$y = 6 + 3x$

8. $5x - 4y = 2$
$-10x + 8y = 7$

Solve the following systems using the substitution method.

9. $3x + y = -4$
$x = \dfrac{2}{3}y$

10. $-5x + 2y = -2$
$x + 6y = 26$

Suppose that two linear equations are graphed on the same set of coordinate axes. Sketch what the graph might look like if the system has the given description.

11. The system has a single solution.

12. The system has no solution.

13. The system has infinitely many solutions.

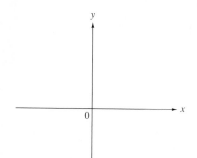

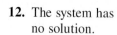

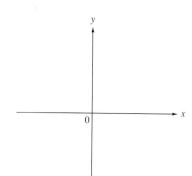

14. Without doing any algebraic work, explain why the system $\begin{array}{l} y = 3x + 2 \\ y = 3x - 4 \end{array}$ has $\emptyset$ as its solution set. Answer based only on your knowledge of the graphs of the two lines.

[4.2] *Solve the following systems of equations using the elimination method.*

15.
$$\begin{array}{rcl} 2x + 3y - z &=& -16 \\ x + 2y + 2z &=& -3 \\ -3x + y + z &=& -5 \end{array}$$

16.
$$\begin{array}{rcl} 3x - y - z &=& -8 \\ 4x + 2y + 3z &=& 15 \\ -6x + 2y + 2z &=& 10 \end{array}$$

17.
$$\begin{array}{rcl} 4x - y &=& 2 \\ 3y + z &=& 9 \\ x + 2z &=& 7 \end{array}$$

18.
$$\begin{array}{rcl} x + 5y - 3z &=& 0 \\ 2x + 6y + z &=& 0 \\ 3x - y + 4z &=& 0 \end{array}$$

[4.3] *Solve the following problems by writing a system of equations.*

19. A rectangular table top is 2 feet longer than it is wide, and its perimeter is 20 feet. Find the length and the width of the table top.

20. On an 8-day business trip, Clay rented a car for $32 per day at weekday rates and $19 per day at weekend rates. If his total rental bill was $217, how many days did he rent at each rate?

21. A plane flies 560 miles in 1.75 hours traveling with the wind. The return trip later against the same wind takes the plane 2 hours. Find the speed of the plane and the speed of the wind.

22. Sweet's Candy Store is offering a special mix for Valentine's Day. Latoya Sweet will mix some $2-a-pound candy with some $1-a-pound candy to get 100 pounds of mix that she will sell at $1.30 a pound. How many pounds of each should she use?

23. The sum of the measures of the angles of a triangle is 180°. One angle measures 10° less than the sum of the other two. The measure of the middle-sized angle is the average of the other two. Find the measures of the three angles.

24. David Zerangue sells real estate. On three recent sales, he made 10% commission, 6% commission, and 5% commission. His total commissions on these sales were $17,000, and he sold property worth $280,000. If the 5% sale amounted to the sum of the other two, what were the three sales' prices?

25. The manager of a candy store wants to feature a special Easter candy mixture of jelly beans, small chocolate eggs, and marshmallow chicks. She plans to make 15 pounds of mix to sell at $1 a pound. Jelly beans sell for $.80 a pound, chocolate eggs for $2 a pound, and marshmallow chicks for $1 a pound. She will use twice as many pounds of jelly beans as eggs and chicks combined, and fives times as many pounds of jelly beans as chocolate eggs. How many pounds of each candy should she use?

26. Gil Troutman has a collection of tropical fish. For each fish, he paid either $20, $40, or $65. The number of $40 fish is one less than twice the number of $20 fish. If there are 29 fish in all worth $1150, how many of each kind of fish is in the collection?

[4.4] *Evaluate each of the following determinants.*

27. $\begin{vmatrix} 2 & -9 \\ 8 & 4 \end{vmatrix}$

28. $\begin{vmatrix} 7 & 0 \\ 5 & -3 \end{vmatrix}$

29. $\begin{vmatrix} 2 & 10 & 4 \\ 0 & 1 & 3 \\ 0 & 6 & -1 \end{vmatrix}$

30. $\begin{vmatrix} 0 & 0 & 0 \\ 0 & 2 & 5 \\ -1 & 3 & 6 \end{vmatrix}$

31. $\begin{vmatrix} 0 & 0 & 2 \\ 2 & 1 & 0 \\ -1 & 0 & 0 \end{vmatrix}$

32. $\begin{vmatrix} 1 & 3 & -2 \\ 2 & 6 & -4 \\ 5 & 0 & 1 \end{vmatrix}$

[4.5]

33. Under what conditions can a system *not* be solved using Cramer's rule?

Use Cramer's rule to solve the following systems of equations.

34. $3x - 4y = 5$
$2x + y = 8$

35. $-4x + 3y = -12$
$2x + 6y = 15$

36. $4x + y + z = 11$
$x - y - z = 4$
$y + 2z = 0$

37. $-x + 3y - 4z = 2$
$2x + 4y + z = 3$
$3x - z = 9$

MIXED REVIEW EXERCISES

Solve by any method.

38. $\dfrac{2}{3}x + \dfrac{1}{6}y = \dfrac{19}{2}$

$\dfrac{1}{3}x - \dfrac{2}{9}y = 2$

39. $2x + 5y - z = 12$
$-x + y - 4z = -10$
$-8x - 20y + 4z = 31$

40. $x = 7y + 10$
$2x + 3y = 3$

41. $x + 4y = 17$
$-3x + 2y = -9$

42. $-7x + 3y = 12$
$5x + 2y = 8$

43. $2x - 5y = 8$
$3x + 4y = 10$

44. To make a 10% acid solution for chemistry class, Xavier wants to mix some 5% solution with 10 liters of 20% solution. How many liters of 5% solution should he use?

45. The sum of the three angles of a triangle is 180°. The largest angle is twice the measure of the smallest, and the third angle measures 10° less than the largest. Find the measures of the three angles.

CHAPTER 4 TEST

1. Use a graph to solve the system

$x + y = 7$
$x - y = 5.$

1. _____

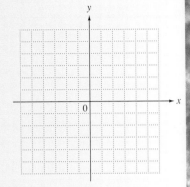

Solve each system using elimination.

2. $3x + y = 12$
$2x - y = 3$

2. _____

3. $-5x + 2y = -4$
$6x + 3y = -6$

3. _____

4. $3x + 4y = 8$
$8y = 7 - 6x$

4. _____

5. $3x + 5y + 3z = 2$
$6x + 5y + z = 0$
$3x + 10y - 2z = 6$

5. _____

Solve each system using substitution.

6. $2x - 3y = 24$
$y = -\dfrac{2}{3}x$

6. _____

7. $12x - 5y = 8$
$3x = \dfrac{5}{4}y + 2$

7. _____

Solve each problem by writing a system of equations.

8. In an election, one candidate received 45 more votes than the other. The total number of votes cast in the election was 405. Find the number of votes received by each candidate.

8. _____

9. _____

9. A chemist needs 12 liters of a 40% alcohol solution. She must mix a 20% solution and a 50% solution. How many liters of each will be required to obtain what she needs?

10. _____

10. A local electronics store will sell 7 AC adaptors and 2 rechargeable flashlights for $86, or 3 AC adaptors and 4 rechargeable flashlights for $84. What is the price of a single AC adaptor and a single rechargeable flashlight?

11. _____

11. The owner of a tea shop wants to mix three kinds of tea to make 100 ounces of a mixture that will sell for $.83 an ounce. He uses Orange Pekoe, which sells for $.80 an ounce, Irish Breakfast, for $.85 an ounce, and Earl Grey, for $.95 an ounce. If he wants to use twice as much Orange Pekoe as Irish Breakfast, how much of each kind of tea should he use?

Evaluate each of the following determinants.

12. _____

12. $\begin{vmatrix} 6 & -3 \\ 5 & -2 \end{vmatrix}$

13. _____

13. $\begin{vmatrix} 4 & 1 & 0 \\ -2 & 7 & 3 \\ 0 & 5 & 2 \end{vmatrix}$

Solve each of the following systems by Cramer's rule.

14. _____

14. $3x - y = -8$
$2x + 6y = 3$

15. _____

15. $x + y + z = 4$
$-2x \quad\quad + z = 5$
$\quad\quad 3y + z = 9$

Evaluate.

1. $(-3)^4$

2. -3^4

3. $-(-3)^4$

4. $\sqrt{.49}$

5. $-\sqrt{.49}$

6. $\sqrt{-.49}$

7. $\sqrt[3]{64}$

8. $\sqrt[3]{-64}$

Evaluate if $x = -4$, $y = 3$, and $z = 6$.

9. $|2x| + 3y - z^3$

10. $-5(x^3 - y^3)$

11. $\dfrac{2x^2 - x + z}{3y - z}$

Solve each equation.

12. $7(2x + 3) - 4(2x + 1) = 2(x + 1)$

13. $|6x - 8| = 4$

14. $ax + by = cx + d$ for x

15. $.04x + .06(x - 1) = 1.04$

Solve each inequality.

16. $\dfrac{2}{3}y + \dfrac{5}{12}y \le 20$

17. $|3x + 2| \le 4$

18. $|12t + 7| \ge 0$

Solve each problem.

19. Two cars start from points 420 miles apart and travel toward each other. They meet after 3.5 hours. Find the average speed of each car if one travels 30 miles per hour slower than the other.

20. A triangle has an area of 42 square meters. The base is 14 meters long. Find the height of the triangle.

21. A jar contains only pennies, nickels, and dimes. The number of dimes is 1 more than the number of nickels, and the number of pennies is 6 more than the number of nickels. How many of each denomination can be found in the jar, if the total value is $4.80?

22. Two angles of a triangle have the same measure. The measure of the third angle is 4° less than twice the measure of each of the equal angles. Find the measures of the three angles.

In Exercises 23–28, point A has coordinates $(-2, 6)$ and point B has coordinates $(4, -2)$.

23. What is the equation of the horizontal line through A?

24. What is the equation of the vertical line through B?

25. What is the slope of AB?

26. What is the slope of a line perpendicular to line AB?

27. What is the standard form of the equation of line AB?

28. Write the equation of the line in the form of a linear function.

29. Find the standard form of the equation of the line with x-intercept $(-3, 0)$ and y-intercept $(0, 5)$.

30. Graph the linear function whose graph has slope $\frac{2}{3}$ and passes through the point $(-1, -3)$.

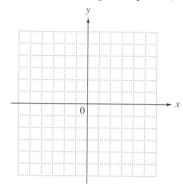

31. Graph the inequality $-3x - 2y \le 6$.

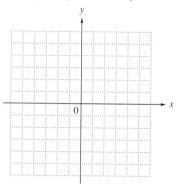

Solve by any method.

32. $-2x + 3y = -15$
$4x - y = 15$

33. $x + y + z = 10$
$x - y - z = 0$
$-x + y - z = -4$

34. Evaluate

$$\begin{vmatrix} 1 & 2 & 3 \\ 0 & 5 & 1 \\ -1 & 0 & 4 \end{vmatrix}$$

In Exercises 35–36, solve the problem using a system of equations. Use the method of your choice: elimination, substitution, or Cramer's rule.

35. Mabel Johnston bought apples and oranges at DeVille's Grocery. She bought 6 pounds of fruit. Oranges cost $.90 per pound, while apples cost $.70 per pound. If she spent a total of $5.20, how many pounds of each kind of fruit did she buy?

36. Alexis has inherited $80,000 from her aunt. She invests part of the money in a video rental firm that produces a return of 7% per year, and divides the rest equally between a tax-free bond at 6% a year and a money market fund at 4% a year. Her annual return on these investments is $5200. How much is invested in each?

The graph shows a company's costs to produce computer parts and the revenue from the sale of computer parts.

37. At what production level does the cost equal the revenue? What is the revenue at that point?

38. Profit is revenue less cost. Estimate the profit on the sale of 1100 parts.

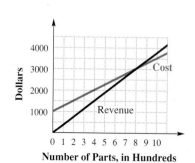

Polynomials 5

5.1 Integer Exponents

In Chapter 1 we showed that two integers whose product is a third number are factors of that number. For example, 2 and 5 are factors of 10, because $2 \cdot 5 = 10$.

OBJECTIVE ▶1▶ Recall that we use exponents to write products of repeated factors. For example, 2^5 is defined as $2 \cdot 2 \cdot 2 \cdot 2 \cdot 2 = 32$. The number 5, the *exponent*, shows that the *base* 2 appears as a factor 5 times. The quantity 2^5 is called an *exponential* or a *power*. We read 2^5 as "2 to the fifth power" or "2 to the fifth."

OBJECTIVES

▶1▶ Review identifying exponents and bases.

▶2▶ Use the product rule.

▶3▶ Define a zero exponent.

▶4▶ Define a negative exponent.

▶5▶ Use the quotient rule.

FOR EXTRA HELP

Tutorial   Tape 6   SSM, Sec. 5.1

E X A M P L E   1   **Identifying Exponents and Bases and Evaluating Exponentials**

Identify the exponent and the base, and evaluate each exponential.

(a) -2^4

The exponent is 4 and the base is 2 (not -2).

$$-2^4 = -(2^4) = -16$$

(b) $(-2)^4$

The exponent is 4, and the base is -2.

$$(-2)^4 = (-2)(-2)(-2)(-2) = 16$$

(c) $3z^7$

The exponent is 7. The base is z (not $3z$).

$$3z^7 = 3z \cdot z \cdot z \cdot z \cdot z \cdot z$$

(d) $(-9y)^3$

Because of the parentheses, the exponent 3 refers to the base $-9y$, so

$$(-9y)^3 = (-9y)(-9y)(-9y) = -729y^3.$$

Caution

As shown in Example 1, when no parentheses are used, the exponent refers only to the factor closest to it. In part (a), think of -2^4 as $(-1)(2)^4$, to see why 2 (not -2) is the factor closest to the exponent.

1. Identify the base of each exponential.

(a) $-2p^8$

(b) $(-2p)^8$

(c) $3y^7$

(d) $-m^{10}$

WORK PROBLEM 1 AT THE SIDE. ▶▶

ANSWERS

1. (a) p **(b)** $-2p$ **(c)** y **(d)** m

2. Find the product.

(a) $m^8 \cdot m^6$

(b) $r^7 \cdot r$

(c) $k^4 k^3 k^6$

(d) $m^5 \cdot p^4$

(e) $(-4a^3)(6a^2)$

(f) $(-5p^4)(-9p^5)$

OBJECTIVE 2 Several useful rules can be developed that simplify the use of exponents. For example, the product $2^5 \cdot 2^3$ can be simplified as follows.

$$2^5 \cdot 2^3 = (2 \cdot 2 \cdot 2 \cdot 2 \cdot 2)(2 \cdot 2 \cdot 2) = 2^8$$

$$5 + 3 = 8$$

This result, that products of exponentials with the same base are found by adding exponents, can be generalized by the product rule for exponents.

Product Rule for Exponents

If m and n are integers and a is any nonzero real number, then

$$a^m \cdot a^n = a^{m+n}.$$

Keep the same base; add the exponents.

E X A M P L E   2   Using the Product Rule

(a) $3^4 \cdot 3^7 = 3^{4+7} = 3^{11}$

(b) $5^3 \cdot 5 = 5^3 \cdot 5^1 = 5^{3+1} = 5^4$

(c) $y^3 \cdot y^8 \cdot y^2 = y^{3+8+2} = y^{13}$

(d) $x^3 \cdot y^5$ cannot be simplified further because the bases x and y are not the same.

E X A M P L E   3   Using the Product Rule

Use the product rule for exponents to find each product.

(a) $(5y^2)(-3y^4)$

We can use the commutative and associative properties to group the constants together and the variables together.

$$(5y^2)(-3y^4) = [5(-3)](y^2y^4)$$
$$= -15y^{2+4} \qquad \text{Product rule}$$
$$= -15y^6$$

(b) $(7p^3q)(2p^5q^2) = [7(2)](p^3p^5qq^2) \qquad \text{Regroup.}$
$$= 14p^8q^3 \qquad \text{Product rule}$$

◀◀ WORK PROBLEM 2 AT THE SIDE.

OBJECTIVE 3 So far we have discussed only positive exponents. Let us consider how we might define a zero exponent. Suppose we multiply 4^2 by 4^0. By the product rule,

$$4^2 \cdot 4^0 = 4^{2+0} = 4^2.$$

For the product rule to hold true, 4^0 must equal 1, and so a^0 is defined as follows for any nonzero real number a.

Zero Exponent

For any nonzero real number a,

$$a^0 = 1.$$

0^0 is undefined.

┌─ E X A M P L E   4   **Using Zero Exponents**

 (a) $12^0 = 1$

 (b) $(-19)^0 = 1$

 (c) $-19^0 = -(19^0) = -1$

 (d) $5^0 + 12^0 = 1 + 1 = 2$

 (e) $(8k)^0 = 1,$ if $k \neq 0$

└─── ■

WORK PROBLEM 3 AT THE SIDE. ▶▶

OBJECTIVE ▶ Now, how should we define a negative exponent? Using the product rule again,

$$8^2 \cdot 8^{-2} = 8^{2+(-2)} = 8^0 = 1.$$

This indicates that 8^{-2} is the reciprocal of 8^2. But $\frac{1}{8^2}$ is the reciprocal of 8^2, and a number can have only one reciprocal. Thus, we must define 8^{-2} to equal $\frac{1}{8^2}$, so negative exponents are defined as follows.

Definition of Negative Exponent

If a is any nonzero number, and n is any whole number,

$$a^{-n} = \frac{1}{a^n}.$$

With this definition, the expression a^n is meaningful for any integer exponent n and any nonzero real number a.

┌─ E X A M P L E   5   **Using Negative Exponents**

 Simplify the following expressions and evaluate, if possible.

 (a) $2^{-3} = \dfrac{1}{2^3} = \dfrac{1}{8}$

 (b) $3^{-2} = \dfrac{1}{3^2} = \dfrac{1}{9}$

 (c) $6^{-1} = \dfrac{1}{6^1} = \dfrac{1}{6}$

 (d) $5^{-3} = \dfrac{1}{5^3} = \dfrac{1}{125}$

 (e) $k^{-7} = \dfrac{1}{k^7}$ $(k \neq 0)$

 (f) $(5z^2)^{-3} = \dfrac{1}{(5z^2)^3}$ $(z \neq 0)$

└─── ■

3. Simplify.

 (a) 46^0

 (b) $(-29)^0$

 (c) -29^0

 (d) $8^0 + 15^0$

 (e) $(-15p^5)^0,$ $p \neq 0$

ANSWERS

3. (a) 1 **(b)** 1 **(c)** −1 **(d)** 2 **(e)** 1

4. Simplify, using the definition of negative exponents.

(a) 8^{-1}

(b) 6^{-3}

(c) -3^{-2}

(d) y^{-3}, $y \neq 0$

(e) $(2m^4)^{-3}$, $m \neq 0$

5. Evaluate each expression.

(a) $3^{-1} + 5^{-1}$

(b) $4^{-1} - 2^{-1}$

(c) $\dfrac{4^{-1}}{2^{-2}}$

(d) $\dfrac{3^{-3}}{9^{-1}}$

4. (a) $\dfrac{1}{8}$ (b) $\dfrac{1}{216}$ (c) $-\dfrac{1}{9}$

  (d) $\dfrac{1}{y^3}$ (e) $\dfrac{1}{(2m^4)^3}$

5. (a) $\dfrac{8}{15}$ (b) $-\dfrac{1}{4}$ (c) 1 (d) $\dfrac{1}{3}$

Caution

A negative exponent *does not indicate* a negative number; negative exponents lead to reciprocals, as shown below.

Expression	*Example*	
$a^{-m} = \dfrac{1}{a^m}$	$3^{-2} = \dfrac{1}{3^2} = \dfrac{1}{9}$	Not negative
$-a^{-m} = -\dfrac{1}{a^m}$	$-3^{-2} = -\dfrac{1}{3^2} = -\dfrac{1}{9}$	Negative

◀◀ **WORK PROBLEM 4 AT THE SIDE.**

E X A M P L E   6   Using Negative Exponents

Evaluate the following expressions.

(a) $3^{-1} + 4^{-1}$

Because $3^{-1} = \dfrac{1}{3}$ and $4^{-1} = \dfrac{1}{4}$,

$$3^{-1} + 4^{-1} = \frac{1}{3} + \frac{1}{4} = \frac{4}{12} + \frac{3}{12} = \frac{7}{12}.$$

(b) $5^{-1} - 2^{-1} = \dfrac{1}{5} - \dfrac{1}{2} = \dfrac{2}{10} - \dfrac{5}{10} = -\dfrac{3}{10}$

Caution

In Example 6, note that

$$3^{-1} + 4^{-1} \neq (3 + 4)^{-1}$$

because $3^{-1} + 4^{-1} = \frac{7}{12}$ but $(3 + 4)^{-1} = 7^{-1} = \frac{1}{7}$.
Also,

$$5^{-1} - 2^{-1} \neq (5 - 2)^{-1}$$

because $5^{-1} - 2^{-1} = -\frac{3}{10}$ but $(5 - 2)^{-1} = 3^{-1} = \frac{1}{3}$.

E X A M P L E   7   Using Negative Exponents

Evaluate each expression.

(a) $\dfrac{1}{2^{-3}} = \dfrac{1}{\dfrac{1}{2^3}} = 2^3 = 8$

(b) $\dfrac{1}{3^{-2}} = \dfrac{1}{\dfrac{1}{3^2}} = 3^2 = 9$

(c) $\dfrac{2^{-3}}{3^{-2}} = \dfrac{\dfrac{1}{2^3}}{\dfrac{1}{3^2}} = \dfrac{1}{2^3} \cdot \dfrac{3^2}{1} = \dfrac{3^2}{2^3} = \dfrac{9}{8}$

Example 7 suggests the following generalizations. If $a \neq 0$, $b \neq 0$,

$$\frac{1}{a^{-n}} = a^n \quad \text{and} \quad \frac{a^{-n}}{b^{-m}} = \frac{b^m}{a^n}.$$

◀◀ **WORK PROBLEM 5 AT THE SIDE.**

Objective 5 A quotient such as $\frac{a^8}{a^3}$ can be simplified in much the same way as a product. (In all quotients of this type, we assume the denominator is not zero.) Using the definition of an exponent,

$$\frac{a^8}{a^3} = \frac{a \cdot a \cdot a \cdot a \cdot a \cdot a \cdot a \cdot a}{a \cdot a \cdot a}$$

$$= a \cdot a \cdot a \cdot a \cdot a$$

$$= a^5.$$

Notice that $8 - 3 = 5$. In the same way,

$$\frac{a^3}{a^8} = \frac{a \cdot a \cdot a}{a \cdot a \cdot a \cdot a \cdot a \cdot a \cdot a \cdot a}$$

$$\frac{a^3}{a^8} = a^{-5} = \frac{1}{a^5}.$$

These examples suggest the following quotient rule for exponents.

Quotient Rule for Exponents

If a is any nonzero real number, and m and n are integers, then

$$\frac{a^m}{a^n} = a^{m-n}.$$

Keep the same base; subtract the exponents.

E X A M P L E   8   Using the Quotient Rule

(a) $\dfrac{3^7}{3^2} = 3^{7-2} = 3^5$

(b) $\dfrac{p^6}{p^2} = p^{6-2} = p^4 \quad (p \neq 0)$

(c) $\dfrac{12^{10}}{12^9} = 12^{10-9} = 12^1 = 12$

(d) $\dfrac{7^4}{7^6} = 7^{4-6} = 7^{-2} = \dfrac{1}{7^2}$

(e) $\dfrac{k^7}{k^{12}} = k^{7-12} = k^{-5} = \dfrac{1}{k^5} \quad (k \neq 0)$

(f) $\dfrac{p^6}{q^4}$ cannot be simplified because the bases p and q are not the same.

WORK PROBLEM 6 AT THE SIDE. ▶▶

Caution
Be very careful when working with quotients that involve negative exponents in the denominator. Always be sure to write the numerator exponent, then a minus sign, and then the denominator exponent.

---

6. Divide. Write all answers with only positive exponents.

(a) $\dfrac{6^9}{6^4}$

(b) $\dfrac{9^{12}}{9^7}$

(c) $\dfrac{15^7}{15^{10}}$

(d) $\dfrac{m^3}{m^8}, \quad m \neq 0$

(e) $\dfrac{x^3}{y^5}, \quad y \neq 0$

Answers

6. (a) 6^5 (b) 9^5 (c) $\dfrac{1}{15^3}$ (d) $\dfrac{1}{m^5}$
   (e) cannot be simplified

7. Simplify. Write all answers with only positive exponents.

(a) $\dfrac{2^{-4}}{2^2}$

(b) $\dfrac{8^{-2}}{8^{-6}}$

(c) $\dfrac{9^{-5}}{9^{-2}}$

(d) $\dfrac{7^{-1}}{7}$

(e) $\dfrac{k^4}{k^{-5}}, \quad k \neq 0$

EXAMPLE 9 Using the Quotient Rule

(a) $\dfrac{8^{-2}}{8^5} = 8^{-2-5} = 8^{-7} = \dfrac{1}{8^7}$

Numerator exponent
Minus sign Denominator exponent

(b) $\dfrac{6^{-5}}{6^{-2}} = 6^{-5-(-2)} = 6^{-3} = \dfrac{1}{6^3}$

Numerator exponent
Minus sign Denominator exponent

(c) $\dfrac{2^7}{2^{-3}} = 2^{7-(-3)} = 2^{10}$

(d) $\dfrac{6}{6^{-1}} = \dfrac{6^1}{6^{-1}} = 6^{1-(-1)} = 6^2$

(e) $\dfrac{z^{-5}}{z^{-8}} = z^{-5-(-8)} = z^3 \quad (z \neq 0)$

◀◀ WORK PROBLEM 7 AT THE SIDE.

ANSWERS

7. (a) $\dfrac{1}{2^6}$ (b) 8^4 (c) $\dfrac{1}{9^3}$ (d) $\dfrac{1}{7^2}$ (e) k^9

5.1 Exercises

1. To use the product rule or the quotient rule with two exponential expressions, what must be true of their bases?

2. Which one of the following is equal to 1 $(a \neq 0)$?
(a) $3a^0$ (b) $-3a^0$ (c) $(3a)^0$
(d) $3(-a)^0$

Identify the exponent and the base. Do not evaluate. See Example 1.

3. -5^4

4. -2^6

5. $(-5)^4$

6. $(-2)^6$

7. $-3p^{-1}$

8. $-5m^{-2}$

Evaluate. Assume all variables are nonzero. See Example 4.

9. 25^0

10. 14^0

11. -7^0

12. -10^0

13. $(-15)^0$

14. $(-20)^0$

15. $-4^0 - m^0$

16. $-8^0 - k^0$

Rewrite using a positive exponent. See Examples 5 and 7.

17. y^{-2}

18. x^{-5}

19. $(-7)^{-2}$

20. $(-5)^{-3}$

21. $\dfrac{1}{x^{-2}}$

22. $\dfrac{1}{z^{-4}}$

Rewrite with a negative exponent.

23. $\dfrac{1}{5^2}$

24. $\dfrac{1}{2^4}$

25. $\dfrac{1}{x^3}$

26. $\dfrac{1}{y^5}$

27. $\dfrac{1}{(-4)^3}$

28. $\dfrac{1}{(-2)^5}$

29. In some cases, $-a^n$ and $(-a)^n$ do give the same result. Using $a = 2$ and $n = 2, 3, 4$, and 5, draw a conclusion as to when they are equal and when they are opposites.

30. Your friend evaluated $4^5 \cdot 4^2$ as 16^7. Explain to him why his answer is incorrect.

Evaluate. See Examples 1 and 4–7.

31. $\left(\dfrac{2}{3}\right)^2$ **32.** $\left(\dfrac{4}{3}\right)^3$ **33.** 4^{-3} **34.** 5^{-2}

35. -4^{-3} **36.** -5^{-2} **37.** $(-4)^{-3}$ **38.** $(-5)^{-2}$

39. $\dfrac{1}{3^{-2}}$ **40.** $\dfrac{1}{6^{-1}}$ **41.** $\dfrac{-3^{-1}}{4^{-2}}$ **42.** $\dfrac{2^{-3}}{-3^{-1}}$

43. $\left(\dfrac{2}{3}\right)^{-3}$ **44.** $\left(\dfrac{5}{4}\right)^{-2}$ **45.** $3^{-1} + 2^{-1}$ **46.** $4^{-1} + 5^{-1}$

Use the product rule or quotient rule to simplify. Write all answers with only positive exponents. Assume that all variables represent nonzero real numbers. See Examples 2–3 and 8–9.

47. $2^6 \cdot 2^{10}$ **48.** $3^5 \cdot 3^7$ **49.** $\dfrac{3^5}{3^2}$ **50.** $\dfrac{5^7}{5^2}$

51. $\dfrac{3^{-5}}{3^{-2}}$ **52.** $\dfrac{2^{-4}}{2^{-3}}$ **53.** $\dfrac{9^{-1}}{9}$ **54.** $\dfrac{12}{12^{-1}}$

55. $t^5 t^{-12}$ **56.** $p^5 p^{-6}$ **57.** $r^4 r$ **58.** $k \cdot k^6$

59. $a^{-3} a^2 a^{-4}$ **60.** $k^{-5} k^{-3} k^4$ **61.** $\dfrac{x^7}{x^{-4}}$ **62.** $\dfrac{p^{-3}}{p^5}$

63. $\dfrac{r^3 r^{-4}}{r^{-2} r^{-5}}$ **64.** $\dfrac{z^{-4} z^{-2}}{z^3 z^{-1}}$ **65.** $7k^2(-2k)(4k^{-5})$ **66.** $3a^2(-5a^{-6})(-2a)$

Many students believe that each pair of expressions in Exercises 67–69 represents the same quantity. This is wrong. Show that each expression in the pair represents a different quantity by replacing x with 2 and y with 3.

67. $(x + y)^{-1}; \; x^{-1} + y^{-1}$ **68.** $(x^{-1} + y^{-1})^{-1}; \; x + y$ **69.** $(x + y)^2; \; x^2 + y^2$

70. Which one of the following does not represent the reciprocal of x $(x \neq 0)$?

     (a) x^{-1} **(b)** $\dfrac{1}{x}$ **(c)** $\left(\dfrac{1}{x^{-1}}\right)^{-1}$ **(d)** $-x$

5.2 *Further Properties of Exponents*

OBJECTIVE 1 ▶ By the product rule for exponents, the expression $(3^4)^2$ can be simplified as

$$(3^4)^2 = 3^4 \cdot 3^4 = 3^{4+4} = 3^8$$

where $4 \cdot 2 = 8$. This example suggests the first of the **power rules for exponents;** the other two parts can be demonstrated with similar examples.

Power Rules for Exponents

If a and b are real numbers and m and n are integers, then

$$(a^m)^n = a^{mn}$$

$$(ab)^m = a^m b^m$$

$$\left(\frac{a}{b}\right)^m = \frac{a^m}{b^m} \quad (b \neq 0).$$

EXAMPLE 1 Using the Power Rules

(a) $(p^8)^3 = p^{8 \cdot 3} = p^{24}$

(b) $\left(\frac{2}{3}\right)^4 = \frac{2^4}{3^4} = \frac{16}{81}$

(c) $(3y)^4 = 3^4 y^4 = 81 y^4$

(d) $(6p^7)^2 = (6^1 p^7)^2 = 6^{1 \cdot 2} p^{7 \cdot 2} = 6^2 p^{14} = 36 p^{14}$

(e) $\left(\frac{2m^5}{z}\right)^3 = \left(\frac{2^1 m^5}{z^1}\right)^3 = \frac{2^{1 \cdot 3} m^{5 \cdot 3}}{z^{1 \cdot 3}} = \frac{2^3 m^{15}}{z^3} = \frac{8 m^{15}}{z^3}$

WORK PROBLEM 1 AT THE SIDE. ▶▶

Caution

Although $(a \cdot b)^m = a^m \cdot b^m$,

$$(a + b)^m \neq a^m + b^m,$$

because the addition in the parentheses must be done before the exponentiation.

In the previous section, we saw that

$$\frac{1}{a^{-n}} = a^n \quad \text{and} \quad \frac{a^{-n}}{b^{-m}} = \frac{b^m}{a^n}.$$

We can extend these ideas as follows.

The reciprocal of a^n is $\frac{1}{a^n} = \left(\frac{1}{a}\right)^n$. Also, by definition, a^n and a^{-n} are reciprocals, because

$$a^n \cdot a^{-n} = a^n \cdot \frac{1}{a^n} = 1.$$

Thus, because both are reciprocals of a^n, $a^{-n} = \left(\frac{1}{a}\right)^n$.

OBJECTIVES

▶ 1 Use the power rules for exponents.

▶ 2 Simplify exponential expressions.

▶ 3 Use the rules for exponents with scientific notation.

FOR EXTRA HELP

Tutorial

Tape 6

SSM, Sec. 5.2

1. Simplify.

(a) $(m^5)^4$

(b) $(x^3)^9$

(c) $\left(\frac{3}{8}\right)^7$

(d) $(2r)^{10}$

(e) $(-3y^5)^2$

(f) $\left(\frac{5p^2}{r^3}\right)^4$, $\quad r \neq 0$

ANSWERS

1. (a) m^{20} (b) x^{27} (c) $\dfrac{3^7}{8^7}$ (d) $2^{10} r^{10}$

(e) $(-3)^2 y^{10}$ or $9y^{10}$ (f) $\dfrac{5^4 p^8}{r^{12}}$

299

2. Write each expression with a positive exponent, and evaluate it.

(a) 4^{-3}

(b) $\left(\dfrac{2}{3}\right)^{-4}$

(c) $\left(\dfrac{1}{6}\right)^{-3}$

That is, any nonzero number raised to the negative nth power equals the *reciprocal* of that number raised to the nth power. For example, using this result,

$$6^{-3} = \left(\frac{1}{6}\right)^{3} \quad \text{and} \quad \left(\frac{1}{3}\right)^{-2} = 3^2$$

because $\frac{1}{6}$ is the reciprocal of 6, and 3 is the reciprocal of $\frac{1}{3}$.

EXAMPLE 2 Using Negative Exponents with Fractions

Write the following expressions with only positive exponents, and then evaluate.

(a) $5^{-4} = \left(\dfrac{1}{5}\right)^{4} = \dfrac{1}{625}$

(b) $\left(\dfrac{3}{7}\right)^{-2} = \left(\dfrac{7}{3}\right)^{2} = \dfrac{49}{9}$

(c) $\left(\dfrac{4}{5}\right)^{-3} = \left(\dfrac{5}{4}\right)^{3} = \dfrac{125}{64}$

◀◀ WORK PROBLEM 2 AT THE SIDE.

Example 2 suggests the following generalizations. If $a \neq 0$ and $b \neq 0$,

$$a^{-n} = \left(\frac{1}{a}\right)^{n} \quad \text{and} \quad \left(\frac{a}{b}\right)^{-n} = \left(\frac{b}{a}\right)^{n}.$$

In the previous section, we expanded the definition of an exponent to include *all* integers—positive, negative, or zero. This was done in such a way that all past rules and definitions for exponents are still valid. These rules and definitions are summarized below.

Definitions and Rules for Exponents

If a and b are real numbers, m and n are integers, and no denominators are zero,

Product rule $a^m \cdot a^n = a^{m+n}$

Quotient rule $\dfrac{a^m}{a^n} = a^{m-n}$

Power rules $(a^m)^n = a^{mn}$

$(ab)^m = a^m b^m$

$\left(\dfrac{a}{b}\right)^m = \dfrac{a^m}{b^m}$

Zero exponent $a^0 = 1$

Negative exponent $a^{-n} = \dfrac{1}{a^n} = \left(\dfrac{1}{a}\right)^{n}$

$\dfrac{a^{-n}}{b^{-m}} = \dfrac{b^m}{a^n}$

$\left(\dfrac{a}{b}\right)^{-n} = \left(\dfrac{b}{a}\right)^{n}.$

ANSWERS

2. (a) $\dfrac{1}{64}$ **(b)** $\dfrac{81}{16}$ **(c)** 216

OBJECTIVE 2 ▶ The next two examples show how these definitions and rules for exponents are used to simplify exponential expressions.

E X A M P L E   3   **Using the Definitions and Rules for Exponents**

Simplify.

(a) $3^2 \cdot 3^{-5} = 3^{2+(-5)} = 3^{-3} = \dfrac{1}{3^3}$

(b) $x^{-3} \cdot x^{-4} \cdot x^2 = x^{-3+(-4)+2} = x^{-5} = \dfrac{1}{x^5}$ $(x \neq 0)$

(c) $(2^5)^{-3} = 2^{5(-3)} = 2^{-15} = \dfrac{1}{2^{15}}$

(d) $(4^{-2})^{-5} = 4^{(-2)(-5)} = 4^{10}$

(e) $(x^{-4})^6 = x^{(-4)6} = x^{-24} = \dfrac{1}{x^{24}}$ $(x \neq 0)$

(f) $(xy)^3 = (x^1y^1)^3 = x^3y^3$

Caution
As shown in part (f) of Example 3, $(xy)^3 = x^3y^3$, so that $(xy)^3 \neq xy^3$. Remember that ab^m is *not* the same as $(ab)^m$.

> **WORK PROBLEM 3 AT THE SIDE.** ▶▶

E X A M P L E   4   **Using the Definitions and Rules for Exponents**

Simplify. Assume that all variables represent nonzero real numbers.

(a) $\dfrac{x^{-4}y^2}{x^2y^{-5}} = \dfrac{y^5y^2}{x^4x^2}$ Use $\dfrac{a^{-n}}{b^{-m}} = \dfrac{b^m}{a^n}$.

$\qquad\qquad = \dfrac{y^7}{x^6}$ Product rule

(b) $(2^3x^{-2})^{-2} = \left(\dfrac{2^3}{x^2}\right)^{-2}$ Use $a^{-n} = \dfrac{1}{a^n}$.

$\qquad\qquad = \left(\dfrac{x^2}{2^3}\right)^2$ Use $\left(\dfrac{a}{b}\right)^{-n} = \left(\dfrac{b}{a}\right)^n$.

$\qquad\qquad = \dfrac{(x^2)^2}{(2^3)^2}$ Use $\left(\dfrac{a}{b}\right)^m = \dfrac{a^m}{b^m}$.

$\qquad\qquad = \dfrac{x^4}{2^6}$ or $\dfrac{x^4}{64}$ Use $(a^m)^n = a^{mn}$.

Note
Expressions like those in Example 4 can be correctly simplified in more than one way. For instance, we could simplify Example 4(a) as follows.

$$\dfrac{x^{-4}y^2}{x^2y^{-5}} = \dfrac{x^{-4}}{x^2} \cdot \dfrac{y^2}{y^{-5}}$$

$$= x^{-4-2} \cdot y^{2-(-5)}$$

$$= x^{-6}y^7$$

$$= \dfrac{y^7}{x^6}$$

> **WORK PROBLEM 4 AT THE SIDE.** ▶▶

3. Simplify each expression, and write the answer with a positive exponent.

(a) $5^2 \cdot 5^{-4}$

(b) $p^{-5} \cdot p^{-3} \cdot p^4$ $(p \neq 0)$

(c) $(4^2)^{-5}$

(d) $(m^{-3})^{-4}$ $(m \neq 0)$

(e) $(4a)^5$

4. Simplify, writing answers with positive exponents. Assume all variables are nonzero.

(a) $\dfrac{a^{-3}b^5}{a^4b^{-2}}$

(b) $(3^2k^{-4})^{-1}$

ANSWERS

3. (a) $\dfrac{1}{5^2}$ or $\dfrac{1}{25}$ **(b)** $\dfrac{1}{p^4}$ **(c)** $\dfrac{1}{4^{10}}$
(d) m^{12} **(e)** 4^5a^5

4. (a) $\dfrac{b^7}{a^7}$ **(b)** $\dfrac{k^4}{3^2}$ or $\dfrac{k^4}{9}$

OBJECTIVE ▶ Many numbers that occur in science are very large. For example, the number of one-celled organisms that will sustain a whale for a few hours is 400,000,000,000,000. Other numbers are very small, such as the wavelength of visible light, which is .0000004 meter. Writing these numbers is simpler when we use *scientific* (sy-en-TIFF-ik) *notation*.

Scientific Notation

In **scientific notation,** a number is written as the product of a number between 1 and 10 (or -1 and -10) and some integer power of 10.

For example, because $8000 = 8 \cdot 1000 = 8 \cdot 10^3$, the number 8000 is written in scientific notation as

$$8000 = 8 \times 10^3.$$

(It is customary to use $\times$ instead of a dot to show multiplication.)
The following numbers are not in scientific notation.

$$.230 \times 10^4 \qquad 46.5 \times 10^{-3}$$
$$\uparrow \qquad\qquad \uparrow$$
.230 is less than 1   46.5 is greater than 10

We can convert a number to scientific notation with the following steps. (If the number is negative, ignore the negative sign, go through these steps, and then attach a negative sign to the result.)

Converting to Scientific Notation

Step 1 Place a caret, $\wedge$, to the right of the first nonzero digit.

Step 2 Count the number of digits from the caret to the decimal point. This number gives the absolute value of the exponent on the ten.

Step 3 Decide whether multiplying by 10^n should make the number larger or smaller. The exponent should be positive to make the product larger; it should be negative to make the product smaller.

EXAMPLE 5   Writing a Number in Scientific Notation

Write in scientific notation.

(a) 820,000

Place a caret after the 8 (the first nonzero digit).

$$8_\wedge 20,000$$

Count from the caret to the decimal point, which is understood to be after the last 0.

$$8_\wedge 20,000. \quad \leftarrow \text{Decimal point}$$
$$\qquad\qquad \text{Count 5 places.}$$

Now decide whether the exponent on 10 should be 5 or -5. The number will be written in scientific notation as 8.2×10^n. Comparing 8.2×10^n with 820,000 shows that n must be 5 so the product 8.2×10^5 will equal 820,000.

CONTINUED ON NEXT PAGE ─

(b) −.000072

$$-.00007\!\!\underset{\curvearrowleft}{\wedge}2$$

5 places

Because .000072 is smaller than 7.2, use a *negative* exponent on 10.

$$-.000072 = -7.2 \times 10^{-5}$$

WORK PROBLEM 5 AT THE SIDE. ▶▶

To convert a number written in scientific notation to standard notation, work in reverse. Multiplying a number by a positive power of 10 makes the number larger; multiplying by a negative power of 10 makes the number smaller.

E X A M P L E   6   Converting from Scientific Notation to Standard Notation

Write the following numbers in standard notation.

(a) 6.93×10^5

$$6\underset{\curvearrowright}{\wedge}93000. \quad \text{5 places}$$

Because the exponent is positive, we moved the decimal point 5 places to the right to get a larger number. (We had to attach 3 zeros.)

$$6.93 \times 10^5 = 693,000$$

(b) $3.52 \times 10^7 = 35,200,000$

(c) 4.7×10^{-6}

$$\text{6 places} \quad .000004\underset{\curvearrowleft}{\wedge}7$$

Because of the negative exponent, we moved the decimal point 6 places to the left to get a smaller number.

WORK PROBLEM 6 AT THE SIDE. ▶▶

With scientific notation, rules for exponents can be used to simplify lengthy computation.

E X A M P L E   7   Using Scientific Notation in Computation

Find $\dfrac{1,920,000 \times .0015}{.000032 \times 45,000}$.

First, express all numbers in scientific notation.

$$\frac{1,920,000 \times .0015}{.000032 \times 45,000} = \frac{1.92 \times 10^6 \times 1.5 \times 10^{-3}}{3.2 \times 10^{-5} \times 4.5 \times 10^4}$$

Next, use the commutative property and the rules for exponents to simplify the expression.

— **CONTINUED ON NEXT PAGE**

5. Write in scientific notation.

(a) 400,000

(b) 29,800,000

(c) −6083

(d) .00172

(e) .0000000503

6. Write in standard notation.

(a) 3.7×10^8

(b) 2.51×10^3

(c) 4.6×10^{-5}

(d) 9.372×10^{-6}

(e) 8.5×10^{-1}

ANSWERS

5. (a) 4×10^5 **(b)** 2.98×10^7
(c) -6.083×10^3 **(d)** 1.72×10^{-3}
(e) 5.03×10^{-8}

6. (a) 370,000,000 **(b)** 2510 **(c)** .000046
(d) .000009372 **(e)** .85

7. Use the rules for exponents to find each value.

(a) $\dfrac{8 \times 10^3}{2 \times 10^2}$

$$\frac{1.92 \times 10^6 \times 1.5 \times 10^{-3}}{3.2 \times 10^{-5} \times 4.5 \times 10^4} = \frac{1.92 \times 1.5 \times 10^6 \times 10^{-3}}{3.2 \times 4.5 \times 10^{-5} \times 10^4}$$

$$= \frac{1.92 \times 1.5}{3.2 \times 4.5} \times 10^4$$

Now perform the calculations and write the answer in standard notation.

$$= .2 \times 10^4$$
$$= 2000$$

▶◀ WORK PROBLEM 7 AT THE SIDE.

(b) $\dfrac{9 \times 10^3}{3 \times 10^{-2}}$

To enter numbers in scientific notation, you can use the $\boxed{\text{EE}}$ or $\boxed{\text{EXP}}$ key on a scientific calculator. For instance, to work Example 7 using a calculator with an EE key, enter the following symbols.

$$1.92 \,\boxed{\text{EE}}\, 6 \times 1.5 \,\boxed{\text{EE}}\, 3 \,\boxed{\pm}\, \div (3.2 \,\boxed{\text{EE}}\, 5 \,\boxed{\pm}\, \times 4.5 \,\boxed{\text{EE}}\, 4) =$$

The $\boxed{\text{EXP}}$ key is used in exactly the same way. The result should be 2. 03, which is interpreted as $2 \times 10^3 = 2000$. Notice that the negative exponent -3 is entered by pressing 3, then $\pm$. (Keystrokes vary among different models of calculators, so you should refer to your owner's guide if this sequence does not apply to your particular model.)

In applied examples, we often leave answers in scientific notation.

(c) $\dfrac{.06}{.003}$

E X A M P L E 8 Applying Scientific Notation

The star Sirius is 2.65 light years away from Earth. A light year is approximately 5.880×10^{12} miles. About how many miles is Sirius from Earth?

Multiply the number of light years times the number of miles in a light year. Use a calculator.

$$2.65(5.880 \times 10^{12}) = 1.5582 \times 10^{13}$$

Sirius is about 1.5582×10^{13} miles from Earth.

(d) $\dfrac{200,000 \times .0003}{.06}$

▶◀ WORK PROBLEM 8 AT THE SIDE.

8. The star Pollux is 10.6 light years from Earth. How far is that in miles?

ANSWERS

7. (a) 40 **(b)** 300,000 **(c)** 20 **(d)** 1000
8. 6.2328×10^{13}

5.2 Exercises

Write true *or false* *for each statement.*

1. $(5^2)^3 = 5^2 \cdot 5^3$

2. $(3x)^2 = 3x^2$

3. $4a^{-1} = \dfrac{4}{a}$

4. $\left(\dfrac{1}{a}\right)^{-2} = a^2$

5. Which one of the following is correct?

  (a) $-\dfrac{3}{4} = \left(\dfrac{3}{4}\right)^{-1}$

  (b) $\dfrac{3^{-1}}{4^{-1}} = \left(\dfrac{4}{3}\right)^{-1}$

  (c) $\dfrac{3^{-1}}{4} = \dfrac{3}{4^{-1}}$

  (d) $\dfrac{3^{-1}}{4^{-1}} = \left(\dfrac{3}{4}\right)^{-1}$

6. Which one of the following is incorrect?

  (a) $(3r)^{-2} = 3^{-2}r^{-2}$

  (b) $3r^{-2} = \dfrac{1}{3r^2}$

  (c) $(3r)^{-2} = \dfrac{1}{(3r)^2}$

  (d) $(3r)^{-2} = \dfrac{r^{-2}}{9}$

Decide whether the following expressions have been simplified correctly. If an expression is simplified incorrectly, correct it.

7. $(ab)^2 = ab^2$

8. $(xy)^4 = xy^4$

9. $(5x)^3 = 5^3x^3$

10. $(2k)^5 = 2^5k^5$

11. $\left(\dfrac{4}{a}\right)^5 = \dfrac{4^5}{a}$

12. $\left(\dfrac{7}{y}\right)^3 = \dfrac{7^3}{y}$

13. $(z^4)^5 = z^9$

14. $(m^3)^4 = m^7$

Write each expression with a positive exponent. See Example 2.

15. $\left(\dfrac{3}{4}\right)^{-2}$

16. $\left(\dfrac{2}{5}\right)^{-3}$

17. $\left(\dfrac{6}{5}\right)^{-1}$

18. $\left(\dfrac{8}{3}\right)^{-2}$

Simplify each expression. Write with only positive exponents. Assume that variables represent nonzero real numbers. See Examples 1–4.

19. $(2^{-3} \cdot 5^{-1})^3$

20. $(5^{-4} \cdot 6^{-2})^3$

21. $(5^{-4} \cdot 6^{-2})^{-3}$

22. $(2^{-5} \cdot 3^{-4})^{-1}$

23. $(k^2)^{-3}k^4$ **24.** $(x^3)^{-4}x^5$ **25.** $-4r^{-2}(r^4)^2$ **26.** $-2m^{-1}(m^3)^2$

27. $(5a^{-1})^4(a^2)^{-3}$ **28.** $(3p^{-4})^2(p^3)^{-1}$ **29.** $(z^{-4}x^3)^{-1}$ **30.** $(y^{-2}z^4)^{-3}$

31. $\dfrac{(p^{-2})^3}{5p^4}$ **32.** $\dfrac{(m^4)^{-1}}{9m^3}$ **33.** $\dfrac{4a^5(a^{-1})^3}{(a^{-2})^{-2}}$ **34.** $\dfrac{12k^{-2}(k^{-3})^{-4}}{6k^5}$

35. $\dfrac{(-y^{-4})^2}{6(y^{-5})^{-1}}$ **36.** $\dfrac{2(-m^{-1})^{-4}}{9(m^{-3})^2}$ **37.** $\dfrac{(2k)^2m^{-5}}{(km)^{-3}}$ **38.** $\dfrac{(3rs)^{-2}}{3^2r^2s^{-4}}$

39. In your own words, describe how to rewrite a fraction raised to a negative power as a fraction raised to a positive power.

40. Explain in your own words how we raise a power to a power.

Write each number in scientific notation. See Example 5.

41. 530

42. 1600

43. .830

44. .0072

45. .00000692

46. .875

47. −38,500

48. −976,000,000

Write each of the following in standard notation. See Example 6.

49. 7.2×10^4

50. 8.91×10^2

51. 2.54×10^{-3}

52. 5.42×10^{-4}

53. -6×10^4

54. -9×10^3

55. 1.2×10^{-5}

56. 2.7×10^{-6}

Use the rules for exponents to find each value. See Example 7.

57. $\dfrac{3 \times 10^{-2}}{12 \times 10^3}$

58. $\dfrac{5 \times 10^{-3}}{25 \times 10^2}$

59. $\dfrac{.05 \times 1600}{.0004}$

60. $\dfrac{.003 \times 40,000}{.00012}$

Use scientific notation to work the following problems. See Example 8.

61. The planet Mercury has a mean distance from the sun of 3.6×10^7 miles, while the mean distance of Venus to the sun is 6.7×10^7 miles. How long would it take a spacecraft traveling at 1.55×10^3 miles per hour to travel from Venus to Mercury?

62. Use the information from the previous exercise to find the number of days it would take the spacecraft to travel from Venus to Mercury. Round your answer to the nearest whole number of days.

63. In a state lottery, a player must choose six numbers from 1 through 40. It can be shown that there are 3.83838×10^6 different ways to do this. Suppose that a group of 1000 persons decides to purchase tickets for all these numbers and each ticket costs $1.00. How much should each person expect to pay?

64. According to Bode's law, the distance d of the nth planet from the sun is described by the function

$$d(n) = \frac{3(2^{n-2}) + 4}{10},$$

in astronomical units. Find the distance of each of the following planets from the sun.
(a) Venus $(n = 2)$ **(b)** Earth $(n = 3)$
(c) Mars $(n = 4)$

The graph shows the number of "800" phone calls made in the AT&T system (in dark blue) plus others made (estimated, in light blue), in billions of calls (1 billion = 1,000,000,000).

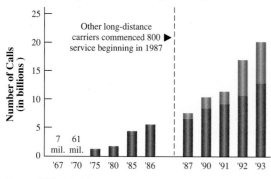

─────── **'800' PHONE CALLS** ───────
AT&T system usage, plus others after 1986
(estimated, in light blue)

Other long-distance
carriers commenced 800 ▶
service beginning in 1987

7   61
mil. mil.

'67 '70 '75 '80 '85 '86   '87 '90 '91 '92 '93

Sources: AT&T; Federal Communications Commission

65. In which year was the number of calls made in the AT&T system approximately 7×10^9?

66. In which year was the total number of calls approximately 1.7×10^{10}?

5.3 Addition and Subtraction of Polynomials

OBJECTIVES

1. Know the basic definitions for polynomials.

2. Identify monomials, binomials, and trinomials.

3. Find the degree of a polynomial.

4. Add and subtract polynomials.

5. Recognize and evaluate polynomial functions.

FOR EXTRA HELP

Tutorial   Tape 7   SSM, Sec. 5.3

Just as whole numbers are the basis of arithmetic, *polynomials* (defined below) are the basis of algebra. Recall that a *term* is a number, a variable, or the product of a number and one or more variables raised to a power. Examples of terms include

$$4x, \quad \frac{1}{2}m^5 \left(\text{or } \frac{m^5}{2}\right), \quad -7z^9, \quad \text{and} \quad 5.$$

Any factor in a term is the **coefficient** of the product of the remaining factors. For example, $3x^2$ is the coefficient of y in the term $3x^2y$, and $3y$ is the coefficient of x^2 in $3x^2y$. However, "coefficient" is often used to mean "numerical coefficient," and from now on we will use it that way in this book. With this understanding, in the term $8x^3$, the coefficient is 8, and in the term $-4p^5$, it is -4. The coefficient of the term k is understood to be 1. The coefficient of $-r$ is -1.

> **WORK PROBLEM 1 AT THE SIDE.**

OBJECTIVE ▶ Recall that any combination of variables or constants joined by the basic operations of addition, subtraction, multiplication, division (except by zero), or extraction of roots is called an *algebraic expression*. The simplest kind of algebraic expression is a **polynomial** (pah-luh-NOH-mee-ul), a term or a finite sum of terms in which all variables have only whole number exponents so no variables appear in denominators. Examples of polynomials include

$$3x - 5, \quad 4m^3 - 5m^2p + 8, \quad \text{and} \quad -5t^3s^3.$$

On the other hand, the expressions

$$5x^2 + 2x + \frac{1}{x} \quad \text{and} \quad x^{-3} - 6x$$

are *not* polynomials because variables appear in a denominator in the first case, and an exponent on a variable is negative in the second case. Even though the expression $3x - 5$ involves subtraction, it is still called a sum of terms because it could be written as $3x + (-5)$.

Although polynomials may have more than one variable, most of the polynomials discussed in this chapter have only one variable. A polynomial containing only the variable x is called a **polynomial in x.** A polynomial in one variable is written in **descending** (dee-SEND-ing) **powers** of the variable if the exponents on the terms of the polynomial decrease from left to right. For example,

$$x^5 - 6x^2 + 12x - 5$$

is a polynomial in descending powers of x (the term -5 can be thought of as $-5x^0$). This is usually the preferred way to write a polynomial.

EXAMPLE 1 Writing a Polynomial in Descending Powers

Write each of the following in descending powers of the variable.

(a) $y - 6y^3 + 8y^5 - 9y^4 + 12$

Write the polynomial as

$$8y^5 - 9y^4 - 6y^3 + y + 12.$$

(b) $-2 + m + 6m^2 - 4m^3$ would be written as

$$-4m^3 + 6m^2 + m - 2.$$

1. Identify each coefficient.

(a) $-9m^5$

(b) $12y^2x$

(c) x

(d) $-y$

2. Write each polynomial in descending powers.

(a) $-4 + 9y + y^3$

(b) $-3z^4 + 2z^3 + z^5 - 6z$

(c) $-12m^{10} + 8m^9 + 10m^{12}$

3. Identify each polynomial as a trinomial, binomial, monomial, or none of these.

(a) $12m^4 - 6m^2$

(b) $-6y^3 + 2y^2 - 8y$

(c) $3a^5$

(d) $-2k^{10} + 2k^9 - 8k^5 + 2k$

2. (a) $y^3 + 9y - 4$
   (b) $z^5 - 3z^4 + 2z^3 - 6z$
   (c) $10m^{12} - 12m^{10} + 8m^9$

3. (a) binomial (b) trinomial
   (c) monomial (d) none of these

◄◄ WORK PROBLEM 2 AT THE SIDE.

OBJECTIVE 2 ▶ Certain types of polynomials are so common that they are given special names based on the number of terms that they have. A polynomial with exactly three terms is a **trinomial** (TRY-noh-mee-ul), while a polynomial with two terms is a **binomial** (BY-noh-mee-ul). A single-term polynomial is a **monomial** (MAH-noh-mee-ul).

E X A M P L E 2   Identifying Types of Polynomials

The list below gives examples of monomials, binomials, and trinomials, as well as polynomials that are none of these.

Type of Polynomial	*Examples*
Monomial	$5x$, $7m^9$, -8
Binomial	$3x^2 - 6$, $11y + 8$, $5k + 15$
Trinomial	$y^2 + 11y + 6$, $8p^3 - 7p + 2m$, $-3 + 2k^5 + 9z^4$
None of the above	$p^3 - 5p^2 + 2p - 5$, $-9z^3 + 5c^3 + 2m^5 + 11r^2 - 7r$

◄◄ WORK PROBLEM 3 AT THE SIDE.

OBJECTIVE 3 ▶ The **degree** (dih-GREE) **of a term** with one variable is the exponent on that variable. For example, the degree of $2x^3$ is 3, the degree of $-x^4$ is 4, and the degree of $17x$ is 1. The degree of a term in more than one variable is defined to be the sum of the exponents of the variables. For example, the degree of $5xy^2z^3$ is 6 because $1 + 2 + 3$ is 6. (Remember, $x = x^1$.)

The greatest degree of any of the terms in a polynomial is called the **degree of the polynomial.** In most cases, we will be interested in finding the degree of a polynomial in one variable.

E X A M P L E 3   Finding the Degree of a Polynomial

Find the degree of each polynomial.

(a) $9x^2 - 5x + 8$ — Largest exponent is 2.
   The largest exponent is 2, so the polynomial is of degree 2.

(b) $17m^9 + 8m^{14} - 9m^3$
   This polynomial is of degree 14.

(c) $5x$
   The degree is 1 because $5x = 5x^1$.

(d) -2
   A constant term, other than 0, has degree 0. (This is because, for example, $-2 = -2 \cdot x^0$.)

(e) $x^3y^9 + 21xy^4 + 7xy$
   The degrees of the terms are 12, 5, and 2. Therefore, the degree of the polynomial is 12, the highest degree of any of the terms.

Note
In Example 3, we saw that the degree of a constant is 0. There is no degree for the constant 0 itself, however, because 0 times a variable to any power is 0.

WORK PROBLEM 4 AT THE SIDE. ▶▶

OBJECTIVE 4 In Chapter 1 we used the distributive property to combine like terms, as follows.

$$4x^2 + 5x^2 = (4 + 5)x^2 = 9x^2.$$

Unlike terms, such as $4x$ and $7x^2$, cannot be combined.

─ **E X A M P L E   4   Combining Terms**

Combine terms.

(a) $-5y^3 + 8y^3 - 6y^3$

These like terms may be combined by using the distributive property.

$$-5y^3 + 8y^3 - 6y^3 = (-5 + 8 - 6)y^3 = -3y^3$$

(b) $6x + 5y - 9x + 2y$

Use the associative and commutative properties to rewrite the expression with all the x's together and all the y's together.

$$6x + 5y - 9x + 2y = \mathbf{6x - 9x + 5y + 2y}$$

Now combine like terms:

$$= -3x + 7y.$$

It is not possible to go further because $-3x$ and $7y$ are unlike terms.

(c) $5x^2y - 6xy^2 + 9x^2y + 13xy^2 = 5x^2y + 9x^2y - 6xy^2 + 13xy^2$
$$= 14x^2y + 7xy^2$$

WORK PROBLEM 5 AT THE SIDE. ▶▶

The following rule is used to add two polynomials.

Adding Polynomials

To add two polynomials, combine like terms.

─ **E X A M P L E   5   Adding Polynomials Horizontally**

Add $4k^2 - 5k + 2$ and $-9k^2 + 3k - 7$.

Use the commutative and associative properties to rearrange the polynomials so that like terms are together.

$$(4k^2 - 5k + 2) + (-9k^2 + 3k - 7)$$
$$= 4k^2 - 9k^2 - 5k + 3k + 2 - 7$$
$$= -5k^2 - 2k - 5$$

4. Give the degree of each polynomial.

(a) $9y^4 + 8y^3 - 6$

(b) $-12m^7 + 11m^3 + m^9$

(c) $-2k$

(d) 10

(e) $3mn^2 + 2m^3n$

5. Combine terms.

(a) $11x + 12x - 7x - 3x$

(b) $11p^5 + 4p^5 - 6p^3 + 8p^3$

(c) $2y^2z^4 + 3y^4 + 5y^4 - 9y^4z^2$

(d) $2x^2y^3 + 5x^3y^2$

ANSWERS
4. (a) 4 **(b)** 9 **(c)** 1 **(d)** 0 **(e)** 4
5. (a) $13x$ **(b)** $15p^5 + 2p^3$
   (c) $2y^2z^4 + 8y^4 - 9y^4z^2$
   (d) cannot be simplified

6. Add.

(a) $-4x^3 + 2x^2$
$\underline{8x^3 - 6x^2}$

EXAMPLE 6 Adding Polynomials Vertically

The two polynomials of Example 5 can also be added vertically by lining up like terms in columns, then adding by columns.

$$4k^2 - 5k + 2$$
$$\underline{-9k^2 + 3k - 7}$$
$$-5k^2 - 2k - 5 \qquad \text{Add columns.}$$

EXAMPLE 7 Adding Polynomials

Add $3a^5 - 9a^3 + 4a^2$ and $-8a^5 + 8a^3 + 2$.
By the first method shown above,

$$(3a^5 - 9a^3 + 4a^2) + (-8a^5 + 8a^3 + 2)$$
$$= 3a^5 - 8a^5 - 9a^3 + 8a^3 + 4a^2 + 2$$
$$= -5a^5 - a^3 + 4a^2 + 2. \qquad \text{Combine like terms.}$$

(b) $(-5p^3 + 6p^2)$
$+ (8p^3 - 12p^2)$

We can also add these two polynomials by placing them in columns, with like terms in the same columns.

$$3a^5 - 9a^3 + 4a^2$$
$$\underline{-8a^5 + 8a^3 + 2}$$
$$-5a^5 - a^3 + 4a^2 + 2 \qquad \text{Add like terms.}$$

For many people, there is less chance of error with vertical addition.

(c) $-6r^5 + 2r^3 - r^2$
$\underline{8r^5 - 2r^3 + 5r^2}$

◀◀ **WORK PROBLEM 6 AT THE SIDE.**

In Chapter 1, subtraction of real numbers was defined for real numbers a and b as

$$a - b = a + (-b).$$

That is, the first number and the negative of the second are added. The definition for subtraction of polynomials is similar. The negative of a polynomial is defined as the polynomial with every sign changed. Thus, polynomials are subtracted as follows.

Subtracting Polynomials

To subtract two polynomials, change the signs of the second polynomial and add.

(d) $(12y^2 - 7y + 9)$
$+ (-4y^2 - 11y + 5)$

EXAMPLE 8 Subtracting Polynomials Horizontally

Subtract: $(-6m^2 - 8m + 5) - (-3m^2 + 7m - 8)$.
  Change every sign in the second polynomial and add.

$$(-6m^2 - 8m + 5) - (-3m^2 + 7m - 8)$$
$$= -6m^2 - 8m + 5 + 3m^2 - 7m + 8$$
$$\uparrow \uparrow \uparrow$$
$$\text{Change every sign.}$$

CONTINUED ON NEXT PAGE

ANSWERS

6. (a) $4x^3 - 4x^2$ **(b)** $3p^3 - 6p^2$
  (c) $2r^5 + 4r^2$
  (d) $8y^2 - 18y + 14$

Now add by combining terms.

$$= -6m^2 + 3m^2 - 8m - 7m + 5 + 8$$
$$= -3m^2 - 15m + 13$$

Check by adding $-3m^2 - 15m + 13$ to the second polynomial. The result should be the first polynomial.

WORK PROBLEM 7 AT THE SIDE. ▶▶

EXAMPLE 9 Subtracting Polynomials Vertically

Use the same polynomials as in Example 8, and subtract in columns.

Write the first polynomial over the second, lining up like terms in columns.

$$-6m^2 - 8m + 5$$
$$-3m^2 + 7m - 8$$

Change all the signs in the second polynomial, and add.

$$\begin{array}{l} -6m^2 - 8m + 5 \\ +3m^2 - 7m + 8 \qquad \text{All signs changed} \\ \hline -3m^2 - 15m + 13 \qquad \text{Add in columns.} \end{array}$$

WORK PROBLEM 8 AT THE SIDE. ▶▶

OBJECTIVE 5 In Chapter 3 we studied linear (first-degree polynomial) functions, defined as $f(x) = mx + b$. Now we can consider more general polynomial functions.

Polynomial Functions

A **polynomial function of degree n** is defined by

$$P(x) = a_n x^n + a_{n-1} x^{n-1} + \cdots + a_1 x + a_0,$$

for real numbers $a_n, a_{n-1}, \ldots, a_1$, and a_0, where $a_n \neq 0$ and n is a whole number.

We use $P(x)$ in this definition instead of $f(x)$ to emphasize that these are polynomial functions. Of course $f(x)$ could be used instead.

As shown in Chapter 3, if $x = -2$ then $P(x) = 3x^2 - 5x + 7$ takes the value

$$P(-2) = 3(-2)^2 - 5(-2) + 7 \qquad \text{Let } x = -2.$$
$$= 3(4) + 10 + 7 = 29.$$

Caution

Note that $P(x)$ does *not* mean P times x. It is a special notation to name a polynomial function with the variable x.

7. Subtract.

(a) $(2a^2 - a) - (5a^2 + 8a)$

(b) $\begin{array}{l}(6y^3 - 9y^2 + 8) \\ -(2y^3 + y^2 + 5)\end{array}$

8. Subtract.

(a) $\begin{array}{l}m^4 - 2m^3 \\ 5m^4 + 8m^3\end{array}$

(b) $\begin{array}{l}6y^3 - 2y^2 + 5y \\ -2y^3 + 8y^2 - 11y\end{array}$

9. Let $P(x) = -x^2 + 5x - 11$.
Find each of the following.

(a) $P(1)$

(b) $P(-4)$

(c) $P(5)$

(d) $P(0)$

E X A M P L E   10   **Evaluating a Polynomial Function**

Let $P(x) = 4x^3 - x^2 + 5$. Find each of the following.

(a) $P(3)$

First, substitute 3 for x.

$$P(x) = 4x^3 - x^2 + 5$$
$$P(3) = 4 \cdot 3^3 - 3^2 + 5 \qquad \text{Let } x = 3.$$

Now use the order of operations from Chapter 1.

$$= 4 \cdot 27 - 9 + 5 \qquad \text{Evaluate exponentials.}$$
$$= 108 - 9 + 5 \qquad \text{Multiply.}$$
$$= 104 \qquad \text{Subtract and add.}$$

(b) $P(-4) = 4 \cdot (-4)^3 - (-4)^2 + 5 \qquad \text{Let } x = -4.$

$$= 4 \cdot (-64) - 16 + 5$$
$$= -267$$

◀◀ **WORK PROBLEM 9 AT THE SIDE.**

Later we will investigate second-degree polynomial functions, called *quadratic functions,* in detail.

ANSWERS

9. (a) -7 **(b)** -47 **(c)** -11 **(d)** -11

5.3 Exercises

1. Which one of the following is *not* a polynomial?
   (a) $2x - 3$ **(b)** $4p^2 + 3p - 1$ **(c)** $5m^{-2} + m^2$ **(d)** 7

2. Which one of the following is *not* written in descending powers of y?
   (a) $4y^3 + 2y - y^6$ **(b)** $y^4 - 3y^2 - 5$ **(c)** $y^3 + y$ **(d)** $y^2 + 2y - 3$

3. Which one of the following is a trinomial in descending powers, having degree 6?
   (a) $5x^6 - 4x^5 + 12$ **(b)** $6x^5 - x^6 + 4$ **(c)** $2x + 4x^2 - x^6$ **(d)** $4x^6 - 6x^4 + 9x^2 - 8$

4. Give an example of a polynomial of four terms in the variable x, having degree 5, written in descending powers, lacking a fourth degree term.

Give the coefficient and the degree of each term. See Example 3.

5. $7z$ **6.** $3r$ **7.** $-15p^2$ **8.** $-27k^3$

9. x^4 **10.** y^6 **11.** $-mn^5$ **12.** $-a^5 b$

13. The exponent in the expression 4^5 is 5. Explain why the degree of 4^5 is not 5. What is its degree?

14. What is the coefficient of 4^5?

We defined a polynomial written in descending powers in the text. See Example 1. Sometimes we write a polynomial in ascending powers, *with the degree of the terms increasing from left to right. Decide whether each polynomial is written in descending powers, ascending powers, or neither.*

15. $2x^3 + x - 3x^2$ **16.** $3y^5 + y^4 - 2y^3 + y$ **17.** $4p^3 - 8p^5 + p^7$

18. $q^2 + 3q^4 - 2q + 1$ **19.** $-m^3 + 5m^2 + 3m + 10$ **20.** $4 - x + 3x^2$

Identify each polynomial as a monomial, binomial, trinomial, *or* none of these. *Give the degree of each. See Examples 2 and 3.*

Polynomial	Type	Degree	Polynomial	Type	Degree
21. 24			**22.** 5		
23. $7m - 21$			**24.** $-x^2 + 3x^5$		
25. $2r^3 + 3r^2 + 5r$			**26.** $5z^2 - 5z + 7$		
27. $-6p^4q - 3p^3q^2 + 2pq^3 - q^4$			**28.** $8s^3t - 3s^2t^2 + 2st^3 + 9$		

Combine terms. See Example 4.

29. $5z^4 + 3z^4$

30. $8r^5 - 2r^5$

31. $-m^3 + 2m^3 + 6m^3$

32. $3p^4 + 5p^4 - 2p^4$

33. $x + x + x + x + x$

34. $z - z - z + z$

35. $m^4 - 3m^2 + m$

36. $5a^5 + 2a^4 - 9a^3$

37. $y^2 + 7y - 4y^2$

38. $2c^2 - 4 + 8 - c^2$

39. $2k + 3k^2 + 5k^2 - 7$

40. $4x^2 + 2x - 6x^2 - 6$

41. $n^4 - 2n^3 + n^2 - 3n^4 + n^3$

42. $2q^3 + 3q^2 - 4q - q^3 + 5q^2$

Add or subtract as indicated. See Examples 5–9.

43. Add.
$$-12p^2 + 4p - 1$$
$$\underline{\quad 3p^2 + 7p - 8}$$

44. Add.
$$-6y^3 + 8y + 5$$
$$\underline{\quad 9y^3 + 4y - 6}$$

45. Subtract.
$$12a + 15$$
$$\underline{\quad 7a - 3}$$

46. Subtract.
$$-3b + 6$$
$$\underline{\quad 2b - 8}$$

47. Subtract.
$$6m^2 - 11m + 5$$
$$\underline{-8m^2 + \quad 2m - 1}$$

48. Subtract.
$$-4z^2 + 2z - 1$$
$$\underline{\quad 3z^2 - 5z + 2}$$

49. Add.
$$12z^2 - 11z + 8$$
$$5z^2 + 16z - 2$$
$$\underline{-4z^2 + \quad 5z - 9}$$

50. Add.
$$-6m^3 + 2m^2 + 5m$$
$$8m^3 + 4m^2 - 6m$$
$$\underline{-3m^3 + 2m^2 - 7m}$$

51. Add.
$$6y^3 - 9y^2 \qquad + 8$$
$$\underline{4y^3 + 2y^2 + 5y \qquad}$$

52. Add.
$$-7r^8 + 2r^6 - r^5$$
$$\underline{\qquad 3r^6 \qquad + 5}$$

53. Subtract.
$$-5a^4 \qquad + 8a^2 - 9$$
$$\underline{\qquad 6a^3 - \quad a^2 + 2}$$

54. Subtract.
$$\qquad - 2m^3 + 8m^2$$
$$\underline{m^4 - \quad m^3 \qquad + 2m}$$

55. $(3r + 8) - (2r - 5)$

56. $(2d + 7) - (3d - 1)$

57. $(5x^2 + 7x - 4) + (3x^2 - 6x + 2)$

58. $(4k^3 + k^2 + k) + (2k^3 - 4k^2 - 3k)$

59. $(2a^2 + 3a - 1) - (4a^2 + 5a + 6)$

60. $(q^4 - 2q^2 + 10) - (3q^4 + 5q^2 - 5)$

61. $(z^5 + 3z^2 + 2z) - (4z^5 + 2z^2 - 5z)$

62. $(5t^3 - 3t^2 + 2t) - (4t^3 + 2t^2 + 3t)$

63. $(9y^4 - 5y^2 + 10) + (2 - 5y^2 - 3y^4)$

64. $(2x^6 + 3x^3 - x) + (4x + 2x^3 - 3x^6)$

65. $[-(4m^2 - 8m + 4m^3) - (3m^2 + 2m + 5m^3)] + m^2$

66. $[-(y^4 - y^2 + 1) - (y^4 + 2y^2 + 1)] + (3y^4 - 3y^2 - 2)$

For each polynomial function, find (a) $P(-1)$ and (b) $P(2)$. See Example 10.

67. $P(x) = 6x - 4$

68. $P(x) = -2x + 5$

69. $P(x) = x^2 - 3x + 4$

70. $P(x) = 3x^2 + x - 5$

71. $P(x) = 5x^4 - 3x^2 + 6$

72. $P(x) = -4x^4 + 2x^2 - 1$

Addition and subtraction of polynomials may be used to add and subtract polynomial functions. Work each of the following problems.

73. Let $P(z) = 2z - 3z^2$, $Q(z) = 3z^2 + 2$, and $R(z) = 6z + 2z^2$. Find $Q(z) + R(z) - P(z)$.

74. Let $P(k) = -5k + 3k^2$, $Q(k) = 2k - 3k^3$, and $R(k) = 4k + 6k^2$. Find $Q(k) - P(k) + R(k)$.

The polynomial function $P(x) = 1.06x^3 - 6.00x^2 + 7.86x + 31.6$ gives sales of recorded rock music as a percent of all recorded music sales for the years 1990 through 1994, where $x = 0$ corresponds to 1990, $x = 1$ corresponds to 1991, and so on. This is an example of a mathematical model. Use a calculator to determine to the nearest tenth of a percent all recorded music sales represented by rock in the following years.

75. 1990

76. 1991

77. 1993

78. 1994

79. Assuming that the polynomial in Exercises 75–78 continues to be an accurate representation, estimate the percent of sales in 1998. Is your answer reasonable? Explain why.

5.4 Multiplication of Polynomials

OBJECTIVES

 Define operations on functions.

 Multiply monomials.

 Multiply any two polynomials.

Multiply two binomials.

Find the product of the sum and difference of two terms.

Find the square of a binomial.

FOR EXTRA HELP

| Tutorial | Tape 7 | SSM, Sec. 5.4 |

OBJECTIVE 1 In the previous section we saw how polynomials are added and subtracted. If each of two polynomials defines a function, then the sum or difference of the polynomials gives the sum or difference of the two functions. For example, if $P(x) = 3x^2 - 5x$ and $Q(x) = 5x^3 - x^2 + 8$, then

$$P(x) + Q(x) = (3x^2 - 5x) + (5x^3 - x^2 + 8)$$
$$= 5x^3 + 2x^2 - 5x + 8.$$

Several operations on functions are defined below.

Operations on Functions

For all values of x for which both functions $f(x)$ and $g(x)$ exist,

$$(f + g)(x) = f(x) + g(x),$$
$$(f - g)(x) = f(x) - g(x),$$
$$(fg)(x) = f(x) \cdot g(x),$$
$$\left(\frac{f}{g}\right)(x) = \frac{f(x)}{g(x)}, \quad \text{where } g(x) \neq 0.$$

In this chapter we will be most interested in operations on polynomial functions. Section 5.3 illustrated addition and subtraction, this section illustrates multiplication, and Section 5.5 illustrates division.

OBJECTIVE 2 A polynomial with just one term is a monomial. To see how to multiply two monomials, recall that the product of the two terms $3x^4$ and $5x^3$ is found by using the commutative and associative properties, along with the rules for exponents.

$$(3x^4)(5x^3) = 3 \cdot 5 \cdot x^4 \cdot x^3$$
$$= 15x^{4+3}$$
$$= 15x^7$$

1. Find each product.

(a) $-6m^5(2m^4)$

EXAMPLE 1 Multiplying Monomials

(a) $(-4a^3)(3a^5) = (-4)(3)a^3 \cdot a^5 = -12a^8$

(b) $(2m^2z^4)(8m^3z^2) = (2)(8)m^2 \cdot m^3 \cdot z^4 \cdot z^2 = 16m^5z^6$

(b) $(8k^3y)(9ky^3)$

WORK PROBLEM 1 AT THE SIDE. ▶▶

OBJECTIVE 3 The distributive property can be used to extend this process to find the product of any two polynomials.

EXAMPLE 2 Multiplying a Monomial and a Polynomial

(a) Find the product of -2 and $8x^3 - 9x^2$.
Use the distributive property.

$$-2(8x^3 - 9x^2) = -2(8x^3) - 2(-9x^2)$$
$$= -16x^3 + 18x^2$$

CONTINUED ON NEXT PAGE

ANSWERS

1. (a) $-12m^9$ **(b)** $72k^4y^4$

2. Find each product.

(a) $-2r(9r - 5)$

(b) $3p^2(5p^3 + 2p^2 - 7)$

3. Find each product.

(a) $2m - 5$
$3m + 4$

(b) $(4a - 5)(3a + 6)$

(c) $(2k - 5m)(3k + 2m)$

(b) Find the product of $5x^2$ and $-4x^2 + 3x - 2$.

$$5x^2(-4x^2 + 3x - 2)$$
$$= 5x^2(-4x^2) + 5x^2(3x) + 5x^2(-2) \quad \text{Distributive property}$$
$$= -20x^4 + 15x^3 - 10x^2$$

◀◀ **WORK PROBLEM 2 AT THE SIDE.**

E X A M P L E   3   Multiplying Two Polynomials

Find the product of $3x - 4$ and $2x^2 + x$.

Use the distributive property to multiply each term of $2x^2 + x$ by $3x - 4$.

$$(3x - 4)(2x^2 + x) = (3x - 4)(2x^2) + (3x - 4)(x)$$

Here $3x - 4$ has been treated as a single expression so that the distributive property could be used. Now use the distributive property twice again.

$$(3x - 4)(2x^2 + x) = 3x(2x^2) + (-4)(2x^2) + (3x)(x) + (-4)(x)$$
$$= 6x^3 - 8x^2 + 3x^2 - 4x$$
$$= 6x^3 - 5x^2 - 4x$$

It is often easier to multiply polynomials by writing them vertically, in much the same way that numbers are multiplied. We proceed as follows to find the product in Example 3 by this method. (Notice how this process is similar to finding the product of two numbers, such as 24×78.)

Step 1 Multiply x and $3x - 4$.

$$\begin{array}{r} 3x - 4 \\ 2x^2 + x \\ \hline x(3x - 4) \rightarrow 3x^2 - 4x \end{array}$$

Step 2 Multiply $2x^2$ and $3x - 4$.
Line up like terms of the products in columns.

$$\begin{array}{r} 3x - 4 \\ 2x^2 + x \\ \hline 3x^2 - 4x \\ 2x^2(3x - 4) \rightarrow 6x^3 - 8x^2 \end{array}$$

Step 3 Combine like terms.

$$6x^3 - 5x^2 - 4x$$

E X A M P L E   4   Multiplying Polynomials Vertically

Find the product of $5a - 2b$ and $3a + b$.

$$\begin{array}{r} 5a - 2b \\ 3a + b \\ \hline 5ab - 2b^2 \quad b(5a - 2b) \\ 15a^2 - 6ab 3a(5a - 2b) \\ \hline 15a^2 - ab - 2b^2 \quad \text{Combine like terms.} \end{array}$$

◀◀ **WORK PROBLEM 3 AT THE SIDE.**

E X A M P L E 5   Multiplying Polynomials Vertically

Find the product of $3m - 5$ and $3m^3 - 2m^2 + 4$.

$$
\begin{array}{r}
3m^3 - 2m^2 + 4 \\
3m - 5 \\
\hline
-15m^3 + 10m^2 \qquad\quad - 20 \\
9m^4 - 6m^3 \qquad\quad + 12m \\
\hline
9m^4 - 21m^3 + 10m^2 + 12m - 20
\end{array}
$$

-5 times $3m^3 - 2m^2 + 4$
$3m$ times $3m^3 - 2m^2 + 4$
Combine like terms.

4. Find each product.

(a) $(r^4 - 2r^3 + 6)(3r - 1)$

> **WORK PROBLEM 4 AT THE SIDE.** ▶▶

OBJECTIVE ▶ In working with polynomials, a special kind of product comes up repeatedly—the product of two binomials. In the rest of this section, we will discuss a shortcut method for finding these products.

Recall that a binomial is a polynomial with just two terms, such as $3x - 4$ or $2x + 3$. We can find the product of the binomials $3x - 4$ and $2x + 3$ by using the distributive property as follows.

$$
\begin{aligned}
(3x - 4)(2x + 3) &= (3x - 4)(2x) + (3x - 4)(3) \\
&= (3x)(2x) + (-4)(2x) + (3x)(3) + (-4)(3) \\
&= 6x^2 - 8x + 9x - 12
\end{aligned}
$$

Before combining like terms to find the simplest form of the answer, let us check the origin of each of the four terms in the sum.

$6x^2$ is the product of the *first* terms.

$$(3x - 4)(2x + 3) \qquad (3x)(2x) = 6x^2 \qquad \text{First terms}$$

$9x$ is the product of the *outside* terms.

$$(3x - 4)(2x + 3) \qquad 3x(3) = 9x \qquad \text{Outside terms}$$

$-8x$ is the product of the *inside terms*.

$$(3x - 4)(2x + 3) \qquad -4(2x) = -8x \qquad \text{Inside terms}$$

-12 is the product of the *last terms*.

$$(3x - 4)(2x + 3) \qquad -4(3) = -12 \qquad \text{Last terms}$$
$$\downarrow$$
$$\text{FOIL}$$

(b) $5a^3 - 6a^2 + 2a - 3$
$\qquad\qquad\qquad 2a - 5$

The product is found by combining these four results.

$$
\begin{aligned}
(3x - 4)(2x + 3) &= 6x^2 + 9x - 8x - 12 \\
&= 6x^2 + x - 12
\end{aligned}
$$

To keep track of the order of multiplying these terms, use the initials FOIL (First-Outside-Inside-Last). The steps of the FOIL (pronounced "foil") method can be done quickly, as follows.

The FOIL method will be very helpful in factoring, which is discussed in the rest of this chapter.

ANSWERS
4. **(a)** $3r^5 - 7r^4 + 2r^3 + 18r - 6$
   (b) $10a^4 - 37a^3 + 34a^2 - 16a + 15$

5. Use the FOIL method to find each product.

(a) $(3z - 2)(z + 1)$

(b) $(5r + 3)(2r - 5)$

6. Find each product.

(a) $(4p + 5q)(3p - 2q)$

(b) $(4y - z)(2y + 3z)$

(c) $(8r + 1)(8r - 1)$

(d) $(3p + 5)(3p + 5)$

E X A M P L E 6   Using the FOIL Method

Use the FOIL method to find $(4m - 5)(3m + 1)$.
   Find the product of the first terms.

$$(4m - 5)(3m + 1) \qquad (4m)(3m) = 12m^2 \qquad \text{F}$$

Multiply the outside terms.

$$(4m - 5)(3m + 1) \qquad (4m)(1) = 4m \qquad \text{O}$$

Find the product of the inside terms.

$$(4m - 5)(3m + 1) \qquad (-5)(3m) = -15m \qquad \text{I}$$

Multiply the last terms.

$$(4m - 5)(3m + 1) \qquad (-5)(1) = -5 \qquad \text{L}$$

Combine the four terms obtained above, and simplify.

$$(4m - 5)(3m + 1) = 12m^2 + 4m - 15m - 5$$
$$= 12m^2 - 11m - 5$$

Practice adding the middle terms mentally so that you can write down the three terms of the answer directly, or use the compact form shown below.

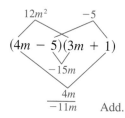

Combine these four results to get $12m^2 - 11m - 5$.

◀◀ **WORK PROBLEM 5 AT THE SIDE.**

E X A M P L E 7   Using the FOIL Method

Find each product.

(a) $(6a - 5b)(3a + 4b) = 18a^2 + 24ab - 15ab - 20b^2$

$$\qquad\qquad\qquad\qquad\qquad \text{F} \qquad \text{O} \qquad \text{I} \qquad \text{L}$$
$$= 18a^2 + 9ab - 20b^2$$

(b) $(2k + 3z)(5k - 3z) = 10k^2 - 6kz + 15kz - 9z^2$
$$= 10k^2 + 9kz - 9z^2$$

◀◀ **WORK PROBLEM 6 AT THE SIDE.**

OBJECTIVE 5 A special type of binomial product that occurs frequently is the product of the sum and difference of the same two terms. By the FOIL method, the product of $(x + y)(x - y)$ is

$$(x + y)(x - y) = x^2 - xy + xy - y^2$$
$$= x^2 - y^2.$$

ANSWERS

5. (a) $3z^2 + z - 2$ **(b)** $10r^2 - 19r - 15$
6. (a) $12p^2 + 7pq - 10q^2$
   (b) $8y^2 + 10yz - 3z^2$
   (c) $64r^2 - 1$ **(d)** $9p^2 + 30p + 25$

The **product of the sum and difference of two terms** is the difference of the squares of the terms, or

$$(x + y)(x - y) = x^2 - y^2.$$

EXAMPLE 8   Multiplying the Sum and Difference of Two Terms

$$(x + y)(x - y) = x^2 - y^2$$

(a) $(p + 7)(p - 7) = p^2 - 7^2 = p^2 - 49$

(b) $(2r + 5)(2r - 5) = (2r)^2 - 5^2$
$$= 2^2 r^2 - 25$$
$$= 4r^2 - 25$$

(c) $(6m + 5n)(6m - 5n) = (6m)^2 - (5n)^2$
$$= 36m^2 - 25n^2$$

(d) $(9y + 2z)(9y - 2z) = (9y)^2 - (2z)^2$
$$= 81y^2 - 4z^2$$

WORK PROBLEM 7 AT THE SIDE. ▶▶

OBJECTIVE 6 ▶ Another special binomial product is the *square of a binomial*. To find the square of $x + y$ or $(x + y)^2$, multiply $x + y$ and $x + y$.

$$(x + y)^2 = (x + y)(x + y) = x^2 + xy + xy + y^2$$
$$= x^2 + 2xy + y^2$$

A similar result is true for the square of a difference, as shown below.

The **square of a binomial** is the sum of the square of the first term, twice the product of the two terms, and the square of the last term.

$$(x + y)^2 = x^2 + 2xy + y^2$$
$$(x - y)^2 = x^2 - 2xy + y^2$$

EXAMPLE 9   Squaring a Binomial

$$(x + y)^2 = x^2 + \underbrace{2xy} + y^2$$

(a) $(m + 7)^2 = m^2 + 2 \cdot m \cdot 7 + 7^2$
$$= m^2 + 14m + 49$$

(b) $(p - 5)^2 = p^2 - 2 \cdot p \cdot 5 + 5^2$
$$= p^2 - 10p + 25$$

(c) $(2p + 3v)^2 = (2p)^2 + 2(2p)(3v) + (3v)^2$
$$= 4p^2 + 12pv + 9v^2$$

(d) $(3r - 5s)^2 = (3r)^2 - 2(3r)(5s) + (5s)^2$
$$= 9r^2 - 30rs + 25s^2$$

7. Find each product.

(a) $(m + 5)(m - 5)$

(b) $(x - 4y)(x + 4y)$

(c) $(7m - 2n)(7m + 2n)$

(d) $(6a + b)(6a - b)$

ANSWERS

7. (a) $m^2 - 25$ **(b)** $x^2 - 16y^2$
(c) $49m^2 - 4n^2$ **(d)** $36a^2 - b^2$

8. Find each product.

(a) $(a + 2)^2$

(b) $(2m - 5)^2$

(c) $(y + 6z)^2$

(d) $(3k - 2n)^2$

9. Find each product.

(a) $[(m - 2n) - 3]$
  $\cdot [(m - 2n) + 3]$

(b) $[(k - 5h) + 2]^2$

(c) $(p + 2q)^3$

◀◀ WORK PROBLEM 8 AT THE SIDE.

Caution

As the products in the definition of the square of a binomial show,
$$(x + y)^2 \neq x^2 + y^2.$$
Also, more generally,
$$(x + y)^n \neq x^n + y^n.$$

The patterns given above for the product of the sum and difference of two terms and for the square of a binomial can be used with more complicated binomial expressions or higher powers of a binomial.

EXAMPLE 10 **Multiplying More Complicated Binomials**

Find each product.

(a) $[(3p - 2) + 5q][(3p - 2) - 5q]$

$$(x + y)(x - y) = x^2 - y^2$$

$$[(3p - 2) + 5q][(3p - 2) - 5q] = (3p - 2)^2 - (5q)^2$$
$$= 9p^2 - 12p + 4 - 25q^2 \quad \text{Square both quantities.}$$

(b) $[(2z + r) + 3]^2$
$$= (2z + r)^2 + 2(2z + r)(3) + 3^2 \quad \text{Square of a binomial}$$
$$= 4z^2 + 4zr + r^2 + 12z + 6r + 9 \quad \text{Square } 2z + r.$$

EXAMPLE 11 **Finding Higher Powers of a Binomial**

Find each product.

(a) $(x + y)^3$
$$(x + y)^3 = (x + y)^2(x + y)$$
$$= (x^2 + 2xy + y^2)(x + y) \quad \text{Square } x + y.$$
$$= x^3 + 2x^2y + xy^2 + x^2y + 2xy^2 + y^3$$
$$= x^3 + 3x^2y + 3xy^2 + y^3$$

(b) $(2a + b)^4$
$$(2a + b)^4 = (2a + b)^2(2a + b)^2$$
$$= (4a^2 + 4ab + b^2)(4a^2 + 4ab + b^2)$$
$$\text{Square } 2a + b.$$
$$= 16a^4 + 16a^3b + 4a^2b^2 + 16a^3b + 16a^2b^2$$
$$\quad + 4ab^3 + 4a^2b^2 + 4ab^3 + b^4$$
$$= 16a^4 + 32a^3b + 24a^2b^2 + 8ab^3 + b^4$$

◀◀ WORK PROBLEM 9 AT THE SIDE.

ANSWERS
8. (a) $a^2 + 4a + 4$
  (b) $4m^2 - 20m + 25$
  (c) $y^2 + 12yz + 36z^2$
  (d) $9k^2 - 12kn + 4n^2$
9. (a) $m^2 - 4mn + 4n^2 - 9$
  (b) $k^2 - 10kh + 25h^2 + 4k - 20h + 4$
  (c) $p^3 + 6p^2q + 12pq^2 + 8q^3$

5.4 Exercises

Complete each statement.

1. The _____ and _____ properties of multiplication are used to rewrite the product $(4m^3)(6m^5)$ as $(4 \cdot 6)(m^3 \cdot m^5)$.

2. The _____ rule for _____ is used to rewrite $m^3 \cdot m^5$ as m^8.

3. The product of any two polynomials is found by multiplying each _____ of the first by each _____ of the second, and then combining _____.

4. The FOIL method is used to multiply two _____.

5. The product of the sum and difference of two terms equals the _____ of the _____ of the terms.

6. The square of a binomial is the sum of the square of the _____, twice the product of the _____, and the square of the _____.

Find each product. See Examples 1–5.

7. $-8m^3(3m^2)$

8. $4p^2(-5p^4)$

9. $3x(-2x + 5)$

10. $5y(-6y - 1)$

11. $-q^3(2 + 3q)$

12. $-3a^4(4 - a)$

13. $6k^2(3k^2 + 2k + 1)$

14. $5r^3(2r^2 - 3r - 4)$

15. $(2m + 3)(3m^2 - 4m - 1)$

16. $(4z - 2)(z^2 + 3z + 5)$

17. $(-d + 6)(d^3 - 2d + 1)$

18. $(-q + 3)(q^3 + 6q + 2)$

19. $2y + 3$
$\underline{3y - 4}$

20. $5m - 3$
$\underline{2m + 6}$

21. $-b^2 + 3b + 3$
$\underline{2b + 4}$

22. $-r^2 - 4r + 8$
$\underline{3r - 2}$

23. $5m - 3n$
$\underline{5m + 3n}$

24. $2k + 6q$
$\underline{2k - 6q}$

25. $2z^3 - 5z^2 + 8z - 1$
$\underline{4z + 3}$

26. $3z^4 - 2z^3 + z - 5$
$\underline{2z - 5}$

27. $2p^2 + 3p + 6$
$\underline{3p^2 - 4p - 1}$

28. $5y^2 - 2y + 4$
$\underline{2y^2 + y + 3}$

Use the FOIL method to find each product. See Examples 6 and 7.

29. $(m + 5)(m - 8)$

30. $(p - 6)(p + 4)$

31. $(4k + 3)(3k - 2)$

32. $(5w + 2)(2w + 5)$

33. $(z - w)(3z + 4w)$

34. $(s + t)(2s - 5t)$

35. $(6c - d)(2c + 3d)$

36. $(2m - n)(3m + 5n)$

37. $(.2x + 1.3)(.5x - .1)$

38. $(.5y - .4)(.1y + 2.1)$

39. $\left(3r + \dfrac{1}{4}y\right)(r - 2y)$

40. $\left(5w - \dfrac{2}{3}z\right)(w + 5z)$

41. Describe the FOIL method in your own words.

42. Explain why the product of the sum and difference of two terms is not a trinomial.

Find each product. See Example 8.

43. $(2p - 3)(2p + 3)$

44. $(3x - 8)(3x + 8)$

45. $(5m - 1)(5m + 1)$

46. $(6y + 3)(6y - 3)$

47. $(3a + 2c)(3a - 2c)$

48. $(5r - 4s)(5r + 4s)$

49. $\left(4x - \dfrac{2}{3}\right)\left(4x + \dfrac{2}{3}\right)$

50. $\left(3t + \dfrac{5}{4}\right)\left(3t - \dfrac{5}{4}\right)$

51. $(4m + 7n^2)(4m - 7n^2)$

52. $(2k^2 + 6h)(2k^2 - 6h)$

53. $(5y^3 + 2)(5y^3 - 2)$

54. $(3x^3 + 4)(3x^3 - 4)$

Find each square. See Example 9.

55. $(y - 5)^2$

56. $(a - 3)^2$

57. $(2p + 7)^2$

58. $(3z + 8)^2$

59. $(4n - 3m)^2$

60. $(5r - 7s)^2$

61. $\left(k - \dfrac{5}{7}p\right)^2$

62. $\left(q - \dfrac{3}{4}r\right)^2$

63. Explain how the expressions $(x + y)^2$ and $x^2 + y^2$ differ.

64. Show how you can find the product $101 \cdot 99$ using the special product $(a + b)(a - b) = a^2 - b^2$.

Find each of the following products. See Example 10.

65. $[(5x + 1) + 6y]^2$

66. $[(3m - 2) + p]^2$

67. $[(2a + b) - 3][(2a + b) + 3]$

68. $[(m + p) + 5][(m + p) - 5]$

69. $[(2h - k) + j][(2h - k) - j]$

70. $[(3m - y) + z][(3m - y) - z]$

71. $(5r - s)^3$

72. $(x + 3y)^3$

73. $(m - p)^4$

74. Let $P(x) = 2x^2 - 3x + 7$ and let $Q(x) = 3x + 5$. Find each of the following.

    (a) $(P + Q)(x)$
    (b) $(P - Q)(x)$
    (c) $(Q - P)(x)$
    (d) $(PQ)(x)$

Write a simplified mathematical expression for the area of each of the following geometric figures.

75.
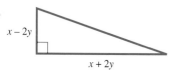
$x - 2y$
$x + 2y$

76.
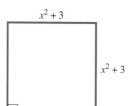
$x^2 + 3$
$x^2 + 3$

77.
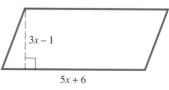
$3x - 1$
$5x + 6$

Show that each of the following is false by replacing x with 2 and y with 3, and then rewrite each statement with the correct product.

78. $(x + y)^2 = x^2 + y^2$

79. $(x + y)^3 = x^3 + y^3$

80. $(x + y)^4 = x^4 + y^4$

5.5 *Division of Polynomials*

OBJECTIVE ▶ Methods for adding, subtracting, and multiplying polynomials have now been shown. We discuss polynomial division in this section, beginning with the division of a polynomial by a monomial. (Recall that a monomial is a single term, such as $8x$, $-9m^4$, or $11y^2$.)

> **Dividing a Polynomial by a Monomial**
>
> To divide a polynomial by a monomial, divide each term in the polynomial by the monomial, and then write each quotient in lowest terms.

E X A M P L E 1 Dividing a Polynomial by a Monomial

(a) Divide $15x^2 - 12x + 6$ by 3.

Divide each term of the polynomial by 3; then write in lowest terms.

$$\frac{15x^2 - 12x + 6}{3} = \frac{15x^2}{3} - \frac{12x}{3} + \frac{6}{3}$$

$$= 5x^2 - 4x + 2$$

Check this answer by multiplying it by the divisor, 3. You should get $15x^2 - 12x + 6$ as the result.

$$3\underbrace{(5x^2 - 4x + 2)}_{\text{Quotient}} = \underbrace{15x^2 - 12x + 6}_{\text{Original polynomial}}$$

Divisor

(b) $\dfrac{5m^3 - 9m^2 + 10m}{5m^2} = \dfrac{5m^3}{5m^2} - \dfrac{9m^2}{5m^2} + \dfrac{10m}{5m^2}$ Divide each term by $5m^2$.

$$= m - \frac{9}{5} + \frac{2}{m}$$ Write in lowest terms.

This result is not a polynomial. The quotient of two polynomials need not be a polynomial.

(c) $\dfrac{8xy^2 - 9x^2y + 6x^2y^2}{x^2y^2} = \dfrac{8xy^2}{x^2y^2} - \dfrac{9x^2y}{x^2y^2} + \dfrac{6x^2y^2}{x^2y^2}$

$$= \frac{8}{x} - \frac{9}{y} + 6$$

WORK PROBLEM 1 AT THE SIDE. ▶▶

OBJECTIVE ▶ The process for dividing one polynomial by a polynomial that is not a monomial is very similar to that for dividing one whole number by another. The following examples show how this is done.

E X A M P L E 2 Dividing One Polynomial by Another

Divide $2m^2 + m - 10$ by $m - 2$.

Write the problem, making sure that both polynomials are written in descending powers of the variables.

$$m - 2\overline{)2m^2 + m - 10}$$

— **CONTINUED ON NEXT PAGE**

1. Divide.

(a) $\dfrac{12p + 30}{6}$

(b) $\dfrac{9y^3 - 4y^2 + 8y}{2y^2}$

(c) $\dfrac{8a^2b^2 - 20ab^3}{4a^3b}$

329

2. Divide.

(a) $\dfrac{2r^2 + r - 21}{r - 3}$

Divide the first term of $2m^2 + m - 10$ by the first term of $m - 2$. Because $2m^2 \div m = 2m$, place this result above the division line.

$$m - 2 \overline{)2m^2 + \;\; m - 10} \qquad \overset{2m}{\longleftarrow} \text{Result of } \dfrac{2m^2}{m}$$

Multiply $m - 2$ and $2m$, and write the result below $2m^2 + m - 10$.

$$
\begin{array}{r}
2m \\
m - 2 \overline{)2m^2 + \;\; m - 10} \\
\underline{2m^2 - 4m} \quad \longleftarrow \quad 2m(m - 2) = 2m^2 - 4m
\end{array}
$$

Now subtract $2m^2 - 4m$ from $2m^2 + m$. Do this by changing the signs on $2m^2 - 4m$ and *adding*:

$$2m^2 + m - (2m^2 - 4m) = 2m^2 + m - 2m^2 + 4m = 5m$$

$$
\begin{array}{r}
2m \\
m - 2 \overline{)2m^2 + \;\; m - 10} \\
\underline{2m^2 - 4m} \\
5m \quad \longleftarrow \quad \text{Subtract.}
\end{array}
$$

Bring down -10, and continue by dividing $5m$ by m.

$$
\begin{array}{r}
2m + \;\; 5 \quad \overset{\leftarrow}{} \dfrac{5m}{m} = 5 \\
m - 2 \overline{)2m^2 + \;\; m - 10} \\
\underline{2m^2 - 4m} \\
5m - 10 \\
\underline{5m - 10} \quad \leftarrow 5(m - 2) = 5m - 10 \\
0 \quad \leftarrow \text{Subtract.}
\end{array}
$$

Finally, $(2m^2 + m - 10) \div (m - 2) = 2m + 5$. To check, multiply $m - 2$ and $2m + 5$.

(b) $\dfrac{2k^2 + 17k + 30}{2k + 5}$

> **Note**
> Another way of stating the problem in Example 2 is: Find the quotient function $\left(\dfrac{P}{Q}\right)(m)$, where $P(m) = 2m^2 + m - 10$ and $Q(m) = m - 2$.

(c) $\dfrac{10y^3 - 39y^2 + 41y - 10}{2y - 5}$

E X A M P L E   3   Dividing Polynomials

Divide $2p^3 + 7p^2 - 5p - 8$ by $2p + 1$.

$$
\begin{array}{r}
p^2 + 3p - 4 \\
2p + 1 \overline{)2p^3 + 7p^2 - 5p - 8} \\
\underline{2p^3 + \;\; p^2} \\
6p^2 - 5p \\
\underline{6p^2 + 3p} \\
- 8p - 8 \\
\underline{- 8p - 4} \\
-4
\end{array}
$$

We write the remainder as a fraction: $\dfrac{-4}{2p + 1}$. The quotient is $p^2 + 3p - 4 + \dfrac{-4}{2p + 1}$. Check by multiplying $2p + 1$ and $p^2 + 3p - 4$ and adding -4 to the result.

◄◄ **WORK PROBLEM 2 AT THE SIDE.**

E X A M P L E 4   **Dividing a Polynomial That Has a Missing Term**

Divide $3x^3 - 2x - 75$ by $x - 3$.

Make sure that $3x^3 - 2x - 75$ is in descending powers of the variable. Add a term with a 0 coefficient for the missing x^2 term.

$$x - 3 \overline{)3x^3 + 0x^2 - 2x - 75} \quad \longleftarrow \text{Missing term}$$

Start with $3x^3 \div x = 3x^2$.

$$\begin{array}{r} 3x^2 \\ x - 3 \overline{)3x^3 + 0x^2 - 2x - 75} \\ \underline{3x^3 - 9x^2} \end{array} \quad \begin{array}{l} \frac{3x^3}{x} = 3x^2 \\ 3x^2(x - 3) \end{array}$$

Subtract by changing the signs on $3x^3 - 9x^2$ and adding.

$$\begin{array}{r} 3x^2 \\ x - 3 \overline{)3x^3 + 0x^2 - 2x - 75} \\ \underline{3x^3 - 9x^2} \\ 9x^2 \end{array} \quad \longleftarrow \text{Subtract.}$$

Bring down the next term.

$$\begin{array}{r} 3x^2 \\ x - 3 \overline{)3x^3 + 0x^2 - 2x - 75} \\ \underline{3x^3 - 9x^2} \\ 9x^2 - 2x \end{array} \quad \longleftarrow \text{Bring down } -2x.$$

In the next step, $9x^2 \div x = 9x$.

$$\begin{array}{r} 3x^2 + 9x \\ x - 3 \overline{)3x^3 + 0x^2 - 2x - 75} \\ \underline{3x^3 - 9x^2} \\ 9x^2 - 2x \\ \underline{9x^2 - 27x} \\ 25x - 75 \end{array} \quad \begin{array}{l} \frac{9x^2}{x} = 9x \\ \\ 9x(x - 3) \\ \leftarrow \text{Subtract and bring down } -75. \end{array}$$

Finally, $25x \div x = 25$.

$$\begin{array}{r} 3x^2 + 9x + 25 \\ x - 3 \overline{)3x^3 + 0x^2 - 2x - 75} \\ \underline{3x^3 - 9x^2} \\ 9x^2 - 2x \\ \underline{9x^2 - 27x} \\ 25x - 75 \\ \underline{25x - 75} \quad \leftarrow 25(x - 3) \\ 0 \quad \leftarrow \text{Subtract.} \end{array}$$

The quotient is

$$\frac{3x^3 - 2x - 75}{x - 3} = 3x^2 + 9x + 25.$$

Check by multiplying $x - 3$ and $3x^2 + 9x + 25$.

WORK PROBLEM 3 AT THE SIDE. ▶▶

3. Divide.

(a) $\dfrac{2r^2 + 2}{2r - 4}$

(b) $\dfrac{3k^3 + 9k - 12}{3k - 3}$

(c) $\dfrac{5x^4 - 13x^3 + 11x^2 - 33x}{5x - 3}$

ANSWERS

3. (a) $r + 2 + \dfrac{10}{2r - 4}$

  (b) $k^2 + k + 4$

  (c) $x^3 - 2x^2 + x - 6 + \dfrac{-18}{5x - 3}$

4. Divide.

(a) $\dfrac{4x^4 - 7x^2 + x + 5}{2x^2 - x}$

(b) $\dfrac{3r^5 - 15r^4 - 2r^3 + 19r^2 - 7}{3r^2 - 2}$

E X A M P L E   5   Dividing by a Polynomial That Has a Missing Term

Divide $6r^4 + 9r^3 + 2r^2 - 8r + 7$ by $3r^2 - 2$.

The polynomial $3r^2 - 2$ has a missing term. Write it as $3r^2 + 0r - 2$, and divide as usual.

$$
\begin{array}{r}
2r^2 + 3r + 2 \\
3r^2 + 0r - 2 \overline{)6r^4 + 9r^3 + 2r^2 - 8r + 7} \\
\underline{6r^4 + 0r^3 - 4r^2} \\
9r^3 + 6r^2 - 8r \\
\underline{9r^3 + 0r^2 - 6r} \\
6r^2 - 2r + 7 \\
\underline{6r^2 + 0r - 4} \\
-2r + 11
\end{array}
$$

Because the degree of the remainder, $-2r + 11$, is less than the degree of $3r^2 - 2$, the division process is now finished. The result is written

$$2r^2 + 3r + 2 + \frac{-2r + 11}{3r^2 - 2}.$$

Check: $(3r^2 - 2)\left[(2r^2 + 3r + 2) + \dfrac{-2r + 11}{3r^2 - 2}\right]$

$= (3r^2 - 2)(2r^2 + 3r + 2) + (-2r + 11)$

$= (6r^4 + 9r^3 + 6r^2 - 4r^2 - 6r - 4) - 2r + 11$

$= 6r^4 + 9r^3 + 2r^2 - 8r + 7$

◄◄ **WORK PROBLEM 4 AT THE SIDE.**

Caution
When dividing a polynomial by a monomial, do not confuse the two methods shown in this section. The long division method is not used to divide by a monomial.

$$\frac{9y^3 - 12y^2}{3y} = \frac{9y^3}{3y} - \frac{12y^2}{3y} = 3y^2 - 4y$$

Remember the following two steps when dividing a polynomial by a nonmonomial.

1. Be sure the terms in both polynomials are in descending powers.
2. Replace any missing terms with a 0 coefficient.

ANSWERS

4. (a) $2x^2 + x - 3 + \dfrac{-2x + 5}{2x^2 - x}$

(b) $r^3 - 5r^2 + 3 + \dfrac{-1}{3r^2 - 2}$

5.5 Exercises

Complete each statement.

1. The quotient of two monomials is found using the _____ rule of _____ .

2. To divide a polynomial by a monomial, we divide _____ of the polynomial by the _____ .

3. When dividing polynomials that are not monomials, the first step is to write them in _____ .

4. If a polynomial in a division problem has a missing term, we should _____ as a place holder.

Divide. See Example 1.

5. $\dfrac{9y^2 + 12y - 15}{3y}$

6. $\dfrac{80r^2 - 40r + 10}{10r}$

7. $\dfrac{15m^3 + 25m^2 + 30m}{5m^2}$

8. $\dfrac{64x^3 - 72x^2 + 12x}{8x^3}$

9. $\dfrac{14m^2n^2 - 21mn^3 + 28m^2n}{14m^2n}$

10. $\dfrac{24h^2k + 56hk^2 - 28hk}{16h^2k^2}$

Complete the division.

11.
$$
\begin{array}{r}
r^2 \\
3r - 1{\overline{\smash{\big)}\,3r^3 - 22r^2 + 25r - 6}} \\
\underline{3r^3 - r^2} \\
-21r^2
\end{array}
$$

12.
$$
\begin{array}{r}
3b^2 \\
2b - 5{\overline{\smash{\big)}\,6b^3 - 7b^2 - 4b - 40}} \\
\underline{6b^3 - 15b^2} \\
8b^2
\end{array}
$$

Divide. See Examples 2–5.

13. $\dfrac{y^2 + 3y - 18}{y + 6}$

14. $\dfrac{q^2 + 4q - 32}{q - 4}$

15. $\dfrac{3t^2 + 17t + 10}{3t + 2}$

16. $\dfrac{2k^2 - 3k - 20}{2k + 5}$

17. $(2z^3 - 5z^2 + 6z - 15) \div (2z - 5)$

18. $(3p^3 + p^2 + 18p + 6) \div (3p + 1)$

19. $(4x^3 + 9x^2 - 10x + 3) \div (4x + 1)$

20. $(10z^3 - 26z^2 + 17z - 13) \div (5z - 3)$

21. $\dfrac{6x^3 - 19x^2 + 14x - 15}{3x^2 - 2x + 4}$

22. $\dfrac{8m^3 - 18m^2 + 37m - 13}{2m^2 - 3m + 6}$

23. $\dfrac{4k^4 + 6k^3 + 3k - 1}{2k^2 + 1}$

24. $\dfrac{6y^4 + 4y^3 + 4y - 6}{3y^2 + 2y - 3}$

25. $(9z^4 - 13z^3 + 23z^2 - 10z + 8) \div (z^2 - z + 2)$

26. $(2q^4 + 5q^3 - 11q^2 + 11q - 20) \div (2q^2 - q + 2)$

27. $\left(2x^2 - \dfrac{7}{3}x - 1\right) \div (3x + 1)$

28. $\left(m^2 + \dfrac{7}{2}m + 3\right) \div (2m + 3)$

29. $\left(3a^2 - \dfrac{23}{4}a - 5\right) \div (4a + 3)$

30. $\left(3q^2 + \dfrac{19}{5}q - 3\right) \div (5q - 2)$

31. If $P(x) = 2x^3 + 15x^2 + 28x$ and if $Q(x) = x^2 + 4x$, find $\left(\dfrac{P}{Q}\right)(x)$.

32. If $P(x) = 2x^3 + 15x^2 + 13x - 63$ and if $Q(x) = 2x + 9$, find $\left(\dfrac{P}{Q}\right)(x)$.

MATHEMATICAL CONNECTIONS (EXERCISES 33–38)

Let $10 = t$. Then $100 = t^2$, $1000 = t^3$, and so on. We can write integers using t as follows: $23 = 20 + 3 = 2t + 3$, $547 = 500 + 40 + 7 = 5t^2 + 4t + 7$, $1080 = 1000 + 80 = t^3 + 8t$. To see the close connection between division of polynomials and division of integers, work the following exercises in order.

33. Write 1654 using t.

34. Write 14 using t.

35. Divide your answer to Exercise 33 by your answer to Exercise 34. What is the quotient?

36. Divide 1654 by 14. (Use a calculator if you like, but write any decimal as a common fraction.)

37. Write your answer to Exercise 35 without t, substituting $t = 10$, $t^2 = 100$, and so on.

38. What do you observe about the quotients in Exercises 36 and 37?

5.6 Synthetic Division

OBJECTIVE Many times when one polynomial is divided by a second, the second polynomial is of the form $x - k$, where the coefficient of the x term is 1 and k is a constant. There is a shortcut way for doing these divisions. To see how this shortcut works, look first at the left below, where the division of $3x^3 - 2x + 5$ by $x - 3$ is shown. (Notice that a 0 was inserted for the missing x^2 term.)

$$
\begin{array}{r}
3x^2 + 9x + 25 \\
x - 3)\overline{3x^3 + 0x^2 - 2x + 5} \\
\underline{3x^3 - 9x^2} \\
9x^2 - 2x \\
\underline{9x^2 - 27x} \\
25x + 5 \\
\underline{25x - 75} \\
80
\end{array}
\qquad
\begin{array}{r}
3 \quad 9 \quad 25 \\
1 - 3)\overline{3 \quad 0 \quad -2 \quad 5} \\
\underline{3 \quad -9} \\
9 \quad -2 \\
\underline{9 \quad -27} \\
25 \quad 5 \\
\underline{25 \quad -75} \\
80
\end{array}
$$

On the right, we show exactly the same division, written without the variables. All the numbers in color on the right are repetitions of the numbers directly above them, so they may be omitted, as shown below.

$$
\begin{array}{r}
3 \quad 9 \quad 25 \\
1 - 3)\overline{3 \quad 0 \quad -2 \quad 5} \\
\underline{-9} \\
9 \quad -2 \\
\underline{-27} \\
25 \quad 5 \\
\underline{-75} \\
80
\end{array}
$$

The numbers in color are again repetitions of the numbers directly above them; they too can be omitted, as shown here.

$$
\begin{array}{r}
3 \quad 9 \quad 25 \\
1 - 3)\overline{3 \quad 0 \quad -2 \quad 5} \\
\underline{-9} \\
9 \\
\underline{-27} \\
25 \\
\underline{-75} \\
80
\end{array}
$$

Now the problem can be condensed. If the 3 in the top row is brought down to the beginning of the bottom row, the top row can be omitted, because it duplicates the bottom row.

$$
\begin{array}{r}
1 - 3)\overline{3 \quad 0 \quad -2 \quad 5} \\
\underline{-9 \quad -27 \quad -75} \\
3 \quad 9 \quad 25 \quad 80
\end{array}
$$

Finally, the number at the upper left can be omitted because that number will always be a 1. Also, to simplify the arithmetic, subtraction in the second row is replaced by addition. Compensate for this by changing the -3

OBJECTIVES

 Use synthetic division to divide by a polynomial of the form $x - k$.

Use the remainder theorem to evaluate a polynomial.

Decide whether a given number is a solution of an equation.

FOR EXTRA HELP

Tutorial   Tape 7   SSM, Sec. 5.6

335

1. Divide, using synthetic division.

(a) $\dfrac{3z^2 + 10z - 8}{z + 4}$

at upper left to its additive inverse, 3. The result of doing all this is shown below.

$$\begin{array}{r@{}r@{\quad}r@{\quad}r@{\quad}r}
\text{Additive} & & & & \\
\text{inverse} \rightarrow 3\overline{)3} & 0 & -2 & 5 & \\
& 9 & 27 & 75 & \leftarrow \text{Signs changed} \\
\hline
3 & 9 & 25 & 80 &
\end{array}$$

Read the quotient from the bottom row. $3x^2 + 9x + 25 + \dfrac{80}{x - 3}$

The first three numbers in the bottom row are used to obtain a polynomial of degree 1 less than the degree of the dividend. The last number gives the remainder.

> ### Synthetic Division
>
> This shortcut procedure is called **synthetic** (sin-THET-ik) **division.** It is used only when dividing a polynomial by a polynomial of the form $x - k$.

E X A M P L E I   Using Synthetic Division

Use synthetic division to divide $5x^2 + 16x + 15$ by $x + 2$.

As mentioned above, synthetic division can be used only when dividing by a polynomial of the form $x - k$. To get $x + 2$ in this form, write it as

$$x + 2 = x - (-2),$$

where $k = -2$. Now write the coefficients of $5x^2 + 16x + 15$, placing -2 to the left.

$$x + 2 \text{ leads to } -2$$

$$-2\overline{)5 \quad 16 \quad 15}$$
$$\qquad\qquad \text{Coefficients}$$

Bring down the 5, and multiply: $-2 \cdot 5 = -10$.

$$\begin{array}{r}
-2\overline{)5 \quad\; 16 \quad 15} \\
\underline{-10 \qquad\qquad} \\
5 \qquad\qquad\;
\end{array}$$

(b) $(2x^2 + 3x - 5)$
$\div (x + 1)$

Add 16 and -10, getting 6. Multiply 6 and -2 to get -12.

$$\begin{array}{r}
-2\overline{)5 \quad\; 16 \quad\; 15} \\
\underline{-10 \quad -12} \\
5 \qquad 6 \qquad\;
\end{array}$$

Add 15 and -12, getting 3.

$$\begin{array}{r}
-2\overline{)5 \quad\; 16 \quad\; 15} \\
-10 \quad -12 \\
\hline
5 \qquad 6 \qquad 3
\end{array}$$

Read the result from the bottom row.

$$\frac{5x^2 + 16x + 15}{x + 2} = 5x + 6 + \frac{3}{x + 2}$$

◀◀ **WORK PROBLEM I AT THE SIDE.**

E X A M P L E 2   **Using Synthetic Division with Missing Terms**

Use synthetic division to find

$$(-4x^5 + x^4 + 6x^3 + 2x^2 + 50) \div (x - 2).$$

Use the steps given above, inserting a 0 for the missing x term.

Insert 0 for missing term.

$$
\begin{array}{r|rrrrrr}
2) & -4 & 1 & 6 & 2 & 0 & 50 \\
   &   & -8 & -14 & -16 & -28 & -56 \\
\hline
   & -4 & -7 & -8 & -14 & -28 & -6
\end{array}
$$

Read the result from the bottom row.

$$\frac{-4x^5 + x^4 + 6x^3 + 2x^2 + 50}{x - 2}$$

$$= -4x^4 - 7x^3 - 8x^2 - 14x - 28 + \frac{-6}{x - 2}$$

WORK PROBLEM 2 AT THE SIDE. ▶▶

OBJECTIVE ▶ We can use synthetic division to evaluate polynomials. For instance, in Example 2, synthetic division was used to divide $-4x^5 + x^4 + 6x^3 + 2x^2 + 50$ by $x - 2$. The remainder in this division was -6. If x is replaced with 2,

$$-4x^5 + x^4 + 6x^3 + 2x^2 + 50 = -4 \cdot 2^5 + 2^4 + 6 \cdot 2^3 + 2 \cdot 2^2 + 50$$
$$= -4 \cdot 32 + 16 + 6 \cdot 8 + 2 \cdot 4 + 50$$
$$= -128 + 16 + 48 + 8 + 50$$
$$= -6,$$

the same number as the remainder; that is, dividing by $x - 2$ produces a remainder equal to the result when x is replaced with 2. This always happens, as the following **remainder theorem** (ree-MAIN-der THEAR-em) states.

Remainder Theorem

If the polynomial $P(x)$ is divided by $x - k$, the remainder equals $P(k)$.

This result is proved in more advanced courses.

E X A M P L E 3   **Using the Remainder Theorem**

Let $P(x) = 2x^3 - 5x^2 - 3x + 11$. Find $P(-2)$.

By the remainder theorem, we can find $P(-2)$ by dividing $P(x)$ by $x - (-2) = x + 2$.

$$
\begin{array}{r}
\text{Value of } x \rightarrow -2)\overline{\begin{array}{rrrr} 2 & -5 & -3 & 11 \end{array}} \\
\begin{array}{rrrr} \quad -4 & 18 & -30 \end{array} \\
\hline
\begin{array}{rrrr} 2 & -9 & 15 & -19 \end{array} \leftarrow \text{Remainder}
\end{array}
$$

By this result, $P(-2) = -19$. Verify the answer by substituting -2 for x in $P(x)$, as we did above.

WORK PROBLEM 3 AT THE SIDE. ▶▶

2. Divide, using synthetic division.

(a) $\dfrac{3a^3 - 2a + 21}{a + 2}$

(b) $(-4x^4 + 3x^3 + 18x + 2) \div (x - 2)$

3. Let $P(x) = x^3 - 5x^2 + 7x - 3$. Use synthetic division to find each of the following.

(a) $P(1)$ (Divide by $x - 1$.)

(b) $P(-2)$

ANSWERS

2. (a) $3a^2 - 6a + 10 + \dfrac{1}{a + 2}$

  (b) $-4x^3 - 5x^2 - 10x - 2 + \dfrac{-2}{x - 2}$

3. (a) 0 **(b)** -45

4. Use synthetic division to decide whether or not 2 is a solution for each of the following.

(a) $3r^3 - 11r^2 + 17r - 14 = 0$

E X A M P L E   4   Using the Remainder Theorem

Show that -5 is a solution of the equation

$$2x^4 + 12x^3 + 6x^2 - 5x + 75 = 0.$$

One way to show that -5 is a solution is to substitute -5 for x in the equation. However, an easier way is to use synthetic division and the remainder theorem given above.

$$
\begin{array}{r|rrrrr}
\text{Proposed} \rightarrow -5) & 2 & 12 & 6 & -5 & 75 \\
\text{solution} & & -10 & -10 & 20 & -75 \\
\hline
& 2 & 2 & -4 & 15 & 0 \leftarrow \text{Remainder} = 0
\end{array}
$$

Because the remainder is 0, the value of the polynomial on the left side of the equation when $x = -5$ is 0, and the number -5 is a solution of the equation.

◀◀ WORK PROBLEM 4 AT THE SIDE.

(b) $4k^5 - 7k^4 - 11k^2 + 2k + 6 = 0$

5.6 Exercises

1. What is the purpose of synthetic division?

2. What type of polynomial divisors may be used with synthetic division?

3. Explain why it is important to insert zeros as place holders for missing terms before performing synthetic division.

4. Explain why a zero remainder in synthetic division of $P(x)$ by k indicates that k is a solution of the equation $P(x) = 0$.

Use synthetic division in each of the following. See Examples 1 and 2.

5. $\dfrac{x^2 - 6x + 5}{x - 1}$

6. $\dfrac{x^2 - 4x - 21}{x + 3}$

7. $\dfrac{4m^2 + 19m - 5}{m + 5}$

8. $\dfrac{3k^2 - 5k - 12}{k - 3}$

9. $\dfrac{2a^2 + 8a + 13}{a + 2}$

10. $\dfrac{4y^2 - 5y - 20}{y - 4}$

11. $(p^2 - 3p + 5) \div (p + 1)$

12. $(z^2 + 4z - 6) \div (z - 5)$

13. $\dfrac{4a^3 - 3a^2 + 2a - 3}{a - 1}$

14. $\dfrac{5p^3 - 6p^2 + 3p + 14}{p + 1}$

15. $(x^5 - 2x^3 + 3x^2 - 4x - 2) \div (x - 2)$

16. $(2y^5 - 5y^4 - 3y^2 - 6y - 23) \div (y - 3)$

17. $(-4r^6 - 3r^5 - 3r^4 + 5r^3 - 6r^2 + 3r) \div (r - 1)$ **18.** $(-3t^5 + 2t^4 - 5t^3 + 6t^2 - 3t - 2) \div (t - 2)$

19. $(-3y^5 + 2y^4 - 5y^3 - 6y^2 - 1) \div (y + 2)$ **20.** $(m^6 + 2m^4 - 5m + 11) \div (m - 2)$

21. $\dfrac{y^3 + 1}{y - 1}$ **22.** $\dfrac{z^4 + 81}{z - 3}$

Use the remainder theorem to find P(k). See Example 3.

23. $P(x) = 2x^3 - 4x^2 + 5x - 3; k = 2$ **24.** $P(y) = y^3 + 3y^2 - y + 5; k = -1$

25. $P(r) = -r^3 - 5r^2 - 4r - 2; k = -4$ **26.** $P(z) = -z^3 + 5z^2 - 3z + 4; k = 3$

27. $P(y) = 2y^3 - 4y^2 + 5y - 33; k = 3$ **28.** $P(x) = x^3 - 3x^2 + 4x - 4; k = 2$

Use synthetic division to decide whether the given number is a solution for each of the following equations. See Example 4.

29. $x^3 - 2x^2 - 3x + 10 = 0; x = -2$ **30.** $x^3 - 3x^2 - x + 10 = 0; x = -2$

31. $m^4 + 2m^3 - 3m^2 + 8m - 8 = 0; m = -2$ **32.** $r^4 - r^3 - 6r^2 + 5r + 10 = 0; r = -2$

33. $3a^3 + 2a^2 - 2a + 11 = 0; a = -2$ **34.** $3z^3 + 10z^2 + 3z - 9 = 0; z = -2$

5.7 Greatest Common Factors; Factoring by Grouping

OBJECTIVES

1 ▶ Factor out the greatest common factor.

2 ▶ Factor by grouping.

FOR EXTRA HELP

Tutorial

Tape 8

SSM, Sec. 5.7

Writing a polynomial as the product of two or more simpler polynomials is called **factoring** (FAK-ter-ing) the polynomial. For example, the product of $3x$ and $5x - 2$ is $15x^2 - 6x$, and $15x^2 - 6x$ can be factored as the product $3x(5x - 2)$.

$$3x(5x - 2) = 15x^2 - 6x \qquad \text{Multiplication}$$
$$15x^2 - 6x = 3x(5x - 2) \qquad \text{Factoring}$$

Notice that both multiplication and factoring are examples of the distributive property, used in opposite directions. Factoring is the reverse of multiplying.

OBJECTIVE 1 ▶ The first step in factoring a polynomial is to find the *greatest common factor* for the terms of the polynomial. The **greatest common factor** is the largest term that is a factor of all the terms in the polynomial. For example, the greatest common factor for $8x + 12$ is 4 because 4 is the largest number that is a factor of both $8x$ and 12. Using the distributive property,

$$8x + 12 = 4(2x) + 4(3) = 4(2x + 3).$$

This process is called **factoring out the greatest common factor.** As a check, multiply 4 and $2x + 3$. The result is $8x + 12$.

E X A M P L E 1 Factoring Out the Greatest Common Factor

Factor out the greatest common factor.

(a) $9z - 18$

Because 9 is the greatest common factor,

$$9z - 18 = 9 \cdot z - 9 \cdot 2 = 9(z - 2).$$

(b) $56m + 35p = 7(8m + 5p)$ 7 is the greatest common factor.

(c) $2y + 5$ There is no common factor other than 1.

(d) $12 + 24z = 12 \cdot 1 + 12 \cdot 2z$
$$= 12(1 + 2z) \qquad \text{12 is the greatest common factor.}$$

Caution
When factoring, it is very common to forget the 1. Be careful to include it as needed. Do not forget that any answer can be checked by multiplication.

WORK PROBLEM 1 AT THE SIDE. ▶▶

E X A M P L E 2 Factoring Out the Greatest Common Factor

Factor out the greatest common factor.

(a) $9x^2 + 12x^3$

The numerical part of the common factor is 3, the largest number that is a factor of both 9 and 12. For the variable portions, x^2 and x^3, use x to the least degree; here the least degree is 2. The greatest common factor is $3x^2$.

$$9x^2 + 12x^3 = 3x^2(3) + 3x^2(4x) = 3x^2(3 + 4x)$$

— **CONTINUED ON NEXT PAGE**

---

1. Factor out the greatest common factor.

(a) $7k + 28$

(b) $32m + 24$

(c) $8a - 9$

(d) $5z + 5$

ANSWERS

1. (a) $7(k + 4)$ **(b)** $8(4m + 3)$
(c) cannot be factored **(d)** $5(z + 1)$

2. Factor out the greatest common factor.

(a) $16y^4 + 8y^3$

(b) $32p^4 - 24p^3 + 40p^5$

The least degree is 3. The greatest common factor is thus $8p^3$.

$$32p^4 - 24p^3 + 40p^5 = \mathbf{8p^3}(4p) + \mathbf{8p^3}(-3) + \mathbf{8p^3}(5p^2)$$
$$= 8p^3(4p - 3 + 5p^2)$$

(c) $3k^4 - 15k^7 + 24k^9 = 3k^4(1 - 5k^3 + 8k^5)$

(b) $14p^2 - 9p^3 + 6p^4$

◀◀ **WORK PROBLEM 2 AT THE SIDE.**

E X A M P L E   3   **Factoring Out the Greatest Common Factor**

Factor out the greatest common factor.

(a) $24m^3n^2 - 18m^2n + 6m^4n^3$

The numerical part of the greatest common factor is 6. Find the variable part by writing each variable with its least degree. Here, 2 is the least exponent that appears on m, while 1 is the least exponent on n. Finally, $6m^2n$ is the greatest common factor.

$$24m^3n^2 - 18m^2n + 6m^4n^3$$
$$= (6m^2n)(4mn) + (6m^2n)(-3) + (6m^2n)(m^2n^2)$$
$$= 6m^2n(4mn - 3 + m^2n^2)$$

(c) $15z^2 + 45z^5 - 60z^6$

(b) $25x^2y^3 + 30xy^5 - 15x^4y^7 = 5xy^3(5x + 6y^2 - 3x^3y^4)$

◀◀ **WORK PROBLEM 3 AT THE SIDE.**

A greatest common factor need not be a monomial. The next example shows a binomial greatest common factor.

E X A M P L E   4   **Factoring Out a Binomial Factor**

Factor out the greatest common factor.

(a) $(x - 5)(x + 6) + (x - 5)(2x + 5)$

The greatest common factor here is $x - 5$.

$$(x - 5)(x + 6) + (x - 5)(2x + 5)$$
$$= (x - 5)[(x + 6) + (2x + 5)]$$
$$= (x - 5)(x + 6 + 2x + 5)$$
$$= (x - 5)(3x + 11)$$

3. Factor out the greatest common factor.

(a) $12y^5x^2 + 8y^3x^3$

(b) $z^2(m + n) + x^2(m + n) = (m + n)(z^2 + x^2)$

(c) $p(r + 2s) - q^2(r + 2s) = (r + 2s)(p - q^2)$

(d) $(p - 5)(p + 2) - (p - 5)(3p + 4)$
$$= (p - 5)[(p + 2) - (3p + 4)]$$
$$= (p - 5)(p + 2 - 3p - 4)$$
$$= (p - 5)(-2p - 2)$$
$$= (p - 5)(-2)(p + 1) \quad \text{or} \quad -2(p - 5)(p + 1)$$

(b) $5m^4x^3 + 15m^5x^6 - 20m^4x^6$

ANSWERS

2. (a) $8y^3(2y + 1)$
   (b) $p^2(14 - 9p + 6p^2)$
   (c) $15z^2(1 + 3z^3 - 4z^4)$
3. (a) $4y^3x^2(3y^2 + 2x)$
   (b) $5m^4x^3(1 + 3mx^3 - 4x^3)$

WORK PROBLEM 4 AT THE SIDE. ▶▶

4. Factor out the greatest common factor.

─ E X A M P L E 5   **Factoring Out a Negative Factor**

Factor out the greatest common factor from

$$-a^3 + 3a^2 - 5a.$$

There are two ways to factor this polynomial, both of which are correct. We could use a as the common factor, giving

$$-a^3 + 3a^2 - 5a = a(-a^2) + a(3a) + a(-5)$$
$$= a(-a^2 + 3a - 5).$$

Alternatively, $-a$ could be used as the common factor.

$$-a^3 + 3a^2 - 5a = -a(a^2) + (-a)(-3a) + (-a)(5)$$
$$= -a(a^2 - 3a + 5)$$

(a) $(a + 2)(a - 3)$
$\qquad + (a + 2)(a + 6)$

(b) $(y - 1)(y + 3)$
$\qquad - (y - 1)(y + 4)$

Note

In Example 5 we showed two ways of factoring a polynomial. Sometimes in a particular problem, one of these forms will be preferable to the other, but both are correct. The answer section in this book will give only one of these forms, usually the one where the common factor has a positive coefficient, but either is correct.

(c) $k^2(a + 5b) + m^2(a + 5b)$

WORK PROBLEM 5 AT THE SIDE. ▶▶

OBJECTIVE 2 Sometimes a polynomial has a greatest common factor of 1, but it still may be possible to write the polynomial as a product of factors. The idea is to look for factors common to *some* of the terms, rearrange the terms accordingly, and then factor. This process is called **factoring by grouping** (GROOP-ing).

For example, to factor the polynomial

$$ax - ay + bx - by,$$

we can group the terms as follows.

(d) $r^2(y + 6) + r^2(y + 3)$

Terms with common factors
↓   ↓
$$(ax - ay) + (bx - by)$$

Next, factor $ax - ay$ as $a(x - y)$, and factor $bx - by$ as $b(x - y)$, to get

$$ax - ay + bx - by = a(x - y) + b(x - y).$$

On the right, the common factor is $x - y$. The final factorization is

$$ax - ay + bx - by = (x - y)(a + b).$$

5. Factor each polynomial in two ways.

(a) $-k^2 + 3k$

(b) $-6r^2 + 5r$

Factoring by Grouping

Step 1 Collect the terms into groups so that each group has a common factor.

Step 2 Factor out the common factor in each group.

Step 3 If each group now has a common factor, factor it out. If not, try a different grouping.

6. Factor.

(a) $mn + 6 + 2n + 3m$

(b) $4y - zx + yx - 4z$

E X A M P L E 6   Factoring by Grouping

Factor $p^2q^2 - 10 - 2q^2 + 5p^2$.

The first two terms have no common factor (nor do the last two terms). We can group the terms as follows, however.

$$(p^2q^2 - 2q^2) + (5p^2 - 10) \qquad \text{Group terms.}$$
$$q^2(p^2 - 2) + 5(p^2 - 2) \qquad \text{Factor out the common factors.}$$
$$(p^2 - 2)(q^2 + 5) \qquad \text{Factor out } p^2 - 2.$$

Caution

It is a common error to stop at the step

$$q^2(p^2 - 2) + 5(p^2 - 2).$$

This expression is *not in factored form* because it is a *sum* of two terms: $q^2(p^2 - 2)$ and $5(p^2 - 2)$.

◄◄ WORK PROBLEM 6 AT THE SIDE.

7. Factor.

(a) $xy + 2y - 4x - 8$

(b) $10p^2 + 15p - 12p - 18$

E X A M P L E 7   Factoring by Grouping

Factor $3x - 3y - ax + ay$.

Grouping terms as we did above and factoring gives

$$(3x - 3y) + (-ax + ay) = 3(x - y) + a(-x + y).$$

There is no simple common factor here. However, if we factor out $-a$ instead of a from the second group, we get

$$3(x - y) - a(x - y).$$

(Be careful with the signs.) Now factor out $x - y$ to get

$$(x - y)(3 - a).$$

E X A M P L E 8   Factoring by Grouping

Factor $6x^2 - 4x - 15x + 10$.

Work as above. Note that we must factor -5 rather than 5 from the second group in order to get a common factor of $3x - 2$.

$$6x^2 - 4x - 15x + 10 = 2x(3x - 2) - 5(3x - 2)$$
$$= (3x - 2)(2x - 5)$$

◄◄ WORK PROBLEM 7 AT THE SIDE.

ANSWERS

6. (a) $(m + 2)(n + 3)$
   (b) $(y - z)(x + 4)$
7. (a) $(x + 2)(y - 4)$
   (b) $(2p + 3)(5p - 6)$

5.7 Exercises

1. Explain in your own words what it means to factor a polynomial.

2. What is the first step in attempting to factor a polynomial?

3. What is the greatest common factor of the following terms? $7z^2(m + n)^4$, $9z^3(m + n)^5$

4. Which one of the following is an example of a polynomial in factored form?
  (a) $3x^2y^3 + 6x^2(2x + y)$ **(b)** $5(x + y)^2 - 10(x + y)^3$
  (c) $(-2 + 3x)(5y^2 + 4y + 3)$ **(d)** $(3x + 4)(5x - y) - (3x + 4)(2x - 1)$

Find the greatest common factor for each list of terms.

5. $9m^3, 3m^2, 15m$

6. $4a^2, 6ab, 2a^3$

7. $6m(r + t)^2, 3p(r + t)^4$

8. $7z^2(m + n)^4, 9z^3(m + n)^5$

9. Which one of the following has the greatest common factor of $6x^3y^4 - 12x^5y^2 + 24x^4y^8$
as one of the factors?
  (a) $6x^3y^2(y^2 - 2x^2 + 4xy^6)$ **(b)** $6xy(x^2y^3 - 2x^4y + 4x^3y^7)$
  (c) $2x^3y^2(3y^2 - 6x^2 + 12xy^6)$ **(d)** $6x^2y^2(xy^2 - 2x^3 + 4x^2y^6)$

10. When directed to factor the polynomial $4x^2y^5 - 8xy^3$ completely, a student responded
with $2xy^3(2xy^2 - 4)$. When the teacher did not give him full credit, he complained
because when his answer is multiplied out, the result is the original polynomial. Was the
teacher justified in her grading? Why or why not?

Factor out the greatest common factor. See Examples 1–4.

11. $8k^3 + 24k$

12. $9z^4 + 72z$

13. $3xy - 5xy^2$

14. $5h^2j + 7hj$

15. $-4p^3q^4 - 2p^2q^5$

16. $-3z^5w^2 - 18z^3w^4$

17. $21x^5 + 35x^4 - 14x^3$

18. $18k^3 - 36k^4 + 48k^5$

19. $15a^2c^3 - 25ac^2 + 5a^2c$

20. $15y^3z^3 + 27y^2z^4 - 36yz^5$

21. $-27m^3p^5 + 5r^4s^3 - 8x^5z^4$

22. $-50r^4t^2 + 81x^3y^3 - 49p^2q^4$

23. $(m - 4)(m + 2) + (m - 4)(m + 3)$

24. $(z - 5)(z + 7) + (z - 5)(z + 9)$

25. $(2z - 1)(z + 6) - (2z - 1)(z - 5)$

26. $(3x + 2)(x - 4) - (3x + 2)(x + 8)$

27. $-y^5(r + w) - y^6(z + k)$

28. $-r^6(m + n) - r^7(p + q)$

29. $5(2 - x)^2 - (2 - x)^3 + 4(2 - x)$

30. $3(5 - x)^4 + 2(5 - x)^3 - (5 - x)^2$

Factor each of the following polynomials twice. First use a common factor with a positive coefficient, and then use a common factor with a negative coefficient. See Example 5.

31. $42z^3 - 56z^4$

32. $-2x^5 + 6x^3 + 4x^2$

33. $-5a^3 + 10a^4 - 15a^5$

Factor by grouping. See Examples 6–8.

34. $mx + 3qx + my + 3qy$

35. $2k + 2h + jk + jh$

36. $10m + 2n + 5mk + nk$

37. $3ma + 3mb + 2ab + 2b^2$

38. $m^2 - 3m - 15 + 5m$

39. $z^2 - 6z - 54 + 9z$

40. $p^2 - 4zq + pq - 4pz$

41. $r^2 - 9tw + 3rw - 3rt$

42. $3a^2 + 15a - 10 - 2a$

43. $7k + 2k^2 - 6k - 21$

44. $-15p^2 + 5pq - 6pq + 2q^2$

45. $-6r^2 + 9rs + 8rs - 12s^2$

46. $-3a^3 - 3ab^2 + 2a^2b + 2b^3$

47. $-16m^3 + 4m^2p^2 - 4mp + p^3$

48. $4 + xy - 2y - 2x$

49. $2ab^2 - 4 - 8b^2 + a$

50. $8 + 9y^4 - 6y^3 - 12y$

51. $x^3y^2 - 3 - 3y^2 + x^3$

52. Which one of the following is not a factored form of $1 - a + ab - b$?
(a) $(a - 1)(b - 1)$
(b) $(-a + 1)(-b + 1)$
(c) $(-1 + a)(-1 + b)$
(d) $(1 - a)(b + 1)$

53. $2ab^2 - 8b^2 + a - 4$ can be factored as $(2b^2 + 1)(a - 4)$. Give two other acceptable factored forms of this polynomial.

Factor out the variable that is raised to the smaller exponent.

54. $3m^{-5} + m^{-3}$

55. $k^{-2} + 2k^{-4}$

56. $3p^{-3} + 2p^{-2}$

5.8 *Factoring Trinomials*

The product of $x + 3$ and $x - 5$ is

$$(x + 3)(x - 5) = x^2 - 5x + 3x - 15$$
$$= x^2 - 2x - 15.$$

By this result, the **factored form** of $x^2 - 2x - 15$ is $(x + 3)(x - 5)$.

$$\text{Factored form} \rightarrow (x + 3)(x - 5) \xrightarrow{\text{Multiplication}} x^2 - 2x - 15 \leftarrow \text{Product}$$
$$\xleftarrow{\text{Factoring}}$$

OBJECTIVE 1 We show how to factor trinomials in this section, beginning with those having 1 as the coefficient of the squared term. Let us start by analyzing the example above, $x^2 - 2x - 15$. Because multiplying and factoring are inverses, factoring trinomials involves using FOIL backwards. As shown below, the x^2 term came from multiplying x and x, and -15 came from multiplying 3 and -5.

$$\text{Product of } x \text{ and } x \text{ is } x^2.$$
$$(x + 3)(x - 5) = x^2 - 2x - 15$$
$$\text{Product of 3 and } -5 \text{ is } -15.$$

The $-2x$ in $x^2 - 2x - 15$ was found by multiplying the outside terms, and then the inside terms, and adding.

Outside terms: $x(-5) = -5x$
$$(x + 3)(x - 5) \qquad \text{Add to get } -2x.$$
Inside terms: $3 \cdot x = 3x$

Based on this example, we can factor a trinomial with 1 as the coefficient of the squared term by using the steps below.

Factoring $x^2 + bx + c$

Step 1 Find all pairs of numbers whose product is the third term of the trinomial.

Step 2 Choose the pair whose sum is the coefficient of the middle term.

Step 3 If there are no such numbers, the polynomial cannot be factored.

A polynomial that cannot be factored is called **prime.**

E X A M P L E 1   Factoring a Trinomial $x^2 + bx + c$

Factor $x^2 + 8x + 12$.

We need to find the correct two numbers to place in the blanks.

$$x^2 + 8x + 12 = (x + \underline{\hspace{1cm}})(x + \underline{\hspace{1cm}})$$

WORK PROBLEM 1 AT THE SIDE. ▶▶

— CONTINUED ON NEXT PAGE

OBJECTIVES

1 ▶ Factor trinomials in which the coefficient of the squared term is 1.

2 ▶ Factor trinomials in which the coefficient of the squared term is not 1.

3 ▶ Use an alternative method to factor trinomials.

4 ▶ Factor by substitution.

FOR EXTRA HELP

Tutorial   Tape 8   SSM, Sec. 5.8

1. Find all pairs of numbers whose product is 12. Find the sum of each pair of numbers.

$$12 \cdot (\underline{\hspace{0.5cm}}); \qquad 12 + \underline{\hspace{0.5cm}} = \underline{\hspace{0.5cm}}$$
$$6 \cdot (\underline{\hspace{0.5cm}}); \qquad 6 + \underline{\hspace{0.5cm}} = \underline{\hspace{0.5cm}}$$
$$3 \cdot (\underline{\hspace{0.5cm}}); \qquad 3 + \underline{\hspace{0.5cm}} = \underline{\hspace{0.5cm}}$$
$$-12 \cdot (\underline{\hspace{0.5cm}}); \qquad -12 + \underline{\hspace{0.5cm}} = \underline{\hspace{0.5cm}}$$
$$-6 \cdot (\underline{\hspace{0.5cm}}); \qquad -6 + \underline{\hspace{0.5cm}} = \underline{\hspace{0.5cm}}$$
$$-3 \cdot (\underline{\hspace{0.5cm}}); \qquad -3 + \underline{\hspace{0.5cm}} = \underline{\hspace{0.5cm}}$$

ANSWERS

1. 1, 1, 13; 2, 2, 8; 4, 4, 7;
   $-1, -1, -13; -2, -2, -8;$
   $-4, -4, -7$

2. Factor.

(a) $p^2 + 6p + 5$

(b) $a^2 + 9a + 20$

As Problem 1 shows, the numbers 6 and 2 have a product of 12 and a sum of 8. Therefore,

$$x^2 + 8x + 12 = (x + 6)(x + 2).$$

Because of the commutative property, it would be equally correct to write $(x + 2)(x + 6)$.

◀◀ WORK PROBLEM 2 AT THE SIDE.

E X A M P L E   2   Factoring a Trinomial $x^2 + bx + c$

Factor $y^2 + 2y - 35$.

Find pairs of numbers whose product is -35.

Pairs of Factors	Sums of the Pairs
$-35(1)$	-34
$35(-1)$	34
$7(-5)$	2 ← Coefficient of middle term
$5(-7)$	-2

Use 7 and -5.

$$y^2 + 2y - 35 = (y + 7)(y - 5)$$

To check, find the product of $y + 7$ and $y - 5$.

3. Factor.

(a) $k^2 - k - 6$

(b) $b^2 - 7b + 10$

(c) $y^2 - 8y + 6$

E X A M P L E   3   Recognizing a Prime Polynomial

Factor $m^2 + 6m + 7$.

Look for two numbers whose product is 7 and whose sum is 6. Only two pairs of integers, 7 and 1 and -7 and -1, give a product of 7. Neither of these pairs has a sum of 6, so $m^2 + 6m + 7$ cannot be factored and is prime.

◀◀ WORK PROBLEM 3 AT THE SIDE.

4. Factor.

(a) $m^2 + 2mn - 8n^2$

E X A M P L E   4   Factoring a Trinomial in Two Variables

Factor $p^2 + 6ap - 16a^2$.

Look for two expressions whose sum is $6a$ and whose product is $-16a^2$. The quantities $8a$ and $-2a$ have the necessary sum and product, so

$$p^2 + 6ap - 16a^2 = (p + 8a)(p - 2a).$$

(b) $z^2 - 7zx + 9x^2$

◀◀ WORK PROBLEM 4 AT THE SIDE.

E X A M P L E   5   Factoring a Trinomial with a Common Factor

Factor $16y^3 - 32y^2 - 48y$.

Start by factoring out the greatest common factor, $16y$.

$$16y^3 - 32y^2 - 48y = 16y(y^2 - 2y - 3)$$

To factor $y^2 - 2y - 3$, look for two integers whose sum is -2 and whose product is -3. The necessary integers are -3 and 1, with

$$16y^3 - 32y^2 - 48y = 16y(y - 3)(y + 1).$$

ANSWERS

2. (a) $(p + 1)(p + 5)$
  (b) $(a + 5)(a + 4)$
3. (a) $(k - 3)(k + 2)$
  (b) $(b - 5)(b - 2)$
  (c) prime
4. (a) $(m - 2n)(m + 4n)$
  (b) prime

Caution
In factoring, always look for a common factor first. Do not forget to write the common factor as part of the answer.

WORK PROBLEM 5 AT THE SIDE. ▶▶

OBJECTIVE 2 ▶ A generalization of the method shown above can be used to factor a trinomial of the form $ax^2 + bx + c$, where the coefficient of the squared term is not equal to 1. In the first step we find all pairs of factors whose product is ac. In Step 2, we choose the pair whose sum is b. To see how this method works, let us factor $3x^2 + 7x + 2$. First, identify the values of a, b, and c.

$$\begin{array}{ccc} ax^2 & + bx & + c \\ \downarrow & \downarrow & \downarrow \\ 3x^2 & + 7x & + 2 \end{array}$$

$$a = 3, \, b = 7, \, c = 2$$

The product ac is $3 \cdot 2 = 6$, so we need integers having a product of 6 and a sum of 7 (because the coefficient of the middle term is 7). By inspection, the necessary integers are 1 and 6. Write $7x$ as $1x + 6x$, or $x + 6x$, giving

$$3x^2 + 7x + 2 = 3x^2 + \underbrace{x + 6x}_{} + 2.$$

$$\quad\quad\quad\quad\quad\quad\quad\quad\quad x + 6x = 7x$$

$$\begin{aligned} &= x(3x + 1) + 2(3x + 1) \quad \text{Factor by grouping.} \\ &= (3x + 1)(x + 2) \quad\quad\quad \text{Factor out the common factor.} \end{aligned}$$

As before, check this result by multiplying $3x + 1$ and $x + 2$.

E X A M P L E   6   Factoring a Trinomial
$$ax^2 + bx + c, \quad a \neq 1$$

Factor $12r^2 - 5r - 2$.
  Because $a = 12$, $b = -5$, and $c = -2$, the product ac is -24. The two integers whose product is -24 and whose sum is -5 are -8 and 3. Write $-5r$ as $3r - 8r$.

$$\begin{aligned} 12r^2 - 5r - 2 &= 12r^2 + 3r - 8r - 2 \\ &= 3r(4r + 1) - 2(4r + 1) \quad \text{Factor by grouping.} \\ &= (4r + 1)(3r - 2) \quad\quad\quad \text{Factor out the common factor.} \end{aligned}$$

WORK PROBLEM 6 AT THE SIDE. ▶▶

OBJECTIVE 3 ▶ An alternative approach, the method of trying repeated combinations, is especially helpful when the product ac is large. Let us examine this method using the same polynomials as above.

E X A M P L E   7   Factoring a Trinomial
$$ax^2 + bx + c, \quad a \neq 1$$

Factor each of the following.

(a) $3x^2 + 7x + 2$
  To factor this polynomial, we need to find the correct numbers to put in the blanks.

$$3x^2 + 7x + 2 = (\underline{\quad\quad} x + \underline{\quad\quad})(\underline{\quad\quad} x + \underline{\quad\quad})$$

CONTINUED ON NEXT PAGE

---

5. Factor $5m^4 - 5m^3 - 100m^2$.

6. Factor.

  (a) $3y^2 - 11y - 4$

  (b) $6k^2 - 19k + 10$

Plus signs were used because all the signs in the polynomial are plus. The first two expressions have a product of $3x^2$, so they must be $3x$ and x.

$$3x^2 + 7x + 2 = (3x + \underline{\hspace{1cm}})(x + \underline{\hspace{1cm}})$$

The product of the two last terms must be 2, so that the numbers must be 2 and 1. We have a choice. We could use the 2 with the $3x$ or with the x. Only one of these choices can give the correct middle term. Try each.

$3x$

$(3x + 2)(x + 1)$

$2x$

$3x + 2x = 5x$
Wrong middle term

$6x$

$(3x + 1)(x + 2)$

x

$6x + x = 7x$
Correct middle term

Therefore, $3x^2 + 7x + 2 = (3x + 1)(x + 2)$.

(b) $12r^2 - 5r - 2$

For factors of 12 we could choose 4 and 3, 6 and 2, or 12 and 1. To reduce the number of trials, we note that the trinomial has no common factor. This means that neither of its factors can have a common factor. We should keep this in mind as we choose factors. Let us try 4 and 3 for the two first terms. If these do not work, we will make another choice.

$$12r^2 - 5r - 2 = (4r\underline{\hspace{1cm}})(3r\underline{\hspace{1cm}})$$

We do not know what signs to use yet. The factors of -2 are -2 and 1 or 2 and -1. Try some possibilities.

$(4r - 2)(3r + 1)$
Wrong: $4r - 2$ has a
common factor.

$8r$

$(4r - 1)(3r + 2)$

$-3r$

$8r - 3r = 5r$
Wrong middle term

The middle term on the right is $5r$, instead of the $-5r$ we need. To get $-5r$, we need only exchange the middle signs.

$-8r$

$(4r + 1)(3r - 2)$

$3r$

$-8r + 3r = -5r$
Correct middle term

Thus, $12r^2 - 5r - 2 = (4r + 1)(3r - 2)$.

Note

As shown in Example 7, if the terms of a polynomial have no common factor (except 1), then none of the terms of its factors have a common factor. Remembering this fact helps us eliminate some potential choices.

┌─
E X A M P L E   8   **Factoring a Trinomial $ax^2 + bx + c$ with $a < 0$**

Factor $-2x^2 - 5x + 3$.

    Although a is negative here, the steps are the same. Factors of $-2x^2$ are $-2x$ and x or $2x$ and $-x$. Factors of 3 are 3 and 1 or -3 and -1. Check the middle term using various combinations. We show just two of the possibilities.

$$\underbrace{(-2x + 3)(x + 1)}_{3x}\ ^{-2x} \qquad \underbrace{(-2x + 1)(x + 3)}_{x}\ ^{-6x}$$

$$-2x + 3x = x \qquad\qquad -6x + x = -5x$$
Wrong middle term $\qquad\qquad$ Correct middle term

Thus, $-2x^2 - 5x + 3 = (-2x + 1)(x + 3)$.
─┘

    The alternative method of factoring a trinomial with the coefficient of the squared term not equal to 1 is summarized here.

Factoring a Trinomial

Step 1 Write all pairs of factors of the coefficient of the squared term.

Step 2 Write all pairs of factors of the last term.

Step 3 Use various combinations of these factors until you find the necessary middle term.

Step 4 If the necessary combination does not exist, the polynomial is prime.

WORK PROBLEM 7 AT THE SIDE. ▶▶

┌─
E X A M P L E   9   **Factoring a Trinomial in Two Variables**

Factor $18m^2 - 19mx - 12x^2$.

    There is no common factor (except 1). Go through the steps to factor the trinomial. There are many possible factors of both 18 and -12. As a general rule, the middle-sized factors should be tried first. Let's try 6 and 3 for 18 and -3 and 4 for -12.

$$(6m - 3x)(3m + 4x) \qquad (6m + 4x)(3m - 3x)$$
Wrong: common factor $\qquad$ Wrong: common factors

We did not get very far with 6 and 3 as factors of 18; so we will try 9 and 2 instead, and try -4 and 3 as factors of -12.

$$(9m + 3x)(2m - 4x) \qquad \underbrace{(9m - 4x)(2m + 3x)}_{-8mx}\ ^{27mx}$$
Common factors

$$27mx + (-8mx) = 19mx$$

The result on the right differs from the correct middle term only in the sign, so we exchange the middle signs on the two factors to get

$$18m^2 - 19mx - 12x^2 = (9m + 4x)(2m - 3x).$$
─┘

WORK PROBLEM 8 AT THE SIDE. ▶▶

7. Factor.

(a) $2p^2 + 5p - 12$

(b) $6k^2 - k - 2$

(c) $8m^2 + 18m - 5$

(d) $-6r^2 + 13r + 5$

8. Factor.

(a) $7p^2 + 15pq + 2q^2$

(b) $6m^2 + 7mn - 5n^2$

(c) $12z^2 - 5zy - 2y^2$

ANSWERS
7. (a) $(p + 4)(2p - 3)$
   (b) $(3k - 2)(2k + 1)$
   (c) $(4m - 1)(2m + 5)$
   (d) $(-2r + 5)(3r + 1)$
8. (a) $(7p + q)(p + 2q)$
   (b) $(3m + 5n)(2m - n)$
   (c) $(3z - 2y)(4z + y)$

9. Factor.

   (a) $2m^3 - 4m^2 - 6m$

   (b) $12r^4 + 6r^3 - 90r^2$

   (c) $30y^5 - 55y^4 - 50y^3$

10. Factor.

   (a) $y^4 + y^2 - 6$

   (b) $2p^4 + 7p^2 - 15$

   (c) $6r^4 - 13r^2 + 5$

11. Factor.

   (a) $6(a - 1)^2 + (a - 1) - 2$

   (b) $8(z + 5)^2 - 2(z + 5) - 3$

   (c) $15(m - 4)^2 - 11(m - 4) + 2$

ANSWERS

9. (a) $2m(m + 1)(m - 3)$
  (b) $6r^2(r + 3)(2r - 5)$
  (c) $5y^3(2y - 5)(3y + 2)$
10. (a) $(y^2 - 2)(y^2 + 3)$
  (b) $(2p^2 - 3)(p^2 + 5)$
  (c) $(3r^2 - 5)(2r^2 - 1)$
11. (a) $(2a - 3)(3a - 1)$
  (b) $(4z + 17)(2z + 11)$
  (c) $(3m - 13)(5m - 22)$

EXAMPLE 10   Factoring a Trinomial with Terms That Have a Common Factor

Factor $16y^3 + 24y^2 - 16y$.

The terms of this polynomial have a greatest common factor of $8y$. Factor this out first.

$$16y^3 + 24y^2 - 16y = 8y(2y^2 + 3y - 2)$$

Factor $2y^2 + 3y - 2$ by either method given above.

$$16y^3 + 24y^2 - 16y = 8y(2y - 1)(y + 2)$$

Remember to write the common factor in front.

◀◀ WORK PROBLEM 9 AT THE SIDE.

EXAMPLE 11   Factoring a Trinomial $ax^4 + bx^2 + c$

Factor $6y^4 + 7y^2 - 20$.

We know that $y^4 = (y^2)^2$, so factors of $6y^4$ might be $6y^2$ and y^2 or $3y^2$ and $2y^2$. Let us try $3y^2$ and $2y^2$ with -4 and 5 as factors of -20. This gives

$$(3y^2 - 4)(2y^2 + 5).$$

Check the middle term: $(3y^2)(5) + (-4)(2y^2) = 15y^2 - 8y^2 = 7y^2$, as required. Thus,

$$6y^4 + 7y^2 - 20 = (3y^2 - 4)(2y^2 + 5).$$

◀◀ WORK PROBLEM 10 AT THE SIDE.

OBJECTIVE 4 In Example 11, we could have substituted another variable for y^2 and then factored the simpler trinomial. This **method of substitution** (sub-stih-TOO-shun) is used in the next example with a more complicated trinomial.

EXAMPLE 12   Factoring a Trinomial Using Substitution

Factor $2(x + 3)^2 + 5(x + 3) - 12$.

Because the binomial $x + 3$ appears in powers of 2 and 1, let the substitution variable represent $x + 3$. We may choose any letter except x. Let us choose y.

$$2(x + 3)^2 + 5(x + 3) - 12 = 2y^2 + 5y - 12 \quad \text{Let } y = x + 3.$$
$$= (2y - 3)(y + 4) \quad \text{Factor.}$$
$$= [2(x + 3) - 3][(x + 3) + 4] \quad \text{Replace } y \text{ with } x + 3.$$
$$= (2x + 6 - 3)(x + 7)$$
$$= (2x + 3)(x + 7)$$

◀◀ WORK PROBLEM 11 AT THE SIDE.

5.8 *Exercises*

Write true *or* false *for each statement.*

_____ **1.** A polynomial that cannot be factored is called a prime polynomial.

_____ **2.** To factor a trinomial, we use FOIL backwards.

_____ **3.** It is not necessary to factor out any common factors when factoring a trinomial.

_____ **4.** The method of substitution is used to simplify a polynomial with a common binomial factor before factoring.

Factor each of the following trinomials. See Examples 1–10.

5. $y^2 + 7y - 30$

6. $z^2 + 2z - 24$

7. $p^2 - p - 56$

8. $k^2 - 11k + 30$

9. $-m^2 + 16m - 60$

10. $-p^2 + 6p + 27$

11. $a^2 - 2ab - 35b^2$

12. $z^2 + 8zw + 15w^2$

13. $y^2 - 3yq - 15q^2$

14. $k^2 - 11hk + 28h^2$

15. $x^2y^2 + 11xy + 18$

16. $p^2q^2 - 5pq - 18$

17. $-6m^2 - 13m + 15$

18. $-15y^2 + 17y + 18$

19. $10x^2 + 3x - 18$

20. $8k^2 + 34k + 35$

21. $20k^2 + 47k + 24$

22. $27z^2 + 42z - 5$

23. $15a^2 - 22ab + 8b^2$

24. $15p^2 + 24pq + 8q^2$

25. $36m^2 - 60m + 25$

26. $25r^2 - 90r + 81$

27. $40x^2 + xy + 6y^2$

28. $14c^2 - 17cd - 6d^2$

29. $6x^2z^2 + 5xz - 4$

30. $8m^2n^2 - 10mn + 3$

31. $24x^2 + 42x + 15$

32. $36x^2 + 18x - 4$

33. $15a^2 + 70a - 120$

34. $12a^2 + 10a - 42$

35. $11x^3 - 110x^2 + 264x$

36. $9k^3 + 36k^2 - 189k$

37. $2x^3y^3 - 48x^2y^4 + 288xy^5$

38. $6m^3n^2 - 24m^2n^3 - 30mn^4$

39. Asked to factor $4x^2 + 2x - 20$ completely, a student gave the answer as $(4x + 10)(x - 2)$. Write a short explanation of why she lost some credit for this answer.

40. Find a polynomial that can be factored as $-9a(a - 5b)(2a + 7b)$.

Factor each of the following trinomials. See Example 11.

41. $2x^4 - 9x^2 - 18$

42. $6z^4 + z^2 - 1$

43. $16x^4 + 16x^2 + 3$

44. $9r^4 + 9r^2 + 2$

45. $12p^6 - 32p^3r + 5r^2$

46. $2y^6 + 7xy^3 + 6x^2$

Factor each of the following trinomials. See Example 12.

47. $10(k + 1)^2 - 7(k + 1) + 1$

48. $4(m - 5)^2 - 4(m - 5) - 15$

49. $3(m + p)^2 - 7(m + p) - 20$

50. $4(x - y)^2 - 23(x - y) - 6$

51. $a^2(a + b)^2 - ab(a + b)^2 - 6b^2(a + b)^2$

52. $m^2(m - p) + mp(m - p) - 2p^2(m - p)$

MATHEMATICAL CONNECTIONS (EXERCISES 53–58)

Refer to the note following Example 7 in this section. Then work Exercises 53–58 in order.

53. Is 2 a factor of the composite number 45?

54. List all positive integer factors of 45. Is 2 a factor of any of these factors?

55. Is 5 a factor of $10x^2 + 29x + 10$?

56. Factor $10x^2 + 29x + 10$. Is 5 a factor of either of its factors?

57. Suppose that k is an odd integer and you are asked to factor $2x^2 + kx + 8$. Why is $2x + 4$ not a possible choice in factoring this polynomial?

58. The polynomial $12y^2 - 11y - 15$ can be factored using the methods of this section. Explain why $3y + 15$ cannot be one of its factors.

5.9 Special Factoring

Certain types of factoring occur so often that they deserve special study.

OBJECTIVE As discussed earlier in this chapter, the product of the sum and difference of two terms is

$$(x + y)(x - y) = x^2 - y^2.$$

This result leads to the **difference of two squares,** a formula that is useful in factoring.

Difference of Two Squares

$$x^2 - y^2 = (x + y)(x - y)$$

E X A M P L E   1   Factoring the Difference of Squares

(a) $4a^2 - 64$

There is a common factor of 4.

$$4a^2 - 64 = 4(a^2 - 16) \qquad \text{Factor out the common factor.}$$
$$= 4(a + 4)(a - 4) \qquad \text{Factor the difference of squares.}$$

(b) $16m^2 - 49p^2$

$$
\begin{array}{cccccc}
A^2 & - & B^2 & = (A & + & B)(A & - & B) \\
\downarrow & & \downarrow & \downarrow & & \downarrow & \downarrow & & \downarrow
\end{array}
$$
$$16m^2 - 49p^2 = (4m)^2 - (7p)^2 = (4m + 7p)(4m - 7p)$$

(c) $81k^2 - (a + 2)^2$

$$
\begin{array}{cccccc}
A^2 & - & B^2 & = (A & + & B) & (A & - & B) \\
\downarrow & & \downarrow & & \downarrow & & \downarrow & & \downarrow
\end{array}
$$
$$81k^2 - (a + 2)^2 = (9k)^2 - (a + 2)^2 = (9k + a + 2)(9k - [a + 2])$$
$$= (9k + a + 2)(9k - a - 2)$$

We could have used the method of substitution here.

(d) $x^4 - 81 = (x^2 + 9)(x^2 - 9) \qquad \text{Factor the difference of squares.}$
$$= (x^2 + 9)(x + 3)(x - 3) \qquad \text{Factor } x^2 - 9.$$

Caution

Assuming no greatest common factor except 1, it is not possible to factor (with real numbers) a *sum* of two squares such as $x^2 + 25$. In particular, $x^2 + y^2 \neq (x + y)^2$, as shown next.

WORK PROBLEM 1 AT THE SIDE. ▶▶

OBJECTIVE Two other special products lead to the following rules for factoring.

Square Trinomials

$$x^2 + 2xy + y^2 = (x + y)^2$$
$$x^2 - 2xy + y^2 = (x - y)^2$$

OBJECTIVES

1 ▶ Factor the difference of two squares.

2 ▶ Factor a square trinomial.

3 ▶ Factor the difference of two cubes.

4 ▶ Factor the sum of two cubes.

FOR EXTRA HELP

Tutorial   Tape 8   SSM, Sec. 5.9

1. Factor.

  (a) $9a^2 - 16b^2$

  (b) $(m + 3)^2 - 49z^2$

ANSWERS

1. (a) $(3a - 4b)(3a + 4b)$
   (b) $(m + 3 + 7z)(m + 3 - 7z)$

2. Identify any square trinomials.

(a) $z^2 + 12z + 36$

(b) $2x^2 - 4x + 4$

(c) $9a^2 + 12ab + 16b^2$

3. Factor.

(a) $49z^2 - 14zk + k^2$

(b) $9a^2 + 48ab + 64b^2$

(c) $(k + m)^2 - 12(k + m) + 36$

(d) $x^2 - 2x + 1 - y^2$

The trinomial $x^2 + 2xy + y^2$ is the square of $x + y$. For this reason $x^2 + 2xy + y^2$ is called a **square trinomial.** In these patterns, both the first and last terms of the trinomial must be perfect squares. In the factored form, twice the product of the first and last terms must give the middle term of the trinomial. It is important to understand these patterns in terms of words because they occur with many different symbols (other than x and y).

$$4m^2 + 20m + 25 \qquad\qquad p^2 - 8p + 64$$

Square trinomial

Not a square trinomial
Middle term should be $16p$.

◀◀ **WORK PROBLEM 2 AT THE SIDE.**

E X A M P L E 2   Factoring a Square Trinomial

Factor.

(a) $144p^2 - 120p + 25$

$$144p^2 - 120p + 25 = (12p)^2 - 120p + 5^2 \qquad \text{Identify the square terms.}$$

$$= (12p - 5)^2 \qquad \text{Use the same middle sign.}$$

To see if this is correct, take twice the product of these two terms,

$$2(12p)(-5) = -120p,$$

which is the middle term of the given trinomial. Thus,

$$144p^2 - 120p + 25 = (12p - 5)^2.$$

(b) $4m^2 + 20mn + 49n^2$

If this is a square trinomial, it will equal $(2m + 7n)^2$. Check the middle term: $2(2m)(7n) = 28mn$, which *does not equal* $20mn$. Verify that we cannot factor this trinomial by the methods of the previous section either. It is prime.

(c) $(r + 5)^2 + 6(r + 5) + 9 = [(r + 5) + 3]^2 \qquad$ Because $2(3)(r + 5)$
$$= (r + 8)^2 \qquad\qquad\qquad\qquad\qquad = 6(r + 5)$$

(d) $m^2 - 8m + 16 - p^2$

The first three terms are a square trinomial; group them together, and factor the trinomial as follows.

$$(m^2 - 8m + 16) - p^2 = (m - 4)^2 - p^2 \qquad \text{Group terms; factor the square trinomial.}$$

$$(m - 4)^2 - p^2 = (m - 4 + p)(m - 4 - p) \qquad \text{Factor the difference of squares.}$$

◀◀ **WORK PROBLEM 3 AT THE SIDE.**

OBJECTIVE ▶ The **difference of two cubes,** $x^3 - y^3$, can be factored as follows.

Difference of Two Cubes

$$x^3 - y^3 = (x - y)(x^2 + xy + y^2)$$

ANSWERS

2. (a) square **(b)** not square
   (c) not square
3. (a) $(7z - k)^2$ **(b)** $(3a + 8b)^2$
   (c) $[(k + m) - 6]^2$ or $(k + m - 6)^2$
   (d) $(x - 1 + y)(x - 1 - y)$

We can check this by finding the product of $x - y$ and $x^2 + xy + y^2$, as follows.

$$
\begin{array}{r}
x^2 + xy + y^2 \\
x - y \\
\hline
-x^2y - xy^2 - y^3 \\
x^3 + x^2y + xy^2 \\
\hline
x^3 \qquad\qquad - y^3
\end{array}
$$

This result shows that

$$x^3 - y^3 = (x - y)(x^2 + xy + y^2).$$

E X A M P L E   3   Factoring the Difference of Cubes

Factor.

(a) $m^3 - 8 = m^3 - 2^3 = (m - 2)(m^2 + 2m + 2^2)$

$$= (m - 2)(m^2 + 2m + 4)$$

Opposite of the product of the roots gives the middle term.

(b) $27x^3 - 8y^3 = (3x)^3 - (2y)^3$

$$= (3x - 2y)[(3x)^2 + (3x)(2y) + (2y)^2]$$

$$= (3x - 2y)(9x^2 + 6xy + 4y^2)$$

(c) $1000k^3 - 27n^3 = (10k)^3 - (3n)^3$

$$= (10k - 3n)[(10k)^2 + (10k)(3n) + (3n)^2]$$

$$= (10k - 3n)(100k^2 + 30kn + 9n^2)$$

WORK PROBLEM 4 AT THE SIDE. ▶▶

Objective ▶ While an expression of the form $x^2 + y^2$ (a sum of two squares) usually cannot be factored with real numbers, the **sum of two cubes** can be factored with a pattern very similar to that for the difference of two cubes.

Sum of Two Cubes

$$x^3 + y^3 = (x + y)(x^2 - xy + y^2)$$

To verify this result, find the product of $x + y$ and $x^2 - xy + y^2$. Compare this pattern with the pattern for the difference of two cubes.

Note
The sign in the first factor of the sum or difference of cubes rule is *always the same* as the sign in the original problem. In the second factor, the first and last terms are *always positive;* the sign of the middle term is *the opposite of* the sign in the first factor.

4. Factor.

(a) $x^3 - 1000$

(b) $8k^3 - y^3$

(c) $27m^3 - 64$

Answers

4. (a) $(x - 10)(x^2 + 10x + 100)$
   (b) $(2k - y)(4k^2 + 2ky + y^2)$
   (c) $(3m - 4)(9m^2 + 12m + 16)$

5. Factor.

(a) $2x^3 + 2000$

E X A M P L E   4   Factoring the Sum of Cubes

Factor.

(a) $3r^4 + 81r$

There is a common factor of $3r$.

$$3r^4 + 81r = 3r(r^3 + 27) \qquad \text{Factor out the common factor.}$$
$$= 3r(r + 3)(r^2 - 3r + 3^2) \qquad \text{Factor the sum of cubes.}$$
$$= 3r(r + 3)(r^2 - 3r + 9)$$

(b) $27z^3 + 125 = (3z)^3 + 5^3 = (3z + 5)[(3z)^2 - (3z)(5) + 5^2]$
$$= (3z + 5)(9z^2 - 15z + 25)$$

(c) $125t^3 + 216s^6 = (5t)^3 + (6s^2)^3$
$$= (5t + 6s^2)[(5t)^2 - (5t)(6s^2) + (6s^2)^2]$$
$$= (5t + 6s^2)(25t^2 - 30ts^2 + 36s^4)$$

(b) $8p^3 + 125$

Caution

A common error is to think there is a 2 as coefficient of xy when factoring the sum or difference of the two cubes, x^3 and y^3. There is no 2, so in general, expressions of the form $x^2 + xy + y^2$ and $x^2 - xy + y^2$ cannot be factored.

◄◄ **WORK PROBLEM 5 AT THE SIDE.**

The special types of factoring discussed in this section are summarized below and should be memorized.

Difference of two squares	$x^2 - y^2 = (x + y)(x - y)$
Perfect square trinomial	$x^2 + 2xy + y^2 = (x + y)^2$
	$x^2 - 2xy + y^2 = (x - y)^2$
Difference of two cubes	$x^3 - y^3 = (x - y)(x^2 + xy + y^2)$
Sum of two cubes	$x^3 + y^3 = (x + y)(x^2 - xy + y^2)$

(c) $27m^3 + 125n^3$

ANSWERS

5. (a) $2(x + 10)(x^2 - 10x + 100)$
   (b) $(2p + 5)(4p^2 - 10p + 25)$
   (c) $(3m + 5n)(9m^2 - 15mn + 25n^2)$

5.9 Exercises

1. Which of the following binomials are differences of squares?
   (a) $64 - m^2$ (b) $2x^2 - 25$ (c) $k^2 + 9$ (d) $4z^4 - 49$

2. Which of the following binomials are sums or differences of cubes?
   (a) $64 + y^3$ (b) $125 - p^6$ (c) $9x^3 + 125$ (d) $(x + y)^3 - 1$

3. Which of the following trinomials are perfect squares?
   (a) $x^2 - 8x - 16$ (b) $4m^2 + 20m + 25$ (c) $9z^4 + 30z^2 + 25$ (d) $25a^2 - 45a + 81$

4. Of the twelve polynomials listed in Exercises 1–3, which ones can be factored using the methods of this section?

5. The binomial $9x^2 + 81$ is an example of the sum of two squares that can be factored. Under what conditions can the sum of two squares be factored?

6. Insert the correct signs in the blanks.
   (a) $8 + t^3 = (2 \_\_ t)(4 \_\_ 2t \_\_ t^2)$ (b) $z^3 - 1 = (z \_\_ 1)(z^2 \_\_ z \_\_ 1)$

Factor each of the following polynomials. See Examples 1–4.

7. $p^2 - 16$

8. $k^2 - 9$

9. $25x^2 - 4$

10. $36m^2 - 25$

11. $9a^2 - 49b^2$

12. $16c^2 - 49d^2$

13. $64m^4 - 4y^4$

14. $243x^4 - 3t^4$

15. $(y + z)^2 - 81$

16. $(h + k)^2 - 9$

17. $16 - (x + 3y)^2$

18. $64 - (r + 2t)^2$

19. $(p + q)^2 - (p - q)^2$

20. $(a + b)^2 - (a - b)^2$

21. $k^2 - 6k + 9$

22. $x^2 + 10x + 25$

23. $4z^2 + 4zw + w^2$

24. $9y^2 + 6yz + z^2$

25. $16m^2 - 8m + 1 - n^2$ **26.** $25c^2 - 20c + 4 - d^2$ **27.** $4r^2 - 12r + 9 - s^2$

28. $9a^2 - 24a + 16 - b^2$ **29.** $x^2 - y^2 + 2y - 1$ **30.** $-k^2 - h^2 + 2kh + 4$

31. $98m^2 + 84mn + 18n^2$ **32.** $80z^2 - 40zw + 5w^2$ **33.** $(p + q)^2 + 2(p + q) + 1$

34. $(x + y)^2 + 6(x + y) + 9$ **35.** $(a - b)^2 + 8(a - b) + 16$ **36.** $(m - n)^2 + 4(m - n) + 4$

37. $8x^3 - y^3$ **38.** $z^3 + 125p^3$ **39.** $64g^3 + 27h^3$

40. $27a^3 - 8b^3$ **41.** $24n^3 + 81p^3$ **42.** $250x^3 - 16y^3$

43. $(y + z)^3 - 64$ **44.** $(p - q)^3 + 125$ **45.** $m^6 - 125$

46. $k^6 + (k + 3)^3$ **47.** $(a + b)^3 - (a - b)^3$

MATHEMATICAL CONNECTIONS (EXERCISES 48–51)

Work Exercises 48–51 in order.

48. Factor $x^6 - y^6$ by first factoring as a difference of squares.

49. Factor $x^6 - y^6$ by first factoring as a difference of cubes.

50. The factorizations you found in Exercises 48 and 49 should be equal, since they both equal $x^6 - y^6$. Both answers include factors of $x - y$ and $x + y$. What must be true of the remaining factors? Verify this.

51. Based on the results in Exercises 48–50, which method should be used to get a complete factorization of such a polynomial?

SUMMARY EXERCISES ON FACTORING METHODS

A polynomial is completely factored when the polynomial is in the form described below.

> **1.** The polynomial is written as a product of prime polynomials with integer coefficients.
>
> **2.** None of the polynomial factors can be factored further, except that a monomial factor need not be factored completely.

For example, $9x^2(x + 2)$ is the factored form of $9x^3 + 18x^2$. As stated in the second rule above, it is not necessary to factor $9x^2$ as $3 \cdot 3 \cdot x \cdot x$. The order of the factors does not matter.

The steps to follow in factoring a polynomial are listed below.

> *Step 1*   Factor out any common factor. See Section 5.7.
>
> *Step 2a*   If the polynomial is a binomial, check to see if it is the difference of two squares, the difference of two cubes, or the sum of two cubes. See Section 5.9.
>
> *Step 2b*   If the polynomial is a trinomial, check to see if it is a square trinomial. If so, factor as in Section 5.9. If it is not, factor as in Section 5.8.
>
> *Step 2c*   If the polynomial has four terms, try to factor by grouping. See Section 5.7.

Factor each of the following polynomials.

1. $100a^2 - 9b^2$ **2.** $10r^2 + 13r - 3$ **3.** $18p^5 - 24p^3 + 12p^6$

4. $15x^2 - 20x$ **5.** $x^2 + 2x - 35$ **6.** $9 - a^2 + 2ab - b^2$

7. $49z^2 - 16$ **8.** $225p^2 + 256$ **9.** $x^3 - 1000$

10. $6b^2 - 17b - 3$

11. $k^2 - 6k + 16$

12. $18m^3n + 3m^2n^2 - 6mn^3$

13. $6t^2 + 19tu - 77u^2$

14. $2p^2 + 11pq + 15q^2$

15. $40p^2 - 32p$

16. $9m^2 - 45m + 18m^3$

17. $4k^2 + 28kr + 49r^2$

18. $54m^3 - 2000$

19. $mn - 2n + 5m - 10$

20. $2a^2 - 7a - 4$

21. $9m^2 - 30mn + 25n^2 - p^2$

22. $x^3 + 3x^2 - 9x - 27$

23. $56k^3 - 875$

24. $9r^2 + 100$

25. $16z^3x^2 - 32z^2x$

26. $8p^3 - 125$

27. $m^2(m - 2) - 4(m - 2)$

28. $6k^2 - k - 1$

29. $27m^2 + 144mn + 192n^2$

30. $x^4 - 625$

31. $125m^6 + 216$

32. $ab + 6b + ac + 6c$

33. $2m^2 - mn - 15n^2$

34. $p^3 + 64$

35. $4y^2 - 8y$

36. $6a^4 - 11a^2 - 10$

37. $14z^2 - 3zk - 2k^2$

38. $12z^3 - 6z^2 + 18z$

39. $256b^2 - 400c^2$

40. $z^2 - zp + 20p^2$

41. $1000z^3 + 512$

42. $64m^2 - 25n^2$

43. $10r^2 + 23rs - 5s^2$

44. $12k^2 - 17kq - 5q^2$

45. $32x^2 + 16x^3 - 24x^5$

46. $48k^4 - 243$

47. $14x^2 - 25xq - 25q^2$

48. $5p^2 - 10p$

49. $y^2 + 3y - 10$

50. $b^2 - 7ba - 18a^2$

51. $2a^3 + 6a^2 - 4a$

52. $12m^2rx + 4mnrx + 40n^2rx$

53. $18p^2 + 53pr - 35r^2$

54. $21a^2 - 5ab - 4b^2$

55. $(x - 2y)^2 - 4$

56. $(3m - n)^2 - 25$

57. $(5r + 2s)^2 - 6(5r + 2s) + 9$

58. $(p + 8q)^2 - 10(p + 8q) + 25$

59. $z^4 - 9z^2 + 20$

60. $21m^4 - 32m^2 - 5$

5.10 Solving Equations by Factoring

OBJECTIVES

1. Learn the zero-factor property.
2. Use the zero-factor property to solve equations.
3. Solve applied problems that need the zero-factor property.

FOR EXTRA HELP

Tutorial   Tape 8   SSM, Sec. 5.10

Up to now in Chapter 5 we have worked with polynomials, which are algebraic *expressions*. Now we discuss polynomial *equations,* so we will be using the addition and multiplication properties of equality from Chapter 2.

OBJECTIVE 1 Some equations that cannot be solved by other methods can be solved by factoring. This process depends on a special property of the number 0, called the **zero-factor property.**

> **Zero-Factor Property**
>
> If two numbers have a product of 0, then at least one of the numbers must be 0. That is, if $ab = 0$, then $a = 0$ or $b = 0$.

To prove the zero-factor property, we first assume $a \neq 0$. (If a does equal 0, then the property is proved already.) If $a \neq 0$, then $\frac{1}{a}$ exists, and we can multiply both sides of $ab = 0$ by $\frac{1}{a}$ to get

$$\frac{1}{a} \cdot ab = \frac{1}{a} \cdot 0 \qquad \text{Multiply by } \tfrac{1}{a}.$$

$$b = 0.$$

Thus, if $a \neq 0$, then $b = 0$, and the result is proved.

OBJECTIVE 2 The next examples show how to use the zero-factor property to solve equations.

E X A M P L E   1   Using the Zero-Factor Property to Solve an Equation

Solve the equation $(x + 6)(2x - 3) = 0$.

Here the product of $x + 6$ and $2x - 3$ is 0. By the zero-factor property, this can be true only if $x + 6$ equals 0 or if $2x - 3$ equals 0:

$$x + 6 = 0 \quad \text{or} \quad 2x - 3 = 0.$$

Solve $x + 6 = 0$ by subtracting 6 from both sides of the equation. Solve $2x - 3 = 0$ by adding 3 on both sides and then dividing by 2.

$$
\begin{array}{ccc}
x + 6 = 0 & \text{or} & 2x - 3 = 0 \\
x + 6 - 6 = 0 - 6 & & 2x - 3 + 3 = 0 + 3 \\
x = -6 & & 2x = 3 \\
& & x = \dfrac{3}{2}
\end{array}
$$

This equation has two solutions that should be checked by substitution in the original equation.

If $x = -6$,

$$(-6 + 6)[2(-6) - 3] = 0 \quad ?$$

$$0(-15) = 0 \quad \text{True}$$

If $x = \frac{3}{2}$,

$$\left(\frac{3}{2} + 6\right)\left(2 \cdot \frac{3}{2} - 3\right) = 0 \quad ?$$

$$\frac{15}{2}(0) = 0 \quad \text{True}$$

Both solutions check; the solution set is $\{-6, \frac{3}{2}\}$.

WORK PROBLEM 1 AT THE SIDE. ▶▶

---

1. Solve.

(a) $(3k + 5)(k + 1) = 0$

(b) $(3r + 11)(5r - 2) = 0$

ANSWERS

1. (a) $\left\{ -\dfrac{5}{3}, -1 \right\}$ (b) $\left\{ -\dfrac{11}{3}, \dfrac{2}{5} \right\}$

2. Solve.

(a) $3r^2 - r = 4$

Because the product $(x + 6)(2x - 3)$ equals $2x^2 + 9x - 18$, the equation in Example 1 has a squared term and is an example of a *quadratic* (kwah-DRAD-ik) *equation*.

> **Quadratic Equation**
>
> An equation that can be written in the form
> $$ax^2 + bx + c = 0,$$
> $(a \neq 0)$, is a **quadratic equation.**

Quadratic equations are discussed in more detail in a later chapter.

The steps involved in solving a polynomial equation by factoring are summarized below.

> **Solving a Polynomial Equation by Factoring**
>
> *Step 1* Rewrite the equation if necessary so that one side is zero.
>
> *Step 2* Factor the polynomial.
>
> *Step 3* Set each factor equal to 0, using the zero-factor property.
>
> *Step 4* Solve each equation in Step 3.
>
> *Step 5* Check each solution in the original equation.

(b) $15m^2 + 7m = 2$

E X A M P L E 2   Solving a Quadratic Equation by Factoring

Solve $2p^2 + 3p = 2$.

First rewrite the equation with zero on one side.

$$
\begin{aligned}
2p^2 + 3p &= 2 \\
2p^2 + 3p - 2 &= 0 && \text{Subtract 2.} \\
(2p - 1)(p + 2) &= 0 && \text{Factor.} \\
2p - 1 = 0 \quad &\text{or} \quad p + 2 = 0 && \text{Zero-factor property} \\
2p = 1 \quad\quad\quad & \quad\quad\quad p = -2 && \text{Solve each equation.} \\
p = \frac{1}{2} &
\end{aligned}
$$

Check these solutions by substituting them into the original equation.

$$\text{If } p = \frac{1}{2},$$

$$2\left(\frac{1}{2}\right)^2 + 3\left(\frac{1}{2}\right) = 2 \quad ?$$

$$2\left(\frac{1}{4}\right) + \frac{3}{2} = 2 \quad ?$$

$$\frac{1}{2} + \frac{3}{2} = 2 \quad ?$$

$$2 = 2 \quad \text{True}$$

$$\text{If } p = -2,$$

$$2(-2)^2 + 3(-2) = 2 \quad ?$$

$$2(4) - 6 = 2 \quad ?$$

$$8 - 6 = 2 \quad ?$$

$$2 = 2 \quad \text{True}$$

Because both solutions check, the solution set is $\{\frac{1}{2}, -2\}$.

◀◀ **WORK PROBLEM 2 AT THE SIDE.**

┌─ E X A M P L E 3   **Solving a Quadratic Equation**
│   **with a Missing Term**

Solve $5z^2 - 25z = 0$.

This quadratic equation has a missing term. Comparing it with the general form $ax^2 + bx + c = 0$ shows that $c = 0$. The zero-factor property still applies, however, because $5z^2 - 25z$ can be factored as

$$5z^2 - 25z = 5z(z - 5).$$

Now use the zero-factor property to solve the equation.

$$5z^2 - 25z = 0$$

$$5z(z - 5) = 0 \qquad \text{Factor.}$$

$$5z = 0 \quad \text{or} \quad z - 5 = 0 \qquad \text{Zero-factor property}$$

$$z = 0 \qquad\qquad z = 5$$

The solution set is $\{0, 5\}$, as can be verified by substituting back in the original equation.
└─

WORK PROBLEM 3 AT THE SIDE. ▶▶

> **Caution**
> It is important to remember that the zero-factor property works only for a product equal to *zero*. If $ab = 0$, then $a = 0$ or $b = 0$. However, if $ab = 6$, for example, we do not know that $a = 6$ or $b = 6$; it is very likely that *neither $a = 6$ nor $b = 6$*.

┌─ E X A M P L E 4   **Solving an Equation Requiring**
│   **Rewriting**

Solve $(2q + 1)(q + 1) = 2(1 - q) + 6$.

Begin by getting zero on one side.

$$(2q + 1)(q + 1) = 2(1 - q) + 6$$

$$2q^2 + 3q + 1 = 2 - 2q + 6 \qquad \text{FOIL; distributive property}$$

$$2q^2 + 3q + 1 = 8 - 2q \qquad \text{Combine terms.}$$

$$2q^2 + 5q - 7 = 0 \qquad \text{Add } -8 + 2q \text{ to both sides.}$$

$$(2q + 7)(q - 1) = 0 \qquad \text{Factor.}$$

$$2q + 7 = 0 \quad \text{or} \quad q - 1 = 0 \qquad \text{Zero-factor property}$$

$$q = -\frac{7}{2} \qquad\qquad q = 1 \qquad \text{Solve for } q.$$

Check that the solution set is $\{-\frac{7}{2}, 1\}$.
└─

WORK PROBLEM 4 AT THE SIDE. ▶▶

OBJECTIVE ▶ The next example shows an application that leads to a quadratic equation.

3. Solve.

(a) $p^2 = -12p$

(b) $k^2 - 16 = 0$

4. Solve.

$(a + 6)(a - 2) = 2 + a - 10$

ANSWERS

3. (a) $\{0, -12\}$ **(b)** $\{4, -4\}$

4. $\{1, -4\}$

5. Carl is planning to build a rectangular deck along the back of his house. He wants the area of the deck to be 60 square meters and the width to be 1 meter less than half the length. What length and width should he use?

E X A M P L E   5   Using a Quadratic Equation in an Application

Surveyors are surveying a lot that is in the shape of a parallelogram. They find that the longer sides of the parallelogram are each 8 meters longer than the distance between them. The area of the lot is 48 square meters. Find the length of the longer sides and the distance between them.

Sketch a parallelogram as shown in Figure 1. (See the list of formulas from geometry inside the covers of this book.) Label your sketch as follows.

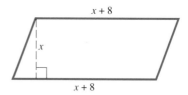

FIGURE 1

Let x be the distance between the longer sides; then $x + 8$ is the length of each longer side.

The area of a parallelogram is given by $A = bh$, where b is the length of the longer side and h is the height, the distance between the longer sides. Here, $b = x + 8$ and $h = x$.

$$A = bh$$
$$48 = (x + 8)x \qquad \text{Let } A = 48, b = x + 8, h = x.$$
$$48 = x^2 + 8x \qquad \text{Distributive property}$$
$$0 = x^2 + 8x - 48 \qquad \text{Subtract 48 on both sides.}$$
$$0 = (x + 12)(x - 4) \qquad \text{Factor.}$$
$$x + 12 = 0 \quad \text{or} \quad x - 4 = 0 \qquad \text{Zero-factor property}$$
$$x = -12 \qquad\qquad x = 4$$

A parallelogram cannot have a negative height, so reject $x = -12$ as a solution. The only solution is 4; the distance between the longer sides is 4 meters. The length of the longer sides is $4 + 8 = 12$ meters.

◄◄ WORK PROBLEM 5 AT THE SIDE.

5.10 Exercises

1. Explain in your own words how the zero-factor property is used in solving a quadratic equation.

2. One of the following equations is *not* in proper form for using the zero-factor property. Which one is it, and why is it not in proper form?
   (a) $(x + 2)(x - 6) = 0$
   (b) $x(3x - 7) = 0$
   (c) $3t(t + 8)(t - 9) = 0$
   (d) $y(y - 3) + 6(y - 3) = 0$

3. In trying to solve $(x + 4)(x - 1) = 1$, a student reasons that because $1 \cdot 1 = 1$, the equation can be solved by solving $x + 4 = 1$ or $x - 1 = 1$. Explain the error in the reasoning. What is the correct way to solve this equation?

4. In solving the equation $4(x - 3)(x + 7) = 0$, a student writes $4 = 0$ or $x - 3 = 0$ or $x + 7 = 0$. Then the student becomes confused about how to handle the equation $4 = 0$. Explain what should be done, and give the solutions of the equation.

Solve each equation using the zero-factor property. See Example 1.

5. $(x - 5)(x + 10) = 0$

6. $(y + 3)(y + 7) = 0$

7. $(2k - 5)(3k + 8) = 0$

8. $(3q - 4)(2q + 5) = 0$

Find all solutions by factoring. See Examples 2–4.

9. $m^2 - 3m - 10 = 0$

10. $x^2 + x - 12 = 0$

11. $z^2 + 9z + 18 = 0$

12. $x^2 - 18x + 80 = 0$

13. $2x^2 = 7x + 4$

14. $2x^2 = 3 - x$

15. $15k^2 - 7k = 4$

16. $3c^2 + 3 = -10c$

17. $2y^2 - 12 - 4y = y^2 - 3y$

18. $3p^2 + 9p + 30 = 2p^2 - 2p$

19. $(5y + 1)(y + 3) = -2(5y + 1)$

20. $(3x + 1)(x - 3) = 2 + 3(x + 5)$

21. $6m^2 - 36m = 0$

22. $-3m^2 + 27m = 0$

23. $-3m^2 + 27 = 0$

24. $-2a^2 + 8 = 0$

25. $4p^2 - 16 = 0$

26. $9x^2 - 81 = 0$

27. Write a quadratic equation with solutions 3 and $\frac{1}{4}$. (*Hint:* Use the zero-factor property in reverse.)

28. Without actually solving each equation, determine which one of the following has 0 in its solution set.
(a) $4x^2 - 25 = 0$ **(b)** $x^2 + 2x - 3 = 0$
(c) $6x^2 + 9x + 1 = 0$
(d) $x^3 + 4x^2 = 3x$

Solve each applied problem by writing a quadratic equation and solving it. See Example 5.

29. A building has a floor area of 140 square meters. The building is rectangular with length 4 meters more than the width. Find the width and length of the building.

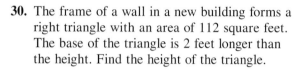

30. The frame of a wall in a new building forms a right triangle with an area of 112 square feet. The base of the triangle is 2 feet longer than the height. Find the height of the triangle.

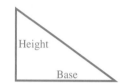

31. A toy rocket is launched vertically upward from ground level with an initial velocity of 128 feet per second. Its height, h, after t seconds is given by $h = -16t^2 + 128t$ (if air resistance is neglected). How long will it take to return to the ground? (When the rocket is at ground level, $h = 0$.)

32. How many seconds will it take for the rocket in Exercise 31 to be 112 feet above the ground?

33. A rancher wants to use 300 feet of fencing to enclose a rectangular paddock of 5000 square feet. What dimensions should the rectangle have?

34. When Europeans arrived in America, many native Americans of the Northeast lived in *longhouses* that sheltered several related families. The rectangular floor area of a typical Huron longhouse was about 2750 square feet. The length was 85 feet greater than the width. What were the dimensions of the floor?

35. A box with no top is to be constructed from a piece of cardboard whose length measures 6 inches more than its width. The box is to be formed by cutting squares that measure 2 inches on each side from the four corners, and then folding up the sides. If the volume of the box will be 110 cubic inches, what are the dimensions of the piece of cardboard?

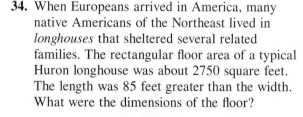

36. The surface area of the open box shown in the figure is 161 square inches. Find the dimensions of the base. (*Hint:* The surface area S is given by the formula $S = x^2 + 4xh$.)

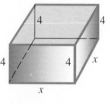

5.2	**scientific notation**	In scientific notation, a number is written as the product of a number between 1 and 10 (or -1 and -10) and some integer power of 10.
5.3	**term**	A term is the product of a number and one or more variables raised to a power.
	coefficient	A coefficient is a factor in a term (usually used for the numerical factor).
	polynomial	A polynomial is a term or a finite sum of terms with only whole number exponents on the variables and no variable denominators.
	descending powers	A polynomial is written in descending powers if the exponents on the variables in the terms decrease from left to right.
	trinomial	A trinomial is a polynomial with exactly three terms.
	binomial	A binomial is a polynomial with exactly two terms.
	monomial	A monomial is a polynomial with exactly one term.
	degree of a term	The degree of a term with one variable is the exponent on that variable.
	degree of a polynomial	The degree of a polynomial is the greatest degree of any of the terms in the polynomial.
	polynomial function of degree n	A function defined by $P(x) = a_n x^n + a_{n-1} x^{n-1} + \cdots + a_1 x + a_0$ is a polynomial function of degree n.
5.6	**synthetic division**	Synthetic division is a shortcut procedure for dividing a polynomial by a polynomial of the form $x - k$.
5.7	**greatest common factor**	The product of the largest numerical factor and the variable factor of least degree of every term in a polynomial is the greatest common factor of the terms of the polynomial.
5.8	**prime polynomial**	A polynomial that cannot be factored is a prime polynomial.
5.9	**square trinomial**	A trinomial that is the square of a binomial is a square trinomial.
5.10	**quadratic equation**	A quadratic equation is one that can be written in the form $ax^2 + bx + c = 0$ $(a \neq 0)$.

$P(x)$ polynomial in x (read "P of x")

Concepts	Examples
5.1 Integer Exponents	
Definitions and Rules for Exponents	
Product Rule: $a^m \cdot a^n = a^{m+n}$	$3^4 \cdot 3^2 = 3^6$
Quotient Rule: $\dfrac{a^m}{a^n} = a^{m-n}$	$\dfrac{2^5}{2^3} = 2^2$
Negative Exponent: $a^{-n} = \dfrac{1}{a^n}$	$5^{-2} = \dfrac{1}{5^2}$
$\dfrac{a^{-n}}{b^{-m}} = \dfrac{b^m}{a^n}$	$\dfrac{5^{-3}}{4^{-6}} = \dfrac{4^6}{5^3}$
Zero Exponent: $a^0 = 1$	$27^0 = 1 \qquad (-5)^0 = 1$

Concepts	*Examples*

5.2 Further Properties of Exponents

Power Rules $(a^m)^n = a^{mn}$

$(ab)^m = a^m b^m$

$\left(\dfrac{a}{b}\right)^n = \dfrac{a^n}{b^n}$

$\left(\dfrac{a}{b}\right)^{-n} = \left(\dfrac{b}{a}\right)^n$

$(6^3)^4 = 6^{12}$

$(5p)^4 = 5^4 p^4$

$\left(\dfrac{2}{3}\right)^5 = \dfrac{2^5}{3^5}$

$\left(\dfrac{4}{7}\right)^{-2} = \left(\dfrac{7}{4}\right)^2$

5.3 Addition and Subtraction of Polynomials

Add or subtract polynomials by combining like terms.

$(x^2 - 2x + 3) + (2x^2 - 8)$
$= 3x^2 - 2x - 5$
$(5x^4 + 3x^2) - (7x^4 + x^2 - x)$
$= -2x^4 + 2x^2 + x$

Function notation, introduced in Chapter 3, is often used for polynomial functions.

If $P(x) = x^3 + 3x^2 - x + 8$, then
$P(2) = 2^3 + 3(2)^2 - 2 + 8 = 26.$

5.4 Multiplication of Polynomials

Multiply each term in one polynomial by each term in the other polynomial, and combine any like terms.

$(x^3 + 3x)(4x^2 - 5x + 2)$
$= 4x^5 + 12x^3 - 5x^4 - 15x^2 + 2x^3 + 6x$
$= 4x^5 - 5x^4 + 14x^3 - 15x^2 + 6x$

The FOIL Method

$$\overset{\text{F} \qquad\qquad \text{L}}{(ax + b)(cx + d)}$$

$$\underset{\text{I} + \text{O}}{\text{O}} \qquad \text{Add (if possible).}$$

$$\overset{6x^2 \qquad\qquad -5}{(2x + 5)(3x - 1)}$$

$$15x$$

$$\underset{13x}{-2x} \qquad \text{Add.}$$

Special Products

$(x + y)(x - y) = x^2 - y^2$

$(x + y)^2 = x^2 + 2xy + y^2$

$(x - y)^2 = x^2 - 2xy + y^2$

$(3m + 8)(3m - 8) = 9m^2 - 64$

$(5a + 3b)^2 = 25a^2 + 30ab + 9b^2$

$(2k - 1)^2 = 4k^2 - 4k + 1$

5.5 Division of Polynomials

To divide a polynomial by a monomial, divide each term of the polynomial by the monomial and write each quotient in lowest terms.

$$\dfrac{5x^4 - 15x^2 + 25x}{5x} = x^3 - 3x + 5$$

Divide any polynomial by another polynomial using a long division process similar to that used for numbers.

$$
\begin{array}{r}
x^3 + 2x^2 - x \\
x - 2 \overline{)x^4 + 0x^3 - 5x^2 + 2x} \\
\underline{x^4 - 2x^3} \\
2x^3 - 5x^2 \\
\underline{2x^3 - 4x^2} \\
-x^2 + 2x \\
\underline{-x^2 + 2x} \\
0
\end{array}
$$

Concepts	Examples

5.6 Synthetic Division

Remainder Theorem

If $P(x)$ is divided by $x - k$, the remainder is $P(k)$.

Find $P(3)$, given $P(x) = 2x^2 - 5x + 6$.

$$
\begin{array}{r}
3\,\overline{)\,2 \quad -5 \quad 6} \\
\quad 6 \quad 3 \\
\hline
2 \quad\ \ 1 \quad 9 \ \leftarrow P(3) = 9
\end{array}
$$

5.7 Greatest Common Factors; Factoring by Grouping

Factoring out the Greatest Common Factor

The product of the largest numerical factor and the variable of least degree of every term in a polynomial is the greatest common factor of the terms of the polynomial.

Factor $4x^2y - 50xy^2 = 2^2x^2y - 2 \cdot 5^2xy^2$.
The greatest common factor is $2xy$.

$$4x^2y - 50xy^2 = 2xy(2x - 25y)$$

Factoring by Grouping

Group the terms so that each group has a common factor. Factor out the common factor in each group. If the groups now have a common factor, factor it out. If not, try a different grouping.

$$
\begin{aligned}
5a - 5b - ax + bx &= (5a - 5b) + (-ax + bx) \\
&= 5(a - b) - x(a - b) \\
&= (a - b)(5 - x)
\end{aligned}
$$

5.8 Factoring Trinomials

Choose factors of the coefficient of the first term and factors of the coefficient of the last term. Use combinations of these factors to find the correct middle term of the trinomial.

Factor $15x^2 + 14x - 8$.
Factors of 15 are 5 and 3, 1 and 15.
Factors of -8 are -4 and 2, 4 and -2, -1 and 8, 8 and -1.

$$15x^2 + 14x - 8 = (5x - 2)(3x + 4)$$

5.9 Special Factoring

Difference of Two Squares

$$x^2 - y^2 = (x + y)(x - y)$$

$$
\begin{aligned}
4m^2 - 25n^2 &= (2m)^2 - (5n)^2 \\
&= (2m + 5n)(2m - 5n)
\end{aligned}
$$

Perfect Square Trinomial

$$x^2 + 2xy + y^2 = (x + y)^2$$
$$x^2 - 2xy + y^2 = (x - y)^2$$

$$9y^2 + 6y + 1 = (3y + 1)^2$$
$$16p^2 - 56p + 49 = (4p - 7)^2$$

Difference of Two Cubes

$$x^3 - y^3 = (x - y)(x^2 + xy + y^2)$$

$$
\begin{aligned}
8 - 27a^3 &= 2^3 - (3a)^3 \\
&= (2 - 3a)(4 + 6a + 9a^2)
\end{aligned}
$$

Sum of Two Cubes

$$x^3 + y^3 = (x + y)(x^2 - xy + y^2)$$

$$
\begin{aligned}
64z^3 + 1 &= (4z)^3 + 1^3 \\
&= (4z + 1)(16z^2 - 4z + 1)
\end{aligned}
$$

Concepts	Examples
5.10 Solving Equations by Factoring If necessary, rewrite the equation so that one side is 0. Factor the polynomial. Set each factor equal to 0 (zero-factor property). Solve each equation. Check each solution.	Solve $2x^2 + 5x = 3$. $$2x^2 + 5x - 3 = 0$$ $$(2x - 1)(x + 3) = 0$$ $$2x - 1 = 0 \quad \text{or} \quad x + 3 = 0$$ $$2x = 1 \qquad\qquad x = -3$$ $$x = \frac{1}{2}$$ A check verifies that the solution set is $\{\frac{1}{2}, -3\}$.

CHAPTER 5 REVIEW EXERCISES

[5.1] *Evaluate.*

1. 4^3

2. $\left(\dfrac{1}{3}\right)^4$

3. $(-.2)^2$

4. $\dfrac{2}{(-3)^{-2}}$

5. $\left(\dfrac{2}{3}\right)^{-4}$

6. Explain the difference between the expressions $(-6)^0$ and -6^0.

Simplify. Write answers with only positive exponents. Assume that all variables represent nonzero real numbers.

7. $7^{12} \cdot 7^{-4}$

8. $\dfrac{3^5}{3^{-3}}$

9. $m^4 m^6 m$

10. $6(3x^4)(5x^{-7})$

11. $\dfrac{4^{-3}}{4^{-6}}$

12. $\dfrac{y^{-3}y^{-2}}{y^{-5}y^{-4}}$

[5.1–5.2]

13. Which one of the following statements is true? Correct each false statement.
   (a) $-3^2 = -9$ **(b)** $-a^4 = (-a)^4$ **(c)** $5^{-2} = -5^2$ **(d)** $(xy)^3 = x^3 y^3$

[5.2] *Write in scientific notation.*

14. 2790

15. .0000085

16. .296

Write without scientific notation.

17. 3.6×10^4

18. 5.71×10^{-3}

19. 9.04×10^{-2}

20. Explain why multiplying by 10^{-a} is the same as dividing by 10^a.

The population of Fresno, California, is approximately 3.45×10^5. According to the 1994 World Almanac, the population density is 5449 per square mile.

21. Write the population density in scientific notation.

22. To the nearest whole number, what is the area of Fresno?

23. We know that $\frac{25}{25} = 1$ and, since $25 = 5^2$, $\frac{25}{25} = \frac{5^2}{5^2}$. Use the quotient rule to simplify $\frac{5^2}{5^2}$. Explain how this supports the definition of a^0.

Simplify. Write answers with only positive exponents. Assume that all variables represent nonzero real numbers.

24. $(4^{-3})^2$

25. $(k^{-2})^3 k^{-4}$

26. $\dfrac{(5z)^3 z^2}{z^{-3} z^{-2}}$

27. $\left(\dfrac{7p^{-3}}{p^{-8}}\right)^{-1} \cdot \dfrac{p^{-3}}{5}$

28. $\left(\dfrac{3a^4}{4a^{-2}}\right)^{-3} \left(\dfrac{9a^{-3}}{8a^{-6}}\right)^4$

29. $\left(\dfrac{q^{-1} r^{-3}}{5q^3}\right)^{-2} \left(\dfrac{r^{-3} \cdot 2q^3}{3r^{-4}}\right)^{-1}$

[5.3] *For each polynomial (a) write in descending powers of the variable, (b) identify as monomial, binomial, trinomial, or none of these, and (c) give the degree of the polynomial.*

30. $9k + 11k^3 - 3k^2$ (a)   (b)   (c)

31. $14m^6 - 9m^7$ (a)   (b)   (c)

32. $-5y^4 + 3y^3 + 7y^2 - 2y$ (a)   (b)   (c)

For each polynomial function, find (a) $P(-2)$ and (b) $P(3)$.

33. $P(x) = -6x + 15$ (a)   (b)

34. $P(x) = -2x^2 + 5x + 7$ (a)   (b)

Add or subtract as indicated.

35. Add.
$$3x^2 - 5x + 6$$
$$-4x^2 + 2x - 5$$

36. Subtract.
$$10m - 4$$
$$-8m + 6$$

37. Subtract.
$$-5y^3 \qquad + 8y - 3$$
$$4y^2 + 2y + 9$$

38. $(4a^3 - 9a + 15) - (-2a^3 + 4a^2 + 7a)$

39. $(3y^2 + 2y - 1) + (5y^2 - 11y + 6)$

[5.4] *Find each product.*

40. $-5b(3b^2 + 10)$

41. $-6k(2k^2 + 7)$

42. $(3m - 2)(5m + 1)$

43. $(7y - 8)(2y + 3)$

44. $(3w - 2t)(2w - 3t)$

45. $(2p^2 + 6p)(5p^2 - 4)$

46. $(3q^2 + 2q - 4)(q - 5)$

47. $(3z^3 - 2z^2 + 4z - 1)(3z - 2)$

48. $(6r^2 - 1)(6r^2 + 1)$

49. $\left(z + \dfrac{3}{5}\right)\left(z - \dfrac{3}{5}\right)$

50. $(4m + 3)^2$

51. $(2n - 10)^2$

[5.5] *Divide.*

52. $\dfrac{18x^2 - 32x + 12}{12x}$

53. $\dfrac{4y^3 - 12y^2 + 5y}{4y}$

54. $\dfrac{6x^2 - 17x + 2}{2x - 3}$

55. $\dfrac{15k^2 + 11k - 17}{3k - 2}$

56. $\dfrac{2p^3 + 9p^2 + 27}{2p - 3}$

57. $\dfrac{12y^4 + 7y^2 - 2y + 1}{3y^2 + 1}$

58. $\dfrac{4x^4 - 28x^3 + 27x^2 - 21x + 18}{4x^2 + 3}$

59. $\dfrac{5p^4 + 15p^3 - 33p^2 - 9p + 18}{5p^2 - 3}$

[5.6] *Use synthetic division for each of the following.*

60. $\dfrac{3p^2 - p - 2}{p - 1}$

61. $\dfrac{10k^2 - 3k - 15}{k + 2}$

62. $(2k^3 - 5k^2 + 12) \div (k - 3)$

63. $(-a^4 + 19a^2 + 18a + 15) \div (a + 4)$

Use synthetic division to decide whether or not -5 is a solution for each of the following equations.

64. $2w^3 + 8w^2 - 14w - 20 = 0$

65. $-3q^4 + 2q^3 + 5q^2 - 9q + 1 = 0$

Use synthetic division to evaluate $P(k)$ for the given value of k.

66. $P(x) = 3x^3 - 5x^2 + 4x - 1; \ k = -1$

67. $P(z) = z^4 - 2z^3 - 9z - 5; \ k = 3$

[5.7–5.8] *Factor completely.*

68. $12p^2 - 6p$

69. $21y^2 + 35y$

70. $12q^2b + 8qb^2 - 20q^3b^2$

71. $6r^3t - 30r^2t^2 + 18rt^3$

72. $(x + 3)(4x - 1) - (x + 3)(3x + 2)$

73. $(z + 1)(z - 4) + (z + 1)(2z + 3)$

74. $4m + nq + mn + 4q$

75. $x^2 + 5y + 5x + xy$

76. $3p^2 - p - 4$

77. $6k^2 + 11k - 10$

78. $12r^2 - 5r - 3$

79. $10m^2 + 37m + 30$

80. $10k^2 - 11kh + 3h^2$

81. $9x^2 + 4xy - 2y^2$

82. $24x - 2x^2 - 2x^3$

83. $6b^3 - 9b^2 - 15b$

84. $y^4 + 2y^2 - 8$

85. $2k^4 - 5k^2 - 3$

86. When asked to factor $x^2y^2 - 6x^2 + 5y^2 - 30$, a student gave $x^2(y^2 - 6) + 5(y^2 - 6)$ as the answer. Why is this answer incorrect? What is the correct answer?

[5.9]

87. Which of the following statements are true? Correct each false statement.
   (a) $x^2 - y^2 = (x + y)(x - y)$ **(b)** $x^2 + y^2 = (x + y)(x + y)$
   (c) $x^3 + y^3 = (x + y)^3$ **(d)** $x^3 - y^3 = (x - y)^3$

Factor completely.

88. $16p^2 - 9$

89. $9z^2 - 100$

90. $36(t + 1)^2 - 64(z + 1)^2$

91. $p^2 + 10p + 100$

92. $16x^2 - 40x + 25$

93. $4z^2 + 12zm + 9m^2$

94. $1 - y^3$

95. $125x^3 - 8$

96. $4r^3 + 108$

[5.10]

97. A student began the solution of $(x + 2)(x - 4) = -5$ by writing $x + 2 = -1$ or $x - 4 = 5$. What is wrong with this step? Find the correct solution of the equation.

Solve each equation.

98. $3y^2 + 10y - 8 = 0$

99. $16r^2 - 3 = 8r$

100. $6x^2 + 2x = 20$

Work the following problems.

101. A triangular wall brace has the shape of a right triangle. One of the perpendicular sides is 1 foot longer than twice the other. The area enclosed by the triangle is 10.5 square feet. Find the shorter of the perpendicular sides.

102. A rectangular parking lot has a length 20 feet more than its width. Its area is 2400 square feet. What are the dimensions of the lot?

103. A rectangular shed has a width that is 5 meters less than its length. The area of the floor of the shed is 50 square meters. Find the length of the shed.

104. The length of a rectangular picture frame is 2 inches longer than its width. The area enclosed by the frame is 48 square inches. What is the width?

MIXED REVIEW EXERCISES

In Exercises 105–114, perform the indicated operations, then simplify. Write answers with only positive exponents. Assume all variables represent nonzero real numbers.

105. $(2x + 1)(4x - 3)$

106. $\left(\dfrac{3}{2}\right)^{-2}$

107. $(y^6)^{-5}(2y^{-3})^{-4}$

108. $(11w^2 - 5w + 6) - (-15 - 8w^2)$

109. $\dfrac{m^2 - 8m + 15}{m - 5}$

110. $-(-3)^2$

111. $\dfrac{(5z^2x^3)^2}{(-10zx^{-3})^{-2}}$

112. $(2k - 1) - (3k^2 - 2k + 6)$

113. $7p^5(3p^4 + p^3 + 2p^2)$

114. $\dfrac{20y^3x^3 + 15y^4x + 25yx^4}{10yx^2}$

115. Find $P(-2)$, if $P(x) = 2x^2 - x + 3$.

Factor completely.

116. $10m^2 - m - 3$

117. $12z - 72z^2$

118. $64 - p^3$

119. $2z^2 - 7xz - 4x^2$

120. $15d^4 - 10d^2$

121. $25c^2 + 30c + 9$

CHAPTER 5 TEST

1. Give an example to show that $(2a)^{-3}$ does not equal $\dfrac{2}{a^3}$ by choosing a specific value for a.

1. _____

Simplify. Write answers with only positive exponents. Assume all variables are nonzero.

2. $\left(\dfrac{3}{2}\right)^{-2}\left(\dfrac{3}{2}\right)^{5}\left(\dfrac{3}{2}\right)^{-6}$

2. _____

3. $(-4m^{-2}n^4)^{-1}$

3. _____

4. $\dfrac{5^{-3}a^{-2}}{2(a^3)^{-3}}$

4. _____

Perform the indicated operations.

5. $(4x^3 - 3x^2 + 2x - 5) - (3x^3 + 11x + 8) + (x^2 - x)$

5. _____

6. $(5x - 3)(2x + 1)$

6. _____

7. $(2m - 5)(3m^2 + 4m - 5)$

7. _____

8. $(6x + y)(6x - y)$

8. _____

9. $(3k + q)^2$

9. _____

10. $\dfrac{16p^3 - 32p^2 + 24p}{4p^2}$

10. _____

11. $\dfrac{9q^4 - 18q^3 + 11q^2 + 10q - 10}{3q - 2}$

11. _____

12. **(a)** If $P(x) = x^4 - 8x^3 + 21x^2 - 14x - 24$, find $P(1)$.
   (b) Use synthetic division to decide whether 4 is a solution of $P(x) = 0$.

12. _____

Factor each polynomial in Exercises 13–20.

13. $11z^2 - 44z$

13. _____

14. $(h - 1)(3h + 4) - (h - 1)(h + 2)$

14. _____

15. $3x + by + bx + 3y$

15. _____

16. $4p^2 + 3pq - q^2$

16. _____

17. $16a^2 + 20ab + 25b^2$

17. _____

18. $y^3 - 216$

18. _____

19. $9k^2 - 121j^2$

19. _____

20. $6k^4 - k^2 - 35$

20. _____

21. Which one of the following is not a factored form of $-x^2 - x + 12$?
    (a) $(3 - x)(x + 4)$ **(b)** $-(x - 3)(x + 4)$
    (c) $(-x + 3)(x + 4)$ **(d)** $(x - 3)(-x + 4)$

21. _____

Solve each equation.

22. $3x^2 + 8x + 4 = 0$

22. _____

23. $10x^2 = 17x - 3$

23. _____

24. $5m(m - 1) = 2(1 - m)$

24. _____

25. A brace forms a right triangle with the floor and one wall of a room. The base of the triangle is 2 feet less than the height. If the area of the triangle is 24 square feet, find the height.

25. _____

Let $R = \left\{-5, -\dfrac{3}{4}, 0, 1.6, \sqrt{7}, 9\right\}$. *List the elements that belong to each set.*

1. Irrational numbers   **2.** Rational numbers   **3.** Integers

4. Write $\{x \mid -1 < x \le 7\}$ in interval notation.

Evaluate.

5. $|\,2 - 5\,| + |\,-3\,| - 7$

6. $\dfrac{3 \cdot 2^2 - 4}{-6 + 2^3 \cdot 3}$

Solve.

7. $7x - (1 + 4x) = 8 + 3x$ **8.** $|\,2r - 7\,| = 5$ **9.** $3k - v^2 = 4p$ for k

10. $\dfrac{x - 6}{4} = 1 + \dfrac{2x + 3}{2}$ **11.** $3 - \dfrac{1}{2}m < 6 + \dfrac{3}{2}m$ **12.** $3p \le -2$ and $p + 1 \ge -4$

In what quadrant is each of the following points located?

13. $(-1, 4)$ **14.** $(3, -9)$ **15.** $(-2, -5)$ **16.** $(0, 6)$

Graph each relation.

17. $f(x) = \dfrac{2}{5}x - 2$

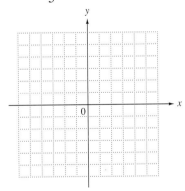

18. $3x + y < 6$

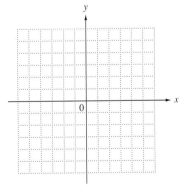

For each of the following, find $f(-2)$.

19. $f(x) = -x + 4$

20. $f(x) = \dfrac{3}{x + 1}$

21. Use the elimination method to solve the system
$$x + 2y = -1$$
$$3y - x = -4$$
$$4x - 2z = -2.$$

22. Using two variables, write a system of equations to solve the following problem. A game board has the shape of a parallelogram. The longer sides are each 2 inches longer than the shorter sides. The perimeter of the board is 68 inches. Find the lengths of all the sides.

23. Use Cramer's rule to solve the system in Exercise 21.

Evaluate.

24. 7^{-3}

25. -3^0

26. $P(3)$, if $P(x) = x^3 - x^2 + x - 5$

Perform the indicated operations. Assume no denominators are zero.

27. $(3x^2y^{-1})^{-2}(2x^{-3}y)^{-1}$

28. $\dfrac{5m^{-2}y^3}{3m^{-3}y^{-1}}$

29. $(2p + 3)(5p^2 - 4p - 8)$

30. $(3x^3 + 4x^2 - 7) - (2x^3 - 8x^2 + 3x)$

Factor.

31. $16w^2 + 50wz - 21z^2$

32. $4x^2 - 4x + 1 - y^2$

33. $4y^2 - 36y + 81$

34. $100x^4 - 81$

35. $8p^3 + 27$

Solve.

36. $(p - 1)(2p + 3)(p + 4) = 0$

37. $9q^2 = 6q - 1$

38. $5x^2 - 45 = 0$

39. Find the period of time for which $3000 must be invested to earn $384 simple interest at an annual interest rate of 3.2%.

40. A sign is to have the shape of a triangle with a height 3 meters greater than the length of the base. How long should the base be if the area is to be 14 square meters?

Rational Expressions 6

6.1 Rational Expressions and Functions; Multiplication and Division

OBJECTIVE 1  A rational number is the quotient of two integers, with the denominator not 0. In algebra, this idea is generalized: A **rational expression** (RA-shun-ul eks-PRESH-un) (or **algebraic fraction**) is the quotient of two polynomials, again with the denominator not 0. For example,

$$\frac{m+4}{m-2}, \qquad \frac{8x^2 - 2x + 5}{4x^2 + 5x}, \qquad \text{and} \qquad x^5 \left(\text{or } \frac{x^5}{1} \right)$$

are all rational expressions. A function defined by a rational expression in a single variable is called a **rational function**.

OBJECTIVE 2 When a number used as a replacement for the variable in a rational function makes the denominator 0, the rational function is undefined. For example, the number 5 cannot be used as a replacement for x in the function

$$f(x) = \frac{2}{x-5}$$

because 5 would make the denominator equal 0.

OBJECTIVES

1. Define rational expressions and rational functions.
2. Find the numbers that make a rational function undefined.
3. Write rational expressions in lowest terms.
4. Multiply rational expressions.
5. Find reciprocals for rational expressions.
6. Divide rational expressions.

FOR EXTRA HELP

Tutorial   Tape 9   SSM, Sec. 6.1

┌─ **E X A M P L E I**   **Determining When Rational Functions Are Undefined**

Find all numbers that make the rational function undefined.

(a) $f(x) = \dfrac{3}{7x - 14}$

The only values that cannot be used are those that make the denominator 0. Find these values by setting the denominator equal to 0 and solving the resulting equation.

$$7x - 14 = 0$$
$$7x = 14 \qquad \text{Add 14.}$$
$$x = 2 \qquad \text{Divide by 7.}$$

The number 2 cannot be used as a replacement for x; 2 makes the rational function undefined.

└─ **CONTINUED ON NEXT PAGE**

1. Find all numbers that make the rational function undefined.

(a) $f(x) = \dfrac{x + 4}{x - 6}$

(b) $f(x) = \dfrac{8x}{5}$

(c) $f(r) = \dfrac{r + 6}{r^2 - r - 6}$

(d) $f(k) = \dfrac{k + 2}{k^2 + 1}$

(b) $f(p) = \dfrac{3 + p}{p^2 - 4p + 3}$

Set the denominator equal to 0, and solve the equation.

$$p^2 - 4p + 3 = 0$$
$$(p - 3)(p - 1) = 0 \qquad \text{Factor.}$$
$$p - 3 = 0 \quad \text{or} \quad p - 1 = 0 \qquad \text{Set each factor equal to 0.}$$
$$p = 3 \quad \text{or} \qquad p = 1$$

Both 3 and 1 make the rational function undefined.

(c) $f(x) = \dfrac{8x + 2}{3}$

Because the denominator is a constant rather than a variable, there is no real number that makes the rational function undefined.

(d) $f(x) = \dfrac{2}{x^2 + 4}$

Setting $x^2 + 4$ equal to 0 leads to $x^2 = -4$. There is no real number whose square is -4. Therefore, any real number can be used, and as in part (c), there is no real number that makes the rational function undefined.

◀◀ WORK PROBLEM I AT THE SIDE.

> **Note**
> The function $f(x) = \dfrac{8x + 2}{3}$ in Example 1(c) is considered a rational function, even though the denominator contains no variable.

OBJECTIVE 3 A fraction is in lowest terms when the numerator and denominator have no factors in common (other than 1). We write a fraction in this form by using the **property of 1.**

> **Property of I**
> A nonzero number divided by itself is equal to 1.

Because any factors of a rational expression represent real numbers, we can use the property of 1 to write rational expressions in lowest terms as follows.

> **Writing Rational Expressions in Lowest Terms**
> *Step 1* Factor the numerator and the denominator completely.
> *Step 2* Replace each pair of factors common to the numerator and denominator with 1.

┌───

E X A M P L E 2   **Writing Rational Expressions in Lowest Terms**

Write each rational expression in lowest terms.

(a) $\dfrac{8k}{16}$

Factor the numerator and denominator. Replace the quotient of common factors with 1.

$$\frac{8k}{16} = \frac{k \cdot 8}{2 \cdot 8} = \frac{k}{2} \cdot \frac{8}{8} = \frac{k}{2} \cdot 1 = \frac{k}{2}$$

Notice that we use the identity property for multiplication in the last step.

(b) $\dfrac{12x^3y^2}{6x^4y} = \dfrac{2y \cdot 6x^3y}{x \cdot 6x^3y} = \dfrac{2y}{x} \cdot 1 = \dfrac{2y}{x}$

(c) $\dfrac{a^2 - a - 6}{a^2 + 5a + 6}$

Start by factoring the numerator and denominator.

$$\frac{a^2 - a - 6}{a^2 + 5a + 6} = \frac{(a - 3)(a + 2)}{(a + 3)(a + 2)} = \frac{a - 3}{a + 3}$$

(d) $\dfrac{y^2 - 4}{2y + 4} = \dfrac{(y + 2)(y - 2)}{2(y + 2)} = \dfrac{y - 2}{2}$

─── ■

Caution

One of the most common errors in algebra involves incorrect use of the property of 1. Only common *factors* may be divided to get 1. For instance, in Example 2(d)

$$\frac{y - 2}{2} \neq y \quad \text{or} \quad y - 1.$$

The 2 in $y - 2$ is not a *factor* of the numerator. Similarly, it would be incorrect to "divide" the a^2 terms in the numerator and denominator in Example 2(c). It is essential to factor before writing in lowest terms.

WORK PROBLEM 2 AT THE SIDE. ▶▶

In the rational expression from Example 2(c),

$$\frac{a^2 - a - 6}{a^2 + 5a + 6}, \quad \text{or} \quad \frac{(a - 3)(a + 2)}{(a + 3)(a + 2)},$$

a can take on any value at all except -3 or -2. In the rational expression

$$\frac{a - 3}{a + 3},$$

a cannot equal -3. Because of this,

$$\frac{a^2 - a - 6}{a^2 + 5a + 6} = \frac{a - 3}{a + 3}$$

for all values of a except -3 or -2. From now on we shall write such statements of equality with the understanding that they apply only for those real numbers that make no denominator equal 0.

2. Write each rational expression in lowest terms.

(a) $\dfrac{18m^2}{9m^5}$

(b) $\dfrac{6y^2z^2}{12yz^3}$

(c) $\dfrac{y^2 + 2y - 3}{y^2 - 3y + 2}$

(d) $\dfrac{3y + 9}{y^2 - 9}$

(e) $\dfrac{y + 2}{y^2 + 4}$

3. Write each rational expression in lowest terms.

(a) $\dfrac{y - 2}{2 - y}$

(b) $\dfrac{8 - b}{8 + b}$

(c) $\dfrac{p - 2}{4 - p^2}$

(d) $\dfrac{1 + p^3}{1 + p}$

(e) $\dfrac{3x + 3y + rx + ry}{5x + 5y - rx - ry}$

The next example illustrates the techniques of factoring out a -1, factoring the difference of cubes, and factoring by grouping.

E X A M P L E   3   **Writing Rational Expressions in Lowest Terms**

Write each rational expression in lowest terms.

(a) $\dfrac{m - 3}{3 - m}$

In this rational expression, the numerator and denominator are exactly opposite. We can write the given expression in lowest terms by writing the denominator as $3 - m = -1(m - 3)$, giving

$$\frac{m - 3}{3 - m} = \frac{m - 3}{-1(m - 3)} = \frac{1}{-1} = -1.$$

The numerator could have been rewritten instead to get the same result.

(b) $\dfrac{r^2 - 16}{4 - r} = \dfrac{(r + 4)(r - 4)}{4 - r}$

$\qquad = \dfrac{(r + 4)(r - 4)}{-1(r - 4)}$ Write $4 - r$ as $-1(r - 4)$.

$\qquad = \dfrac{r + 4}{-1}$ Lowest terms

$\qquad = -(r + 4)$ or $-r - 4$

(c) $\dfrac{x^3 - 27}{x - 3} = \dfrac{(x - 3)(x^2 + 3x + 9)}{x - 3}$ Factor the difference of cubes.

$\qquad = x^2 + 3x + 9$ Lowest terms

(d) $\dfrac{pr + qr + ps + qs}{pr + qr - ps - qs} = \dfrac{(pr + qr) + (ps + qs)}{(pr + qr) - (ps + qs)}$

$\qquad = \dfrac{r(p + q) + s(p + q)}{r(p + q) - s(p + q)}$

$\qquad = \dfrac{(p + q)(r + s)}{(p + q)(r - s)}$ Factor by grouping.

$\qquad = \dfrac{r + s}{r - s}$ Lowest terms

As shown in Examples 3(a) and 3(b), the quotient $\dfrac{a}{-a}$ $(a \neq 0)$ can be simplified as $\dfrac{a}{-a} = \dfrac{a}{-1(a)} = \dfrac{1}{-1} = -1$. The following generalization applies.

> The quotient of two quantities that differ only in sign is -1.

Based on this result, $\dfrac{q - 7}{7 - q} = -1$ and $\dfrac{-5a + 2b}{5a - 2b} = -1$, but $\dfrac{r - 2}{r + 2}$ cannot be simplified further.

ANSWERS

3. **(a)** -1 **(b)** already in lowest terms

(c) $\dfrac{-1}{2 + p}$ **(d)** $1 - p + p^2$ **(e)** $\dfrac{3 + r}{5 - r}$

◀◀ WORK PROBLEM 3 AT THE SIDE.

Caution
Rational expressions often can be written in lowest terms in *seemingly* different ways. For example,

$$\frac{y - 3}{-5} \quad \text{and} \quad \frac{-y + 3}{5}$$

look different, but the second quotient is obtained by multiplying the first by -1 in both the numerator and denominator. If your answer does not exactly match the one given in the text, check to see if it is equivalent to the one given.

OBJECTIVE 4▶ To multiply two rational expressions, we multiply the numerators and multiply the denominators. The product should be simplified by writing it in lowest terms. In practice, we simplify before performing any multiplication.

Multiplying Rational Expressions

Step 1 Factor all numerators and denominators as completely as possible.

Step 2 Replace quotients of factors common to the numerator and denominator with 1.

Step 3 Multiply remaining factors in the numerator, and multiply remaining factors in the denominator.

Step 4 Be certain that the product is in lowest terms.

┌─ **E X A M P L E 4** **Multiplying Rational Expressions**

Multiply.

(a) $\dfrac{3x^2}{5} \cdot \dfrac{10}{x^3} = \dfrac{3x^2 \cdot 5 \cdot 2}{5 \cdot x^2 \cdot x} = \dfrac{3 \cdot 1 \cdot 1 \cdot 2}{x} = \dfrac{6}{x}$

(b) $\dfrac{5p - 5}{p} \cdot \dfrac{3p^2}{10p - 10}$

Factor where possible.

$$\frac{5p - 5}{p} \cdot \frac{3p^2}{10p - 10} = \frac{5(p - 1)}{p} \cdot \frac{3p^2}{5 \cdot 2(p - 1)}$$

$$= \frac{5 \cdot 3 \cdot p \cdot p(p - 1)}{5 \cdot 2 \cdot p(p - 1)} = \frac{3p}{2}$$

(c) $\dfrac{k^2 + 2k - 15}{k^2 - 4k + 3} \cdot \dfrac{k^2 - k}{k^2 + k - 20}$

$$= \frac{(k + 5)(k - 3)}{(k - 3)(k - 1)} \cdot \frac{k(k - 1)}{(k + 5)(k - 4)} = \frac{k}{k - 4}$$

└──■

WORK PROBLEM 4 AT THE SIDE. ▶▶

OBJECTIVE 5▶ Recall that the reciprocal of a nonzero real number $\frac{a}{b}$ is the real number $\frac{b}{a}$.

4. Multiply.

(a) $\dfrac{12r}{7} \cdot \dfrac{14}{r^2}$

(b) $\dfrac{6z^2}{5} \cdot \dfrac{4z}{3}$

(c) $\dfrac{2r + 4}{5r} \cdot \dfrac{3r}{5r + 10}$

(d) $\dfrac{m^2 - 16}{m + 2} \cdot \dfrac{1}{m + 4}$

(e) $\dfrac{c^2 + 2c}{c^2 - 4} \cdot \dfrac{c^2 - 4c + 4}{c^2 - c}$

ANSWERS

4. (a) $\dfrac{24}{r}$ **(b)** $\dfrac{8z^3}{5}$ **(c)** $\dfrac{6}{25}$

(d) $\dfrac{m - 4}{m + 2}$ **(e)** $\dfrac{c - 2}{c - 1}$

5. Find the reciprocal.

(a) $\dfrac{-3}{r}$

(b) $\dfrac{7}{y + 8}$

(c) $\dfrac{a^2 + 7a}{2a - 1}$

(d) $\dfrac{0}{-5}$

6. Divide.

(a) $\dfrac{16k^2}{5} \div \dfrac{3k}{10}$

(b) $\dfrac{5p + 2}{6} \div \dfrac{15p + 6}{5}$

(c) $\dfrac{y^2 - 2y - 3}{y^2 + 4y + 4}$

$\div \dfrac{y^2 - 1}{y^2 + y - 2}$

> **Procedure for Finding a Reciprocal**
>
> To find the *reciprocal* of a nonzero rational expression, invert the rational expression.

EXAMPLE 5 Finding the Reciprocal of a Rational Expression

Find the reciprocal of each rational expression.

Rational Expression	Reciprocal
$\dfrac{5}{k}$	$\dfrac{k}{5}$
$\dfrac{m^2 - 9m}{2}$	$\dfrac{2}{m^2 - 9m}$
$\dfrac{0}{4}$	Undefined—no reciprocal for 0

◀◀ **WORK PROBLEM 5 AT THE SIDE.**

OBJECTIVE 6 ▶ Division of rational expressions follows the rule for division of rational numbers.

> **Dividing Rational Expressions**
>
> To divide two rational expressions, *multiply* the first by the reciprocal of the second (the divisor).

EXAMPLE 6 Dividing Rational Expressions

Divide.

(a) $\dfrac{2z}{9} \div \dfrac{5z^2}{18} = \dfrac{2z}{9} \cdot \dfrac{\mathbf{18}}{\mathbf{5z^2}} = \dfrac{4}{5z}$ Multiply by the reciprocal of the divisor.

(b) $\dfrac{8k - 16}{3k} \div \dfrac{6 - 3k}{4k^2} = \dfrac{8k - 16}{3k} \cdot \dfrac{\mathbf{4k^2}}{\mathbf{6 - 3k}}$ Multiply by the reciprocal of the divisor.

$= \dfrac{8(k - 2)}{3k} \cdot \dfrac{4k^2}{3(2 - k)} = -\dfrac{32k}{9}$ Factor.

The negative sign appears in the quotient because the quotient of $k - 2$ in the numerator and $2 - k$ in the denominator is -1.

(c) $\dfrac{m^2 - 1}{25m^2 - 9} \cdot \dfrac{5m^2 + 17m - 12}{3m^2 + 7m - 20} \div \dfrac{5m^2 + 2m - 3}{15m^2 - 34m + 15}$

$= \dfrac{m^2 - 1}{25m^2 - 9} \cdot \dfrac{5m^2 + 17m - 12}{3m^2 + 7m - 20} \cdot \dfrac{15m^2 - 34m + 15}{5m^2 + 2m - 3}$

$= \dfrac{(m + 1)(m - 1)}{(5m + 3)(5m - 3)} \cdot \dfrac{(5m - 3)(m + 4)}{(m + 4)(3m - 5)} \cdot \dfrac{(3m - 5)(5m - 3)}{(5m - 3)(m + 1)}$

$= \dfrac{m - 1}{5m + 3}$

ANSWERS

5. (a) $\dfrac{r}{-3}$ (b) $\dfrac{y + 8}{7}$ (c) $\dfrac{2a - 1}{a^2 + 7a}$
(d) There is no reciprocal.

6. (a) $\dfrac{32k}{3}$ (b) $\dfrac{5}{18}$ (c) $\dfrac{y - 3}{y + 2}$

◀◀ **WORK PROBLEM 6 AT THE SIDE.**

6.1 Exercises

Recall in the Caution at the end of Objective 3, two rational expressions that are equivalent may look different. To prepare yourself for recognizing equivalent rational expressions, match the expressions in Exercises 1–6 with their equivalents in choices A–F.

_____ **1.** $\dfrac{x-3}{x+4}$ **A.** $\dfrac{-x-3}{4-x}$

_____ **2.** $\dfrac{x+3}{x-4}$ **B.** $\dfrac{-x-3}{-x-4}$

_____ **3.** $\dfrac{x-3}{x-4}$ **C.** $\dfrac{3-x}{-x-4}$

_____ **4.** $\dfrac{x+3}{x+4}$ **D.** $\dfrac{x+3}{-x-4}$

_____ **5.** $\dfrac{-x-3}{x+4}$ **E.** $\dfrac{-x-3}{x+4}$

_____ **6.** $\dfrac{x+3}{-x-4}$ **F.** $\dfrac{3-x}{4-x}$

Find all real numbers that make the rational function undefined. See Example 1.

7. $f(x)=\dfrac{x+1}{x-7}$ **8.** $f(x)=\dfrac{x+9}{x+3}$ **9.** $f(x)=\dfrac{3x-6}{7x+1}$

10. $f(x)=\dfrac{8x-3}{2x+7}$ **11.** $f(t)=\dfrac{3t+1}{t}$ **12.** $f(t)=\dfrac{t^2+1}{t}$

13. $f(x)=\dfrac{3x+1}{2x^2+x-6}$ **14.** $f(x)=\dfrac{2x+4}{3x^2+11x-42}$ **15.** $f(x)=\dfrac{x+2}{14}$

16. $f(x) = \dfrac{x + 9}{4}$

17. $f(x) = \dfrac{2x^2 - 3x + 4}{3x^2 + 8}$

18. $f(x) = \dfrac{9x^2 - 8x + 3}{4x^2 + 1}$

19. (a) Identify the two *terms* in the numerator and the two *terms* in the denominator of the rational expression $\dfrac{x^2 + 4x}{x + 4}$.

    (b) Explain the process of reducing this rational expression to lowest terms.

20. Only one of the following rational expressions is not expressed in lowest terms. Which one is it?

   (a) $\dfrac{x^2 + 5}{x^2}$ (b) $\dfrac{x}{x - 5}$ (c) $\dfrac{x + 5}{x - 5}$ (d) $\dfrac{x^2 - 5x}{x}$

Write each rational expression in lowest terms. See Examples 2 and 3.

21. $\dfrac{24x^2y^4}{18xy^5}$

22. $\dfrac{36m^4n^3}{24m^2n^5}$

23. $\dfrac{(x + 4)(x - 3)}{(x + 5)(x + 4)}$

24. $\dfrac{(2x + 7)(x - 1)}{(2x + 3)(2x + 7)}$

25. $\dfrac{4x(x + 3)}{8x^2(x - 3)}$

26. $\dfrac{5y^2(y + 8)}{15y(y - 8)}$

27. $\dfrac{3x + 7}{3x}$

28. $\dfrac{4x - 9}{4x}$

29. $\dfrac{7x - 7}{8x - 8}$

30. $\dfrac{9t + 9}{11t + 11}$

31. $\dfrac{6m + 18}{7m + 21}$

32. $\dfrac{5r - 20}{3r - 12}$

33. $\dfrac{3z^2 + z}{18z + 6}$

34. $\dfrac{2x^2 - 5x}{16x - 40}$

35. $\dfrac{2t + 6}{t^2 - 9}$

36. $\dfrac{5s - 25}{s^2 - 25}$

37. $\dfrac{4b^2 - 4}{8b - 8}$

38. $\dfrac{3k^2 - 3}{6k - 6}$

39. $\dfrac{x^2 + 2x - 15}{x^2 + 6x + 5}$

40. $\dfrac{y^2 - 5y - 14}{y^2 + y - 2}$

41. $\dfrac{8x^2 - 10x - 3}{8x^2 - 6x - 9}$

42. $\dfrac{12x^2 - 4x - 5}{8x^2 - 6x - 5}$

43. $\dfrac{a^3 + b^3}{a + b}$

44. $\dfrac{r^3 - s^3}{r - s}$

45. $\dfrac{2c^2 + 2cd - 60d^2}{2c^2 - 12cd + 10d^2}$

46. $\dfrac{3s^2 - 9st - 54t^2}{3s^2 - 6st - 72t^2}$

47. $\dfrac{ac - ad + bc - bd}{ac - ad - bc + bd}$

48. $\dfrac{2xy + 2xw + y + w}{2xy + y - 2xw - w}$

Write each rational expression in lowest terms. See Example 3.

49. $\dfrac{7 - b}{b - 7}$

50. $\dfrac{r - 13}{13 - r}$

51. $\dfrac{x^2 - y^2}{y - x}$

52. $\dfrac{m^2 - n^2}{n - m}$

53. $\dfrac{(a - 3)(x + y)}{(3 - a)(x - y)}$

54. $\dfrac{(8 - p)(x + 2)}{(p - 8)(x - 2)}$

55. $\dfrac{5k - 10}{20 - 10k}$

56. $\dfrac{7x - 21}{63 - 21x}$

57. $\dfrac{a^2 - b^2}{a^2 + b^2}$

58. $\dfrac{p^2 + q^2}{p^2 - q^2}$

59. Which of the following rational expressions equals -1?

(a) $\dfrac{2x + 3}{2x - 3}$ (b) $\dfrac{2x - 3}{3 - 2x}$

(c) $\dfrac{2x + 3}{3 + 2x}$ (d) $\dfrac{2x + 3}{-2x - 3}$

60. Only one of the following rational expressions is *not* equivalent to

$$\frac{x - 3}{4 - x}.$$

Which one is it?

(a) $\dfrac{3 - x}{x - 4}$ (b) $\dfrac{x + 3}{4 + x}$

(c) $-\dfrac{3 - x}{4 - x}$ (d) $-\dfrac{x - 3}{x - 4}$

Multiply or divide as indicated. See Examples 4 and 6.

61. $\dfrac{x^3}{3y} \cdot \dfrac{9y^2}{x^5}$

62. $\dfrac{a^4}{5b^2} \cdot \dfrac{25b^4}{a^3}$

63. $\dfrac{5a^4b^2}{16a^2b} \div \dfrac{25a^2b}{60a^3b^2}$

64. $\dfrac{s^3t^2}{10s^2t^4} \div \dfrac{8s^4t^2}{5t^6}$

65. $\dfrac{4x}{8x + 4} \cdot \dfrac{14x + 7}{6}$

66. $\dfrac{12x - 20}{5x} \cdot \dfrac{6}{9x - 15}$

67. $\dfrac{7t + 7}{-6} \div \dfrac{4t + 4}{15}$

68. $\dfrac{8z - 16}{-20} \div \dfrac{3z - 6}{40}$

69. $\dfrac{a^2 - 1}{4a} \cdot \dfrac{2}{1 - a}$

70. $\dfrac{p^2 - 25}{4p} \cdot \dfrac{2}{5 - p}$

71. $\dfrac{m^2 - 49}{m + 1} \div \dfrac{7 - m}{m}$

72. $\dfrac{k^2 - 4}{3k^2} \div \dfrac{2 - k}{11k}$

73. $\dfrac{12x - 10y}{3x + 2y} \cdot \dfrac{6x + 4y}{10y - 12x}$

74. $\dfrac{9s - 12t}{2s + 2t} \cdot \dfrac{3s + 3t}{4t - 3s}$

75. $\dfrac{x^2 - 25}{x^2 + x - 20} \cdot \dfrac{x^2 + 7x + 12}{x^2 - 2x - 15}$

76. $\dfrac{t^2 - 49}{t^2 + 4t - 21} \cdot \dfrac{t^2 + 8t + 15}{t^2 - 2t - 35}$

77. $\dfrac{15x^2 - xy - 2y^2}{15x^2 + 11xy + 2y^2} \cdot \dfrac{15x^2 + xy - 2y^2}{15x^2 + 4xy - 4y^2}$

78. $\dfrac{6p^2 + 19pq + 10q^2}{3p^2 + 11pq + 6q^2} \cdot \dfrac{18p^2 + 3pq - 10q^2}{6p^2 + 11pq - 10q^2}$

79. $\dfrac{3k^2 + 17kp + 10p^2}{6k^2 + 13kp - 5p^2} \div \dfrac{6k^2 + kp - 2p^2}{6k^2 - 5kp + p^2}$

80. $\dfrac{16c^2 + 24cd + 9d^2}{16c^2 - 16cd + 3d^2} \div \dfrac{16c^2 - 9d^2}{16c^2 - 24cd + 9d^2}$

81. $\left(\dfrac{6k^2 - 13k - 5}{k^2 + 7k} \div \dfrac{2k - 5}{k^3 + 6k^2 - 7k} \right) \cdot \dfrac{k^2 - 5k + 6}{3k^2 - 8k - 3}$

82. $\left(\dfrac{2x^3 + 3x^2 - 2x}{3x - 15} \div \dfrac{2x^3 - x^2}{x^2 - 3x - 10} \right) \cdot \dfrac{5x^2 - 10x}{3x^2 + 12x + 12}$

83. $\dfrac{a^2(2a + b) + 6a(2a + b) + 5(2a + b)}{3a^2(2a + b) - 2a(2a + b) + (2a + b)} \div \dfrac{a + 1}{a - 1}$

84. $\dfrac{2x^2(x - 3z) - 5x(x - 3z) + 2(x - 3z)}{4x^2(x - 3z) - 11x(x - 3z) + 6(x - 3z)} \div \dfrac{4x - 3}{4x + 1}$

The word rational *comes from the root word* ratio. *A* ratio *is a quotient of two quantities. Ratios provide a way to compare two quantities. Use the graphs to write the ratios in Exercises 85–88. Then express the quotient in decimal form to the nearest hundredth using a calculator.*

JAPANESE INVESTMENT IN U.S. REAL ESTATE

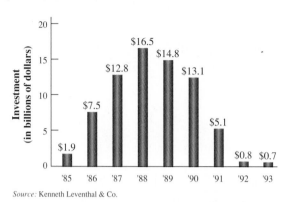

Source: Kenneth Leventhal & Co.

JAPANESE INVESTMENT BY STATE

For 1993, in millions of dollars

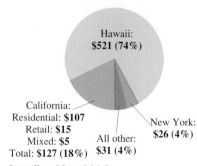

Source: Kenneth Leventhal & Co.

85. The greatest number of dollars (in billions) invested in the U.S. by the Japanese in one year compared to the total investment in the years shown.

86. The least amount invested in one year compared to the greatest amount invested in one year

87. The 1993 investment in New York compared to the total investment in 1993

88. The 1993 investment in California compared to the amount invested that year in Hawaii

6.2 Addition and Subtraction of Rational Expressions

We call fractions with the same denominators, such as $\frac{2}{3}$ and $\frac{4}{3}$, **like fractions.** Fractions with different denominators, such as $\frac{3}{4}$ and $\frac{5}{6}$, are called **unlike fractions.**

The following steps are used in adding or subtracting rational expressions.

Adding or Subtracting Rational Expressions

Step 1(a) If the denominators are the same, add or subtract the numerators. Place the result over the common denominator.

Step 1(b) If the denominators are different, first find the least common denominator. Write all rational expressions with this least common denominator, and then add or subtract the numerators. Place the result over the common denominator.

Step 2 Write all answers in lowest terms.

OBJECTIVE 1 The first example shows how to add and subtract like fractions.

EXAMPLE 1 Adding and Subtracting Like Fractions

(a) $\dfrac{5}{m} + \dfrac{7}{m} = \dfrac{5 + 7}{m} = \dfrac{12}{m}$

Because the denominators are the same, just add the numerators, and place the sum over the common denominator.

(b) $\dfrac{7}{2r^2} - \dfrac{11}{2r^2} = \dfrac{7 - 11}{2r^2} = \dfrac{-4}{2r^2} = -\dfrac{2}{r^2}$

The rule for subtraction is similar to that for addition. Because the denominators are the same, subtract the numerators. Keep the common denominator, and write the answer in lowest terms.

(c) $\dfrac{m}{m^2 - p^2} + \dfrac{p}{m^2 - p^2} = \dfrac{m + p}{m^2 - p^2}$

$\qquad\qquad = \dfrac{m + p}{(m + p)(m - p)}$ Factor the denominator.

$\qquad\qquad = \dfrac{1}{m - p}$ Lowest terms

WORK PROBLEM 1 AT THE SIDE. ▶▶

OBJECTIVE 2 If rational expressions to be added or subtracted have different denominators, we first find their **least common denominator (LCD),** an expression divisible by the denominator of each of the rational expressions.

OBJECTIVES

1. Add and subtract rational expressions with the same denominators.

2. Find the least common denominator.

3. Add and subtract rational expressions with different denominators.

FOR EXTRA HELP

Tutorial   Tape 9   SSM, Sec. 6.2

1. Add or subtract.

(a) $\dfrac{3m}{8} + \dfrac{5n}{8}$

(b) $\dfrac{7}{3a} + \dfrac{10}{3a}$

(c) $\dfrac{2}{y^2} - \dfrac{5}{y^2}$

(d) $\dfrac{a}{a + b} + \dfrac{b}{a + b}$

(e) $\dfrac{2y - 1}{y^2 + y - 2} - \dfrac{y}{y^2 + y - 2}$

ANSWERS

1. (a) $\dfrac{3m + 5n}{8}$ **(b)** $\dfrac{17}{3a}$ **(c)** $-\dfrac{3}{y^2}$ **(d)** 1

(e) $\dfrac{1}{y + 2}$

2. Find the least common denominator for each pair of rational expressions.

(a) $\dfrac{9}{5k^3s}$, $\dfrac{7}{10ks^4}$

(b) $\dfrac{10}{3 - x}$, $\dfrac{5}{9 - x^2}$

(c) $\dfrac{8}{z}$, $\dfrac{8}{z + 6}$

(d) $\dfrac{y - 4}{2y^2 - 3y - 2}$, $\dfrac{2y + 7}{2y^2 + 3y + 1}$

Finding the Least Common Denominator

Step 1 Factor each denominator.

Step 2 The LCD is the product of all the different factors from each denominator, with each factor raised to the *greatest* power that occurs in any denominator.

E X A M P L E 2   Finding the LCD

Find the LCD for each pair of rational expressions.

(a) $\dfrac{6}{5xy^2}$, $\dfrac{3}{2x^3y}$

Each denominator is already factored.

$$5xy^2 = 5 \cdot x \cdot y^2$$
$$2x^3y = 2 \cdot x^3 \cdot y$$

Greatest exponent on x is 3.

$$\text{LCD} = 5 \cdot 2 \cdot x^3 \cdot y^2 \quad \leftarrow \text{Greatest exponent on } y \text{ is 2.}$$
$$= 10x^3y^2$$

(b) $\dfrac{9}{p^2 - 16}$, $\dfrac{2}{4 - p}$

$$p^2 - 16 = (p + 4)(p - 4) \qquad \text{Factor denominators.}$$
$$4 - p = -1(p - 4)$$
$$\text{LCD} = -(p + 4)(p - 4)$$

(c) $\dfrac{8}{k - 3}$, $\dfrac{7}{k}$

The LCD is $k(k - 3)$. It is usually best to leave the LCD in factored form in the event the expression can be simplified.

(d) $\dfrac{y}{y^2 - 2y - 8}$, $\dfrac{3y + 1}{y^2 + 3y + 2}$

$$y^2 - 2y - 8 = (y + 2)(y - 4) \qquad \text{Factor denominators.}$$
$$y^2 + 3y + 2 = (y + 2)(y + 1)$$
$$\text{LCD} = (y + 2)(y - 4)(y + 1)$$

■ ━━━━━━━━━━━━━━

◀◀ **WORK PROBLEM 2 AT THE SIDE.**

OBJECTIVE 3▶ As mentioned earlier, if the rational expressions to be added or subtracted have different denominators, it is necessary to rewrite each rational expression with the LCD. This is done by multiplying both the numerator and the denominator of each rational expression by the factor required to get the LCD. This procedure is valid because each rational expression is being multiplied by a form of 1, the identity element for multiplication.

ANSWERS

2. **(a)** $10k^3s^4$ **(b)** $(3 + x)(3 - x)$
   (c) $z(z + 6)$
   (d) $(y - 2)(2y + 1)(y + 1)$

EXAMPLE 3   Adding and Subtracting Unlike Fractions

Add or subtract as indicated.

(a) $\dfrac{7}{15} + \dfrac{5}{12}$

The LCD for 15 and 12 is 60. Multiply $\frac{7}{15}$ by $\frac{4}{4}$ (a form of 1) and multiply $\frac{5}{12}$ by $\frac{5}{5}$ so that each fraction has denominator 60, and then add the numerators.

$$\dfrac{7}{15} + \dfrac{5}{12} = \dfrac{7 \cdot 4}{15 \cdot 4} + \dfrac{5 \cdot 5}{12 \cdot 5} \qquad \text{Get a common denominator.}$$

$$= \dfrac{28}{60} + \dfrac{25}{60}$$

$$= \dfrac{28 + 25}{60} \qquad \text{Add numerators.}$$

$$= \dfrac{53}{60}$$

(b) $\dfrac{5}{2p} + \dfrac{3}{8p}$

The LCD for $2p$ and $8p$ is $8p$. To write the first rational expression with a denominator of $8p$, multiply by $\frac{4}{4}$, a form of 1.

$$\dfrac{5}{2p} + \dfrac{3}{8p} = \dfrac{5 \cdot 4}{2p \cdot 4} + \dfrac{3}{8p}$$

$$= \dfrac{20}{8p} + \dfrac{3}{8p}$$

$$= \dfrac{20 + 3}{8p} = \dfrac{23}{8p}$$

(c) $\dfrac{6}{r} - \dfrac{5}{r-3}$

The LCD is $r(r-3)$. Rewrite each rational expression with this denominator.

$$\dfrac{6}{r} - \dfrac{5}{r-3} = \dfrac{6(r-3)}{r(r-3)} - \dfrac{5r}{r(r-3)}$$

$$= \dfrac{6(r-3) - 5r}{r(r-3)} \qquad \text{Subtract numerators.}$$

$$= \dfrac{6r - 18 - 5r}{r(r-3)} \qquad \text{Distributive property}$$

$$= \dfrac{r - 18}{r(r-3)} \qquad \text{Combine terms in numerator.}$$

■

WORK PROBLEM 3 AT THE SIDE. ▶▶

3. Add or subtract.

(a) $\dfrac{6}{7} + \dfrac{1}{5}$

(b) $\dfrac{8}{3k} - \dfrac{2}{9k}$

(c) $\dfrac{2}{y} - \dfrac{1}{y+4}$.

4. Subtract.

(a) $\dfrac{5x + 7}{2x + 7} - \dfrac{-x - 14}{2x + 7}$

Caution

One of the most common sign errors in algebra occurs when a rational expression with two or more terms in its numerator is being subtracted. Remember that in this situation, the subtraction sign must be distributed to *every* term in the numerator of the fraction that follows.

The next example illustrates this situation.

E X A M P L E 4   **Subtracting a Rational Expression with a Binomial Numerator**

Subtract.

(a) $\dfrac{7x}{3x + 1} - \dfrac{x - 2}{3x + 1}$

The denominator is the same for both rational expressions. The subtraction sign must be applied to *both* terms in the numerator of the second rational expression. Notice the careful use of parentheses here.

$$\dfrac{7x}{3x + 1} - \dfrac{x - 2}{3x + 1} = \dfrac{7x - (x - 2)}{3x + 1} \qquad \text{Write as a single rational expression.}$$

$$= \dfrac{7x - x + 2}{3x + 1} \qquad \text{Subtract: } -(x - 2) = -x + 2.$$

$$= \dfrac{6x + 2}{3x + 1} \qquad \text{Combine terms in numerator.}$$

$$= \dfrac{2(3x + 1)}{3x + 1} \qquad \text{Factor numerator.}$$

$$= 2 \qquad \text{Reduce to lowest terms.}$$

(b) $\dfrac{2}{r - 2} - \dfrac{r + 3}{r - 1}$

(b) $\dfrac{1}{q - 1} - \dfrac{1}{q + 1}$

$$= \dfrac{1(q + 1)}{(q - 1)(q + 1)} - \dfrac{1(q - 1)}{(q + 1)(q - 1)} \qquad \text{Get a common denominator.}$$

$$= \dfrac{(q + 1) - (q - 1)}{(q - 1)(q + 1)} \qquad \text{Subtract.}$$

$$= \dfrac{q + 1 - q + 1}{(q - 1)(q + 1)} \qquad \text{Distributive property}$$

$$= \dfrac{2}{(q - 1)(q + 1)} \qquad \text{Combine terms in numerator.}$$

◄◄ **WORK PROBLEM 4 AT THE SIDE.**

In some problems, rational expressions to be added or subtracted have denominators that are opposites of each other. The next example illustrates how to proceed in such a problem.

┌ **E X A M P L E   5** **Adding Rational Expressions with Denominators That Are Opposites.**

Add.

$$\frac{y}{y - 2} + \frac{8}{2 - y}$$

To get a common denominator of $y - 2$, multiply the second expression by -1 in both the numerator and the denominator.

$$\frac{y}{y - 2} + \frac{8}{2 - y} = \frac{y}{y - 2} + \frac{8(-1)}{(2 - y)(-1)}$$

$$= \frac{y}{y - 2} + \frac{-8}{y - 2}$$

$$= \frac{y - 8}{y - 2} \qquad \text{Add numerators.}$$

WORK PROBLEM 5 AT THE SIDE. ▶▶

The next example illustrates addition and subtraction involving more than two rational expressions.

┌ **E X A M P L E   6** **Adding and Subtracting Three Rational Expressions**

Add and subtract as indicated.

$$\frac{3}{x - 2} + \frac{5}{x} - \frac{6}{x^2 - 2x}$$

The denominator of the third rational expression is $x(x - 2)$, which is the LCD for the three rational expressions.

$$\frac{3}{x - 2} + \frac{5}{x} - \frac{6}{x^2 - 2x}$$

$$= \frac{3x}{x(x - 2)} + \frac{5(x - 2)}{x(x - 2)} - \frac{6}{x(x - 2)} \qquad \text{Get a common denominator.}$$

$$= \frac{3x + 5(x - 2) - 6}{x(x - 2)} \qquad \text{Add and subtract numerators.}$$

$$= \frac{3x + 5x - 10 - 6}{x(x - 2)} \qquad \text{Distributive property}$$

$$= \frac{8x - 16}{x(x - 2)} \qquad \text{Combine terms in numerator.}$$

$$= \frac{8(x - 2)}{x(x - 2)} \qquad \text{Factor numerator.}$$

$$= \frac{8}{x} \qquad \text{Lowest terms}$$

WORK PROBLEM 6 AT THE SIDE. ▶▶

5. Add or subtract as indicated.

(a) $\dfrac{8}{x - 4} + \dfrac{2}{4 - x}$

(b) $\dfrac{9}{2x - 9} - \dfrac{4}{9 - 2x}$

6. Add and subtract as indicated.

$$\frac{4}{x - 5} + \frac{-2}{x} - \frac{10}{x^2 - 5x}$$

ANSWERS

5. (a) $\dfrac{6}{x - 4}$ or $\dfrac{-6}{4 - x}$

(b) $\dfrac{13}{2x - 9}$ or $\dfrac{-13}{9 - 2x}$

6. $\dfrac{2}{x - 5}$

7. Subtract.

$$\frac{-a}{a^2 + 3a - 4} - \frac{4a}{a^2 + 7a + 12}$$

E X A M P L E   7   **Subtracting Rational Expressions**

Subtract: $\dfrac{m + 4}{m^2 - 2m - 3} - \dfrac{2m - 3}{m^2 - 5m + 6}$.

$$\frac{m + 4}{m^2 - 2m - 3} - \frac{2m - 3}{m^2 - 5m + 6}$$

$$= \frac{m + 4}{(m - 3)(m + 1)} - \frac{2m - 3}{(m - 3)(m - 2)} \qquad \text{Factor each denominator.}$$

The LCD is $(m - 3)(m + 1)(m - 2)$.

$$= \frac{(m + 4)(m - 2)}{(m - 3)(m + 1)(m - 2)} - \frac{(2m - 3)(m + 1)}{(m - 3)(m - 2)(m + 1)}$$

Get a common denominator.

$$= \frac{(m + 4)(m - 2) - (2m - 3)(m + 1)}{(m - 3)(m + 1)(m - 2)} \qquad \text{Subtract.}$$

$$= \frac{m^2 + 2m - 8 - (2m^2 - m - 3)}{(m - 3)(m + 1)(m - 2)} \qquad \text{Multiply in numerator.}$$

$$= \frac{m^2 + 2m - 8 - 2m^2 + m + 3}{(m - 3)(m + 1)(m - 2)} \qquad \text{Distributive property}$$

$$= \frac{-m^2 + 3m - 5}{(m - 3)(m + 1)(m - 2)} \qquad \text{Combine terms in numerator.}$$

◀◀ **WORK PROBLEM 7 AT THE SIDE.**

6.2 Exercises

1. Write a step-by-step method of adding or subtracting rational expressions that have a common denominator.

2. Write a step-by-step method of adding or subtracting rational expressions that have different denominators.

Add or subtract as indicated. Write all answers in lowest terms. See Example 1.

3. $\dfrac{7}{t} + \dfrac{2}{t}$

4. $\dfrac{5}{r} + \dfrac{9}{r}$

5. $\dfrac{11}{5x} - \dfrac{1}{5x}$

6. $\dfrac{7}{4y} - \dfrac{3}{4y}$

7. $\dfrac{5x + 4}{6x + 5} + \dfrac{x + 1}{6x + 5}$

8. $\dfrac{6y + 12}{4y + 3} + \dfrac{2y - 6}{4y + 3}$

9. $\dfrac{x^2}{x + 5} - \dfrac{25}{x + 5}$

10. $\dfrac{y^2}{y + 6} - \dfrac{36}{y + 6}$

11. $\dfrac{4}{p^2 + 7p + 12} + \dfrac{p}{p^2 + 7p + 12}$

12. $\dfrac{5}{x^2 + x - 20} + \dfrac{x}{x^2 + x - 20}$

13. $\dfrac{a^3}{a^2 + ab + b^2} - \dfrac{b^3}{a^2 + ab + b^2}$

14. $\dfrac{p^3}{p^2 - pq + q^2} + \dfrac{q^3}{p^2 - pq + q^2}$

Find the least common denominator for each group of rational expressions. See Example 2.

15. $\dfrac{5}{18x^2y^3}, \ \dfrac{-3}{24x^4y^5}$

16. $\dfrac{-7}{24a^3b^4}, \ \dfrac{13}{18a^5b^2}$

17. $\dfrac{7}{z - 2}, \ \dfrac{5}{z}$

18. $\dfrac{9}{k + 3}, \ \dfrac{3}{k}$

19. $\dfrac{9}{2y + 8}, \ \dfrac{-8}{y + 4}$

20. $\dfrac{17}{3r - 21}, \ \dfrac{9}{r - 7}$

21. $\dfrac{x}{x^2 - 81}, \ \dfrac{3x}{x^2 + 18x + 81}$

22. $\dfrac{y}{y^2 - 16}, \ \dfrac{9y}{y^2 - 8y + 16}$

23. $\dfrac{3m}{m + n}, \ \dfrac{2n}{m - n}$

24. $\dfrac{5r}{r+s}, \quad \dfrac{8s}{r-s}$

25. $\dfrac{x+8}{x^2-3x-4}, \quad \dfrac{-9}{x+x^2}$

26. $\dfrac{y+1}{y^2-8y+12}, \quad \dfrac{-3}{y^2-6y}$

27. $\dfrac{t}{2t^2+7t-15}, \quad \dfrac{t}{t^2+3t-10}$

28. $\dfrac{s}{s^2-3s-4}, \quad \dfrac{s}{3s^2+s-2}$

29. $\dfrac{y}{2y+6}, \quad \dfrac{3}{y^2-9}, \quad \dfrac{6}{y}$

30. $\dfrac{x}{9x+18}, \quad \dfrac{-1}{x^2-4}, \quad \dfrac{5}{x}$

31. One student added two rational expressions and obtained the answer $\dfrac{3}{5-y}$. Another student obtained the answer $\dfrac{-3}{y-5}$ for the same problem. Is it possible that both answers are correct? Explain.

32. What is *wrong* with the following work?

$$\frac{x}{x+2}-\frac{4x-1}{x+2}=\frac{x-4x-1}{x+2}=\frac{-3x-1}{x+2}$$

Add or subtract as indicated. Write all answers in lowest terms. See Examples 3–7.

33. $\dfrac{8}{t}+\dfrac{7}{3t}$

34. $\dfrac{5}{x}+\dfrac{9}{4x}$

35. $\dfrac{5}{12x^2y}-\dfrac{11}{6xy}$

36. $\dfrac{7}{18a^3b^2}-\dfrac{2}{9ab}$

37. $\dfrac{1}{x-1}-\dfrac{1}{x}$

38. $\dfrac{3}{x-3}-\dfrac{1}{x}$

39. $\dfrac{3a}{a+1}+\dfrac{2a}{a-3}$

40. $\dfrac{2x}{x+4}+\dfrac{3x}{x-7}$

41. $\dfrac{17y+3}{9y+7}-\dfrac{-10y-18}{9y+7}$

42. $\dfrac{7x + 8}{3x + 2} - \dfrac{x + 4}{3x + 2}$

43. $\dfrac{2}{4 - x} + \dfrac{5}{x - 4}$

44. $\dfrac{3}{2 - t} + \dfrac{1}{t - 2}$

45. $\dfrac{w}{w - z} - \dfrac{z}{z - w}$

46. $\dfrac{a}{a - b} - \dfrac{b}{b - a}$

47. $\dfrac{5}{12 + 4x} - \dfrac{7}{9 + 3x}$

48. $\dfrac{3}{10x + 15} - \dfrac{8}{12x + 18}$

49. $\dfrac{4x}{x - 1} - \dfrac{2}{x + 1} - \dfrac{4}{x^2 - 1}$

50. $\dfrac{4}{x + 3} - \dfrac{x}{x - 3} - \dfrac{18}{x^2 - 9}$

51. $\dfrac{15}{y^2 + 3y} + \dfrac{2}{y} + \dfrac{5}{y + 3}$

52. $\dfrac{7}{t - 2} - \dfrac{6}{t^2 - 2t} - \dfrac{3}{t}$

53. $\dfrac{5}{x - 2} + \dfrac{1}{x} + \dfrac{2}{x^2 - 2x}$

54. $\dfrac{5x}{x - 3} + \dfrac{2}{x} + \dfrac{6}{x^2 - 3x}$

55. $\dfrac{3x}{x + 1} + \dfrac{4}{x - 1} - \dfrac{6}{x^2 - 1}$

56. $\dfrac{5x}{x + 3} + \dfrac{x + 2}{x} - \dfrac{6}{x^2 + 3x}$

57. $\dfrac{4}{x + 1} + \dfrac{1}{x^2 - x + 1} - \dfrac{12}{x^3 + 1}$

58. $\dfrac{5}{x + 2} + \dfrac{2}{x^2 - 2x + 4} - \dfrac{60}{x^3 + 8}$

59. $\dfrac{2x + 4}{x + 3} + \dfrac{3}{x} - \dfrac{6}{x^2 + 3x}$

60. $\dfrac{4x + 1}{x + 5} - \dfrac{2}{x} + \dfrac{10}{x^2 + 5x}$

61. $\dfrac{5x}{x^2 + xy - 2y^2} - \dfrac{3x}{x^2 + 5xy - 6y^2}$

62. $\dfrac{6x}{6x^2 + 5xy - 4y^2} - \dfrac{2y}{9x^2 - 16y^2}$

63. $\dfrac{r+s}{3r^2+2rs-s^2} - \dfrac{s-r}{6r^2-5rs+s^2}$

64. $\dfrac{3y}{y^2+yz-2z^2} + \dfrac{4y-1}{y^2-z^2}$

MATHEMATICAL CONNECTIONS (EXERCISES 65–70)

In Example 6 we showed that

$$\frac{3}{x-2} + \frac{5}{x} - \frac{6}{x^2-2x}$$

is equal to $\dfrac{8}{x}$*. Algebra is, in a sense, a generalized form of arithmetic. In the following exercises, we show how the algebra in this example is related to the arithmetic of common fractions. Work Exercises 65–70 in order.*

65. Perform the following operations, and express your answer in lowest terms.

$$\frac{3}{7} + \frac{5}{9} - \frac{6}{63}$$

66. Substitute 9 for x in the given problem from Example 6. Compare this problem to the one given in Exercise 65. What do you notice?

67. Perform the operations in the problem you wrote in Exercise 66. Now substitute 9 for x in the answer given in Example 6. Do your results agree?

68. Replace x in the problem from Example 6 with the number of letters in your last name (assuming that this number is not 2). If your last name has 2 letters, let $x = 3$. Now predict the answer to your problem. Verify that your prediction is correct.

69. Why will $x = 2$ not work for the problem from Example 6?

70. What other value for x is not allowed in the problem given from Example 6?

6.3 Complex Fractions

A **complex fraction** is a fraction that has a fraction in the numerator, the denominator, or both. Examples of complex fractions include

$$\frac{1 + \frac{1}{x}}{2}, \quad \frac{\frac{4}{y}}{6 - \frac{3}{y}}, \quad \text{and} \quad \frac{\frac{m^2 - 9}{m + 1}}{\frac{m + 3}{m^2 - 1}}.$$

OBJECTIVE 1 There are two different methods for simplifying complex fractions.

Simplifying Complex Fractions

Method 1 Simplify the numerator and denominator separately, as much as possible. Then multiply the numerator by the reciprocal of the denominator. Simplify the resulting fraction, if possible.

E X A M P L E 1   Simplifying Complex Fractions Using Method 1

Use Method 1 to simplify each complex fraction.

(a) $\dfrac{\dfrac{x + 1}{x}}{\dfrac{x - 1}{2x}}$

Both the numerator and denominator are already simplified, so multiply the numerator by the reciprocal of the denominator.

$$\frac{\frac{x + 1}{x}}{\frac{x - 1}{2x}} = \frac{x + 1}{x} \div \frac{x - 1}{2x}$$

$$= \frac{x + 1}{x} \cdot \frac{2x}{x - 1} \qquad \text{Reciprocal of } \frac{x - 1}{2x}$$

$$= \frac{2(x + 1)}{x - 1} \qquad \text{Multiply and simplify.}$$

(b) $\dfrac{2 + \dfrac{1}{y}}{3 - \dfrac{2}{y}} = \dfrac{\dfrac{2y}{y} + \dfrac{1}{y}}{\dfrac{3y}{y} - \dfrac{2}{y}} = \dfrac{\dfrac{2y + 1}{y}}{\dfrac{3y - 2}{y}}$ Simplify numerator and denominator.

$$= \frac{2y + 1}{y} \cdot \frac{y}{3y - 2} \qquad \text{Reciprocal of } \frac{3y - 2}{y}$$

$$= \frac{2y + 1}{3y - 2}$$

WORK PROBLEM 1 AT THE SIDE. ▶▶

OBJECTIVE 2 Next we show the second method for simplifying complex fractions.

OBJECTIVES

1 Simplify complex fractions by simplifying numerator and denominator.

2 Simplify complex fractions by multiplying by a common denominator.

3 Compare the two methods of simplifying complex fractions.

4 Simplify rational expressions written with negative exponents.

FOR EXTRA HELP

Tutorial   Tape 9   SSM, Sec. 6.3

1. Use Method 1 to simplify each complex fraction.

(a) $\dfrac{\dfrac{a + 2}{5a}}{\dfrac{a - 3}{7a}}$

(b) $\dfrac{2 + \dfrac{1}{k}}{2 - \dfrac{1}{k}}$

(c) $\dfrac{\dfrac{r^2 - 4}{4}}{1 + \dfrac{2}{r}}$

ANSWERS

1. (a) $\dfrac{7(a + 2)}{5(a - 3)}$ (b) $\dfrac{2k + 1}{2k - 1}$ (c) $\dfrac{r(r - 2)}{4}$

2. Use Method 2 to simplify each complex fraction.

(a) $\dfrac{\dfrac{5}{y} + 6}{\dfrac{8}{3y} - 1}$

(b) $\dfrac{\dfrac{1}{y} + \dfrac{1}{y-1}}{\dfrac{1}{y} - \dfrac{2}{y-1}}$

Simplifying Complex Fractions

Method 2 Multiply the numerator and denominator of the complex fraction by the LCD of all fractions appearing in the numerator or denominator of the complex fraction. Simplify the resulting fraction, if possible.

E X A M P L E   2   Simplifying a Complex Fraction Using Method 2

Use Method 2 to simplify the following complex fraction.

$$\dfrac{2 + \dfrac{1}{y}}{3 - \dfrac{2}{y}}$$

For Method 2, multiply the numerator and denominator by the LCD of all the fractions appearing in either part of the complex fraction.

$$\dfrac{2 + \dfrac{1}{y}}{3 - \dfrac{2}{y}} = \dfrac{\left(2 + \dfrac{1}{y}\right) \cdot y}{\left(3 - \dfrac{2}{y}\right) \cdot y} \qquad \text{Multiply by the LCD, } y.$$

$$= \dfrac{2 \cdot y + \dfrac{1}{y} \cdot y}{3 \cdot y - \dfrac{2}{y} \cdot y} \qquad \text{Use the distributive property.}$$

$$= \dfrac{2y + 1}{3y - 2}$$

Compare this method of solution with that used for this same complex fraction in Example 1(b).

◀◀ **WORK PROBLEM 2 AT THE SIDE.**

OBJECTIVE 3▶ Choosing whether to use Method 1 or Method 2 to simplify a complex fraction is usually a matter of preference. Some students prefer one method over the other, while other students feel comfortable with both methods, and rely on practice with many examples to determine which method they will use on a particular problem. In the next example, we illustrate how to simplify a complex fraction using both methods. In so doing, you can observe the processes and decide for yourself the pros and cons of each method.

E X A M P L E   3   Simplifying Complex Fractions Using Both Methods

Use both Method 1 and Method 2 to simplify each complex fraction.

(a) $\dfrac{\dfrac{2}{x - 3}}{\dfrac{5}{x^2 - 9}}$

CONTINUED ON NEXT PAGE

Method 1

$$\dfrac{\dfrac{2}{x-3}}{\dfrac{5}{x^2-9}}$$

$$= \dfrac{\dfrac{2}{x-3}}{\dfrac{5}{(x-3)(x+3)}}$$

$$= \dfrac{2}{x-3} \div \dfrac{5}{(x-3)(x+3)}$$

$$= \dfrac{2}{x-3} \cdot \dfrac{(x-3)(x+3)}{5}$$

$$= \dfrac{2(x+3)}{5}$$

Method 2

$$\dfrac{\dfrac{2}{x-3}}{\dfrac{5}{x^2-9}}$$

$$= \dfrac{\dfrac{2}{x-3}}{\dfrac{5}{(x-3)(x+3)}} \cdot \dfrac{(x-3)(x+3)}{(x-3)(x+3)}$$

$$= \dfrac{2(x+3)}{5}$$

(b) $\dfrac{\dfrac{1}{x}+\dfrac{1}{y}}{\dfrac{1}{x^2}-\dfrac{1}{y^2}}$

Method 1

$$\dfrac{\dfrac{1}{x}+\dfrac{1}{y}}{\dfrac{1}{x^2}-\dfrac{1}{y^2}}$$

$$= \dfrac{\dfrac{y}{xy}+\dfrac{x}{xy}}{\dfrac{y^2}{x^2y^2}-\dfrac{x^2}{x^2y^2}}$$

$$= \dfrac{\dfrac{y+x}{xy}}{\dfrac{y^2-x^2}{x^2y^2}}$$

$$= \dfrac{y+x}{xy} \div \dfrac{y^2-x^2}{x^2y^2}$$

$$= \dfrac{y+x}{xy} \cdot \dfrac{x^2y^2}{(y-x)(y+x)}$$

$$= \dfrac{xy}{y-x}$$

Method 2

$$\dfrac{\dfrac{1}{x}+\dfrac{1}{y}}{\dfrac{1}{x^2}-\dfrac{1}{y^2}}$$

$$= \dfrac{\left(\dfrac{1}{x}+\dfrac{1}{y}\right)\cdot x^2y^2}{\left(\dfrac{1}{x^2}-\dfrac{1}{y^2}\right)\cdot x^2y^2}$$

$$= \dfrac{xy^2+x^2y}{y^2-x^2}$$

$$= \dfrac{xy(y+x)}{(y+x)(y-x)}$$

$$= \dfrac{xy}{y-x}$$

WORK PROBLEM 3 AT THE SIDE. ▶▶

OBJECTIVE 4 Rational expressions and complex fractions often involve negative exponents, as in the following example.

3. Use both methods to simplify each complex fraction.

(a) $\dfrac{\dfrac{5}{y+2}}{\dfrac{-3}{y^2-4}}$

(b) $\dfrac{\dfrac{1}{a}-\dfrac{1}{b}}{\dfrac{1}{a^2}-\dfrac{1}{b^2}}$

ANSWERS

3. (Both methods give the same answer.)

(a) $\dfrac{5(y-2)}{-3}$

(b) $\dfrac{ab}{b+a}$

4. Simplify each of the following, using only positive exponents in your answers.

(a) $\dfrac{1}{a^{-1} + b^{-1}}$

E X A M P L E   4   **Simplifying a Rational Expression with Negative Exponents**

Simplify the following rational expression.

$$\frac{m^{-1} + p^{-2}}{2m^{-2} - p^{-1}}$$

Begin by using the definition of a negative exponent to write the expression without negative exponents.

$$\frac{m^{-1} + p^{-2}}{2m^{-2} - p^{-1}} = \frac{\dfrac{1}{m} + \dfrac{1}{p^2}}{\dfrac{2}{m^2} - \dfrac{1}{p}}$$

We use Method 2 to simplify this complex fraction, multiplying each term in the numerator and denominator by the LCD of all terms in the complex fraction, $m^2 p^2$.

$$\frac{\dfrac{1}{m} + \dfrac{1}{p^2}}{\dfrac{2}{m^2} - \dfrac{1}{p}} = \frac{m^2 p^2 \cdot \dfrac{1}{m} + m^2 p^2 \cdot \dfrac{1}{p^2}}{m^2 p^2 \cdot \dfrac{2}{m^2} - m^2 p^2 \cdot \dfrac{1}{p}} = \frac{mp^2 + m^2}{2p^2 - m^2 p}$$

◀◀ **WORK PROBLEM 4 AT THE SIDE.**

(b) $\dfrac{k^{-1}}{3k^{-2} + 1}$

Caution

The first term in the denominator in Example 4 was rewritten as $\dfrac{2}{m^2}$, not $\dfrac{1}{2m^2}$. The exponent -2 applies only to m, and not to 2.

6.3 Exercises

1. Explain in your own words Method 1 of simplifying complex fractions.

2. Method 2 of simplifying complex fractions says that we can multiply both the numerator and the denominator of the complex fraction by the same nonzero expression. What property of real numbers from Section 1.5 justifies this method?

Use either method to simplify each complex fraction. See Examples 1–3.

3. $\dfrac{\dfrac{12}{x-1}}{\dfrac{6}{x}}$

4. $\dfrac{\dfrac{24}{t+4}}{\dfrac{6}{t}}$

5. $\dfrac{\dfrac{k+1}{2k}}{\dfrac{3k-1}{4k}}$

6. $\dfrac{\dfrac{1-r}{4r}}{\dfrac{-1-r}{8r}}$

7. $\dfrac{\dfrac{4z^2x^4}{9}}{\dfrac{12x^2z^5}{15}}$

8. $\dfrac{\dfrac{3y^2x^3}{8}}{\dfrac{9y^3x^4}{16}}$

9. $\dfrac{\dfrac{1}{x}+1}{-\dfrac{1}{x}+1}$

10. $\dfrac{\dfrac{2}{k}-1}{\dfrac{2}{k}+1}$

11. $\dfrac{\dfrac{3}{x}+\dfrac{3}{y}}{\dfrac{3}{x}-\dfrac{3}{y}}$

12. $\dfrac{\dfrac{4}{t}-\dfrac{4}{s}}{\dfrac{4}{t}+\dfrac{4}{s}}$

13. $\dfrac{\dfrac{8x-24y}{10}}{\dfrac{x-3y}{5x}}$

14. $\dfrac{\dfrac{10x-5y}{12}}{\dfrac{2x-y}{6y}}$

15. $\dfrac{\dfrac{x^2-16y^2}{xy}}{\dfrac{1}{y}-\dfrac{4}{x}}$

16. $\dfrac{\dfrac{2}{s}-\dfrac{3}{t}}{\dfrac{4t^2-9s^2}{st}}$

17. $\dfrac{y-\dfrac{y-3}{3}}{\dfrac{4}{9}+\dfrac{2}{3y}}$

18. $\dfrac{p - \dfrac{p+2}{4}}{\dfrac{3}{4} - \dfrac{5}{2p}}$

19. $\dfrac{\dfrac{x+2}{x} + \dfrac{1}{x+2}}{\dfrac{5}{x} + \dfrac{x}{x+2}}$

20. $\dfrac{\dfrac{y+3}{y} - \dfrac{4}{y-1}}{\dfrac{y}{y-1} + \dfrac{1}{y}}$

MATHEMATICAL CONNECTIONS (EXERCISES 21–26)

Simplifying a complex fraction by Method 1 is a good way to review the methods of adding, subtracting, multiplying, and dividing rational expressions. Method 2 gives a good review of the fundamental theorem of rational expressions. Refer to the complex fraction below and work the following exercises in order.

$$\dfrac{\dfrac{4}{m} + \dfrac{m+2}{m-1}}{\dfrac{m+2}{m} - \dfrac{2}{m-1}}$$

21. Add the fractions in the numerator.

22. Subtract as indicated in the denominator.

23. Divide your answer from Exercise 21 by your answer from Exercise 22.

24. Go back to the original complex fraction and find the least common denominator of all denominators.

25. Multiply the numerator and denominator of the complex fraction by your answer from Exercise 24.

26. Your answers for Exercises 23 and 25 should be the same. Write an explanation comparing the two methods. Which method do you prefer? Explain why.

Simplify each of the following, using only positive exponents in your answer. See Example 4.

27. $\dfrac{1}{x^{-2} + y^{-2}}$

28. $\dfrac{1}{p^{-2} - q^{-2}}$

29. $\dfrac{x^{-2} + y^{-2}}{x^{-1} + y^{-1}}$

30. $\dfrac{x^{-1} - y^{-1}}{x^{-2} - y^{-2}}$

31. $(r^{-1} + s^{-1})^{-1}$

32. $((2k)^{-1} + (4s)^{-1})^{-1}$

6.4 Equations Involving Rational Expressions

OBJECTIVES

1 ▶ Solve equations with rational expressions.

2 ▶ Know when potential solutions must be checked.

FOR EXTRA HELP

Tutorial   Tape 9   SSM, Sec. 6.4

OBJECTIVE **1** ▶ The easiest way to solve most equations involving rational expressions is to multiply all the terms in the equation by the LCD, as the next examples show. This process can only be used with equations, not expressions.

E X A M P L E 1 **Solving an Equation with Rational Expressions**

Solve $\frac{2x}{5} - \frac{x}{3} = 2$.

The LCD for $\frac{2x}{5}$ and $\frac{x}{3}$ is 15, so multiply both sides of the equation by 15.

$$15\left(\frac{2x}{5} - \frac{x}{3}\right) = 15(2)$$

$$15\left(\frac{2x}{5}\right) - 15\left(\frac{x}{3}\right) = 15(2) \quad \text{Distributive property}$$

$$6x - 5x = 30 \quad \text{Simplify.}$$

$$x = 30 \quad \text{Combine terms.}$$

Substitute 30 for x in the original equation to check this solution.

$$\frac{2 \cdot 30}{5} - \frac{30}{3} = 2 \quad ? \quad \text{Let } x = 30.$$

$$\frac{60}{5} - 10 = 2 \quad ?$$

$$12 - 10 = 2 \quad ?$$

$$2 = 2 \quad \text{True}$$

This check shows that the solution set is $\{30\}$.

WORK PROBLEM 1 AT THE SIDE. ▶▶

1. Solve.

(a) $\dfrac{2k}{3} - \dfrac{k}{6} = -3$

(b) $\dfrac{3p}{2} - \dfrac{p}{4} = 1$

(c) $\dfrac{7k}{12} - \dfrac{5k}{8} = \dfrac{1}{4}$

E X A M P L E 2 **Solving an Equation with Rational Expressions**

Solve $\frac{2}{y} - \frac{3}{2} = \frac{7}{2y}$.

Multiply both sides by the LCD, $2y$.

$$2y\left(\frac{2}{y} - \frac{3}{2}\right) = 2y\left(\frac{7}{2y}\right)$$

$$2y\left(\frac{2}{y}\right) - 2y\left(\frac{3}{2}\right) = 2y\left(\frac{7}{2y}\right) \quad \text{Distributive property}$$

$$4 - 3y = 7 \quad \text{Simplify.}$$

$$-3y = 3 \quad \text{Subtract 4.}$$

$$y = -1 \quad \text{Divide by } -3.$$

To see if -1 is a solution for the equation, replace y with -1 in the original equation:

$$\frac{2}{-1} - \frac{3}{2} = \frac{7}{2(-1)} \quad ? \quad \text{Let } y = -1.$$

$$-\frac{4}{2} - \frac{3}{2} = -\frac{7}{2}. \quad \text{True}$$

The solution set is $\{-1\}$.

2. Solve.

(a) $\dfrac{3}{p} - \dfrac{7}{10} = \dfrac{8}{5p}$

(b) $\dfrac{3x}{7} + \dfrac{1}{2} = \dfrac{5x}{14}$

ANSWERS

1. (a) $\{-6\}$ (b) $\left\{\dfrac{4}{5}\right\}$ (c) $\{-6\}$

2. (a) $\{2\}$ (b) $\{-7\}$

WORK PROBLEM 2 AT THE SIDE. ▶▶

3. Solve.

$$\frac{3}{a+1} = \frac{1}{a-1} - \frac{2}{a^2-1}$$

4. Solve.

(a) $\dfrac{10}{m^2} - \dfrac{3}{m} = 1$

(b) $\dfrac{x}{x-3} + \dfrac{2}{x+3} = \dfrac{-12}{x^2-9}$

OBJECTIVE 2 To solve the equation in Example 2, we multiplied both sides by $2y$. Be careful, though, when multiplying by a variable expression.

> **Caution**
> When both sides of an equation are multiplied by an expression containing a variable, it is possible that the resulting "solutions" are not actually solutions of the given equation.

E X A M P L E 3 Solving an Equation with Rational Expressions

Solve $\dfrac{2}{m-3} - \dfrac{3}{m+3} = \dfrac{12}{m^2-9}$.

Multiply both sides by the LCD, which is $(m+3)(m-3)$.

$$(m+3)(m-3) \cdot \frac{2}{m-3} - (m+3)(m-3) \cdot \frac{3}{m+3}$$

$$= (m+3)(m-3) \cdot \frac{12}{m^2-9}$$

$$2(m+3) - 3(m-3) = 12$$

$$2m + 6 - 3m + 9 = 12 \qquad \text{Distributive property}$$

$$-m + 15 = 12 \qquad \text{Combine terms.}$$

$$-m = -3 \qquad \text{Subtract 15.}$$

$$m = 3$$

Because both sides were multiplied by a term containing a variable, we must check the potential solution.

$$\frac{2}{3-3} - \frac{3}{3+3} = \frac{12}{3^2-9} \qquad \text{?} \quad \text{Let } m = 3.$$

$$\frac{2}{0} - \frac{3}{6} = \frac{12}{0} \qquad \text{?}$$

Division by 0 is not possible; the given equation has no solution, and the solution set is $\emptyset$.

◀◀ **WORK PROBLEM 3 AT THE SIDE.**

E X A M P L E 4 Solving an Equation That Leads to a Quadratic Equation

Solve $\dfrac{2}{3x+1} = \dfrac{1}{x} - \dfrac{6x}{3x+1}$.

Multiply both sides by $x(3x+1)$. The resulting equation is

$$2x = (3x+1) - 6x^2.$$

$$6x^2 - 3x + 2x - 1 = 0 \qquad \text{Get 0 on the right.}$$

$$6x^2 - x - 1 = 0 \qquad \text{Combine terms.}$$

$$(3x+1)(2x-1) = 0 \qquad \text{Factor.}$$

$$3x + 1 = 0 \quad \text{or} \quad 2x - 1 = 0 \qquad \text{Zero-factor property}$$

$$x = -\frac{1}{3} \quad \text{or} \qquad x = \frac{1}{2}$$

Using $-\frac{1}{3}$ in the original equation causes the denominator $3x+1$ to equal 0, so it is not a solution. The solution set is $\{\frac{1}{2}\}$.

◀◀ **WORK PROBLEM 4 AT THE SIDE.**

6.4 Exercises

As explained in this section, any values that would cause a denominator to equal zero must be excluded from possible solutions of equations that have variable expressions in their denominators. Without actually solving the equation, list all possible numbers that would have to be rejected if they appeared as potential solutions.

1. $\dfrac{1}{x + 1} - \dfrac{1}{x - 2} = 0$

2. $\dfrac{3}{x + 4} - \dfrac{2}{x - 9} = 0$

3. $\dfrac{5}{3x + 5} - \dfrac{1}{x} = \dfrac{1}{2x + 3}$

4. $\dfrac{6}{4x + 7} - \dfrac{3}{x} = \dfrac{5}{6x - 13}$

5. $\dfrac{1}{3x} + \dfrac{1}{2x} = \dfrac{x}{3}$

6. $\dfrac{5}{6x} - \dfrac{8}{2x} = \dfrac{x}{4}$

7. $\dfrac{3x + 1}{x - 4} = \dfrac{6x + 5}{2x - 7}$

8. $\dfrac{4x - 1}{2x + 3} = \dfrac{12x - 25}{6x - 2}$

9. $\dfrac{x + 5}{10} - \dfrac{2x + 3}{5} = \dfrac{x}{20}$

10. Is it possible that any potential solutions to the equation

$$\frac{x + 7}{4} - \frac{x + 3}{3} = \frac{x}{12}$$

would have to be rejected? Explain why or why not.

Solve each of the following equations. See Examples 1–4.

11. $\dfrac{x}{4} - \dfrac{x}{6} = \dfrac{2}{3}$

12. $\dfrac{y}{10} + \dfrac{3y}{5} = -\dfrac{7}{2}$

13. $\dfrac{x + 8}{5} = \dfrac{6 + x}{3}$

14. $\dfrac{r + 1}{4} = \dfrac{1 + 2r}{5}$

15. $\dfrac{x - 4}{x + 6} = \dfrac{2x + 3}{2x - 1}$

16. $\dfrac{5x - 8}{x + 2} = \dfrac{5x - 1}{x + 3}$

17. $\dfrac{3x + 1}{x - 4} = \dfrac{6x + 5}{2x - 7}$

18. $\dfrac{4x - 1}{2x + 3} = \dfrac{12x - 25}{6x - 2}$

19. $\dfrac{-5}{2x} + \dfrac{3}{4x} = \dfrac{-7}{4}$

20. $\dfrac{6}{5x} - \dfrac{2}{3x} = \dfrac{-8}{45}$

21. $x - \dfrac{24}{x} = -2$

22. $p + \dfrac{15}{p} = -8$

23. $\dfrac{1}{y - 1} + \dfrac{5}{12} = \dfrac{-2}{3y - 3}$

24. $\dfrac{4}{m + 2} - \dfrac{11}{9} = \dfrac{1}{3m + 6}$

25. $\dfrac{-2}{3t - 6} - \dfrac{1}{36} = \dfrac{-3}{4t - 8}$

26. $\dfrac{3}{4m + 2} = \dfrac{17}{2} - \dfrac{7}{2m + 1}$

27. $\dfrac{3}{k + 2} - \dfrac{2}{k^2 - 4} = \dfrac{1}{k - 2}$

28. $\dfrac{3}{x - 2} + \dfrac{21}{x^2 - 4} = \dfrac{14}{x + 2}$

29. $\dfrac{1}{y + 2} + \dfrac{3}{y + 7} = \dfrac{5}{y^2 + 9y + 14}$

30. $\dfrac{1}{t + 3} + \dfrac{4}{t + 5} = \dfrac{2}{t^2 + 8t + 15}$

31. $\dfrac{9}{x} + \dfrac{4}{6x - 3} = \dfrac{2}{6x - 3}$

32. $\dfrac{5}{n} + \dfrac{4}{6 - 3n} = \dfrac{2n}{6 - 3n}$

33. $\dfrac{6}{w + 3} + \dfrac{-7}{w - 5} = \dfrac{-48}{w^2 - 2w - 15}$

34. $\dfrac{2}{r - 5} + \dfrac{3}{2r + 1} = \dfrac{22}{2r^2 - 9r - 5}$

35. $\dfrac{x}{x - 3} + \dfrac{4}{x + 3} = \dfrac{18}{x^2 - 9}$

36. $\dfrac{2x}{x-3} + \dfrac{4}{x+3} = \dfrac{-24}{x^2-9}$

37. $\dfrac{6}{x-4} + \dfrac{5}{x} = \dfrac{-20}{x^2-4x}$

38. $\dfrac{7}{x-4} + \dfrac{3}{x} = \dfrac{-12}{x^2-4x}$

39. $\dfrac{2}{4x+7} + \dfrac{x}{3} = \dfrac{6}{12x+21}$

40. $\dfrac{5x+14}{x^2-9} = \dfrac{-2x^2-5x+2}{x^2-9} + \dfrac{2x+4}{x-3}$

41. $\dfrac{4x-7}{4x^2-9} = \dfrac{-2x^2+5x-4}{4x^2-9} + \dfrac{x+1}{2x+3}$

42. What is wrong with the following problem? "Solve $\dfrac{2x+1}{3x-4} + \dfrac{1}{2x+3}$."

MATHEMATICAL CONNECTIONS (EXERCISES 43–46)

An equation of the form

$$\frac{A}{x+B} + \frac{x}{x-B} = \frac{C}{x^2-B^2}$$

will have one rejected solution if the relationship $C = -2AB$ holds true. (This can be proved using methods not covered in intermediate algebra.) For example, if $A = 1$ and $B = 2$, then $C = -2AB = -2(1)(2) = -4$, and the equation becomes

$$\frac{1}{x+2} + \frac{x}{x-2} = \frac{-4}{x^2-4}.$$

This equation has solution set $\{-1\}$; the potential solution -2 must be rejected.

To further illustrate this idea, work Exercises 43–46 in order.

43. Show that the equation shown above does indeed have solution set $\{-1\}$ and -2 must be rejected.

44. Let $A = 2$ and let $B = 1$. What is the corresponding value of C? Solve the equation determined by A, B, and C. What is the solution set? What value must be rejected?

45. Let $A = 4$ and let $B = -3$. What is the corresponding value of C? Solve the equation determined by A, B, and C. What is the solution set? What value must be rejected?

46. Choose two numbers of your own, letting one be A and the other be B. Repeat the process described in Exercises 44 and 45.

---

47. Make up an equation similar to the one in Exercise 11, and then solve it. (*Hint:* Start with the answer, and work backward.)

48. Explain the difference between *simplifying the expression* $\dfrac{2}{x+1} + \dfrac{3}{x-2} - \dfrac{6}{x^2 - x - 2}$

and *solving the equation* $\dfrac{2}{x+1} + \dfrac{3}{x-2} = \dfrac{6}{x^2 - x - 2}$.

SUMMARY EXERCISES OPERATIONS AND EQUATIONS INVOLVING RATIONAL EXPRESSIONS

A common student error is to confuse an *equation,* such as $\frac{x}{2} + \frac{x}{3} = -5$, with an *expression involving an operation,* such as $\frac{x}{2} + \frac{x}{3}$. Look for the equals sign to distinguish between them. Equations are solved for a numerical answer, while problems involving operations result in simplified *expressions,* as shown below.

Solving an Equation

Solve: $\dfrac{x}{2} + \dfrac{x}{3} = -5$.

Multiply both sides by the LCD, 6.

$$6\left(\frac{x}{2} + \frac{x}{3}\right) = 6(-5)$$
$$3x + 2x = -30$$
$$5x = -30$$
$$x = -6$$

Check that the solution is -6, and the solution set is $\{-6\}$.

Simplifying an Expression

Add: $\dfrac{x}{2} + \dfrac{x}{3}$.

Write both fractions with LCD, 6.

$$\frac{x}{2} + \frac{x}{3} = \frac{x \cdot 3}{2 \cdot 3} + \frac{x \cdot 2}{3 \cdot 2}$$
$$= \frac{3x}{6} + \frac{2x}{6}$$
$$= \frac{3x + 2x}{6}$$
$$= \frac{5x}{6}$$

In each exercise, identify as an equation *or an* expression. *Then perform the indicated operation or solve the given equation, as appropriate.*

1. $\dfrac{x}{2} - \dfrac{x}{4} = 5$

2. $\dfrac{8x^4 z}{12x^3 z^2} \cdot \dfrac{7x}{3x^5}$

3. $\dfrac{4x - 20}{x^2 - 25} \cdot \dfrac{(x + 5)^2}{10}$

4. $\dfrac{6}{7x} - \dfrac{4}{x}$

5. $\dfrac{\dfrac{1}{x} + \dfrac{1}{y}}{\dfrac{1}{x} - \dfrac{1}{y}}$

6. $\dfrac{5}{7t} = \dfrac{52}{7} - \dfrac{3}{t}$

7. $\dfrac{x - 5}{3} + \dfrac{1}{3} = \dfrac{x - 2}{5}$

8. $\dfrac{7}{6x} + \dfrac{5}{8x}$

9. $\dfrac{4}{x} - \dfrac{8}{x + 1} = 0$

10. $\dfrac{\dfrac{6}{x + 1} - \dfrac{1}{x}}{\dfrac{2}{x} - \dfrac{4}{x + 1}}$

11. $\dfrac{8}{r + 2} - \dfrac{7}{4r + 8}$

12. $\dfrac{x}{x + y} + \dfrac{2y}{x - y}$

13. $\dfrac{3p^2 - 6p}{p + 5} \div \dfrac{p^2 - 4}{8p + 40}$

14. $\dfrac{x - 2}{9} \cdot \dfrac{5}{8 - 4x}$

15. $\dfrac{a - 4}{3} + \dfrac{11}{6} = \dfrac{a + 1}{2}$

16. $\dfrac{b^2 + b - 6}{b^2 + 2b - 8} \cdot \dfrac{b^2 + 8b + 16}{3b + 12}$

17. $\dfrac{10z^2 - 5z}{3z^3 - 6z^2} \div \dfrac{2z^2 + 5z - 3}{z^2 + z - 6}$

18. $\dfrac{5}{x^2 - 2x} - \dfrac{3}{x^2 - 4}$

19. $\dfrac{6}{t + 1} + \dfrac{4}{5t + 5} = \dfrac{34}{15}$

20. $\dfrac{x^{-1} + y^{-1}}{y^{-1} - x^{-1}}$

21. $\dfrac{\dfrac{5}{x} - \dfrac{3}{y}}{\dfrac{9x^2 - 25y^2}{x^2y}}$

22. $\dfrac{-2}{a^2 + 2a - 3} - \dfrac{5}{3 - 3a} = \dfrac{4}{3a + 9}$

23. $\dfrac{2r^{-1} + 5s^{-1}}{\dfrac{4s^2 - 25r^2}{3rs}}$

24. $\dfrac{4y^2 - 13y + 3}{2y^2 - 9y + 9} \div \dfrac{4y^2 + 11y - 3}{6y^2 - 5y - 6}$

25. $\dfrac{8}{3k + 9} - \dfrac{8}{15} = \dfrac{2}{5k + 15}$

26. $\dfrac{3r}{r - 2} = 1 + \dfrac{6}{r - 2}$

27. $\dfrac{6z^2 - 5z - 6}{6z^2 + 5z - 6} \cdot \dfrac{12z^2 - 17z + 6}{12z^2 - z - 6}$

28. $\dfrac{-1}{3 - x} - \dfrac{2}{x - 3}$

29. $\dfrac{\dfrac{t}{4} - \dfrac{1}{t}}{1 + \dfrac{t + 4}{t}}$

30. $\dfrac{2}{y + 1} - \dfrac{3}{y^2 - y - 2} = \dfrac{3}{y - 2}$

31. $\dfrac{7}{2x^2 - 8x} + \dfrac{3}{x^2 - 16}$

32. $\dfrac{3}{y - 3} - \dfrac{3}{y^2 - 5y + 6} = \dfrac{2}{y - 2}$

33. $\dfrac{2k + \dfrac{5}{k - 1}}{3k - \dfrac{2}{k}}$

6.5 Formulas Involving Rational Expressions

OBJECTIVES

1 ▸ Find the value of an unknown variable.

2 ▸ Solve a formula for a given variable.

FOR EXTRA HELP

Tutorial   Tape 10   SSM, Sec. 6.5

Many common formulas involve rational expressions. Methods of working with these formulas are shown in this section.

OBJECTIVE 1 ▸ The first example shows how to find the value of an unknown variable in a formula.

E X A M P L E   1   Finding a Value Using a Formula

In physics, the focal length, f, of a lens is given by the formula

$$\frac{1}{f} = \frac{1}{p} + \frac{1}{q},$$

where p is the distance from the object to the lens and q is the distance from the lens to the image. Find q if $p = 20$ centimeters and $f = 10$ centimeters.

Replace f with 10 and p with 20.

$$\frac{1}{f} = \frac{1}{p} + \frac{1}{q}$$

$$\frac{1}{10} = \frac{1}{20} + \frac{1}{q} \qquad \text{Let } f = 10 \text{ and } p = 20.$$

Multiply both sides by the LCD, $20q$.

$$20q \cdot \frac{1}{10} = 20q \cdot \frac{1}{20} + 20q \cdot \frac{1}{q}$$

$$2q = q + 20$$

$$q = 20$$

The distance q from the lens to the image is 20 centimeters.

WORK PROBLEM 1 AT THE SIDE. ▶▶

1. Use the formula given in Example 1 to answer each part.

 (a) Find p if $f = 15$ and $q = 25$.

 (b) Find f if $p = 6$ and $q = 9$.

OBJECTIVE 2 ▸ The next example shows how to solve a formula for a particular variable.

E X A M P L E   2   Solving a Formula for a Particular Variable

Solve $\frac{1}{f} = \frac{1}{p} + \frac{1}{q}$ for p.

To solve the formula for p, begin by multiplying both sides by the LCD, fpq. Then get p alone on one side of the equals sign by subtracting fp from both sides, so that both terms with p are on one side.

$$fpq \cdot \frac{1}{f} = fpq\left(\frac{1}{p} + \frac{1}{q}\right)$$

$$pq = fq + fp \qquad \text{Distributive property}$$

$$pq - fp = fq \qquad \text{Subtract } fp.$$

$$p(q - f) = fq \qquad \text{Distributive property}$$

$$p = \frac{fq}{q - f} \qquad \text{Divide by } q - f.$$

The last step requires that $q \neq f$.

 (c) Find q if $f = 12$ and $p = 16$.

ANSWERS

1. (a) $\dfrac{75}{2}$ (b) $\dfrac{18}{5}$ (c) 48

2. (a) Solve $\dfrac{1}{f} = \dfrac{1}{p} + \dfrac{1}{q}$ for q.

◄◄ WORK PROBLEM 2 AT THE SIDE.

E X A M P L E   3   **Solving a Formula for a Particular Variable**

Solve $I = \dfrac{nE}{R + nr}$ for n.

First, multiply both sides by $R + nr$.

$$(R + nr)I = (R + nr)\dfrac{nE}{R + nr}$$

$$RI + nrI = nE$$

Get the terms with n (the specified variable) together on one side of the equals sign. To do this, subtract nrI from both sides.

$$RI = nE - nrI$$

$$RI = n(E - rI) \qquad \text{Distributive property}$$

Finally, divide both sides by $E - rI$.

$$\dfrac{RI}{E - rI} = n$$

(b) Solve $\dfrac{8}{7x} - \dfrac{9}{5y} = \dfrac{2}{z}$ for y.

3. (a) Solve for R.

$$A = \dfrac{Rr}{R + r}$$

◄◄ WORK PROBLEM 3 AT THE SIDE.

Caution

Refer to the steps in Examples 2 and 3 that use the distributive property. This is a step that often gives students difficulty. Remember that the variable for which you are solving *must* be a factor on one side of the equation so that in the last step, both sides are divided by the remaining factor there. The *distributive property* allows us to perform this factorization.

(b) Solve for r.

$$I = \dfrac{nE}{R + nr}$$

ANSWERS

2. (a) $q = \dfrac{fp}{p - f}$ **(b)** $y = \dfrac{63xz}{40z - 70x}$

3. (a) $R = \dfrac{-Ar}{A - r}$ or $R = \dfrac{Ar}{r - A}$

(b) $r = \dfrac{nE - IR}{In}$

6.5 Exercises

In Exercises 1–4, a familiar formula is given. Give the letter of the choice that is an equivalent form of the given formula (involving a rational expression).

1. $d = rt$ (motion)

    (a) $r = \dfrac{d}{t}$ **(b)** $t = \dfrac{r}{d}$ **(c)** $r = \dfrac{t}{d}$ **(d)** $d = \dfrac{t}{r}$

2. $I = prt$ (simple interest)

    (a) $p = \dfrac{r}{It}$ **(b)** $r = \dfrac{It}{p}$ **(c)** $p = \dfrac{I}{rt}$ **(d)** $t = \dfrac{pr}{I}$

3. $A = \dfrac{1}{2}bh$ (area of a triangle)

    (a) $h = \dfrac{2A}{b}$ **(b)** $h = \dfrac{b}{2A}$ **(c)** $b = \dfrac{2}{Ah}$ **(d)** $b = \dfrac{2h}{A}$

4. $PVT = pvt$ (chemistry)

    (a) $t = \dfrac{pV}{PvT}$ **(b)** $p = \dfrac{vt}{PVT}$ **(c)** $v = \dfrac{pt}{PVT}$ **(d)** $p = \dfrac{PVT}{vt}$

Solve for the unknown quantity in each problem. See Example 1.

5. A law from physics says that

$$m = \frac{F}{a}.$$

Find F if $m = 30$ and $a = 9$.

6. Ohm's law in electricity says that

$$I = \frac{E}{R}.$$

Find R if $I = 20$ and $E = 8$.

7. The gravitational force between two masses is given by

$$F = \frac{GMm}{d^2}.$$

Find M if $F = 10$, $G = 6.67 \times 10^{-11}$, $m = 1$, and $d = 3 \times 10^{-6}$.

8. A gas law in chemistry says that

$$\frac{PV}{T} = \frac{pv}{t}.$$

Suppose that $T = 300$, $t = 350$, $V = 9$, $P = 50$, and $v = 8$. Find p.

9. In work with electric circuits, the formula

$$\frac{1}{a} = \frac{1}{b} + \frac{1}{c}$$

occurs. Find b if $a = 8$ and $c = 12$.

10. A formula from anthropology says that

$$c = \frac{100b}{L}.$$

Find L if $c = 80$ and $b = 5$.

MATHEMATICAL CONNECTIONS (EXERCISES 11–14)

Solving the "formula" $bx = c + d(x + 2)$ for x is very similar to solving the equation $5x = 4 + 2(x + 2)$. In Exercises 11–14, we illustrate these similarities. Work them in order.

11. To solve $5x = 4 + 2(x + 2)$ for x, perform these steps.
    (a) Distribute 2 on the right side.
    (b) Subtract $2x$ from both sides. Leave the left side as a difference.
    (c) Factor out x on the left side, and combine terms on the right side.
    (d) Simplify the coefficient of x on the left.
    (e) Divide both sides by the coefficient of x.

12. To solve the "formula" $bx = c + d(x + 2)$ for x, perform these steps.
 (a) Distribute d on the right side.
 (b) Subtract dx from both sides.
 (c) Factor out x from the two terms on the left side, and combine terms on the right side.
 (d) Divide both sides by the coefficient of x.

13. In Exercise 11, there were two instances where we combined terms. Yet in Exercise 12, this was not done. Why not?

14. Comment on the following statement: When solving a formula for a particular variable, treat that variable as if it was the only one, and treat all other variables as if they were constants.

Solve each formula for the specified variable. See Examples 2 and 3.

15. $F = \dfrac{GMm}{d^2}$ for G (physics)

16. $F = \dfrac{GMm}{d^2}$ for M (physics)

17. $\dfrac{1}{a} = \dfrac{1}{b} + \dfrac{1}{c}$ for a (electricity)

18. $\dfrac{1}{a} = \dfrac{1}{b} + \dfrac{1}{c}$ for b (electricity)

19. $\dfrac{PV}{T} = \dfrac{pv}{t}$ for v (chemistry)

20. $\dfrac{PV}{T} = \dfrac{pv}{t}$ for T (chemistry)

21. $A = P + Prt$ for P (finance)

22. $A = \dfrac{V - v}{t}$ for V (physics)

23. $A = \dfrac{1}{2}h(B + b)$ for b (mathematics)

24. $S = \dfrac{n}{2}(a + \ell)d$ for n (mathematics)

6.6 Applications Involving Rational Expressions

OBJECTIVES

1 ▶ Solve problems using proportions.

2 ▶ Solve problems about distance, rate, and time.

3 ▶ Solve problems about work.

FOR EXTRA HELP

Tutorial   Tape 10   SSM, Sec. 6.6

A **ratio** (RAY-show) is a comparison of two quantities with the same units. The ratio of a to b may be written in any of the following ways.

$$a \text{ to } b, \quad a : b, \quad \text{or} \quad \frac{a}{b}$$

A bar graph with double bars, such as the one in Figure 1, is useful in interpreting ratios. For example, we see that in 1984, the ratio of box office revenue to home video revenue was about 4 billion to 2 billion. In 1986, home video revenue overtook box office revenue. In 1990, home video revenue was twice that of box office revenue, a ratio of 2 to 1. In the exercises for this section, we will ask other questions regarding this figure.

ENLARGING THE PIE

New distribution centers don't have to hurt old ones. Hollywood quickly learned to stop worrying and love home video.

Sources: Motion Picures Association of America; Paul Kagan Associates Inc.

FIGURE 1

OBJECTIVE 1 ▶ Ratios are usually written as quotients in algebra. A **proportion** (proh-POR-shun) is a statement that two ratios are equal. Proportions are a useful and important type of rational expression. The first example shows how to solve a problem that leads to a proportion.

┌─ **E X A M P L E   I   Solving a Proportion**

In a recent year, 15 of every 100 Americans had no health insurance coverage. The population at that time was about 246 million. How many million had no health insurance?

We continue to use the problem-solving steps given in Chapter 2.

Let $x =$ the number who had no health insurance. To get an equation, we set up a proportion. The ratio x to 246 should equal the ratio 15 to 100. Write the proportion and solve the equation.

└─ **CONTINUED ON NEXT PAGE**

1. Several years ago, the average American family spent 11.1 of every 100 dollars on health care. This amounted to $4296 per family. To the nearest dollar, what was the average family income at that time?

$$\frac{15}{100} = \frac{x}{246}$$

$$24{,}600\left(\frac{15}{100}\right) = 24{,}600\left(\frac{x}{246}\right) \quad \text{Multiply by a common denominator.}$$

$$246(15) = 100x \quad \text{Simplify.}$$

$$3690 = 100x$$

$$x = 36.9$$

There were 36.9 million Americans with no health insurance. Check that this number compared to 246 million is equivalent to 15/100.

◀◀ **WORK PROBLEM 1 AT THE SIDE.**

A comparison of two quantities with different units is called a **rate.** Two equal rates can be expressed as a proportion. It is important to be sure the two rates are expressed with the units in the same order.

EXAMPLE 2 **Solving a Proportion Involving Rates**

Marissa's car uses 10 gallons of gas to travel 210 miles. She has 5 gallons of gas in the car, and she wants to know how much more gas she will need to drive 640 miles. If we assume the car uses gas at the same rate, how many more gallons will she need?

We can set up a proportion.

Let $x =$ the additional amount of gas needed.

$$\frac{\text{gallons}}{\text{miles}} \quad \frac{10}{210} = \frac{5 + x}{640} \quad \frac{\text{gallons}}{\text{miles}}$$

The LCD is $10 \cdot 21 \cdot 64$.

$$10 \cdot 21 \cdot 64\left(\frac{10}{210}\right) = 10 \cdot 21 \cdot 64\left(\frac{5 + x}{640}\right)$$

$$64 \cdot 10 = 21(5 + x)$$

$$30.5 = 5 + x \quad \text{Divide 640 by 21; round to}$$

$$25.5 = x \quad \text{nearest tenth.}$$

2. Americans pay an average tax of $.34 on gasoline priced at $1.22 per gallon. If this ratio were maintained and the cost of gas rose to $3.65 per gallon, the price paid in France and Germany, what additional tax would Americans pay?

Marissa will need 25.5 more gallons of gas. Check the answer in the words of the problem. The 25.5 gallons plus the 5 gallons equals 30.5 gallons.

$$\frac{30.5}{640} = .0476 \text{ (approximately)}$$

$$\frac{10}{210} = .0476 \text{ (approximately)}$$

Because the rates are equal, the solution is correct.

◀◀ **WORK PROBLEM 2 AT THE SIDE.**

OBJECTIVE 2 A familiar example of a rate is speed, which is the ratio of distance to time. The next examples use the distance formula $d = rt$ introduced in Chapter 2.

┌─
EXAMPLE 3   **Solving a Problem about Distance,
               Rate, and Time**

A tour boat goes 10 miles against the current in a small river in the same time that it goes 15 miles with the current. If the speed of the current is 3 miles per hour, find the speed of the boat in still water.

Use the distance formula:

$$\text{distance} = \text{rate} \times \text{time}, \quad \text{or} \quad d = rt.$$

Let $x =$ the speed of the boat in still water;
     $x - 3 =$ the speed of the boat against the current;
     $x + 3 =$ the speed of the boat with the current.

Because the time is the same going against the current as with the current, find time in terms of distance and rate (speed) for each situation.

Start with $d = rt$, and divide both sides by r to get

$$t = \frac{d}{r}.$$

Going against the current, the distance is 10 miles and the rate is $x - 3$, giving

$$t = \frac{d}{r} = \frac{10}{x - 3}.$$

Going with the current, the distance is 15 miles and the rate is $x + 3$, so that

$$t = \frac{d}{r} = \frac{15}{x + 3}.$$

This information is summarized in the following chart.

Direction	Distance	Rate	Time
Against current	10	$x - 3$	$\dfrac{10}{x - 3}$
With current	15	$x + 3$	$\dfrac{15}{x + 3}$

Times are equal.

Because the times are equal, $\dfrac{10}{x - 3} = \dfrac{15}{x + 3}$. The LCD is $(x - 3)(x + 3)$.

Multiplying on both sides by the LCD gives

$10(x + 3) = 15(x - 3)$	
$10x + 30 = 15x - 45$	Distributive property
$30 = 5x - 45$	Subtract $10x$.
$75 = 5x$	Add 45.
$15 = x.$	Divide by 5.

The speed of the boat in still water is 15 miles per hour.
└─

WORK PROBLEM 3 AT THE SIDE. ▶▶

3. A plane travels 100 miles against the wind in the same time that it takes to travel 120 miles with the wind. The wind speed is 20 miles per hour.

(a) Complete this table.

Direction	d	r	t
Against wind	100	$x - 20$	
With wind	120	$x + 20$	

(b) Find the speed of the plane in still air.

E X A M P L E 4 **Solving a Problem about Distance, Rate, and Time**

At the airport, Cheryl and Bill are walking to the gate (at the same speed) to catch their flight to Akron, Ohio. Bill wants a window seat, so he steps onto the moving sidewalk and continues to walk while Cheryl uses the stationary sidewalk. If the sidewalk moves at 1 meter per second and Bill saves 50 seconds covering the 300-meter distance, what is their walking speed?

Let x represent their walking speed. Then Cheryl travels at x meters per second and Bill travels at $x + 1$ meters per second. Because Bill's time is 50 seconds less than Cheryl's time, express their times in terms of the known distances and the variable rates. Start with $d = rt$, and divide both sides by r to get

$$t = \frac{d}{r}.$$

For Cheryl, distance is 300 meters, and the rate is x. Cheryl's time is

$$t = \frac{d}{r} = \frac{300}{x}.$$

Bill goes 300 meters at a rate of $x + 1$, so his time is

$$t = \frac{d}{r} = \frac{300}{x + 1}.$$

This information is summarized in the following chart.

Passenger	d	r	t
Cheryl	300	x	$\dfrac{300}{x}$
Bill	300	$x + 1$	$\dfrac{300}{x + 1}$

Now use the information given in the problem about the times to write an equation.

Bill's time	is	Cheryl's time	less 50 seconds.
$\dfrac{300}{x + 1}$	$=$	$\dfrac{300}{x}$	$- 50$

The common denominator is $x(x + 1)$. Multiply both sides by $x(x + 1)$.

CONTINUED ON NEXT PAGE

$$x(x + 1)\left(\frac{300}{x + 1}\right) = x(x + 1)\left(\frac{300}{x} - 50\right)$$

$$300x = 300(x + 1) - 50x(x + 1)$$

$$300x = 300x + 300 - 50x^2 - 50x \quad \text{Distributive property}$$

$$0 = 50x^2 + 50x - 300 \quad \begin{array}{l}\text{Subtract } 300x;\\ \text{multiply by } -1.\end{array}$$

$$0 = x^2 + x - 6 \quad \begin{array}{l}\text{Divide both sides by}\\ 50.\end{array}$$

$$0 = (x + 3)(x - 2) \quad \text{Factor.}$$

$$x + 3 = 0 \quad \text{or} \quad x - 2 = 0 \quad \text{Zero-factor property}$$

$$x = -3 \quad \text{or} \quad x = 2$$

Discard the negative answer since speed cannot be negative. Their walking speed is 2 meters per second. Check the solution in the words of the original problem.

WORK PROBLEM 4 AT THE SIDE. ▶▶

OBJECTIVE 3 People and machines work at different rates. Rates of work are another common example of rates. Let the letters r, t, and A represent the rate at which the work is done, the time required, and the amount of work accomplished, respectively. Then $A = rt$. Notice the similarity to the distance formula, $d = rt$. The amount of work is often measured in terms of jobs accomplished. Thus, if 1 job is completed, $A = 1$, and the formula gives

$$1 = rt$$

$$r = \frac{1}{t}$$

as the rate. If a job can be accomplished in t units of time, then the rate of work is

$$\frac{1}{t} \text{ job per unit of time.}$$

In solving a work problem, we begin by using this fact to express each rate of work. Example 5 shows one possible approach in solving a work problem.

E X A M P L E   5   Solving a Problem about Work Rates

Lindsay and Michael are working on a neighborhood cleanup. Michael can clean up all the trash in the area in 7 hours, while Lindsay can do the same job in 5 hours. How long will it take them if they work together?

Let x = the number of hours it will take the two people working together. Just as we made a chart for the distance formula, $d = rt$, we can make a chart here for $A = rt$, with $A = 1$. Because $A = 1$, the rate for each person will be $1/t$, where t is the time it takes each person to complete the job alone. For example, Michael can clean up all the trash in 7 hours, so his rate is $1/7$ of the job per hour. Similarly, Lindsay's rate is $1/5$ of the job per hour. Fill in the chart as shown.

CONTINUED ON NEXT PAGE

4. Luz drove 300 miles north from San Antonio, mostly on the freeway. She usually averaged 55 miles per hour, but an accident slowed her speed through Dallas to 15 miles per hour. If her trip took 6 hours, how many miles did she drive at reduced speed?

ANSWERS

4. $11\frac{1}{4}$ miles

5. (a) Stan needs 45 minutes to do the dishes, while Bobbie can do them in 30 minutes. How long will it take them if they work together?

Worker	Rate	Time Working Together	Fractional Part of the Job Done
Michael	$\dfrac{1}{7}$	x	$\dfrac{1}{7}x$
Lindsay	$\dfrac{1}{5}$	x	$\dfrac{1}{5}x$

Because together they complete 1 job, the sum of the fractional parts accomplished by each of them should equal 1.

$$\underset{\substack{\text{part done} \\ \text{by Michael}}}{\frac{1}{7}x} \quad + \quad \underset{\substack{\text{part done} \\ \text{by Lindsay}}}{\frac{1}{5}x} \quad = \quad \underset{\substack{\text{1 whole} \\ \text{job}}}{1}$$

Solve this equation. The least common denominator is 35.

$$35\left(\frac{1}{7}x + \frac{1}{5}x\right) = 35 \cdot 1$$
$$5x + 7x = 35 \qquad (*)$$
$$12x = 35$$
$$x = \frac{35}{12}$$

Working together, Michael and Lindsay can do the entire job in $\frac{35}{12} = 2\frac{11}{12} = 2\frac{55}{60}$ hours or 2 hours and 55 minutes. Check this result in the original problem.

(b) Suppose it takes Stan 35 minutes to do the dishes, and together they can do them in 15 minutes. How long will it take Bobbie to do them alone?

There is another way to approach the problem in Example 5. Once again, let x represent the number of hours it will take the two people working together. In one hour, $\frac{1}{x}$ of the entire job will be completed. Michael completes $\frac{1}{7}$ of the job in one hour, and Lindsay completes $\frac{1}{5}$ of the job, so the sum of their rates should equal $\frac{1}{x}$.

$$\frac{1}{7} + \frac{1}{5} = \frac{1}{x}$$

Multiplying both sides of this equation by $35x$ leads to $5x + 7x = 35$. Notice that this is the same equation obtained in Example 5 in the line marked (*). Thus, the solution of the equation is the same in both approaches.

◀◀ **WORK PROBLEM 5 AT THE SIDE.**

6.6 Exercises

Refer to Figure 1 to answer the following questions.

1. In 1986, what was the ratio of the box office revenue to the home video revenue?

2. In 1992, what was the ratio of the home video revenue to the box office revenue?

3. In 1993, what was the ratio of the box office revenue to the home video revenue?

4. In a particular year, the ratio of home video revenue to box office revenue was about 11 to 5. What year was it?

Use proportions to solve each of the following problems mentally.

5. In a mathematics class, 3 of every 4 students is a girl. If there are 20 students in the class, how many are girls? How many are boys?

6. In a certain southern state, sales tax on a purchase of $1.50 is $.12. What is the sales tax on a purchase of $6.00?

7. If Juanita can mow her yard in 2 hours, what is her rate (in job per hour)?

8. A bus traveling from Atlanta to Detroit averages 50 miles per hour and takes 14 hours to make the trip. How far is it from Atlanta to Detroit?

Solve each of the following problems. See Examples 1 and 2.

9. In a recent year, 25.5 of every 100 households in the United States was composed of a married couple with children under 18. If there were 95,700,000 households altogether, how many fell into this category? Round your answer to the nearest hundred thousand.

10. The Office of the Surgeon General reported several years ago that 6.2 of every 100 students in the 7th through 12th grades drink wine. How many of the 10,600,000 students in these grades drink wine? Round your answer to the nearest ten thousand.

11. In a recent year, a single person could expect to pay $4500 in income taxes on an adjusted gross income of $30,000. How much more tax should the person expect to pay if her adjusted gross income increased $2000, knowing that this would not move her into a higher tax bracket?

12. In a recent year, the top-performing mutual fund would have produced income of $44,300 on an investment of $100,000. If this investment had been increased to $260,000, how much more income would this fund have earned?

13. Biologists tagged 500 fish in a lake on January 1. On February 1 they returned and collected a random sample of 400 fish, 8 of which had been previously tagged. Approximately how many fish does the lake have based on this experiment?

14. Suppose that in the experiment of Exercise 13, 10 of the previously tagged fish were collected on February 1. What would be the estimated fish population?

Solve each of the following problems. See Examples 3 and 4.

15. Kellen's boat goes 12 miles per hour. Find the speed of the current of the river if he can go 6 miles upstream in the same amount of time he can go 10 miles downstream.

	Distance	*Rate*	*Time*
Downstream	10	$12 + x$	
Upstream	6	$12 - x$	

16. A canal has a current of 2 miles per hour. Find the speed of Kasey's boat in still water if it goes 11 miles downstream in the same time that it goes 8 miles upstream.

	Distance	*Rate*	*Time*
Downstream	11	$x + 2$	
Upstream	8	$x - 2$	

17. If Dr. Dawson rides his bike to his office, he averages 12 miles per hour. If he drives his car, he averages 36 miles per hour. His time driving is $\frac{1}{4}$ hour less than his time riding his bike. How far is his office from home?

18. Driving from Tulsa to Detroit, Jeff averaged 50 miles per hour. He figured that if he had averaged 60 miles per hour, his driving time would have decreased 3 hours. How far is it from Tulsa to Detroit?

19. On the first part of a trip to Carmel traveling on the freeway, Marge averaged 60 miles per hour. On the rest of the trip, which was 10 miles longer than the first part, she averaged 50 miles per hour. Find the total distance to Carmel if the second part of the trip took 30 minutes more than the first part.

20. While on vacation, Jim and Annie decided to drive all day. During the first part of their trip on the highway, they averaged 60 miles per hour. When they got to Houston, traffic caused them to average only 30 miles per hour. The distance they drove in Houston was 100 miles less than their distance on the highway. What was their total driving distance if they spent 50 minutes more on the highway than they did in Houston?

21. Explain the similarities between the methods of solving problems about distance, rate, and time and problems about work.

22. If one person takes 3 hours to do a job and another takes 4 hours to do the same job, why is "$3\frac{1}{2}$ hours" *not* a valid answer to the problem "How long will it take them to do the job working together?"?

Solve each of the following problems. See Example 5.

23. Lou can groom Joe's dogs in 8 hours, but it takes his business partner, Janet, only 5 hours to groom the same dogs. How long will it take them to groom Joe's dogs if they work together?

24. Butch and Peggy want to pick up the mess that their grandson, William, has made in his playroom. Butch could do it in 15 minutes working alone. Peggy, working alone, could clean it in 12 minutes. How long will it take them if they work together?

25. Sandi and Cary Goldstein are refinishing a table. Working alone, Cary could do the job in 7 hours. If the two work together, the job takes 5 hours. How long will it take Sandi to refinish the table if she works alone?

26. Ron can paint a room in 6 hours working alone. If Dee helps him, the job takes 4 hours. How long would it take Dee to do the job if she worked alone?

27. A winery has a vat to hold chardonnay. An inlet pipe can fill the vat in 9 hours, while an outlet pipe can empty it in 12 hours. How long will it take to fill the vat if both the outlet and the inlet pipes are open?

28. If a vat of acid can be filled by an inlet pipe in 10 hours, and emptied by an outlet pipe in 20 hours, how long will it take to fill the vat if both pipes are open?

29. An inlet pipe can fill an artificial lily pond in 60 minutes, while an outlet pipe can empty it in 80 minutes. Through an error, both pipes are left on. How long will it take for the pond to fill?

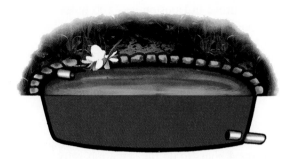

30. Suppose that Hortense and Mort can clean their entire house in 7 hours, while their toddler, Mimi, just by being around, can completely mess it up in only 2 hours. If Hortense and Mort clean the house while Mimi is at her grandma's, and then start cleaning up after Mimi the minute she gets home, how long does it take from the time Mimi gets home until the whole place is a shambles?

In geometry, it is shown that two triangles with corresponding angles equal, called similar triangles, *have corresponding sides proportional. For example, in the figure, angle A = angle D, angle B = angle E, and angle C = angle F, so the triangles are similar. Then the following ratios of corresponding sides are equal.*

$$\frac{4}{6} = \frac{6}{9} = \frac{2x + 1}{2x + 5}$$

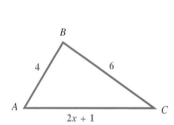

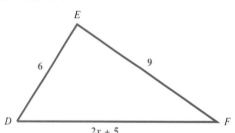

31. Solve for x using the proportion given above to find the lengths of the third sides of the triangles.

32. Suppose the triangles shown below are similar. Find y and the lengths of the two longest sides of each triangle.

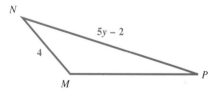

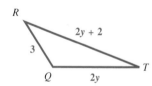

6.1	**rational expression**	A rational expression (or algebraic fraction) is the quotient of two polynomials with denominator not 0.
	rational function	A rational function is a function defined by a rational expression in a single variable.
6.2	**like fractions**	Like fractions are fractions with the same denominators.
	unlike fractions	Unlike fractions are fractions with different denominators.
	least common denominator (LCD)	A least common denominator of several denominators is an expression that is divisible by each of these denominators.
6.3	**complex fraction**	A fraction that has a fraction in the numerator, the denominator, or both, is called a complex fraction.
6.6	**ratio**	A ratio is a comparison of two quantities with the same units.
	proportion	A proportion is a statement that two ratios are equal.
	rate	A comparison of two quantities with different units is called a rate.

Concepts	*Examples*

6.1 Rational Expressions and Functions; Multiplication and Division

Finding Values for which a Rational Function Is Undefined

Set the denominator equal to zero and solve. The solutions of this equation are the values for which the rational function is undefined.

For what real numbers is the rational function

$$f(x) = \frac{x}{x^2 - 25} \text{ undefined?}$$
$$x^2 - 25 = 0$$
$$(x + 5)(x - 5) = 0$$
$$x = \pm 5$$

This function is undefined for -5 and 5.

Writing Rational Expressions in Lowest Terms

Step 1 Factor the numerator and the denominator completely.

Step 2 Replace each pair of factors common to the numerator and denominator with 1.

Write in lowest terms.

$$\frac{2x + 8}{x^2 - 16} = \frac{2(x + 4)}{(x - 4)(x + 4)}$$
$$= \frac{2}{x - 4}$$

Multiplying Rational Expressions

Factor numerators and denominators. Replace all pairs of common factors in numerators and denominators with 1. Multiply remaining factors in the numerator and in the denominator.

Multiply.

$$\frac{x^2 + 2x + 1}{x^2 - 1} \cdot \frac{5}{3x + 3}$$
$$= \frac{(x + 1)^2}{(x - 1)(x + 1)} \cdot \frac{5}{3(x + 1)}$$
$$= \frac{5}{3(x - 1)}$$

Dividing Rational Expressions

Multiply the first by the reciprocal of the second.

Divide.

$$\frac{2x + 5}{x - 3} \div \frac{2x^2 + 3x - 5}{x^2 - 9}$$
$$= \frac{2x + 5}{x - 3} \cdot \frac{(x + 3)(x - 3)}{(2x + 5)(x - 1)}$$
$$= \frac{x + 3}{x - 1}$$

Concepts	Examples
6.2 Addition and Subtraction of Rational Expressions **Adding or Subtracting Rational Expressions** If the denominators are the same, add or subtract the numerators. Place the result over the common denominator. If the denominators are different, write all rational expressions with the LCD. Then add or subtract the like fractions. Be sure answer is in lowest terms.	Subtract. $\dfrac{1}{x+6} - \dfrac{3}{x+2}$ $= \dfrac{x+2}{(x+6)(x+2)} - \dfrac{3(x+6)}{(x+6)(x+2)}$ $= \dfrac{x+2-3x-18}{(x+6)(x+2)}$ $= \dfrac{-2x-16}{(x+6)(x+2)}$
6.3 Complex Fractions **Simplifying Complex Fractions** *Method 1* Simplify the numerator and denominator separately, as much as possible. Then multiply the numerator by the reciprocal of the denominator. Simplify the resulting fraction, if possible.	Simplify the complex fraction. *Method 1* $\dfrac{\dfrac{1}{2}+\dfrac{1}{3}}{\dfrac{1}{4}-\dfrac{1}{2}} = \dfrac{\dfrac{3}{6}+\dfrac{2}{6}}{\dfrac{1}{4}-\dfrac{2}{4}}$ $= \dfrac{\dfrac{5}{6}}{\dfrac{-1}{4}} = \dfrac{5}{6}\cdot\dfrac{4}{-1}$ $= \dfrac{20}{-6} = -\dfrac{10}{3}$
Method 2 Multiply the numerator and denominator of the complex fraction by the least common denominator of all fractions appearing in the numerator or denominator of the complex fraction. Simplify the resulting fraction, if possible.	*Method 2* $\dfrac{\dfrac{1}{2}+\dfrac{1}{3}}{\dfrac{1}{4}-\dfrac{1}{2}} = \dfrac{12\left(\dfrac{1}{2}\right)+12\left(\dfrac{1}{3}\right)}{12\left(\dfrac{1}{4}\right)-12\left(\dfrac{1}{2}\right)}$ $= \dfrac{6+4}{3-6} = \dfrac{10}{-3} = -\dfrac{10}{3}$
6.4 Equations Involving Rational Expressions To solve an equation involving rational expressions, multiply all the terms in the equation by the least common denominator. Each potential solution must be checked to make sure that no denominator in the original equation is 0.	Solve for x: $\dfrac{1}{x} + x = \dfrac{26}{5}.$ $5 + 5x^2 = 26x \qquad \text{Multiply by } 5x.$ $5x^2 - 26x + 5 = 0$ $(5x-1)(x-5) = 0$ $x = \dfrac{1}{5} \quad \text{or} \quad x = 5$ Both check. The solution set is $\{\frac{1}{5}, 5\}$.

Concepts	Examples
6.5 Formulas Involving Rational Expressions To solve a formula for a particular variable, get that variable alone on one side.	Solve for L. $$c = \frac{100b}{L}$$ $cL = 100b$ Multiply by L. $L = \dfrac{100b}{c}$ Divide by c.
6.6 Applications Involving Rational Expressions If an applied problem translates into an equation with rational expressions, solve the equation using the method described in Section 6.4.	If a 6.4-ounce tube of toothpaste costs \$1.89, what should an 8.2-ounce tube cost? Let x represent the unknown cost. Use a proportion. $$\frac{6.4}{1.89} = \frac{8.2}{x}$$ $$1.89x\left(\frac{6.4}{1.89}\right) = 1.89x\left(\frac{8.2}{x}\right)$$ $$6.4x = 15.498$$ $$x = 2.42$$ The 8.2-ounce tube should cost \$2.42. Check by comparing the two ratios.

Numbers in the Real World

a graphing calculator minicourse

Lesson 5: Matrices

Graphing calculators are capable of working with matrices and associated topics such as their determinants. At the beginning of Section 4.4, we briefly introduced matrices, and then showed how determinants of square matrices can be used to solve systems of equations with Cramer's rule. This can be done efficiently with a graphing calculator.

Suppose that we wish to solve the system
$$5x + 7y = -1$$
$$6x + 8y = 1.$$

Let $D = \begin{bmatrix} 5 & 7 \\ 6 & 8 \end{bmatrix}$ be the matrix of coefficients, $A = D_x = \begin{bmatrix} -1 & 7 \\ 1 & 8 \end{bmatrix}$ be the matrix obtained by substituting the constants for the coefficients of x, and $B = D_y = \begin{bmatrix} 5 & -1 \\ 6 & 1 \end{bmatrix}$ be the matrix obtained by substituting the constants for the coefficients of y. These matrices can be entered into a graphing calculator. (See the owner's manual for instructions.) The calculator can calculate the determinants and the correct quotients. As seen in the first three figures at the left, the quotient for x is 7.5 and the quotient for y is -5.5. Thus the solution set is $\{(7.5, -5.5)\}$.

In more advanced courses such as college algebra, matrices are studied in a more theoretical way. A method of solving systems called *row reduction* employing properties of matrices is very efficient and is used by computers to solve large systems. Arithmetic operations with matrices are defined, and identities and inverses for multiplication are studied. For example, two matrices with the same dimensions are added or subtracted by adding or subtracting the corresponding entries. The next-to-last screen at the left shows how the matrices A and B defined above are added and subtracted. Multiplication of matrices is defined in what might seem to be a rather strange way. (Consult a college algebra text for the definition.) A very interesting property of matrix multiplication is that it is not commutative. The bottom screen shows that the matrix products AB and BA are not equal.

GRAPHING CALCULATOR EXPLORATIONS

1. Use a graphing calculator to apply Cramer's rule to find the solution of each system.

   (a) $x + y = 4$
   $2x - y = 2$

   (b) $x + y - z = -2$
   $2x - y + z = -5$
   $x - 2y + 3z = 4$

2. In Section 4.4 we saw that if the determinant of the coefficient matrix is 0, the system cannot be solved by Cramer's rule. Use a graphing calculator to show that this is the case for the following systems.

   (a) $3x + 2y = 9$
   $6x + 4y = 3$

   (b) $2x - 3y + 4z = 10$
   $6x - 9y + 12z = 24$
   $x + 2y - 3z = 5$

[6.1] *Find all real numbers that make the rational function undefined.*

1. $f(x) = \dfrac{9}{x+3}$

2. $f(x) = \dfrac{8}{x^2 - 36}$

3. $f(x) = \dfrac{x+9}{x^2 - 7x + 10}$

Write in lowest terms.

4. $\dfrac{55m^4n^3}{10m^5n}$

5. $\dfrac{12x^2 + 6x}{24x + 12}$

6. $\dfrac{y^2 + 3y - 10}{y^2 - 5y + 6}$

7. $\dfrac{25m^2 - n^2}{25m^2 - 10mn + n^2}$

8. $\dfrac{r-2}{4-r^2}$

Multiply or divide. Write all answers in lowest terms.

9. $\dfrac{y^5}{6} \cdot \dfrac{9}{y^4}$

10. $\dfrac{25p^3q^2}{8p^4q} \div \dfrac{15pq^2}{16p^5}$

11. $\dfrac{3m + 12}{-8} \div \dfrac{5m + 20}{4}$

12. $\dfrac{(2y+3)^2}{5y} \cdot \dfrac{15y^3}{4y^2 - 9}$

13. $\dfrac{w^2 - 16}{w} \cdot \dfrac{3}{4-w}$

14. $\dfrac{y^2 + 2y}{y^2 + y - 2} \div \dfrac{y-5}{y^2 + 4y - 5}$

15. $\dfrac{z^2 - z - 6}{z - 6} \cdot \dfrac{z^2 - 6z}{z^2 + 2z - 15}$

16. $\dfrac{2p^2 - 5p - 12}{5p^2 - 18p - 8} \cdot \dfrac{25p^2 - 4}{30p - 12}$

17. $\dfrac{m^3 - n^3}{m^2 - n^2} \div \dfrac{m^2 + mn + n^2}{m + n}$

18. What is *wrong* with the following work?

$$\frac{x^2 + 5x}{x + 5} = \frac{x^2}{x} + \frac{5x}{5} = x + x = 2x$$

What is the *correct* simplified form?

[6.2] *Find the least common denominator for each group of rational expressions.*

19. $\dfrac{7x}{12}, \quad \dfrac{8}{15x}$

20. $\dfrac{5a}{32b^3}, \quad \dfrac{31}{24b^5}$

21. $\dfrac{17}{9r^2}, \quad \dfrac{5r - 3}{3r + 1}$

22. $\dfrac{14}{6k + 3}, \quad \dfrac{7k^2 + 2k + 1}{10k + 5}, \quad \dfrac{-11k}{18k + 9}$

23. $\dfrac{4x - 9}{6x^2 + 13x - 5}, \quad \dfrac{x + 15}{9x^2 + 9x - 4}$

Add or subtract as indicated.

24. $\dfrac{2}{y} + \dfrac{3}{8y}$

25. $\dfrac{8}{z} - \dfrac{3}{2z^2}$

26. $\dfrac{3}{t-2} - \dfrac{5}{2-t}$

27. $\dfrac{5y + 13}{y + 1} - \dfrac{1 - 7y}{y + 1}$

28. $\dfrac{6}{5a + 10} + \dfrac{7}{6a + 12}$

29. $\dfrac{3x}{x - 5} + \dfrac{20}{x^2 - 25} + \dfrac{2}{x + 5}$

30. $\dfrac{3r}{10r^2 - 3rs - s^2} + \dfrac{2r}{2r^2 + rs - s^2}$

[6.3] *Simplify each complex fraction.*

31. $\dfrac{\dfrac{3}{t} + 2}{\dfrac{4}{t} - 7}$

32. $\dfrac{\dfrac{4m^5n^6}{mn}}{\dfrac{8m^7n^3}{m^4n^2}}$

33. $\dfrac{\dfrac{r + 2s}{20}}{\dfrac{8r + 16s}{5}}$

34. $\dfrac{3x^{-1} - 5}{6 + x^{-1}}$

35. $\dfrac{\dfrac{3}{p} - \dfrac{2}{q}}{\dfrac{9q^2 - 4p^2}{qp}}$

36. $\dfrac{\dfrac{1}{x} + \dfrac{2}{x^2} + \dfrac{1}{x^3}}{\dfrac{1}{x} - \dfrac{1}{x^3}}$

[6.4] *Solve each equation involving rational expressions.*

37. $\dfrac{4}{3} - \dfrac{1}{x} = \dfrac{1}{3x}$

38. $\dfrac{1}{t + 4} + \dfrac{1}{2} = \dfrac{3}{2t + 8}$

39. $\dfrac{-5m}{m + 1} + \dfrac{m}{3m + 3} = \dfrac{56}{6m + 6}$

40. $\dfrac{2}{k - 1} - \dfrac{4k + 1}{k^2 - 1} = \dfrac{-1}{k + 1}$

41. $\dfrac{x + 3}{x^2 - 5x + 4} - \dfrac{1}{x} = \dfrac{2}{x^2 - 4x}$

42. $\dfrac{5}{x + 2} + \dfrac{3}{x + 3} = \dfrac{x}{x^2 + 5x + 6}$

43. After solving the equation

$$\frac{3}{x-3} - \frac{2}{x-2} = \frac{3}{x^2 - 5x + 6},$$

a student got $x = 3$ as her final step. She could not understand why the answer in the back of the book was "∅," because she checked her algebra several times and was sure that all her algebraic work was correct. Was she wrong or was the answer in the back of the book wrong? Explain.

44. Explain the difference between simplifying the expression $\frac{4}{x} + \frac{1}{2} - \frac{1}{3}$ and solving the equation $\frac{4}{x} + \frac{1}{2} = \frac{1}{3}$.

[6.5] *Work each problem.*

45. According to a law from physics, $\frac{1}{A} = \frac{1}{B} + \frac{1}{C}$. Find A if $B = 30$ and $C = 10$.

46. Using the law given in Exercise 45, find B if $A = 10$ and $C = 15$.

Solve each formula for the specified variable.

47. $F = \dfrac{GMm}{d^2}$ for m (physics)

48. $S = \dfrac{n}{2}(a + \ell)$ for ℓ (mathematics)

49. $\mu = \dfrac{Mv}{M + m}$ for m (electronics)

50. $I = \dfrac{nE}{R + nr}$ for R (electricity)

[6.6] *Solve each problem.*

51. At a certain gasoline station, 3 gallons of unleaded gasoline cost $4.86. How much would 13 gallons of the same gasoline cost?

52. In a sample of 2000 registered voters, 1430 responded that they were in favor of increasing funding to the local animal shelter. If it is predicted that 15,000 people will vote in the election, how many will vote for this funding if the survey proves to be accurate?

53. A sink can be filled by a cold-water tap in 8 minutes, and filled by the hot-water tap in 12 minutes. How long would it take to fill the sink with both taps open?

54. Jane Ann Lindstedt and Greg Tobin need to sort a pile of bottles at the recycling center. Working alone, Jane could do the entire job in 9 hours, while Greg could do the entire job in 6 hours. How long will it take them if they work together?

55. A bus can travel 80 miles in the same time that a train goes 96 miles. The speed of the train is 10 miles per hour faster than the speed of the bus. Find both speeds.

56. A river has a current of 4 kilometers per hour. Find the speed of Lynn McTernan's boat in still water if it goes 40 kilometers downstream in the same time that it takes to go 24 kilometers upstream.

MIXED REVIEW EXERCISES

Perform the indicated operations.

57. $\dfrac{2}{m} + \dfrac{5}{3m^2}$

58. $\dfrac{k^2 - 6k + 9}{1 - 216k^3} \cdot \dfrac{6k^2 + 17k - 3}{9 - k^2}$

59. $\dfrac{\dfrac{-3}{x} + \dfrac{x}{2}}{1 + \dfrac{x + 1}{x}}$

60. $\dfrac{9x^2 + 46x + 5}{3x^2 - 2x - 1} \div \dfrac{x^2 + 11x + 30}{x^3 + 5x^2 - 6x}$

61. $\dfrac{9}{3 - x} - \dfrac{2}{x - 3}$

62. $\dfrac{4y + 16}{30} \div \dfrac{2y + 8}{5}$

63. $\dfrac{t^{-2} + s^{-2}}{t^{-1} - s^{-1}}$

64. $\dfrac{4a}{a^2 - ab - 2b^2} - \dfrac{6b}{a^2 + 4ab + 3b^2}$

Solve.

65. $A = \dfrac{Rr}{R + r}$ for r (mathematics)

66. $1 - \dfrac{5}{r} = \dfrac{-4}{r^2}$

67. $\dfrac{3x}{x - 4} + \dfrac{2}{x} = \dfrac{48}{x^2 - 4x}$

68. The hot-water tap can fill a tub in 20 minutes. The cold-water tap takes 15 minutes to fill the tub. How long would it take to fill the tub with both taps open?

1. Find all real numbers that make the following rational function undefined: $f(x) = \dfrac{2x - 1}{3x^2 + 2x - 8}$.

   1. _____

2. Write the following rational expression in lowest terms: $\dfrac{6x^2 - 13x - 5}{9x^3 - x}$.

   2. _____

Multiply or divide.

3. $\dfrac{4x^2y^5}{7xy^8} \div \dfrac{8xy^6}{21xy}$

   3. _____

4. $\dfrac{y^2 - 16}{y^2 - 25} \cdot \dfrac{y^2 + 2y - 15}{y^2 - 7y + 12}$

   4. _____

5. $\dfrac{x^2 - 9}{x^3 + 3x^2} \div \dfrac{x^2 + x - 12}{x^3 + 9x^2 + 20x}$

   5. _____

6. Find the least common denominator for the following group of rational expressions: $\dfrac{t}{t^2 + t - 6}, \dfrac{17t^2}{t^2 + 3t}, \dfrac{-1}{t^2}$.

   6. _____

Add or subtract as indicated.

7. $\dfrac{7}{6t^2} - \dfrac{1}{3t}$

   7. _____

8. $\dfrac{9}{x - 7} + \dfrac{4}{x + 7}$

   8. _____

9. $\dfrac{6}{x + 4} + \dfrac{1}{x + 2} - \dfrac{3x}{x^2 + 6x + 8}$

   9. _____

Simplify each complex fraction.

10. $\dfrac{\dfrac{12}{r + 4}}{\dfrac{11}{6r + 24}}$

    10. _____

11. $\dfrac{\dfrac{1}{a} - \dfrac{1}{b}}{\dfrac{a}{b} - \dfrac{b}{a}}$

    11. _____

12. (a) _____

(b) _____

12. One of the following is an expression to be simplified by algebraic operations, and the other is an equation to be solved. Simplify the expression, and solve the equation.

(a) $\dfrac{2x}{3} + \dfrac{x}{4} - \dfrac{11}{2}$ **(b)** $\dfrac{2x}{3} + \dfrac{x}{4} = \dfrac{11}{2}$

Solve each of the following equations involving rational expressions.

13. _____

13. $1 - \dfrac{3}{y} = \dfrac{1}{2}$

14. _____

14. $\dfrac{1}{x} - \dfrac{4}{3x} = \dfrac{1}{x - 2}$

15. _____

15. $\dfrac{y}{y + 2} - \dfrac{1}{y - 2} = \dfrac{8}{y^2 - 4}$

16. _____

16. Solve for the variable r in this formula that comes from the field of electronics: $I = \dfrac{En}{Rn + r}$.

17. _____

17. A formula involving pulleys is $P = \dfrac{W(R - r)}{2R}$. Find R if $W = 120$, $r = 8$, and $P = 10$.

Solve each of the following problems.

18. _____

18. Wayne can do a job in 9 hours, while Susan can do the same job in 5 hours. How long would it take them to do the job if they worked together?

19. _____

19. The current of a river runs at 3 miles per hour. Nana's boat can go 36 miles downstream in the same time that it takes to go 24 miles upstream. Find the speed of the boat in still water.

20. _____

20. Biologists collected a sample of 600 fish from Lake Linda on May 1 and tagged each of them. When they returned on June 1, they collected a new sample of 800 fish, and 10 of these had been previously tagged. Use this experiment to determine the approximate fish population of Lake Linda.

1. Solve for x: $3x + 2(x - 4) = 5x - 8$.

2. Solve the inequality, and graph its solution set.

$$-3 \le \frac{2}{3}x - 1 \le 1$$

3. Otis Taylor invested some money at 4% interest and twice as much at 3% interest. His interest for the first year was $400. How much did he invest at each rate?

4. A student must have an average grade of at least 80 on the four tests in a course to earn a grade of B. David Hingle had grades of 79, 75, and 88 on the first three tests. What possible scores can he make on the fourth test so that he can get a B in the course?

5. Solve for t: $|6t - 4| + 8 = 3$.

6. Evaluate $-4^2 + (-4)^2$.

7. Solve for x: $3x - 5 \ge 1$ or $2x + 7 \le 9$.

8. Solve for t: $.04t + .06(t - 1) = 1.04$.

9. Solve and graph the solution set.

$$4 - 7(q + 4) < -3q$$

10. Solve the compound inequality and graph the solution set.

$$3x - 2 < 10 \text{ and } -2x < 10$$

Find the intercepts of each line.

11. $-4x + 5y = 20$

12. $y = -4$

13. $x = 3y$

14. Solve: $5x + 2y = 16$
$3x - 3y = 18.$

15. If $f(x) = \dfrac{x + 1}{3 - x}$, find $f(2)$.

Write in scientific notation.

16. .000076

Write without scientific notation.

17. 5.6×10^9

Simplify. Write answers with only positive exponents. Assume that all variables represent nonzero real numbers.

18. $\dfrac{3x^{-4}y^3}{7^{-1}x^5y^{-4}}$

19. $\left(\dfrac{a^{-3}b^4}{a^2b^{-1}}\right)^{-2}$

20. $\left(\dfrac{m^{-4}n^2}{m^2n^{-3}}\right) \cdot \left(\dfrac{m^5n^{-1}}{m^{-2}n^5}\right)$

Perform the indicated operations.

21. $9(-3x^3 - 4x + 12) + 2(x^3 - x^2 + 3)$

22. $(4f + 3)(3f - 1)$

23. $(x + y)(x^2 - xy + y^2)$

24. $(7t^3 + 8)(7t^3 - 8)$

25. $\left(\dfrac{1}{4}x + 5\right)^2$

26. If $P(x) = -4x^3 + 2x - 8$, find $P(-2)$.

27. Use synthetic division to divide $(2x^4 + 3x^3 - 8x^2 + x + 2)$ by $(x - 1)$.

Factor each of the following polynomials completely.

28. $2x^2 - 13x - 45$

29. $100t^4 - 25$

30. Solve the equation $3x^2 + 4x = 7$.

31. For what values of x is

$$f(x) = \dfrac{x + 8}{x^2 - 3x - 4}$$

undefined?

32. Write in lowest terms: $\dfrac{8x^2 - 18}{8x^2 + 4x - 12}$.

Perform the indicated operations. Express answers in lowest terms.

33. $\dfrac{x + 4}{x - 2} + \dfrac{2x - 10}{x - 2}$

34. $\dfrac{2}{a + b} - \dfrac{3}{a - b}$

35. $\dfrac{2x}{2x - 1} + \dfrac{4}{2x + 1} + \dfrac{8}{4x^2 - 1}$

36. $\dfrac{5}{x^3 - y^3} + \dfrac{3}{x^2 + xy + y^2}$

37. Solve the equation $\dfrac{-3x}{x+1} + \dfrac{4x+1}{x} = \dfrac{-3}{x^2+x}$.

38. Solve the formula for q: $\dfrac{1}{f} = \dfrac{1}{p} + \dfrac{1}{q}$.

Solve each problem.

39. Lucinda can fly her plane 200 miles against the wind in the same time it takes her to fly 300 miles with the wind. The wind blows at 30 miles per hour. Find the speed of her plane in still air.

40. Machine A can complete a certain job in 2 hours. To speed up the work, Machine B, which could complete the job alone in 3 hours, is brought in to help. How long will it take the two machines to complete the job working together?

Roots and Radicals

7.1 Radicals

OBJECTIVE 1 We found roots of numbers in Chapter 1. Recall that to find the square root of a number a, we must find the number that was squared to get a.

Square Root

The number b is a **square root** of a if $b^2 = a$.

E X A M P L E I   Finding Square Roots of a Number

Find the real square roots of each number.

(a) 36

Since $6^2 = 36$, and $(-6)^2 = 36$, both 6 and -6 are square roots of 36.

(b) 100

Both 10 and -10 are square roots of 100.

(c) -4

There is no real number that can be squared to give -4, so -4 has no real number square root.

WORK PROBLEM I AT THE SIDE. ▶▶

A square root can be written with a **radical** (RAD-ih-kul) **sign**, $\sqrt{}$. For example, $\sqrt{64} = 8$, since $8^2 = 64$. While 0 has only one square root ($\sqrt{0} = 0$), a given positive number has two square roots, one positive and one negative.

The symbol $\sqrt{a}$ is used only for the *nonnegative* square root of a; the negative square root of a is written $-\sqrt{a}$.

As an abbreviation, the two square roots of the positive number a are sometimes written together as $\pm\sqrt{a}$, with the sign $\pm$ read "plus or minus."

www.mathnotes.com

1. Find all real square roots.

(a) 64

(b) 169

(c) -9

ANSWERS

1. (a) 8, -8 **(b)** 13, -13
(c) no real number square roots

2. Simplify.

(a) $\sqrt[3]{8}$

(b) $\sqrt[3]{1000}$

(c) $\sqrt[3]{-1}$

(d) $\sqrt[4]{81}$

(e) $\sqrt[4]{-1}$

(f) $\sqrt[6]{64}$

3. Find each root.

(a) $\sqrt{4}$

(b) $\sqrt[3]{27}$

(c) $-\sqrt{36}$

(d) $\sqrt[4]{625}$

(e) $\sqrt[5]{-32}$

ANSWERS
2. (a) 2 (b) 10 (c) −1 (d) 3
(e) not a real number (f) 2
3. (a) 2 (b) 3 (c) −6 (d) 5 (e) −2

OBJECTIVE 2 Radical signs also can be used with higher roots, such as **cube roots, fourth roots,** and so on. By definition,

$$\sqrt[n]{a} = b \quad \text{means} \quad b^n = a.$$

n is called the **index,** a is called the **radicand,** and the entire expression $\sqrt[n]{a}$ is called a **radical.**

Both 2 and −2 are fourth roots of 16; however, the symbol $\sqrt[4]{16}$ is used for the *positive* root.

EXAMPLE 2 Simplifying Higher Roots

Simplify.
(a) $\sqrt[3]{27} = 3$, because $3^3 = 27$.
(b) $\sqrt[3]{125} = 5$, because $5^3 = 125$.
(c) $\sqrt[3]{-216} = -6$, because $(-6)^3 = -216$.
(d) $\sqrt[4]{16} = 2$, because $2^4 = 16$.
(e) $\sqrt[4]{-16}$ is not a real number.
(f) $\sqrt[5]{32} = 2$, because $2^5 = 32$.

◄◄ WORK PROBLEM 2 AT THE SIDE.

OBJECTIVE 3 When a number has two roots, the positive root is called the **principal root.**

$\sqrt[n]{a}$ represents the *principal* nth root of a, and $-\sqrt[n]{a}$ is the negative nth root of a.
If a is positive, then $\sqrt[n]{a}$ is positive.
If a is negative, then

$\sqrt[n]{a}$ is negative when n is odd,
$\sqrt[n]{a}$ is not a real number when n is even.

EXAMPLE 3 Finding Principal Roots

Find each root.
(a) $\sqrt{100} = 10$
Here the radicand 100 is positive. There are two square roots, 10 and −10, but 10 is the principal root.
(b) $\sqrt[4]{81} = 3$
(c) $-\sqrt{16} = -4$
(d) $\sqrt[6]{-8}$
The index is even and the radicand is negative, so the principal root is not a real number.
(e) $\sqrt[3]{-8} = -2$ because $(-2)^3 = -8$.

◄◄ WORK PROBLEM 3 AT THE SIDE.

A square root of a^2 (where $a \neq 0$) is either a or $-a$. Since one root is negative and one is positive, and since the symbol $\sqrt{a^2}$, represents the *nonnegative* square root, this root is written with absolute value bars as $|a|$.

EXAMPLE 4 Simplifying Square Roots Using Absolute Value

Find each square root that is a real number.

(a) $\sqrt{7^2} = |7| = 7$

(b) $\sqrt{(-7)^2} = |-7| = 7$

(c) $\sqrt{k^2} = |k|$

(d) $\sqrt{(-k)^2} = |-k| = |k|$

WORK PROBLEM 4 AT THE SIDE. ▶▶

As shown above, $\sqrt{a^2} = |a|$. The fourth root of a^4, the sixth root of a^6, and other even nth roots of a^n also must be written as absolute values. For example, $\sqrt[4]{a^4} = |a|$ and $\sqrt[6]{a^6} = |a|$. On the other hand, the cube (or third) root of a positive number is positive and the cube root of a negative number is negative, so $\sqrt[3]{a^3} = a$ whether a is positive or negative. All odd roots behave like cube roots. These examples suggest the following rule.

Rules for $\sqrt[n]{a^n}$

If n is an **even** positive integer, then $\sqrt[n]{a^n} = |a|$,
and if n is an **odd** positive integer, then $\sqrt[n]{a^n} = a$.

Use absolute value when n is even; do not use absolute value when n is odd.

EXAMPLE 5 Simplifying Higher Roots Using Absolute Value

Simplify each root.

(a) $\sqrt[6]{(-3)^6} = |-3| = 3$

(b) $\sqrt[5]{(-4)^5} = -4$

(c) $-\sqrt[4]{(-9)^4} = -|-9| = -9$

(d) $\sqrt{\dfrac{9}{16}} = \dfrac{3}{4}$

(e) $-\sqrt{m^4} = -|m^2| = -m^2$

No absolute value bars are needed here, because m^2 is nonnegative for any real number value of m.

(f) $\sqrt[3]{a^{12}} = a^4$ (because $a^{12} = (a^4)^3$)

(g) $\sqrt[4]{x^{12}} = |x^3|$

We use absolute value bars to guarantee that the result is not negative (because x^3 can be either positive or negative, depending on x). If desired, $|x^3|$ can be written as $x^2 \cdot |x|$.

WORK PROBLEM 5 AT THE SIDE. ▶▶

4. Simplify.

(a) $\sqrt{49}$

(b) $-\sqrt{\dfrac{36}{25}}$

(c) $\sqrt{(-6)^2}$

(d) $\sqrt{r^2}$

5. Simplify.

(a) $\sqrt[6]{64}$

(b) $-\sqrt[4]{16}$

(c) $\sqrt[3]{\dfrac{216}{125}}$

(d) $\sqrt[5]{-243}$

(e) $\sqrt[6]{(-p)^6}$

(f) $-\sqrt[6]{y^{24}}$

ANSWERS

4. (a) 7 (b) $-\dfrac{6}{5}$ (c) 6 (d) $|r|$

5. (a) 2 (b) -2 (c) $\dfrac{6}{5}$ (d) -3

(e) $|p|$ (f) $-y^4$

6. Find a decimal approximation for each of the following to three decimal places.

(a) $\sqrt{10}$

(b) $\sqrt{51}$

(c) $-\sqrt{99}$

(d) $\sqrt{950}$

(e) $-\sqrt{670}$

A function defined by

$$f(x) = \sqrt[n]{x}$$

for natural number values of n is the simplest example of a **root function** or **radical function.** We will investigate certain kinds of root functions further in Chapter 9. If n is even, the domain is $\{x \mid x \geq 0\}$. If n is odd, the domain is the set of real numbers.

OBJECTIVE ▶**4** Not all square roots are rational numbers. For example, there is no rational number whose square is 15, so $\sqrt{15}$ is not a rational number. (Recall from Chapter 1 that $\sqrt{15}$ is an *irrational* number.) We can find an approximation of $\sqrt{15}$ by using a calculator.

To use a calculator to approximate $\sqrt{15}$, enter the number 15 and press the key marked $\sqrt{}$. (This procedure may not apply to some more sophisticated calculators.) The display will then read 3.8729833. (There may be fewer or more decimal places, depending upon the model of calculator used.) In this book we will show approximations correct to three decimal places. Therefore,

$$\sqrt{15} \approx 3.873,$$

where $\approx$ means "is approximately equal to."

There is a simple way to check that a calculator approximation is "in the ballpark." Because 16 is a little larger than 15, $\sqrt{16} = 4$ should be a little larger than $\sqrt{15}$. Thus, 3.873 is a reasonable approximation of $\sqrt{15}$.

E X A M P L E   6   **Finding Approximations for Square Roots**

Find a decimal approximation for each of the following, using a calculator.

(a) $\sqrt{39} \approx 6.245$

(b) $\sqrt{83} \approx 9.110$

(c) $-\sqrt{72} \approx -8.485$

To find the negative square root, we first find the positive square root and then its negative.

(d) $\sqrt{770} \approx 27.749$

(e) $-\sqrt{420} \approx -20.494$

◀◀ **WORK PROBLEM 6 AT THE SIDE.**

As mentioned earlier, fourth roots can be found by taking the square root twice. Some calculators have cube root keys or *n*th root keys. Scientific calculators have a key marked $\boxed{y^x}$ or $\boxed{x^y}$ for finding exponentials. With many scientific calculators, this key can be used with the key marked $\boxed{\text{INV}}$, or 2nd , to find roots. For example, the sequence of keystrokes

$$\boxed{4}\boxed{6}\boxed{7}\boxed{\text{INV}}\boxed{y^x}\boxed{5}\boxed{=}$$

will display 3.4187188 as an approximation of $\sqrt[5]{467}$.

ANSWERS

6. (a) 3.162 **(b)** 7.141
(c) −9.950 **(d)** 30.822 **(e)** −25.884

7.1 Exercises

Fill in the blanks with the correct responses.

1. The principal square root of 36 is _____ .

2. The negative square root of 36 is _____ .

3. The principal cube root of 27 is _____ .

4. The negative fourth root of 81 is _____ .

5. If a whole number is a four-digit perfect square and its units digit is 5, what is the units digit of its real square roots?

6. If a whole number is a four-digit perfect square and its units digit is 9, what are the possibilities for the units digit of its real square roots?

Find all real square roots of each number. Use a calculator as needed. See Example 1.

7. 4   8. 9   9. 121   10. 144

11. 1764   12. 3364   13. -225 14. -6241

Use estimation to choose the closest approximation of each square root.

15. $\sqrt{123.5}$
    (a) 9   (b) 10   (c) 11   (d) 12

16. $\sqrt{67.8}$
    (a) 7   (b) 8   (c) 9   (d) 10

Suppose that a rectangle has length $\sqrt{98}$ and width $\sqrt{26}$.

17. Which one of the following is the best estimate of its area?
    (a) 2500   (b) 250   (c) 50   (d) 100

18. Which one of the following is the best estimate of its perimeter?
    (a) 15   (b) 250   (c) 100   (d) 30

Find each square root that is a real number. Use a calculator as needed. See Example 1.

19. $\sqrt{36}$ 20. $\sqrt{100}$ 21. $\sqrt{2209}$ 22. $\sqrt{3721}$ 23. $\sqrt{-81}$

24. $\sqrt{-25}$ 25. $-\sqrt{1444}$ 26. $-\sqrt{4624}$ 27. $-\sqrt{-169}$ 28. $-\sqrt{-400}$

29. Consider the expression $-\sqrt{-a}$. Decide whether it is positive, negative, zero, or not a real number if
    (a) $a > 0$ (b) $a < 0$ (c) $a = 0$.

30. If n is odd, under what conditions is $\sqrt[n]{a}$
    (a) positive   (b) negative   (c) zero?

Find each root that is a real number. Use a calculator as necessary. See Examples 2 and 3.

31. $-\sqrt{81}$

32. $-\sqrt{121}$

33. $\sqrt[3]{216}$

34. $\sqrt[3]{343}$

35. $\sqrt[3]{-64}$

36. $\sqrt[3]{-125}$

37. $-\sqrt[3]{512}$

38. $-\sqrt[3]{1000}$

39. $\sqrt[4]{1296}$

40. $\sqrt[4]{625}$

41. $-\sqrt[4]{81}$

42. $-\sqrt[4]{256}$

43. $\sqrt[4]{-16}$

44. $\sqrt[4]{-81}$

45. $\sqrt[6]{(-2)^6}$

46. $\sqrt[6]{(-4)^6}$

47. $\sqrt[5]{(-9)^5}$

48. $\sqrt[5]{(-8)^5}$

49. $\sqrt{\dfrac{64}{81}}$

50. $\sqrt{\dfrac{100}{9}}$

Simplify each root. See Examples 4 and 5.

51. $\sqrt{x^2}$

52. $-\sqrt{x^2}$

53. $\sqrt[3]{x^3}$

54. $-\sqrt[3]{x^3}$

55. $\sqrt[3]{x^{15}}$

56. $\sqrt[4]{k^{20}}$

Find a decimal approximation for each of the following. Round answers to three decimal places. See Example 6.

57. $\sqrt{9483}$

58. $\sqrt{6825}$

59. $\sqrt{284.361}$

60. $\sqrt{846.104}$

MATHEMATICAL CONNECTIONS (EXERCISES 61–70)

Every positive number a has two even nth roots, the principal root $\sqrt[n]{a}$, which is positive, and a negative root $-\sqrt[n]{a}$. The following exercises, which should be worked in order, explore connections between these roots.

61. Find the square roots of 16.

62. Find the principal square root of 16.

63. Find $\sqrt{16}$ and $-\sqrt{16}$. Which of these is the principal square root?

64. What is the solution set of $x^2 = 16$?

65. Find the fourth roots of 81.

66. Find the principal fourth root of 81.

67. Find $\sqrt[4]{81}$ and $-\sqrt[4]{81}$. Which one is the principal fourth root?

68. Give the solution set of $x^4 = 81$.

69. Explain what is meant by $\pm\sqrt{25}$.

70. Explain why $\sqrt[4]{x^4}$ is simplified as $|x|$, but $\sqrt[3]{x^3}$ is simply x.

7.2 *Rational Exponents*

OBJECTIVE ▶ In this section we extend the definitions and rules for exponents to include rational exponents as well as integer exponents. If the rules for exponents are still to be valid, then we should find the product $(3^{1/2})^2 = 3^{1/2} \cdot 3^{1/2}$ by adding exponents.

$$(3^{1/2})^2 = 3^{1/2} \cdot 3^{1/2}$$
$$= 3^{1/2+1/2}$$
$$= 3^1 = 3$$

However, by definition $(\sqrt{3})^2 = \sqrt{3} \cdot \sqrt{3} = 3$. Since both $(3^{1/2})^2$ and $(\sqrt{3})^2$ are equal to 3, we must have

$$3^{1/2} = \sqrt{3}.$$

Generalizing from this example, $a^{1/n}$ is defined as the principal nth root of a.

When $\sqrt[n]{a}$ is real,

$$a^{1/n} = \sqrt[n]{a}.$$

E X A M P L E 1 **Evaluating Exponentials of the Form $a^{1/n}$**

(a) $64^{1/3} = \sqrt[3]{64} = 4$

(b) $100^{1/2} = \sqrt{100} = 10$

(c) $-256^{1/4} = -\sqrt[4]{256} = -4$

(d) $(-25)^{1/2} = \sqrt{-25}$ is not real because $\sqrt{-25}$ is not a real number.

(e) $m^{1/5} = \sqrt[5]{m}$

> **WORK PROBLEM 1 AT THE SIDE.** ▶▶

OBJECTIVE ▶ What about the symbol $8^{2/3}$? For past rules of exponents to be valid,

$$8^{2/3} = 8^{(1/3)2} = (8^{1/3})^2.$$

Because $8^{1/3} = \sqrt[3]{8}$,

$$8^{2/3} = (\sqrt[3]{8})^2 = 2^2 = 4.$$

Generalizing, $a^{m/n}$ is defined as follows.

If m and n are positive integers with m/n in lowest terms, then

$$a^{m/n} = (a^{1/n})^m$$

provided that $a^{1/n}$ is a real number. If $a^{1/n}$ is not a real number, then $a^{m/n}$ is not real.

1. Simplify in parts (a)–(d). In part (e), write as a radical.

(a) $8^{1/3}$

(b) $9^{1/2}$

(c) $-81^{1/4}$

(d) $(-16)^{1/4}$

(e) $a^{1/7}$

ANSWERS

1. (a) 2 **(b)** 3 **(c)** −3
  (d) not a real number **(e)** $\sqrt[7]{a}$

2. Evaluate.

(a) $64^{2/3}$

(b) $100^{3/2}$

(c) $-16^{3/4}$

(d) $(-16)^{3/4}$

E X A M P L E   2   **Evaluating Exponentials of the Form $a^{m/n}$**

(a) $36^{3/2} = (36^{1/2})^3 = 6^3 = 216$

(b) $125^{2/3} = (125^{1/3})^2 = 5^2 = 25$

(c) $-4^{5/2} = -(4^{5/2}) = -(4^{1/2})^5 = -(2)^5 = -32$

(d) $(-27)^{2/3} = [(-27)^{1/3}]^2 = (-3)^2 = 9$

Notice how the $-$ sign is used in (c) and in (d). In (c), we first evaluate the exponential and then find its negative. In (d), the $-$ sign is part of the base, -27.

(e) $(-100)^{3/2}$ is not a real number, since $(-100)^{1/2}$ is not a real number.

◀◀ **WORK PROBLEM 2 AT THE SIDE.**

E X A M P L E   3   **Evaluating Exponentials with Negative Rational Exponents**

Simplify each of the following.

(a) $16^{-3/4}$

By the definition of a negative exponent,

$$16^{-3/4} = \frac{1}{16^{3/4}}.$$

Since $16^{3/4} = (\sqrt[4]{16})^3 = 2^3 = 8,$

$$16^{-3/4} = \frac{1}{16^{3/4}} = \frac{1}{8}.$$

(b) $25^{-3/2} = \frac{1}{25^{3/2}} = \frac{1}{(\sqrt{25})^3} = \frac{1}{5^3} = \frac{1}{125}$

(c) $\left(\frac{8}{27}\right)^{-2/3} = \frac{1}{\left(\frac{8}{27}\right)^{2/3}} = \frac{1}{\left(\frac{2}{3}\right)^2} = \frac{1}{\frac{4}{9}} = \frac{9}{4}$

We could also use the rule $\left(\frac{b}{a}\right)^{-m} = \left(\frac{a}{b}\right)^m$ here as follows.

$$\left(\frac{8}{27}\right)^{-2/3} = \left(\frac{27}{8}\right)^{2/3} = \left(\frac{3}{2}\right)^2 = \frac{9}{4}$$

---

3. Evaluate.

(a) $36^{-3/2}$

(b) $32^{-4/5}$

(c) $\left(\frac{4}{9}\right)^{-5/2}$

Caution

When using the rule in Example 3(c), we take the reciprocal only of the base, *not* the exponent. Also, be careful to distinguish between exponential expressions like $-16^{1/4}$, $16^{-1/4}$, and $-16^{-1/4}$.

$$-16^{1/4} = -2, \quad 16^{-1/4} = 1/2, \quad \text{and} \quad -16^{-1/4} = -1/2.$$

◀◀ **WORK PROBLEM 3 AT THE SIDE.**

The exponential expression $a^{m/n}$ was defined as $(a^{1/n})^m$. An alternative definition can be obtained by using the power rule for exponents in another way. If all indicated roots are real,

$$a^{m/n} = a^{m(1/n)}$$
$$= (a^m)^{1/n},$$

ANSWERS

2. (a) 16 **(b)** 1000 **(c)** -8
    (d) not a real number

3. (a) $\frac{1}{216}$ **(b)** $\frac{1}{16}$ **(c)** $\frac{243}{32}$

so that

$$a^{m/n} = (a^m)^{1/n}.$$

With this result, $a^{m/n}$ can be defined in either of two ways.

If all indicated roots are real, then

$$a^{m/n} = (a^{1/n})^m = (a^m)^{1/n}.$$

We can now evaluate an expression such as $27^{2/3}$ in two ways:

$$27^{2/3} = (27^{1/3})^2 = 3^2 = 9$$
$$27^{2/3} = (27^2)^{1/3} = 729^{1/3} = 9.$$

In most cases, it is easier to use $(a^{1/n})^m$.

This rule can also be expressed with radicals as follows.

If all indicated roots are real, then

$$a^{m/n} = \sqrt[n]{a^m} = (\sqrt[n]{a})^m.$$

We can raise to a power and then take the root, or take the root and then raise to a power.

For example, $8^{2/3} = \sqrt[3]{8^2} = \sqrt[3]{64} = 4,$
and $\qquad\quad 8^{2/3} = (\sqrt[3]{8})^2 = 2^2 = 4,$
so $\qquad\qquad 8^{2/3} = \sqrt[3]{8^2} = (\sqrt[3]{8})^2.$

OBJECTIVE 3 With the definition of a rational exponent given above, all the rules for exponents are still valid. They are repeated here for reference.

Definitions and Rules for Exponents

If a and b are real numbers, all powers are real, and m and n are rational numbers:

Product rule	$a^m \cdot a^n = a^{m+n}$	
Quotient rule	$\dfrac{a^m}{a^n} = a^{m-n}$	$(a \neq 0)$
Power rules	$(a^m)^n = a^{mn}$	
	$(ab)^m = a^m b^m$	
	$\left(\dfrac{a}{b}\right)^m = \dfrac{a^m}{b^m}$	$(b \neq 0)$
Zero exponent	$a^0 = 1$	$(a \neq 0)$
Negative exponent	$a^{-n} = \dfrac{1}{a^n}$	$(a \neq 0)$

The next example shows applications of these rules.

4. Simplify.

    (a) $11^{3/4} \cdot 11^{5/4}$

    (b) $\dfrac{7^{3/4}}{7^{7/4}}$

    (c) $\dfrac{9^{2/3}}{9^{-1/3}}$

    (d) $(x^{3/2})^4,\ x > 0$

5. Simplify. Assume all variables represent positive real numbers.

    (a) $\sqrt{y^{10}}$

    (b) $\sqrt[6]{27y^9}$

    (c) $\sqrt[5]{y} \cdot \sqrt[10]{y^7}$

    (d) $\dfrac{\sqrt[3]{m^2}}{\sqrt[5]{m}}$

E X A M P L E 4   **Applying Exponent Rules to Rational Exponents**

(a) $3^{3/2} \cdot 3^{1/2} = 3^{3/2+1/2} = 3^2 = 9$ Product rule

(b) $\dfrac{5^{2/3}}{5^{5/3}} = 5^{2/3-5/3} = 5^{-1} = \dfrac{1}{5}$ Quotient rule

(c) $6^{5/8} \cdot 6^{-3/8} = 6^{5/8+(-3/8)} = 6^{1/4}$ Product rule

(d) $(3^{4/3})^5 = 3^{(4/3)\cdot5} = 3^{20/3}$ Power rule

(e) $\dfrac{(y^{2/3})^4}{y} = \dfrac{y^{8/3}}{y^1} = y^{(8/3)-1} = y^{5/3}$ $(y \neq 0)$ Power rule; quotient rule

---

Caution

Errors often occur in exercises like the ones in Example 4 because students try to convert the expressions to radical form. Remember that the *rules of exponents* apply here.

◀◀ **WORK PROBLEM 4 AT THE SIDE.**

OBJECTIVE ▶ Using the definition of rational exponents, we can simplify many problems involving radicals by converting the radicals to numbers with rational exponents. After simplifying, we convert the answer back to radical form.

    In the next example, we assume that all variables represent positive real numbers. We make this assumption in order to eliminate the need to use absolute value symbols when the root index is even.

E X A M P L E 5   **Converting between Radicals and Rational Exponents**

Simplify. Assume all variables represent positive real numbers.

(a) $\sqrt{m^4} = m^{4/2} = m^2$

(b) $\sqrt[4]{25y^2} = \sqrt[4]{5^2y^2}$
$$= (5^2y^2)^{1/4} = 5^{1/2}y^{1/2} = (5y)^{1/2} = \sqrt{5y}$$

(c) $\sqrt[3]{w} \cdot \sqrt[4]{w} = w^{1/3} \cdot w^{1/4}$
$$= w^{1/3+1/4} = w^{7/12} = \sqrt[12]{w^7}$$

(d) $\sqrt[8]{k^5} \cdot \sqrt{k} \cdot \sqrt[4]{k^3} = k^{5/8} \cdot k^{1/2} \cdot k^{3/4}$
$$= k^{15/8} = k^{1+7/8} = k \cdot k^{7/8} = k\sqrt[8]{k^7}$$

(e) $\dfrac{\sqrt{x^3}}{\sqrt[3]{x^2}} = \dfrac{x^{3/2}}{x^{2/3}} = x^{3/2-2/3}$
$$= x^{5/6} = \sqrt[6]{x^5}$$

◀◀ **WORK PROBLEM 5 AT THE SIDE.**

7.2 Exercises

Decide whether each statement is true or false. If the statement is false, explain why.

1. To evaluate $8^{2/3}$, we first find the cube root of 8, then square that result to get $2^2 = 4$.

2. The only way to evaluate $8^{2/3}$ is to proceed as in Exercise 1.

3. $\sqrt{3^2 + 4^2} = 3 + 4$

4. In general, $\sqrt{m^2 + n^2} \neq m + n$.

5. Which one of the following is a positive number?
    (a) $(-27)^{2/3}$ **(b)** $(-64)^{5/3}$ **(c)** $(-100)^{1/2}$ **(d)** $(-32)^{1/5}$

6. Explain why $(-64)^{1/2}$ is not a real number, while $-64^{1/2}$ is a real number.

Simplify each expression involving rational exponents. See Examples 1–3.

7. $169^{1/2}$ **8.** $121^{1/2}$ **9.** $729^{1/3}$ **10.** $512^{1/3}$

11. $16^{1/4}$ **12.** $625^{1/4}$ **13.** $\left(\dfrac{64}{81}\right)^{1/2}$ **14.** $\left(\dfrac{8}{27}\right)^{1/3}$

15. $(-27)^{1/3}$ **16.** $(-32)^{1/5}$ **17.** $100^{3/2}$ **18.** $64^{3/2}$

19. $-144^{1/2}$ **20.** $-32^{1/5}$ **21.** $(-144)^{1/2}$ **22.** $(-36)^{1/2}$

23. $64^{-3/2}$ **24.** $81^{-3/2}$ **25.** $\left(\dfrac{625}{16}\right)^{-1/4}$ **26.** $\left(\dfrac{36}{25}\right)^{-3/2}$

27. $\left(-\dfrac{8}{27}\right)^{-2/3}$ **28.** $\left(-\dfrac{64}{125}\right)^{-2/3}$ **29.** $\left(-\dfrac{4}{9}\right)^{-1/2}$ **30.** $\left(-\dfrac{16}{25}\right)^{-1/2}$

    The exponential key on a scientific calculator, usually marked $\boxed{y^x}$ or $\boxed{x^y}$, can often be used to evaluate expressions involving rational exponents. For example, to evaluate $32^{4/5}$, a typical sequence of keystrokes would be as follows:

$\boxed{3}\,\boxed{2}\,\boxed{y^x}\,\boxed{(}\,\boxed{4}\,\boxed{\div}\,\boxed{5}\,\boxed{)}\,\boxed{=}$ $\boxed{16}$ Displayed answer

Or, because $4/5 = .8$ as a decimal number, the following keystrokes would give the same result:

$\boxed{3}\,\boxed{2}\,\boxed{y^x}\,\boxed{.}\,\boxed{8}\,\boxed{=}$ $\boxed{16}$ Displayed answer

Use a calculator and one of the methods described above to evaluate each of the following in Exercises 31–36. Compare your answer to the answer found in the exercise or example noted in parentheses. (In Example 3(a), convert the answer in the example to a decimal to check.)

31. $169^{1/2}$ (Exercise 7)   **32.** $121^{1/2}$ (Exercise 8)   **33.** $100^{3/2}$ (Exercise 17)

34. $64^{3/2}$ (Exercise 18)   **35.** $125^{2/3}$ (Example 2(b))   **36.** $16^{-3/4}$ (Example 3(a))

Use the rules of exponents to simplify each of the following. Write all answers with positive exponents. Assume that all variables represent positive real numbers. See Example 4.

37. $3^{1/2} \cdot 3^{3/2}$ **38.** $6^{4/3} \cdot 6^{2/3}$ **39.** $\dfrac{64^{5/3}}{64^{4/3}}$ **40.** $\dfrac{125^{7/3}}{125^{5/3}}$

41. $y^{7/3} \cdot y^{-4/3}$ **42.** $r^{-8/9} \cdot r^{17/9}$ **43.** $\dfrac{k^{1/3}}{k^{2/3} \cdot k^{-1}}$ **44.** $\dfrac{z^{3/4}}{z^{5/4} \cdot z^{-2}}$

45. $(27x^{12}y^{15})^{2/3}$ **46.** $(64p^4q^6)^{3/2}$ **47.** $\dfrac{(x^{2/3})^2}{(x^2)^{7/3}}$ **48.** $\dfrac{(p^3)^{1/4}}{(p^{5/4})^2}$

49. $\dfrac{m^{3/4}n^{-1/4}}{(m^2n)^{1/2}}$

50. $\dfrac{(a^2b^5)^{-1/4}}{(a^{-3}b^2)^{1/6}}$

51. $\left(\dfrac{b^{-3/2}}{c^{-5/3}}\right)^2 \cdot (b^{-1/4}c^{-1/3})^{-1}$

52. $\left(\dfrac{m^{-2/3}}{a^{-3/4}}\right)^4 \cdot (m^{-3/8}a^{1/4})^{-2}$

53. Observe the following simplification:
$$\sqrt[3]{5^{12}} = 5^{12/3} = 5^4 = 625.$$

Now fill in the blanks with the correct responses: One way to evaluate $\sqrt[3]{5^{12}}$ is to raise _____ to the power obtained when _____ is divided by _____. The resulting exponent is _____, which gives a final answer of _____.

54. Which one of the following is *not* equal to $\sqrt[4]{9^2}$?
   (a) 3   **(b)** $\sqrt{9}$ **(c)** $9^{1/2}$ **(d)** -3

Simplify each of the following by first converting to rational exponents. Assume that all variables represent positive real numbers. See Example 5.

55. $\sqrt{2^{12}}$

56. $\sqrt{5^{10}}$

57. $\sqrt[3]{4^9}$

58. $\sqrt[4]{6^8}$

59. $\sqrt{x^{20}}$

60. $\sqrt{r^{50}}$

61. $\sqrt[3]{x} \cdot \sqrt{x}$

62. $\sqrt[4]{y} \cdot \sqrt[5]{y^2}$

63. $\sqrt[4]{49y^6}$

64. $\sqrt[4]{100y^{10}}$

65. $\dfrac{\sqrt[3]{t^4}}{\sqrt[5]{t^4}}$

66. $\dfrac{\sqrt[4]{w^3}}{\sqrt[6]{w}}$

MATHEMATICAL CONNECTIONS (EXERCISES 67–74)

Earlier, we factored expressions like $x^4 - x^5$ by factoring out the greatest common factor, so that $x^4 - x^5 = x^4(1 - x)$. We can adapt this approach to factor expressions with rational exponents. When one or more of the exponents is negative or a fraction, we use the ordering on the number line discussed in Chapter 1 to decide on the common factor. In this type of factoring, we want the binomial factor to have only positive exponents, so we always factor out the variable with the smallest exponent. A positive exponent is greater than a negative exponent, so in $7z^{5/8} + z^{-3/4}$, we factor out $z^{-3/4}$, because $-3/4$ is smaller than $5/8$.

Find the appropriate common factor in Exercises 67–69.

67. $3x^{-1/2} - 4x^{1/2}$

68. $m^3 - 3m^{5/2}$

69. $9k^{-3/4} + 2k^{-1/4}$

70. Factor $3x^{-1/2} - 4x^{1/2}$. **71.** Factor $m^3 - 3m^{5/2}$. **72.** Factor $9k^{-3/4} + 2k^{-1/4}$.

Use the method discussed earlier to factor the following.

73. $\dfrac{4}{\sqrt{t}} + 7\sqrt{t^3}$ **74.** $8\sqrt[3]{x^2} - \dfrac{5}{\sqrt[3]{x}}$

Solve each problem.

75. Meteorologists can determine the duration of a storm by using the function defined by $T(D) = .07D^{3/2}$, where D is the diameter of the storm in miles and T is the time in hours. Find the duration of a storm with a diameter of 16 miles. Round your answer to the nearest tenth of an hour.

76. The threshold weight T, in pounds, for a person is the weight above which the risk of death increases greatly. The threshold weight in pounds for men aged 40–49 is related to height in inches by the function defined by

$$h(T) = 12.3T^{1/3}.$$

What height corresponds to a threshold weight of 216 pounds for a 43-year-old man? Round your answer to the nearest inch, and then to the nearest tenth of a foot.

INTERPRETING TECHNOLOGY (EXERCISES 77–78)

77. The screen suggests that $\sqrt[4]{\dfrac{2143}{22}} = \pi$. Is this conclusion correct? Why or why not?

```
4ˣ√(2143/22)
         3.14159265
π
         3.14159265
```

78. The screen shows that $\sqrt{\sqrt{81}} = 3$. Explain why.

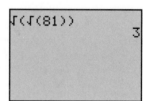

7.3 Simplifying Radicals

Objective ► The product of $\sqrt{36}$ and $\sqrt{4}$ is

$$\sqrt{36} \cdot \sqrt{4} = 6 \cdot 2 = 12.$$

Multiplying 36 and 4 and then taking the square root gives

$$\sqrt{36 \cdot 4} = \sqrt{144} = 12,$$

the same answer. This result is an example of the **product rule for radicals.**

> **Product Rule for Radicals**
>
> If a and b are real numbers, not both negative, all roots are real, and n is a natural number,
>
> $$\sqrt[n]{a} \cdot \sqrt[n]{b} = \sqrt[n]{ab}.$$
>
> The product of two radicals is the radical of the product.

The product rule can be justified using rational exponents. Since $\sqrt[n]{a} = a^{1/n}$ and $\sqrt[n]{b} = b^{1/n}$,

$$\sqrt[n]{a} \cdot \sqrt[n]{b} = a^{1/n} \cdot b^{1/n} = (ab)^{1/n} = \sqrt[n]{ab}.$$

In this section we assume all radicands are positive real numbers. Examples with negative radicands are discussed in Section 7 of this chapter.

EXAMPLE 1 Using the Product Rule

Multiply. Assume that all variables represent positive real numbers.

(a) $\sqrt{5} \cdot \sqrt{7} = \sqrt{5 \cdot 7} = \sqrt{35}$

(b) $\sqrt{2} \cdot \sqrt{19} = \sqrt{2 \cdot 19} = \sqrt{38}$

(c) $\sqrt{7} \cdot \sqrt{11xyz} = \sqrt{77xyz}$

(d) $\sqrt{\dfrac{7}{y}} \cdot \sqrt{\dfrac{3}{p}} = \sqrt{\dfrac{21}{yp}}$

WORK PROBLEM 1 AT THE SIDE. ►►

EXAMPLE 2 Using the Product Rule

Multiply. Assume that all variables represent positive real numbers.

(a) $\sqrt[3]{3} \cdot \sqrt[3]{12} = \sqrt[3]{3 \cdot 12} = \sqrt[3]{36}$

(b) $\sqrt[4]{8y} \cdot \sqrt[4]{3r^2} = \sqrt[4]{24yr^2}$

(c) $\sqrt[6]{10m^4} \cdot \sqrt[6]{5m} = \sqrt[6]{50m^5}$

(d) $\sqrt[4]{5} \cdot \sqrt[5]{2}$ cannot be simplified by the product rule, because the two indexes (4 and 5) are different.

WORK PROBLEM 2 AT THE SIDE. ►►

Objective ► The **quotient rule for radicals** follows from a power rule of exponents: $\left(\dfrac{a}{b}\right)^{1/n} = \dfrac{a^{1/n}}{b^{1/n}}.$

1. Multiply. Assume all variables represent positive real numbers.

(a) $\sqrt{5} \cdot \sqrt{13}$

(b) $\sqrt{10y} \cdot \sqrt{3k}$

(c) $\sqrt{\dfrac{5}{a}} \cdot \sqrt{\dfrac{11}{z}}$

2. Multiply. Assume all variables represent positive real numbers.

(a) $\sqrt[3]{2} \cdot \sqrt[3]{7}$

(b) $\sqrt[6]{8r^2} \cdot \sqrt[6]{2r^3}$

(c) $\sqrt[5]{9y^2x} \cdot \sqrt[5]{8xy^2}$

(d) $\sqrt{7} \cdot \sqrt[3]{5}$

ANSWERS

1. (a) $\sqrt{65}$ (b) $\sqrt{30yk}$
   (c) $\sqrt{\dfrac{55}{az}}$

2. (a) $\sqrt[3]{14}$ (b) $\sqrt[6]{16r^5}$ (c) $\sqrt[5]{72y^4x^2}$
   (d) cannot be simplified by the product rule

3. Simplify. Assume that all variables represent positive real numbers.

(a) $\sqrt{\dfrac{100}{81}}$

(b) $\sqrt{\dfrac{11}{25}}$

(c) $\sqrt[3]{\dfrac{18}{125}}$

(d) $\sqrt{\dfrac{y^8}{16}}$

(e) $\sqrt[3]{\dfrac{x^2}{r^{12}}}$

Quotient Rule for Radicals

If a and b are real numbers, not both negative, all roots are real, if $b \neq 0$, and if n is a natural number, then

$$\sqrt[n]{\frac{a}{b}} = \frac{\sqrt[n]{a}}{\sqrt[n]{b}}.$$

The radical of a quotient is the quotient of the radicals.

E X A M P L E   3   Using the Quotient Rule

Simplify. Assume that all variables represent positive real numbers.

(a) $\sqrt{\dfrac{16}{25}} = \dfrac{\sqrt{16}}{\sqrt{25}} = \dfrac{4}{5}$

(b) $\sqrt{\dfrac{7}{36}} = \dfrac{\sqrt{7}}{\sqrt{36}} = \dfrac{\sqrt{7}}{6}$

(c) $\sqrt[3]{-\dfrac{8}{125}} = \sqrt[3]{\dfrac{-8}{125}} = \dfrac{\sqrt[3]{-8}}{\sqrt[3]{125}} = \dfrac{-2}{5} = -\dfrac{2}{5}$

(d) $\sqrt[3]{\dfrac{7}{216}} = \dfrac{\sqrt[3]{7}}{\sqrt[3]{216}} = \dfrac{\sqrt[3]{7}}{6}$

(e) $\sqrt{\dfrac{m^4}{25}} = \dfrac{\sqrt{m^4}}{\sqrt{25}} = \dfrac{m^2}{5}$

◀◀ **WORK PROBLEM 3 AT THE SIDE.**

OBJECTIVE 3 ▶ One of the main uses of the product and quotient rules is in simplifying radicals. A radical is **simplified** if the following four conditions are met.

Simplified Radical

1. The radicand has no factor raised to a power greater than or equal to the index.
2. The radicand has no fractions.
3. No denominator contains a radical.
4. Exponents in the radicand and the index of the radical have no common factors (except 1).

E X A M P L E   4   Simplifying Radicals Involving Numbers

Simplify each radical.

(a) $\sqrt{24}$

Check to see if 24 is divisible by one of the perfect squares 4, 9, . . . , and choose the largest perfect square (square of a natural number) that divides into 24. The largest such number is 4. Write 24 as the product of 4 and 6, and then use the product rule.

$$\sqrt{24} = \sqrt{4 \cdot 6} = \sqrt{4} \cdot \sqrt{6} = 2\sqrt{6}$$

CONTINUED ON NEXT PAGE

ANSWERS

3. (a) $\dfrac{10}{9}$ (b) $\dfrac{\sqrt{11}}{5}$ (c) $\dfrac{\sqrt[3]{18}}{5}$

(d) $\dfrac{y^4}{4}$ (e) $\dfrac{\sqrt[3]{x^2}}{r^4}$

(b) $\sqrt{108}$

The number 108 is divisible by the perfect square 36. If this is not obvious, try factoring 108 into its prime factors.

$$\sqrt{108} = \sqrt{2^2 \cdot 3^3}$$
$$= \sqrt{2^2 \cdot 3^2 \cdot 3}$$
$$= 2 \cdot 3\sqrt{3} \qquad \text{Product rule}$$
$$= 6\sqrt{3}$$

(c) $\sqrt{500} = \sqrt{100 \cdot 5} = \sqrt{100} \cdot \sqrt{5} = 10\sqrt{5}$

(d) $\sqrt{10}$

No perfect square (other than 1) divides into 10, so $\sqrt{10}$ cannot be simplified further.

(e) $\sqrt[3]{16}$

Look for the largest perfect *cube* that divides into 16. The number 8 satisfies this condition, so we write 16 as $8 \cdot 2$ (or factor 16 into prime factors).

$$\sqrt[3]{16} = \sqrt[3]{8 \cdot 2} = \sqrt[3]{8} \cdot \sqrt[3]{2} = 2\sqrt[3]{2}$$

(f) $\sqrt[4]{162} = \sqrt[4]{81 \cdot 2} = \sqrt[4]{3^4 \cdot 2} \qquad \text{Factor.}$
$$= \sqrt[4]{3^4} \cdot \sqrt[4]{2} \qquad \text{Product rule}$$
$$= 3\sqrt[4]{2}$$

> **WORK PROBLEM 4 AT THE SIDE.** ▶▶

┌ **E X A M P L E 5** **Simplifying Radicals Involving Variables**

Simplify. Assume that all variables represent positive real numbers.

(a) $\sqrt{16m^3} = \sqrt{16m^2 \cdot m}$
$$= \sqrt{16m^2} \cdot \sqrt{m}$$
$$= 4m\sqrt{m}$$

No absolute value bars are needed around the m because of the assumption that all the variables represent *positive* real numbers.

(b) $\sqrt{200k^7q^8} = \sqrt{10^2 \cdot 2 \cdot (k^3)^2 \cdot k \cdot (q^4)^2} \qquad \text{Factor.}$
$$= 10k^3q^4\sqrt{2k} \qquad \begin{array}{l}\text{Remove perfect square} \\ \text{factors.}\end{array}$$

(c) $\sqrt{75p^6q^9} = \sqrt{(25p^6q^8)(3q)} = \sqrt{25p^6q^8} \cdot \sqrt{3q} = 5p^3q^4\sqrt{3q}$

(d) $\sqrt[3]{8x^4y^5} = \sqrt[3]{(8x^3y^3)(xy^2)} = \sqrt[3]{8x^3y^3} \cdot \sqrt[3]{xy^2} = 2xy\sqrt[3]{xy^2}$

(e) $\sqrt[4]{32y^9} = \sqrt[4]{(16y^8)(2y)} = \sqrt[4]{16y^8} \cdot \sqrt[4]{2y} = 2y^2\sqrt[4]{2y}$

> **WORK PROBLEM 5 AT THE SIDE.** ▶▶

OBJECTIVE ▶ The conditions for a simplified radical given earlier state that an exponent in the radicand and the index of the radical should have no common factor (except 1). The next example shows how to simplify radicals with such common factors.

---

4. Simplify.

(a) $\sqrt{32}$

(b) $\sqrt{45}$

(c) $\sqrt{300}$

(d) $\sqrt{35}$

(e) $\sqrt[3]{54}$

(f) $\sqrt[4]{243}$

5. Simplify. Assume all variables represent positive real numbers.

(a) $\sqrt{25p^7}$

(b) $\sqrt{72y^3x}$

(c) $\sqrt[3]{y^7x^5z^6}$

(d) $\sqrt[4]{32a^5b^6}$

ANSWERS

4. (a) $4\sqrt{2}$ (b) $3\sqrt{5}$ (c) $10\sqrt{3}$
(d) cannot be further simplified
(e) $3\sqrt[3]{2}$ (f) $3\sqrt[4]{3}$
5. (a) $5p^3\sqrt{p}$ (b) $6y\sqrt{2yx}$
(c) $y^2xz^2\sqrt[3]{yx^2}$ (d) $2ab\sqrt[4]{2ab^2}$

6. Simplify. Assume all variables represent positive real numbers.

(a) $\sqrt[12]{2^3}$

(b) $\sqrt[6]{t^2}$

E X A M P L E   6   Simplifying Radicals by Using Smaller Indexes

Simplify. Assume that all variables represent positive real numbers.

(a) $\sqrt[9]{5^6}$

We can write this radical using rational exponents and then write the exponent in lowest terms. Express the answer as a radical.

$$\sqrt[9]{5^6} = 5^{6/9} = 5^{2/3} = \sqrt[3]{5^2} \quad \text{or} \quad \sqrt[3]{25}$$

(b) $\sqrt[4]{p^2} = p^{2/4} = p^{1/2} = \sqrt{p}$ (Recall the assumption $p > 0$.)

These examples suggest the following rule.

> If m is an integer, n is a positive integer, k is a positive integer, and a is a positive real number,
> $$\sqrt[kn]{a^{km}} = \sqrt[n]{a^m}.$$

◀◀ **WORK PROBLEM 6 AT THE SIDE.**

The next example shows how to simplify the product of two radicals having different indexes.

E X A M P L E   7   Multiplying Radicals with Different Indexes

Simplify $\sqrt{7} \cdot \sqrt[3]{2}$.

Because the indexes, 2 and 3, have a least common index of 6, we can use rational exponents to write each radical as a sixth root.

$$\sqrt{7} = 7^{1/2} = 7^{3/6} = \sqrt[6]{7^3} = \sqrt[6]{343}$$
$$\sqrt[3]{2} = 2^{1/3} = 2^{2/6} = \sqrt[6]{2^2} = \sqrt[6]{4}$$
$$\sqrt{7} \cdot \sqrt[3]{2} = \sqrt[6]{343} \cdot \sqrt[6]{4} = \sqrt[6]{1372} \quad \text{Product rule}$$

◀◀ **WORK PROBLEM 7 AT THE SIDE.**

7. Simplify $\sqrt{5} \cdot \sqrt[3]{4}$.

Objective ▶5 One useful application of radicals occurs when using the **Pythagorean formula** (puh-THAG-uh-REE-un FORM-yoo-luh) from geometry to find the length of one side of a right triangle when the lengths of the other sides are known.

> ### Pythagorean Formula
>
> If c is the length of the longest side of a right triangle and a and b are the lengths of the shorter sides, then
>
> $$c^2 = a^2 + b^2.$$
>
>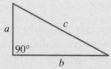
>
> The longest side is the **hypotenuse** and the two shorter sides are the **legs** of the triangle. The hypotenuse is the side opposite the right angle.

From the formula $c^2 = a^2 + b^2$, the length of the hypotenuse (hy-POT-en-oos) is given by $c = \sqrt{a^2 + b^2}$.

E X A M P L E 8   Using the Pythagorean Formula

Use the Pythagorean formula to find the length of the hypotenuse in the triangle with legs of lengths 4 feet and 6 feet.

By the formula, the length of the hypotenuse is

$$c = \sqrt{a^2 + b^2} = \sqrt{4^2 + 6^2} \qquad \text{Let } a = 4 \text{ and } b = 6.$$
$$= \sqrt{16 + 36} = \sqrt{52} = \sqrt{4 \cdot 13} \qquad \text{Factor.}$$
$$= \sqrt{4} \cdot \sqrt{13} = 2\sqrt{13} \text{ feet.} \qquad \text{Product rule}$$

WORK PROBLEM 8 AT THE SIDE. ▶▶

OBJECTIVE ▶6 The Pythagorean formula is used to derive a very important result in algebra, the distance formula. With the distance formula, we can determine algebraically the distance between two points in a rectangular coordinate system. For example, Figure 1 shows the points $(3, -4)$ and $(-5, 3)$. From the Pythagorean formula,

$$d^2 = a^2 + b^2.$$

As shown in Figure 1,

$$a = 3 - (-5) = 8 \quad \text{and} \quad b = 3 - (-4) = 7.$$

By the Pythagorean formula,

$$d^2 = 7^2 + 8^2 = 49 + 64 = 113,$$

and $$d = \sqrt{113}.$$

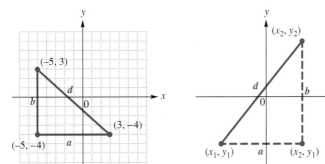

FIGURE 1   **FIGURE 2**

Now this result can be generalized. Figure 2 shows the two points (x_1, y_1) and (x_2, y_2). We want to find a general formula for the distance d between these two points. The distance between (x_1, y_1) and (x_2, y_1) is given by

$$a = x_2 - x_1,$$

and the distance between (x_2, y_2) and (x_2, y_1) is given by

$$b = y_2 - y_1.$$

From the Pythagorean formula,

$$d^2 = a^2 + b^2$$
$$= (x_2 - x_1)^2 + (y_2 - y_1)^2.$$

Taking square roots on both sides gives the distance formula.

8. Find the length of the missing side in each triangle.

(a)

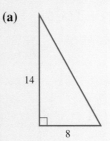

14

8

(b)

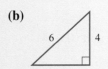

6   4

9. Find the distance between the points in each pair.

(a) $(2, -1)$ and $(5, 3)$

Distance Formula

The distance between (x_1, y_1) and (x_2, y_2) is

$$d = \sqrt{(x_2 - x_1)^2 + (y_2 - y_1)^2}.$$

E X A M P L E 9   Using the Distance Formula

Find the distance between $(-3, 5)$ and $(6, 4)$.

Using the distance formula,

$$d = \sqrt{[6 - (-3)]^2 + (4 - 5)^2} = \sqrt{9^2 + (-1)^2} = \sqrt{82}.$$

◀◀ WORK PROBLEM 9 AT THE SIDE.

(b) $(-3, 2)$ and $(0, -4)$

| Name | Date | Class |

7.3 Exercises

Fill in the blanks to complete each statement. Assume that m and n represent positive numbers.

1. By the product rule, $\sqrt{m} \cdot \sqrt{n} = $ _____ .

2. By the quotient rule, $\dfrac{\sqrt{m}}{\sqrt{n}} = $ _____ .

3. A simplified radical cannot have a(n) _____ in its _____ .

4. To multiply radicals with different _____ , write each radical in _____ form, add the _____ , and simplify the answer, if possible. Write the result in _____ form.

5. Explain why $\sqrt[3]{x} \cdot \sqrt[3]{x}$ is not equal to x. What is it equal to?

6. Explain why $\sqrt[4]{x} \cdot \sqrt[4]{x}$ is not equal to x, but *is* equal to $\sqrt{x}$, for $x \geq 0$.

Multiply. See Examples 1 and 2.

7. $\sqrt{5} \cdot \sqrt{6}$

8. $\sqrt{10} \cdot \sqrt{3}$

9. $\sqrt[3]{7x} \cdot \sqrt[3]{2y}$

10. $\sqrt[3]{9x} \cdot \sqrt[3]{4y}$

11. $\sqrt[4]{12} \cdot \sqrt[4]{3}$

12. $\sqrt[4]{6} \cdot \sqrt[4]{9}$

Simplify each radical. Assume that all variables represent positive real numbers. See Example 3.

13. $\sqrt{\dfrac{64}{121}}$

14. $\sqrt{\dfrac{16}{49}}$

15. $\sqrt{\dfrac{3}{25}}$

16. $\sqrt{\dfrac{13}{49}}$

17. $\sqrt{\dfrac{x}{25}}$

18. $\sqrt{\dfrac{k}{100}}$

19. $\sqrt{\dfrac{p^6}{81}}$

20. $\sqrt{\dfrac{w^{10}}{36}}$

21. $\sqrt[3]{\dfrac{27}{64}}$

22. $\sqrt[3]{\dfrac{216}{125}}$

23. $\sqrt[3]{-\dfrac{r^2}{8}}$

24. $\sqrt[3]{-\dfrac{t}{125}}$

Express each of the following in simplified form. See Example 4.

25. $\sqrt{12}$

26. $\sqrt{18}$

27. $\sqrt{288}$

28. $\sqrt{72}$

29. $-\sqrt{32}$

30. $-\sqrt{48}$

31. $-\sqrt{28}$

32. $-\sqrt{24}$

33. $\sqrt{-300}$

34. $\sqrt{-150}$

35. $\sqrt[3]{128}$

36. $\sqrt[3]{24}$

37. $\sqrt[3]{-16}$

38. $\sqrt[3]{-250}$

39. $\sqrt[3]{40}$

40. $\sqrt[3]{375}$

41. $-\sqrt[4]{512}$

42. $-\sqrt[4]{1250}$

43. $\sqrt[5]{64}$

44. $\sqrt[5]{128}$

45. A student claimed that $\sqrt[3]{14}$ is not in simplified form, since $14 = 8 + 6$, and 8 is a perfect cube. Was his reasoning correct? Why or why not?

46. Explain in your own words why $\sqrt[3]{k^4}$ is not a simplified radical.

Express each of the following in simplified form. Assume that all variables represent positive real numbers. See Example 5.

47. $\sqrt{72k^2}$

48. $\sqrt{18m^2}$

49. $\sqrt[3]{\dfrac{81}{64}}$

50. $\sqrt[3]{\dfrac{32}{216}}$

51. $\sqrt{121x^6}$

52. $\sqrt{256z^{12}}$

53. $-\sqrt[3]{27t^{12}}$

54. $-\sqrt[3]{64y^{18}}$

55. $-\sqrt{100m^8z^4}$

56. $-\sqrt{25t^6s^{20}}$

57. $-\sqrt[3]{-125a^6b^9c^{12}}$

58. $-\sqrt[3]{-216y^{15}x^6z^3}$

59. $\sqrt[4]{\dfrac{1}{16}r^8t^{20}}$

60. $\sqrt[4]{\dfrac{81}{256}t^{12}u^8}$

61. $\sqrt{50x^3}$

62. $\sqrt{300z^3}$

63. $-\sqrt{500r^{11}}$

64. $-\sqrt{200p^{13}}$

65. $\sqrt{13x^7y^8}$

66. $\sqrt{23k^9p^{14}}$

67. $\sqrt[3]{8z^6w^9}$

68. $\sqrt[3]{64a^{15}b^{12}}$

69. $\sqrt[3]{-16z^5t^7}$

70. $\sqrt[3]{-81m^4n^{10}}$

71. $\sqrt[4]{81x^{12}y^{16}}$

72. $\sqrt[4]{81t^8u^{28}}$

73. $-\sqrt[4]{162r^{15}s^{10}}$

74. $-\sqrt[4]{32k^5m^{10}}$

75. $\sqrt{\dfrac{y^{11}}{36}}$

76. $\sqrt{\dfrac{v^{13}}{49}}$

77. $\sqrt[3]{\dfrac{x^{16}}{27}}$

78. $\sqrt[3]{\dfrac{y^{17}}{125}}$

Simplify each of the following. Assume that all variables represent positive real numbers. See Example 6.

79. $\sqrt[4]{48^2}$

80. $\sqrt[4]{50^2}$

81. $\sqrt[10]{x^{25}}$

82. $\sqrt[12]{x^{44}}$

Simplify each by first writing all radicals as radicals with the same index. Then multiply. See Example 7.

83. $\sqrt[3]{4} \cdot \sqrt{3}$

84. $\sqrt[3]{5} \cdot \sqrt{6}$

85. $\sqrt[4]{3} \cdot \sqrt[3]{4}$

86. $\sqrt[5]{7} \cdot \sqrt[7]{5}$

Find the missing length in the right triangle. Simplify the answer if necessary. (Hint: If one leg is unknown, write the formula as a $= \sqrt{c^2 - b^2}$ or $b = \sqrt{c^2 - a^2}$.) See Example 8.

87.

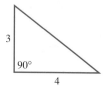

88.

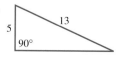

89.

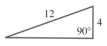

90.

Find the distance between each pair of points. See Example 9.

91. $(5, 3)$ and $(-1, 2)$

92. $(-1, 4)$ and $(5, 3)$

93. $(-1, 5)$ and $(5, 3)$

94. $(4, 5)$ and $(-8, 4)$

95. $(\sqrt{2}, \sqrt{6})$ and $(-2\sqrt{2}, 4\sqrt{6})$

96. $(\sqrt{7}, 9\sqrt{3})$ and $(-\sqrt{7}, 4\sqrt{3})$

97. $(x + y, y)$ and $(x - y, x)$

98. $(c, c - d)$ and $(d, c + d)$

In Exercises 99 and 100, give each answer first as a simplified radical, then as an approximation correct to the place value specified.

99. The illumination I, in footcandles, produced by a light source is related to the distance d, in feet, from the light source by the equation

$$d = \sqrt{\frac{k}{I}}$$

where k is a constant. If $k = 640$, how far from the light source will the illumination be 2 footcandles? Give the exact value, and then round to the nearest tenth of a foot.

100. The length of the diagonal of a box is given by

$$D = \sqrt{L^2 + W^2 + H^2}$$

where L, W, and H are the length, the width, and the height of the box. Find the length of the diagonal, D, of a box that is 4 feet long, 3 feet high, and 2 feet wide. Give the exact value, then round to the nearest tenth of a foot.

7.4 Addition and Subtraction of Radical Expressions

OBJECTIVES

1. Define a radical expression.

2. Simplify sums and differences of radical expressions by using the distributive property.

3. Use radicals in an applied problem.

FOR EXTRA HELP

Tutorial   Tape 11   SSM, Sec. 7.4

OBJECTIVE 1   A **radical expression** is an algebraic expression that contains radicals. For example,

$$\sqrt[4]{3} + \sqrt{6}, \qquad \sqrt{2} - 1, \qquad \text{and} \qquad \sqrt{8} - \sqrt{2}$$

are radical expressions. The examples in the preceding section required simplifying radical expressions involving only multiplication and division. Now we will simplify radical expressions that involve addition and subtraction.

OBJECTIVE 2   An expression such as $4\sqrt{2} + 3\sqrt{2}$ can be simplified by using the distributive property.

$$4\sqrt{2} + 3\sqrt{2} = (4 + 3)\sqrt{2} = 7\sqrt{2}$$

As another example, $2\sqrt{3} - 5\sqrt{3} = (2 - 5)\sqrt{3} = -3\sqrt{3}$. This is much like simplifying $2x + 3x$ to $5x$ or $5y - 8y$ to $-3y$.

> **Caution**
> Only radical expressions with the same index and the same radicand may be combined. Expressions such as $5\sqrt{3} + 2\sqrt{2}$ or $3\sqrt{3} + 2\sqrt[3]{3}$ cannot be simplified.

EXAMPLE 1   **Adding and Subtracting Radicals**

Add or subtract the following radical expressions.

(a) $3\sqrt{24} + \sqrt{54}$

We begin by simplifying each radical; then we use the distributive property.

$$3\sqrt{24} + \sqrt{54} = 3\sqrt{4} \cdot \sqrt{6} + \sqrt{9} \cdot \sqrt{6}$$
$$= 3 \cdot 2\sqrt{6} + 3\sqrt{6}$$
$$= 6\sqrt{6} + 3\sqrt{6}$$
$$= 9\sqrt{6}$$

(b) $-3\sqrt{8} + 4\sqrt{18} = -3\sqrt{4}\sqrt{2} + 4\sqrt{9}\sqrt{2}$ Product rule
$$= -3 \cdot 2\sqrt{2} + 4 \cdot 3\sqrt{2}$$
$$= -6\sqrt{2} + 12\sqrt{2} = 6\sqrt{2} \quad \text{Combine terms.}$$

(c) $2\sqrt{20x} - \sqrt{45x} = 2\sqrt{4}\sqrt{5x} - \sqrt{9}\sqrt{5x}$ Product rule
$$= 2 \cdot 2\sqrt{5x} - 3\sqrt{5x}$$
$$= 4\sqrt{5x} - 3\sqrt{5x} = \sqrt{5x} \quad \text{Combine terms.}$$

Because the radicand is $5x$, we must have $x \geq 0$.

(d) $2\sqrt{125} - 3\sqrt{180} + 2\sqrt{500}$
$$= 2\sqrt{25} \cdot \sqrt{5} - 3\sqrt{36} \cdot \sqrt{5} + 2\sqrt{100} \cdot \sqrt{5}$$
$$= 2 \cdot 5\sqrt{5} - 3 \cdot 6\sqrt{5} + 2 \cdot 10\sqrt{5}$$
$$= 10\sqrt{5} - 18\sqrt{5} + 20\sqrt{5} = 12\sqrt{5}$$

(e) $2\sqrt{3} - 4\sqrt{5}$

Here the radicals differ and are already simplified, so $2\sqrt{3} - 4\sqrt{5}$ cannot be simplified further.

---

1. Simplify.

(a) $3\sqrt{5} + 7\sqrt{5}$

(b) $2\sqrt{11} - \sqrt{11} + 3\sqrt{44}$

(c) $5\sqrt{12y} + 6\sqrt{75y}, \; y \geq 0$

(d) $3\sqrt{8} - 6\sqrt{50} + 2\sqrt{200}$

(e) $9\sqrt{5} - 4\sqrt{10}$

ANSWERS

1. **(a)** $10\sqrt{5}$
    (b) $7\sqrt{11}$ **(c)** $40\sqrt{3y}$
    (d) $-4\sqrt{2}$
    (e) cannot be further simplified

WORK PROBLEM 1 AT THE SIDE. ▶▶

2. Simplify each radical expression. Assume that all variables represent positive real numbers.

(a) $7\sqrt[3]{81} + 3\sqrt[3]{24}$

(b) $-2\sqrt[4]{32} - 7\sqrt[4]{162}$

(c) $\sqrt[3]{p^4q^7} - \sqrt[3]{64pq}$

3. Find the area of a trapezoid with bases $\sqrt{45}$ and $\sqrt{80}$ inches and height $\sqrt{18}$ inches.

Caution
Do not confuse the product rule with combining like terms. The root of a sum **does not equal** the sum of the roots. That is,
$$\sqrt{25} \neq \sqrt{9} + \sqrt{16}.$$
$$\sqrt{25} = 5, \text{ but } \sqrt{9} + \sqrt{16} = 3 + 4 = 7.$$

E X A M P L E 2 Adding and Subtracting Radicals with Higher Indexes

Add or subtract the following radical expressions. Assume that all variables represent positive real numbers.

(a) $2\sqrt[3]{16} - 5\sqrt[3]{54} = 2\sqrt[3]{8 \cdot 2} - 5\sqrt[3]{27 \cdot 2}$ Factor.
$$= 2\sqrt[3]{8} \cdot \sqrt[3]{2} - 5\sqrt[3]{27} \cdot \sqrt[3]{2}$$ Product rule
$$= 2 \cdot 2 \cdot \sqrt[3]{2} - 5 \cdot 3 \cdot \sqrt[3]{2}$$ Simplify.
$$= 4\sqrt[3]{2} - 15\sqrt[3]{2}$$ Multiply.
$$= -11\sqrt[3]{2}$$ Combine terms.

(b) $2\sqrt[3]{x^2y} + \sqrt[3]{8x^5y^4} = 2\sqrt[3]{x^2y} + \sqrt[3]{(8x^3y^3)x^2y}$ Factor.
$$= 2\sqrt[3]{x^2y} + 2xy\sqrt[3]{x^2y}$$ Product rule
$$= (2 + 2xy)\sqrt[3]{x^2y}$$ Distributive property

◀◀ **WORK PROBLEM 2 AT THE SIDE.**

Caution
It is a common error to forget to write the index when working with cube roots, fourth roots, and so on. Always remember to write these indexes.

OBJECTIVE 3▶ Applied problems may involve radical expressions that can be simplified using the techniques introduced in this chapter.

E X A M P L E 3 Solving an Applied Problem Involving Radicals

Find the area of a trapezoid with bases measuring $\sqrt{28}$ feet and $\sqrt{63}$ feet and height $\sqrt{14}$ feet.

The formula for the area of a trapezoid is $A = .5(B + b)h$. We have $B = \sqrt{63}$ feet, $b = \sqrt{28}$ feet, and $h = \sqrt{14}$ feet.

$$A = .5(B + b)h$$
$$= .5(\sqrt{63} + \sqrt{28})\sqrt{14}$$ Let $B = \sqrt{63}$, $b = \sqrt{28}$, $h = \sqrt{14}$.
$$= .5(3\sqrt{7} + 2\sqrt{7})\sqrt{2} \cdot \sqrt{7}$$ Product rule; simplify.
$$= .5(5\sqrt{7})\sqrt{2} \cdot \sqrt{7}$$ Combine terms.
$$= .5(5)7\sqrt{2}$$ Multiply.
$$= 17.5\sqrt{2}$$

The area is $17.5\sqrt{2}$ square feet.

◀◀ **WORK PROBLEM 3 AT THE SIDE.**

ANSWERS
2. (a) $27\sqrt[3]{3}$
(b) $-25\sqrt[4]{2}$
(c) $(pq^2 - 4)\sqrt[3]{pq}$
3. $10.5\sqrt{10}$ square inches

7.4 Exercises

1. Which one of the following sums could we simplify without first simplifying the individual radical expressions?

    (a) $\sqrt{50} + \sqrt{32}$ (b) $3\sqrt{6} + 9\sqrt{6}$ (c) $\sqrt[3]{32} - \sqrt[3]{108}$ (d) $\sqrt[5]{6} - \sqrt[5]{192}$

2. Let $a = 1$, and let $b = 64$.

    (a) Evaluate $\sqrt{a} + \sqrt{b}$. Then find $\sqrt{a+b}$. Are they equal?

    (b) Evaluate $\sqrt[3]{a} + \sqrt[3]{b}$. Then find $\sqrt[3]{a+b}$. Are they equal?

    (c) Complete the following: In general, $\sqrt[n]{a} + \sqrt[n]{b} \neq$ _____ , based on the observations in parts (a) and (b) of this exercise.

3. Even though the root indexes of the terms are not equal, the sum $\sqrt{64} + \sqrt[3]{125} + \sqrt[4]{16}$ can be simplified quite easily. What is this sum? Why can we add them so easily?

4. "You can't add apples and oranges." How does this old saying apply to the expression $\sqrt{a} + \sqrt[3]{a}$?

Simplify. Assume that all variables represent positive real numbers. See Examples 1 and 2.

5. $\sqrt{36} - \sqrt{100}$

6. $\sqrt{25} - \sqrt{81}$

7. $-2\sqrt{48} + 3\sqrt{75}$

8. $4\sqrt{32} - 2\sqrt{8}$

9. $6\sqrt{18} - \sqrt{32} + 2\sqrt{50}$

10. $5\sqrt{8} + 3\sqrt{72} - 3\sqrt{50}$

11. $-2\sqrt{63} + 2\sqrt{28} + 2\sqrt{7}$

12. $-\sqrt{27} + 2\sqrt{48} - \sqrt{75}$

13. $2\sqrt{5} + 3\sqrt{20} + 4\sqrt{45}$

14. $5\sqrt{54} - 2\sqrt{24} - 2\sqrt{96}$

15. $8\sqrt{2x} - \sqrt{8x} + \sqrt{72x}$

16. $4\sqrt{18k} - \sqrt{72k} + \sqrt{50k}$

17. $3\sqrt{72m^2} - 5\sqrt{32m^2} - 3\sqrt{18m^2}$

18. $9\sqrt{27p^2} - 14\sqrt{108p^2} + 2\sqrt{48p^2}$

19. $-\sqrt[3]{54} + 2\sqrt[3]{16}$

20. $15\sqrt[3]{81} - 4\sqrt[3]{24}$

21. $2\sqrt[3]{27x} - 2\sqrt[3]{8x}$

22. $6\sqrt[3]{128m} + 3\sqrt[3]{16m}$

23. $5\sqrt[4]{32} + 3\sqrt[4]{162}$

24. $2\sqrt[4]{512} + 4\sqrt[4]{32}$

25. $3\sqrt[4]{x^5y} - 2x\sqrt[4]{xy}$

26. $2\sqrt[4]{m^9p^6} - 3m^2p\sqrt[4]{mp^2}$

27. $\sqrt[3]{64xy^2} + \sqrt[3]{27x^4y^5}$

28. $\sqrt[4]{625s^3t} - \sqrt[4]{81s^7t^5}$

29. $\sqrt{\dfrac{8}{9}} + \sqrt{\dfrac{18}{36}}$ **30.** $\sqrt{\dfrac{12}{16}} + \sqrt{\dfrac{48}{64}}$ **31.** $\dfrac{\sqrt{32}}{3} + \dfrac{2\sqrt{2}}{3} - \dfrac{\sqrt{2}}{\sqrt{9}}$ **32.** $\dfrac{\sqrt{27}}{2} - \dfrac{3\sqrt{3}}{2} + \dfrac{\sqrt{3}}{\sqrt{4}}$

In Example 1(a) we show that $3\sqrt{24} + \sqrt{54} = 9\sqrt{6}$. All of these terms are expressed as square roots using radical symbols. To support this result, we may find a calculator approximation of $3\sqrt{24}$, then find a calculator approximation of $\sqrt{54}$, and add these two approximations. Then, we find a calculator approximation of $9\sqrt{6}$. It should correspond to the sum that we just found. (For this example, both approximations are 22.04540769. Due to rounding procedures, there may be a discrepancy in the final digit if you try to duplicate this work.) Follow this procedure to support the statements in Exercises 33–36.

33. $3\sqrt{32} - 2\sqrt{8} = 8\sqrt{2}$ **34.** $4\sqrt{12} - 7\sqrt{27} = -13\sqrt{3}$

35. $2\sqrt{40} + 6\sqrt{90} - 3\sqrt{160} = 10\sqrt{10}$ **36.** $5\sqrt{28} - 3\sqrt{63} + 2\sqrt{112} = 9\sqrt{7}$

37. A rectangular yard has a length of $\sqrt{192}$ meters and a width of $\sqrt{48}$ meters. Choose the best estimate of its dimensions. (All measures are in meters.)
  (a) 14 by 7   **(b)** 5 by 7   **(c)** 14 by 8
  (d) 15 by 8

38. If the base of a triangle is $\sqrt{65}$ inches and its height is $\sqrt{26}$ inches, which one of the following is the best estimate for its area?
  (a) 20 square inches   **(b)** 26 square inches
  (c) 40 square inches   **(d)** 52 square inches

Work each of the following problems. Give answers as simplified radical expressions. See Example 3.

39. Find the perimeter of the triangle shown here.

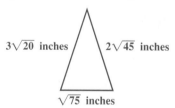

$3\sqrt{20}$ inches $2\sqrt{45}$ inches $\sqrt{75}$ inches

40. Find the perimeter of the rectangle shown here.

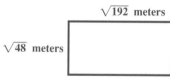

$\sqrt{192}$ meters $\sqrt{48}$ meters

41. What is the perimeter of a computer graphic with sides measuring $4\sqrt{18}$, $4\sqrt{32}$, $6\sqrt{50}$, and $5\sqrt{12}$ centimeters?

42. Find the area of a trapezoid with bases $\sqrt{288}$ and $\sqrt{72}$ inches, and height $\sqrt{24}$ inches.

7.5 *Multiplication and Division of Radical Expressions*

OBJECTIVES

1. Multiply binomial expressions involving radicals.
2. Rationalize denominators involving a single square root.
3. Rationalize denominators involving cube roots.
4. Rationalize denominators with binomials involving radicals.
5. Write radical expressions in lowest terms.

FOR EXTRA HELP

Tutorial   Tape 11   SSM, Sec. 7.5

OBJECTIVE 1 We multiply binomial expressions involving radicals by using the FOIL (First, Outside, Inside, Last) method. For example, we find the product of the binomials $\sqrt{5} + 3$ and $\sqrt{6} + 1$ as follows.

$$(\sqrt{5} + 3)(\sqrt{6} + 1) = \overbrace{\sqrt{5} \cdot \sqrt{6}}^{\text{First}} + \overbrace{\sqrt{5} \cdot 1}^{\text{Outside}} + \overbrace{3 \cdot \sqrt{6}}^{\text{Inside}} + \overbrace{3 \cdot 1}^{\text{Last}}$$
$$= \sqrt{30} + \sqrt{5} + 3\sqrt{6} + 3$$

This result cannot be simplified further.

EXAMPLE 1   Multiplying Binomials Involving Radical Expressions

Multiply.

(a) $(7 - \sqrt{3})(\sqrt{5} + \sqrt{2}) = 7\sqrt{5} + 7\sqrt{2} - \sqrt{3} \cdot \sqrt{5} - \sqrt{3} \cdot \sqrt{2}$
$$= 7\sqrt{5} + 7\sqrt{2} - \sqrt{15} - \sqrt{6}$$

(b) $(\sqrt{10} + \sqrt{3})(\sqrt{10} - \sqrt{3})$
$$= \sqrt{10}\sqrt{10} - \sqrt{10}\sqrt{3} + \sqrt{10}\sqrt{3} - \sqrt{3}\sqrt{3}$$
$$= 10 - 3$$
$$= 7$$

Notice that this is the kind of product that results in the difference of two squares:
$$(a + b)(a - b) = a^2 - b^2.$$
Here, $a = \sqrt{10}$ and $b = \sqrt{3}$.

(c) $(\sqrt{7} - 3)^2 = (\sqrt{7} - 3)(\sqrt{7} - 3)$
$$= \sqrt{7} \cdot \sqrt{7} - 3\sqrt{7} - 3\sqrt{7} + 3 \cdot 3$$
$$= 7 - 6\sqrt{7} + 9$$
$$= 16 - 6\sqrt{7}$$

(d) $(5 - \sqrt[3]{3})(5 + \sqrt[3]{3}) = 5 \cdot 5 + 5\sqrt[3]{3} - 5\sqrt[3]{3} - \sqrt[3]{3} \cdot \sqrt[3]{3}$
$$= 25 - \sqrt[3]{3^2}$$
$$= 25 - \sqrt[3]{9}$$

> **WORK PROBLEM 1 AT THE SIDE.** ▶▶

OBJECTIVE 2 In some circumstances it is useful to write a radical expression such as $\frac{1}{\sqrt{2}}$ so that there is no radical in the denominator. The process of removing radicals from the denominator is called **rationalizing** (RA-shun-ul-eye-zing) **the denominator** because the result is a rational number denominator.

---

1. Multiply.

(a) $(2 + \sqrt{3})(1 + \sqrt{5})$

(b) $(2\sqrt{3} + \sqrt{5})(\sqrt{6} - 3\sqrt{5})$

(c) $(4 + \sqrt{3})(4 - \sqrt{3})$

(d) $(\sqrt{6} - \sqrt{5})^2$

(e) $(4 + \sqrt[3]{7})(4 - \sqrt[3]{7})$

ANSWERS

1. **(a)** $2 + 2\sqrt{5} + \sqrt{3} + \sqrt{15}$
   (b) $6\sqrt{2} - 6\sqrt{15} + \sqrt{30} - 15$
   (c) 13 **(d)** $11 - 2\sqrt{30}$ **(e)** $16 - \sqrt[3]{49}$

477

2. Rationalize the denominators.

(a) $\dfrac{8}{\sqrt{3}}$

(b) $\dfrac{9}{\sqrt{6}}$

(c) $\dfrac{\sqrt{3}}{\sqrt{7}}$

E X A M P L E   2   **Rationalizing Denominators with Square Roots**

Rationalize each denominator.

(a) $\dfrac{3}{\sqrt{7}}$

We want to make the denominator a rational number. Because $\sqrt{7} \cdot \sqrt{7} = \sqrt{7 \cdot 7} = 7$, using the property of 1, we multiply by $\dfrac{\sqrt{7}}{\sqrt{7}}$.

$$\frac{3}{\sqrt{7}} = \frac{3 \cdot \sqrt{7}}{\sqrt{7} \cdot \sqrt{7}} = \frac{3\sqrt{7}}{7}.$$

The radical is now removed from the denominator, so the denominator is rational.

(b) $\dfrac{5\sqrt{2}}{\sqrt{5}}$

Use the property of 1. Multiply by $\dfrac{\sqrt{5}}{\sqrt{5}}$.

$$\frac{5\sqrt{2}}{\sqrt{5}} = \frac{5\sqrt{2} \cdot \sqrt{5}}{\sqrt{5} \cdot \sqrt{5}} = \frac{5\sqrt{10}}{5} = \sqrt{10}$$

◀◀ **WORK PROBLEM 2 AT THE SIDE.**

E X A M P L E   3   **Rationalizing Denominators with Square Roots**

Rationalize each denominator.

(a) $\dfrac{6}{\sqrt{12}}$

Less work is involved if we simplify the radical in the denominator first.

$$\frac{6}{\sqrt{12}} = \frac{6}{\sqrt{4 \cdot 3}} = \frac{6}{2\sqrt{3}} = \frac{3}{\sqrt{3}}$$

Now rationalize the denominator by multiplying by $\dfrac{\sqrt{3}}{\sqrt{3}}$.

$$\frac{3 \cdot \sqrt{3}}{\sqrt{3} \cdot \sqrt{3}} = \frac{3\sqrt{3}}{3} = \sqrt{3}$$

Another approach is to notice that the denominator can be rationalized directly by multiplying by $\dfrac{\sqrt{3}}{\sqrt{3}}$ as follows.

$$\frac{6}{\sqrt{12}} = \frac{6}{\sqrt{12}} \cdot \frac{\sqrt{3}}{\sqrt{3}} = \frac{6\sqrt{3}}{\sqrt{36}} = \frac{6\sqrt{3}}{6} = \sqrt{3}$$

(b) $\dfrac{9}{\sqrt{72}}$

First simplify $\sqrt{72}$ in the denominator.

$$\frac{9}{\sqrt{72}} = \frac{9}{\sqrt{36 \cdot 2}} = \frac{9}{6\sqrt{2}} = \frac{3}{2\sqrt{2}}$$

ANSWERS

2. (a) $\dfrac{8\sqrt{3}}{3}$ (b) $\dfrac{3\sqrt{6}}{2}$ (c) $\dfrac{\sqrt{21}}{7}$

CONTINUED ON NEXT PAGE

Now multiply by $\dfrac{\sqrt{2}}{\sqrt{2}}$.

$$\frac{3}{2\sqrt{2}} = \frac{3 \cdot \sqrt{2}}{2\sqrt{2} \cdot \sqrt{2}} = \frac{3\sqrt{2}}{2 \cdot 2} = \frac{3\sqrt{2}}{4}$$

WORK PROBLEM 3 AT THE SIDE. ▶▶

E X A M P L E 4   **Rationalizing Denominators in Roots of Fractions**

Simplify each of the following.

(a) $\sqrt{\dfrac{18}{125}}$

$$\sqrt{\frac{18}{125}} = \frac{\sqrt{18}}{\sqrt{125}} \qquad \text{Quotient rule}$$

$$= \frac{\sqrt{9 \cdot 2}}{\sqrt{25 \cdot 5}} \qquad \text{Factor.}$$

$$= \frac{3\sqrt{2}}{5\sqrt{5}} \qquad \text{Product rule}$$

$$= \frac{3\sqrt{2} \cdot \sqrt{5}}{5\sqrt{5} \cdot \sqrt{5}} \qquad \text{Multiply by } \frac{\sqrt{5}}{\sqrt{5}}.$$

$$= \frac{3\sqrt{10}}{5 \cdot 5} \qquad \text{Product rule}$$

$$= \frac{3\sqrt{10}}{25}$$

(b) $\sqrt{\dfrac{50m^4}{p^5}}, \quad p > 0$

$$\sqrt{\frac{50m^4}{p^5}} = \frac{\sqrt{50m^4}}{\sqrt{p^5}} \qquad \text{Quotient rule}$$

$$= \frac{5m^2\sqrt{2}}{p^2\sqrt{p}} \qquad \text{Product rule}$$

$$= \frac{5m^2\sqrt{2} \cdot \sqrt{p}}{p^2\sqrt{p} \cdot \sqrt{p}} \qquad \text{Multiply by } \frac{\sqrt{p}}{\sqrt{p}}.$$

$$= \frac{5m^2\sqrt{2p}}{p^2 \cdot p} \qquad \text{Product rule}$$

$$= \frac{5m^2\sqrt{2p}}{p^3}$$

WORK PROBLEM 4 AT THE SIDE. ▶▶

OBJECTIVE ▶ In the next example we show how to rationalize denominators in expressions with higher roots, such as cube roots.

3. Rationalize the denominators.

(a) $\dfrac{3}{\sqrt{48}}$

(b) $\dfrac{9}{\sqrt{20}}$

(c) $\dfrac{-16}{\sqrt{32}}$

4. Simplify. Assume all variables represent positive real numbers.

(a) $\sqrt{\dfrac{8}{45}}$

(b) $\sqrt{\dfrac{72}{y}}$

(c) $\sqrt{\dfrac{200k^6}{y^7}}$

ANSWERS

3. (a) $\dfrac{\sqrt{3}}{4}$ **(b)** $\dfrac{9\sqrt{5}}{10}$ **(c)** $-2\sqrt{2}$

4. (a) $\dfrac{2\sqrt{10}}{15}$ **(b)** $\dfrac{6\sqrt{2y}}{y}$ **(c)** $\dfrac{10k^3\sqrt{2y}}{y^4}$

5. Simplify.

(a) $\sqrt[3]{\dfrac{15}{32}}$

(b) $\sqrt[3]{\dfrac{m^{12}}{n}}, \quad n \neq 0$

E X A M P L E 5   **Rationalizing a Denominator Involving a Cube Root**

Simplify $\sqrt[3]{\dfrac{27}{16}}$.

Use the quotient rule first.

$$\sqrt[3]{\frac{27}{16}} = \frac{\sqrt[3]{27}}{\sqrt[3]{16}} = \frac{3}{\sqrt[3]{16}}$$

Simplify the denominator: $\sqrt[3]{16} = \sqrt[3]{8 \cdot 2} = 2\sqrt[3]{2}$. To get a rational denominator, multiply numerator and denominator by a number that will produce a perfect cube under the radical in the denominator. Because

$$\sqrt[3]{2} \cdot \sqrt[3]{2^2} = \sqrt[3]{2^3} = 2,$$

we should multiply numerator and denominator by $\sqrt[3]{2^2}$ or $\sqrt[3]{4}$.

$$\sqrt[3]{\frac{27}{16}} = \frac{3}{2\sqrt[3]{2}} = \frac{3 \cdot \sqrt[3]{4}}{2\sqrt[3]{2} \cdot \sqrt[3]{4}} = \frac{3\sqrt[3]{4}}{2\sqrt[3]{8}} = \frac{3\sqrt[3]{4}}{2 \cdot 2} = \frac{3\sqrt[3]{4}}{4}$$

> **Caution**
>
> It is easy to make mistakes in problems like the one in Example 5. It is incorrect to multiply numerator and denominator by $\sqrt[3]{2}$, because
>
> $$\sqrt[3]{2} \cdot \sqrt[3]{2} \neq 2.$$
>
> You need *three* factors of 2 to get 2^3 under the radical. As shown in Example 5,
>
> $$\sqrt[3]{2} \cdot \sqrt[3]{2^2} = 2.$$

◀◀ **WORK PROBLEM 5 AT THE SIDE.**

Objective ▶ Recall the special product

$$(a + b)(a - b) = a^2 - b^2.$$

The expressions $a + b$ and $a - b$ are **conjugates** (KAHN-juh-guts). In order to rationalize a denominator that contains a binomial expression (one that contains exactly two terms) involving radicals, such as

$$\frac{3}{1 + \sqrt{2}},$$

we must use conjugates. The conjugate of $1 + \sqrt{2}$ is $1 - \sqrt{2}$.

> Whenever a radical expression has a sum or difference with square root radicals in the denominator, we rationalize by multiplying both the numerator and the denominator by the conjugate of the denominator.

For the expression $\dfrac{3}{1 + \sqrt{2}}$, we rationalize the denominator by multiplying both the numerator and denominator by the conjugate of the denominator.

$$\frac{3}{1 + \sqrt{2}} = \frac{3(1 - \sqrt{2})}{(1 + \sqrt{2})(1 - \sqrt{2})}$$

Answers

5. (a) $\dfrac{\sqrt[3]{30}}{4}$

(b) $\dfrac{m^4\sqrt[3]{n^2}}{n}$

According to the special product mentioned earlier, $(1 + \sqrt{2})(1 - \sqrt{2}) = 1^2 - (\sqrt{2})^2 = 1 - 2 = -1$. Placing -1 in the denominator gives

$$= \frac{3(1 - \sqrt{2})}{-1}$$

$$= \frac{3}{-1}(1 - \sqrt{2})$$

$$= -3(1 - \sqrt{2}), \quad \text{or} \quad -3 + 3\sqrt{2}.$$

E X A M P L E 6   Rationalizing a Binomial Denominator

Rationalize the denominator of $\dfrac{5}{4 - \sqrt{3}}$.

Multiply numerator and denominator by $4 + \sqrt{3}$.

$$\frac{5}{4 - \sqrt{3}} = \frac{5(4 + \sqrt{3})}{(4 - \sqrt{3})(4 + \sqrt{3})}$$

$$= \frac{5(4 + \sqrt{3})}{16 - 3}$$

$$= \frac{5(4 + \sqrt{3})}{13}$$

Notice that the numerator is left in factored form. Doing this makes it easier to determine whether the expression can be reduced to lowest terms.

WORK PROBLEM 6 AT THE SIDE. ▶▶

E X A M P L E 7   Rationalizing Binomial Denominators

Rationalize each denominator.

(a) $\dfrac{\sqrt{2} - \sqrt{3}}{\sqrt{5} + \sqrt{3}} = \dfrac{(\sqrt{2} - \sqrt{3})(\sqrt{5} - \sqrt{3})}{(\sqrt{5} + \sqrt{3})(\sqrt{5} - \sqrt{3})}$

$$= \frac{\sqrt{10} - \sqrt{6} - \sqrt{15} + 3}{5 - 3}$$

$$= \frac{\sqrt{10} - \sqrt{6} - \sqrt{15} + 3}{2}$$

(b) $\dfrac{3}{\sqrt{5m} - \sqrt{p}} = \dfrac{3(\sqrt{5m} + \sqrt{p})}{(\sqrt{5m} - \sqrt{p})(\sqrt{5m} + \sqrt{p})}$

$$= \frac{3(\sqrt{5m} + \sqrt{p})}{5m - p} \quad (5m \neq p, m > 0, p > 0)$$

WORK PROBLEM 7 AT THE SIDE. ▶▶

OBJECTIVE 5 ▶ The final example shows how to write radical expressions in lowest terms. We will use this technique in the next chapter.

---

6. Rationalize the denominators.

(a) $\dfrac{-4}{\sqrt{5} + 2}$

(b) $\dfrac{18}{9 + \sqrt{18}}$

7. Rationalize the denominators.

(a) $\dfrac{15}{\sqrt{7} + \sqrt{2}}$

(b) $\dfrac{\sqrt{3} + \sqrt{5}}{\sqrt{2} - \sqrt{7}}$

(c) $\dfrac{2}{\sqrt{k} + \sqrt{z}}$, $k \neq z, k > 0$, $z > 0$

ANSWERS

6. (a) $-4(\sqrt{5} - 2)$ (b) $\dfrac{6(3 - \sqrt{2})}{7}$

7. (a) $3(\sqrt{7} - \sqrt{2})$

(b) $\dfrac{-(\sqrt{6} + \sqrt{21} + \sqrt{10} + \sqrt{35})}{5}$

(c) $\dfrac{2(\sqrt{k} - \sqrt{z})}{k - z}$

8. Write in lowest terms.

(a) $\dfrac{15 - 5\sqrt{3}}{5}$

E X A M P L E   8   **Writing Radical Expressions in Lowest Terms**

Write in lowest terms.

(a) $\dfrac{6 + 2\sqrt{5}}{4}$

Factor the numerator and denominator, then write in lowest terms.

$$\frac{6 + 2\sqrt{5}}{4} = \frac{2(3 + \sqrt{5})}{2 \cdot 2}$$

$$= \frac{3 + \sqrt{5}}{2}$$

Here is an alternative method to write this expression in lowest terms.

$$\frac{6 + 2\sqrt{5}}{4} = \frac{6}{4} + \frac{2\sqrt{5}}{4}$$

$$= \frac{3}{2} + \frac{\sqrt{5}}{2}$$

$$= \frac{3 + \sqrt{5}}{2}$$

(b) $\dfrac{5y - \sqrt{8y^2}}{6y} = \dfrac{5y - 2y\sqrt{2}}{6y}$ Product rule

$$= \frac{y(5 - 2\sqrt{2})}{6y}$$ Factor the numerator.

$$= \frac{5 - 2\sqrt{2}}{6} \quad (y > 0)$$

(b) $\dfrac{24 - 36\sqrt{7}}{16}$

Note that the final fraction cannot be reduced further.

$$\frac{5 - 2\sqrt{2}}{6} \neq \frac{5 - \sqrt{2}}{3},$$

because there is no common factor of 2 in the numerator.

Caution

Refer to part (a) in Example 8. It is incorrect to try to write in lowest terms *before* factoring. Always factor first before attempting to write in lowest terms.

◀◀ **WORK PROBLEM 8 AT THE SIDE.**

7.5 Exercises

MATHEMATICAL CONNECTIONS (EXERCISES 1–6)

The following operations apply to the exercises of this section. You should be able to perform them without writing intermediate steps. Fill in the blank with the correct response. Assume all variables represent positive real numbers.

1. $\sqrt{a} \cdot \sqrt{b} =$ _____

2. $(x + y)(x - y) =$ _____

3. $(x + \sqrt{y})(x - \sqrt{y}) =$ _____

4. $(\sqrt{x} + \sqrt{y})(\sqrt{x} - \sqrt{y}) =$ _____

5. $(x + y)^2 =$ _____

6. $(\sqrt{x} + \sqrt{y})^2 =$ _____

Multiply, then simplify the products. Assume that all variables represent positive real numbers. See Example 1.

7. $\sqrt{3}(\sqrt{12} - 4)$

8. $\sqrt{5}(\sqrt{125} - 6)$

9. $\sqrt{2}(\sqrt{18} - \sqrt{3})$

10. $\sqrt{5}(\sqrt{15} + \sqrt{5})$

11. $(\sqrt{6} + 2)(\sqrt{6} - 2)$

12. $(\sqrt{7} + 8)(\sqrt{7} - 8)$

13. $(\sqrt{12} - \sqrt{3})(\sqrt{12} + \sqrt{3})$

14. $(\sqrt{18} + \sqrt{8})(\sqrt{18} - \sqrt{8})$

15. $(\sqrt{3} + 2)(\sqrt{6} - 5)$

16. $(\sqrt{7} + 1)(\sqrt{2} - 4)$

17. $(\sqrt{3x} + 2)(\sqrt{3x} - 2)$

18. $(\sqrt{6y} - 4)(\sqrt{6y} + 4)$

19. $(2\sqrt{x} + \sqrt{y})(2\sqrt{x} - \sqrt{y})$

20. $(\sqrt{p} + 5\sqrt{s})(\sqrt{p} - 5\sqrt{s})$

21. $(4\sqrt{x} + 3)^2$

22. $(5\sqrt{p} - 6)^2$

23. $(9 - \sqrt[3]{2})(9 + \sqrt[3]{2})$

24. $(7 + \sqrt[3]{6})(7 - \sqrt[3]{6})$

25. The correct answer to Exercise 7 is $6 - 4\sqrt{3}$. Explain why this is not equal to $2\sqrt{3}$.

26. When we rationalize the denominator in the radical expression $\frac{1}{\sqrt{2}}$, we multiply both the numerator and the denominator by $\sqrt{2}$. What property of real numbers covered in Section 1.5 justifies this procedure?

Rationalize the denominator in each expression. Assume that all variables represent positive real numbers. See Examples 2 and 3.

27. $\dfrac{7}{\sqrt{7}}$

28. $\dfrac{11}{\sqrt{11}}$

29. $\dfrac{15}{\sqrt{3}}$

30. $\dfrac{12}{\sqrt{6}}$

31. $\dfrac{\sqrt{3}}{\sqrt{2}}$

32. $\dfrac{\sqrt{7}}{\sqrt{6}}$

33. $\dfrac{9\sqrt{3}}{\sqrt{5}}$

34. $\dfrac{3\sqrt{2}}{\sqrt{11}}$

35. $\dfrac{-6}{\sqrt{18}}$

36. $\dfrac{-5}{\sqrt{24}}$

37. $\dfrac{-8\sqrt{3}}{\sqrt{k}}$

38. $\dfrac{-4\sqrt{13}}{\sqrt{m}}$

39. $\dfrac{6\sqrt{3y}}{\sqrt{y^3}}$

40. $\dfrac{-8\sqrt{5y}}{\sqrt{y^5}}$

41. Look again at the expression in Exercise 39. Start by multiplying both the numerator and the denominator by $\sqrt{y}$, and obtain the final answer. Then start over, multiplying both the numerator and the denominator by $\sqrt{y^3}$, to obtain the same answer. Which method do you prefer? Why?

42. Explain why you cannot write $\dfrac{1}{\sqrt[3]{2}}$ with the denominator rationalized if you begin by multiplying both the numerator and the denominator by $\sqrt[3]{2}$. By what can you multiply them both to achieve the desired result?

Simplify. Assume that all variables represent positive real numbers. See Examples 4 and 5.

43. $\sqrt{\dfrac{7}{2}}$

44. $\sqrt{\dfrac{10}{3}}$

45. $-\sqrt{\dfrac{7}{50}}$

46. $-\sqrt{\dfrac{13}{75}}$

47. $\sqrt{\dfrac{24}{x}}$

48. $\sqrt{\dfrac{52}{y}}$

49. $-\sqrt{\dfrac{98r^3}{s}}$

50. $-\sqrt{\dfrac{150m^5}{n}}$

51. $\sqrt{\dfrac{288x^7}{y^9}}$

52. $\sqrt{\dfrac{242t^9}{u^{11}}}$

53. $\sqrt[3]{\dfrac{2}{3}}$ **54.** $\sqrt[3]{\dfrac{4}{5}}$ **55.** $\sqrt[3]{\dfrac{4}{9}}$ **56.** $\sqrt[3]{\dfrac{5}{16}}$ **57.** $-\sqrt[3]{\dfrac{2p}{r^2}}$

58. $-\sqrt[3]{\dfrac{6x}{y^2}}$ **59.** $\sqrt[4]{\dfrac{16}{x}}$ **60.** $\sqrt[4]{\dfrac{81}{y}}$

61. Explain the procedure you will use to rationalize the denominator of the expression in Exercise 63: $\dfrac{2}{4 + \sqrt{3}}$.

62. Would multiplying both the numerator and the denominator of $\dfrac{2}{4 + \sqrt{3}}$ by $4 + \sqrt{3}$ lead to a rationalized denominator? Why or why not?

Rationalize the denominator in each expression. Assume that all variables represent positive real numbers and that no denominators are zero. See Examples 6 and 7.

63. $\dfrac{2}{4 + \sqrt{3}}$ **64.** $\dfrac{6}{5 + \sqrt{2}}$ **65.** $\dfrac{6}{\sqrt{5} + \sqrt{3}}$ **66.** $\dfrac{12}{\sqrt{6} + \sqrt{3}}$

67. $\dfrac{-4}{\sqrt{3} - \sqrt{7}}$ **68.** $\dfrac{-3}{\sqrt{2} + \sqrt{5}}$ **69.** $\dfrac{1 - \sqrt{2}}{\sqrt{7} + \sqrt{6}}$ **70.** $\dfrac{-1 - \sqrt{3}}{\sqrt{6} + \sqrt{5}}$

71. $\dfrac{4\sqrt{x}}{\sqrt{x} - 2\sqrt{y}}$ **72.** $\dfrac{5\sqrt{r}}{3\sqrt{r} + \sqrt{s}}$ **73.** $\dfrac{\sqrt{x} - \sqrt{y}}{\sqrt{x} + \sqrt{y}}$ **74.** $\dfrac{\sqrt{a} + \sqrt{b}}{\sqrt{a} - \sqrt{b}}$

75. If a and b are both positive numbers and $a^2 = b^2$, then $a = b$. Use this fact to show that
$$\dfrac{\sqrt{6} - \sqrt{2}}{4} = \dfrac{\sqrt{2} - \sqrt{3}}{2}.$$

76. Use a calculator approximation to support the result in Exercise 75.

Write each expression in lowest terms. Assume that all variables represent positive real numbers. See Example 8.

77. $\dfrac{25 + 10\sqrt{6}}{20}$

78. $\dfrac{12 - 6\sqrt{2}}{24}$

79. $\dfrac{16 + 4\sqrt{8}}{12}$

80. $\dfrac{12 + 9\sqrt{72}}{18}$

81. $\dfrac{6x + \sqrt{24x^3}}{3x}$

82. $\dfrac{11y + \sqrt{242y^5}}{22y}$

83. The following expression occurs in a certain standard problem in trigonometry:

$$\frac{1}{\sqrt{2}} \cdot \frac{\sqrt{3}}{2} - \frac{1}{\sqrt{2}} \cdot \frac{1}{2}.$$

Show that it simplifies to $\dfrac{\sqrt{6} - \sqrt{2}}{4}$. Then verify using a calculator approximation.

84. The following expression occurs in a certain standard problem in trigonometry:

$$\frac{\sqrt{3} + 1}{1 - \sqrt{3}}.$$

Show that it simplifies to $-2 - \sqrt{3}$. Then verify using a calculator approximation.

MATHEMATICAL CONNECTIONS (EXERCISES 85–88)

Sometimes it is desirable to rationalize the *numerator* in an expression. The procedure is similar to rationalizing the denominator. For example, to rationalize the numerator of

$$\frac{6 - \sqrt{2}}{4},$$

we multiply by the conjugate of the numerator, $6 + \sqrt{2}$.

$$\frac{6 - \sqrt{2}}{4} = \frac{(6 - \sqrt{2})(6 + \sqrt{2})}{4(6 + \sqrt{2})} = \frac{36 - 2}{4(6 + \sqrt{2})} = \frac{34}{4(6 + \sqrt{2})}$$

In the final expression, the numerator is rationalized.

85. Rationalize the numerator of $\dfrac{8\sqrt{5} - 1}{6}$.

86. Rationalize the numerator of $\dfrac{3\sqrt{a} + \sqrt{b}}{\sqrt{b} - \sqrt{a}}$.

87. Rationalize the denominator of the expression in Exercise 86.

88. Describe the difference in the procedures used in Exercises 86 and 87.

7.6 Equations with Radicals

OBJECTIVE 1 ▶ The equation $x = 1$ has only one solution. Its solution set is $\{1\}$. If we square both sides of this equation, we get $x^2 = 1$. This new equation has two solutions: -1 and 1. Notice that the solution of the original equation is also a solution of the squared equation. However, the squared equation has another solution, -1, that is *not* a solution of the original equation. When solving equations with radicals, we will use this idea of raising both sides to a power. It is an application of the *power rule*.

Power Rule

If both sides of an equation are raised to the same power, all solutions of the original equation are also solutions of the new equation.

Read the power rule carefully; it does *not* say that all solutions of the new equation are solutions of the original equation. They may or may not be.

Note

When the power rule is used to solve an equation, every solution of the new equation must be checked in the original equation.

Solutions that do not satisfy the original equation are called **extraneous** (eks-TRAIN-ee-us) **solutions;** they must be discarded.

OBJECTIVE 2 ▶ The first example shows how to use the power rule in solving an equation.

E X A M P L E I   **Using the Power Rule**

Solve $\sqrt{3x + 4} = 8$.

We use the power rule and square both sides to get

$$(\sqrt{3x + 4})^2 = 8^2$$
$$3x + 4 = 64$$
$$3x = 60$$
$$x = 20.$$

To check, substitute the proposed solution in the original equation.

$$\sqrt{3x + 4} = 8$$
$$\sqrt{3 \cdot 20 + 4} = 8 \quad ? \quad \text{Let } x = 20.$$
$$\sqrt{64} = 8 \quad ?$$
$$8 = 8 \quad \text{True}$$

Since 20 satisfies the *original* equation, the solution set is $\{20\}$.

WORK PROBLEM I AT THE SIDE. ▶▶

The solution of the equation in Example 1 can be generalized to give a method for solving equations with radicals.

Solving Equations with Radicals

Step 1 Isolate one radical on one side of the equals sign.
Step 2 Raise each side of the equation to a power that is the same as the index of the radical.
Step 3 If possible, solve the resulting equation; if it still contains a radical, repeat Steps 1 and 2.
Step 4 Check all potential solutions in the *original equation*.

---

1. Solve.

(a) $\sqrt{r} = 3$

(b) $\sqrt{5x + 1} = 4$

487

2. Solve.

(a) $\sqrt{k} + 4 = -3$

Be careful not to skip Step 4, or you may get an incorrect solution set.

EXAMPLE 2 Using the Power Rule

Solve $\sqrt{5q - 1} + 3 = 0$.

Step 1 To get the radical alone on one side, we subtract 3 from both sides.

$$\sqrt{5q - 1} = -3$$

Step 2 Now we square both sides.

$$(\sqrt{5q - 1})^2 = (-3)^2$$

Step 3
$$5q - 1 = 9$$
$$5q = 10$$
$$q = 2$$

Step 4 The potential solution, 2, must be checked by substituting it in the original equation.

$$\sqrt{5q - 1} + 3 = 0$$
$$\sqrt{5 \cdot 2 - 1} + 3 = 0 \qquad ? \quad \text{Let } q = 2.$$
$$\mathbf{3 + 3 = 0} \qquad \text{False}$$

This false result shows that 2 is *not* a solution of the original equation; it is extraneous. The solution set is $\emptyset$.

Note
We could have determined after the first step that the equation in Example 2 has no solution. The equation $\sqrt{5q - 1} = -3$ has no solution because the expression on the left cannot be negative.

(b) $\sqrt{x - 9} - 3 = 0$

◀◀ **WORK PROBLEM 2 AT THE SIDE.**

OBJECTIVE 3 The next examples involve finding the square of a binomial. Recall that $(x + y)^2 = x^2 + 2xy + y^2$.

EXAMPLE 3 Using the Power Rule; Squaring a Binomial

Solve $\sqrt{4 - x} = x + 2$.

Square both sides; the square of $x + 2$ is $(x + 2)^2 = x^2 + 4x + 4$.

$$(\sqrt{4 - x})^2 = (x + 2)^2$$
$$4 - x = x^2 + 4x + 4$$
$$\qquad\qquad \llcorner \text{ Twice the product of 2 and } x$$
$$0 = x^2 + 5x \qquad \text{Subtract 4 and add } x.$$
$$0 = x(x + 5) \qquad \text{Factor.}$$
$$x = 0 \quad \text{or} \quad x + 5 = 0 \qquad \text{Zero-factor property}$$
$$x = -5$$

CONTINUED ON NEXT PAGE

Check each potential solution in the original equation.

If $x = 0$,

$$\sqrt{4 - x} = x + 2$$
$$\sqrt{4 - 0} = 0 + 2 \quad ?$$
$$\sqrt{4} = 2 \quad ?$$
$$2 = 2. \quad \text{True}$$

If $x = -5$,

$$\sqrt{4 - x} = x + 2$$
$$\sqrt{4 - (-5)} = -5 + 2 \quad ?$$
$$\sqrt{9} = -3 \quad ?$$
$$3 = -3. \quad \text{False}$$

The solution set is $\{0\}$. The other potential solution, -5, is extraneous.

Caution

When a radical equation requires squaring a binomial as in Example 3, the middle term is often omitted in error. Remember:

$$(x + 2)^2 \neq x^2 + 4y \qquad (x + 2)^2 = x^2 + 4x + 4$$

INCORRECT **CORRECT**

WORK PROBLEM 3 AT THE SIDE.

3. Solve $\sqrt{3z - 5} = z - 1$.

EXAMPLE 4 Using the Power Rule; Squaring a Binomial

Solve $\sqrt{m^2 - 4m + 9} = m - 1$.

Square both sides. The square of the binomial $m - 1$ is $(m - 1)^2 = m^2 - 2(m)(1) + 1^2$.

$$(\sqrt{m^2 - 4m + 9})^2 = (m - 1)^2$$
$$m^2 - 4m + 9 = m^2 - \mathbf{2m} + 1$$

(Twice the product of m and -1)

Subtract m^2 and 1 from both sides, then add $4m$ to both sides to get

$$8 = 2m$$
$$4 = m.$$

Check this potential solution in the original equation.

$$\sqrt{m^2 - 4m + 9} = m - 1$$
$$\sqrt{4^2 - 4 \cdot 4 + 9} = 4 - 1 \quad ? \quad \text{Let } m = 4.$$
$$3 = 3 \quad \text{True}$$

The solution set of the given equation is $\{4\}$.

WORK PROBLEM 4 AT THE SIDE.

4. Solve $\sqrt{4a^2 + 2a - 3} = 2a + 7$.

OBJECTIVE 4 In the next example, we solve an equation in which both sides must be squared twice. The solution requires an extra step.

EXAMPLE 5 Using the Power Rule; Squaring Twice

Solve $\sqrt{5m + 6} + \sqrt{3m + 4} = 2$.

Start by isolating one radical on one side of the equals sign. We can do this by subtracting $\sqrt{3m + 4}$ from both sides.

$$\sqrt{5m + 6} = 2 - \sqrt{3m + 4}$$

Now square both sides.

$$(\sqrt{5m + 6})^2 = (2 - \sqrt{3m + 4})^2$$
$$5m + 6 = 4 - \mathbf{4\sqrt{3m + 4}} + (3m + 4)$$

(Twice the product of 2 and $-\sqrt{3m + 4}$)

CONTINUED ON NEXT PAGE

ANSWERS

3. $\{2, 3\}$
4. $\{-2\}$

5. Solve $\sqrt{p + 1} - \sqrt{p - 4} = 1$.

This equation still contains a radical, so we will have to square both sides again. Before doing this, we combine terms on the right and simplify the equation.

$$5m + 6 = 8 + 3m - 4\sqrt{3m + 4}$$
$$2m - 2 = -4\sqrt{3m + 4} \qquad \text{Subtract 8 and } 3m.$$
$$m - 1 = -2\sqrt{3m + 4} \qquad \text{Divide by 2.}$$

Now we square both sides again.

$$(m - 1)^2 = (-2\sqrt{3m + 4})^2$$
$$m^2 - 2m + 1 = (-2)^2(\sqrt{3m + 4})^2 \qquad (ab)^2 = a^2b^2$$
$$m^2 - 2m + 1 = 4(3m + 4)$$
$$m^2 - 2m + 1 = 12m + 16 \qquad \text{Distributive property}$$

This equation is quadratic and may be solved with the zero-factor property. Start by getting 0 on one side of the equation, then factor.

$$m^2 - 14m - 15 = 0$$
$$(m - 15)(m + 1) = 0$$

By the zero-factor property,

$$m - 15 = 0 \quad \text{or} \quad m + 1 = 0$$
$$m = 15 \qquad\qquad m = -1.$$

6. Solve each equation.

(a) $\sqrt[3]{p^2 + 3p + 12} = \sqrt[3]{p^2}$

Check each of these potential solutions in the original equation. Only -1 works, so the equation has solution set $\{-1\}$. The other potential solution, 15, is extraneous.

◀◀ **WORK PROBLEM 5 AT THE SIDE.**

OBJECTIVE 5 ▶ The power rule also works for powers higher than 2, as the next example shows.

EXAMPLE 6   Using the Power Rule; Higher Powers

Solve $\sqrt[4]{m + 5} = \sqrt[4]{2m - 6}$.

We must raise both sides to the fourth power.

$$(\sqrt[4]{m + 5})^4 = (\sqrt[4]{2m - 6})^4$$
$$m + 5 = 2m - 6$$
$$11 = m$$

(b) $\sqrt[4]{2k + 5} + 1 = 0$

Verify that the solution set is $\{11\}$.

◀◀ **WORK PROBLEM 6 AT THE SIDE.**

7.6 Exercises

1. Is 9 a solution of the equation $\sqrt{x} = -3$? If not, what is the solution of this equation?

2. Before even attempting to solve $\sqrt{3x + 18} = x$, how can you be sure that the equation cannot have a negative solution?

Solve each equation. See Examples 1 and 2.

3. $\sqrt{x - 3} = 4$

4. $\sqrt{y + 2} = 5$

5. $\sqrt{3k - 2} = 6$

6. $\sqrt{4t + 7} = 9$

7. $\sqrt{x} + 9 = 0$

8. $\sqrt{w} + 4 = 0$

9. $\sqrt{3x - 6} - 3 = 0$

10. $\sqrt{7y + 11} - 5 = 0$

11. $\sqrt{6x + 2} - \sqrt{5x + 3} = 0$

12. $\sqrt{3 + 5x} - \sqrt{x + 11} = 0$

13. $3\sqrt{x} = \sqrt{8x + 9}$

14. $6\sqrt{p} = \sqrt{30p + 24}$

15. What is wrong with this first step in the solution process for $\sqrt{3x + 4} = 8 - x$?

$$3x + 4 = 64 + x^2$$

16. What is wrong with this first step in the solution process for $\sqrt{5y + 6} - \sqrt{y + 3} = 3$?

$$(5y + 6) + (y + 3) = 9$$

Solve each of the following equations. See Examples 3 and 4.

17. $\sqrt{3x + 4} = 8 - x$

18. $\sqrt{5x + 1} = 2x - 2$

19. $\sqrt{13 + 4t} = t + 4$

20. $\sqrt{50 + 7k} = k + 8$

21. $\sqrt{r^2 - 15r + 15} + 5 = r$

22. $\sqrt{p^2 + 12p - 4} + 4 = p$

Solve each equation. See Example 5.

23. $\sqrt{r + 4} - \sqrt{r - 4} = 2$

24. $\sqrt{m + 1} - \sqrt{m - 2} = 1$

25. $\sqrt{11 + 2q} + 1 = \sqrt{5q + 1}$

26. $\sqrt{6 + 5y} - 3 = \sqrt{y + 3}$

27. $\sqrt{3 - 3p} - \sqrt{3p + 2} = 3$

28. $\sqrt{3x + 4} - \sqrt{2x - 4} = 2$

29. What is the smallest power to which you can raise both sides of the radical equation $\sqrt[3]{x + 3} = \sqrt[3]{5 + 4x}$ so that the radicals are eliminated?

30. What is the smallest power to which you can raise both sides of the radical equation $\sqrt{x + 3} = \sqrt[3]{10x + 14}$ so that the radicals are eliminated?

Solve each equation. See Example 6.

31. $\sqrt[3]{2x^2 + 3x - 7} = \sqrt[3]{2x^2 + 4x + 6}$

32. $\sqrt[3]{3y^2 - 4y + 6} = \sqrt[3]{3y^2 - 2y + 8}$

33. $\sqrt[3]{1 - 2k} - \sqrt[3]{-k - 13} = 0$

34. $\sqrt[3]{11 - 2t} - \sqrt[3]{-1 - 5t} = 0$

35. $\sqrt[4]{x-1} + 2 = 0$

36. $\sqrt[4]{2k+3} + 1 = 0$

37. $\sqrt[4]{x+7} = \sqrt[4]{2x}$

38. $\sqrt[4]{y+8} = \sqrt[4]{3y}$

For the following equations, rewrite the expressions with rational exponents as radical expressions. Then solve using the procedures explained in this section.

39. $(5r-6)^{1/2} = 2 + (3r-6)^{1/2}$

40. $(3w+7)^{1/2} = 1 + (w+2)^{1/2}$

41. $(2w-1)^{2/3} - w^{1/3} = 0$

42. $(x^2-2x)^{1/3} - x^{1/3} = 0$

INTERPRETING TECHNOLOGY (EXERCISES 43–46)

Earlier, we saw two methods for using a graphing calculator to solve equations. The x-intercept method requires getting 0 on one side of the equation, calling the expression on the other side f(x), graphing y = f(x), and using the capability of the calculator to determine the x-intercept, which gives the solution of the equation. Solve the following equations using this method.

43. The screen shows the graph of the function $f(x) = \sqrt{x^2 - 4x + 9} - x + 1$. The x-intercept is given at the bottom of the screen. Use the screen to solve the equation $\sqrt{x^2 - 4x + 9} = x - 1$.

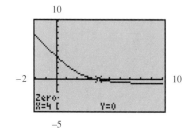

44. The screen shows the graph of $f(x) = (x+5)^{1/4} - (2x-6)^{1/4}$. Use it to solve $(x+5)^{1/4} = (2x-6)^{1/4}$.

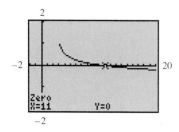

With the intersection-of-graphs method we call the expression on one side of the equation y_1, the expression on the other side y_2, graph both functions, and determine the x-value of the intersection, which is the solution of the equation. Solve the following equations using this method.

45. The screen shows the graphs of $f(x) = \sqrt{x + 4}$ and $g(x) = 2 + \sqrt{x - 4}$. Use it to solve $\sqrt{x + 4} = 2 + \sqrt{x - 4}$.

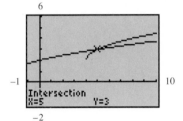

46. Use the screen, which shows the graphs of $f(x) = \sqrt{11 + 2x} + 1$ and $g(x) = \sqrt{5x + 1}$, to solve $\sqrt{11 + 2x} + 1 = \sqrt{5x + 1}$.

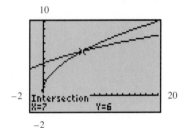

Work each problem, using a calculator to obtain an approximation for the answer. Give answers as decimals to the nearest tenth.

47. The period of a pendulum in seconds depends on its length, L, in feet, and is given by the formula

$$P = 2\pi\sqrt{\frac{L}{32}}.$$

How long is a pendulum with a period of 2.5 seconds? Use 3.14 for π.

48. The length, L, of an animal, in centimeters, is related to its surface area, S, by the formula

$$L = \sqrt{\frac{S}{a}},$$

where a is a constant that depends on the type of animal. If an animal has $a = .51$ and is 52 centimeters long, what is the surface area of the animal?

7.7 Complex Numbers

As we saw in Chapter 1, the set of real numbers includes many other number sets (the rational numbers, integers, and natural numbers, for example). In this section a new set of numbers is introduced that includes the set of real numbers, as well as numbers that are even roots of negative numbers, like $\sqrt{-2}$.

OBJECTIVE The equation $x^2 + 1 = 0$ has no real number solutions, since any solution must be a number whose square is -1. In the set of real numbers all squares are nonnegative numbers, because multiplication is defined in such a way that the product of two positive numbers or two negative numbers is always positive. To provide a solution for the equation $x^2 + 1 = 0$, a new number i, called the **imaginary** (ih-MAJ-uh-nair-ee) **unit,** is defined so that

$$i^2 = -1.$$

That is, i is a number whose square is -1. This definition of i makes it possible to define any square root of a negative number as follows.

> For any positive number b,
> $$\sqrt{-b} = i\sqrt{b}.$$

E X A M P L E   1   Simplifying Square Roots of Negative Numbers

Write each number as a product of a real number and i.

(a) $\sqrt{-100} = i\sqrt{100} = 10i$

(b) $\sqrt{-2} = i\sqrt{2}$

Caution
It is easy to mistake $\sqrt{2}i$ for $\sqrt{2i}$, with the i under the radical. For this reason, we often write $\sqrt{2}i$ as $i\sqrt{2}$.

WORK PROBLEM 1 AT THE SIDE. ▶▶

When finding a product such as $\sqrt{-4} \cdot \sqrt{-9}$, we cannot use the product rule for radicals because that rule applies only when no more than one radicand is negative. For this reason, we change $\sqrt{-b}$ to the form $i\sqrt{b}$ before performing any multiplications or divisions. For example,

$$\sqrt{-4} \cdot \sqrt{-9} = i\sqrt{4} \cdot i\sqrt{9}$$
$$= i \cdot 2 \cdot i \cdot 3$$
$$= 6i^2$$
$$= 6(-1) \qquad \text{Let } i^2 = -1.$$
$$= -6.$$

An *incorrect* use of the product rule for radicals would give a *wrong* answer.

$$\sqrt{-4} \cdot \sqrt{-9} = \sqrt{(-4)(-9)}$$
$$= \sqrt{36}$$
$$= 6 \qquad \text{INCORRECT}$$

OBJECTIVES

1 ▶ Simplify numbers of the form $\sqrt{-b}$, where $b > 0$.

2 ▶ Recognize imaginary and complex numbers.

3 ▶ Add and subtract complex numbers.

4 ▶ Find products of complex numbers.

5 ▶ Find quotients of complex numbers.

6 ▶ Find powers of i.

FOR EXTRA HELP

Tutorial   Tape 12   SSM, Sec. 7.7

1. Write each number as a product of a real number and i.

(a) $\sqrt{-16}$

(b) $-\sqrt{-81}$

(c) $\sqrt{-7}$

ANSWERS

1. (a) $4i$ **(b)** $-9i$ **(c)** $i\sqrt{7}$

2. Multiply.

(a) $\sqrt{-7} \cdot \sqrt{-7}$

(b) $\sqrt{-5} \cdot \sqrt{-10}$

(c) $\sqrt{-15} \cdot \sqrt{2}$

3. Divide.

(a) $\dfrac{\sqrt{-32}}{\sqrt{-2}}$

(b) $\dfrac{\sqrt{-27}}{\sqrt{-3}}$

(c) $\dfrac{\sqrt{-40}}{\sqrt{10}}$

E X A M P L E   2   Multiplying Square Roots of Negative Numbers

Multiply.

(a) $\sqrt{-3} \cdot \sqrt{-7} = i\sqrt{3} \cdot i\sqrt{7}$
$$= i^2\sqrt{3 \cdot 7}$$
$$= (-1)\sqrt{21}$$
$$= -\sqrt{21}$$

(b) $\sqrt{-2} \cdot \sqrt{-8} = i\sqrt{2} \cdot i\sqrt{8}$
$$= i^2\sqrt{2 \cdot 8}$$
$$= (-1)\sqrt{16}$$
$$= (-1)4 = -4$$

(c) $\sqrt{-5} \cdot \sqrt{6} = i\sqrt{5} \cdot \sqrt{6} = i\sqrt{30}$

◄◄ **WORK PROBLEM 2 AT THE SIDE.**

The methods used to find the products in Example 2 also apply to quotients, as the next example shows.

E X A M P L E   3   Dividing Square Roots of Negative Numbers

Divide.

(a) $\dfrac{\sqrt{-75}}{\sqrt{-3}} = \dfrac{i\sqrt{75}}{i\sqrt{3}} = \sqrt{\dfrac{75}{3}} = \sqrt{25} = 5$

(b) $\dfrac{\sqrt{-32}}{\sqrt{8}} = \dfrac{i\sqrt{32}}{\sqrt{8}} = i\sqrt{\dfrac{32}{8}} = i\sqrt{4} = 2i$

◄◄ **WORK PROBLEM 3 AT THE SIDE.**

OBJECTIVE 2 With the new number i and the real numbers, a new set of numbers can be formed that includes the real numbers as a subset. The *complex numbers* are defined as follows.

If a and b are real numbers, then any number of the form $a + bi$ is called a **complex number.**

In the complex number $a + bi$, the number a is called the **real part** and b is called the **imaginary part.*** When $b = 0$, $a + bi$ is a real number, so the real numbers are a subset of the complex numbers. Complex numbers with $b \neq 0$ are called **imaginary numbers.**** In spite of their name, imaginary numbers are very useful in applications, particularly in work with electricity.

_____
* In some texts, bi is called the imaginary part.
** Imaginary numbers are sometimes defined as complex numbers with $a = 0$ and $b \neq 0$.

ANSWERS
2. (a) -7 (b) $-5\sqrt{2}$ (c) $i\sqrt{30}$
3. (a) 4 (b) 3 (c) $2i$

The relationships among the various sets of numbers discussed in this book are shown in Figure 3.

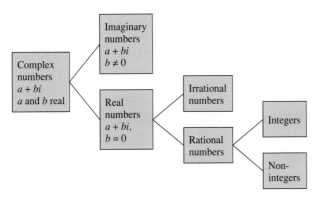

FIGURE 3

OBJECTIVE 3▶ The commutative, associative, and distributive properties for real numbers are also valid for complex numbers. Thus, to add complex numbers, we add their real parts and add their imaginary parts.

┌─**E X A M P L E   4**   **Adding Complex Numbers**
Add.
(a) $(2 + 3i) + (6 + 4i)$

   $= (2 + 6) + (3 + 4)i$ Commutative and associative properties

   $= 8 + 7i$

(b) $5 + (9 - 3i) = (5 + 9) - 3i$

   $= 14 - 3i$
└──■

WORK PROBLEM 4 AT THE SIDE. ▶▶

We subtract complex numbers by subtracting their real parts and subtracting their imaginary parts.

┌─**E X A M P L E   5**   **Subtracting Complex Numbers**
Subtract.
(a) $(6 + 5i) - (3 + 2i) = (6 - 3) + (5 - 2)i$

   $= 3 + 3i$

(b) $(7 - 3i) - (8 - 6i) = (7 - 8) + [-3 - (-6)]i$

   $= -1 + 3i$

(c) $(-9 + 4i) - (-9 + 8i) = (-9 + 9) + (4 - 8)i$

   $= 0 - 4i = -4i$
└──■

WORK PROBLEM 5 AT THE SIDE. ▶▶

4. Add.

   (a) $(4 + 6i) + (-3 + 5i)$

   (b) $(-1 + 8i) + (9 - 3i)$

5. Subtract.

   (a) $(7 + 3i) - (4 + 2i)$

   (b) $(-6 - i) - (-5 - 4i)$

   (c) $8 - (3 - 2i)$

ANSWERS
4. (a) $1 + 11i$ **(b)** $8 + 5i$
5. (a) $3 + i$ **(b)** $-1 + 3i$ **(c)** $5 + 2i$

6. Multiply.

(a) $6i(4 + 3i)$

(b) $(6 - 4i)(2 + 4i)$

(c) $(3 - 2i)(3 + 2i)$

Objective ▶4 Complex numbers of the form $a + bi$ have the same form as binomials, so we multiply two complex numbers by using the FOIL method for multiplying binomials. (Recall that FOIL stands for *First-Outside-Inside-Last*.)

EXAMPLE 6 Multiplying Complex Numbers

(a) $4i(2 + 3i) = (4i)(2) + (4i)(3i) = 8i + 12i^2 = 8i + 12(-1)$

$$= -12 + 8i$$

(b) $(3 + 5i)(4 - 2i) = \underbrace{3(4)}_{\text{First}} + \underbrace{3(-2i)}_{\text{Outside}} + \underbrace{5i(4)}_{\text{Inside}} + \underbrace{5i(-2i)}_{\text{Last}}$

Now simplify. (Remember that $i^2 = -1$.)

$$= 12 - 6i + 20i - 10i^2$$
$$= 12 + 14i - 10(-1) \qquad \text{Let } i^2 = -1.$$
$$= 12 + 14i + 10$$
$$= 22 + 14i$$

(c) $(2 + 3i)(1 - 5i) = 2(1) + 2(-5i) + 3i(1) + 3i(-5i)$

$$= 2 - 10i + 3i - 15i^2$$
$$= 2 - 7i - 15(-1)$$
$$= 2 - 7i + 15$$
$$= 17 - 7i$$

◀◀ **WORK PROBLEM 6 AT THE SIDE.**

The two complex numbers $a + bi$ and $a - bi$ are called **conjugates** of each other. The product of a complex number and its conjugate is always a real number, as shown here.

$$(a + bi)(a - bi) = a \cdot a - abi + abi - b^2i^2$$
$$= a^2 - b^2(-1)$$
$$(a + bi)(a - bi) = a^2 + b^2$$

For example, $(3 + 7i)(3 - 7i) = 3^2 + 7^2 = 9 + 49 = 58$.

Objective ▶5 The quotient of two complex numbers should be a complex number. To write the quotient as a complex number, we need to eliminate i in the denominator. We use conjugates to do this.

EXAMPLE 7 Dividing Complex Numbers

Find the quotients.

(a) $\dfrac{4 - 3i}{5 + 2i}$

Using the property of 1, we multiply the numerator and denominator by the conjugate of the denominator. The conjugate of $5 + 2i$ is $5 - 2i$.

CONTINUED ON NEXT PAGE

$$\frac{4 - 3i}{5 + 2i} = \frac{(4 - 3i)(5 - 2i)}{(5 + 2i)(5 - 2i)}$$

$$= \frac{20 - 8i - 15i + 6i^2}{5^2 + 2^2} \qquad \text{FOIL method}$$

$$= \frac{14 - 23i}{29} \quad \text{or} \quad \frac{14}{29} - \frac{23}{29}i \qquad \text{Write as } a + bi.$$

Notice that this is just like rationalizing the denominator.

(b) $\dfrac{1 + i}{i}$

The conjugate of i is $-i$. Multiply the numerator and denominator by $-i$.

$$\frac{1 + i}{i} = \frac{(1 + i)(-i)}{i(-i)}$$

$$= \frac{-i - i^2}{-i^2}$$

$$= \frac{-i - (-1)}{-(-1)}$$

$$= \frac{-i + 1}{1} = 1 - i$$

WORK PROBLEM 7 AT THE SIDE. ▶▶

OBJECTIVE 6 ▶ The fact that i^2 is equal to -1 can be used to find higher powers of i, as shown below.

$$i^3 = i \cdot i^2 = i(-1) = -i \qquad i^6 = i^2 \cdot i^4 = (-1) \cdot 1 = -1$$
$$i^4 = i^2 \cdot i^2 = (-1)(-1) = 1 \qquad i^7 = i^3 \cdot i^4 = (-i) \cdot 1 = -i$$
$$i^5 = i \cdot i^4 = i \cdot 1 = i \qquad i^8 = i^4 \cdot i^4 = 1 \cdot 1 = 1$$

As these examples show, the powers of i rotate through four numbers: i, -1, $-i$, and 1. Larger powers of i can be simplified by using the fact that $i^4 = 1$. For example, $i^{75} = (i^4)^{18} \cdot i^3 = 1^{18} \cdot i^3 = 1 \cdot i^3 = i^3 = -i$. This example suggests a quick method for simplifying large powers of i.

EXAMPLE 8 Simplifying Powers of i

Find each power of i.

(a) $i^{12} = (i^4)^3 = 1^3 = 1$

(b) $i^{39} = i^{36} \cdot i^3$
$$= (i^4)^9 \cdot i^3$$
$$= 1^9 \cdot (-i)$$
$$= -i$$

(c) $i^{-2} = \dfrac{1}{i^2} = \dfrac{1}{-1} = -1$

CONTINUED ON NEXT PAGE

7. Find the quotients.

(a) $\dfrac{2 + i}{3 - i}$

(b) $\dfrac{6 + 2i}{4 - 3i}$

(c) $\dfrac{5}{3 - 2i}$

(d) $\dfrac{5 - i}{i}$

ANSWERS

7. (a) $\dfrac{1}{2} + \dfrac{1}{2}i$ **(b)** $\dfrac{18}{25} + \dfrac{26}{25}i$
(c) $\dfrac{15}{13} + \dfrac{10}{13}i$ **(d)** $-1 - 5i$

8. Find each power of i.

(a) i^{21}

(b) i^{36}

(c) i^{50}

(d) i^{-9}

(d) $i^{-1} = \dfrac{1}{i}$

To simplify this quotient, multiply numerator and denominator by $-i$, the conjugate of i.

$$\frac{1}{i} = \frac{1(-i)}{i(-i)}$$

$$= \frac{-i}{-i^2}$$

$$= \frac{-i}{-(-1)}$$

$$= \frac{-i}{1}$$

$$= -i$$

◀◀ **WORK PROBLEM 8 AT THE SIDE.**

7.7 *Exercises*

Complete each statement.

1. _____ is the imaginary unit.

2. $i^2 =$ _____ .

3. $\sqrt{-b} = i$ _____ , for $b > 0$.

4. The real part of the complex number $a + bi$ is _____ .

5. Every real number is a complex number. Explain why this is so.

6. Not every complex number is a real number. Give an example of this, and explain why this statement is true.

Write each as a product of a real number and i. Simplify all radical expressions. See Example 1.

7. $\sqrt{-169}$

8. $\sqrt{-225}$

9. $-\sqrt{-144}$

10. $-\sqrt{-196}$

11. $\sqrt{-5}$

12. $\sqrt{-21}$

13. $\sqrt{-48}$

14. $\sqrt{-96}$

Multiply or divide as indicated. See Examples 2 and 3.

15. $\sqrt{-15} \cdot \sqrt{-15}$

16. $\sqrt{-19} \cdot \sqrt{-19}$

17. $\sqrt{-4} \cdot \sqrt{-25}$

18. $\sqrt{-9} \cdot \sqrt{-81}$

19. $\dfrac{\sqrt{-300}}{\sqrt{-100}}$

20. $\dfrac{\sqrt{-40}}{\sqrt{-10}}$

21. $\dfrac{\sqrt{-75}}{\sqrt{3}}$

22. $\dfrac{\sqrt{-160}}{\sqrt{10}}$

Add or subtract as indicated. Write your answers in the form $a + bi$. See Examples 4 and 5.

23. $(3 + 2i) + (-4 + 5i)$

24. $(7 + 15i) + (-11 + 14i)$

25. $(5 - i) + (-5 + i)$

26. $(-2 + 6i) + (2 - 6i)$

27. $(4 + i) - (-3 - 2i)$

28. $(9 + i) - (3 + 2i)$

29. $(-3 - 4i) - (-1 - 4i)$

30. $(-2 - 3i) - (-5 - 3i)$

31. $(-4 + 11i) + (-2 - 4i) + (7 + 6i)$

32. $(-1 + i) + (2 + 5i) + (3 + 2i)$

33. $[(7 + 3i) - (4 - 2i)] + (3 + i)$

34. $[(7 + 2i) + (-4 - i)] - (2 + 5i)$

35. Fill in the blank with the correct response: Because $(4 + 2i) - (3 + i) = 1 + i$, using the definition of subtraction we can check this to find that $(1 + i) + (3 + i) = $ _____ .

36. Fill in the blank with the correct response: Because $\frac{-5}{2-i} = -2 - i$, using the definition of division we can check this to find that $(-2 - i)(2 - i) = $ _____ .

Multiply. See Example 6.

37. $(3i)(27i)$

38. $(5i)(125i)$

39. $(-8i)(-2i)$

40. $(-32i)(-2i)$

41. $5i(-6 + 2i)$

42. $3i(4 + 9i)$

43. $(4 + 3i)(1 - 2i)$

44. $(7 - 2i)(3 + i)$

45. $(4 + 5i)^2$

46. $(3 + 2i)^2$

47. $(12 + 3i)(12 - 3i)$

48. $(6 + 7i)(6 - 7i)$

49. (a) What is the conjugate of $a + bi$?
(b) If we multiply $a + bi$ by its conjugate, we get _____ + _____ , which is always a real number.

50. Explain the procedure you would use to find the quotient

$$\frac{-1 + 5i}{3 + 2i}.$$

Write in the form a + bi. See Example 7.

51. $\dfrac{2}{1 - i}$

52. $\dfrac{29}{5 + 2i}$

53. $\dfrac{-7 + 4i}{3 + 2i}$

54. $\dfrac{-38 - 8i}{7 + 3i}$

55. $\dfrac{8i}{2 + 2i}$

56. $\dfrac{-8i}{1 + i}$

57. $\dfrac{2 - 3i}{2 + 3i}$

58. $\dfrac{-1 + 5i}{3 + 2i}$

MATHEMATICAL CONNECTIONS (EXERCISES 59–64)

Consider the following expressions:

Binomials	**Complex Numbers**
$x + 2, \quad 3x - 1$	$1 + 2i, \quad 3 - i.$

When we add, subtract, or multiply complex numbers in standard form, the rules are the same as those for the corresponding operations on binomials. That is, we add or subtract like terms, and we use FOIL *to multiply. Division, however, is comparable to division by the sum or difference of radicals, where we multiply by the conjugate to get a rational denominator. To express the quotient of two complex numbers in standard form, we also multiply by the conjugate of the denominator.*

The following exercises illustrate these ideas. Work them in order.

59. (a) Add the two binomials.   **(b)** Add the two complex numbers.

60. (a) Subtract the second binomial from the first.
   (b) Subtract the second complex number from the first.

61. (a) Multiply the two binomials.   **(b)** Multiply the two complex numbers.

62. (a) Rationalize the denominator: $\dfrac{\sqrt{3} - 1}{1 + \sqrt{2}}$.   **(b)** Write in standard form: $\dfrac{3 - i}{1 + 2i}$.

63. Explain why the answers for parts (a) and (b) in Exercise 61 do not correspond as the answers in Exercises 59 and 60 do.

64. Explain why the answers for parts (a) and (b) in Exercise 62 do not correspond as the answers in Exercises 59 and 60 do.

65. Recall that if $a \neq 0$, $\frac{1}{a}$ is called the reciprocal of a. Use this definition to express the reciprocal of $5 - 4i$ in the form $a + bi$.

66. Recall that if $a \neq 0$, a^{-1} is defined to be $\frac{1}{a}$. Use this definition to express $(4 - 3i)^{-1}$ in the form $a + bi$.

Find each power of i. See Example 8.

67. i^{18}

68. i^{26}

69. i^{89}

70. i^{45}

71. i^{96}

72. i^{48}

73. i^{-5}

74. i^{-17}

75. A student simplified i^{-18} as follows:

$$i^{-18} = i^{-18} \cdot i^{20} = i^{-18+20} = i^2 = -1.$$

Explain the mathematical justification for this correct work.

76. Add: $3(2 - i)^{-1} + 5(1 + i)^{-1}$.

Ohm's law for the current I in a circuit with voltage E, resistance R, capacitance reactance X_c, and inductive reactance X_L is

$$I = \frac{E}{R + (X_L - X_c)i}.$$

77. Find I if $E = 2 + 3i$, $R = 5$, $X_L = 4$, and $X_c = 3$.

78. Using the law given for Exercise 77, find E if $I = 1 - i$, $R = 2$, $X_L = 3$, and $X_c = 1$.

79. Show that $1 + 5i$ is a solution of

$$x^2 - 2x + 26 = 0.$$

80. Show that $3 + 2i$ is a solution of

$$x^2 - 6x + 13 = 0.$$

CHAPTER 7 SUMMARY

7.1	square root	The number b is a square root of a if $b^2 = a$.
	cube root, fourth root	These are defined in a manner similar to square roots (see above).
	radicand, index	In the expression $\sqrt[n]{a}$, a is the radicand and n is the index.
	principal root	For a positive number a and even value of n, the principal nth root of a is the positive nth root of a.
7.4	radical expression	A radical expression is an algebraic expression that contains radicals.
7.5	rationalizing the denominator	The process of removing radicals from the denominator so that the denominator contains only rational quantities is called rationalizing the denominator.
	conjugate	The conjugate of $a + bi$ is $a - bi$.
7.6	extraneous solution	An extraneous solution of a radical equation is a solution of $x = a^2$ that is not a solution of $\sqrt{x} = a$.
7.7	imaginary unit	The number i is called the imaginary unit.
	complex number	A complex number is a number that can be written in the form $a + bi$, where a and b are real numbers.
	real part	The real part of $a + bi$ is a.
	imaginary part	The imaginary part of $a + bi$ is b.
	imaginary number	A complex number $a + bi$ with $b \neq 0$ is called an imaginary number.

$\sqrt{}$	radical sign
$\pm$	plus or minus
$\sqrt[n]{a}$	radical; principal nth root of a
$\approx$	is approximately equal to
i	a number whose square is -1
$a^{1/n}$	a to the power $\dfrac{1}{n}$
$a^{m/n}$	a to the power $\dfrac{m}{n}$

Concepts	Examples
7.1 Radicals	
$\sqrt[n]{a} = b$ means $b^n = a$. $\sqrt[n]{a}$ is the principal nth root of a. $\sqrt[n]{a^n} = \lvert a \rvert$ if n is even. $\sqrt[n]{a^n} = a$ if n is odd.	The two square roots of 64 are $\sqrt{64} = 8$ and $-\sqrt{64} = -8$. Of these, 8 is the principal square root of 64. $\sqrt[3]{-27} = -3 \qquad \sqrt[4]{(-2)^4} = \lvert -2 \rvert = 2$

Concepts	Examples
7.2 Rational Exponents $a^{1/n} = \sqrt[n]{a}$ whenever $\sqrt[n]{a}$ exists. If m and n are positive integers with m/n in lowest terms, then $$a^{m/n} = (a^{1/n})^m$$ provided that $a^{1/n}$ is a real number. All of the usual rules for exponents are valid for rational exponents.	$$25^{1/2} = \sqrt{25} = 5$$ $$(-64)^{1/3} = \sqrt[3]{-64} = -4$$ $$8^{5/3} = (8^{1/3})^5 = 2^5 = 32$$ $$5^{-1/2} \cdot 5^{1/4} = 5^{-1/2+1/4} = 5^{-1/4}$$ $$(y^{2/5})^{10} = y^4$$ $$\frac{x^{-1/3}}{x^{-1/2}} = x^{-1/3-(-1/2)}$$ $$= x^{-1/3+1/2} = x^{1/6} \quad (x > 0)$$
7.3 Simplifying Radicals **Product and Quotient Rules for Radicals** If a and b are real numbers, not both negative, all roots are real, and n is a natural number, $$\sqrt[n]{a} \cdot \sqrt[n]{b} = \sqrt[n]{ab}$$ and $$\sqrt[n]{\frac{a}{b}} = \frac{\sqrt[n]{a}}{\sqrt[n]{b}} \quad (b \ne 0).$$ **Simplified Radical** 1. The radicand has no factor raised to a power greater than or equal to the index. 2. The radicand has no fractions. 3. No denominator contains a radical. 4. Exponents in the radicand and the index of the radical have no common factors (except 1).	$$\sqrt{3} \cdot \sqrt{7} = \sqrt{21}$$ $$\sqrt[5]{x^3 y} \cdot \sqrt[5]{xy^2} = \sqrt[5]{x^4 y^3}$$ $$\frac{\sqrt{x^5}}{\sqrt{x^4}} = \sqrt{\frac{x^5}{x^4}} = \sqrt{x} \quad (x > 0)$$ $$\sqrt{18} = \sqrt{9 \cdot 2} = 3\sqrt{2}$$ $$\sqrt[3]{54x^5 y^3} = \sqrt[3]{27x^3 y^3 \cdot 2x^2} = 3xy\sqrt[3]{2x^2}$$ $$\sqrt{\frac{7}{4}} = \frac{\sqrt{7}}{\sqrt{4}} = \frac{\sqrt{7}}{2}$$ $$\sqrt[9]{x^3} = x^{3/9} = x^{1/3} \quad \text{or} \quad \sqrt[3]{x}$$
Pythagorean Formula If c is the length of the longest side of a right triangle and a and b are the lengths of the shorter sides, then $c^2 = a^2 + b^2$. The longest side is the hypotenuse and the two shorter sides are the legs of the triangle. The hypotenuse is opposite the right angle.	Find b for the triangle in the figure. 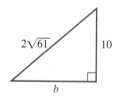 $$10^2 + b^2 = (2\sqrt{61})^2$$ $$b^2 = 4(61) - 100 = 144$$ $$\mathbf{b = 12}$$
Distance Formula The distance between (x_1, y_1) and (x_2, y_2) is $$d = \sqrt{(x_2 - x_1)^2 + (y_2 - y_1)^2}.$$	The distance between $(3, -2)$ and $(-1, 1)$ is given by $$\sqrt{(-1 - 3)^2 + [1 - (-2)]^2}$$ $$= \sqrt{(-4)^2 + 3^2} = \sqrt{16 + 9}$$ $$= \sqrt{25} = 5.$$

Concepts	Examples
7.4 Addition and Subtraction of Radical Expressions Only radical expressions with the same index and the same radicand may be combined.	$3\sqrt{17} + 2\sqrt{17} - 8\sqrt{17} = (3 + 2 - 8)\sqrt{17}$ $= -3\sqrt{17}$ $\sqrt[3]{2} - \sqrt[3]{250} = \sqrt[3]{2} - 5\sqrt[3]{2}$ $= -4\sqrt[3]{2}$ $\left.\begin{array}{l}\sqrt{15} + \sqrt{30}\\ \sqrt{3} + \sqrt[3]{9}\end{array}\right\}$ cannot be further simplified
7.5 Multiplication and Division of Radical Expressions Multiply radical expressions by using the FOIL method. Special products from Section 5.4 may apply.	$(\sqrt{2} + \sqrt{7})(\sqrt{3} - \sqrt{6})$ $= \sqrt{6} - \sqrt{12} + \sqrt{21} - \sqrt{42}$ $= \sqrt{6} - 2\sqrt{3} + \sqrt{21} - \sqrt{42}$ $(\sqrt{5} - \sqrt{10})(\sqrt{5} + \sqrt{10}) = 5 - 10 = -5$ $(\sqrt{3} - \sqrt{2})^2 = 3 - 2\sqrt{3}\cdot\sqrt{2} + 2$ $= 5 - 2\sqrt{6}$
Rationalize the denominator by multiplying both the numerator and denominator by the same expression (using the property of 1).	$\dfrac{\sqrt{7}}{\sqrt{5}} = \dfrac{\sqrt{7}}{\sqrt{5}}\cdot\dfrac{\sqrt{5}}{\sqrt{5}} = \dfrac{\sqrt{35}}{5}$ $\dfrac{\sqrt[3]{2}}{\sqrt[3]{4}} = \dfrac{\sqrt[3]{2}}{\sqrt[3]{4}}\cdot\dfrac{\sqrt[3]{2}}{\sqrt[3]{2}} = \dfrac{\sqrt[3]{4}}{\sqrt[3]{8}} = \dfrac{\sqrt[3]{4}}{2}$ $\dfrac{4}{\sqrt{5} - \sqrt{2}} = \dfrac{4}{\sqrt{5} - \sqrt{2}}\cdot\dfrac{\sqrt{5} + \sqrt{2}}{\sqrt{5} + \sqrt{2}}$ $= \dfrac{4(\sqrt{5} + \sqrt{2})}{5 - 2}$ $= \dfrac{4(\sqrt{5} + \sqrt{2})}{3}$
To write a quotient involving radicals, such as $$\dfrac{5 + 15\sqrt{6}}{10}$$ in lowest terms, factor the numerator and denominator, and then replace their common factors by 1.	$\dfrac{5 + 15\sqrt{6}}{10} = \dfrac{5(1 + 3\sqrt{6})}{5\cdot 2}$ $= 1\cdot\dfrac{1 + 3\sqrt{6}}{2} = \dfrac{1 + 3\sqrt{6}}{2}$

Concepts	Examples
7.6 Equations with Radicals	Solve $\sqrt{2x + 3} - x = 0$.
Solving Equations with Radicals	
Step 1 Isolate one radical on one side of the equal sign.	$$\sqrt{2x + 3} = x$$
Step 2 Raise each side of the equation to a power that is the same as the index of the radical.	$$2x + 3 = x^2$$
Step 3 If possible, solve the resulting equation; if it still contains a radical, repeat Steps 1 and 2.	$$x^2 - 2x - 3 = 0$$ $$(x - 3)(x + 1) = 0$$
Step 4 Check all potential solutions in the *original* equation.	$x - 3 = 0$ or $x + 1 = 0$ $\qquad x = 3 \qquad\qquad x = -1$
Potential solutions that do not check are *extraneous;* they are not part of the solution set.	A check shows that 3 is a solution, but -1 is extraneous. The solution set is $\{3\}$.
7.7 Complex Numbers $$i^2 = -1$$ For any positive number b, $\sqrt{-b} = i\sqrt{b}$. To multiply $\sqrt{-3} \cdot \sqrt{-27}$, first change each factor to the form $i\sqrt{b}$, then multiply.	$$\sqrt{-3} \cdot \sqrt{-27} = i\sqrt{3} \cdot i\sqrt{27}$$ $$= i^2\sqrt{81}$$ $$= -1 \cdot 9 = -9$$
Quotients such as $$\frac{\sqrt{-18}}{\sqrt{-2}}$$ are found similarly.	$$\frac{\sqrt{-18}}{\sqrt{-2}} = \frac{i\sqrt{18}}{i\sqrt{2}} = \frac{\sqrt{18}}{\sqrt{2}} = \sqrt{9} = 3$$
Adding and Subtracting Complex Numbers Add (or subtract) the real parts and add (or subtract) the imaginary parts.	$$(5 + 3i) + (8 - 7i) = 13 - 4i$$ $$(5 + 3i) - (8 - 7i)$$ $$= (5 - 8) + [3 - (-7)]i$$ $$= -3 + 10i$$
Multiplying and Dividing Complex Numbers Multiply complex numbers by using the FOIL method. Divide complex numbers by multiplying the numerator and the denominator by the conjugate of the denominator.	$$(2 + i)(5 - 3i) = 10 - 6i + 5i - 3i^2$$ $$= 10 - i - 3(-1)$$ $$= 10 - i + 3$$ $$= 13 - i$$ $$\frac{2}{3 + i} = \frac{2}{3 + i} \cdot \frac{3 - i}{3 - i}$$ $$= \frac{2(3 - i)}{9 - i^2}$$ $$= \frac{2(3 - i)}{10}$$ $$= \frac{3 - i}{5}$$

CHAPTER 7 REVIEW EXERCISES

[7.1] *Find each of the following real number roots. Use a calculator as necessary.*

1. $\sqrt{1764}$ **2.** $-\sqrt{289}$ **3.** $-\sqrt{-841}$ **4.** $\sqrt[3]{216}$

5. $\sqrt[3]{-125}$ **6.** $-\sqrt[3]{27z^{12}}$ **7.** $\sqrt[5]{-32}$

8. Under what conditions is $\sqrt[n]{a}$ not a real number?

Find decimal approximations for each of the following. Round to the nearest thousandth.

9. $\sqrt{40}$ **10.** $\sqrt{77}$ **11.** $\sqrt{310}$

[7.2] *Simplify. Assume that all variables represent positive real numbers.*

12. $16^{5/4}$ **13.** $-8^{2/3}$ **14.** $-\left(\dfrac{36}{25}\right)^{3/2}$ **15.** $\left(-\dfrac{1}{8}\right)^{-5/3}$

16. $\left(\dfrac{81}{10{,}000}\right)^{-3/4}$ **17.** $7^{1/3} \cdot 7^{5/3}$ **18.** $\dfrac{96^{2/3}}{96^{-1/3}}$ **19.** $\dfrac{k^{2/3}k^{-1/2}k^{3/4}}{2(k^2)^{-1/4}}$

20. Explain the relationship between the expressions $a^{m/n}$ and $\sqrt[n]{a^m}$.

Simplify by first converting to rational exponents. Convert all answers back to radical form in Exercises 22–24. Assume all variables represent positive real numbers.

21. $\sqrt{3^{18}}$ **22.** $\sqrt{7^9}$ **23.** $\sqrt[3]{m^5} \cdot \sqrt[3]{m^8}$ **24.** $\sqrt[4]{k^2} \cdot \sqrt[4]{k^7}$

[7.3] *Simplify each of the following. Assume that all variables represent positive real numbers.*

25. $\sqrt{6} \cdot \sqrt{11}$ **26.** $\sqrt{5} \cdot \sqrt{r}$ **27.** $\sqrt[3]{6} \cdot \sqrt[3]{5}$ **28.** $\sqrt[4]{7} \cdot \sqrt[4]{3}$

29. $\sqrt{20}$ **30.** $\sqrt{75}$ **31.** $-\sqrt{125}$ **32.** $\sqrt[3]{-108}$

33. $\sqrt{100y^7}$

34. $\sqrt[3]{64p^4q^6}$

35. $\sqrt{\dfrac{49}{81}}$

36. $\sqrt{\dfrac{y^3}{144}}$

37. $\sqrt[3]{\dfrac{m^{15}}{27}}$

38. $\sqrt[3]{\dfrac{r^2}{8}}$

Find the distance between each pair of points.

39. $(2, 7)$ and $(-1, -4)$

40. $(-3, -5)$ and $(4, -3)$

[7.4] *Perform the indicated operations. Assume that all variables represent positive real numbers.*

41. $2\sqrt{8} - 3\sqrt{50}$

42. $8\sqrt{80} - 3\sqrt{45}$

43. $-\sqrt{27y} + 2\sqrt{75y}$

44. $2\sqrt{54m^3} + 5\sqrt{96m^3}$

45. $3\sqrt[3]{54} + 5\sqrt[3]{16}$

46. $-6\sqrt[4]{32} + \sqrt[4]{512}$

[7.5] *Multiply, then simplify the products.*

47. $(\sqrt{3} + 1)(\sqrt{3} - 2)$

48. $(\sqrt{7} + \sqrt{5})(\sqrt{7} - \sqrt{5})$

49. $(3\sqrt{2} + 1)(2\sqrt{2} - 3)$

50. $(\sqrt{11} + 3\sqrt{5})(\sqrt{11} + 5\sqrt{5})$

51. $(\sqrt{13} - \sqrt{2})^2$

52. $(\sqrt{5} - \sqrt{7})^2$

53. Use a calculator to show that the answer to Exercise 52, $12 - 2\sqrt{35}$, is not equal to $10\sqrt{35}$.

Rationalize the denominators. Assume that all variables represent positive real numbers.

54. $\dfrac{\sqrt{6}}{\sqrt{5}}$

55. $\dfrac{-6\sqrt{3}}{\sqrt{2}}$

56. $\dfrac{3\sqrt{7p}}{\sqrt{y}}$

57. $\sqrt{\dfrac{11}{8}}$

58. $-\sqrt[3]{\dfrac{9}{25}}$

59. $\sqrt[3]{\dfrac{108m^3}{n^5}}$

60. $\dfrac{1}{\sqrt{2} + \sqrt{7}}$

61. $\dfrac{-5}{\sqrt{6} - \sqrt{3}}$

[7.6] *Solve each equation.*

62. $\sqrt{8y + 9} = 5$

63. $\sqrt{2z - 3} - 3 = 0$

64. $\sqrt{3m + 1} = -1$

65. $\sqrt{7z + 1} = z + 1$

66. $3\sqrt{m} = \sqrt{10m - 9}$

67. $\sqrt{p^2 + 3p + 7} = p + 2$

68. $\sqrt{a + 2} - \sqrt{a - 3} = 1$

69. $\sqrt[3]{5m - 1} = \sqrt[3]{3m - 2}$

70. $\sqrt[4]{b + 6} = \sqrt[4]{2b}$

[7.7] *Write as a product of a real number and i.*

71. $\sqrt{-25}$

72. $\sqrt{-200}$

73. If a is a positive real number, is $-\sqrt{-a}$ a real number?

Perform the indicated operations. Write each imaginary number answer in the form $a + bi$.

74. $(-2 + 5i) + (-8 - 7i)$

75. $(5 + 4i) - (-9 - 3i)$

76. $\sqrt{-5} \cdot \sqrt{-7}$

77. $\sqrt{-25} \cdot \sqrt{-81}$

78. $\dfrac{\sqrt{-72}}{\sqrt{-8}}$

79. $(2 + 3i)(1 - i)$

80. $(6 - 2i)^2$

81. $\dfrac{3 - i}{2 + i}$

82. $\dfrac{5 + 14i}{2 + 3i}$

Find each power of i.

83. i^{11}

84. i^{52}

85. i^{-13}

MIXED REVIEW EXERCISES

Simplify. Assume all variables represent positive real numbers.

86. $-\sqrt[4]{256}$

87. $-\sqrt{169a^2b^4}$

88. $1000^{-2/3}$

89. $\dfrac{y^{-1/3} \cdot y^{5/6}}{y}$

90. $\dfrac{z^{-1/4} x^{1/2}}{z^{1/2} x^{-1/4}}$

91. $\sqrt[4]{k^{24}}$

92. $\sqrt[3]{54 z^9 t^8}$

93. $-5\sqrt{18} + 12\sqrt{72}$

94. $8\sqrt[3]{x^3 y^2} - 2x\sqrt[3]{y^2}$

95. $(\sqrt{5} - \sqrt{3})(\sqrt{7} + \sqrt{3})$

96. $\dfrac{-1}{\sqrt{12}}$

97. $\sqrt[3]{\dfrac{12}{25}}$

98. $\dfrac{2\sqrt{z}}{\sqrt{z} - 2}$

99. $\sqrt{-49}$

100. $(4 - 9i) + (-1 + 2i)$

101. $\dfrac{\sqrt{50}}{\sqrt{-2}}$

102. i^{-1000}

Solve each equation.

103. $\sqrt{x + 4} = x - 2$

104. $\sqrt{6 + 2y} - 1 = \sqrt{7 - 2y}$

105. Find the distance between the points $(6, -2)$ and $(5, 8)$.

Work the following problems. Express your answers to the nearest tenth.

106. Carpenters stabilize wall frames with a diagonal brace as shown in the figure. The length of the brace is given by $L = \sqrt{H^2 + W^2}$. If the bottom of the brace is attached 9 feet from the corner and the brace is 12 feet long, how far up the corner post should it be nailed?

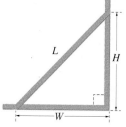

107. According to an article in *The World Scanner Report* (August 1991), the distance, D, in miles, to the horizon from an observer's point of view over water or "flat" earth is given by the function with

$$D(H) = (2H)^{1/2},$$

where H is the height of the point of view, in feet. If a person whose eyes are 6 feet above ground level is standing at the top of a hill 28 feet above the "flat" earth, approximately how far to the horizon will she be able to see?

CHAPTER 7 TEST

Find the following roots. Use a calculator as necessary.

1. $-\sqrt{841}$

1. _____

2. $\sqrt[3]{3375}$

2. _____

3. For $\sqrt{146.25}$, find the closest estimate of the choices given.
   (a) 10   **(b)** 11   **(c)** 12   **(d)** 13

3. _____

4. Give a calculator approximation of $\sqrt{146.25}$ to the nearest hundredth.

4. _____

Simplify each of the following. Assume that all variables represent positive real numbers.

5. $(-64)^{-4/3}$

5. _____

6. $\dfrac{3^{2/5}x^{-1/4}y^{2/5}}{3^{-8/5}x^{7/4}y^{1/10}}$

6. _____

Simplify each of the following. Assume that all variables represent positive real numbers.

7. $\sqrt{54x^5y^6}$

7. _____

8. $\sqrt[4]{32a^7b^{13}}$

8. _____

9. $\sqrt{2} \cdot \sqrt[3]{5}$

9. _____

10. $3\sqrt{20} - 5\sqrt{80} + 4\sqrt{500}$

10. _____

11. $(7\sqrt{5} + 4)(2\sqrt{5} - 1)$

11. _____

12. _____

$$12. \quad \frac{-4}{\sqrt{7} + \sqrt{5}}$$

13. _____

$$13. \quad \frac{-5}{\sqrt{40}}$$

14. _____

$$14. \quad \frac{2}{\sqrt[3]{5}}$$

15. _____

15. Find the distance between the points $(-3, 8)$ and $(2, 7)$.

16. _____

16. The function defined below is used in physics to relate the velocity of sound, V, to the temperature, T.

$$V(T) = \frac{V_0}{\sqrt{1 - kT}}$$

Find an approximation of V to the nearest tenth if $V_0 = 50$, $k = .01$, and $T = 30$.

17. _____

17. Solve for x: $x - 5 = \sqrt{7 - x}$.

Perform the indicated operations. Express answers in the form $a + bi$.

18. _____

18. $(-2 + 5i) - (3 + 6i) - 7i$

19. _____

$$19. \quad \frac{7 + i}{1 - i}$$

20. _____

20. Simplify i^{35}.

Solve the following equations.

1. $7 - (4 + 3t) + 2t = -6(t - 2) - 5$

2. $|6x - 9| = |-4x + 2|$

Solve the following inequalities.

3. $-5 - 3(m - 2) < 11 - 2(m + 2)$ **4.** $1 + 4x > 5$ and $-2x > -6$ **5.** $-2 < 1 - 3y < 7$

6. Write an equation of the line through the points $(-4, 6)$ and $(7, -6)$.

7. The lines with equations $2x + 3y = 8$ and $6y = 4x + 16$ are **(a)** parallel **(b)** perpendicular **(c)** neither.

8. For the graph of $f(x) = -3x + 6$,
   (a) what is the y-intercept?
   (b) what is the x-intercept?

9. For many items, the cost per item to manufacture the item varies inversely as the number made. It costs \$200 each to manufacture 1500 widgets. How much will it cost per item to make 2500 widgets, if widgets are that type of item?

10. Graph the inequality $-2x + y < -6$.

11. Find the measures of the marked angles.

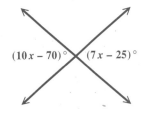

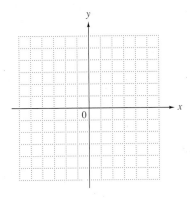

Solve each system.

12. $5x + 2y = 7$
   $10x + 4y = 12$

13. $2x + y - z = 5$
   $3x + 2y + z = 8$
   $4x + 2y - 2z = 10$

Evaluate each determinant.

14. $\begin{vmatrix} 1 & 5 & 2 \\ 2 & 7 & 4 \\ 3 & -3 & 6 \end{vmatrix}$

15. $\begin{vmatrix} 5 & 6 \\ 7 & 3 \end{vmatrix}$

Perform the indicated operations.

16. $(3k^3 - 5k^2 + 8k - 2) - (4k^3 + 11k + 7) + (2k^2 - 5k)$

17. $(8x - 7)(x + 3)$

18. $\dfrac{8z^3 - 16z^2 + 24z}{8z^2}$

19. $\dfrac{6y^4 - 3y^3 + 5y^2 + 6y - 9}{2y + 1}$

Factor each polynomial completely.

20. $2p^2 - 5pq + 3q^2$

21. $18k^4 + 9k^2 - 20$

22. $x^3 + 512$

Perform each operation and express answers in lowest terms.

23. $\dfrac{y^2 + y - 12}{y^3 + 9y^2 + 20y} \div \dfrac{y^2 - 9}{y^3 + 3y^2}$

24. $\dfrac{1}{x + y} + \dfrac{3}{x - y}$

Simplify each complex fraction.

25. $\dfrac{\dfrac{-6}{x - 2}}{\dfrac{8}{3x - 6}}$

26. $\dfrac{\dfrac{1}{a} - \dfrac{1}{b}}{\dfrac{a}{b} - \dfrac{b}{a}}$

Solve by factoring.

27. $2x^2 + 11x + 15 = 0$

28. $5t(t - 1) = 2(1 - t)$

Simplify.

29. $27^{-5/3}$

30. $\dfrac{x^{-2/3}}{x^{-3/4}}$ $(x \neq 0)$

31. $8\sqrt{20} + 3\sqrt{80} - 2\sqrt{500}$

32. $\dfrac{-9}{\sqrt{80}}$

33. $\dfrac{4}{\sqrt{6} - \sqrt{5}}$

34. $\dfrac{12}{\sqrt[3]{2}}$

35. Find the distance between the points $(-4, 4)$ and $(-2, 9)$.

36. Solve $\sqrt{8x - 4} - \sqrt{7x + 2} = 0$.

Solve each problem.

37. The current of a river runs at 3 miles per hour. Brent's boat can go 36 miles downstream in the same time that it takes to go 24 miles upstream. Find the speed of the boat in still water.

38. Brenda rides her bike 4 miles per hour faster than her husband, Chuck. If Brenda can ride 48 miles in the same time that Chuck can ride 24 miles, what are their speeds?

39. A jar containing only dimes and quarters has 29 coins with a face value of $4.70. How many of each denomination are there?

40. How many liters of pure alcohol must be mixed with 40 liters of 18% alcohol to obtain a 22% alcohol solution?

Quadratic Equations and Inequalities

8.1 The Square Root Property and Completing the Square

In an earlier chapter, we solved quadratic (kwah-DRAD-ik) equations by using the zero-factor property. In this section, we introduce additional methods for solving quadratic equations.

Recall that a *quadratic equation* is defined as follows.

Quadratic Equation

An equation that can be written in the form

$$ax^2 + bx + c = 0,$$

where a, b, and c are real numbers, with $a \neq 0$, is a **quadratic equation.**

A quadratic equation written in the form $ax^2 + bx + c = 0$ is in **standard form.** For example, $3q^2 - 4q - 8 = 0$ and $2x^2 - 9 = 0$ are quadratic equations in standard form. The quadratic equation $3x^2 = 6x$ is not in standard form.

OBJECTIVE 1 The zero-factor property stated earlier was restricted to real number factors. Now that property can be extended to complex number factors.

Zero-Factor Property

If a and b are complex numbers and if $ab = 0$, then either $a = 0$ or $b = 0$, or both.

For example, the equation $x^2 + 4 = 0$ can now be solved with the zero-factor property as follows.

$$x^2 + 4 = 0$$
$$x^2 - (-4) = 0 \qquad \text{Write as a difference.}$$
$$(x + \sqrt{-4})(x - \sqrt{-4}) = 0 \qquad \text{Factor.}$$
$$x + \sqrt{-4} = 0 \quad \text{or} \quad x - \sqrt{-4} = 0 \qquad \text{Zero-factor property}$$
$$x = -\sqrt{-4} \qquad\qquad x = \sqrt{-4}$$
$$x = -2i \qquad\qquad x = 2i \qquad \sqrt{-4} = 2i$$

OBJECTIVE 2 Although factoring is the simplest way to solve many quadratic equations, not every quadratic equation can be solved easily by factoring. In this section and the next, other methods of solving quadratic equations are developed based on the following property.

www.mathnotes.com

OBJECTIVES

1 Extend the zero-factor property to complex number factors.

2 Learn the square root property.

3 Solve quadratic equations of the form $(ax - k)^2 = n$ by using the square root property.

4 Solve quadratic equations by completing the square.

FOR EXTRA HELP

Tutorial

Tape 12

SSM, Sec. 8.1

517

1. Solve.

(a) $m^2 = 64$

(b) $p^2 = 7$

Square Root Property

If x and b are complex numbers and if $x^2 = b$, then $x = \sqrt{b}$ or $x = -\sqrt{b}$.

To see why this property works, write $x^2 = b$ as $x^2 - b = 0$. Then factor to get $(x - \sqrt{b})(x + \sqrt{b}) = 0$. Set each factor equal to zero to get $x = \sqrt{b}$ or $x = -\sqrt{b}$. For example, by this property, $x^2 = 16$ implies that $x = 4$ or $x = -4$.

Caution
Remember that if $b \neq 0$, using the square root property always produces *two* square roots, one positive and one negative.

E X A M P L E 1   Using the Square Root Property

Solve $r^2 = 5$.

From the square root property,

$$r = \sqrt{5} \quad \text{or} \quad r = -\sqrt{5}$$

and the solution set is $\{-\sqrt{5}, \sqrt{5}\}$.

◄◄ WORK PROBLEM 1 AT THE SIDE.

Recall that solutions such as those in Example 1 are sometimes abbreviated with the symbol $\pm$ (read "plus or minus"); with this symbol the solutions in Example 1 would be written $\pm\sqrt{5}$.

Objective 3 We can use the square root property to solve more complicated equations, such as

$$(x - 5)^2 = 36,$$

by substituting $(x - 5)^2$ for x^2 and 36 for b in the square root property to get

$$x - 5 = 6 \quad \text{or} \quad x - 5 = -6$$
$$x = 11 \qquad\qquad x = -1.$$

Check: $(x - 5)^2 = 36$

$(11 - 5)^2 = 36 \quad ? \qquad\qquad (-1 - 5)^2 = 36 \quad ?$

$6^2 = 36 \qquad \text{True} \qquad (-6)^2 = 36 \qquad \text{True}$

E X A M P L E 2   Using the Square Root Property

Solve $(2a - 3)^2 = 18$.

By the square root property,

$$2a - 3 = \sqrt{18} \quad \text{or} \quad 2a - 3 = -\sqrt{18}$$

from which

$$2a = 3 + \sqrt{18} \qquad 2a = 3 - \sqrt{18}$$
$$a = \frac{3 + \sqrt{18}}{2} \qquad a = \frac{3 - \sqrt{18}}{2}.$$

Because $\sqrt{18} = \sqrt{9 \cdot 2} = 3\sqrt{2}$, the solution set can be written

$$\left\{ \frac{3 + 3\sqrt{2}}{2}, \frac{3 - 3\sqrt{2}}{2} \right\}.$$

Answers

1. **(a)** $\{-8, 8\}$ **(b)** $\{-\sqrt{7}, \sqrt{7}\}$

CONTINUED ON NEXT PAGE

We show the check for the first solution. The check for the second solution is very similar.

Check:
$$(2a - 3)^2 = 18$$
$$\left[2\left(\frac{3 + 3\sqrt{2}}{2}\right) - 3\right]^2 = 18 \quad ?$$
$$(3 + 3\sqrt{2} - 3)^2 = 18 \quad ?$$
$$(3\sqrt{2})^2 = 18 \quad ?$$
$$9 \cdot 2 = 18 \quad \text{True}$$

WORK PROBLEM 2 AT THE SIDE. ▶▶

The next example shows an equation with imaginary number solutions.

E X A M P L E   3   Using the Square Root Property

Solve $(b + 2)^2 = -16$.

By the square root property,
$$b + 2 = \sqrt{-16} \quad \text{or} \quad b + 2 = -\sqrt{-16}.$$
Because $\sqrt{-16} = 4i$,
$$b + 2 = 4i \qquad \text{or} \quad b + 2 = -4i$$
$$b = -2 + 4i \qquad\qquad b = -2 - 4i.$$

Check these solutions to verify that the solution set is $\{-2 + 4i, -2 - 4i\}$.

WORK PROBLEM 3 AT THE SIDE. ▶▶

OBJECTIVE 4▶ The square root property can be used to solve any quadratic equation by writing it in the form $(x + k)^2 = n$. That is, the left side of the equation must be rewritten as a trinomial that is a perfect square (one that can be factored as $(x + k)^2$) and the right side must be a constant. Rewriting the equation in this form is called **completing the square.**

For example,
$$m^2 + 8m + 10 = 0$$
is a quadratic equation that cannot be solved easily by factoring. To get a perfect square trinomial on the left side of the equation, we first subtract 10 from both sides:
$$m^2 + 8m = -10.$$

The left side should be a perfect square, say $(m + k)^2$. Since $(m + k)^2 = m^2 + 2mk + k^2$, comparing $m^2 + 8m$ with $m^2 + 2mk$ shows that
$$2mk = 8m$$
$$k = 4.$$

If $k = 4$, then $(m + k)^2$ becomes $(m + 4)^2$, or $m^2 + 8m + 16$. To get the necessary $+16$, add 16 on both sides:
$$m^2 + 8m + 16 = -10 + 16.$$
Factor the left side and add on the right to get
$$(m + 4)^2 = 6.$$

2. Solve.

(a) $(a - 3)^2 = 25$

(b) $(3k + 1)^2 = 2$

(c) $(2r + 3)^2 = 8$

3. Solve $(k + 5)^2 = -100$.

4. Determine the number that will complete each perfect square trinomial.

(a) $x^2 + 6x +$ _____

(b) $m^2 - 14m +$ _____

(c) $r^2 + 3r +$ _____

This equation can be solved with the square root property:

$$m + 4 = \sqrt{6} \quad \text{or} \quad m + 4 = -\sqrt{6},$$

leading to the solution set $\{-4 + \sqrt{6}, -4 - \sqrt{6}\}$.

Based on the work of this example, we can convert an equation of the form $x^2 + px = q$ into an equation of the form $(x + k)^2 = n$ by adding the square of half the coefficient of the first degree term to both sides of the equation.

◀◀ WORK PROBLEM 4 AT THE SIDE.

In summary, to find the solutions of a quadratic equation by completing the square, proceed as follows.

Completing the Square

To solve $ax^2 + bx + c = 0$ $(a \neq 0)$, use the following steps.

Step 1 If $a \neq 1$, divide both sides by a.

Step 2 Rewrite the equation so that both terms containing variables are on one side of the equals sign and the constant is on the other side.

Step 3 Take half the coefficient of x (the first-degree term) and square it.

Step 4 Add the square to both sides.

Step 5 One side is now a perfect square. Write it as the square of a binomial.

Step 6 Use the square root property to complete the solution.

E X A M P L E   4   **Solving a Quadratic Equation by Completing the Square**

Solve $k^2 + 5k - 1 = 0$ by completing the square.

Follow the steps listed above. Since $a = 1$, Step 1 is not needed here. Begin by adding 1 to both sides.

$$k^2 + 5k = 1 \qquad \text{Step 2}$$

Take half of 5 (the coefficient of the first-degree term) and square the result.

$$\frac{1}{2} \cdot 5 = \frac{5}{2} \quad \text{and} \quad \left(\frac{5}{2}\right)^2 = \frac{25}{4} \qquad \text{Step 3}$$

Add the square to each side of the equation to get

$$k^2 + 5k + \left(\frac{5}{2}\right)^2 = 1 + \frac{25}{4}. \qquad \text{Step 4}$$

Write the left side as a square and add the terms on the right. Then use the square root property.

CONTINUED ON NEXT PAGE

$$\left(k + \frac{5}{2}\right)^2 = \frac{29}{4} \qquad \text{Step 5}$$

$$k + \frac{5}{2} = \sqrt{\frac{29}{4}} \qquad \text{or} \quad k + \frac{5}{2} = -\sqrt{\frac{29}{4}} \qquad \text{Step 6}$$

$$k + \frac{5}{2} = \frac{\sqrt{29}}{2} \qquad \text{or} \quad k + \frac{5}{2} = \frac{-\sqrt{29}}{2} \qquad \text{Simplify.}$$

$$k = -\frac{5}{2} + \frac{\sqrt{29}}{2} \qquad\qquad k = -\frac{5}{2} - \frac{\sqrt{29}}{2} \qquad \text{Solve.}$$

$$k = \frac{-5 + \sqrt{29}}{2} \qquad \text{or} \qquad k = \frac{-5 - \sqrt{29}}{2}. \qquad \text{Combine terms.}$$

Check that the solution set is $\left\{ \dfrac{-5 + \sqrt{29}}{2}, \dfrac{-5 - \sqrt{29}}{2} \right\}$.

WORK PROBLEM 5 AT THE SIDE. ▶▶

EXAMPLE 5 Solving a Quadratic Equation by Completing the Square

Solve $2x^2 - 4x - 5 = 0$.

Go through the steps for completing the square. First divide both sides of the equation by 2 to make the coefficient of the second-degree term equal to 1, getting

$$x^2 - 2x - \frac{5}{2} = 0. \qquad \text{Step 1}$$

$$x^2 - 2x = \frac{5}{2} \qquad \text{Step 2}$$

$$\frac{1}{2}(-2) = -1 \text{ and } (-1)^2 = 1 \qquad \text{Step 3}$$

$$x^2 - 2x + 1 = \frac{5}{2} + 1 \qquad \text{Step 4}$$

$$(x - 1)^2 = \frac{7}{2} \qquad \text{Step 5}$$

$$x - 1 = \sqrt{\frac{7}{2}} \qquad \text{or} \quad x - 1 = -\sqrt{\frac{7}{2}} \qquad \text{Step 6}$$

$$x = 1 + \sqrt{\frac{7}{2}} \qquad\qquad x = 1 - \sqrt{\frac{7}{2}} \qquad \text{Solve.}$$

$$x = 1 + \frac{\sqrt{14}}{2} \qquad \text{or} \qquad x = 1 - \frac{\sqrt{14}}{2} \qquad \begin{array}{l}\text{Rationalize}\\\text{denominators.}\end{array}$$

— **CONTINUED ON NEXT PAGE**

5. Solve by completing the square.

(a) $y^2 + 6y + 4 = 0$

(b) $x^2 + 2x - 10 = 0$

6. Solve by completing the square.

(a) $2r^2 - 4r + 1 = 0$

Combine the two terms in each solution as follows:

$$1 + \frac{\sqrt{14}}{2} = \frac{2}{2} + \frac{\sqrt{14}}{2} = \frac{2 + \sqrt{14}}{2}$$

$$1 - \frac{\sqrt{14}}{2} = \frac{2}{2} - \frac{\sqrt{14}}{2} = \frac{2 - \sqrt{14}}{2}.$$

Verify by checking that the solution set is

$$\left\{ \frac{2 + \sqrt{14}}{2}, \frac{2 - \sqrt{14}}{2} \right\}.$$

◀◀ **WORK PROBLEM 6 AT THE SIDE.**

(b) $5t^2 - 15t + 12 = 0$

ANSWERS

6. (a) $\left\{ \dfrac{2 + \sqrt{2}}{2}, \dfrac{2 - \sqrt{2}}{2} \right\}$

(b) $\left\{ \dfrac{15 + i\sqrt{15}}{10}, \dfrac{15 - i\sqrt{15}}{10} \right\}$

8.1 Exercises

1. What would be your first step in solving $2x^2 + 8x = 9$ by completing the square?

2. Which one of the following is equal to $\sqrt{-5}$?
   (a) $5i$ **(b)** $i\sqrt{5}$ **(c)** $-\sqrt{5}$ **(d)** $-5i$

3. Of the two equations
   $$(2x + 1)^2 = 5 \quad \text{and} \quad x^2 + 4x = 12,$$
   one is more suitable for solving by the square root property, and the other by completing the square. Which method do you think most students would use for each equation?

4. Why would most students find the equation $x^2 + 4x = 20$ easier to solve by completing the square than the equation $5x^2 + 2x = 3$?

Use the square root property to solve each equation. (All solutions for these equations are real numbers.) See Examples 1 and 2.

5. $x^2 = 81$

6. $y^2 = 225$

7. $t^2 = 17$

8. $k^2 = 19$

9. $(x + 2)^2 = 25$

10. $(8 - y)^2 = 9$

11. $(1 - 3k)^2 = 7$

12. $(2x + 4)^2 = 10$

13. $(4p + 1)^2 = 24$

14. $(5k - 2)^2 = 12$

Find the imaginary number solutions of each of the following equations. See Example 3.

15. $x^2 = -12$

16. $y^2 = -18$

17. $(r - 5)^2 = -3$

18. $(t + 6)^2 = -5$

19. $(6k - 1)^2 = -8$

20. $(4m - 7)^2 = -27$

Solve each equation by completing the square. (All solutions for these equations are real numbers.) See Examples 4 and 5.

21. $x^2 - 2x - 24 = 0$

22. $m^2 - 4m - 32 = 0$

23. $3y^2 + y = 24$

24. $4z^2 - z = 39$

25. $2k^2 + 5k - 2 = 0$

26. $3r^2 + 2r - 2 = 0$

Solve each equation by completing the square. (Some solutions for these equations are imaginary numbers.) See Examples 4 and 5.

27. $m^2 + 4m + 13 = 0$

28. $t^2 + 6t + 10 = 0$

29. $9x^2 - 24x = -13$

30. $25n^2 - 20n = 1$

31. $z^2 - \frac{4}{3}z = -\frac{1}{9}$

32. $p^2 - \frac{8}{3}p = -1$

33. $3r^2 + 4r + 4 = 0$

34. $4x^2 + 5x + 5 = 0$

35. $.1x^2 - .2x - .1 = 0$

36. $.1p^2 - .4p + .1 = 0$

37. $-m^2 - 6m - 12 = 0$

38. $-k^2 - 5k - 10 = 0$

MATHEMATICAL CONNECTIONS (EXERCISES 39–44)

The Greeks had a method of completing the square geometrically in which they literally changed a figure into a square. For example, to complete the square for $x^2 + 6x$, we begin with a square of side x, as in the figure. We add three rectangles of width 1 to the right side and the bottom to get a region with area $x^2 + 6x$. To fill in the corner (complete the square) we must add 9 1-by-1 squares as shown.

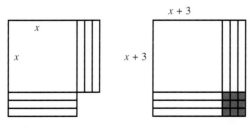

Work Exercises 39–44 in order.

_____ **39.** What is the area of the original square?

_____ **40.** What is the area of each strip?

_____ **41.** What is the total area of the six strips?

_____ **42.** What is the area of each small square in the corner of the second figure?

_____ **43.** What is the total area of the small squares?

_____ **44.** What is the area of the new larger square?

8.2 *The Quadratic Formula*

The examples in the previous section showed that we can solve any quadratic equation by completing the square; however, completing the square is often tedious and time consuming. Later in this section, we will complete the square on the general quadratic equation to develop a formula that can be used to solve any quadratic equation. For now, we state the formula and show how it is used.

> **Quadratic Formula**
>
> The solutions of $ax^2 + bx + c = 0$ ($a \neq 0$) are
> $$x = \frac{-b \pm \sqrt{b^2 - 4ac}}{2a}.$$

OBJECTIVE 1 To use the quadratic formula, first write the given equation in standard form $ax^2 + bx + c = 0$; then identify the values of a, b, and c and substitute them into the quadratic formula, as shown in the next examples.

WORK PROBLEM 1 AT THE SIDE. ▶▶

EXAMPLE 1 Using the Quadratic Formula

Solve $6x^2 - 5x - 4 = 0$.

First, identify the letters a, b, and c of the general quadratic equation, $ax^2 + bx + c = 0$. Here a, the coefficient of the second-degree term, is 6, while b, the coefficient of the first-degree term, is -5, and the constant c is -4. Substitute these values into the quadratic formula.

$$x = \frac{-b \pm \sqrt{b^2 - 4ac}}{2a}$$

$$x = \frac{-(-5) \pm \sqrt{(-5)^2 - 4(6)(-4)}}{2(6)} \qquad a = 6, b = -5, c = -4$$

$$= \frac{5 \pm \sqrt{25 + 96}}{12}$$

$$= \frac{5 \pm \sqrt{121}}{12}$$

$$x = \frac{5 \pm 11}{12}$$

This last statement leads to two solutions, one from $+$ and one from $-$, giving

$$x = \frac{5 + 11}{12} = \frac{16}{12} = \frac{4}{3} \quad \text{or} \quad x = \frac{5 - 11}{12} = \frac{-6}{12} = -\frac{1}{2}.$$

Check these solutions by substituting each one in the original equation. The solution set is $\{-\frac{1}{2}, \frac{4}{3}\}$.

> **Caution**
>
> Notice in the quadratic formula that the square root is added to or subtracted from the value of $-b$ *before* dividing by $2a$.

OBJECTIVES

1 ▶ Solve quadratic equations using the quadratic formula.

2 ▶ Solve applied problems using the quadratic formula.

3 ▶ Use the discriminant to determine the number and type of solutions.

4 ▶ Use the discriminant to decide whether a quadratic trinomial can be factored.

FOR EXTRA HELP

Tutorial   Tape 12   SSM, Sec. 8.2

1. Identify the letters a, b, and c. (*Hint:* If necessary, write the equation in standard form first.) *Do not solve.*

(a) $-3q^2 + 9q - 4 = 0$

(b) $3y^2 = 6y + 2$

2. Solve $6x^2 + 4x - 1 = 0$ using the quadratic formula.

◀◀ WORK PROBLEM 2 AT THE SIDE.

EXAMPLE 2 Using the Quadratic Formula

Solve $4r^2 = 8r - 1$.

Rewrite the equation as $4r^2 - 8r + 1 = 0$, and identify $a = 4$, $b = -8$, and $c = 1$. Now use the quadratic formula.

$$r = \frac{-b \pm \sqrt{b^2 - 4ac}}{2a}$$

$$r = \frac{-(-8) \pm \sqrt{(-8)^2 - 4(4)(1)}}{2(4)} \qquad a = 4, b = -8, c = 1$$

$$= \frac{8 \pm \sqrt{64 - 16}}{8}$$

$$= \frac{8 \pm \sqrt{48}}{8} = \frac{8 \pm 4\sqrt{3}}{8}$$

$$r = \frac{4(2 \pm \sqrt{3})}{8} = \frac{2 \pm \sqrt{3}}{2}$$

The solution set for $4r^2 = 8r - 1$ is $\left\{ \dfrac{2 + \sqrt{3}}{2}, \dfrac{2 - \sqrt{3}}{2} \right\}$.

The solutions to the equation in the next example are imaginary numbers.

EXAMPLE 3 Using the Quadratic Formula

Solve $(9q + 3)(q - 1) = -8$.

Every quadratic equation must be in standard form before we begin to solve it, whether we are factoring or using the quadratic formula. To put this equation in standard form, we first multiply on the left, then collect all nonzero terms on the left.

$$(9q + 3)(q - 1) = -8$$
$$9q^2 - 6q - 3 = -8$$
$$9q^2 - 6q + 5 = 0$$

After writing the equation in the required form, we can identify $a = 9$, $b = -6$, and $c = 5$. Then we use the quadratic formula.

$$q = \frac{-(-6) \pm \sqrt{(-6)^2 - 4(9)(5)}}{2(9)}$$

$$= \frac{6 \pm \sqrt{-144}}{18} = \frac{6 \pm 12i}{18} \qquad i = \sqrt{-1}$$

$$= \frac{6(1 \pm 2i)}{18} = \frac{1 \pm 2i}{3}$$

The solution set is $\left\{ \dfrac{1 + 2i}{3}, \dfrac{1 - 2i}{3} \right\}$.

We could have written the solutions in Example 3 as $a + bi$, the standard form for complex numbers, as follows:

$$\frac{1 \pm 2i}{3} = \frac{1}{3} \pm \frac{2}{3}i.$$

> **WORK PROBLEM 3 AT THE SIDE.** ▶▶

OBJECTIVE 2 The next example shows how the quadratic formula is used to solve applied problems that cannot be solved by factoring.

EXAMPLE 4 Solving an Applied Problem Using the Quadratic Formula

If a rock is thrown upward from the top of a 144-foot building, its position (in feet above the ground) is given by $s(t) = -16t^2 + 112t + 144$, where t is time in seconds after it was dropped. When does it hit the ground?

When the rock hits the ground, its distance above the ground is zero. Find t when s is zero by solving the equation

$$0 = -16t^2 + 112t + 144. \qquad \text{Let } s(t) = 0.$$
$$0 = t^2 - 7t - 9 \qquad \text{Divide both sides by } -16.$$
$$t = \frac{7 \pm \sqrt{49 + 36}}{2} \qquad \text{Quadratic formula}$$
$$t = \frac{7 \pm \sqrt{85}}{2}$$

In applied problems, we often prefer approximations to exact values. Using a calculator, we find $\sqrt{85} \approx 9.2$, so

$$t \approx \frac{7 \pm 9.2}{2},$$

giving the solutions $t \approx 8.1$ or $t \approx -1.1$. Discard the negative solution. The rock will hit the ground about 8.1 seconds after it is thrown.

> **WORK PROBLEM 4 AT THE SIDE.** ▶▶

As mentioned earlier, the quadratic formula is developed by completing the square to solve the general quadratic equation $ax^2 + bx + c = 0$ ($a \neq 0$). We show the steps here for the case where $a > 0$. First divide both sides by a (using the fact that $a > 0$).

$$x^2 + \frac{b}{a}x + \frac{c}{a} = 0 \qquad \text{Step 1}$$

Now subtract $\frac{c}{a}$ from each side.

$$x^2 + \frac{b}{a}x = -\frac{c}{a} \qquad \text{Step 2}$$

Take half of $\frac{b}{a}$ and square it.

$$\frac{1}{2}\left(\frac{b}{a}\right) = \frac{b}{2a} \qquad \text{Step 3}$$
$$\left(\frac{b}{2a}\right)^2 = \frac{b^2}{4a^2}$$

---

3. Solve each equation using the quadratic formula.

(a) $2k^2 + 19 = 14k$

(b) $z^2 = 4z - 5$

4. A ball is thrown vertically upward from the ground. Its distance in feet from the ground at t seconds is $s(t) = -16t^2 + 64t$. At what times will the ball be 32 feet from the ground? Use a calculator and round answers to the nearest tenth. (*Hint:* There are two answers.)

ANSWERS

3. (a) $\left\{ \dfrac{7 + \sqrt{11}}{2}, \dfrac{7 - \sqrt{11}}{2} \right\}$
   (b) $\{2 + i, 2 - i\}$
4. at .6 second and at 3.4 seconds

Add the square to both sides.

$$x^2 + \frac{b}{a}x + \frac{b^2}{4a^2} = -\frac{c}{a} + \frac{b^2}{4a^2} \qquad \text{Step 4}$$

Write the left side as a perfect square and rearrange the right side.

$$\left(x + \frac{b}{2a}\right)^2 = \frac{b^2}{4a^2} + \frac{-c}{a} \qquad \text{Step 5}$$

$$\left(x + \frac{b}{2a}\right)^2 = \frac{b^2}{4a^2} + \frac{-4ac}{4a^2} \qquad \text{Write with a common denominator.}$$

$$\left(x + \frac{b}{2a}\right)^2 = \frac{b^2 - 4ac}{4a^2} \qquad \text{Add fractions.}$$

By the square root property,

$$x + \frac{b}{2a} = \sqrt{\frac{b^2 - 4ac}{4a^2}} \quad \text{or} \quad x + \frac{b}{2a} = -\sqrt{\frac{b^2 - 4ac}{4a^2}}. \qquad \text{Step 6}$$

Because

$$\sqrt{\frac{b^2 - 4ac}{4a^2}} = \frac{\sqrt{b^2 - 4ac}}{\sqrt{4a^2}} = \frac{\sqrt{b^2 - 4ac}}{2a},$$

the result above can be expressed as

$$x + \frac{b}{2a} = \frac{\sqrt{b^2 - 4ac}}{2a} \quad \text{or} \quad x + \frac{b}{2a} = \frac{-\sqrt{b^2 - 4ac}}{2a}$$

$$x = -\frac{b}{2a} + \frac{\sqrt{b^2 - 4ac}}{2a} \quad \text{or} \quad x = -\frac{b}{2a} - \frac{\sqrt{b^2 - 4ac}}{2a}$$

$$x = \frac{-b + \sqrt{b^2 - 4ac}}{2a} \quad \text{or} \quad x = \frac{-b - \sqrt{b^2 - 4ac}}{2a}.$$

This result agrees with the solution in the quadratic formula, $x = \frac{-b \pm \sqrt{b^2 - 4ac}}{2a}$. If $a < 0$, going through the same steps gives the same two solutions.

OBJECTIVE 3 The quadratic formula gives the solutions of the quadratic equation $ax^2 + bx + c = 0$ as

$$x = \frac{-b \pm \sqrt{b^2 - 4ac}}{2a}.$$

If a, b, and c are integers, the type of solutions of a quadratic equation (that is, rational, irrational, or imaginary) is determined by the quantity under the square root sign, $b^2 - 4ac$. Because it distinguishes among the three types of solutions, the quantity $b^2 - 4ac$ is called the **discriminant** (dis-KRIM-ih-nunt). By calculating the discriminant before solving a quadratic equation, we can predict whether the solutions will be rational numbers, irrational numbers, or imaginary numbers. This can be useful in an applied problem, for example, where imaginary number solutions are not acceptable. If the discriminant is a perfect square (including 0), we can solve the equation by factoring. Otherwise, the quadratic formula should be used.

Discriminant

The discriminant of $ax^2 + bx + c = 0$ is given by $b^2 - 4ac$. If a, b, and c are integers, then the type of solution is determined as follows.

Discriminant	Solutions
Positive, and the square of an integer	Two different rational solutions
Positive, but not the square of an integer	Two different irrational solutions
Zero	One rational solution
Negative	Two different imaginary solutions

E X A M P L E 5   Using the Discriminant

Given the equation $6x^2 - x - 15 = 0$, find the discriminant and determine whether the solutions of the equation will be rational, irrational, or imaginary.

We find the discriminant by evaluating $b^2 - 4ac$. In this example, $a = 6$, $b = -1$, $c = -15$, and the discriminant is

$$b^2 - 4ac = (-1)^2 - 4(6)(-15) = 1 + 360 = 361.$$

A calculator shows that $361 = 19^2$. Because the discriminant is a perfect square and a, b, and c are integers, the solutions to the given equation will be two different rational numbers, and the equation can be solved by factoring.

WORK PROBLEM 5 AT THE SIDE. ▶▶

E X A M P L E 6   Using the Discriminant

Predict the number and type of solutions for each of the following equations.

(a) $4y^2 + 9 = 12y$

Rewrite the equation as $4y^2 - 12y + 9 = 0$ to find $a = 4$, $b = -12$, and $c = 9$. The discriminant is

$$b^2 - 4ac = (-12)^2 - 4(4)(9) = 144 - 144 = 0.$$

Because the discriminant is 0, the quantity under the radical in the quadratic formula is 0, and there is only one rational solution. Again, we can solve the equation by factoring.

(b) $3m^2 - 4m = 5$

Rewrite the equation as $3m^2 - 4m - 5 = 0$. Then $a = 3$, $b = -4$, $c = -5$, and the discriminant is

$$b^2 - 4ac = (-4)^2 - 4(3)(-5) = 16 + 60 = 76.$$

Because 76 is not a perfect square, $\sqrt{76}$ is irrational, and since a, b, and c are integers, the given equation will have two different irrational solutions, one from using $\sqrt{76}$ and one from using $-\sqrt{76}$.

CONTINUED ON NEXT PAGE

5. Find the discriminant and decide whether it is a perfect square.

(a) $6m^2 - 13m - 28 = 0$

(b) $4y^2 + 2y + 1 = 0$

(c) $15k^2 + 11k = 14$

6. Predict the number and type of solutions for each equation.

(a) $2x^2 + 3x = 4$

(b) $2x^2 + 3x + 4 = 0$

(c) $x^2 + 20x + 100 = 0$

(d) $3x^2 + 7x = 0$

7. Use the discriminant to decide whether each trinomial can be factored; then factor it, if possible.

(a) $2y^2 + 13y - 7$

(b) $6z^2 - 11z + 18$

(c) $4x^2 + x + 1 = 0$
   Here $a = 4$, $b = 1$, $c = 1$, and the discriminant is
$$b^2 - 4ac = 1^2 - 4(4)(1) = 1 - 16 = -15.$$

Because the discriminant is negative, the equation $4x^2 + x + 1 = 0$ will have two imaginary number solutions.

◀◀ WORK PROBLEM 6 AT THE SIDE.

OBJECTIVE 4 As mentioned earlier, a quadratic trinomial can be factored with rational coefficients only if the corresponding quadratic equation has rational solutions. Thus, we can use the discriminant to decide whether a given trinomial is factorable.

EXAMPLE 7 **Deciding Whether a Trinomial Is Factorable**

Decide whether the following trinomials can be factored.

(a) $24x^2 + 7x - 5$
   To decide whether the solutions of $24x^2 + 7x - 5 = 0$ are rational numbers, evaluate the discriminant.
$$b^2 - 4ac = 7^2 - 4(24)(-5) = 49 + 480 = \mathbf{529} = \mathbf{23^2}$$

Because 529 is a perfect square, the solutions are rational and the trinomial can be factored. It factors as
$$24x^2 + 7x - 5 = (3x - 1)(8x + 5).$$

(b) $11m^2 - 9m + 12$
   The discriminant is $b^2 - 4ac = (-9)^2 - 4(11)(12) = -447$. This number is negative, so the corresponding quadratic equation has imaginary number solutions and therefore the trinomial cannot be factored.

◀◀ WORK PROBLEM 7 AT THE SIDE.

ANSWERS

6. (a) two; irrational **(b)** two; imaginary
   (c) one; rational **(d)** two; rational
7. (a) $(2y - 1)(y + 7)$
   (b) cannot be factored

8.2 *Exercises*

Decide whether the statement is true or false based on the discussion in Sections 8.1 and 8.2.

1. The equation $(x + 4)^2 = -8$ has no real solutions.

2. The discriminant of $x^2 + x + 1 = 0$ is -3, and so the equation has two imaginary solutions.

3. The equations $x^2 + 3x - 4 = 0$ and $-x^2 - 3x + 4 = 0$ have the same solution set.

4. An equation of the form $x^2 + kx = 0$ will have two real solutions, one of which is 0.

5. If p is a prime number, $x^2 = p$ will have two irrational solutions.

6. The equation $x^2 - 5 = 0$ cannot be solved using the quadratic formula, since there is no value for b.

Use the quadratic formula to solve each equation. (All solutions for these equations are real numbers.) See Examples 1 and 2.

7. $m^2 - 8m + 15 = 0$ **8.** $x^2 + 3x - 28 = 0$ **9.** $2k^2 + 4k + 1 = 0$

10. $2y^2 + 3y - 1 = 0$ **11.** $2x^2 - 2x = 1$ **12.** $9t^2 + 6t = 1$

13. $x^2 + 18 = 10x$ **14.** $x^2 - 4 = 2x$ **15.** $-2t(t + 2) = -3$

16. $-3x(x + 2) = -4$ **17.** $(r - 3)(r + 5) = 2$ **18.** $(k + 1)(k - 7) = 1$

Use the quadratic formula to solve each equation. (All solutions for these equations are imaginary numbers.) See Example 3.

19. $k^2 + 47 = 0$

20. $x^2 + 19 = 0$

21. $r^2 - 6r + 14 = 0$

22. $t^2 + 4t + 11 = 0$

23. $4x^2 - 4x = -7$

24. $9x^2 - 6x = -7$

25. $x(3x + 4) = -2$

26. $y(2y + 3) = -2$

27. $\dfrac{x + 5}{2x - 1} = \dfrac{x - 4}{x - 6}$

28. $\dfrac{3x - 4}{2x - 5} = \dfrac{x + 5}{x + 2}$

29. $\dfrac{1}{x^2} + 1 = -\dfrac{1}{x}$

30. $\dfrac{4}{r^2} + 3 = \dfrac{1}{r}$

Solve each problem. See Example 4.

31. A ball is thrown vertically upward from the ground. Its distance in feet from the ground in t seconds is $s(t) = -16t^2 + 128t$. At what times will the ball be 240 feet from the ground?

32. A toy rocket is launched from ground level. Its distance in feet from the ground in t seconds is $s(t) = -16t^2 + 208t$. At what times will the rocket be 640 feet from the ground?

A rock is thrown upward from ground level, and its distance from the ground in t seconds is $s(t) = -16t^2 + 160t$. Use algebra and a short explanation to answer Exercises 33 and 34.

33. After how many seconds does it reach a height of 400 feet? How would you describe in words its position at this height?

34. After how many seconds does it reach a height of 425 feet? How would you interpret the mathematical result here?

MATHEMATICAL CONNECTIONS (EXERCISES 35–40)

In earlier chapters we saw how linear functions can be used to model certain data. In some cases, quadratic functions provide better models, as the data may not be linear in nature. For example, during the years 1988–1990, the quadratic function $f(x) = -6.5x^2 + 132.5x + 2117$ modeled the number of prisoners in the United States under the sentence of death, according to statistics from the U.S. Justice Department. In this model, $x = 0$ corresponds to 1988, $x = 1$ to 1989, and $x = 2$ to 1990. The accompanying graph models these data in another way.

PRISONERS ON DEATH ROW

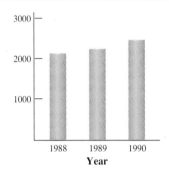

Year

Now work Exercises 35–40 in order, using the quadratic function f above.

35. How many such prisoners were there in 1988?

36. How many such prisoners were there in 1989?

37. How many such prisoners were there in 1990?

38. The equation given was based only on information for the three years named. If we wanted to make a prediction for the number of prisoners in 1991, how would we do it based on this quadratic function?

39. Use this function to predict the year in which the number of prisoners was 2543. (*Hint:* Replace $f(x)$ by 2543 and solve using the quadratic formula.)

40. Suppose that the number of people in a small town grew according to the quadratic function $f(x) = 5x^2 + 45x + 350$, where $x = 0$ corresponds to 1994, $x = 1$ to 1995, and so on. In what year would the population be 700?

Use the discriminant to determine whether the solutions to each of the following equations are
(**a**) *two distinct rational numbers,* (**b**) *exactly one rational number,*
(**c**) *two distinct irrational numbers,* (**d**) *two distinct imaginary numbers.*

Do not solve. See Examples 5 and 6.

41. $25x^2 + 70x + 49 = 0$

42. $4k^2 - 28k + 49 = 0$

43. $x^2 + 4x + 2 = 0$

44. $9x^2 - 12x - 1 = 0$

45. $3x^2 = 5x + 2$

46. $4x^2 = 4x + 3$

47. $3y^2 - 10y + 15 = 0$

48. $18x^2 + 60x + 82 = 0$

Use the discriminant to determine whether each polynomial is factorable. If it can be factored, do so. See Example 7.

49. $24x^2 - 34x - 45$

50. $36y^2 + 69y + 28$

51. $36x^2 + 21x - 24$

52. $18k^2 + 13k - 12$

53. $12x^2 - 83x - 7$

54. $16y^2 - 61y - 12$

55. Find all values of k such that $2x^2 + kx + 2 = 0$ has exactly one rational number solution.

56. Find all values of k such that $3x^2 - 2x + k = 0$ has two distinct imaginary number solutions.

8.3 Equations Quadratic in Form

Four methods have now been introduced for solving quadratic equations written in the form $ax^2 + bx + c = 0$. The chart below gives some advantages and disadvantages of each method.

Methods for Solving Quadratic Equations

Method	Advantages	Disadvantages
Factoring	Usually the fastest method	Not all polynomials are factorable; some factorable polynomials are hard to factor
Completing the square	None for solving equations (the procedure is useful in other areas of mathematics)	Requires more steps than other methods
Quadratic formula	Can always be used	More difficult than factoring because of the square root
Square root method	Simplest method for solving equations of the form $(ax - k)^2 = n$	Few equations are given in this form

OBJECTIVE **1** A variety of nonquadratic equations can be written in the form of a quadratic equation and solved by using these methods. For example, some equations with fractions lead to quadratic equations. As you solve the equations in this section, try to decide which is the best method for each equation.

EXAMPLE 1   **Writing an Equation in Quadratic Form**

Solve $\dfrac{1}{x} + \dfrac{1}{x-1} = \dfrac{7}{12}$.

To clear the equation of fractions, multiply each term by the common denominator, $12x(x - 1)$.

$$12x(x-1)\frac{1}{x} + 12x(x-1)\frac{1}{x-1} = 12x(x-1)\frac{7}{12}$$

$$12(x-1) + 12x = 7x(x-1)$$

$$12x - 12 + 12x = 7x^2 - 7x \qquad \text{Distributive property}$$

$$24x - 12 = 7x^2 - 7x \qquad \text{Combine terms.}$$

A quadratic equation must be in the form $ax^2 + bx + c = 0$ before we can solve it. Combine terms and arrange them so that one side of the equation is zero.

$$0 = 7x^2 - 31x + 12 \qquad \text{Subtract } 24x \text{ and add } 12.$$

$$0 = (7x - 3)(x - 4) \qquad \text{Factor.}$$

Setting each factor equal to 0 and solving the two linear equations gives the solutions $\frac{3}{7}$ and 4. Check by substituting these solutions in the original equation. The solution set is $\{\frac{3}{7}, 4\}$.

1. Solve.

(a) $\dfrac{5}{m} + \dfrac{12}{m^2} = 2$

◀◀ **WORK PROBLEM 1 AT THE SIDE.**

OBJECTIVE 2▶ Earlier we solved distance-rate-time (or motion) problems that led to linear equations or rational equations. Now we can extend that work to motion problems that lead to quadratic equations. We can write an equation to solve the problem just as we did earlier. Distance-rate-time applications often lead to equations with fractions, as in the next example.

E X A M P L E 2   Solving a Motion Problem

A riverboat for tourists averages 12 miles per hour in still water. It takes the boat 1 hour, 4 minutes to go 6 miles upstream and return. Find the speed of the current. See Figure 1.

FIGURE I

(b) $\dfrac{2}{x} + \dfrac{1}{x-2} = \dfrac{5}{3}$

For a problem about rate (or speed), we use the distance formula, $d = rt$.

Let $x =$ the speed of the current;

$12 - x =$ the rate upstream;

$12 + x =$ the rate downstream.

The rate upstream is the difference of the speed of the boat in still water and the speed of the current, or $12 - x$. The speed downstream is, in the same way, $12 + x$. To find the time, rewrite the formula $d = rt$ as

$$t = \dfrac{d}{r}.$$

This information was used to complete the following chart.

(c) $\dfrac{4}{m-1} + 9 = -\dfrac{7}{m}$

	d	r	t
Upstream	6	$12 - x$	$\dfrac{6}{12-x}$
Downstream	6	$12 + x$	$\dfrac{6}{12+x}$

Times in hours

The total time, 1 hour and 4 minutes, can be written as

$$1 + \dfrac{4}{60} = 1 + \dfrac{1}{15} = \dfrac{16}{15} \text{ hours.}$$

Because the time upstream plus the time downstream equals $\frac{16}{15}$ hours,

$$\dfrac{6}{12-x} + \dfrac{6}{12+x} = \dfrac{16}{15}.$$

ANSWERS

1. (a) $\left\{-\dfrac{3}{2}, 4\right\}$ (b) $\left\{\dfrac{4}{5}, 3\right\}$ (c) $\left\{\dfrac{7}{9}, -1\right\}$

CONTINUED ON NEXT PAGE

Now multiply both sides of the equation by the common denominator $15(12 - x)(12 + x)$ and solve the resulting quadratic equation.

$$15(12 + x)6 + 15(12 - x)6 = 16(12 - x)(12 + x)$$

$$90(12 + x) + 90(12 - x) = 16(144 - x^2)$$

$$1080 + 90x + 1080 - 90x = 2304 - 16x^2 \quad \text{Distributive property}$$

$$2160 = 2304 - 16x^2 \quad \text{Combine terms.}$$

$$16x^2 = 144$$

$$x^2 = 9$$

Solve $x^2 = 9$ by using the square root property to get the two solutions

$$x = 3 \quad \text{or} \quad x = -3.$$

The speed of the current cannot be -3, so the solution is $x = 3$ miles per hour.

Caution

As shown in Example 2, when a quadratic equation is used to solve an applied problem, sometimes only *one* answer satisfies the application. It is *always necessary* to check each answer in the words of the stated problem.

WORK PROBLEM 2 AT THE SIDE. ▶▶

In an earlier chapter we solved problems about work rates, using the formula $r = 1/t$ for the rate at which a person completes 1 job in t time units. Now we extend this idea to problems which produce quadratic equations. The method of solution is the same.

E X A M P L E   3   Solving a Work Problem

It takes two carpet layers 4 hours to carpet a room. If each worked alone, one of them could do the job in one hour less time than the other. How long would it take the slower one to complete the job alone?

Let x represent the number of hours for the slower carpet layer to complete the job alone. Then the faster carpet layer could do the entire job in $x - 1$ hours. Together, they do the job in 4 hours. The slower person's rate is $\frac{1}{x}$, and the faster person's rate is $\frac{1}{x-1}$. Fill in a chart as shown.

Worker	Rate	Time Working Together	Fractional Part of the Job Done
Slower	$\dfrac{1}{x}$	4	$\dfrac{1}{x}(4)$
Faster	$\dfrac{1}{x-1}$	4	$\dfrac{1}{x-1}(4)$

Sum is 1 whole job.

Together they complete one whole job, so the sum of the two fractional parts is 1.

$$\frac{1}{x}(4) + \frac{1}{x-1}(4) = 1$$

CONTINUED ON NEXT PAGE

2. (a) In 4 hours Kerrie can go 15 miles upriver and come back. The speed of the current is 5 miles per hour. Complete this chart.

	d	r	t
Up			
Down			

(b) Find the speed of the boat from part (a) in still water.

(c) In $1\frac{3}{4}$ hours Ken rows his boat 5 miles upriver and comes back. The speed of the current is 3 miles per hour. How fast does Ken row?

ANSWERS

2. (a) row 1: 15; $x - 5$; $\dfrac{15}{x - 5}$;

   row 2: 15; $x + 5$; $\dfrac{15}{x + 5}$

(b) 10 miles per hour

(c) 7 miles per hour

3. Carlos can complete a certain lab test in 2 hours less time than Jaime can. If they can finish the job together in 2 hours, how long would it take each of them working alone? Round answers to the nearest tenth.

Multiply both sides by the common denominator, $x(x - 1)$.

$$4(x - 1) + 4x = x(x - 1)$$
$$4x - 4 + 4x = x^2 - x \qquad \text{Distributive property}$$
$$0 = x^2 - 9x + 4 \qquad \text{Standard form}$$

Now use the quadratic formula.

$$x = \frac{9 \pm \sqrt{81 - 16}}{2} = \frac{9 \pm \sqrt{65}}{2} \qquad a = 1, b = -9, c = 4$$

From a calculator, $\sqrt{65} \approx 8.062$, so

$$x \approx \frac{9 \pm 8.062}{2}.$$

Using the $+$ sign gives $x \approx 8.5$, while the $-$ sign leads to $x \approx .5$. (Here we rounded to the nearest tenth.) Only the solution 8.5 makes sense in the original problem. (Why?) Thus, the slower carpet layer can do the job in about 8.5 hours and the faster in about $8.5 - 1 = 7.5$ hours.

◄◄ **WORK PROBLEM 3 AT THE SIDE.**

OBJECTIVE 3▶ In the previous chapter we saw that some equations with radicals lead to quadratic equations.

E X A M P L E   4   Writing an Equation in Quadratic Form

Solve each equation.

(a) $k = \sqrt{6k - 8}$

This equation is not quadratic. However, squaring both sides of the equation gives $k^2 = 6k - 8$, which is a quadratic equation that we can solve by factoring.

$$k^2 = 6k - 8$$
$$k^2 - 6k + 8 = 0$$
$$(k - 4)(k - 2) = 0$$
$$k = 4 \quad \text{or} \quad k = 2 \qquad \text{Potential solutions}$$

Check both of these numbers in the original (and *not* the squared) equation to be sure they are solutions. (This check is an essential step.)

Check: If $k = 4$,   If $k = 2$,

$$4 = \sqrt{6(4) - 8} \quad ? \qquad\qquad 2 = \sqrt{6(2) - 8} \quad ?$$
$$4 = \sqrt{16} \qquad\quad ? \qquad\qquad\quad 2 = \sqrt{4} \qquad\qquad ?$$
$$4 = 4. \qquad\qquad \text{True} \qquad\qquad 2 = 2. \qquad\qquad \text{True}$$

Both numbers check, so the solution set is $\{2, 4\}$.

(b) $x + \sqrt{x} = 6$

$$\sqrt{x} = 6 - x \qquad\qquad \text{Isolate the radical on one side.}$$
$$x = 36 - 12x + x^2 \qquad \text{Square both sides.}$$
$$0 = x^2 - 13x + 36 \qquad \text{Write in standard form.}$$
$$0 = (x - 4)(x - 9) \qquad \text{Factor.}$$
$$x - 4 = 0 \quad \text{or} \quad x - 9 = 0 \qquad \text{Set each factor equal to 0.}$$
$$x = 4 \quad \text{or} \qquad x = 9$$

CONTINUED ON NEXT PAGE

Check both potential solutions.

Check: If $x = \mathbf{4}$,   If $x = \mathbf{9}$,

$$4 + \sqrt{4} = 6. \quad \text{True} \qquad 9 + \sqrt{9} = 6. \quad \text{False}$$

The solution set is $\{4\}$.

WORK PROBLEM 4 AT THE SIDE. ▶▶

OBJECTIVE 4▶ An equation that can be written in the form $au^2 + bu + c = 0$, for $a \neq 0$ and an algebraic expression u, is called **quadratic in form**.

┌─ **E X A M P L E 5**   **Solving an Equation That Is Quadratic in Form**

Solve each of the following.

(a) $x^4 - 13x^2 + 36 = 0$

Because $x^4 = (x^2)^2$, we can write this equation as $(x^2)^2 - 13(x^2) + 36 = 0$, so it is in quadratic form with $u = x^2$ and can be solved by factoring. Since $u = x^2$, $u^2 = x^4$, and the equation becomes

$$u^2 - 13u + 36 = 0 \qquad \text{Let } u = x^2.$$
$$(u - 4)(u - 9) = 0 \qquad \text{Factor.}$$
$$u - 4 = 0 \quad \text{or} \quad u - 9 = 0 \qquad \text{Set each factor equal to 0.}$$
$$u = 4 \qquad\qquad u = 9. \qquad \text{Solve.}$$

To find x, substitute x^2 for u.

$$x^2 = 4 \quad \text{or} \quad x^2 = 9$$
$$x = \pm 2 \qquad x = \pm 3 \qquad \text{Square root property}$$

The equation $x^4 - 13x^2 + 36 = 0$, a fourth-degree equation, has four solutions.* The solution set is $\{-3, -2, 2, 3\}$, as can be verified by substituting into the equation.

(b) $4x^4 + 1 = 5x^2$

Use the fact that $x^4 = (x^2)^2$ again, and let $u = x^2$ and $u^2 = x^4$.

$$4u^2 + 1 = 5u \qquad \text{Let } u = x^2.$$
$$4u^2 - 5u + 1 = 0 \qquad \text{Write in standard form.}$$
$$(4u - 1)(u - 1) = 0 \qquad \text{Factor.}$$
$$4u - 1 = 0 \quad \text{or} \quad u - 1 = 0 \qquad \text{Set each factor equal to 0.}$$
$$u = \frac{1}{4} \quad \text{or} \qquad u = 1 \qquad \text{Solve.}$$
$$x^2 = \frac{1}{4} \quad \text{or} \qquad x^2 = 1 \qquad u = x^2$$
$$x = \pm \frac{1}{2} \quad \text{or} \qquad x = \pm 1 \qquad \text{Square root property}$$

The solution set is $\{-1, -\frac{1}{2}, \frac{1}{2}, 1\}$.

WORK PROBLEM 5 AT THE SIDE. ▶▶

---

*In general, an equation in which an nth-degree polynomial equals 0 has n solutions, although some of them may be repeated, and thus are the same.

4. Solve. Check each answer.

  (a) $x = \sqrt{7x - 10}$

  (b) $2x = \sqrt{x} + 1$

5. Solve.

  (a) $m^4 - 10m^2 + 9 = 0$

  (b) $9k^4 - 37k^2 + 4 = 0$

ANSWERS

4. (a) $\{2, 5\}$ **(b)** $\{1\}$

5. (a) $\{-3, -1, 1, 3\}$

  (b) $\left\{-2, -\dfrac{1}{3}, \dfrac{1}{3}, 2\right\}$

6. Solve.

(a) $5(r + 3)^2 + 9(r + 3) = 2$

EXAMPLE 6 Solving an Equation That Is Quadratic in Form

Solve each equation.

(a) $2(4m - 3)^2 + 7(4m - 3) + 5 = 0$

Because of the repeated quantity $4m - 3$, this equation is quadratic in form with $u = 4m - 3$. (Any letter except m could be used instead of u.)

$$2(4m - 3)^2 + 7(4m - 3) + 5 = 0$$

$$2u^2 + 7u + 5 = 0 \qquad \text{Let } 4m - 3 = u.$$

$$(2u + 5)(u + 1) = 0 \qquad \text{Factor.}$$

$$u = -\frac{5}{2} \quad \text{or} \quad u = -1 \qquad \text{Zero-factor property}$$

To find m, substitute $4m - 3$ for u.

$$4m - 3 = -\frac{5}{2} \quad \text{or} \quad 4m - 3 = -1$$

$$4m = \frac{1}{2} \quad \text{or} \quad 4m = 2$$

$$m = \frac{1}{8} \quad \text{or} \quad m = \frac{1}{2}$$

The solution set of the original equation is $\{\frac{1}{8}, \frac{1}{2}\}$.

(b) $2a^{2/3} - 11a^{1/3} + 12 = 0$

Let $a^{1/3} = u$; then $a^{2/3} = u^2$. Substitute into the given equation.

$$2u^2 - 11u + 12 = 0 \qquad \text{Let } a^{1/3} = u, a^{2/3} = u^2.$$

$$(2u - 3)(u - 4) = 0 \qquad \text{Factor.}$$

$$2u - 3 = 0 \quad \text{or} \quad u - 4 = 0$$

$$u = \frac{3}{2} \quad \text{or} \quad u = 4$$

$$a^{1/3} = \frac{3}{2} \quad \text{or} \quad a^{1/3} = 4 \qquad u = a^{1/3}$$

$$a = \left(\frac{3}{2}\right)^3 = \frac{27}{8} \quad \text{or} \quad a = 4^3 = 64 \qquad \text{Cube both sides.}$$

(Recall that in the previous chapter we solved equations with radicals by raising both sides to the same power.) Check that the solution set is $\{\frac{27}{8}, 64\}$.

(b) $4m^{2/3} = 3m^{1/3} + 1$

◀◀ **WORK PROBLEM 6 AT THE SIDE.**

ANSWERS

6. (a) $\left\{-5, -\frac{14}{5}\right\}$ **(b)** $\left\{-\frac{1}{64}, 1\right\}$

8.3 Exercises

Based on the discussion and examples of this section, write a sentence describing the first step you would take in solving each of the following equations. Do not actually solve the equation.

1. $\dfrac{14}{x} = x - 5$

2. $\sqrt{1 + x} + x = 5$

3. $(r^2 + r)^2 - 8(r^2 + r) + 12 = 0$

4. $3t = \sqrt{16 - 10t}$

Solve by first clearing each equation of fractions. Check your answers. See Example 1.

5. $1 - \dfrac{3}{x} - \dfrac{28}{x^2} = 0$

6. $4 - \dfrac{7}{r} - \dfrac{2}{r^2} = 0$

7. $3 - \dfrac{1}{t} = \dfrac{2}{t^2}$

8. $1 + \dfrac{2}{k} = \dfrac{3}{k^2}$

9. $\dfrac{1}{x} + \dfrac{2}{x + 2} = \dfrac{17}{35}$

10. $\dfrac{2}{m} + \dfrac{3}{m + 9} = \dfrac{11}{4}$

11. $\dfrac{2}{x + 1} + \dfrac{3}{x + 2} = \dfrac{7}{2}$

12. $\dfrac{4}{3 - y} + \dfrac{2}{5 - y} = \dfrac{26}{15}$

13. $\dfrac{3}{2x} - \dfrac{1}{2(x + 2)} = 1$

14. $\dfrac{4}{3x} - \dfrac{1}{2(x + 1)} = 1$

15. If it takes m hours to grade a set of papers, what is the grader's rate (in job per hour)?

16. A boat goes 20 miles per hour in still water, and the rate of the current is t miles per hour.
(a) What is the rate of the boat when it travels upstream?
(b) What is the rate of the boat when it travels downstream?

Solve each problem by writing an equation with fractions and solving it. See Examples 2 and 3.

17. On a windy day Yoshiaki found that he could go 16 miles downstream and then 4 miles back upstream at top speed in a total of 48 minutes. What was the top speed of Yoshiaki's boat if the current was 15 miles per hour?

18. Lekesha flew her plane for 6 hours at a constant speed. She traveled 810 miles with the wind, then turned around and traveled 720 miles against the wind. The wind speed was a constant 15 miles per hour. Find the speed of the plane.

19. Albuquerque and Amarillo are 300 miles apart. Steve rides his Honda 20 miles per hour faster than Paula rides her Yamaha. Find Steve's average speed if he travels from Albuquerque to Amarillo in $1\frac{1}{4}$ hours less time than Paula.

20. The distance from Jackson to Lodi is about 40 miles, as is the distance from Lodi to Manteca. Rico drove from Jackson to Lodi during the rush hour, stopped in Lodi for a root beer, and then drove on to Manteca at 10 miles per hour faster. Driving time for the entire trip was 88 minutes. Find his speed from Jackson to Lodi.

21. A washing machine can be filled in 6 minutes if both the hot and cold water taps are fully opened. Filling the washer with hot water alone takes 9 minutes longer than filling it with cold water alone. How long does it take to fill the washer with cold water?

22. Two pipes together can fill a large tank in 2 hours. One of the pipes, used alone, takes 3 hours longer than the other to fill the tank. How long would each pipe take to fill the tank alone?

Find all solutions by first squaring. Check your answers. See Example 4.

23. $2x = \sqrt{11x + 3}$

24. $4x = \sqrt{6x + 1}$

25. $3y = \sqrt{16 - 10y}$

26. $4t = \sqrt{8t + 3}$

27. $p - 2\sqrt{p} = 8$

28. $k + \sqrt{k} = 12$

29. $m = \sqrt{\dfrac{6 - 13m}{5}}$

30. $r = \sqrt{\dfrac{20 - 19r}{6}}$

Find all solutions to the following equations. Check your answers. See Examples 5 and 6.

31. $t^4 - 18t^2 + 81 = 0$

32. $y^4 - 8y^2 + 16 = 0$

33. $4k^4 - 13k^2 + 9 = 0$

34. $9x^4 - 25x^2 + 16 = 0$

35. $(x + 3)^2 + 5(x + 3) + 6 = 0$

36. $(k - 4)^2 + (k - 4) - 20 = 0$

37. $(t + 5)^2 + 6 = 7(t + 5)$

38. $3(m + 4)^2 - 8 = 2(m + 4)$

39. $2 + \dfrac{5}{3k - 1} = \dfrac{-2}{(3k - 1)^2}$

40. $3 - \dfrac{7}{2p + 2} = \dfrac{6}{(2p + 2)^2}$

41. $2 - 6(m - 1)^{-2} = (m - 1)^{-1}$

42. $3 - 2(x - 1)^{-1} = (x - 1)^{-2}$

Use substitution to solve the following equations. Check your answers. See Example 6.

43. $x^{2/3} + x^{1/3} - 2 = 0$

44. $3x^{2/3} - x^{1/3} - 24 = 0$

45. $2(1 + \sqrt{y})^2 = 13(1 + \sqrt{y}) - 6$

46. $(k^2 + k)^2 + 12 = 8(k^2 + k)$

MATHEMATICAL CONNECTIONS (EXERCISES 47–52)

Consider the following equation, which contains variable expressions in the denominators.

$$\frac{x^2}{(x-3)^2} + \frac{3x}{x-3} - 4 = 0$$

Work Exercises 47–52 in order. They all pertain to this equation.

47. Why can 3 not possibly be a solution for this equation?

48. Multiply both sides of the equation by the LCD, $(x-3)^2$, and solve. There is only one solution—what is it?

49. Write the equation in a different manner so that it is quadratic in form using the expression $\dfrac{x}{x-3}$.

50. In your own words, explain why the expression $\dfrac{x}{x-3}$ cannot equal 1.

51. Solve the equation from Exercise 49 by making the substitution $t = \dfrac{x}{x-3}$. You should get two values for t. Why is one of them impossible for this equation?

52. Solve the equation $x^2(x-3)^{-2} + 3x(x-3)^{-1} - 4 = 0$ by letting $s = (x-3)^{-1}$. You should get two values for s. Why is this impossible for this equation?

8.4 *Formulas and Applications Involving Quadratic Equations*

OBJECTIVE 1 Many useful formulas have a second-degree term. We can use the methods presented earlier in this chapter to solve a formula for a variable that is squared.

The formula in Example 1 is the Pythagorean theorem of geometry.

> **Pythagorean Theorem**
>
> If c is the length of the longest side of a right triangle and a and b are the lengths of the shorter sides, then
> $$c^2 = a^2 + b^2.$$
>
> See the figure.
>
>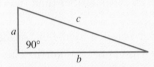
>
> The longest side is the **hypotenuse** and the two shorter sides are the **legs** of the triangle.

OBJECTIVES

 1 Solve second-degree formulas for a specified variable.

2 Solve applied problems using the Pythagorean theorem.

3 Solve applied problems using formulas for area.

4 Solve applied problems about work.

5 Solve applied problems using quadratic functions as models.

FOR EXTRA HELP

Tutorial | Tape 13 | SSM, Sec. 8.4

1. Solve $5y^2 = z^2 + 9x^2$ for x. Assume the variables represent the lengths of the sides of a right triangle.

EXAMPLE 1 Solving for a Squared Variable

Solve the Pythagorean theorem $c^2 = a^2 + b^2$ for b.

Think of c^2 and a^2 as constants. We solve for b by first getting b^2 alone on one side of the equation. Begin by subtracting a^2 from both sides.

$$c^2 = a^2 + b^2$$
$$c^2 - a^2 = b^2$$

Now use the square root property.

$$b = \sqrt{c^2 - a^2} \quad \text{or} \quad b = -\sqrt{c^2 - a^2}$$

Because b represents the side of a triangle, b must be positive. Because of this,

$$b = \sqrt{c^2 - a^2}$$

is the only solution. The solution cannot be simplified further.

WORK PROBLEM 1 AT THE SIDE. ▶▶

EXAMPLE 2 Solving for a Squared Variable

Solve $s = 2t^2 + kt$ for t.

Because the equation has terms with t^2 and t, first put it in the quadratic form $ax^2 + bx + c = 0$ with t as the variable instead of x.

$$s = 2t^2 + kt$$
$$0 = 2t^2 + kt - s$$

Now use the quadratic formula with $a = 2$, $b = k$, and $c = -s$.

$$t = \frac{-k \pm \sqrt{k^2 - 4(2)(-s)}}{2(2)}$$
$$= \frac{-k \pm \sqrt{k^2 + 8s}}{4}$$

The solutions are $t = \dfrac{-k + \sqrt{k^2 + 8s}}{4}$ and $t = \dfrac{-k - \sqrt{k^2 + 8s}}{4}$.

ANSWERS

1. $x = \dfrac{\sqrt{5y^2 - z^2}}{3}$

2. Solve $2t^2 - 5t + k = 0$ for t.

◀◀ WORK PROBLEM 2 AT THE SIDE.

Caution
The following examples show that it is important to check all proposed solutions of applied problems against the information given in the original problem. Numbers that are valid solutions of the equation may not satisfy the physical conditions of the problem.

OBJECTIVE 2   The Pythagorean theorem is used again in the solution of the next example.

E X A M P L E   3   Using the Pythagorean Theorem

Two cars left an intersection at the same time, one heading due north, the other due west. Some time later, they were exactly 100 miles apart. The car headed north had gone 20 miles farther than the car headed west. How far had each car traveled?

Let x be the distance traveled by the car headed west. Then $x + 20$ is the distance traveled by the car headed north. These distances are shown in Figure 2. The cars are 100 miles apart, so the hypotenuse of the right triangle equals 100 and the two legs are equal to x and $x + 20$.

3. A 13-foot ladder is leaning against a house. The distance from the bottom of the ladder to the house is 7 feet less than the distance from the top of the ladder to the ground. How far is the bottom of the ladder from the house?

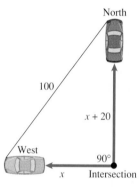

FIGURE 2

By the Pythagorean theorem,

$$c^2 = a^2 + b^2$$
$$100^2 = x^2 + (x + 20)^2$$
$$10{,}000 = x^2 + x^2 + 40x + 400$$
$$0 = 2x^2 + 40x - 9600$$
$$0 = 2(x^2 + 20x - 4800). \qquad \text{Factor out the common factor.}$$

Divide both sides by 2 to get $x^2 + 20x - 4800 = 0$. Although this equation can be solved by factoring, we use the quadratic formula to find x. Here, $a = 1$, $b = 20$, and $c = -4800$.

$$x = \frac{-20 \pm \sqrt{400 - 4(-4800)}}{2} = \frac{-20 \pm \sqrt{19{,}600}}{2}$$

From a calculator, $\sqrt{19{,}600} = 140$, so $x = \dfrac{-20 \pm 140}{2}$.

The solutions are $x = 60$ or $x = -80$. Since distance cannot be negative, we reject -80. Thus 60 and $60 + 20 = 80$ are the required distances in miles.

■

◀◀ WORK PROBLEM 3 AT THE SIDE.

OBJECTIVE 3   Formulas for area may also result in quadratic equations, as the next example shows.

ANSWERS

2. $t = \dfrac{5 + \sqrt{25 - 8k}}{4}$,

$t = \dfrac{5 - \sqrt{25 - 8k}}{4}$

3. 5 feet

EXAMPLE 4   Solving an Area Problem

A rectangular reflecting pool in a park is 20 feet wide and 30 feet long. The park gardener wants to plant a strip of grass of uniform width around the edge of the pool. She has enough seed to cover 336 square feet. How wide will the strip be?

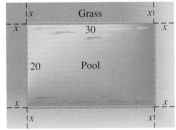

FIGURE 3

The pool is shown in Figure 3. If x represents the unknown width of the grass strip, the width of the large rectangle is given by $20 + 2x$ (the width of the pool plus two grass strips), and the length is given by $30 + 2x$. The area of the large rectangle is given by the product of its length and width, $(20 + 2x)(30 + 2x)$. The area of the pool is $20 \cdot 30 = 600$ square feet. The area of the large rectangle, minus the area of the pool, should equal the area of the grass strip. Since the area of the grass strip is to be 336 square feet, the equation is

$$\underset{\text{Area of rectangle}}{(20 + 2x)(30 + 2x)} - \underset{\substack{\text{Area of} \\ \text{pool}}}{600} = \underset{\substack{\text{Area of} \\ \text{strip}}}{336}.$$

$$600 + 100x + 4x^2 - 600 = 336 \qquad \text{Multiply.}$$

$$4x^2 + 100x - 336 = 0 \qquad \text{Collect terms.}$$

$$x^2 + 25x - 84 = 0 \qquad \text{Divide by 4.}$$

$$(x + 28)(x - 3) = 0 \qquad \text{Factor.}$$

$$x = -28 \quad \text{or} \quad x = 3$$

The width of a grass strip cannot be -28 feet, so 3 feet is the desired width of the strip.

WORK PROBLEM 4 AT THE SIDE. ▶▶

OBJECTIVE 4 ▶   As shown earlier, applied problems about work may result in quadratic equations.

EXAMPLE 5   Solving a Work Problem

A janitorial service provides two people to clean an office building. Working together, the two can clean the building in 5 hours. One person is new to the job and would take 2 hours longer than the other person to clean the building working alone. How long would it take the experienced worker to clean the building working alone?

Let $x =$ time in hours for the experienced worker to clean the building;

$x + 2 =$ time in hours for the new worker to clean the building;

$5 =$ time in hours for the two workers to clean the building together.

Then $\dfrac{1}{x} =$ the rate for the experienced worker;

$\dfrac{1}{x + 2} =$ the rate for the new worker.

─── **CONTINUED ON NEXT PAGE**

4. Suppose the pool in Example 4 is 20 feet by 40 feet and there is enough seed to cover 700 square feet. How wide should the grass strip be?

5. If the new worker in Example 5 takes 1 hour longer than the experienced worker does to clean the building, and together they complete the job in 5 hours, how long would the experienced worker require to complete the job working alone? Round the answer to the nearest tenth of an hour.

The parts of the job done by each are $\frac{1}{x}(5)$ and $\frac{1}{x+2}(5)$, so

$$\frac{1}{x}(5) + \frac{1}{x+2}(5) = 1.$$

$5(x+2) + 5x = x(x+2)$	Multiply by the common denominator.
$5x + 10 + 5x = x^2 + 2x$	Distributive property
$0 = x^2 - 8x - 10$	Collect terms.
$x = \dfrac{8 \pm \sqrt{64+40}}{2}$	Quadratic formula
$= \dfrac{8 \pm \sqrt{104}}{2}$	
$\approx \dfrac{8 \pm 10.20}{2}$	From a calculator
$x = 9.1$	Discard negative solution.

To the nearest tenth, the experienced worker requires 9.1 hours to clean the building working alone.

◀◀ **WORK PROBLEM 5 AT THE SIDE.**

OBJECTIVE 5▶   Quadratic functions can be used to model data, as shown in the following example.

EXAMPLE 6   Using a Quadratic Function as a Model

Union membership in the United States rose from 1930 to 1960 and then declined. The quadratic function

$$f(x) = -.011x^2 + 1.22x - 8.5$$

approximates the number of union members, in millions, in the years 1950 through 1990, where x is the number of years since 1930 (when unions were in their formative stages). (*Source:* U.S. Union Membership, 1930–1990 Table. Bureau of Labor Statistics, U.S. Department of Labor)

(a) Approximate the union membership for the year 1950.
    The year 1950 is 20 years from 1930, so we let $x = 20$ and find $f(20)$:

$$f(20) = -.011(20)^2 + 1.22(20) - 8.5 = 11.5 \text{ (million)}.$$

(b) In what year did union membership reach 15.125 million?
    Here we must find the value of x that makes $f(x) = 15.125$.

$$f(x) = -.011x^2 + 1.22x - 8.5$$
$$15.125 = -.011x^2 + 1.22x - 8.5 \quad \text{Let } f(x) = 15.125.$$
$$.011x^2 - 1.22x + 23.625 = 0 \quad \text{Standard form}$$

Now use $a = .011$, $b = -1.22$, and $c = 23.625$ in the quadratic formula to find the smaller positive solution.

◀◀ **WORK PROBLEM 6 AT THE SIDE.**

    The smaller positive solution is $x = 25$, so the year is $1930 + 25 = 1955$. (We reject the larger positive solution since it leads to a later year than 1990.)

6. Evaluate

$$\frac{1.22 \pm \sqrt{(-1.22)^2 - 4(.011)(23.625)}}{2(.011)}$$

for both solutions. Which one is the smaller positive solution?

8.4 Exercises

In Exercises 1 and 2, solve for m in terms of the other variables (m > 0).

1.

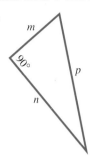

2.

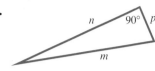

Solve each equation for the indicated variable. (While in practice we would often reject a negative value due to the physical nature of the quantity represented by the variable, leave ± in your answers here.) See Examples 1 and 2.

3. $d = kt^2$ for t

4. $s = kwd^2$ for d

5. $I = \dfrac{ks}{d^2}$ for d

6. $R = \dfrac{k}{d^2}$ for d

7. $F = \dfrac{kA}{v^2}$ for v

8. $L = \dfrac{kd^4}{h^2}$ for h

9. $V = \dfrac{1}{3}\pi r^2 h$ for r

10. $V = \pi(r^2 + R^2)h$ for r

11. $At^2 + Bt = -C$ for t

12. $S = 2\pi rh + \pi r^2$ for r

13. $D = \sqrt{kh}$ for h

14. $F = \dfrac{k}{\sqrt{d}}$ for d

15. $p = \sqrt{\dfrac{k\ell}{g}}$ for ℓ

16. $p = \sqrt{\dfrac{k\ell}{g}}$ for g

17. In the formula of Exercise 15, if g is a positive number, explain why k and l cannot have different signs if the equation is to have a real value for p.

18. In Example 2 of this section, suppose that k and s are both positive numbers. Which one of the two solutions given is positive? Which one is negative?

Solve the following problems. See Example 3.

19. Two ships leave port at the same time, one heading due south and the other heading due east. Several hours later, they are 170 miles apart. If the ship traveling south traveled 70 miles farther than the other, how many miles did they each travel?

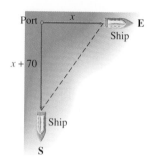

20. Rita is flying a kite that is 30 feet farther above her hand than its horizontal distance from her. The string from her hand to the kite is 150 feet long. How high is the kite?

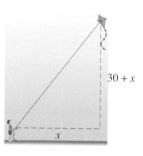

21. The longer leg of a right triangle is 1 meter longer than the shorter leg, while the hypotenuse is 8 meters longer than the longer leg. Find the lengths of the three sides.

22. The hypotenuse of a right triangle is 1 meter longer than twice the length of the shorter leg, and the longer leg is 9 meters shorter than 3 times the length of the shorter leg. Find the lengths of the three sides of the triangle.

Solve the following problems. See Example 4.

23. A square has an area of 256 square centimeters. If the same amount is removed from one dimension and added to the other, the resulting rectangle has an area 16 square centimeters less. Find the dimensions of the rectangle.

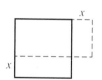

24. A rectangular piece of sheet metal is 2 inches longer than it is wide. A square piece 3 inches on a side is cut from each corner. The sides are then turned up to form an uncovered box of volume 765 cubic inches. Find the dimensions of the original piece of metal.

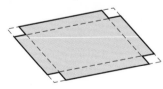

25. A couple wants to buy a rug for a room that is 20 feet long and 15 feet wide. They want to leave an even strip of flooring uncovered around the edges of the room. How wide a strip will they have if they buy a rug with an area of 234 square feet?

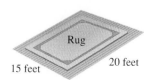

26. A club swimming pool is 30 feet wide and 40 feet long. The club members want an exposed aggregate border in a strip of uniform width around the pool. They have enough material for 296 square feet. How wide can the strip be?

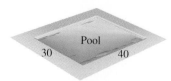

Use the quadratic formula to solve the following problems. Round answers to the nearest tenth. See Example 5.

27. Carmen and Paul can clean the house together in 2 hours. Working alone, it takes Carmen $\frac{1}{2}$ hour longer than it takes Paul to do the job. How long would it take Carmen alone?

28. Mashari and Jamal are distributing brochures for a fund-raising campaign. Together they can complete the job in 3 hours. If Mashari could do the job alone in one hour more than Jamal, how long would it take Jamal working alone?

29. Two pipes can fill a tank in 4 hours when used together. Alone, one can fill the tank in .5 hour more than the other. How long will it take each pipe to fill the tank alone?

30. Charlie Dawkins can process a stack of invoices 1 hour faster than Arnold Parker can. Working together, they take 1.5 hours. How long would it take each person working alone?

For Exercises 31–42, refer to Example 6.

The adjusted poverty threshold for a single person from the year 1984 to the year 1990 is approximated by the quadratic model $f(x) = 18.7x^2 + 105.3x + 4814.1$, where $x = 0$ corresponds to 1984, and $f(x)$ is in dollars. (Source: Congressional Budget Office.) Use this model to answer the questions in Exercises 31–34.

POVERTY THRESHOLD FOR A SINGLE PERSON

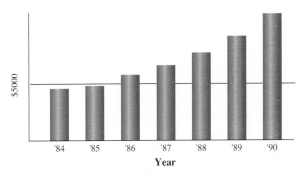

31. What was the threshold in 1984?

32. What was the threshold in 1986?

33. In what year during this period was the threshold $f(x)$ approximately $5300?

34. In what year during this period was the threshold $f(x)$ approximately $6119?

35. The function $D(t) = 13t^2 - 100t$ gives the distance a car going approximately 68 miles per hour will skid in t seconds. Find the time it would take for the car to skid 180 feet.

36. The function in Exercise 35 becomes $D(t) = 13t^2 - 73t$ for a car going 50 miles per hour. Find the time for this car to skid 218 feet.

37. An object is projected directly upward from the ground. After t seconds its distance above the ground is given by the function $f(t) = 144t - 16t^2$ feet. After how many seconds will it be 128 feet above the ground?

38. Refer to Exercise 37. When does the object strike the ground?

39. The formula $A = P(1 + r)^2$ gives the amount A in dollars that P dollars will grow to in 2 years at interest rate r (where r is given as a decimal), using compound interest. What interest rate will cause $2000 to grow to $2142.25 in 2 years?

40. If a square piece of cardboard has 3-inch squares cut from its corners and then has the flaps folded up to form an open-top box, the volume of the box is given by the function $V(x) = 3(x - 6)^2$, where x is the length of each side of the original piece of cardboard in inches. What original length would yield a box with a volume of 432 cubic inches?

William Froude was a 19th century naval architect who used the expression

$$\frac{v^2}{g\ell}$$

in shipbuilding. This expression, known as the Froude number, was also used by R. McNeill Alexander in his research on dinosaurs. (See "How Dinosaurs Ran" in Scientific American, *April, 1991, pp. 130–136.) For each of the following, ℓ is given, as well as the value of the Froude number. Find the value of v (in meters per second). It is known that $g = 9.8$ meters per second squared.*

41. rhinoceros: $\ell = 1.2$; Froude number $= 2.57$

42. triceratops: $\ell = 2.8$; Froude number $= .16$

Recall from Chapter 6 that the corresponding sides of similar triangles are proportional. Use this fact to find the lengths of the indicated sides of the pair of similar triangles. Check all possible solutions in both triangles. Sides of a triangle cannot be negative.

43. side AC

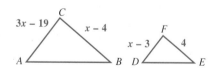

44. side RQ

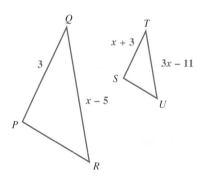

8.5 Nonlinear and Fractional Inequalities

OBJECTIVES

1 ▶ Recognize a quadratic inequality.

2 ▶ Solve quadratic inequalities.

3 ▶ Solve polynomial inequalities of degree 3 or more.

4 ▶ Solve fractional inequalities.

5 ▶ Solve fractional inequalities using an alternative method.

6 ▶ Understand special cases of quadratic inequalities.

FOR EXTRA HELP

Tutorial

Tape 13

SSM, Sec. 8.5

OBJECTIVE 1 ▶ We have discussed methods of solving linear inequalities (earlier) and methods of solving quadratic equations (in this chapter). Now this work can be extended to include solving *quadratic inequalities.*

Quadratic Inequality

A **quadratic inequality** can be written in the form

$$ax^2 + bx + c < 0 \quad \text{or} \quad ax^2 + bx + c > 0$$

where a, b, and c are real numbers, with $a \neq 0$.

As before, $<$ and $>$ can be replaced with $\leq$ and $\geq$ as necessary.

OBJECTIVE 2 ▶ A method for solving quadratic inequalities is shown in the next example.

EXAMPLE 1 Solving a Quadratic Inequality

Solve $x^2 - x - 12 > 0$.

First solve the quadratic equation $x^2 - x - 12 = 0$ by factoring.

$$(x - 4)(x + 3) = 0$$
$$x - 4 = 0 \quad \text{or} \quad x + 3 = 0$$
$$x = 4 \quad \text{or} \quad x = -3$$

The numbers 4 and -3 divide the number line into the three regions shown in Figure 4. (Be careful to put the smaller number on the left.)

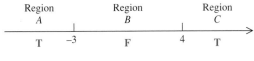

FIGURE 4

The numbers 4 and -3 are the only numbers that make the expression $x^2 - x - 12$ equal to zero. All other numbers make the expression either positive or negative. The sign of the expression can change from positive to negative or from negative to positive only at a number that makes it zero. Therefore, if one number in a region satisfies the inequality, then all the numbers in that region will satisfy the inequality. Choose any number from Region A in Figure 4 (any number less than -3). Substitute this number for x in the inequality $x^2 - x - 12 > 0$. If the result is *true,* then all numbers in Region A satisfy the original inequality.

Let us choose -5 from Region A. Substitute -5 into $x^2 - x - 12 > 0$, getting

$$(-5)^2 - (-5) - 12 > 0 \quad ?$$
$$25 + 5 - 12 > 0 \quad ?$$
$$18 > 0. \quad \text{True}$$

Because -5 from Region A satisfies the inequality, all numbers from Region A are solutions.

Try 0 from Region B. If $x = 0$, then

$$0^2 - 0 - 12 > 0 \quad ?$$
$$-12 > 0. \quad \text{False}$$

The numbers in Region B are *not* solutions.

CONTINUED ON NEXT PAGE

1. Does the number 5 from Region *C* satisfy $x^2 - x - 12 > 0$?

◀◀ **WORK PROBLEM 1 AT THE SIDE.**

In Problem 1 at the side, the number 5 satisfies the inequality, so the numbers in Region *C* are also solutions to the inequality.

Based on these results (shown by the colored letters in Figure 4), the solution set includes the numbers in Regions *A* and *C*, as shown on the graph in Figure 5. The solution set is written in interval notation as

$$(-\infty, -3) \cup (4, \infty).$$

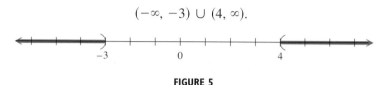

FIGURE 5

In summary, a quadratic inequality is solved by following these steps.

2. Solve. Graph each solution.

(a) $x^2 + x - 6 > 0$

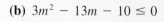

Solving a Quadratic Inequality

Step 1 Replace the inequality symbol with = and solve the equation.

Step 2 Place the numbers found in Step 1 on a number line. These numbers divide the number line into regions.

Step 3 Substitute a number from each region into the inequality to determine the intervals that make the inequality true. If one number in a region satisfies the inequality, then all the numbers in that region will satisfy the inequality. The numbers in those intervals that make the inequality true are in the solution set.

Step 4 The numbers found in Step 1 are included in the solution set if the symbol is ≤ or ≥; they are not included if it is < or >.

(b) $3m^2 - 13m - 10 \leq 0$

◀◀ **WORK PROBLEM 2 AT THE SIDE.**

OBJECTIVE 3 Higher-degree polynomial inequalities that are factorable can be solved in the same way as quadratic inequalities.

E X A M P L E   2   Solving a Third-Degree Polynomial Inequality

Solve $(x - 1)(x + 2)(x - 4) \leq 0$.

This is a *cubic* (third-degree) inequality rather than a quadratic inequality, but it can be solved by the method shown above and by extending the zero-factor property to more than two factors. Begin by setting the factored polynomial *equal* to 0 and solving the equation.

$$(x - 1)(x + 2)(x - 4) = 0$$

$$x - 1 = 0 \quad \text{or} \quad x + 2 = 0 \quad \text{or} \quad x - 4 = 0$$

$$x = 1 \quad \text{or} \quad x = -2 \quad \text{or} \quad x = 4$$

Locate the numbers -2, 1, and 4 on a number line as in Figure 6 to determine the regions *A*, *B*, *C*, and *D*. The numbers -2, 1, and 4 are in the solution set because of the "or equal to" part of the inequality symbol.

ANSWERS

1. yes

2. (a) $(-\infty, -3) \cup (2, \infty)$

$-3 \quad 0 \quad 2$

(b) $\left[-\dfrac{2}{3}, 5\right]$

$-\dfrac{2}{3}$

$-1\,0 \quad 3 \quad 5 \quad 7$

CONTINUED ON NEXT PAGE

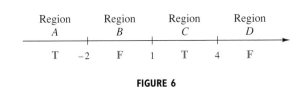

FIGURE 6

Substitute a number from each region into the original inequality to determine which regions satisfy the inequality. These results are shown below the number line in Figure 6. For example, in Region A, using $x = -3$ gives

$$(-3 - 1)(-3 + 2)(-3 - 4) \leq 0$$
$$(-4)(-1)(-7) \leq 0$$
$$-28 \leq 0. \quad \text{True}$$

The numbers in Region A are in the solution set. Verify that the numbers in Region C are also in the solution set, which is written

$$(-\infty, -2] \cup [1, 4].$$

The solution set is graphed in Figure 7.

FIGURE 7

WORK PROBLEM 3 AT THE SIDE. ▶▶

OBJECTIVE 4 ▶ Inequalities involving fractions are solved in a similar manner, by going through the following steps.

Solving Inequalities with Fractions

Step 1 Rewrite the inequality as an equation and solve the equation.

Step 2 Set the denominator equal to zero and solve that equation.

Step 3 Use the solutions from Steps 1 and 2 to divide a number line into regions.

Step 4 Test a number from each region by substitution in the inequality to determine the intervals that satisfy the inequality.

Step 5 Be sure to exclude any values that make the denominator equal to zero.

Caution

Don't forget Step 2. Any number that makes the denominator zero *must* separate two regions on the number line because there will be no point on the number line for that value of the variable.

3. Solve. Graph each solution.

(a) $(y - 3)(y + 2)(y + 1) > 0$

(b) $(k - 5)(k + 1)(k - 3) \leq 0$

ANSWERS

3. (a) $(-2, -1) \cup (3, \infty)$

-2 -1 0 1 2 3 4

(b) $(-\infty, -1] \cup [3, 5]$

-1 0 1   3   5

4. Solve. Graph each solution.

(a) $\dfrac{2}{x-4} < 3$

E X A M P L E 3   Solving an Inequality with a Fraction

Solve the inequality $\dfrac{-1}{p-3} > 1$.

Step 1 Write the corresponding equation and solve it.

$$\frac{-1}{p-3} = 1$$

$$-1 = p - 3 \qquad \text{Multiply by the common denominator.}$$

$$2 = p$$

Step 2 Find the number that makes the denominator 0.

$$p - 3 = 0$$

$$p = 3$$

Step 3 These two numbers, 2 and 3, divide a number line into three regions. (See Figure 8.)

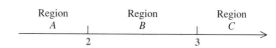

FIGURE 8

Step 4 Testing one number from each region in the given inequality shows that the solution set is the interval (2, 3).

Step 5 This interval does not include any value that might make the denominator of the original inequality equal to zero. A graph of this solution set is given in Figure 9.

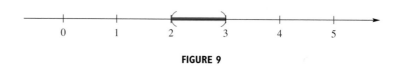

FIGURE 9

(b) $\dfrac{5}{y+1} > 4$

◀◀ **WORK PROBLEM 4 AT THE SIDE.**

E X A M P L E 4   Solving an Inequality with a Fraction

Solve $\dfrac{m-2}{m+2} \le 2$.

Write the corresponding equation and solve it. (Step 1)

$$\frac{m-2}{m+2} = 2$$

$$m - 2 = 2(m+2) \qquad \text{Multiply by the common denominator.}$$

$$m - 2 = 2m + 4 \qquad \text{Distributive property}$$

$$-6 = m$$

Set the denominator equal to zero and solve the equation. (Step 2)

$$m + 2 = 0$$

$$m = -2$$

ANSWERS

4. (a) $(-\infty, 4) \cup \left(\dfrac{14}{3}, \infty\right)$

$$\underset{\substack{\uparrow \\ \tfrac{14}{3}}}{}$$

$$\xleftarrow{} \!\!\!\! \underset{0\ 1\ 2\ 3\ 4\ 5\ 6}{|\ |\ |\ |\ )(\ |} \!\!\!\! \xrightarrow{}$$

(b) $\left(-1, \dfrac{1}{4}\right)$

$$\underset{\tfrac{1}{4}}{}$$

$$\xleftarrow{} \!\!\!\! \underset{-2\ -1\ \ 0\ \ 1\ \ 2}{|\ (\ )\ |\ |} \!\!\!\! \xrightarrow{}$$

CONTINUED ON NEXT PAGE

The numbers -6 and -2 determine three regions (Step 3). Test one number from each region (Step 4) to see that the solution set is the interval

$$(-\infty, -6] \cup (-2, \infty).$$

The number -6 satisfies the equality in $\leq$, but -2 cannot be used as a solution since it makes the denominator equal to zero (Step 5). The graph of the solution set is shown in Figure 10.

FIGURE 10

WORK PROBLEM 5 AT THE SIDE. ▶▶

OBJECTIVE 5 There is an alternative method for solving fractional inequalities like those in Examples 3 and 4. It requires writing the inequality with 0 on one side. The other side is then expressed as a single fraction. The values that cause the numerator or the denominator to equal 0 are then used to divide a number line into regions, and then a test value is used from each region to determine the solution set.

E X A M P L E 5 Using an Alternative Method to Solve a Fractional Inequality

Use the alternative method described above to solve the inequality from Example 3, $\dfrac{-1}{p-3} > 1$.

$$\frac{-1}{p-3} > 1$$

$$\frac{-1}{p-3} - 1 > 0 \qquad \text{Subtract 1 so that one side is equal to 0.}$$

$$\frac{-1}{p-3} - \frac{p-3}{p-3} > 0 \qquad \text{Use } p-3 \text{ as the common denominator.}$$

$$\frac{-1-p+3}{p-3} > 0 \qquad \text{Write as a single fraction.}$$

$$\frac{-p+2}{p-3} > 0 \qquad \text{Combine terms.}$$

The number 2 makes the numerator 0, and 3 makes the denominator 0. Notice that these determine the same regions as indicated in Figure 8. Testing one number from each region shows that only Region B, $(2, 3)$, makes the final inequality true. The solution set is graphed in Figure 9.

WORK PROBLEM 6 AT THE SIDE. ▶▶

OBJECTIVE 6 Special cases of quadratic inequalities may occur, such as those discussed in the next example.

5. Solve $\dfrac{k+2}{k-1} \leq 5$, and graph the solution.

6. Use the alternative method to solve the inequality in Example 4,

$$\frac{m-2}{m+2} \leq 2.$$

7. Solve.

(a) $(3k - 2)^2 > -2$

(b) $(5z + 3)^2 < -3$

E X A M P L E 6   Solving Special Cases

Solve $(2y - 3)^2 > -1$.

Because $(2y - 3)^2$ is never negative, it is always greater than -1. Thus, the solution is the set of all real numbers, $(-\infty, \infty)$. In the same way, there is no solution for $(2y - 3)^2 < -1$ and the solution set is $\emptyset$.

◀◀ WORK PROBLEM 7 AT THE SIDE.

8.5 Exercises

1. Explain how you determine whether to include or exclude endpoints when solving a quadratic or higher-degree inequality.

2. Explain why the number 7 cannot possibly be a solution of a fractional inequality that has $x - 7$ as the denominator of a fraction.

3. The solution set of the inequality $x^2 + x - 12 < 0$ is the interval $(-4, 3)$. Without actually performing any work, give the solution set of the inequality $x^2 + x - 12 \geq 0$.

4. Without actually performing any work, give the solution set of the fractional inequality $\frac{3}{x^2 + 1} > 0$. (*Hint:* Determine the sign of the numerator. Determine what the sign of the denominator *must* be. Then consider the inequality symbol.)

Solve the following inequalities, and graph each solution. See Example 1. (Hint: In Exercises 17 and 18, use the quadratic formula.)

5. $(x + 1)(x - 5) > 0$

6. $(m + 6)(m - 2) > 0$

7. $(r + 4)(r - 6) < 0$

8. $(y + 4)(y - 8) < 0$

9. $x^2 - 4x + 3 \geq 0$

10. $m^2 - 3m - 10 \geq 0$

11. $10a^2 + 9a \geq 9$

12. $3r^2 + 10r \geq 8$

13. $9p^2 + 3p < 2$

14. $2y^2 + y < 15$

15. $6x^2 + x \geq 1$

16. $4y^2 + 7y \geq -3$

17. $y^2 - 6y + 6 \geq 0$

18. $3k^2 - 6k + 2 \leq 0$

Solve the following inequalities and graph each solution. See Example 2.

19. $(p - 1)(p - 2)(p - 4) < 0$

20. $(2r + 1)(3r - 2)(4r + 7) < 0$

21. $(a - 4)(2a + 3)(3a - 1) \geq 0$

22. $(z + 2)(4z - 3)(2z + 7) \geq 0$

Solve the following inequalities and graph each solution. See Examples 3–5.

23. $\dfrac{x - 1}{x - 4} > 0$

24. $\dfrac{x + 1}{x - 5} > 0$

25. $\dfrac{2y + 3}{y - 5} \leq 0$

26. $\dfrac{3t + 7}{t - 3} \le 0$

27. $\dfrac{8}{x - 2} \ge 2$

28. $\dfrac{20}{y - 1} \ge 1$

29. $\dfrac{3}{2t - 1} < 2$

30. $\dfrac{6}{m - 1} < 1$

31. $\dfrac{a}{a + 2} \ge 2$

32. $\dfrac{m}{m + 5} \ge 2$

33. $\dfrac{4k}{2k - 1} < k$

34. $\dfrac{r}{r + 2} < 2r$

Solve the following inequalities. See Example 6.

35. $(4 - 3x)^2 \ge -2$

36. $(6y + 7)^2 \ge -1$

37. $(3x + 5)^2 \le -4$

38. $(8t + 5)^2 \le -5$

MATHEMATICAL CONNECTIONS (EXERCISES 39–42)

A rock is projected vertically upward from the ground. Its distance s in feet above the ground after t seconds is given by the quadratic function $s(t) = -16t^2 + 256t$. Work Exercises 39–42 in order to see how quadratic equations and inequalities are connected.

39. At what times will the rock be 624 feet above the ground? (*Hint:* Set $s(t) = 624$ and solve the quadratic *equation*.)

40. At what times will the rock be more than 624 feet above the ground? (*Hint:* Set $s(t) > 624$ and solve the quadratic *inequality*.)

41. At what times will the rock be at ground level? (*Hint:* Set $s(t) = 0$ and solve the quadratic *equation*.)

42. At what times will the rock be less than 624 feet above the ground? (*Hint:* Set $s(t) < 624$, solve the quadratic *inequality,* and observe the solutions in Exercises 40 and 41 to determine the smallest and largest possible values of t.)

8.1	**standard form**	A quadratic equation written in the form $ax^2 + bx + c = 0$ is in standard form.
8.2	**quadratic formula**	The quadratic formula is a formula for solving quadratic equations.
	discriminant	The discriminant is the quantity under the radical in the quadratic formula.
8.3	**quadratic in form**	An equation that can be written as a quadratic equation is called quadratic in form.
8.4	**Pythagorean theorem**	The Pythagorean theorem states that in a right triangle the square of the length of the hypotenuse equals the sum of the squares of the lengths of the legs.
	hypotenuse	The hypotenuse is the longest side in a right triangle.
	leg	The two shorter sides of a right triangle are called the legs.
8.5	**quadratic inequality**	A quadratic inequality is an inequality that can be written in the form $ax^2 + bx + c < 0$ or $ax^2 + bx + c > 0$, or with $\leq$ or $\geq$.

Concepts	*Examples*

8.1 The Square Root Property and Completing the Square

Square Root Property
If x and b are complex numbers and if $x^2 = b$, then

$$x = \sqrt{b} \quad \text{or} \quad x = -\sqrt{b}.$$

Solve $(x - 1)^2 = 8$.

$$x - 1 = \pm\sqrt{8} = \pm 2\sqrt{2}$$
$$x = 1 \pm 2\sqrt{2}$$

Solution set: $\{1 \pm 2\sqrt{2}\}$

Solve $x^2 = -11$.

$$x = \sqrt{-11} \quad \text{or} \quad x = -\sqrt{-11}$$
$$x = i\sqrt{11} \quad \text{or} \quad x = -i\sqrt{11}$$

Solution set: $\{\pm i\sqrt{11}\}$

Completing the Square
To solve $ax^2 + bx + c = 0$: If $a \neq 1$, divide both sides by a. Write the equation with the variable terms on one side and the constant on the other.

Solve $2x^2 - 4x - 18 = 0$.
$$x^2 - 2x - 9 = 0$$
$$x^2 - 2x = 9$$

Find half the coefficient of x and square it.

$$\left[\frac{1}{2}(-2)\right]^2 = (-1)^2 = 1$$

Add the square to both sides. One side should now be a square trinomial. Write it as the square of a binomial.

$$x^2 - 2x + 1 = 9 + 1$$
$$(x - 1)^2 = 10$$

Use the square root property to complete the solution.

$$x - 1 = \pm\sqrt{10}$$
$$x = 1 \pm \sqrt{10}$$

Solution set: $\{1 \pm \sqrt{10}\}$

Concepts	*Examples*

8.2 The Quadratic Formula

Quadratic Formula

The solutions of $ax^2 + bx + c = 0$ $(a \neq 0)$ are

$$x = \frac{-b \pm \sqrt{b^2 - 4ac}}{2a}.$$

Solve $3x^2 + 5x + 2 = 0$.

$$x = \frac{-5 \pm \sqrt{5^2 - 4(3)(2)}}{2(3)}$$

$$x = -1 \quad \text{or} \quad x = -\frac{2}{3}$$

Solution set: $\left\{ -1, -\frac{2}{3} \right\}$

The Discriminant

If a, b, and c are integers, then the discriminant, $b^2 - 4ac$, of $ax^2 + bx + c = 0$ determines the type of solutions as follows.

Discriminant	*Solutions*
Positive square of an integer	2 rational solutions
Positive, not square of an integer	2 irrational solutions
Zero	1 rational solution
Negative	2 imaginary solutions

For $x^2 + 3x - 10 = 0$, the discriminant is

$$3^2 - 4(1)(-10) = 49.$$

There are **2 rational** solutions.

For $2x^2 + 5x + 1 = 0$, the discriminant is

$$5^2 - 4(2)(1) = 17.$$

There are **2 irrational** solutions.

For $9x^2 - 6x + 1 = 0$, the discriminant is

$$(-6)^2 - 4(9)(1) = 0.$$

There is **1 rational** solution.

For $4x^2 + x + 1 = 0$, the discriminant is

$$1^2 - 4(4)(1) = -15.$$

There are **2 imaginary** solutions.

If the discriminant of $ax^2 + bx + c$ is a perfect square, the trinomial is factorable.

The discriminant of $30x^2 - 13x - 10$ is $b^2 - 4ac = (-13)^2 - 4(30)(-10) = 1369 = 37^2$, so it is factorable. It factors as $(5x + 2)(6x - 5)$.

8.3 Equations Quadratic in Form

An equation that can be written in the form $au^2 + bu + c = 0$, for $a \neq 0$ and an algebraic expression u, is called quadratic in form. Substitute u for the expression, solve for u, and then solve for the variable in the expression.

Solve $3(x + 5)^2 + 7(x + 5) + 2 = 0$.

Here $u = x + 5$, so the equation can be written as $3u^2 + 7u + 2 = 0$. Solve for u by factoring.

$$(3u + 1)(u + 2) = 0$$

$$u = -\frac{1}{3} \quad \text{or} \quad u = -2$$

Now solve for x.

$$u = x + 5 \qquad\qquad u = x + 5$$

$$-\frac{1}{3} = x + 5 \qquad -2 = x + 5$$

$$x = -\frac{16}{3} \qquad\qquad x = -7$$

Solution set: $\left\{ -7, -\frac{16}{3} \right\}$

Concepts	Examples

8.4 Formulas and Applications Involving Quadratic Equations

Formulas

(a) If the variable appears only to the second degree, isolate the squared variable on one side of the equation. Then use the square root property.

Solve $A = \dfrac{2mp}{r^2}$ for r.

$$r^2 A = 2mp \qquad \text{Multiply by } r^2.$$

$$r^2 = \frac{2mp}{A} \qquad \text{Divide by } A.$$

$$r = \pm\sqrt{\frac{2mp}{A}} \qquad \text{Take square roots.}$$

$$r = \pm\frac{\sqrt{2mpA}}{A} \qquad \text{Rationalize.}$$

(b) If the variable appears to the first and second degree, write the equation in standard quadratic form. Then use the quadratic formula to solve.

Solve $m^2 + rm = t$ for m.

$$m^2 + rm - t = 0 \qquad \text{Standard form}$$

$$m = \frac{-r \pm \sqrt{r^2 - 4(1)(-t)}}{2(1)} \quad \begin{matrix} a = 1, b = r, \\ c = -t \end{matrix}$$

$$m = \frac{-r \pm \sqrt{r^2 + 4t}}{2}$$

Applications

Quadratic functions can often be used to model data that are not linear.

The quadratic function $f(x) = .0234x^2 - .5029x + 12.5$ can be used to model the number of infant deaths per 1000 live births between 1980 and 1989, where $x = 0$ corresponds to 1980, $x = 1$ to 1981, and so on.

To find the number of deaths in 1985, let $x = 5$ and find $f(5)$. Since $f(5) \approx 10.6$, there were about 10.6 deaths per 1000 live births in 1985.

To find the year in which the number of deaths was 11.6, solve $11.6 = .0234x^2 - .5029x + 12.5$ to find the approximate solutions 2 and 19.5. Only 2 applies here, and $x = 2$ corresponds to 1982.

8.5 Nonlinear and Fractional Inequalities

Solving a Quadratic Inequality

Step 1 Rewrite the inequality as an equation and solve.

Solve $2x^2 + 5x + 2 < 0$.

$$2x^2 + 5x + 2 = 0$$

$$x = -\frac{1}{2}, x = -2$$

Step 2 Place the numbers found in Step 1 on a number line. These numbers divide the line into regions.

Step 3 Substitute a number from each region into the inequality to determine the intervals that belong in the solution set—those intervals containing numbers that make the inequality true.

$x = -3$ makes it false; $x = -1$ makes it true; $x = 0$ makes it false.

Solution: $\left(-2, -\dfrac{1}{2}\right)$

This method can be extended to inequalities having more than two factors.

Concepts	Examples
8.5 Nonlinear and Fractional Inequalities (Continued)	Solve $\dfrac{x}{x+2} > 4$.
Solving Inequalities with Fractions	Solving $\dfrac{x}{x+2} = 4$ yields $x = -\dfrac{8}{3}$.
Step 1 Rewrite the inequality as an equation and solve the equation.	$x + 2 = 0$
Step 2 Set the denominator equal to zero and solve that equation.	$x = -2$
Step 3 Use the solutions from Steps 1 and 2 to divide a number line into regions.	$\xleftarrow{\hspace{2cm}}\xrightarrow{\hspace{2cm}}$ $-\dfrac{8}{3}$ $\quad$ -2
Step 4 Test a number from each region in the inequality to determine the regions that satisfy the inequality.	-4 makes it false; $-\dfrac{7}{3}$ makes it true; 0 makes it false.
Step 5 Exclude any values that make the denominator zero.	The solution is $\left(-\dfrac{8}{3}, -2\right)$, since -2 makes the denominator 0 and $-\dfrac{8}{3}$ gives a false sentence.
An alternative method of solving inequalities with fractions is discussed in Section 8.5.	

[8.1] *Solve each equation using either the square root property or the method of completing the square.*

1. $t^2 = 121$

2. $p^2 = 3$

3. $(2x + 5)^2 = 100$

*4. $(3k - 2)^2 = -25$

5. $x^2 + 4x = 15$

6. $2m^2 - 3m = -1$

[8.2] *Solve each equation using the quadratic formula.*

7. $2y^2 + y - 21 = 0$

8. $k^2 + 5k = 7$

9. $(t + 3)(t - 4) = -2$

10. $9p^2 = 42p - 49$

*11. $3p^2 = 2(2p - 1)$

12. $m(2m - 7) = 3m^2 + 3$

*13. $2x^2 + 3x + 4 = 0$

14. A student wrote the following as the quadratic formula for solving $ax^2 + bx + c = 0$, $a \neq 0$:
$x = -b \pm \dfrac{\sqrt{b^2 - 4ac}}{2a}$. Was this correct? If not, what is wrong with it?

_____

* Exercises identified with asterisks have imaginary number solutions.

15. A rock is projected vertically upward from the ground. Its distance in feet from the ground in t seconds is $s(t) = -16t^2 + 256t$. At what times will the rock be 768 feet from the ground?

16. Explain why the problem in Exercise 15 has two answers.

Use the quadratic equation $x^2 + 2x + k = 0$ for Exercises 17 and 18.

17. What value(s) of k will give only one real solution?

18. What value(s) of k will give two real solutions?

Use the discriminant to predict whether the solutions to the following equations are
(a) two distinct rational numbers;   *(b) exactly one rational number;*
(c) two distinct irrational numbers;   *(d) two distinct imaginary numbers.*

19. $a^2 + 5a + 2 = 0$ **20.** $4c^2 = 3 - 4c$ **21.** $4x^2 = 6x - 8$ **22.** $9z^2 + 30z + 25 = 0$

Use the discriminant to tell which polynomials can be factored. If a polynomial can be factored, factor it.

23. $24x^2 - 74x + 45$

24. $36x^2 + 69x - 34$

[8.3] *Solve each equation.*

25. $\dfrac{15}{x} = 2x - 1$

26. $\dfrac{1}{y} + \dfrac{2}{y+1} = 2$

27. $8(3x + 5)^2 + 2(3x + 5) - 1 = 0$

28. $-2r = \sqrt{\dfrac{48 - 20r}{2}}$

29. $2x^{2/3} - x^{1/3} - 28 = 0$

30. $(x^2 + x)^2 = 8(x^2 + x) - 12$

31. Lisa Wunderle drove 8 miles to pick up her friend Laurie, and then drove 11 miles to a mall at a speed 15 miles per hour faster. If Lisa's total travel time was 24 minutes, what was her speed on the trip to pick up Laurie?

32. It takes Linda Youngman 2 hours longer to write a report to her boss than it takes Ed Moura. Working together, it would take them 3 hours. How long would it take each one to do the job alone? Round your answer to the nearest tenth of an hour.

33. Why can't the equation $x = \sqrt{2x + 4}$ have a negative solution?

34. If you were to use the quadratic formula to solve $x^4 - 5x^2 + 6 = 0$, what would you have to remember after you applied the formula?

[8.4] *Solve each formula for the indicated variable. (Give answers with* $\pm$.*)*

35. $S = \dfrac{Id^2}{k}$ for d

36. $k = \dfrac{rF}{wv^2}$ for v

37. $S = 2\pi rh + 2\pi r^2$ for r

38. $mt^2 = 3mt + 6$ for t

Solve each of the following problems.

39. The Mart Hotel in Dallas, Texas, is 400 feet high. Suppose that a ball is projected upward from the top of the Mart, and its position in feet above the ground is given by the quadratic function $f(t) = -16t^2 + 45t + 400$, where t is the number of seconds elapsed. How long will it take for the ball to reach a height of 200 feet above the ground?

40. The Toronto Dominion Center in Winnipeg, Manitoba, is 407 feet high. Suppose that a ball is projected upward from the top of the Center, and its position in feet above the ground is given by the quadratic function $s(t) = -16t^2 + 75t + 407$, where t is the number of seconds elapsed. How long will it take for the ball to reach a height of 450 feet above the ground?

41. A rectangle has a length 2 meters more than its width. If one meter is cut from the length and one meter is added to the width, the resulting figure is a square with an area of 121 square meters. Find the dimensions of the original rectangle.

42. The hypotenuse of a right triangle is 9 feet shorter than twice the length of the longer leg. The shorter leg is 3 feet shorter than the longer leg. Find the lengths of the three sides of the triangle.

43. The product of the page numbers of two pages facing each other in this book is 4692. What are the page numbers?

44. Nancy wants to buy a mat for a photograph that measures 14 inches by 20 inches. She wants to have an even border around the picture when it is mounted on the mat. If the area of the mat she chooses is 352 square inches, how wide will the border be?

45. A search light moves horizontally back and forth along a wall with the distance of the light from a starting point at t minutes given by the quadratic function $f(t) = 100t^2 - 300t$. How long will it take before the light returns to the starting point?

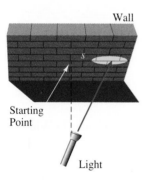

46. The manager of a fast-food outlet has determined that the demand for frozen yogurt is $\frac{25}{p}$ units per day, where p is the price (in dollars) per unit. The supply is $70p + 15$ units per day. Find the price at which supply and demand are equal.

47. Use the formula $A = P(1 + r)^2$ to find the interest rate r at which a principal P of \$10,000 will increase to \$10,920.25 in 2 years.

48. Jim can complete a job alone in one hour less than his brother, Jake. Together they can complete the job in 4 hours. How long does the job take each brother working alone?

[8.5] *Solve the following inequalities and graph the solutions.*

49. $(x - 4)(2x + 3) > 0$

50. $x^2 + x \leq 12$

51. $2k^2 > 5k + 3$

52. $\dfrac{3y + 4}{y - 2} \leq 1$

53. $(x + 2)(x - 3)(x + 5) \leq 0$

54. $(4m + 3)^2 \leq -4$

MIXED REVIEW EXERCISES

Solve.

55. $V = r^2 + R^2 h$ for R

*56. $3t^2 - 6t = -4$

57. $(b^2 - 2b)^2 = 11(b^2 - 2b) - 24$

58. $(r - 1)(2r + 3)(r + 6) < 0$

59. $(3k + 11)^2 = 7$

60. $p = \sqrt{\dfrac{yz}{6}}$ for y

61. $-5x^2 = -8x + 3$

62. $6 + \dfrac{15}{s^2} = -\dfrac{19}{s}$

63. $\dfrac{-2}{x + 5} \leq -5$

64. Two pipes together can fill a large tank in 2 hours. One of the pipes, used alone, takes 3 hours longer than the other does to fill the tank. How long would each pipe take to fill the tank alone?

65. Phong paddled his canoe 20 miles upstream, then paddled back. If the speed of the current was 3 miles per hour and the total trip took 7 hours, what was Phong's speed?

66. $\dfrac{2}{x - 4} + \dfrac{1}{x} = \dfrac{11}{5}$

*67. $y^2 = -242$

68. $(8k - 7)^2 \geq -1$

CHAPTER 8 TEST

*Items marked * require knowledge of imaginary numbers.*

Solve using the square root property.

1. $t^2 = 54$

1. _____

2. $(7x + 3)^2 = 25$

2. _____

Solve by completing the square.

3. $x^2 + 2x = 1$

3. _____

Solve using the quadratic formula.

4. $2x^2 - 3x - 1 = 0$

4. _____

*5. $3t^2 - 4t = -5$

5. _____

6. $3x = \sqrt{\dfrac{9x + 2}{2}}$

6. _____

7. Maretha and Lillaana are typesetters. For a certain report, Lillaana can set the type 2 hours faster than Maretha can. If they work together, they can do the entire report in 5 hours. How long will it take each of them working alone to prepare the report? Round your answers to the nearest tenth of an hour.

7. _____

*8. If k is a negative number, then which one of the following equations will have two imaginary solutions?
(a) $x^2 = 4k$ **(b)** $x^2 = -4k$
(c) $(x + 2)^2 = -k$ **(d)** $x^2 + k = 0$

8. _____

9. What is the discriminant for $2x^2 - 8x - 3 = 0$? How many and what type of solutions does this equation have? (Do not actually solve.)

9. _____

Solve by any method.

10. $3 - \dfrac{16}{x} - \dfrac{12}{x^2} = 0$

10. _____

11. $4x^2 + 7x - 3 = 0$

11. _____

12. _____

12. $9x^4 + 4 = 37x^2$

13. _____

13. $12 = (2d + 1)^2 + (2d + 1)$

14. _____

14. Solve for r: $S = 4\pi r^2$ (Leave $\pm$ in your answer.)

Solve each problem.

15. _____

15. The quadratic function $f(x) = 3.23x^2 - 1.89x + 1.06$ approximates the number of AIDS cases diagnosed in the United States between 1982 and 1993. Here the number of cases is in thousands, and $x = 0$ corresponds to 1982, $x = 1$ to 1983, and so on. In what year did the number of cases diagnosed reach 24.64 thousand?

16. _____

16. Sandi Goldstein paddled her canoe 10 miles upstream, and then paddled back to her starting point. If the rate of the current was 3 miles per hour and the entire trip took $3\frac{1}{2}$ hours, what was Sandi's rate?

17. _____

17. Adam Bryer has a pool 24 feet long and 10 feet wide. He wants to construct a concrete walk around the pool. If he plans for the walk to be of uniform width and cover 152 square feet, what will the width of the walk be?

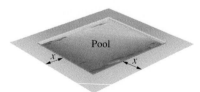

18. _____

18. At a point 30 meters from the base of a tower, the distance to the top of the tower is 2 meters more than twice the height of the tower. Find the height of the tower.

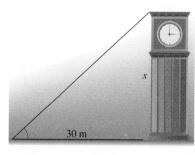

Solve. Graph each solution set.

19. ————————————→

19. $2x^2 + 7x > 15$

20. ————————————→

20. $\dfrac{5}{t - 4} \le 1$

Let $S = \{-\frac{7}{3}, -2, -\sqrt{3}, 0, .7, \sqrt{12}, \sqrt{-8}, 7, \frac{32}{3}\}$. List the elements of S that are elements of the following sets.

1. Integers   **2.** Rational numbers   **3.** Real numbers   **4.** Complex numbers

Simplify each of the following.

5. $|-3| + 8 - |-9| - (-7 + 3)$

6. $2(-3)^2 + (-8)(-5) + (-17)$

Solve the following.

7. $-2x + 4 = 5(x - 4) + 17$ **8.** $|3y - 7| \leq 1$ **9.** $|4z + 2| > 7$

10. Find the slope and y-intercept of the line with equation $2x - 4y = 7$.

11. Write an equation in standard form of the line through $(2, -1)$ and perpendicular to $-3x + y = 5$.

Write with positive exponents only. Assume variables represent positive real numbers.

12. $\left(\dfrac{x^{-3}y^2}{x^5y^{-2}}\right)^{-1}$

13. $\dfrac{(4x^{-2})^2(2y^3)}{8x^{-3}y^5}$

14. If $f(x) = \sqrt{2x - 5}$, find $f(10)$. What is the domain of this function?

15. Solve the system $5x - 3y = 17$
$2y = 4x - 12$.

16. For a certain system of equations, $D_x = 4$, $D_y = 3$, $D_z = -6$, and $D = 12$. Find the solution set of the system.

Perform the indicated operations.

17. $\left(\dfrac{2}{3}t + 9\right)^2$

18. $(3t^3 + 5t^2 - 8t + 7) - (6t^3 + 4t - 8)$

19. Divide $4x^3 + 2x^2 - x + 26$ by $x + 2$.

Factor completely.

20. $16x - x^3$

21. $24m^2 + 2m - 15$

22. $9x^2 - 30xy + 25y^2$

23. $8x^3 + 27y^3$

Perform the operations, and express answers in lowest terms. Assume denominators are nonzero.

24. $\dfrac{x^2 - 3x - 10}{x^2 + 3x + 2} \cdot \dfrac{x^2 - 2x - 3}{x^2 + 2x - 15}$

25. $\dfrac{5t + 2}{-6} \div \dfrac{15t + 6}{5}$

26. $\dfrac{3}{2 - k} - \dfrac{5}{k} + \dfrac{6}{k^2 - 2k}$

27. $\dfrac{\dfrac{r}{s} - \dfrac{s}{r}}{\dfrac{r}{s} + 1}$

Simplify the following radical expressions.

28. $\sqrt[3]{\dfrac{27}{16}}$

29. $\dfrac{2}{\sqrt{7} - \sqrt{5}}$

Solve the following equations.

30. $2x = \sqrt{\dfrac{5x + 2}{3}}$

31. $\dfrac{3}{x - 3} - \dfrac{2}{x - 2} = \dfrac{3}{x^2 - 5x + 6}$

32. $(r - 5)(2r + 3) = 1$

33. $b^4 - 5b^2 + 4 = 0$

Solve the following problems.

34. The perimeter of a rectangle is 20 inches, and its area is 21 square inches. What are the dimensions of the rectangle?

35. Two cars left an intersection at the same time, one heading due south and the other due east. Later they were exactly 95 miles apart. The car heading east had gone 38 miles less than twice as far as the car heading south. How far had each car traveled?

36. A knitting shop orders yarn from three suppliers, I, II, and III. One month the shop ordered a total of 100 units of yarn from these suppliers. The delivery costs were $80, $50, and $65 per unit for the orders from suppliers I, II, and III, respectively, with total delivery costs of $5990. The shop ordered the same amount from suppliers I and III. How many units were ordered from each supplier?

37. An object is propelled upward. Its height in feet is given by the function

$$f(t) = -16t^2 + 80t + 50.$$

At what times t is its height 70 feet? Round to the nearest tenth of a second.

Solve each equation by the method stated.

38. $z^2 - 2z = 15$ (completing the square)

39. $2x^2 - 4x - 3 = 0$ (quadratic formula)

40. Solve the fractional inequality $\dfrac{x + 3}{x - 5} > 2.$

Graphs of Nonlinear Functions and Conic Sections

9.1 Graphs of Quadratic Functions; Vertical Parabolas

In Chapter 3, we saw that the graphs of first-degree equations are straight lines. In this chapter we discuss the graphs of nonlinear equations. In particular, we will investigate graphs of second-degree equations, which are equations with one or more second-degree terms. These graphs are the intersections of an infinite cone and a plane, as shown in Figure 1. Because of this, the graphs are called **conic sections.**

www.mathnotes.com

OBJECTIVES

1. Graph parabolas that are examples of quadratic functions.

2. Graph parabolas with horizontal and vertical shifts.

3. Predict the shape and direction of the graph of a parabola from the coefficient of x^2.

FOR EXTRA HELP

Tutorial   Tape 13   SSM, Sec. 9.1

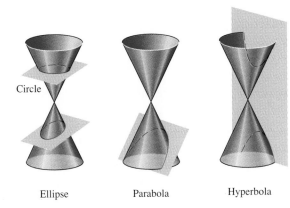

Circle

Ellipse   Parabola   Hyperbola

FIGURE 1

OBJECTIVE 1 Let us begin by graphing the equation $y = x^2$. First, we make a table of ordered pairs satisfying the equation.

x	-2	$-\frac{3}{2}$	-1	$-\frac{1}{2}$	0	$\frac{1}{2}$	1	$\frac{3}{2}$	2
y	4	$\frac{9}{4}$	1	$\frac{1}{4}$	0	$\frac{1}{4}$	1	$\frac{9}{4}$	4

We then plot these points and draw a smooth curve through them to get the graph shown in Figure 2. This graph is called a **parabola** (puh-RAB-uh-luh). The point $(0, 0)$, with the smallest y-value of any point on the curve, is the **vertex** of this parabola. The vertical line through the vertex is the **axis** of this parabola. The parabola is **symmetric with respect to its axis;** that is, if the graph were folded along the axis, the two portions of the curve would coincide.

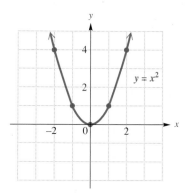

FIGURE 2

Because the graph of $y = x^2$ satisfies the conditions of the graph of a function (Section 3.5), we may write its equation as $f(x) = x^2$. This is the simplest example of a quadratic function.

> **Quadratic Function**
>
> A function defined by an equation that can be written in the form
>
> $$f(x) = ax^2 + bx + c$$
>
> for real numbers, a, b, and c, with $a \neq 0$, is a **quadratic function.**

The graph of any quadratic function is a parabola with a vertical axis.

Parabolas have many applications. If an object is thrown upward, then (disregarding air resistance) the path it follows is a parabola. The large disks seen on the sidelines of televised football games, which are used by television crews to pick up the shouted signals of the players on the field, have cross sections that are parabolas. Cross sections of radar dishes and automobile headlights also form parabolas. Additional applications of parabolas are given in the next section.

For the rest of this section, we shall use the function notation $f(x)$ in discussing parabolas.

OBJECTIVE 2 Parabolas need not have their vertices at the origin, as does $f(x) = x^2$. For example, to graph a parabola of the form $f(x) = x^2 + k$, we start by selecting the sample values of x that were used to graph $f(x) = x^2$. The corresponding values of $f(x)$ in $f(x) = x^2 + k$ differ by k from those of $f(x) = x^2$. For this reason, the graph of $f(x) = x^2 + k$ is shifted, or translated, k units vertically compared with that of $f(x) = x^2$.

EXAMPLE 1 Graphing a Parabola with a Vertical Shift

Graph $f(x) = x^2 - 2$.

As we mentioned before, this graph has the same shape as $f(x) = x^2$, but since k here is -2, the graph is shifted 2 units downward, with vertex

CONTINUED ON NEXT PAGE

at $(0, -2)$. Every function value is 2 less than the corresponding function value of $f(x) = x^2$. Plotting points gives the graph in Figure 3. The graph of $f(x) = x^2$ is also shown for comparison.

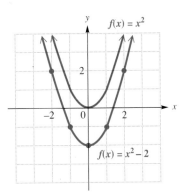

FIGURE 3

Vertical Shift

The graph of $f(x) = x^2 + k$ is a parabola with the same shape as the graph of $f(x) = x^2$. The parabola is shifted k units upward if $k > 0$, and $|k|$ units downward if $k < 0$. The vertex is $(0, k)$.

> **WORK PROBLEM 1 AT THE SIDE.** ▶▶

The graph of $f(x) = (x - h)^2$ is also a parabola with the same shape as $f(x) = x^2$. Because $(x - h)^2 \geq 0$ for all x, the vertex of the parabola $f(x) = (x - h)^2$ should be the lowest point on the parabola. The lowest point occurs here when $f(x)$ is 0. To get $f(x)$ equal to 0, let $x = h$ so the vertex of $f(x) = (x - h)^2$ is at $(h, 0)$. Based on this, the graph of $f(x) = (x - h)^2$ is shifted h units horizontally compared with that of $f(x) = x^2$.

E X A M P L E 2 Graphing a Parabola with a Horizontal Shift

Graph $f(x) = (x - 2)^2$.

When $x = 2$, then $f(x) = 0$, giving the vertex $(2, 0)$. The parabola $f(x) = (x - 2)^2$ has the same shape as $f(x) = x^2$ but is shifted 2 units to the right, as shown in Figure 4. Again, we show the graph of $f(x) = x^2$ for comparison.

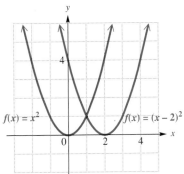

FIGURE 4

1. Graph each parabola.

(a) $f(x) = x^2 + 3$

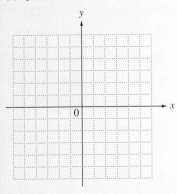

(b) $f(x) = x^2 - 1$

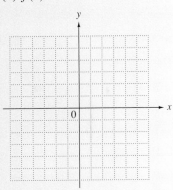

ANSWERS

1. (a)

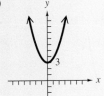

(b)

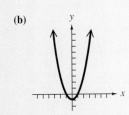

2. Graph each parabola.

(a) $f(x) = (x - 3)^2$

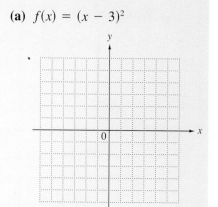

(b) $f(x) = (x + 2)^2$

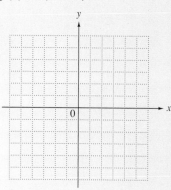

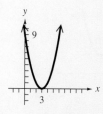
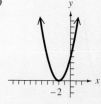

Horizontal Shift

The graph of $f(x) = (x - h)^2$ is a parabola with the same shape as the graph of $f(x) = x^2$. The parabola is shifted h units horizontally: h units to the right if $h > 0$, and $|h|$ units to the left if $h < 0$. The vertex is $(h, 0)$.

Caution
Errors frequently occur when horizontal shifts are involved. In order to determine the direction and magnitude of horizontal shifts, find the value that would cause the expression $x - h$ to equal 0. For example, the graph of $f(x) = (x - 5)^2$ would be shifted 5 units to the *right*, because $+5$ would cause $x - 5$ to equal 0. On the other hand, the graph of $f(x) = (x + 4)^2$ would be shifted 4 units to the *left*, because -4 would cause $x + 4$ to equal 0.

◀◀ WORK PROBLEM 2 AT THE SIDE.

A parabola can have both a horizontal and a vertical shift, as in Example 3.

EXAMPLE 3 Graphing a Parabola with Horizontal and Vertical Shifts

Graph $f(x) = (x + 3)^2 - 2$.

This graph has the same shape as $f(x) = x^2$, but is shifted 3 units to the left (since $x + 3 = 0$ if $x = -3$) and 2 units downward (because of the -2). As shown in Figure 5, the vertex is at $(-3, -2)$.

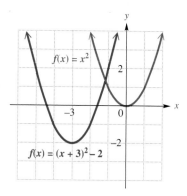

FIGURE 5

EXAMPLE 4 Graphing a Parabola with Horizontal and Vertical Shifts

Graph $f(x) = (x - 1)^2 + 3$.

The graph is shifted one unit to the right and three units up, so the vertex is at $(1, 3)$. See Figure 6.

CONTINUED ON NEXT PAGE

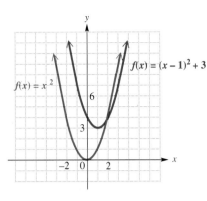

FIGURE 6

The characteristics of the graph of a function of the form $f(x) = (x - h)^2 + k$ are summarized as follows.

> The graph of $f(x) = (x - h)^2 + k$ is a parabola with the same shape as $f(x) = x^2$ and with vertex at (h, k). The axis is the vertical line $x = h$.

WORK PROBLEM 3 AT THE SIDE. ▶▶

OBJECTIVE 3 ▶ Not all parabolas open upward, and not all parabolas have the same shape as $f(x) = x^2$. In the next example we show how to identify parabolas opening downward and having a different shape from that of $f(x) = x^2$.

─**E X A M P L E 5**   **Graphing a Parabola That Opens Downward**

Graph $f(x) = -\frac{1}{2}x^2$.

This parabola is shown in Figure 7. Some ordered pairs that satisfy the equation are $(0, 0)$, $(1, -\frac{1}{2})$, $(2, -2)$, $(-1, -\frac{1}{2})$, and $(-2, -2)$. A table with these ordered pairs is given next to the graph. The coefficient $-\frac{1}{2}$ affects the shape of the graph; the $\frac{1}{2}$ makes the parabola wider (since the values of $f(x)$ grow more slowly than they would for $f(x) = x^2$), and the negative sign makes the parabola open downward. The graph is not shifted in any direction; the vertex is still at $(0, 0)$. Here, the vertex has the *largest* function value (y-value) of any point on the graph.

x	y
-2	-2
-1	$-\frac{1}{2}$
0	0
1	$-\frac{1}{2}$
2	-2

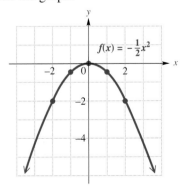

FIGURE 7

3. (a) Graph $f(x) = (x + 2)^2 - 1$. Identify the vertex.

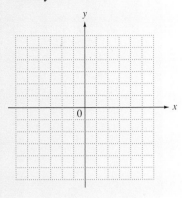

(b) Identify the vertex of the graph of $f(x) = (x - 2)^2 + 5$.

ANSWERS

3. (a) vertex: $(-2, -1)$

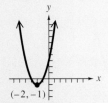

(b) vertex: $(2, 5)$

4. Decide whether each parabola opens upward or downward.

(a) $f(x) = -\dfrac{2}{3}x^2$

(b) $f(x) = \dfrac{3}{4}x^2 + 1$

(c) $f(x) = -2x^2 - 3$

(d) $f(x) = 3x^2 + 2$

5. Decide whether each parabola in Problem 4 is wider or narrower than $f(x) = x^2$.

6. Graph $f(x) = \dfrac{1}{2}(x - 2)^2 + 1$.

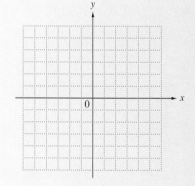

Some general principles concerning the graph of $f(x) = a(x - h)^2 + k$ are summarized as follows.

> **1.** The graph of the quadratic function defined by
> $$f(x) = a(x - h)^2 + k, \qquad a \neq 0$$
> is a parabola with vertex at (h, k) and the vertical line $x = h$ as axis.
> **2.** The graph opens upward if a is positive and downward if a is negative.
> **3.** The graph is wider than $f(x) = x^2$ if $0 < |a| < 1$. The graph is narrower than $f(x) = x^2$ if $|a| > 1$.

◀◀ **WORK PROBLEMS 4 AND 5 AT THE SIDE.**

E X A M P L E   6   Using the General Principles to Graph a Parabola

Graph $f(x) = -2(x + 3)^2 + 4$.

The parabola opens downward (because $a < 0$), and is narrower than the graph of $f(x) = x^2$, since $|-2| = 2 > 1$, causing the values of $f(x)$ to grow more quickly than they would for $f(x) = x^2$. This parabola has its vertex at $(-3, 4)$, as shown in Figure 8. To complete the graph, we plotted the ordered pairs $(-4, 2)$ and $(-2, 2)$, which are shown in the table next to the graph. Notice that these two points are symmetric with respect to the axis of the parabola. This symmetry is very useful for finding additional ordered pairs that satisfy the equation.

x	y
-4	2
-3	4
-2	2

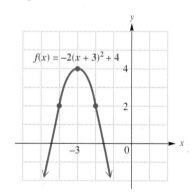

FIGURE 8

◀◀ **WORK PROBLEM 6 AT THE SIDE.**

ANSWERS

4. **(a)** downward **(b)** upward
    (c) downward **(d)** upward
5. **(a)** wider **(b)** wider **(c)** narrower
    (d) narrower
6.

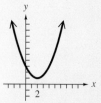

Note

In this section we have seen how the graph of $f(x) = x^2$ can be shifted horizontally or vertically, and how the shape of the graph can be altered. The principles discussed here can be generalized to graphs of other kinds of functions, as we shall see later in this chapter.

9.1 Exercises

1. Match each quadratic function with its graph from choices A–D.

_____ **(a)** $f(x) = (x + 2)^2 - 1$

_____ **(b)** $f(x) = (x + 2)^2 + 1$

_____ **(c)** $f(x) = (x - 2)^2 - 1$

_____ **(d)** $f(x) = (x - 2)^2 + 1$

A.
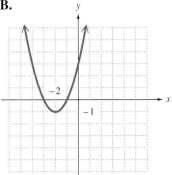

B.
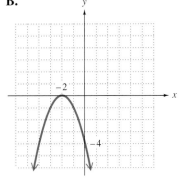

C.
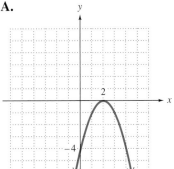

D.

2. Match each quadratic function with its graph from choices A–D.

_____ **(a)** $f(x) = -x^2 + 2$

_____ **(b)** $f(x) = -x^2 - 2$

_____ **(c)** $f(x) = -(x + 2)^2$

_____ **(d)** $f(x) = -(x - 2)^2$

A.
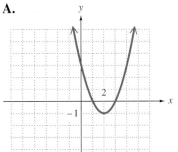

B.
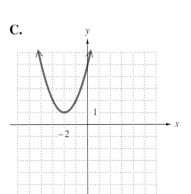

C.
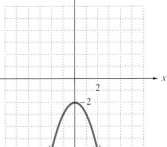

D.
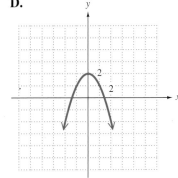

Identify the vertex of each parabola. See Examples 1–4.

3. $f(x) = -3x^2$

4. $f(x) = .5x^2$

5. $f(x) = x^2 + 4$

6. $f(x) = x^2 - 4$

7. $f(x) = (x - 1)^2$

8. $f(x) = (x + 3)^2$

9. $f(x) = (x + 3)^2 - 4$

10. $f(x) = (x - 5)^2 - 8$

For each quadratic function, tell whether the graph opens upward or downward and whether the graph is wider, narrower, or the same as $f(x) = x^2$. See Examples 5 and 6.

11. $f(x) = -.4x^2$

12. $f(x) = -2x^2$

13. $f(x) = 3x^2 + 1$

14. $f(x) = \frac{2}{3}x^2 - 4$

15. Describe how each of the parabolas in Exercises 9 and 10 is shifted compared to the graph of $y = x^2$.

16. What does the value of a in $y = a(x - h)^2 + k$ tell you about the graph of the equation compared to the graph of $y = x^2$?

17. For $f(x) = a(x - h)^2 + k$, in what quadrant is the vertex if
(a) $h > 0, k > 0$;   **(b)** $h > 0, k < 0$;   **(c)** $h < 0, k > 0$;   **(d)** $h < 0, k < 0$?

18. Think of how the graph of the linear function $f(x) = x + 5$ compares to the graph of $g(x) = x$, and fill in the blank: To graph $f(x) = x + 5$, shift the graph of $g(x) = x$ _____ units in the _____ direction.

Sketch the graph of each parabola. Plot at least two points in addition to the vertex.
See Examples 1–6.

19. $f(x) = -2x^2$

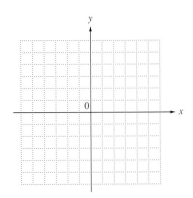

20. $f(x) = \frac{1}{3}x^2$

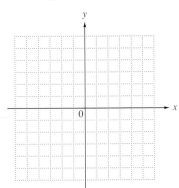

21. $f(x) = x^2 - 1$

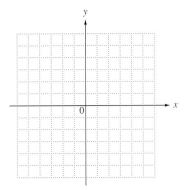

22. $f(x) = x^2 + 3$

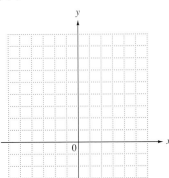

23. $f(x) = -x^2 + 2$

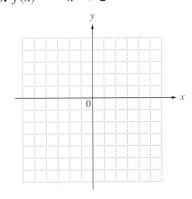

24. $f(x) = 2x^2 - 2$

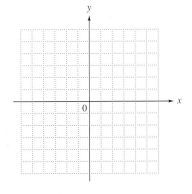

25. $f(x) = .5(x - 4)^2$

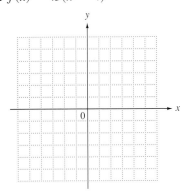

26. $f(x) = -2(x + 1)^2$

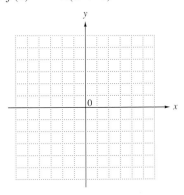

27. $f(x) = (x + 2)^2 - 1$

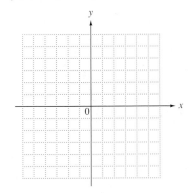

28. $f(x) = -2(x + 3)^2 + 4$

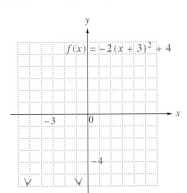

29. $f(x) = -.5(x + 1)^2 + 2$

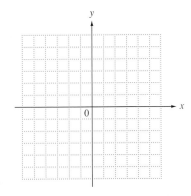

30. $f(x) = -\frac{2}{3}(x + 2)^2 + 1$

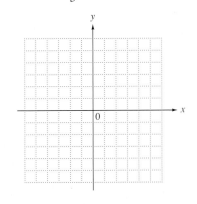

MATHEMATICAL CONNECTIONS (EXERCISES 31–36)

The procedures described in this section that allow the graph of $f(x) = x^2$ to be shifted vertically and horizontally are applicable to other types of functions as well. In Section 3.5 we introduced linear functions (functions of the form $f(x) = ax + b$). Consider the graph of the simplest linear function, $f(x) = x$, shown here, and then work through Exercises 31–36 in order.

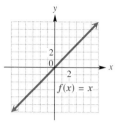

31. Based on the concepts of this section, how does the graph of $f(x) = x^2 + 6$ compare to the graph of $g(x) = x^2$ if a *vertical* shift is considered?

32. Graph the linear function $f(x) = x + 6$.

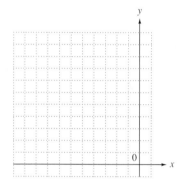

33. Based on the concepts of Chapter 3, how does the graph of $f(x) = x + 6$ compare to the graph of $g(x) = x$ if a vertical shift is considered? (*Hint:* Look at the y-intercept.)

34. Based on the concepts of this section, how does the graph of $f(x) = (x - 6)^2$ compare to the graph of $g(x) = x^2$ if a *horizontal* shift is considered?

35. Graph the linear function $f(x) = x - 6$.

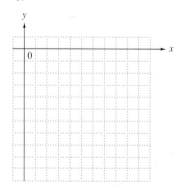

36. Based on the concepts of Chapter 3, how does the graph of $f(x) = x - 6$ compare to the graph of $g(x) = x$ if a horizontal shift is considered? (*Hint:* Look at the x-intercept.)

9.2 More about Quadratic Functions; Horizontal Parabolas

OBJECTIVES

1. Find the vertex of a vertical parabola.
2. Graph a quadratic function.
3. Use the discriminant to find the number of x-intercepts of a vertical parabola.
4. Use quadratic functions to solve problems involving maximum or minimum value.
5. Graph horizontal parabolas.

FOR EXTRA HELP

Tutorial   Tape 13   SSM, Sec. 9.2

OBJECTIVE ▶ When the equation of a parabola is given in the form $f(x) = ax^2 + bx + c$, we need to locate the vertex in order to sketch an accurate graph. This can be done in two ways. The first is by completing the square, as shown in Examples 1 and 2. The second is by using a formula that may be derived by completing the square.

E X A M P L E   1   Completing the Square to Find the Vertex

Find the vertex of the graph of $f(x) = x^2 - 4x + 5$.

To find the vertex, we need to express $x^2 - 4x + 5$ in the form $(x - h)^2 + k$. This is done by completing the square on $x^2 - 4x$, as in Section 8.1. The process is a little different here because we want to keep $f(x)$ alone on one side of the equation. Instead of adding the appropriate number to both sides, *add and subtract* it on the right. This is equivalent to adding 0.

$$f(x) = x^2 - 4x + 5$$
$$= (x^2 - 4x \quad) + 5$$
Half of -4 is -2; $(-2)^2 = 4$.
$$= (x^2 - 4x + 4 - 4) + 5 \qquad \text{Add and subtract 4.}$$
$$= (x^2 - 4x + 4) - 4 + 5 \qquad \text{Bring } -4 \text{ outside the parentheses.}$$
$$f(x) = (x - 2)^2 + 1 \qquad \text{Factor; combine terms.}$$

As we saw in the previous section, the vertex of this parabola is $(2, 1)$.

WORK PROBLEM 1 AT THE SIDE. ▶▶

1. Find the vertex of each parabola by completing the square.

   (a) $f(x) = x^2 - 6x + 7$

   (b) $f(x) = x^2 + 4x - 9$

E X A M P L E   2   Completing the Square to Find the Vertex When $a \neq 1$

Find the vertex of the graph of $f(x) = -3x^2 + 6x - 1$.

We must complete the square on $-3x^2 + 6x$. Because the x^2 term has a coefficient other than 1, factor that coefficient out of the first two terms, and then proceed as in Example 1.

$$f(x) = -3x^2 + 6x - 1$$
$$= -3(x^2 - 2x) - 1$$
Half of -2 is -1, and $(-1)^2 = 1$.
$$= -3(x^2 - 2x + 1 - 1) - 1 \qquad \text{Add and subtract 1.}$$
$$= -3(x^2 - 2x + 1) + (-3)(-1) - 1 \qquad \text{Distributive property}$$
$$= -3(x^2 - 2x + 1) + 3 - 1$$
$$f(x) = -3(x - 1)^2 + 2 \qquad \text{Factor; combine terms.}$$

The vertex is at $(1, 2)$.

WORK PROBLEM 2 AT THE SIDE. ▶▶

2. Find the vertex of each parabola.

   (a) $f(x) = 2x^2 - 4x + 1$

   (b) $f(x) = -\dfrac{1}{2}x^2 + 2x - 3$

ANSWERS

1. **(a)** $(3, -2)$ **(b)** $(-2, -13)$
2. **(a)** $(1, -1)$ **(b)** $(2, -1)$

589

3. Use the formula to find the vertex of the graph of each quadratic function.

(a) $f(x) = -2x^2 + 3x - 1$

(b) $f(x) = 4x^2 - x + 5$

A formula for the vertex of the graph of the quadratic function defined by $f(x) = ax^2 + bx + c$ can be found by completing the square for the standard form of the equation.

$$f(x) = ax^2 + bx + c \quad (a \neq 0)$$
Standard form

$$f(x) = a\left(x^2 + \frac{b}{a}x\right) + c$$
Factor a from the first two terms.

$$f(x) = a\left(x^2 + \frac{b}{a}x + \frac{b^2}{4a^2} - \frac{b^2}{4a^2}\right) + c$$
Add and subtract $\frac{b^2}{4a^2}$ within the parentheses.

$$f(x) = a\left(x^2 + \frac{b}{a}x + \frac{b^2}{4a^2}\right) - \frac{b^2}{4a} + c$$
Distributive property

$$f(x) = a\left(x + \frac{b}{2a}\right)^2 + \frac{4ac - b^2}{4a}$$
Factor and combine terms.

$$f(x) = a\left[x - \left(\frac{-b}{2a}\right)\right]^2 + \underbrace{\frac{4ac - b^2}{4a}}_{k}$$

(with h under the $\frac{-b}{2a}$ term)

The final equation shows that the vertex (h, k) can be expressed in terms of a, b, and c. However, it is not necessary to memorize k, since $k = f(h)$.

Vertex Formula

The graph of the quadratic function defined by $f(x) = ax^2 + bx + c$ has its vertex at

$$\left(\frac{-b}{2a}, f\left(\frac{-b}{2a}\right)\right),$$

and the axis of the parabola is the line $x = \frac{-b}{2a}$.

E X A M P L E   3   Using the Formula to Find the Vertex

Use the vertex formula to find the vertex of the graph of

$$f(x) = x^2 - x - 6.$$

For this function, $a = 1$, $b = -1$, and $c = -6$. The x-coordinate of the vertex of the parabola is given by

$$\frac{-b}{2a} = \frac{-(-1)}{2(1)} = \frac{1}{2}.$$

The y-coordinate is $f\left(\frac{-b}{2a}\right) = f\left(\frac{1}{2}\right)$.

$$f\left(\frac{1}{2}\right) = \left(\frac{1}{2}\right)^2 - \frac{1}{2} - 6 = \frac{1}{4} - \frac{1}{2} - 6 = -\frac{25}{4}$$

Finally, the vertex is $\left(\frac{1}{2}, -\frac{25}{4}\right)$.

◀◀ WORK PROBLEM 3 AT THE SIDE.

OBJECTIVE 2 Parabolas were graphed in the previous section. A more general approach involving finding intercepts and the vertex is given here.

ANSWERS

3. (a) $\left(\frac{3}{4}, \frac{1}{8}\right)$ (b) $\left(\frac{1}{8}, \frac{79}{16}\right)$

Graphing a Quadratic Function f

Step 1 Find the *y*-intercept by evaluating $f(0)$.

Step 2 Find any *x*-intercepts by solving $f(x) = 0$.

Step 3 Find the vertex either by using the formula or by completing the square.

Step 4 Find and plot additional points as needed, using the symmetry about the axis.

Step 5 Verify that the graph opens upward (if $a > 0$) or opens downward (if $a < 0$).

The domain of a quadratic function is $(-\infty, \infty)$, unless otherwise specified. The range is determined by the *y*-value of the vertex, and whether the parabola opens upward or downward.

E X A M P L E 4   Using the Steps for Graphing a Quadratic Function

Graph the quadratic function with $f(x) = x^2 - x - 6$. Give the domain and the range.

Begin by finding the *y*-intercept.

$$f(x) = x^2 - x - 6$$
$$f(0) = 0^2 - 0 - 6 \qquad \text{Find } f(0).$$
$$f(0) = -6$$

The *y*-intercept is $(0, -6)$. Now find any *x*-intercepts.

$$f(x) = x^2 - x - 6$$
$$0 = x^2 - x - 6 \qquad \text{Let } f(x) = 0.$$
$$0 = (x - 3)(x + 2) \qquad \text{Factor.}$$
$$x - 3 = 0 \quad \text{or} \quad x + 2 = 0 \qquad \begin{array}{l}\text{Set each factor}\\\text{equal to 0 and}\\\text{solve.}\end{array}$$
$$x = 3 \quad \text{or} \qquad x = -2$$

The *x*-intercepts are $(3, 0)$ and $(-2, 0)$. The vertex, found in Example 3, is $\left(\frac{1}{2}, -\frac{25}{4}\right)$. Plot the points found so far, and plot any additional points as needed. The symmetry of the graph is helpful here. The graph is shown in Figure 9. The domain is $(-\infty, \infty)$ and the range is $\left[-\frac{25}{4}, \infty\right)$.

x	y
-2	0
-1	-4
0	-6
$\frac{1}{2}$	$-\frac{25}{4}$
2	-4
3	0

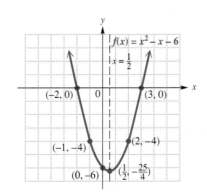

FIGURE 9

4. Graph the quadratic function

$$f(x) = x^2 - 6x + 5.$$

Give the domain and the range.

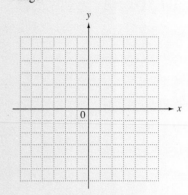

4.

domain: $(-\infty, \infty)$;
range: $[-4, \infty)$

WORK PROBLEM 4 AT THE SIDE. ▶▶

5. Use the discriminant to determine the number of x-intercepts for each graph.

(a) $f(x) = 4x^2 - 20x + 25$

(b) $f(x) = 2x^2 + 3x + 5$

(c) $f(x) = -3x^2 - x + 2$

OBJECTIVE ▶ **3** The graph of a quadratic function may have two x-intercepts, one x-intercept, or no x-intercepts, as shown in Figure 10.

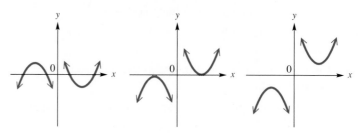

FIGURE 10

Recall from Section 8.2 that the value of $b^2 - 4ac$ is called the *discriminant* of the quadratic equation $ax^2 + bx + c = 0$. It can be used to determine the number of real solutions of a quadratic equation. In a similar way, the discriminant of a quadratic *function* can be used to determine the number of x-intercepts of its graph. If the discriminant is positive, the parabola will have two x-intercepts. If the discriminant is 0, there will be only one x-intercept, and it will be the vertex of the parabola. If the discriminant is negative, the graph will have no x-intercepts.

E X A M P L E   5   Using the Discriminant to Determine the Number of *x*-Intercepts

Determine the number of x-intercepts of the graph of each quadratic function. Use the discriminant.

(a) $f(x) = 2x^2 + 3x - 5$

The discriminant is $b^2 - 4ac$. Here $a = 2$, $b = 3$, and $c = -5$, so

$$b^2 - 4ac = 9 - 4(2)(-5) = 49.$$

Since the discriminant is positive, the parabola has two x-intercepts.

(b) $f(x) = -3x^2 - 1$

In this equation, $a = -3$, $b = 0$, and $c = -1$. The discriminant is

$$b^2 - 4ac = 0 - 4(-3)(-1) = -12.$$

The discriminant is negative, so the graph has no x-intercepts.

(c) $f(x) = 9x^2 + 6x + 1$

Here, $a = 9$, $b = 6$, and $c = 1$. The discriminant is

$$b^2 - 4ac = 36 - 4(9)(1) = 0.$$

The parabola has only one x-intercept (its vertex) because the value of the discriminant is 0.

◀◀ WORK PROBLEM 5 AT THE SIDE.

OBJECTIVE ▶ **4** As we have seen, the vertex of a parabola is either the highest or the lowest point on the parabola. The y-value of the vertex gives the maximum or minimum value of y, while the x-value tells where that maximum or minimum occurs. In many practical problems we want to know the largest or smallest value of some quantity. When that quantity can be expressed as a quadratic function with $y = ax^2 + bx + c$, as in the next example, the vertex can be used to find the desired value.

ANSWERS

5. (a) discriminant is 0; one x-intercept
   (b) discriminant is -31;
       no x-intercepts
   (c) discriminant is 25;
       two x-intercepts

EXAMPLE 6   **Finding the Maximum Area of a Rectangular Region**

A farmer has 120 feet of fencing. He wants to put a fence around a rectangular field next to a river. Find the maximum area he can enclose.

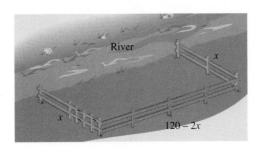

FIGURE 11

Figure 11 shows the field. Let x represent the width of the field. Then, since there are 120 feet of fencing,

$x + x + \text{length} = 120$ Sum of the sides is 120 feet.

$2x + \text{length} = 120$ Combine terms.

$\text{length} = 120 - 2x.$ Subtract $2x$.

The area is given by the product of the length and width, or

$$A = x(120 - 2x) = 120x - 2x^2.$$

To make the area (and thus $120x - 2x^2$) as large as possible, first find the vertex of the parabola $A = 120x - 2x^2$. Writing the equation in standard form as $A = -2x^2 + 120x$ shows that $a = -2$, $b = 120$, and $c = 0$, so

$$h = -\frac{b}{2a} = -\frac{120}{2(-2)} = -\frac{120}{-4} = 30$$

$$k = f(30) = -2(30)^2 + 120(30) = -2(900) + 3600 = 1800.$$

The graph is a parabola that opens downward, and its vertex is $(30, 1800)$. The vertex of the graph shows that the maximum area will be 1800 square feet. This area will occur if x, the width of the field, is 30 feet.

WORK PROBLEM 6 AT THE SIDE. ▶▶

OBJECTIVE 5 ▶ If x and y are exchanged in the equation $y = ax^2 + bx + c$, the equation becomes $x = ay^2 + by + c$. Because of the interchange of the roles of x and y, these parabolas are horizontal (with horizontal lines as axes), compared with the vertical ones graphed previously.

EXAMPLE 7   **Graphing a Horizontal Parabola**

Graph $x = (y - 2)^2 - 3$.

This graph has its vertex at $(-3, 2)$, because the roles of x and y are reversed. It opens to the right, the positive x-direction, and has the same shape as $y = x^2$. Plotting a few additional points gives the graph shown in Figure 12.

CONTINUED ON NEXT PAGE

6. Solve Example 6 if the farmer has only 100 feet of fencing.

ANSWERS

6. The field should be 25 feet by 50 feet with a maximum area of 1250 square feet.

7. Graph $x = (y + 1)^2 - 4$.

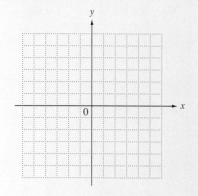

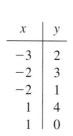

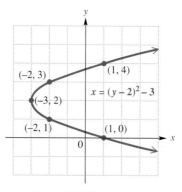

FIGURE 12

Horizontal Parabola

The graph of $x = a(y - k)^2 + h$ is a horizontal parabola with vertex at (h, k) and the horizontal line $y = k$ as axis. The graph opens to the right if a is positive and to the left if a is negative.

◀◀ WORK PROBLEM 7 AT THE SIDE.

When a quadratic equation is given in the form $x = ay^2 + by + c$, completing the square on y will allow us to find the vertex.

8. Find the vertex of each parabola. Tell whether the graph opens to the right or to the left. Give the domain and the range.

(a) $x = 2y^2 - 6y + 5$

E X A M P L E   8   Completing the Square to Graph a Horizontal Parabola

Graph $x = -2y^2 + 4y - 3$. Give the domain and the range of the relation.

$$x = -2y^2 + 4y - 3$$
$$= -2(y^2 - 2y) - 3 \qquad \text{Factor out } -2.$$
$$= -2(y^2 - 2y + 1 - 1) - 3 \qquad \text{Add } 0 \ (1 - 1 = 0).$$
$$= -2(y^2 - 2y + 1) + 2 - 3 \qquad \text{Distributive property}$$
$$x = -2(y - 1)^2 - 1 \qquad \text{Factor.}$$

Because of the negative coefficient (-2), the graph opens to the left (the negative x direction) and is narrower than $y = x^2$. As shown in Figure 13, the vertex is $(-1, 1)$. The domain is $(-\infty, -1]$ and the range is $(-\infty, \infty)$.

(b) $x = -y^2 + 2y + 5$

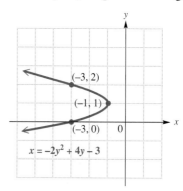

FIGURE 13

7.

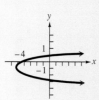

8. (a) $\left(\dfrac{1}{2}, \dfrac{3}{2}\right)$; right; domain: $\left[\dfrac{1}{2}, \infty\right)$; range: $(-\infty, \infty)$ **(b)** $(6, 1)$; left; domain: $(-\infty, 6]$; range: $(-\infty, \infty)$

◀◀ WORK PROBLEM 8 AT THE SIDE.

Caution
Only quadratic equations that are solved for y define functions. The graphs of the equations in Examples 7 and 8 are not graphs of functions. They do not satisfy the conditions of the vertical line test.

9.2 Exercises

1. How can you determine just by looking at the equation of a parabola whether it has a vertical or a horizontal axis?

2. Why can't the graph of a quadratic function be a horizontal parabola?

3. How can you determine the number of x-intercepts of the graph of a quadratic function without graphing the function?

4. If the vertex of the graph of a quadratic function is $(1, -3)$, and the graph opens downward, how many x-intercepts does the graph have?

Find the vertex of each parabola. For each equation, decide whether the graph opens upward, downward, to the left, or to the right; and whether it is wider, narrower, or the same shape as the graph of $y = x^2$. If it is a vertical parabola, use the discriminant to determine the number of x-intercepts. See Examples 1–3, 5, 7, and 8.

5. $y = 2x^2 + 4x + 5$

6. $y = 3x^2 - 6x + 4$

7. $y = -x^2 + 5x + 3$

8. $x = -y^2 + 7y - 2$

9. $x = \dfrac{1}{3}y^2 + 6y + 24$

10. $x = .5y^2 + 10y - 5$

Graph each parabola using the techniques described in this section. In Exercises 11–14 give the domain and range. See Examples 4, 7, and 8.

11. $f(x) = x^2 + 4x + 3$

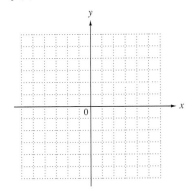

12. $f(x) = x^2 + 2x - 2$

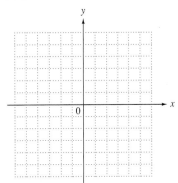

13. $f(x) = -2x^2 + 4x - 5$

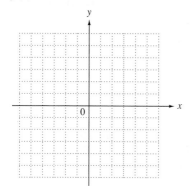

14. $f(x) = -3x^2 + 12x - 8$

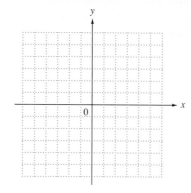

15. $x = -\dfrac{1}{5}y^2 + 2y - 4$

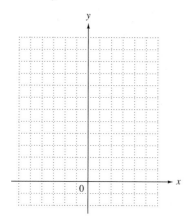

16. $x = -.5y^2 - 4y - 6$

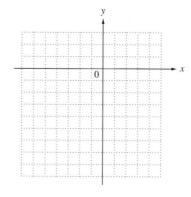

17. $x = 3y^2 + 12y + 5$

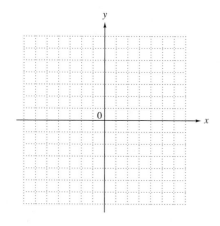

18. $x = 4y^2 + 16y + 11$

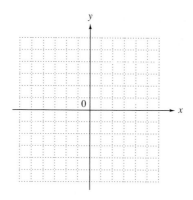

INTERPRETING TECHNOLOGY (EXERCISES 19–22)

Graphing calculators are capable of determining the coordinates of "peaks" and "valleys" of graphs. In the case of quadratic functions, these peaks and valleys are the vertices, and are called maximum and minimum points. For example, the vertex of the graph of
$f(x) = -x^2 - 6x - 13$ *is* $(-3, -4)$, *as indicated in the display at the bottom of the screen. In this case, the vertex is a maximum point.*

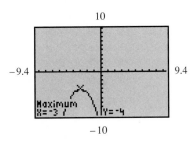

In Exercises 19–22, match the function with its calculator-generated graph by determining the vertex and using the display at the bottom of the screen.

19. $f(x) = x^2 - 8x + 18$

A.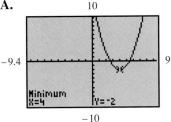

B.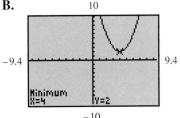

20. $f(x) = x^2 + 8x + 18$

C.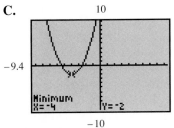

D.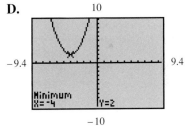

21. $f(x) = x^2 - 8x + 14$

22. $f(x) = x^2 + 8x + 14$

Solve each of the following applied problems. See Example 6.

23. Palo Alto College is planning to construct a rectangular parking lot on land bordered on one side by a highway. The plan is to use 640 feet of fencing to fence off the other three sides. What should the dimensions of the lot be if the enclosed area is to be a maximum?

24. Keisha Hughes has 100 meters of fencing material to enclose a rectangular exercise run for her dog. What width will give the enclosure the maximum area?

25. If an object is thrown upward with an initial velocity of 32 feet per second, then its height (in feet) after t seconds is given by

$$h(t) = 32t - 16t^2.$$

Find the maximum height attained by the object and the number of seconds it takes to hit the ground.

26. A projectile is fired straight upward so that its distance (in feet) above the ground t seconds after firing is

$$s(t) = -16t^2 + 400t.$$

Find the maximum height it reaches and the number of seconds it takes to reach that height.

27. Find the pair of numbers with a sum of 60 whose product is the maximum. (*Hint:* Let x and $60 - x$ represent the two numbers.)

28. For a trip to a resort, a charter bus company charges a fare of $48 per person, plus $2 per person for each unsold seat on the bus. If the bus has 42 seats and x represents the number of unsold seats, find the following:
   (a) a function defined by $R(x)$ that describes the total revenue from the trip (*Hint:* Multiply the total number riding, $42 - x$, by the price per ticket, $48 + 2x$);
   (b) the number of unsold seats that produces the maximum revenue;
   (c) the maximum revenue.

The accompanying bar graph shows the annual average number of nonfarm payroll jobs in California for the years 1988 through 1992. If the tops of the bars were joined by a smooth curve, the curve would resemble the graph of a quadratic function (that is, a parabola). Using a technique from statistics it can be determined that this function can be described approximately as

$$f(x) = -.10x^2 + .42x + 11.90,$$

where x = 0 corresponds to 1988, x = 1 corresponds to 1989, and so on, and f(x) represents the number of payroll jobs in millions. (Source: California Employment Development Department)

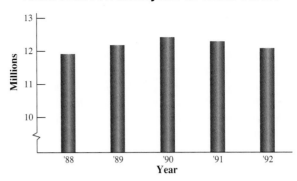

NONFARM PAYROLL JOBS IN CALIFORNIA

29. Explain why the coefficient of x^2 in the function is negative, based on the graph formed by joining the tops of the bars.

30. Determine the coordinates of the vertex of the graph using algebraic methods.

31. How does the x-coordinate of the vertex of the parabola indicate that during the time period under consideration, the maximum number of payroll jobs was in 1990?

32. What does the y-coordinate of the vertex of the parabola indicate?

In Example 1 of Section 8.5, we determined the solution set of the quadratic inequality $x^2 - x - 12 > 0$ by using regions on a number line and testing values in the inequality. If we graph $f(x) = x^2 - x - 12$, the x-intercepts will determine the solutions of the quadratic equation $x^2 - x - 12 = 0$. The solution set is $\{-3, 4\}$. The x-values of the points on the graph that are above the x-axis form the solution set of $x^2 - x - 12 > 0$. As seen in the figure, this solution set is $(-\infty, -3) \cup (4, \infty)$, which supports the result found in Section 8.5. Similarly, the solution set of the quadratic inequality $x^2 - x - 12 < 0$ is found by locating the points on the graph that lie below the x-axis. Those x-values belong to the open interval $(-3, 4)$.

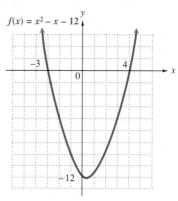

$f(x) = x^2 - x - 12$

The graph is above the x-axis for $(-\infty, -3) \cup (4, \infty)$.

In Exercises 33–36, the graph of a quadratic function f is given. Use only the graph to find the solution set of the equation or inequality. Work through parts (a)–(c) in order each time.

33. $f(x) = x^2 - 4x + 3$

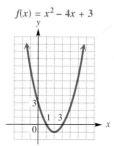

(a) $x^2 - 4x + 3 = 0$
(b) $x^2 - 4x + 3 > 0$
(c) $x^2 - 4x + 3 < 0$

34. $f(x) = 3x^2 + 10x - 8$

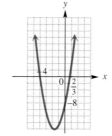

(a) $3x^2 + 10x - 8 = 0$
(b) $3x^2 + 10x - 8 \geq 0$
(c) $3x^2 + 10x - 8 < 0$

35. $f(x) = -x^2 + 3x + 10$

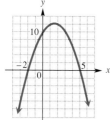

(a) $-x^2 + 3x + 10 = 0$
(b) $-x^2 + 3x + 10 \geq 0$
(c) $-x^2 + 3x + 10 \leq 0$

36. $f(x) = -2x^2 - x + 15$

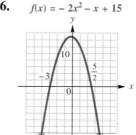

(a) $-2x^2 - x + 15 = 0$
(b) $-2x^2 - x + 15 \geq 0$
(c) $-2x^2 - x + 15 \leq 0$

9.3 Graphs of Elementary Functions and Circles

In the first two sections of this chapter, we introduced an important function, $f(x) = x^2$. This quadratic function is sometimes called the **squaring function,** and it is one of the most important elementary functions in algebra.

OBJECTIVE 1 Three other elementary functions are those defined by $|x|$, $\frac{1}{x}$, and $\sqrt{x}$. The first of these, with $f(x) = |x|$, is called the **absolute value function.** Its graph, along with a table of selected ordered pairs, is shown in Figure 14. Its domain is $(-\infty, \infty)$ and its range is $[0, \infty)$.

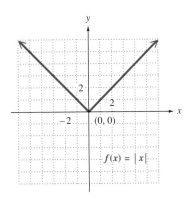

x	y
0	0
± 1	1
± 2	2
± 3	3

FIGURE 14

Another important elementary function is the **reciprocal function,** defined by $f(x) = \frac{1}{x}$. Its graph is shown in Figure 15, along with a table of selected ordered pairs. Notice that x can never equal zero for this function, and as a result, as x gets closer and closer to 0, the graph either approaches ∞ or $-\infty$. Also, $\frac{1}{x}$ can never equal 0, and as x approaches ∞ or $-\infty$, $\frac{1}{x}$ approaches 0. The axes are called **asymptotes** (ASS-im-tohts) for the function. (Asymptotes are studied in more detail in college algebra courses.) For the reciprocal function, the domain and the range are both $(-\infty, 0) \cup (0, \infty)$.

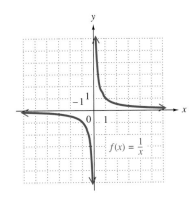

x	y	x	y
$\frac{1}{3}$	3	$-\frac{1}{3}$	-3
$\frac{1}{2}$	2	$-\frac{1}{2}$	-2
1	1	-1	-1
2	$\frac{1}{2}$	-2	$-\frac{1}{2}$
3	$\frac{1}{3}$	-3	$-\frac{1}{3}$

FIGURE 15

OBJECTIVES

1. ▶ Recognize the graphs of functions defined by $|x|$, $\frac{1}{x}$, and $\sqrt{x}$, and graph their translations.

2. ▶ Find the equation of a circle given the center and radius.

3. ▶ Find the center and radius of a circle given its equation.

4. ▶ Graph semicircles defined by functions involving a square root.

FOR EXTRA HELP

Tutorial   Tape 14   SSM, Sec. 9.3

The function with $f(x) = \sqrt{x}$ is called the **square root function.** Its graph is shown in Figure 16. Notice that since we restrict function values to be real numbers, x cannot take on negative values. Thus, the domain of the square root function is $[0, \infty)$. Because the principal square root is always nonnegative, the range is also $[0, \infty)$. A partial table of values is shown along with the graph.

x	y
0	0
1	1
4	2

FIGURE 16

Just as the graph of $f(x) = x^2$ can be translated, as seen in Section 9.1, so can the graphs of these other elementary functions. Example 1 shows how this is done.

E X A M P L E   1   Graphing Translations of Other Elementary Functions

Sketch the graph of each function, using translations as discussed in Section 9.1.

(a) $f(x) = |x - 2|$

The graph of $y = (x - 2)^2$ is obtained by shifting the graph of $y = x^2$ two units to the right. In a similar manner, the graph of $f(x) = |x - 2|$ is found by shifting the graph of $y = |x|$ two units to the right, as shown in Figure 17. The table of ordered pairs accompanying the graph supports this.

x	y
0	2
1	1
2	0
3	1
4	2

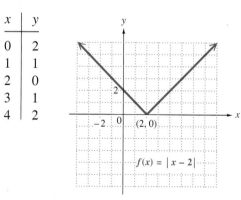

FIGURE 17

CONTINUED ON NEXT PAGE

(b) $f(x) = \dfrac{1}{x} + 3$

The graph of this function is found by translating the graph of $y = \dfrac{1}{x}$ three units upward. See Figure 18.

x	y
$\frac{1}{3}$	6
$\frac{1}{2}$	5
1	4
2	3.5

x	y
$-\frac{1}{3}$	0
$-\frac{1}{2}$	1
-1	2
-2	2.5

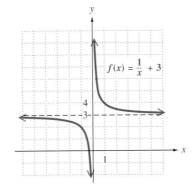

FIGURE 18

(c) $f(x) = \sqrt{x + 1} - 4$

The graph of $y = (x + 1)^2 - 4$ is obtained by shifting the graph of $y = x^2$ one unit to the left and four units downward. Following this pattern, we shift the graph of $y = \sqrt{x}$ one unit to the left and four units downward to get the graph of $f(x) = \sqrt{x + 1} - 4$. See Figure 19.

x	y
-1	-4
0	-3
3	-2

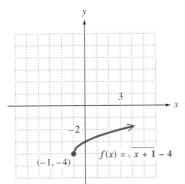

FIGURE 19

WORK PROBLEM 1 AT THE SIDE. ▶▶

OBJECTIVE 2 The distance formula introduced in Chapter 7 is essential in developing the equation for the graph of a circle. Notice that a circle is not the graph of a function; it does not pass the vertical line test.

1. Graph each of the following functions.

(a) $f(x) = \sqrt{x + 4}$

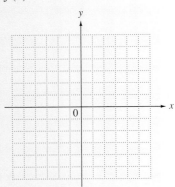

(b) $f(x) = \dfrac{1}{x} - 2$

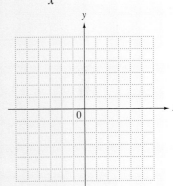

(c) $f(x) = |x + 2| + 1$

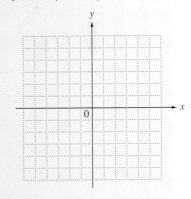

ANSWERS

1. (a) **(b)**

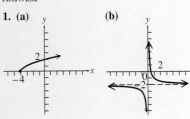

(c)

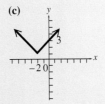

2. Find the equation of the circle with radius 4 and center $(0, 0)$. Sketch its graph.

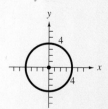

A **circle** is the set of all points in a plane that lie a fixed distance from a fixed point. The fixed point is called the **center** and the fixed distance is called the **radius.** The distance formula can be used to find the equation of a circle.

EXAMPLE 2 Finding the Equation of a Circle and Graphing It

Find the equation of the circle with radius 3 and center at $(0, 0)$. Draw the graph.

If the point (x, y) is on the circle, the distance from (x, y) to the center $(0, 0)$ is 3. By the distance formula,

$$\sqrt{(x_2 - x_1)^2 + (y_2 - y_1)^2} = d$$
$$\sqrt{(x - 0)^2 + (y - 0)^2} = 3$$
$$x^2 + y^2 = 9. \qquad \text{Square both sides.}$$

The equation of this circle is $x^2 + y^2 = 9$. Its graph is given in Figure 20.

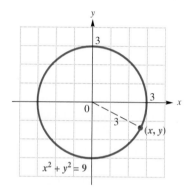

FIGURE 20

◀◀ WORK PROBLEM 2 AT THE SIDE.

EXAMPLE 3 Graphing a Circle Given Its Center and Radius

Find an equation for the circle that has its center at $(4, -3)$ and radius 5, and draw the graph.

Again, use the distance formula with the points (x, y) and $(4, -3)$. The graph of this circle is shown in Figure 21. The equation is

$$\sqrt{(x - 4)^2 + (y + 3)^2} = 5$$
$$(x - 4)^2 + (y + 3)^2 = 25.$$

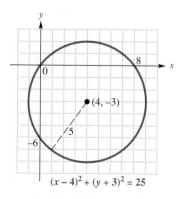

FIGURE 21

ANSWERS

2. $x^2 + y^2 = 16$

The results of Examples 2 and 3 can be generalized to derive an equation of a circle with radius r and center at (h, k). If (x, y) is a point on the circle, the distance from the center (h, k) to the point (x, y) is r. Then by the distance formula with $d = r$, $x_1 = x$, $x_2 = h$, $y_1 = y$, and $y_2 = k$,

$$\sqrt{(x - h)^2 + (y - k)^2} = r.$$

Squaring both sides of the equation gives the following result.

Equation of a Circle

The circle with radius r and center at (h, k) has an equation of the form

$$(x - h)^2 + (y - k)^2 = r^2.$$

WORK PROBLEMS 3 AND 4 AT THE SIDE. ▶▶

OBJECTIVE 3▶ In the equation found in Example 3, multiplying out $(x - 4)^2$ and $(y + 3)^2$ and then combining like terms gives

$$(x - 4)^2 + (y + 3)^2 = 25$$
$$x^2 - 8x + 16 + y^2 + 6y + 9 = 25$$
$$x^2 + y^2 - 8x + 6y = 0.$$

This result suggests that an equation that has both x^2 and y^2 terms with the same coefficient may represent a circle. In many cases it does, and the next example shows how to determine the center and radius of the circle.

E X A M P L E 4   Finding the Center and Radius of a Circle

Find the center and radius of the circle whose equation is

$$x^2 + y^2 + 2x + 4y - 4 = 0.$$

Since the equation has x^2 and y^2 terms with equal coefficients, its graph might be that of a circle. To find the center and radius, we complete the square on x and the square on y as follows. Keep only the terms with the variables on the left side and group the x terms and the y terms.

$$(x^2 + 2x \quad) + (y^2 + 4y \quad) = 4$$

Add the appropriate constants to complete both squares on the left.

$$(x^2 + 2x + 1 - 1) + (y^2 + 4y + 4 - 4) = 4 \qquad \text{Add 0 twice.}$$

$$(x^2 + 2x + 1) - 1 + (y^2 + 4y + 4) - 4 = 4 \qquad \text{Associative property}$$

$$(x^2 + 2x + 1) + (y^2 + 4y + 4) = 4 + 5 \qquad \text{Add 5 on both sides.}$$

$$(x + 1)^2 + (y + 2)^2 = 9 \qquad \text{Factor.}$$

The last equation shows that the center of the circle is $(-1, -2)$ and the radius is 3.

Caution
If the procedure of Example 4 leads to an equation of the form $(x - h)^2 + (y - k)^2 = 0$, the graph is the single point (h, k). If the constant on the right side is negative, the equation has no graph.

---

3. Find the equation of the circle with center at $(3, -2)$ and radius 4. Graph the circle.

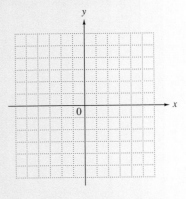

4. Determine the center and radius of $(x - 5)^2 + (y + 2)^2 = 9$ and graph the circle.

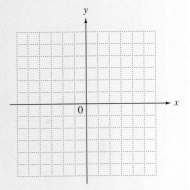

ANSWERS

3. $(x - 3)^2 + (y + 2)^2 = 16$

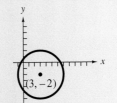

4. center at $(5, -2)$; radius 3

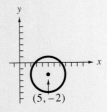

5. Find the center and radius of the circle with equation

$$x^2 + y^2 - 6x + 8y - 11 = 0.$$

◀◀ **WORK PROBLEM 5 AT THE SIDE.**

OBJECTIVE ▶ Earlier in this section we examined the graph of the square root function with $f(x) = \sqrt{x}$. By replacing x with more complicated expressions, other types of functions can be obtained. In particular, if we let the radicand be of the form $k - x^2$, where k is a positive constant, the graph is a semicircle.

E X A M P L E 5   **Graphing a Semicircle**

Graph $f(x) = \sqrt{25 - x^2}$.

Replace $f(x)$ with y and square both sides to get the equation

$$y^2 = 25 - x^2, \quad \text{or} \quad x^2 + y^2 = 25.$$

This is the graph of a circle with center at $(0, 0)$ and radius 5. Since $f(x)$, or y, represents a principal square root in the original equation, $f(x)$ must be nonnegative. This restricts the graph to the upper half of the circle, as shown in Figure 22. Use the graph and the vertical line test to verify that it is indeed a function.

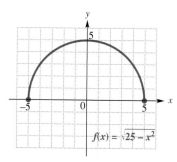

FIGURE 22

6. Graph the function with $f(x) = \sqrt{36 - x^2}$.

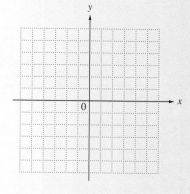

◀◀ **WORK PROBLEM 6 AT THE SIDE.**

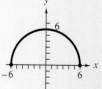

9.3 Exercises

Fill in the blank with the correct response.

1. For the reciprocal function $f(x) = \dfrac{1}{x}$, _____ is the only real number not in the domain.

2. The range of the square root function, $f(x) = \sqrt{x}$, is _____.

3. The lowest point on the graph of $f(x) = |x|$ has coordinates (_____ , _____).

4. The vertical line with equation $x =$ _____ is the axis of symmetry of the graph of $f(x) = |x|$, since the left and right halves of the graph are mirror images of each other across this line.

Without actually plotting points, match the function defined by the absolute value expression with its graph. See Example 1.

5. $f(x) = |x - 2| + 2$ **A.**

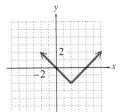

B.

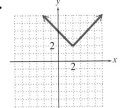

6. $f(x) = |x + 2| + 2$

7. $f(x) = |x - 2| - 2$ **C.**

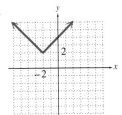

D.

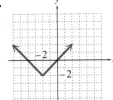

8. $f(x) = |x + 2| - 2$

Graph each of the following functions. See Example 1.

9. $f(x) = |x + 1|$

10. $f(x) = |x - 1|$

11. $f(x) = \dfrac{1}{x} + 1$

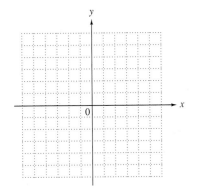

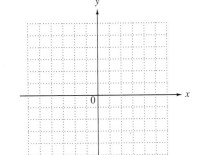

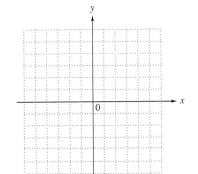

12. $f(x) = \dfrac{1}{x} - 1$

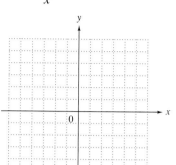

13. $f(x) = \sqrt{x - 2}$

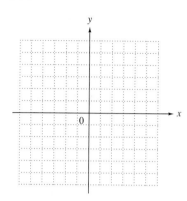

14. $f(x) = \sqrt{x + 5}$

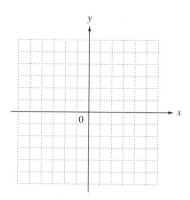

15. $f(x) = \dfrac{1}{x - 2}$

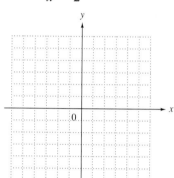

16. $f(x) = \dfrac{1}{x + 2}$

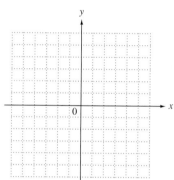

17. $f(x) = \sqrt{x + 3} - 3$

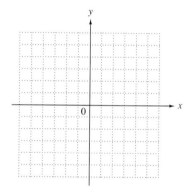

18. What is the center of a circle with equation $x^2 + y^2 = r^2$ $(r > 0)$?

Match the equation with the correct graph. See Example 3.

19. $(x - 3)^2 + (y - 2)^2 = 25$

A.

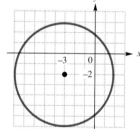

B.

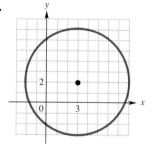

20. $(x - 3)^2 + (y + 2)^2 = 25$

21. $(x + 3)^2 + (y - 2)^2 = 25$

C.

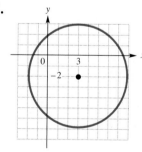

D.

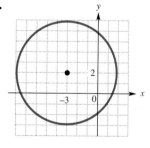

22. $(x + 3)^2 + (y + 2)^2 = 25$

23. Explain why a set of points defined by a circle does not satisfy the definition of a function.

Find the center and radius of each circle. (Hint: In Exercises 28 and 29, divide both sides by a common factor.) See Example 4.

24. $x^2 + y^2 + 4x + 6y + 9 = 0$

25. $x^2 + y^2 - 8x - 12y + 3 = 0$

26. $x^2 + y^2 + 10x - 14y - 7 = 0$

27. $x^2 + y^2 - 2x + 4y - 4 = 0$

28. $3x^2 + 3y^2 - 12x - 24y + 12 = 0$

29. $2x^2 + 2y^2 + 20x + 16y + 10 = 0$

30. A circle can be drawn on a piece of posterboard by fastening one end of a string with a thumbtack, pulling the string taut with a pencil, and tracing a curve as shown in the figure. Explain why this method works.

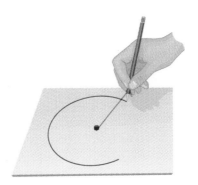

Graph the following. See Examples 2–4.

31. $x^2 + y^2 = 9$

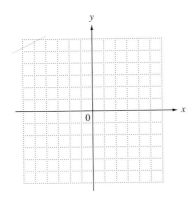

32. $x^2 + y^2 = 4$

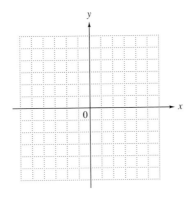

33. $2y^2 = 10 - 2x^2$

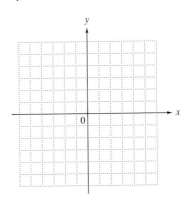

34. $3x^2 = 48 - 3y^2$

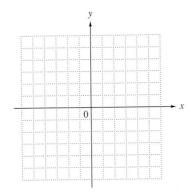

35. $(x + 3)^2 + (y - 2)^2 = 9$

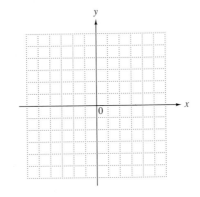

36. $(x - 1)^2 + (y + 3)^2 = 16$

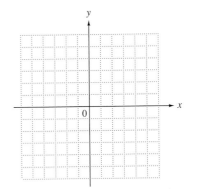

37. $x^2 + y^2 - 4x - 6y + 9 = 0$

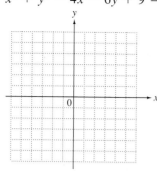

38. $x^2 + y^2 + 8x + 2y - 8 = 0$

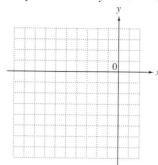

Graph each function involving a square root. See Example 5.

39. $f(x) = \sqrt{16 - x^2}$

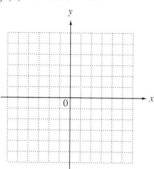

40. $f(x) = \sqrt{9 - x^2}$

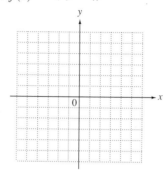

41. $f(x) = -\sqrt{36 - x^2}$

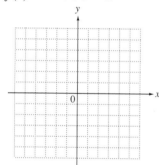

42. $f(x) = -\sqrt{25 - x^2}$

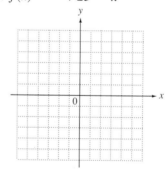

43. A parabola can be defined as the set of all points in a plane equally distant from a given point and a given line not containing the point. See the graph.

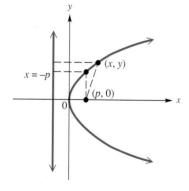

(a) Suppose (x, y) is to be on the parabola. Suppose the line mentioned in the definition is given by $x = -p$. Find the distance between (x, y) and the line. (The distance from a point to a line is the length of the perpendicular from the point to the line.)

(b) If $x = -p$ is the line mentioned in the definition, why should the given point have coordinates $(p, 0)$? (*Hint:* See the graph.)

(c) Find an expression for the distance from (x, y) to $(p, 0)$.

(d) Find an equation for the parabola in the graph. (*Hint:* Use the results of parts (a) and (c) and the fact that (x, y) is equally distant from the point and the line.)

9.4 Ellipses and Hyperbolas

OBJECTIVES

1. Recognize the equation of an ellipse.
2. Graph ellipses.
3. Recognize the equation of a hyperbola.
4. Graph hyperbolas by using the asymptotes.
5. Identify conic sections by name from their equations.
6. Graph portions of conic sections defined by functions involving square roots.

FOR EXTRA HELP

Tutorial   Tape 14   SSM, Sec. 9.4

OBJECTIVE 1 An **ellipse** (ee-LIPS) is the set of all points in a plane the sum of whose distances from two fixed points is constant. These fixed points are called **foci** (singular: *focus*). Figure 23 shows an ellipse centered at the origin, with foci at $(c, 0)$ and $(-c, 0)$, x-intercepts $(a, 0)$ and $(-a, 0)$, and y-intercepts $(0, b)$ and $(0, -b)$. From the definition above, it can be shown by the distance formula that an ellipse has the following equation.

Equation of an Ellipse

The ellipse whose x-intercepts are $(a, 0)$ and $(-a, 0)$ and whose y-intercepts are $(0, b)$ and $(0, -b)$ has an equation of the form

$$\frac{x^2}{a^2} + \frac{y^2}{b^2} = 1.$$

Note that a circle is a special case of an ellipse, where $a^2 = b^2$.

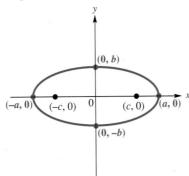

FIGURE 23

The paths of the earth and other planets around the sun are approximately ellipses; the sun is at one focus and a point in space is at the other. The orbits of communication satellites and other space vehicles are elliptical. Elliptical bicycle gears are designed to respond to the legs' natural strengths and weaknesses. At the top and bottom of the powerstroke, where the legs have the least leverage, the gear offers little resistance, but as the gear rotates, the resistance increases. This allows the legs to apply more power where it is most naturally available. See Figure 24.

FIGURE 24

OBJECTIVE 2 To graph an ellipse, plot the four intercepts $(a, 0)$, $(-a, 0)$, $(0, b)$, and $(0, -b)$, and sketch an ellipse through the intercepts.

EXAMPLE 1 Graphing an Ellipse

Graph $\dfrac{x^2}{49} + \dfrac{y^2}{36} = 1$.

The x-intercepts of this ellipse are $(7, 0)$ and $(-7, 0)$. The y-intercepts are $(0, 6)$ and $(0, -6)$. Additional points can be found by choosing a value

CONTINUED ON NEXT PAGE

for x (or y) and substituting into the equation to find the corresponding value for y (or x). For example, let's choose $y = 3$.

$$\frac{x^2}{49} + \frac{y^2}{36} = 1$$

$$\frac{x^2}{49} + \frac{9}{36} = 1 \qquad \text{If } y = 3, \text{ then } y^2 = 9.$$

$$\frac{x^2}{49} + \frac{1}{4} = 1 \qquad \text{Reduce.}$$

$$196\left(\frac{x^2}{49}\right) + 196\left(\frac{1}{4}\right) = 196 \qquad \text{Multiply by } 49 \cdot 4 = 196.$$

$$4x^2 + 49 = 196$$

$$4x^2 = 147 \qquad \text{Subtract 49.}$$

$$x^2 = 36.75 \qquad \text{Divide by 4.}$$

$$x \approx \pm 6.1$$

Thus, the points $(6.1, 3)$ and $(-6.1, 3)$ are on the graph. Choosing $y = -3$ leads to the same x-values so $(6.1, -3)$ and $(-6.1, -3)$ are also on the graph. Plotting these points and the intercepts and sketching the ellipse through them gives the graph in Figure 25.

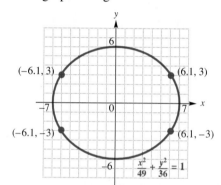

FIGURE 25

E X A M P L E 2 **Graphing an Ellipse**

Graph $\dfrac{x^2}{36} + \dfrac{y^2}{121} = 1$.

The x-intercepts for this ellipse are $(6, 0)$ and $(-6, 0)$, and the y-intercepts are $(0, 11)$ and $(0, -11)$. Additional points could be found, as shown in Example 1. The graph has been sketched in Figure 26.

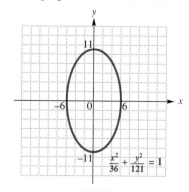

FIGURE 26

WORK PROBLEM I AT THE SIDE. ▶▶

1. Graph.

(a) $\dfrac{x^2}{4} + \dfrac{y^2}{25} = 1$

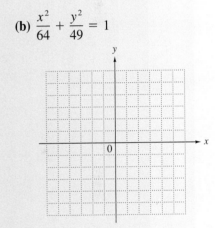

OBJECTIVE 3 A **hyperbola** (hy-PUR-buh-luh) is the set of all points in a plane the *difference* of whose distances from two foci is constant. Figure 27 shows a hyperbola; it can be shown, using this definition and the distance formula, that this hyperbola has equation

$$\frac{x^2}{16} - \frac{y^2}{12} = 1.$$

The x-intercepts are $(4, 0)$ and $(-4, 0)$. When $x = 0$ the equation becomes

$$-\frac{y^2}{12} = 1 \quad \text{or} \quad y^2 = -12.$$

This equation has no real number solutions, so there is no real number value for y corresponding to $x = 0$; that is, there are no y-intercepts.

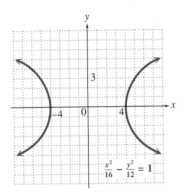

FIGURE 27

(b) $\dfrac{x^2}{64} + \dfrac{y^2}{49} = 1$

Figure 28 shows the graph of the hyperbola

$$\frac{y^2}{25} - \frac{x^2}{9} = 1.$$

Here the y-intercepts are $(0, 5)$ and $(0, -5)$, and there are no x-intercepts.

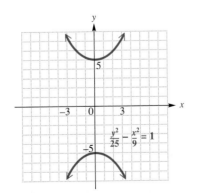

FIGURE 28

ANSWERS

1. (a)

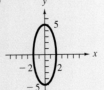

(b)

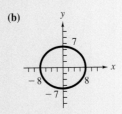

This discussion about hyperbolas is summarized as follows.

Equations of Hyperbolas

A hyperbola with x-intercepts $(a, 0)$ and $(-a, 0)$ has an equation of the form

$$\frac{x^2}{a^2} - \frac{y^2}{b^2} = 1,$$

and a hyperbola with y-intercepts $(0, b)$ and $(0, -b)$ has an equation of the form

$$\frac{y^2}{b^2} - \frac{x^2}{a^2} = 1.$$

A cross section of a large microwave antenna system consists of a parabola and a hyperbola, with the focus of the parabola coinciding with one focus of the hyperbola. See Figure 29. The incoming microwaves that are parallel to the axis of the parabola are reflected from the parabola up toward the hyperbola and back to the other focus of the hyperbola, where the cone of the antenna is located to capture the signal.

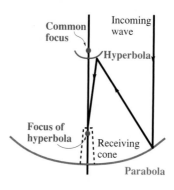

FIGURE 29

OBJECTIVE 4 The two branches of the graph of a hyperbola approach a pair of intersecting straight lines. As mentioned in the previous section when the graph of $f(x) = \dfrac{1}{x}$ was studied, these lines are called *asymptotes*. (See Figure 30.) These lines are useful for sketching the graph of the hyperbola. We find the asymptotes as follows.

Asymptotes of Hyperbolas

The extended diagonals of the rectangle with corners at the points (a, b), $(-a, b)$, $(-a, -b)$, and $(a, -b)$ are the asymptotes of either of the hyperbolas

$$\frac{x^2}{a^2} - \frac{y^2}{b^2} = 1 \quad \text{or} \quad \frac{y^2}{b^2} - \frac{x^2}{a^2} = 1.$$

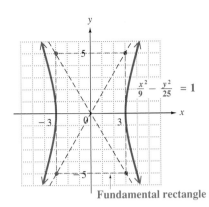

FIGURE 30

This rectangle is called the **fundamental rectangle.**

E X A M P L E 3   Graphing a Hyperbola

Graph $\dfrac{x^2}{9} - \dfrac{y^2}{25} = 1$.

Here $a = 3$ and $b = 5$. The x-intercepts are $(3, 0)$ and $(-3, 0)$. The four points $(3, 5)$, $(3, -5)$, $(-3, 5)$, and $(-3, -5)$ are the corners of the fundamental rectangle that determines the asymptotes shown in Figure 30. The hyperbola approaches these lines as x and y get larger and larger in absolute value.

In summary, to graph either of the two forms of hyperbolas,

$$\frac{x^2}{a^2} - \frac{y^2}{b^2} = 1 \quad \text{or} \quad \frac{y^2}{b^2} - \frac{x^2}{a^2} = 1,$$

follow these steps.

Graphing a Hyperbola

Step 1 Locate the intercepts. They are at $(a, 0)$ and $(-a, 0)$ if the x^2 term has a positive coefficient, or at $(0, b)$ and $(0, -b)$ if the y^2 term has a positive coefficient.

Step 2 Locate the corners of a rectangle at (a, b), $(a, -b)$, $(-a, -b)$, and $(-a, b)$.

Step 3 Sketch the asymptotes (the extended diagonals of the rectangle).

Step 4 Sketch each branch of the hyperbola through an intercept and approaching the asymptotes.

WORK PROBLEM 2 AT THE SIDE. ▶▶

OBJECTIVE 5 By rewriting a second-degree equation in one of the forms given for ellipses, hyperbolas, circles, or parabolas, we can determine when the graph is one of these figures. A summary of the equations and graphs of the conic sections is given here.

2. Use asymptotes to graph the following.

(a) $\dfrac{x^2}{4} - \dfrac{y^2}{25} = 1$

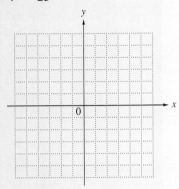

(b) $\dfrac{y^2}{81} - \dfrac{x^2}{64} = 1$

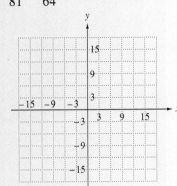

ANSWERS

2. (a)

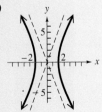

(b)

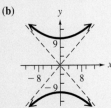

Equation	*Graph*	*Description*
$y = a(x - h)^2 + k$	 Parabola	Opens upward if $a > 0$ or downward if $a < 0$; vertex is at (h, k).
$x = a(y - k)^2 + h$	 Parabola	Opens to right if $a > 0$ or to left if $a < 0$; vertex is at (h, k).
$(x - h)^2 + (y - k)^2 = r^2$	 Circle	Center is at (h, k), radius is r.
$\dfrac{x^2}{a^2} + \dfrac{y^2}{b^2} = 1$	 Ellipse	x-intercepts are $(a, 0)$ and $(-a, 0)$; y-intercepts are $(0, b)$ and $(0, -b)$.
$\dfrac{x^2}{a^2} - \dfrac{y^2}{b^2} = 1$	 Hyperbola	x-intercepts are $(a, 0)$ and $(-a, 0)$.
$\dfrac{y^2}{b^2} - \dfrac{x^2}{a^2} = 1$	 Hyperbola	y-intercepts are $(0, b)$ and $(0, -b)$.

EXAMPLE 4   Identifying the Graph of a
                     Given Equation

Identify the graph of each equation.

(a) $9x^2 = 108 + 12y^2$

Both variables are squared, so the graph is either an ellipse or a hyperbola. (This situation also occurs for a circle, which may be considered a special case of the ellipse.) To see which one it is, rewrite the equation so that the x and y terms are on one side of the equation and 1 is on the other.

$$9x^2 = 108 + 12y^2$$
$$9x^2 - 12y^2 = 108 \qquad \text{Subtract } 12y^2.$$
$$\frac{x^2}{12} - \frac{y^2}{9} = 1 \qquad \text{Divide by 108.}$$

Because of the minus sign, the graph of this equation is a hyperbola.

(b) $x^2 = y - 3$

Only one of the two variables is squared, x, so this is the vertical parabola $y = x^2 + 3$.

(c) $x^2 = 9 - y^2$

Get the variable terms on the same side of the equation.

$$x^2 = 9 - y^2$$
$$x^2 + y^2 = 9 \qquad \text{Add } y^2.$$

This equation represents a circle with center at the origin and radius 3.

WORK PROBLEM 3 AT THE SIDE. ▶▶

OBJECTIVE ▶6▶ In the previous section we saw that semicircles can be defined by functions involving square root radicals. Similarly, portions of parabolas, ellipses, and hyperbolas can be defined by functions involving square root radicals. The final example illustrates such a function.

EXAMPLE 5   Graphing a Portion of an Ellipse

Graph $\dfrac{y}{6} = \sqrt{1 - \dfrac{x^2}{16}}$.

Square both sides to get an equation whose form is known.

$$\frac{y^2}{36} = 1 - \frac{x^2}{16}$$
$$\frac{x^2}{16} + \frac{y^2}{36} = 1 \qquad \text{Add } \tfrac{x^2}{16}.$$

This is the equation of an ellipse with x-intercepts $(4, 0)$ and $(-4, 0)$ and y-intercepts $(0, 6)$ and $(0, -6)$. Since $\tfrac{y}{6}$ equals a principal square root in the original equation, y must be nonnegative, restricting the graph to the upper

CONTINUED ON NEXT PAGE

---

3. Identify the graph of each equation.

(a) $3x^2 = 27 - 4y^2$

(b) $6x^2 = 100 + 2y^2$

(c) $3x^2 = 27 - 4y$

(d) $3x^2 = 27 - 3y^2$

4. Graph

$$\frac{y}{3} = -\sqrt{1 - \frac{x^2}{4}}.$$

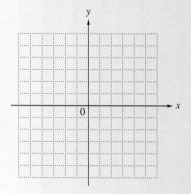

half of the ellipse, as shown in Figure 31. Verify that this is the graph of a function, using the vertical line test.

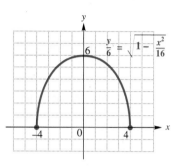

FIGURE 31

◀◀ WORK PROBLEM 4 AT THE SIDE.

ANSWERS

4.

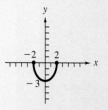

9.4 Exercises

Match the equation with the correct graph. See Examples 1–3.

1. $\dfrac{x^2}{25} + \dfrac{y^2}{9} = 1$

A.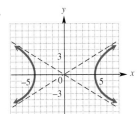

B.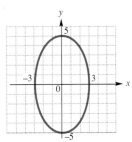

2. $\dfrac{x^2}{9} + \dfrac{y^2}{25} = 1$

3. $\dfrac{x^2}{9} - \dfrac{y^2}{25} = 1$

C.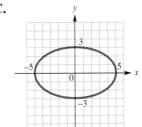

D.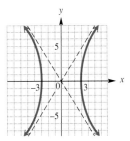

4. $\dfrac{x^2}{25} - \dfrac{y^2}{9} = 1$

5. Write an explanation of how you can tell from the equation whether the branches of a hyperbola open up and down or left and right.

6. Explain why a set of points that form an ellipse does not satisfy the definition of a function.

Graph each ellipse. See Examples 1 and 2.

7. $\dfrac{x^2}{9} + \dfrac{y^2}{25} = 1$

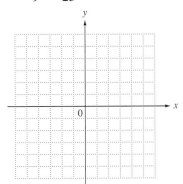

8. $\dfrac{x^2}{9} + \dfrac{y^2}{16} = 1$

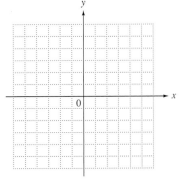

9. $\dfrac{x^2}{36} + \dfrac{y^2}{16} = 1$

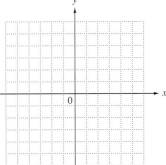

10. $\dfrac{x^2}{9} + \dfrac{y^2}{4} = 1$

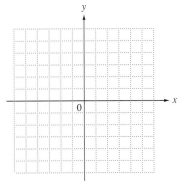

11. $\dfrac{x^2}{49} + \dfrac{y^2}{25} = 1$

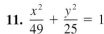
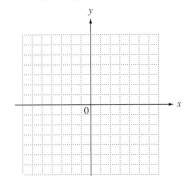

12. $\dfrac{x^2}{16} + \dfrac{y^2}{9} = 1$

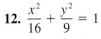
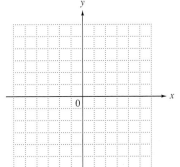

13. Describe how the fundamental rectangle is used to sketch a hyperbola.

14. Explain why the graph of a hyperbola does not satisfy the conditions for the graph of a function.

Graph each hyperbola. See Example 3.

15. $\dfrac{x^2}{16} - \dfrac{y^2}{9} = 1$

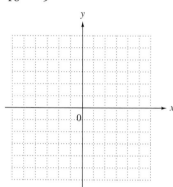

16. $\dfrac{y^2}{4} - \dfrac{x^2}{25} = 1$

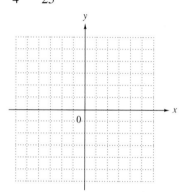

17. $\dfrac{y^2}{9} - \dfrac{x^2}{9} = 1$

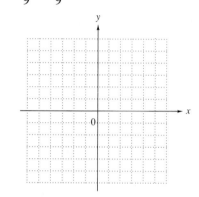

18. $\dfrac{x^2}{49} - \dfrac{y^2}{16} = 1$

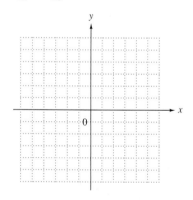

19. $\dfrac{x^2}{25} - \dfrac{y^2}{36} = 1$

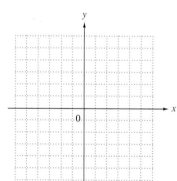

20. $\dfrac{y^2}{9} - \dfrac{x^2}{4} = 1$

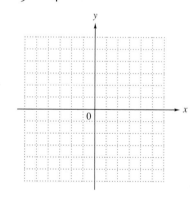

Identify each of the following as a parabola, circle, ellipse, or hyperbola. Sketch the graph. See Example 4.

21. $x^2 - y^2 = 16$

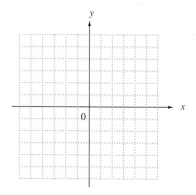

22. $x^2 + y^2 = 16$

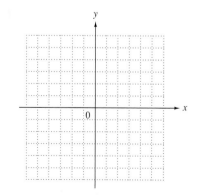

23. $4x^2 + y^2 = 16$

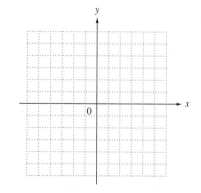

24. $x^2 - 2y = 0$

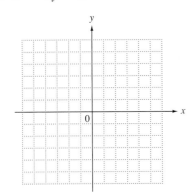

25. $y^2 = 36 - x^2$

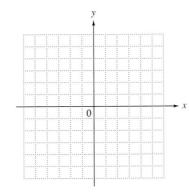

26. $9x^2 + 25y^2 = 225$

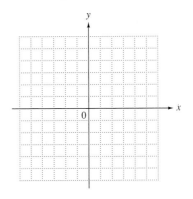

27. $9x^2 = 144 + 16y^2$

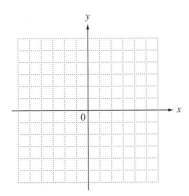

28. $x^2 + 9y^2 = 9$

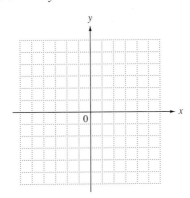

29. $y^2 = 4 + x^2$

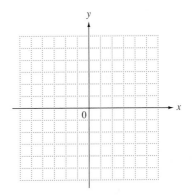

Each of the following is a portion of a conic section. Sketch each graph. See Example 5.

30. $\dfrac{y}{3} = \sqrt{1 + \dfrac{x^2}{9}}$

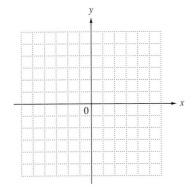

31. $y = \sqrt{\dfrac{x + 4}{2}}$

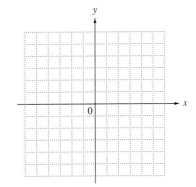

32. $y = -2\sqrt{\dfrac{9 - x^2}{9}}$

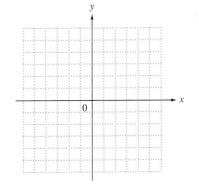

33. It is possible to sketch an ellipse on a piece of posterboard by fastening two ends of a length of string with thumbtacks, pulling the string taut with a pencil, and tracing a curve, as shown in the drawing. Explain why this method works.

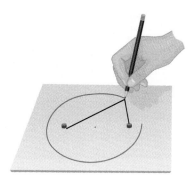

34. A pair of buildings in a sports complex are shaped and positioned like a portion of the branches of the hyperbola $400x^2 - 625y^2 = 250{,}000$ where x and y are in meters.
   (a) How far apart are the buildings at their closest point?
   (b) Find the distance d in the figure.

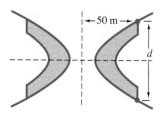

35. The orbit of Venus around the sun (one of the foci) is an ellipse with equation

$$\frac{x^2}{5013} + \frac{y^2}{4970} = 1$$

where x and y are measured in millions of miles.
   (a) Find the farthest distance between Venus and the sun.
   (b) Find the smallest distance between Venus and the sun. (*Hint:* See Figure 23 and use the fact that $c^2 = a^2 - b^2$.)

36. The graph shown here resembles a portion of a hyperbola. It shows the number of gun-related deaths per 100,000 for young African-American males starting in 1985. Here, 1985 is represented by $x = 0$, 1986 by $x = 1$, and so on. It can be shown that the function

$$f(x) = 23.8\sqrt{4.41 + x^2}$$

provides a good model for these data.

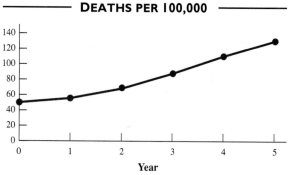

Source: Health and Human Services Department

   (a) According to the graph, what is the number of deaths for the year 1988?
   (b) According to the function, what is the number of deaths for the year 1988?

9.5 Nonlinear Systems of Equations

OBJECTIVES

 Solve a nonlinear system by substitution.

 Use the elimination method to solve a system with two second-degree equations.

 Solve a system that requires a combination of methods.

FOR EXTRA HELP

Tutorial   Tape 14   SSM, Sec. 9.5

An equation in which some terms have more than one variable, or a variable of degree two or higher, is called a **nonlinear** (non-LIN-ee-er) **equation.** A **nonlinear system of equations** includes at least one nonlinear equation. Nonlinear systems can be solved by the elimination method, the substitution method, or a combination of the two. The following examples illustrate the use of these methods for solving nonlinear systems.

OBJECTIVE ▶ The substitution method usually is most useful when one of the equations is linear. The first two examples illustrate this kind of system.

EXAMPLE 1   Using Substitution When One Equation Is Linear

Solve the system

$$x^2 + y^2 = 9 \tag{1}$$
$$2x - y = 3. \tag{2}$$

Solve the linear equation for one of the two variables, then substitute the resulting expression into the nonlinear equation to obtain an equation in one variable. Let us solve equation (2) for y.

$$2x - y = 3 \tag{2}$$
$$y = 2x - 3 \tag{3}$$

Substituting $2x - 3$ for y in equation (1) gives

$$x^2 + (2x - 3)^2 = 9 \qquad \text{Replace } y \text{ with } 2x - 3.$$
$$x^2 + 4x^2 - 12x + 9 = 9 \qquad \text{Square the binomial.}$$
$$5x^2 - 12x = 0 \qquad \text{Standard form}$$
$$x(5x - 12) = 0 \qquad \text{Factor.}$$
$$x = 0 \quad \text{or} \quad x = \frac{12}{5}. \qquad \begin{array}{l}\text{Set each factor equal} \\ \text{to 0; solve.}\end{array}$$

Let $x = 0$ in the equation $y = 2x - 3$ to get $y = -3$. If $x = \frac{12}{5}$, then $y = \frac{9}{5}$. The solution set of the system is $\{(0, -3), (\frac{12}{5}, \frac{9}{5})\}$. The graph of the system, shown in Figure 32, confirms the solution.

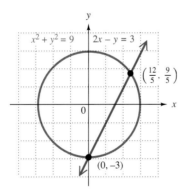

FIGURE 32

WORK PROBLEM 1 AT THE SIDE. ▶▶

1. Solve each system.

(a) $x^2 + y^2 = 10$
$\qquad x = y + 2$

(b) $x^2 - 2y^2 = 8$
$\qquad y + x = 6$

623

2. Solve each system.

(a) $xy = 8$
$x + y = 6$

(b) $xy + 10 = 0$
$4x + 9y = -2$

E X A M P L E 2 Using Substitution When One Equation Is Linear

Solve the system

$$6x - y = 5 \tag{4}$$

$$xy = 4. \tag{5}$$

Although we could solve the linear equation for either x or y and then substitute into the other equation, let us instead solve $xy = 4$ for x, to get $x = \frac{4}{y}$. Substituting $\frac{4}{y}$ for x in equation (4) gives

$$6\left(\frac{4}{y}\right) - y = 5. \qquad \text{Replace } x \text{ with } \tfrac{4}{y}.$$

$$\frac{24}{y} - y = 5$$

$$24 - y^2 = 5y \qquad \text{Multiply by } y. \text{ (Assume } y \neq 0.)$$

$$0 = y^2 + 5y - 24 \qquad \text{Standard form}$$

$$0 = (y - 3)(y + 8) \qquad \text{Factor.}$$

$$y = 3 \quad \text{or} \quad y = -8 \qquad \begin{array}{l}\text{Set each factor equal}\\\text{to 0; solve.}\end{array}$$

Substitute these results into $x = \frac{4}{y}$ to obtain the corresponding values for x.

$$\text{If } y = 3, \text{ then } x = \frac{4}{3}.$$

$$\text{If } y = -8, \text{ then } x = -\frac{1}{2}.$$

The solution set is $\{(\frac{4}{3}, 3), (-\frac{1}{2}, -8)\}$.

◄◄ **WORK PROBLEM 2 AT THE SIDE.**

OBJECTIVE ▶ The elimination method is often useful when both equations are second-degree equations. This method is used in the following example.

E X A M P L E 3 Solving a Nonlinear System by Elimination

Solve the system

$$x^2 + y^2 = 9 \tag{6}$$

$$2x^2 - y^2 = -6. \tag{7}$$

Adding the two equations will eliminate y, leaving an equation that can be solved for x.

$$\begin{array}{rl} x^2 + y^2 = & 9 \qquad (6) \\ \underline{2x^2 - y^2 = -6} & \qquad (7) \\ 3x^2 = & 3 \end{array}$$

$$x^2 = 1$$

$$x = 1 \quad \text{or} \quad x = -1$$

CONTINUED ON NEXT PAGE —

ANSWERS

2. (a) $\{(4, 2), (2, 4)\}$

(b) $\left\{(-5, 2), \left(\frac{9}{2}, -\frac{20}{9}\right)\right\}$

Each value of x gives corresponding values for y when substituted into one of the original equations. Using equation (6) gives the following results.

$$\text{If } x = 1, \qquad\qquad \text{If } x = -1,$$
$$(1)^2 + y^2 = 9 \qquad\qquad (-1)^2 + y^2 = 9$$
$$y^2 = 8 \qquad\qquad\qquad y^2 = 8$$
$$y = \sqrt{8} \quad \text{or} \quad -\sqrt{8} \qquad y = 2\sqrt{2} \quad \text{or} \quad -2\sqrt{2}.$$
$$y = 2\sqrt{2} \quad \text{or} \quad -2\sqrt{2}.$$

The solution set is $\{(1, 2\sqrt{2}), (1, -2\sqrt{2}), (-1, 2\sqrt{2}), (-1, -2\sqrt{2})\}$. The graph in Figure 33 shows the four points of intersection.

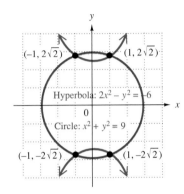

FIGURE 33

WORK PROBLEM 3 AT THE SIDE. ▶▶

OBJECTIVE 3▶ The next example shows a system of second-degree equations that can be solved only by a combination of methods.

EXAMPLE 4   Solving a Nonlinear System by a Combination of Methods

Solve the system

$$x^2 + 2xy - y^2 = 7 \qquad\qquad (8)$$
$$x^2 - y^2 = 3. \qquad\qquad (9)$$

The elimination method is used here in combination with the substitution method. To begin, we eliminate the squared terms by multiplying both sides of equation (9) by -1 and then adding the result to (8).

$$\begin{array}{rcl} x^2 + 2xy - y^2 &=& 7 \\ -x^2 \qquad + y^2 &=& -3 \\ \hline 2xy &=& 4 \end{array}$$

Next, we solve $2xy = 4$ for either variable. Let us solve for y.

$$2xy = 4$$
$$y = \frac{2}{x} \qquad\qquad (10)$$

CONTINUED ON NEXT PAGE

---

3. Solve each system.

(a) $x^2 + y^2 = 41$
$x^2 - y^2 = 9$

(b) $x^2 + 3y^2 = 40$
$4x^2 - y^2 = 4$

4. Solve each system.

(a) $x^2 + xy + y^2 = 3$
$x^2 \qquad + y^2 = 5$

(b) $x^2 + 7xy - 2y^2 = -8$
$-2x^2 \qquad + 4y^2 = 16$

Now substitute $y = \frac{2}{x}$ into one of the original equations. It is easier to do this with (9).

$$x^2 - y^2 = 3 \qquad \textbf{(9)}$$

$$x^2 - \left(\frac{2}{x}\right)^2 = 3$$

$$x^2 - \frac{4}{x^2} = 3$$

To clear the equation of fractions, we multiply both sides by x^2.

$$x^4 - 4 = 3x^2$$
$$x^4 - 3x^2 - 4 = 0$$
$$(x^2 - 4)(x^2 + 1) = 0$$
$$x^2 - 4 = 0 \quad \text{or} \quad x^2 + 1 = 0$$
$$x^2 = 4 \qquad\qquad x^2 = -1$$
$$x = 2 \quad \text{or} \quad x = -2 \qquad x = i \quad \text{or} \quad x = -i$$

By substituting the four values of x from above into equation (10), we get the corresponding values of y.

If $x = 2$, then $y = 1$. 　　　If $x = i$, then $y = -2i$.

If $x = -2$, then $y = -1$. 　　If $x = -i$, then $y = 2i$.

There are four solutions, two ordered pairs of real numbers and two ordered pairs of imaginary numbers, in the solution set: $\{(2, 1), (-2, -1), (i, -2i), (-i, 2i)\}$.

Note that if you substitute the x-values found above into equation (8) or (9) instead of into equation (10), you get extraneous solutions. It is always wise to check all solutions in both of the given equations. The graph of the system, shown in Figure 34, shows only the two real intersection points because the graph is in the real number plane. The two ordered pairs with imaginary components are solutions of the system but do not show up on the graph.

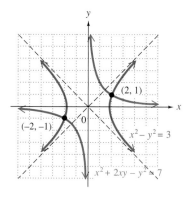

FIGURE 34

◀◀ **WORK PROBLEM 4 AT THE SIDE.**

ANSWERS

4. (a) $\{(1, -2), (-1, 2), (2, -1), (-2, 1)\}$
(b) $\{(0, 2), (0, -2), (2i\sqrt{2}, 0), (-2i\sqrt{2}, 0)\}$

9.5 *Exercises*

1. Write an explanation of the steps you would use to solve the system

$$x^2 + y^2 = 25$$
$$y = x - 1$$

by the substitution method.

2. Write an explanation of the steps you would use to solve the system

$$x^2 + y^2 = 12$$
$$x^2 - y^2 = 13$$

by the elimination method.

3. Is it possible for a nonlinear system consisting of a line and a circle to have three ordered pairs in its solution set? Explain.

4. Suppose that a nonlinear system consists of two ellipses whose graphs look like this:

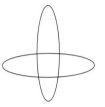

How many ordered pairs are there in the solution set of this system? Explain.

Solve the following systems by the substitution method. See Examples 1 and 2.

5. $y = 4x^2 - x$
   $y = x$

6. $y = x^2 + 6x$
   $3y = 12x$

7. $y = x^2 + 6x + 9$
   $x + y = 3$

8. $y = x^2 + 8x + 16$
   $x - y = -4$

9. $x^2 + y^2 = 2$
   $2x + y = 1$

10. $2x^2 + 4y^2 = 4$
    $x = 4y$

11. $xy = 4$
$3x + 2y = -10$

12. $xy = -5$
$2x + y = 3$

13. $xy = -3$
$x + y = -2$

14. $xy = 12$
$x + y = 8$

15. $y = 3x^2 + 6x$
$y = x^2 - x - 6$

16. $y = 2x^2 + 1$
$y = 5x^2 + 2x - 7$

17. $2x^2 - y^2 = 6$
$y = x^2 - 3$

18. $x^2 + y^2 = 4$
$y = x^2 - 2$

Solve the following systems using the elimination method or a combination of the elimination and substitution methods. See Examples 3 and 4.

19. $3x^2 + 2y^2 = 12$
$x^2 + 2y^2 = 4$

20. $2x^2 + y^2 = 28$
$4x^2 - 5y^2 = 28$

21. $xy = 6$
$3x^2 - y^2 = 12$

22. $xy = 5$
$2y^2 - x^2 = 5$

23. $2x^2 + 2y^2 = 8$
$\quad\ 3x^2 + 4y^2 = 24$

24. $5x^2 + 5y^2 = 20$
$\qquad\ x^2 + 2y^2 = 2$

25. $x^2 + xy + y^2 = 15$
$\qquad\quad x^2 + y^2 = 10$

26. $2x^2 + 3xy + 2y^2 = 21$
$\qquad\qquad\ x^2 + y^2 = 6$

27. $3x^2 + 2xy - 3y^2 = 5$
$\quad\ -x^2 - 3xy + y^2 = 3$

28. $-2x^2 + 7xy - 3y^2 = 4$
$\qquad 2x^2 - 3xy + 3y^2 = 4$

INTERPRETING TECHNOLOGY (EXERCISES 29–32)

If the two equations making up a nonlinear system are graphed with a graphing calculator, the calculator can be used to solve the system by finding the coordinates of the points of intersection of the two graphs. For example, the nonlinear system

$$y = x^2 - 3$$
$$2x + y = 0$$

can be solved by substitution to find that the solution set is $\{(-3, 6), (1, -2)\}$. The graphs of $y_1 = x^2 - 3$ and $y_2 = -2x$ shown in the two screens here indicate that these are indeed the points of intersection.

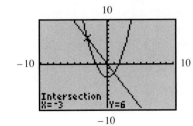
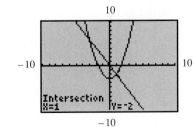

In Exercises 29–32, nonlinear systems are given, along with a screen showing the coordinates of one of the points of intersection of the two graphs. Solve the system using substitution or elimination to find the coordinates of the other *point of intersection.*

29. $y = x^2 + 1$
$x + y = 1$

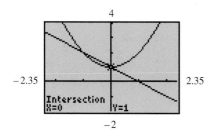

30. $y = -x^2$
$x + y = 0$

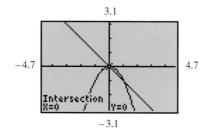

31. $y = \dfrac{1}{2}x^2$
$x + y = 4$

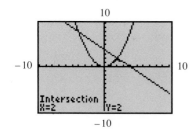

32. $y = -\dfrac{1}{3}x^2$
$2x - y = 9$

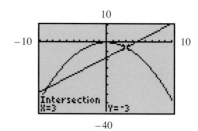

Write a nonlinear system of equations and solve.

33. The area of a rectangular rug is 84 square feet, and its perimeter is 38 feet. Find the length and width of the rug.

34. Find the length and width of a rectangular room whose perimeter is 50 meters and whose area is 100 square meters.

35. A company has found that the price p (in dollars) of its scientific calculator is related to the supply x (in thousands) by the equation $px = 16$. The price is related to the demand x (in thousands) for the calculator by the equation $p = 10x + 12$. The *equilibrium price* is the value of p where demand equals supply. Find the equilibrium price and the supply/demand at that price by solving a system of equations.

36. The calculator company in Exercise 35 has also determined that the cost y to make x (thousand) calculators is $y = 4x^2 + 36x + 20$, while the revenue y from the sale of x (thousand) calculators is $36x^2 - 3y = 0$. Find the *break-even point,* where cost just equals revenue, by solving a system of equations.

9.6 Second-Degree Inequalities; Systems of Inequalities

OBJECTIVES

  1 Graph second-degree inequalities.

  2 Graph a system of inequalities.

FOR EXTRA HELP

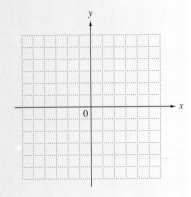

Tutorial   Tape 14   SSM, Sec. 9.6

OBJECTIVE ▶1▶ Recall from Section 3.4 that a linear inequality such as $3x + 2y \leq 5$ is graphed by first graphing the boundary line $3x + 2y = 5$. **Second-degree inequalities** such as $x^2 + y^2 \leq 36$ are graphed in much the same way. The boundary of $x^2 + y^2 \leq 36$ is the graph of the equation $x^2 + y^2 = 36$, a circle with radius 6 and center at the origin, as shown in Figure 35. As with linear inequalities, the inequality $x^2 + y^2 \leq 36$ will include either the points outside the circle or the points inside the circle. Decide which region to shade by substituting any point not on the circle, such as $(0, 0)$, into the inequality. Since $0^2 + 0^2 < 36$ is a true statement, the inequality includes the points inside the circle, the shaded region in Figure 35.

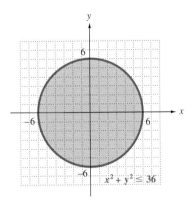

FIGURE 35

1. Graph $y \geq (x + 1)^2 - 5$.

┌─────────────────────────────────┐
│ EXAMPLE 1   **Graphing a Second-Degree Inequality**

Graph $y < -2(x - 4)^2 - 3$.

The boundary, $y = -2(x - 4)^2 - 3$, is a parabola opening downward with vertex at $(4, -3)$. Using the point $(0, 0)$ as a test point gives

$$0 < -2(0 - 4)^2 - 3$$
$$0 < -32 - 3$$
$$0 < -35. \qquad \text{False}$$

This is a false statement, so the points in the region containing $(0, 0)$ do not satisfy the inequality. Figure 36 shows the final graph; the parabola is drawn with a dashed line since the points of the parabola itself do not satisfy the inequality. The solution includes all points inside (below) the parabola.

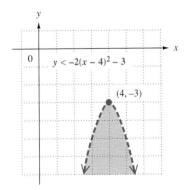

FIGURE 36

WORK PROBLEM 1 AT THE SIDE. ▶▶

ANSWERS

1.

2. Graph $x^2 + 4y^2 > 36$.

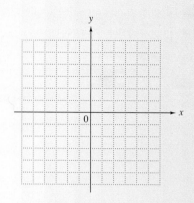

E X A M P L E 2   Graphing a Second-Degree Inequality

Graph $16y^2 \le 144 + 9x^2$.

First rewrite the inequality as follows.

$$16y^2 - 9x^2 \le 144$$

$$\frac{y^2}{9} - \frac{x^2}{16} \le 1 \qquad \text{Divide both sides by 144.}$$

This form of the inequality shows that the boundary is the hyperbola

$$\frac{y^2}{9} - \frac{x^2}{16} = 1.$$

Since the test point $(0, 0)$ satisfies the inequality $16y^2 \le 144 + 9x^2$, the region containing $(0, 0)$, is shaded, as shown in Figure 37.

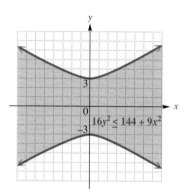

FIGURE 37

◀◀ WORK PROBLEM 2 AT THE SIDE.

Objective 2 The solution set of a **system of inequalities,** such as

$$2x + 3y > 6$$
$$x^2 + y^2 < 16,$$

is the intersection of the solution sets of the individual inequalities.

The graph of the solution set of this system is the set of all points on the plane that belong to the graphs of both inequalities in the system. Graph this system by graphing both inequalities on the same coordinate axes, as shown in Figure 38. The heavily shaded region containing those points that belong to both graphs is the graph of the system. In this case the points of the boundary lines are not included.

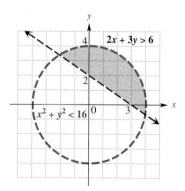

FIGURE 38

Answers

2.

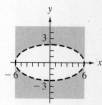

EXAMPLE 3 Graphing a System of Inequalities

Graph the solution set of the system

$$x + y < 1$$
$$y \leq 2x + 3$$
$$y \geq -2.$$

Graph each inequality separately, on the same axes. The graph of the system is the triangular region enclosed by the three lines in Figure 39. It contains all points that satisfy all three inequalities. One boundary line ($x + y = 1$) is dashed, while the other two are solid.

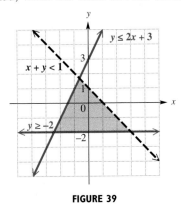

FIGURE 39

> **WORK PROBLEM 3 AT THE SIDE.** ▶▶

EXAMPLE 4 Graphing a System of Inequalities

Graph the solution set of the system

$$y \geq x^2 - 2x + 1$$
$$2x^2 + y^2 > 4$$
$$y < 4.$$

The graph of $y = x^2 - 2x + 1$ is a parabola with vertex at $(1, 0)$. Those points that are above the parabola satisfy the condition $y > x^2 - 2x + 1$. Thus points on the parabola or above it are in the solution set of $y \geq x^2 - 2x + 1$. The graph of $2x^2 + y^2 = 4$ is an ellipse. To satisfy the inequality $2x^2 + y^2 > 4$, a point must lie outside the ellipse. The graph of $y < 4$ includes all points below the line $y = 4$. Finally, the graph of the system is the shaded region in Figure 40 that lies outside the ellipse, above or on the boundary of the parabola, and below the line $y = 4$.

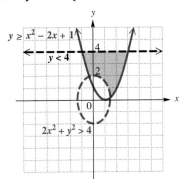

FIGURE 40

> **WORK PROBLEM 4 AT THE SIDE.** ▶▶

3. Graph the solution set of the system

$$3x - 4y \geq 12$$
$$x + 3y \geq 6$$
$$y \leq 2.$$

4. Graph the solution set of the system

$$y \geq x^2 + 1$$
$$\frac{x^2}{9} + \frac{y^2}{4} \geq 1$$
$$y \leq 5.$$

ANSWERS

3.

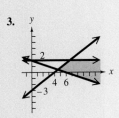

4.

Numbers in the *Real World*   *a graphing calculator minicourse*

Lesson 6: Solution of Quadratic Equations and Inequalities

The equation solved in Example 2 of Section 5.10 is equivalent to the quadratic equation $2x^2 + 3x - 2 = 0$. The graph of $y = 2x^2 + 3x - 2$ is a parabola, as discussed in Section 9.1. Recall that the x-intercept of the line $y = mx + b$ is a solution of the equation $mx + b = 0$. Extending this idea to quadratic equations, we may say that the x-intercept(s), if any, of the graph of $y = ax^2 + bx + c$, are the real solutions of the equation $ax^2 + bx + c = 0$. Therefore, based on the graph and displays shown below, the solutions of $2x^2 + 3x - 2 = 0$ are .5 (or $\frac{1}{2}$) and –2. This is graphical support for the result given in the aforementioned example.

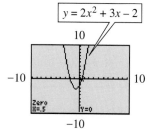
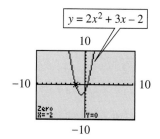

In Section 8.5 we examined algebraic methods for solving quadratic inequalities. We can use graphs of quadratic functions to solve such inequalities graphically. Based on the figures above, we see that the graph of $y = 2x^2 + 3x - 2$ lies *below* the x-axis for x-values between –2 and .5. Thus the solution set of $2x^2 + 3x - 2 < 0$ is the interval $(-2, .5)$. Because the graph of $y = 2x^2 + 3x - 2$ lies *above* the x-axis to the left of –2 and to the right of .5, the solution set of $2x^2 + 3x - 2 > 0$ is $(-\infty, -2) \cup (.5, \infty)$.

GRAPHING CALCULATOR EXPLORATIONS

1. Graph the function $y = x^2 - 4x - 5$ in a window that shows both x-intercepts.

   (a) Solve the equations $x^2 - 4x - 5 = 0$ algebraically. How do the solutions compare to the x-intercepts of the graph?

   (b) Use the graph to give the solution set of $x^2 - 4x - 5 < 0$.

   (c) Use the graph to give the solution set of $x^2 - 4x - 5 > 0$.

2. The figures below show the graph of $y = -x^2 + x + 12$. Use the graph to solve each of the following:

   (a) $-x^2 + x + 12 = 0$ (b) $-x^2 + x + 12 \leq 0$ (c) $-x^2 + x + 12 \geq 0$

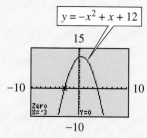
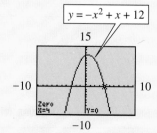

9.6 Exercises

1. Match the nonlinear inequality with its graph.

    (a) $y \geq x^2 + 4$

    (b) $y \leq x^2 + 4$

    (c) $y < x^2 + 4$

    (d) $y > x^2 + 4$

A.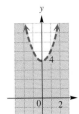
B.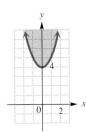
C.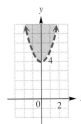
D.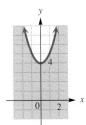

Graph each of the following inequalities. See Examples 1 and 2.

2. $y > x^2 - 1$

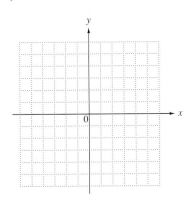

3. $y^2 > 4 + x^2$

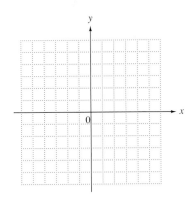

4. $y^2 \leq 4 - 2x^2$

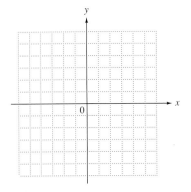

5. $y + 2 \geq x^2$

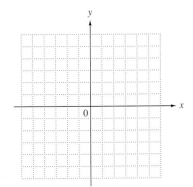

6. $x^2 \leq 16 - y^2$

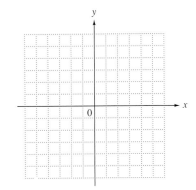

7. $2y^2 \geq 8 - x^2$

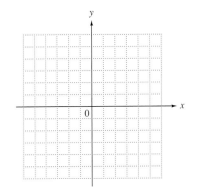

8. $x^2 \le 16 + 4y^2$

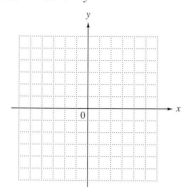

9. $y \le x^2 + 4x + 2$

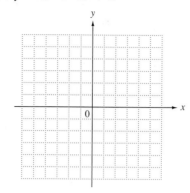

10. $9x^2 < 16y^2 - 144$

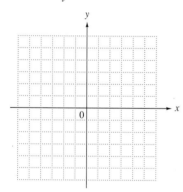

11. $9x^2 > 16y^2 + 144$

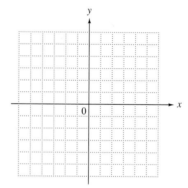

12. $4y^2 \le 36 - 9x^2$

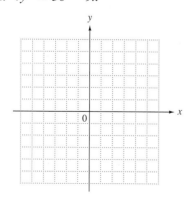

13. $x^2 - 4 \ge -4y^2$

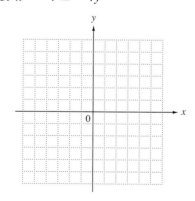

14. $x \geq y^2 - 8y + 14$

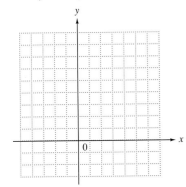

15. $x \leq -y^2 + 6y - 7$

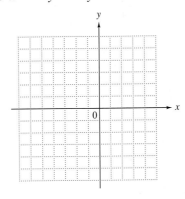

16. $y^2 - 16x^2 \leq 16$

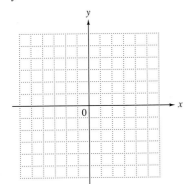

17. $25x^2 \leq 9y^2 + 225$

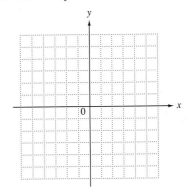

18. Explain how to graph the solution set of a system of nonlinear inequalities.

19. Which one of the following is a description of the graph of the solution set of the system below?

$$x^2 + y^2 < 25$$
$$y > -2$$

  (a) all points outside the circle $x^2 + y^2 = 25$ and above the line $y = -2$
  (b) all points outside the circle $x^2 + y^2 = 25$ and below the line $y = -2$
  (c) all points inside the circle $x^2 + y^2 = 25$ and above the line $y = -2$
  (d) all points inside the circle $x^2 + y^2 = 25$ and below the line $y = -2$

20. Fill in the blank with the appropriate response: The graph of the system

$$y > x^2 + 1$$
$$\frac{x^2}{9} + \frac{y^2}{4} > 1$$
$$y < 5$$

consists of all points _____ the
                        (above/below)
parabola $y = x^2 + 1$, _____ the
                        (inside/outside)
ellipse $x^2/9 + y^2/4 = 1$, and _____
                                  (above/below)
the line $y = 5$.

Graph the following systems of inequalities. See Examples 3 and 4.

21. $2x + 5y < 10$
$\quad x - 2y < 4$

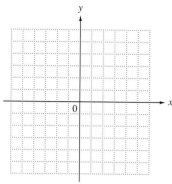

22. $3x - y > -6$
$\quad 4x + 3y > 12$

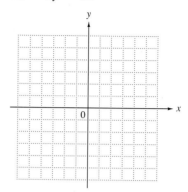

23. $5x - 3y \leq 15$
$\quad 4x + y \geq 4$

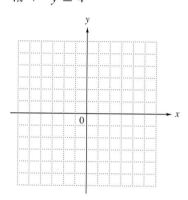

24. $4x - 3y \leq 0$
$\quad x + y \leq 5$

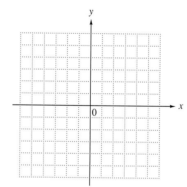

25. $x \leq 5$
$\quad y \leq 4$

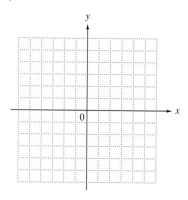

26. $x \geq -2$
$\quad y \leq 4$

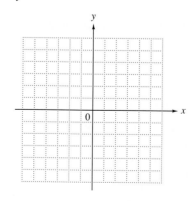

27. $y > x^2 - 4$
   $y < -x^2 + 3$

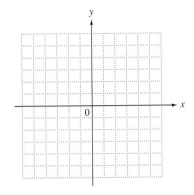

28. $x^2 - y^2 \geq 9$
   $\dfrac{x^2}{16} + \dfrac{y^2}{9} \leq 1$

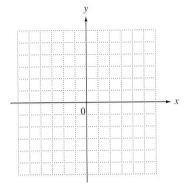

29. $y^2 - x^2 \geq 4$
   $-5 \leq y \leq 5$

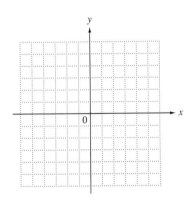

30. $x \geq 0$
         $y \geq 0$
   $x^2 + y^2 \geq 4$
   $x + y \leq 5$

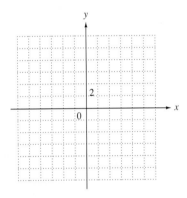

31. $y \leq -x^2$
   $y \geq x - 3$
   $y \leq -1$
   $x < 1$

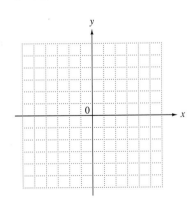

32. $y < x^2$
         $y > -2$
   $x + y < 3$
   $3x - 2y > -6$

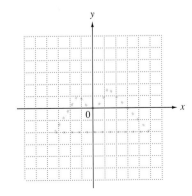

INTERPRETING TECHNOLOGY (EXERCISES 33–36)

Graphing calculators have the capability of shading above or below graphs, thus allowing them to illustrate nonlinear inequalities and systems of nonlinear inequalities. For example, the inequality discussed in Example 1 and graphed in Figure 36,

$$y < -2(x - 4)^2 - 3$$

is shown in the accompanying screen.

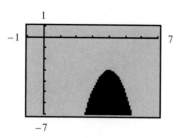

Match the nonlinear inequality or system of nonlinear inequalities with its calculator-generated graph.

33. $y > x^2 + 2$ **A.**

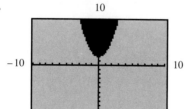

34. $y < x^2 + 2$

35. $y > x^2 + 2$
    $y < 5$ **C.**

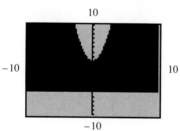

B.

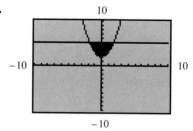

D.

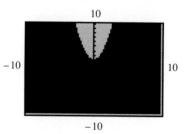

36. $y < x^2 + 2$
    $y > -5$

CHAPTER 9 SUMMARY

9.1	**conic sections**	Graphs that result from cutting an infinite cone with a plane are called conic sections.
	parabola	The graph of a second-degree function is a parabola.
	vertex	The point on a parabola that has the smallest y-value (if the parabola opens upward) or the largest y-value (if the parabola opens downward) is called the vertex of the parabola.
	axis	The vertical or horizontal line through the vertex of a parabola is its axis.
	quadratic function	A function defined by $f(x) = ax^2 + bx + c$, for real numbers a, b, and c, with $a \neq 0$, is a quadratic function.
9.3	**squaring function**	The function defined by $f(x) = x^2$ is called the squaring function.
	absolute value function	The function defined by $f(x) = |x|$ is called the absolute value function.
	reciprocal function	The function defined by $f(x) = \dfrac{1}{x}$ is called the reciprocal function.
	asymptotes	Lines that a graph approaches, such as the x- and y-axes for the reciprocal function, are called asymptotes of the graph.
	square root function	The function defined by $f(x) = \sqrt{x}$ is called the square root function.
	circle	A circle is the set of all points in a plane that lie a fixed distance from a fixed point.
	center	The fixed point discussed in the definition of a circle is the center of the circle.
	radius	The radius of a circle is the fixed distance between the center and any point on the circle.
9.4	**ellipse**	An ellipse is the set of all points in a plane the sum of whose distances from two fixed points is constant.
	hyperbola	A hyperbola is the set of all points in a plane the difference of whose distances from two fixed points is constant.
	asymptotes of a hyperbola	The two intersecting lines that the branches of a hyperbola approach are called asymptotes of the hyperbola.
	fundamental rectangle	The asymptotes of a hyperbola are the extended diagonals of its fundamental rectangle.
9.5	**nonlinear equation**	An equation that cannot be written in the form $Ax + By = C$, for real numbers A, B, and C, is a nonlinear equation.
	nonlinear system of equations	A nonlinear system of equations is a system with at least one nonlinear equation.
9.6	**second-degree inequality**	A second-degree inequality is an inequality with at least one variable of degree two and no variable with degree greater than two.
	system of inequalities	A system of inequalities consists of two or more inequalities to be solved at the same time.

QUICK REVIEW

Concepts	*Examples*

9.1 Graphs of Quadratic Functions;
Vertical Parabolas

The graph of the quadratic function defined by $f(x) = a(x - h)^2 + k$, $a \neq 0$, is a parabola with vertex at (h, k) and the vertical line $x = h$ as axis.

The graph opens upward if a is positive and downward if a is negative.

The graph is wider than $f(x) = x^2$ if $0 < |a| < 1$ and narrower if $|a| > 1$.

Graph $f(x) = -(x + 3)^2 + 1$.

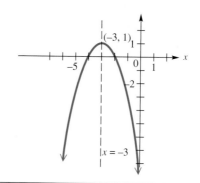

9.2 More about Quadratic Functions;
Horizontal Parabolas

The vertex of the graph of $f(x) = ax^2 + bx + c$, $a \neq 0$, may be found by completing the square. The vertex has coordinates

$$\left(-\frac{b}{2a}, f\left(-\frac{b}{2a}\right)\right).$$

Steps in graphing a quadratic function:

Step 1 Find the y-intercept by evaluating $f(0)$.
Step 2 Find any x-intercepts by solving $f(x) = 0$.
Step 3 Find the vertex either by using the formula or by completing the square.
Step 4 Find and plot any additional points as needed, using the symmetry about the axis.
Step 5 Verify that the graph opens upward (if $a > 0$) or opens downward (if $a < 0$).

If the discriminant, $b^2 - 4ac$, is positive, the graph of $f(x) = ax^2 + bx + c$ has two x-intercepts; if zero, one x-intercept; if negative, no x-intercepts.

The graph of $x = ay^2 + by + c$ is a horizontal parabola, opening to the right if $a > 0$, or to the left if $a < 0$.

Graph $f(x) = x^2 + 4x + 3$.

The vertex is $(-2, -1)$.
Since $f(0) = 3$, the y-intercept is $(0, 3)$. The solutions of $x^2 + 4x + 3 = 0$ are -1 and -3, so the x-intercepts are $(-1, 0)$ and $(-3, 0)$. The domain is $(-\infty, \infty)$ and the range is $[-1, \infty)$.

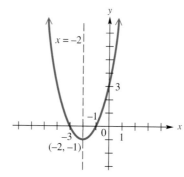

Graph $x = 2y^2 + 6y + 5$.

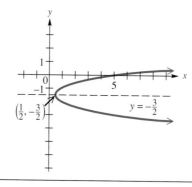

Concepts	Examples
9.3 Graphs of Elementary Functions and Circles In addition to the squaring function, some other important elementary functions in algebra are the absolute value function, defined by $f(x) = |x|$; the reciprocal function, defined by $f(x) = \dfrac{1}{x}$; and the square root function, defined by $f(x) = \sqrt{x}$.	$f(x) = |x| - 2 \qquad f(x) = \dfrac{1}{x+1}$ 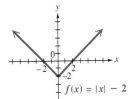 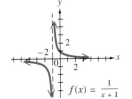 $f(x) = \sqrt{x-2} + 1$ 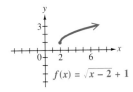
The circle with radius r and center at (h, k) has an equation of the form $$(x - h)^2 + (y - k)^2 = r^2.$$	The circle $(x + 2)^2 + (y - 3)^2 = 25$ has center $(-2, 3)$ and radius 5. 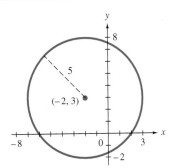
9.4 Ellipses and Hyperbolas The ellipse whose x-intercepts are $(a, 0)$ and $(-a, 0)$ and whose y-intercepts are $(0, b)$ and $(0, -b)$ has an equation of the form $$\frac{x^2}{a^2} + \frac{y^2}{b^2} = 1.$$	Graph $\dfrac{x^2}{9} + \dfrac{y^2}{4} = 1.$ 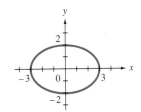
A hyperbola with x-intercepts $(a, 0)$ and $(-a, 0)$ has an equation of the form $$\frac{x^2}{a^2} - \frac{y^2}{b^2} = 1,$$ and a hyperbola with y-intercepts $(0, b)$ and $(0, -b)$ has an equation of the form $$\frac{y^2}{b^2} - \frac{x^2}{a^2} = 1.$$	Graph $\dfrac{x^2}{4} - \dfrac{y^2}{4} = 1.$ 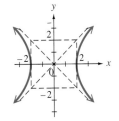
The extended diagonals of the fundamental rectangle with corners at the points (a, b), $(-a, b)$, $(-a, -b)$, and $(a, -b)$ are the asymptotes of these hyperbolas.	The fundamental rectangle has corners at $(2, 2)$, $(-2, 2)$, $(-2, -2)$, and $(2, -2)$.

Concepts	*Examples*
9.5 Nonlinear Systems of Equations	
Nonlinear systems can be solved by the substitution method, the elimination method, or a combination of the two.	Solve the system $$\begin{aligned} x^2 + 2xy - y^2 &= 14 \\ x^2 \qquad\;\; - y^2 &= -16. \end{aligned} \qquad (*)$$ Multiply equation $(*)$ by -1 and use elimination. $$\begin{aligned} x^2 + 2xy - y^2 &= 14 \\ -x^2 \qquad\quad + y^2 &= 16 \\ \hline 2xy \qquad\qquad &= 30 \\ xy &= 15 \end{aligned}$$ Solve for y to obtain $y = \frac{15}{x}$, and substitute into equation $(*)$. $$x^2 - \left(\frac{15}{x}\right)^2 = -16$$ This simplifies to $$x^2 - \frac{225}{x^2} = -16.$$ Multiply by x^2 and get one side equal to 0. $$x^4 + 16x^2 - 225 = 0$$ Factor and solve. $$(x^2 - 9)(x^2 + 25) = 0$$ $$x = \pm 3 \qquad x = \pm 5i$$ Find corresponding y values to get the solution set $\{(3, 5), (-3, -5), (5i, -3i), (-5i, 3i)\}$.
9.6 Second-Degree Inequalities; Systems of Inequalities	
To graph a second-degree inequality, graph the corresponding equation as a boundary and use test points to determine which region(s) form the solution. Shade the appropriate region(s).	Graph $y \geq x^2 - 2x + 3$. 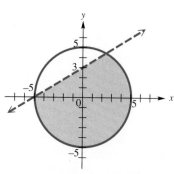
The solution set of a system of inequalities is the intersection of the solution sets of the individual inequalities.	Graph the solution set of the system $$3x - 5y > -15$$ $$x^2 + y^2 \leq 25.$$

[9.1–9.2] *Identify the vertex of each parabola.*

1. $f(x) = 3x^2 - 2$ **2.** $f(x) = 6 - 2x^2$ **3.** $f(x) = (x + 2)^2$ **4.** $f(x) = -(x - 1)^2$

5. $f(x) = (x - 3)^2 + 7$ **6.** $f(x) = \dfrac{4}{3}(x - 2)^2 + 1$ **7.** $x = (y + 2)^2 + 3$ **8.** $x = (y - 3)^2 - 4$

Graph each of the following. In Exercises 9 and 10 give the domain and range.

9. $f(x) = 4x^2 + 4x - 2$ **10.** $f(x) = -2x^2 + 8x - 5$

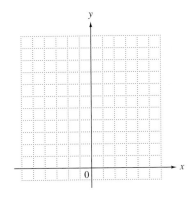   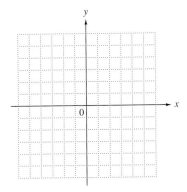

11. $x = -\dfrac{1}{2}y^2 + 6y - 14$ **12.** $x = 2y^2 + 8y + 3$

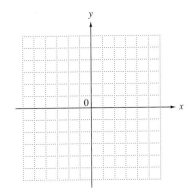

13. Explain how the discriminant can be used to determine the number of x-intercepts of the graph of a quadratic function.

14. Which of the following would most closely resemble the graph of $f(x) = a(x - h)^2 + k$, if $a < 0$, $h > 0$, and $k < 0$?

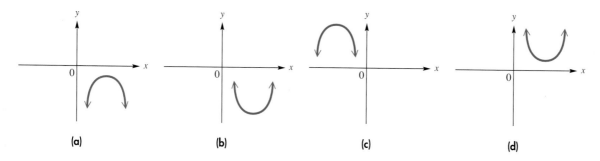

(a)   (b)   (c)   (d)

15. From 1982 to 1990, sales (in billions of dollars) of video games were approximated by the function

$$f(x) = .2x^2 - 1.6x + 3.3,$$

where x is the number of years since 1982. In what year during that period were sales a minimum? What were the minimum sales?

16. The height (in feet) of a projectile t seconds after being fired into the air is given by $f(t) = -16t^2 + 160t$. Find the number of seconds required for the projectile to reach maximum height. What is the maximum height?

17. Find the length and width of a rectangle having a perimeter of 200 meters if the area is to be a maximum.

18. Find the two numbers whose sum is 10 and whose product is a maximum.

[9.3] *Graph each of the following.*

19. $f(x) = |x + 4|$

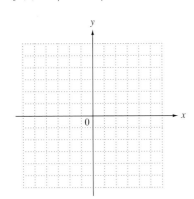

20. $f(x) = \dfrac{1}{x - 4}$

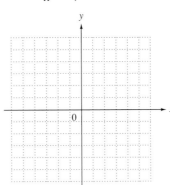

21. $f(x) = \sqrt{x} + 3$

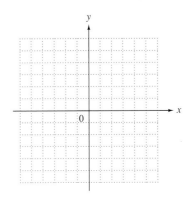

Write an equation for each of the following circles.

22. Center $(-2, 4)$, $r = 3$

23. Center $(-1, -3)$, $r = 5$

24. Center $(4, 2)$, $r = 6$

Find the center and radius of each circle.

25. $x^2 + y^2 + 6x - 4y - 3 = 0$

26. $x^2 + y^2 - 8x - 2y + 13 = 0$

27. $2x^2 + 2y^2 + 4x + 20y = -34$

28. $4x^2 + 4y^2 - 24x + 16y = 48$

[9.4] *Graph each of the following.*

29. $\dfrac{x^2}{16} + \dfrac{y^2}{9} = 1$

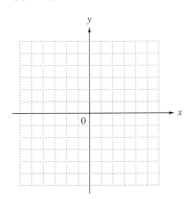

30. $\dfrac{x^2}{49} + \dfrac{y^2}{25} = 1$

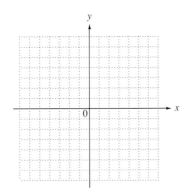

31. $\dfrac{x^2}{16} - \dfrac{y^2}{25} = 1$

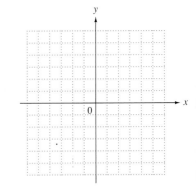

32. $\dfrac{y^2}{25} - \dfrac{x^2}{4} = 1$

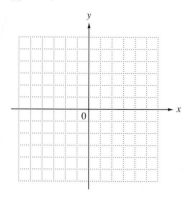

33. $x^2 + 9y^2 = 9$

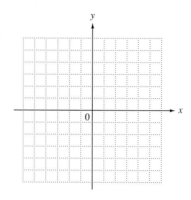

34. $f(x) = -\sqrt{16 - x^2}$

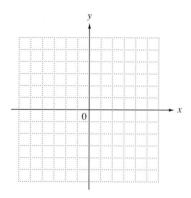

Identify each of the following as a parabola, circle, ellipse, or hyperbola.

35. $x^2 + y^2 = 64$

36. $y = 2x^2 - 3$

37. $y^2 = 2x^2 - 8$

38. $y^2 = 8 - 2x^2$

39. $x = y^2 + 4$

40. $x^2 - y^2 = 64$

[9.5] *Solve each system.*

41. $\begin{aligned} 2y &= 3x - x^2 \\ x + 2y &= -12 \end{aligned}$

42. $\begin{aligned} y + 1 &= x^2 + 2x \\ y + 2x &= 4 \end{aligned}$

43. $\begin{aligned} x^2 + 3y^2 &= 28 \\ y - x &= -2 \end{aligned}$

44. $\begin{aligned} xy &= 8 \\ x - 2y &= 6 \end{aligned}$

45. $\begin{aligned} x^2 + y^2 &= 6 \\ x^2 - 2y^2 &= -6 \end{aligned}$

46. $\begin{aligned} 3x^2 - 2y^2 &= 12 \\ x^2 + 4y^2 &= 18 \end{aligned}$

47. How many solutions are possible for a system of two equations whose graphs are a circle and a line?

48. How many solutions are possible for a system of two equations whose graphs are a parabola and a hyperbola?

[9.6] *Graph each inequality.*

49. $9x^2 \geq 16y^2 + 144$

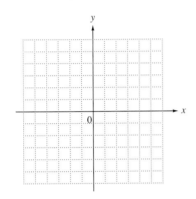

50. $4x^2 + y^2 \geq 16$

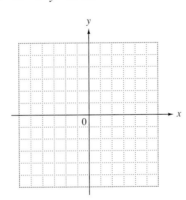

51. $y < -(x + 2)^2 + 1$

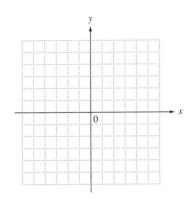

Graph each system of inequalities.

52. $2x + 5y \leq 10$
$3x - y \leq 6$

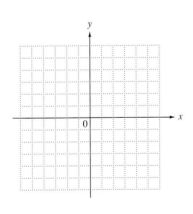

53. $|x| \leq 2$
$|y| > 1$
$4x^2 + 9y^2 \leq 36$

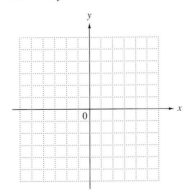

54. $9x^2 \leq 4y^2 + 36$
$x^2 + y^2 \leq 16$

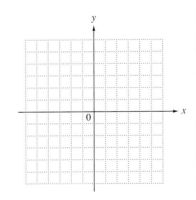

Graph.

55. $\dfrac{x^2}{64} + \dfrac{y^2}{25} = 1$

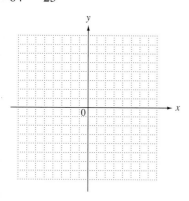

56. $\dfrac{y^2}{4} - 1 = \dfrac{x^2}{9}$

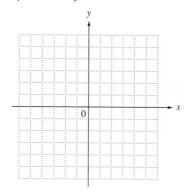

57. $x^2 + y^2 = 25$

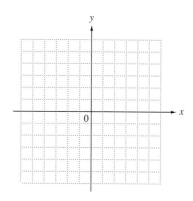

58. $y = 2(x - 2)^2 - 3$

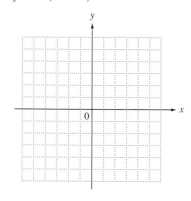

59. $x^2 - 9y^2 = 9$

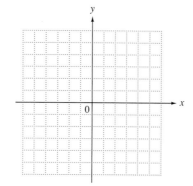

60. $f(x) = \sqrt{4 - x}$

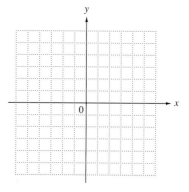

61. $3x + 2y \geq 0$
$\quad\quad y \leq 4$
$\quad\quad x \leq 4$

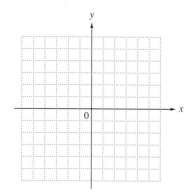

62. $4y > 3x - 12$
$\quad x^2 < 16 - y^2$

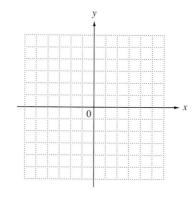

1. Graph the quadratic function defined by $f(x) = \frac{1}{2}x^2 - 2$. Identify the vertex.

1. _____

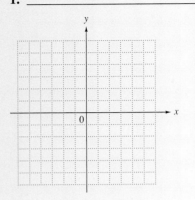

2. Identify the vertex of the graph of $f(x) = -x^2 + 4x - 1$. Sketch the graph. Give the domain and the range.

2. _____

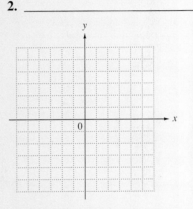

3. The number of chinook salmon that returned each year through the Lower Granite Dam in the state of Washington during the period from 1982 to 1987 can be approximated by the quadratic function with

$$f(x) = -2.6x^2 + 11.7x + 22.5,$$

where $x = 0$ corresponds to 1982, $x = 1$ corresponds to 1983, and so on, and $f(x)$ is in thousands. (*Source:* Idaho Department of Fish and Game)

(a) Based on this model, how many returned in 1983?
(b) In what year did the return reach a maximum? To the nearest thousand, how many salmon returned that year?

3. (a) _____

(b) _____

4. _____

4. Which one of these graphs most closely resembles the graph of $f(x) = (x - 2)^2 + 2$?

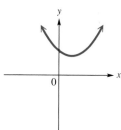

5.

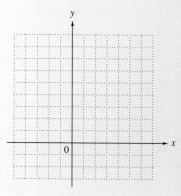

5. Sketch the graph of $f(x) = |x - 3| + 4$.

6. _____

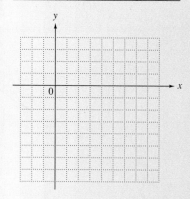

6. Find the center and radius of the circle whose equation is $(x - 2)^2 + (y + 3)^2 = 16$. Sketch the graph.

7. _____

7. Find the center and radius of the circle whose equation is $x^2 + y^2 + 8x - 2y = 8$.

8. _____

8. What characteristics of the equation of a parabola enable you to identify it?

Graph each of the following.

9. $f(x) = \sqrt{9 - x^2}$

9.

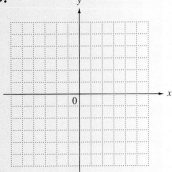

10. $4x^2 + 9y^2 = 36$

10.

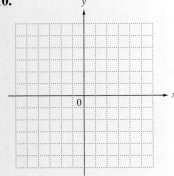

11. $16y^2 - 4x^2 = 64$

11.

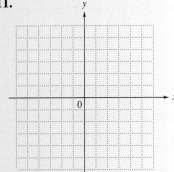

12. $x = -(y - 2)^2 + 2$

12.

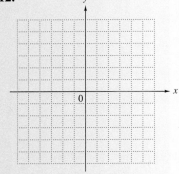

Identify each of the following as the equation of a parabola, hyperbola, ellipse, or circle.

13. _____

13. $6x^2 + 4y^2 = 12$

14. _____

14. $16x^2 = 144 + 9y^2$

15. _____

15. $4y^2 + 4x = 9$

Solve each nonlinear system.

16. _____

16. $2x - y = 9$
$xy = 5$

17. _____

17. $x - 4 = 3y$
$x^2 + y^2 = 8$

18. _____

18. $x^2 + y^2 = 25$
$x^2 - 2y^2 = 16$

19.

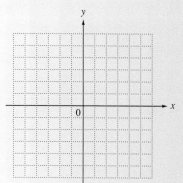

19. Graph the inequality $y \leq x^2 - 2$.

20.

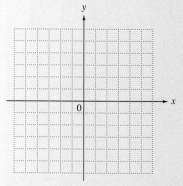

20. Graph the system

$$x^2 + 25y^2 \leq 25$$
$$x^2 + y^2 \leq 9.$$

1. Simplify $-10 + |-5| - |3| + 4$.

Solve.

2. $4 - (2x + 3) + x = 5x - 3$

3. $-4k + 7 \geq 6k + 1$

4. $|5m| - 6 = 14$

5. $|2p - 5| > 15$

6. Find the slope of the line through $(2, 5)$ and $(-4, 1)$.

7. Find the equation of the line through $(-3, -2)$ and perpendicular to $2x - 3y = 7$.

Perform the indicated operations.

8. $(5y - 3)^2$

9. $(2r + 7)(6r - 1)$

10. $(8x^4 - 4x^3 + 2x^2 + 13x + 8) \div (2x + 1)$

Factor.

11. $12x^2 - 7x - 10$ **12.** $2y^4 + 5y^2 - 3$ **13.** $z^4 - 1$ **14.** $a^3 - 27b^3$

Simplify.

15. $\dfrac{5x - 15}{24} \cdot \dfrac{64}{3x - 9}$ **16.** $\dfrac{y^2 - 4}{y^2 - y - 6} \div \dfrac{y^2 - 2y}{y - 1}$

17. $\dfrac{5}{c + 5} - \dfrac{2}{c + 3}$ **18.** $\dfrac{p}{p^2 + p} + \dfrac{1}{p^2 + p}$

19. Kareem and Jamal want to clean their office. Kareem can do the job alone in 3 hours, while Jamal can do it alone in 2 hours. How long will it take them if they work together?

Simplify. Assume all variables represent positive real numbers.

20. $\left(\dfrac{4}{3}\right)^{-1}$ **21.** $\dfrac{(2a)^{-2}a^4}{a^{-3}}$ **22.** $4\sqrt[3]{16} - 2\sqrt[3]{54}$

23. $\dfrac{3\sqrt{5x}}{\sqrt{2x}}$ **24.** $\dfrac{5 + 3i}{2 - i}$

Solve.

25. $2\sqrt{k} = \sqrt{5k + 3}$

26. $10q^2 + 13q = 3$

27. $(4x - 1)^2 = 8$

28. $3k^2 - 3k - 2 = 0$

29. $2(x^2 - 3)^2 - 5(x^2 - 3) = 12$

30. $F = \dfrac{kwv^2}{r}$ for v

Graph.

31. $f(x) = -3x + 5$

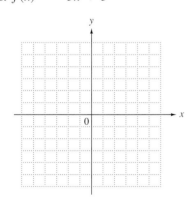

32. $f(x) = -2(x - 1)^2 + 3$

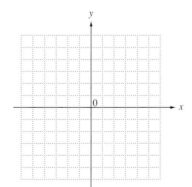

33. $f(x) = \sqrt{x - 2}$

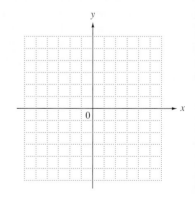

34. $\dfrac{x^2}{4} - \dfrac{y^2}{16} = 1$

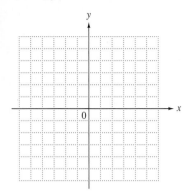

35. $\dfrac{x^2}{25} + \dfrac{y^2}{16} \leq 1$

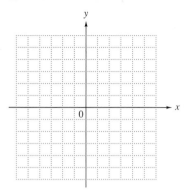

Solve each system.

36. $3x - y = 12$
$2x + 3y = -3$

37. $x + y - 2z = 9$
$2x + y + z = 7$
$3x - y - z = 13$

38. $xy = -5$
$2x + y = 3$

Solve each problem.

39. Al and Bev traveled from their apartment to a picnic 20 miles away. Al traveled on his bike; Bev, who left later, took her car. Al's average speed was half of Bev's average speed. The trip took Al 1/2 hour longer than Bev. What was Bev's average speed?

40. A cash drawer contains only fives and twenties. There are eight more fives than twenties. The total value of the money is $215. How many of each type of bill are in the drawer?

Exponential and Logarithmic Functions

10.1 Inverse Functions

In this chapter we will study two important types of functions, *exponential* and *logarithmic*. These functions are related in a special way. They are *inverses* of one another. We begin by discussing inverse functions in general.

> **Note**
> A calculator with the following keys will be very helpful in this chapter.
>
> $\boxed{y^x}$, $\boxed{10^x}$ or $\boxed{\log x}$, $\boxed{e^x}$ or $\boxed{\ln x}$
>
> We will explain how these keys are used at appropriate places in the chapter.

OBJECTIVE 1 Suppose G is the function $\{(-2, 2), (-1, 1), (0, 0), (1, 3), (2, 5)\}$. We can form another set of ordered pairs from G by exchanging the x- and y-values of each pair in G. Call this set F, with

$$F = \{(2, -2), (1, -1), (0, 0), (3, 1), (5, 2)\}.$$

To show that these two sets are related, F is called the *inverse* of G. For a function to have an inverse that is also a function, the given function must be *one-to-one*. In a **one-to-one function** each x-value corresponds to only one y-value and each y-value corresponds to only one x-value.

The function shown in Figure 1(a) is not one-to-one because the y-value 7 corresponds to *two* x-values, 2 and 3. That is, the ordered pairs $(2, 7)$ and $(3, 7)$ both appear in the function. The function in Figure 1(b) is one-to-one.

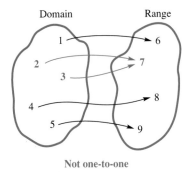
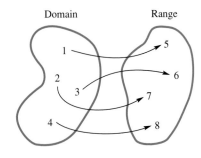

(a)   (b)

FIGURE 1

OBJECTIVES

1. ▶ Decide whether a function is one-to-one and, if it is, find its inverse.

2. ▶ Use the horizontal line test to determine whether a function is one-to-one.

3. ▶ Find the equation of the inverse of a function.

4. ▶ Graph the inverse f^{-1} from the graph of f.

FOR EXTRA HELP

Tutorial   Tape 15   SSM, Sec. 10.1

1. Decide whether or not each function is one-to-one. If it is, find the inverse.

(a) {(1, 2), (2, 4), (3, 3), (4, 5)}

(b) {(0, 3), (−1, 2), (1, 3)}

(c) How Far Can You See from a Plane?

Height	Miles
5,000	87
15,000	149
20,000	172
35,000	228

The *inverse* (IN-vers) of any one-to-one function f is found by exchanging the components of the ordered pairs of f. The inverse of f is written f^{-1}. Read f^{-1} as "the inverse of f" or "f-inverse." The definition of the inverse of a function follows.

> The **inverse** of a one-to-one function f, written f^{-1}, is the set of all ordered pairs of the form (y, x), where (x, y) belongs to f.

Caution
The symbol $f^{-1}(x)$ does not represent $\dfrac{1}{f(x)}$.

Because we form the inverse by interchanging x and y, the domain of f becomes the range of f^{-1}, and the range of f becomes the domain of f^{-1}.

E X A M P L E   1   Deciding Whether a Function Is One-to-One

Decide whether each function is one-to-one. If it is, find its inverse.

(a) $F = \{(-2, 1), (-1, 0), (0, 1), (1, 2), (2, 2)\}$

Each x-value in F corresponds to just one y-value. However, the y-value 2 corresponds to two x-values, 1 and 2. Also, the y-value 1 corresponds to both -2 and 0. Because some y-values correspond to more than one x-value, F is not one-to-one and does not have an inverse.

(b) $G = \{(3, 1), (0, 2), (2, 3), (4, 0)\}$

Every x-value in G corresponds to only one y-value, and every y-value corresponds to only one x-value, so G is a one-to-one function. The inverse function is found by exchanging the numbers in each ordered pair.

$$G^{-1} = \{(1, 3), (2, 0), (3, 2), (0, 4)\}$$

Notice how the domain and range of G become the range and domain, respectively, of G^{-1}.

(c) Let f be the function defined by the correspondence in the table. Then f is not one-to-one because 16 minutes corresponds to vacuuming and bicycling. Also, 13 minutes corresponds to jogging and skiing.

Minutes Needed to Burn 100 Calories by Exercising	
vacuuming	16
walking	27
jogging	13
running	6
skiing	13
bicycling	16

◀◀ **WORK PROBLEM 1 AT THE SIDE.**

ANSWERS

1. (a) {(2, 1), (4, 2), (3, 3), (5, 4)}
  (b) not a one-to-one function
  (c)

Miles	Height
87	5,000
149	15,000
172	20,000
228	35,000

OBJECTIVE 2 It may be difficult to decide whether a function is one-to-one just by looking at the equation that defines the function. However, by graphing the function and observing the graph, we can use the following *horizontal line test* to tell whether the function is one-to-one.

Horizontal Line Test

A function is one-to-one if every horizontal line intersects the graph of the function at most once.

The horizontal line test follows from the definition of a one-to-one function. Any two points that lie on the same horizontal line have the same y-coordinate. No two ordered pairs that belong to a one-to-one function may have the same y-coordinate, and therefore no horizontal line will intersect the graph of a one-to-one function more than once.

E X A M P L E 2   Using the Horizontal Line Test

Use the horizontal line test to determine whether the graphs in Figures 2 and 3 are graphs of one-to-one functions.

(a)

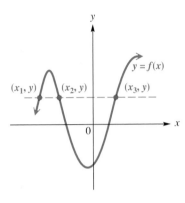

FIGURE 2

Because the horizontal line shown in Figure 2 intersects the graph in more than one point (actually three points in this case), the function is not one-to-one.

(b)

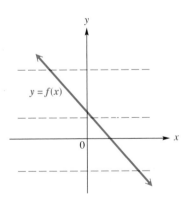

FIGURE 3

Every horizontal line will intersect the graph in Figure 3 in exactly one point. This function is one-to-one.

WORK PROBLEM 2 AT THE SIDE. ▶▶

OBJECTIVE 3 By definition, the inverse of a function is found by exchanging the x- and y-values of the ordered pairs of the function, which reverses the correspondence. The equation of the inverse of a function defined by $y = f(x)$ is found in the same way.

2. Use the horizontal line test to determine whether each graph is the graph of a one-to-one function.

(a)

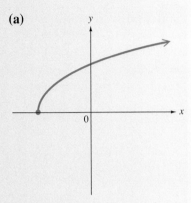

(b)

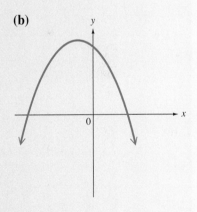

3. Decide whether each equation defines a one-to-one function. If so, find the equation of the inverse.

(a) $f(x) = 3x - 4$

(b) $f(x) = x^3 + 1$

(c) $f(x) = (x - 3)^2$

Inverse of $y = f(x)$

For a one-to-one function f defined by an equation $y = f(x)$, find the defining equation of the inverse as follows.

Step 1 Exchange x and y.

Step 2 Solve for y.

Step 3 Replace y with $f^{-1}(x)$.

This procedure is illustrated in the following example.

E X A M P L E 3 Finding the Equation of the Inverse

Decide whether each of the following defines a one-to-one function. If so, find the equation of the inverse.

(a) $f(x) = 2x + 5$

This linear function has a straight line graph, so by the horizontal line test it is a one-to-one function. Use the steps given above to find the inverse. Let $y = f(x)$ so that

$$y = 2x + 5.$$

Step 1 $x = 2y + 5$ Exchange x and y.

Step 2 $2y = x - 5$

$$y = \frac{x - 5}{2}$$ Solve for y.

Step 3 $f^{-1}(x) = \frac{x - 5}{2}.$ Replace y with $f^{-1}(x)$.

In the function defined by $y = 2x + 5$, we find the value of y by starting with a value for x, multiplying by 2, and adding 5. The inverse function has us *subtract* 5 and then *divide* by 2. This shows how an inverse "undoes" what a function does to the variable x.

(b) $f(x) = (x - 2)^3$

Because of the cube, each value of x produces just one value of y, so this is a one-to-one function. We find the inverse by replacing $f(x)$ with y and then exchanging x and y.

$$y = (x - 2)^3$$
$$x = (y - 2)^3$$ Exchange x and y.
$$\sqrt[3]{x} = \sqrt[3]{(y - 2)^3}$$ Take cube roots.
$$\sqrt[3]{x} = y - 2$$
$$\sqrt[3]{x} + 2 = y$$ Solve for y.
$$f^{-1}(x) = \sqrt[3]{x} + 2$$ Replace y with $f^{-1}(x)$.

(c) $f(x) = x^2 + 2$

Both $x = 3$ and $x = -3$ correspond to $y = 11$. Because of the x^2 term, there are many pairs of x-values that correspond to the same y-value. This means that the function defined by $f(x) = x^2 + 2$ is not one-to-one.

◀◀ WORK PROBLEM 3 AT THE SIDE.

┌───

E X A M P L E 4   **Finding $f(a)$ and $f^{-1}(a)$**
Where a Is a Constant

Let $f(x) = x^3$. Find the following.

(a) $f(-2) = (-2)^3 = -8$

(b) $f(3) = 3^3 = 27$

(c) $f^{-1}(-8)$

From part (a), $(-2, -8)$ belongs to the function f. Because f is one-to-one, $(-8, -2)$ belongs to f^{-1}, with $f^{-1}(-8) = -2$.

(d) $f^{-1}(27)$

Because $(3, 27)$ belongs to f, it follows that $(27, 3)$ belongs to f^{-1} and $f^{-1}(27) = 3$.

───

WORK PROBLEM 4 AT THE SIDE. ▶▶

OBJECTIVE ▶ Suppose the point (a, b) shown in Figure 4 belongs to a one-to-one function f. Then the point (b, a) belongs to f^{-1}. The line segment connecting (a, b) and (b, a) is perpendicular to, and cut in half by, the line $y = x$. The points (a, b) and (b, a) are "mirror images" of each other with respect to $y = x$. For this reason we can find the graph of f^{-1} from the graph of f by locating the mirror image of each point of f with respect to the line $y = x$.

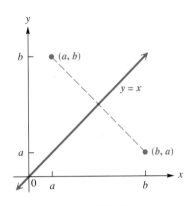

FIGURE 4

4. Find the following for
$$f(x) = \sqrt[3]{1 - x}.$$

(a) $f(2)$

(b) $f(9)$

(c) $f^{-1}(-1)$

(d) $f^{-1}(-2)$

5. Use the given graphs to graph each inverse.

(a)

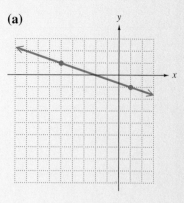

(b)

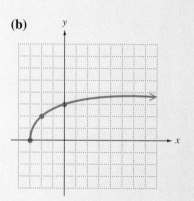

(c)

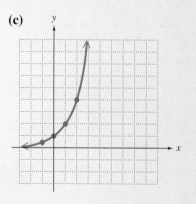

ANSWERS

5. (a)   **(b)**

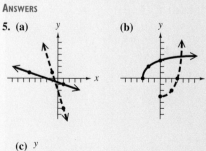

(c)

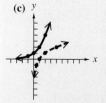

E X A M P L E 5   Finding the Equation of the Inverse

Graph the inverses of the functions shown in Figure 5.

In Figure 5 the graphs of two functions are shown in **blue**. Their inverses are shown in **red**.

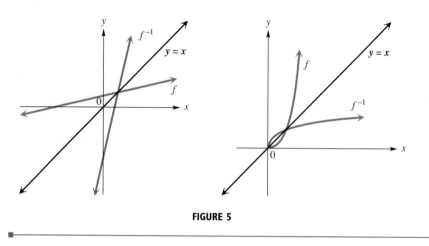

FIGURE 5

◀◀ WORK PROBLEM 5 AT THE SIDE.

10.1 Exercises

1. Is the function defined by $f(x) = x^2$ a one-to-one function?

2. Does the function defined by $f(x) = x^3$ have an inverse? If so, is it defined by $f^{-1}(x) = \sqrt[3]{x}$?

3. If a function f is one-to-one and the point (a, b) lies on the graph of f, which one of the following *must* lie on the graph of f^{-1}?
  (a) (b, a) **(b)** $(-a, b)$ **(c)** $(-b, -a)$ **(d)** $(a, -b)$

4. Suppose that f is a one-to-one function.
  (a) If $f(3) = 5$, then $f^{-1}(5) = $ _____ .
  (b) If $f(-2) = 4$, then $f^{-1}(4) = $ _____ .
  (c) If $f(-19) = 3$, then $f^{-1}(3) = $ _____ .
  (d) If $f(a) = b$, then $f^{-1}(b) = $ _____ .

5. Suppose you consider the set of ordered pairs (x, y) such that x represents a person in your mathematics class and y represents that person's mother. Explain how this function might not be a one-to-one function.

6. The road mileage between Denver, Colorado, and several selected U.S. cities is shown in the table below.

City	Distance to Denver in Miles
Atlanta	1398
Dallas	781
Indianapolis	1058
Kansas City, MO	600
Los Angeles	1059
San Francisco	1235

If we consider this as a function that pairs a city with a distance, is it a one-to-one function? How could we change the answer to this question by adding 1 mile to one of the distances shown?

If the function is one-to-one, find the inverse. See Examples 1–3.

7. $\{(3, 6), (2, 10), (5, 12)\}$

8. $\{(-1, 3), (0, 5), (5, 0), (7, -\frac{1}{2})\}$

9. $\{(-1, 3), (2, 7), (4, 3), (5, 8)\}$

10. $\{(-8, 6), (-4, 3), (0, 6), (5, 10)\}$

11. $f(x) = 2x + 4$

12. $f(x) = 3x + 1$

13. $g(x) = \sqrt{x - 3}, x \geq 3$

14. $g(x) = \sqrt{x + 2}, x \geq -2$ **15.** $f(x) = 3x^2 + 2$ **16.** $f(x) = -4x^2 - 1$

17. $f(x) = x^3 - 4$ **18.** $f(x) = x^3 - 3$

19. If $m \neq 0$, does the linear function defined by $f(x) = mx + b$ have an inverse? Explain.

20. Does the function defined by $f(x) = x^2$ have an inverse? Explain.

Let $f(x) = 2^x$. We will see in the next section that this function is one-to-one. Find each of the following, always working part (a) before part (b). See Example 4.

21. (a) $f(3)$ **22. (a)** $f(4)$ **23. (a)** $f(0)$ **24. (a)** $f(-2)$
    (b) $f^{-1}(8)$ **(b)** $f^{-1}(16)$ **(b)** $f^{-1}(1)$ **(b)** $f^{-1}(\frac{1}{4})$

The graphs of some functions are given in Exercises 25–30. (a) Use the horizontal line test to determine whether the function is one-to-one. (b) If the function is one-to-one, graph the inverse of the function with a dashed line (or curve) on the same set of axes. (Remember that if f is one-to-one and $f(a) = b$, then $f^{-1}(b) = a$.) See Examples 2 and 5.

25.

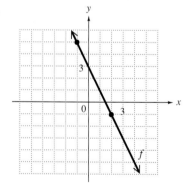

26.

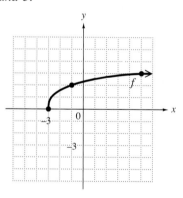

27.

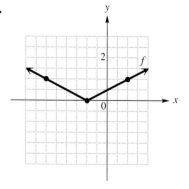

28.

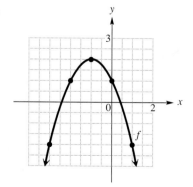

29.

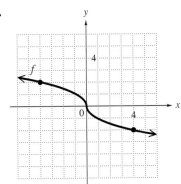

30.

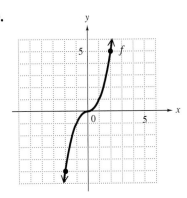

Each function in Exercises 31–38 is a one-to-one function. Graph the function as a solid line (or curve); then graph its inverse on the same set of axes as a dashed line (or curve). In Exercises 35–38 you are given a table to complete so that graphing the function will be a bit easier. See Example 5.

31. $f(x) = 2x - 1$

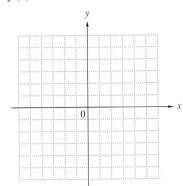

32. $f(x) = 2x + 3$

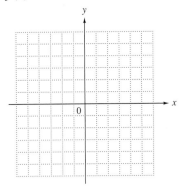

33. $g(x) = -4x$

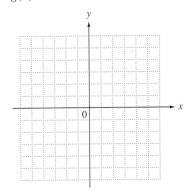

34. $g(x) = -2x$

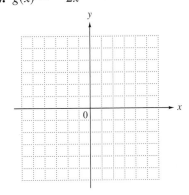

35. $f(x) = \sqrt{x}, x \geq 0$

x	$f(x)$
0	
1	
4	

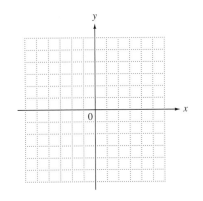

36. $f(x) = -\sqrt{x}, x \geq 0$

x	$f(x)$
0	
1	
4	

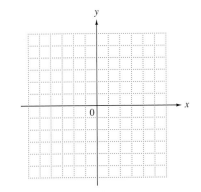

37. $y = x^3 - 2$

x	y
-1	
0	
1	
2	

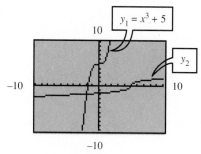

38. $y = x^3 + 3$

x	y
-2	
-1	
0	
1	

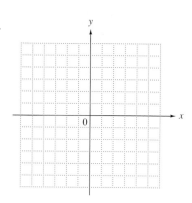

INTERPRETING TECHNOLOGY (EXERCISES 39–41)

The screens show the graphs of a pair of functions. In each case, decide whether the relations are inverse functions, and if they are, determine the expression that defines y_2.

39.

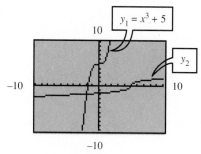

$y_1 = x^3 + 5$

y_2

40.

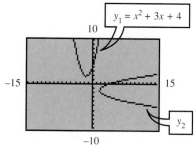

$y_1 = x^2 + 3x + 4$

y_2

41. Explain why the "inverse" of the function in Exercise 40 does not actually satisfy the definition of inverse as given in this section.

---

42. Which one of the following graphs does not have an inverse as defined in this section?

(a)

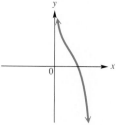

(b)

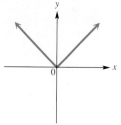

10.2 Exponential Functions

OBJECTIVE Earlier, expressions such as 2^x were evaluated for rational values of x. For example,

$$2^3 = 8$$
$$2^{-1} = \frac{1}{2}$$
$$2^{1/2} = \sqrt{2}$$
$$2^{3/4} = \sqrt[4]{2^3} = \sqrt[4]{8}.$$

In more advanced courses it is shown that 2^x exists for all real-number values of x, both rational and irrational. (Later in the chapter, we will see how to approximate the value of 2^x for irrational x.) With this assumption, we can now define an exponential function.

OBJECTIVES

 Identify exponential functions.

Graph exponential functions.

Solve exponential equations of the form $a^x = a^k$ for x.

Use exponential functions in applications.

FOR EXTRA HELP

Tutorial   Tape 15   SSM, Sec. 10.2

Exponential Function

For $a > 0$, $a \neq 1$, and all real numbers x,

$$f(x) = a^x$$

defines an **exponential function.**

> **Note**
> The two restrictions on a in the definition of an exponential function are important. The restriction that a must be positive is necessary so that the function can be defined for all real numbers x. For example, letting a be negative ($a = -2$, for instance) and letting $x = 1/2$ would give the expression $(-2)^{1/2}$, which is not real. The other restriction, $a \neq 1$, is necessary because 1 raised to any power is equal to 1, and the function would then be the linear function $f(x) = 1$.

OBJECTIVE We can graph an exponential function as we do other functions, by finding several ordered pairs that belong to the function. Plotting these points and connecting them with a smooth curve gives the graph.

EXAMPLE 1   Graphing an Exponential Function

Graph the exponential function $f(x) = 2^x$.

We choose values of x and find the corresponding values of y.

x	-3	-2	-1	0	1	2	3	4
$y = 2^x$	$\frac{1}{8}$	$\frac{1}{4}$	$\frac{1}{2}$	1	2	4	8	16

Plotting these points and drawing a smooth curve through them gives the graph shown in Figure 6. This graph is typical of the graphs of exponential functions of the form $f(x) = a^x$, where $a > 1$. The larger the value of a, the faster the graph rises. To see this, compare the graph of $F(x) = 5^x$ with the graph of $f(x) = 2^x$ in Figure 6.

CONTINUED ON NEXT PAGE

1. Graph.

(a) $f(x) = 10^x$

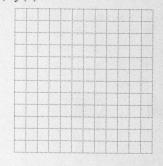

(b) $g(x) = \left(\dfrac{1}{4}\right)^x$

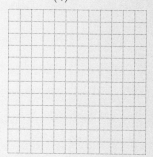

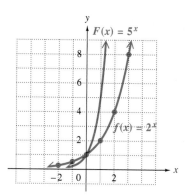

FIGURE 6

By the vertical line test, the graphs in Figure 6 represent functions. As these graphs suggest, the domain of an exponential function includes all real numbers. Because y is always positive, the range is $(0, \infty)$. Figure 6 also shows an important characteristic of exponential functions where $a > 1$; as x gets larger, y increases at a faster and faster rate.

Caution
Be sure to plot enough points to see how rapidly the graph rises.

E X A M P L E   2   Graphing an Exponential Function

Graph $g(x) = \left(\dfrac{1}{2}\right)^x$.

Again, find some points on the graph.

x	-3	-2	-1	0	1	2	3
$y = \left(\dfrac{1}{2}\right)^x$	8	4	2	1	$\dfrac{1}{2}$	$\dfrac{1}{4}$	$\dfrac{1}{8}$

The graph, shown in Figure 7, is the graph of a function. The graph is very similar to that of $f(x) = 2^x$, shown in Figure 6, with the same domain and range, except that here as x gets larger, y *decreases*. This graph is typical of the graph of a function of the form $f(x) = a^x$, where $0 < a < 1$.

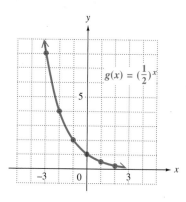

FIGURE 7

ANSWERS

1. (a)

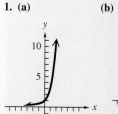

(b)

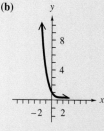

Characteristics of the Graph of $f(x) = a^x$

1. The graph contains the point $(0, 1)$.
2. When $a > 1$, the graph will *rise* from left to right. When $0 < a < 1$, the graph will *fall* from left to right. In both cases, the graph goes from the second quadrant to the first.
3. The graph will approach the x-axis, but never touch it. (It is an asymptote.)
4. The domain is $(-\infty, \infty)$, and the range is $(0, \infty)$.

To graph a more complicated exponential function, we plot carefully selected points, as shown in the next example.

2. Graph $y = 2^{4x-3}$.

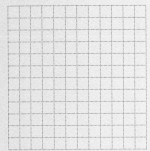

E X A M P L E 3 Graphing an Exponential Function

Graph $f(x) = 3^{2x-4}$.

Find some ordered pairs.

$$\text{If } x = 0, \quad y = 3^{2(0)-4} = 3^{-4} = \frac{1}{81}.$$

$$\text{If } x = 2, \quad y = 3^{2(2)-4} = 3^0 = 1.$$

These ordered pairs, $(0, \frac{1}{81})$ and $(2, 1)$, and other ordered pairs, are shown in the table next to Figure 8. The graph is similar to the graph of $f(x) = 2^x$ except that it is shifted to the right and rises more rapidly.

x	y
0	$\frac{1}{81}$
1	$\frac{1}{9}$
2	1
3	9

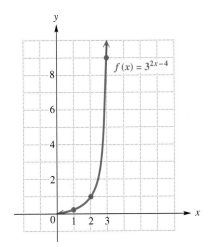

FIGURE 8

WORK PROBLEM 2 AT THE SIDE. ▶▶

OBJECTIVE 3 Until now, we have solved only equations that had the variable as a base; all exponents have been constants. An **exponential equation** is an equation that has a variable in an exponent, such as

$$9^x = 27.$$

Because the exponential function $f(x) = a^x$ is a one-to-one function, we can use the following property to solve many exponential equations.

For $a > 0$ and $a \neq 1$, if $a^x = a^y$ then $x = y$.

This property would not necessarily be true if $a = 1$.

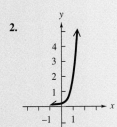

3. Solve. Check your answers.

(a) $25^x = 125$

(b) $4^x = 32$

(c) $81^p = 27$

To solve an exponential equation using this property, go through the following steps.

Solving Exponential Equations

Step 1 Write each side as a power of the same base.

Step 2 If necessary, use the rules of exponents to simplify the exponents.

Step 3 Set the exponents equal.

Step 4 Solve the equation from Step 3.

Note

The steps above cannot be applied to an exponential equation like

$$3^x = 12$$

because Step 1 cannot easily be done. A method for solving such equations is given in Section 10.6.

E X A M P L E   4   Solving an Exponential Equation

Solve $5^x = 125$.

$$5^x = 125$$

Step 1	$5^x = 5^3$	Get the same base.
Step 3	$x = 3$	Set the exponents equal.

Steps 2 and 4 were not needed here. Check by substituting in the given equation:

$$5^x = 125$$
$$5^3 = 125 \quad ?$$
$$125 = 125 \quad \text{True}$$

The solution set is {3}.

E X A M P L E   5   Solving an Exponential Equation

Solve $9^x = 27$.

Step 1	$(3^2)^x = 3^3$	Get the same base.
Step 2	$3^{2x} = 3^3$	Simplify exponents.
Step 3	$2x = 3$	Set exponents equal.
Step 4	$x = \dfrac{3}{2}$	Solve.

Check that the solution set is $\{\frac{3}{2}\}$ by substituting $\frac{3}{2}$ for x in the given equation.

◄◄ **WORK PROBLEM 3 AT THE SIDE.**

OBJECTIVE ▶ Exponential equations frequently occur in applications describing growth or decay of some quantity. In particular, they are used to describe the growth and decay of populations.

ANSWERS

3. (a) $\left\{\frac{3}{2}\right\}$ **(b)** $\left\{\frac{5}{2}\right\}$ **(c)** $\left\{\frac{3}{4}\right\}$

EXAMPLE 6 Solving a Growth and Decay Problem

The air pollution, y, in appropriate units, in a large industrial city has been growing according to the equation

$$y = 1000(2)^{.3x}$$

where x is time in years from 1985. That is, $x = 0$ represents 1985, $x = 2$ represents 1987, and so on.

(a) Find the amount of pollution in 1985.

Let $x = 0$, and solve for y.

$$\begin{aligned} y &= 1000(2)^{.3x} \\ &= 1000(2)^{(.3)(0)} \qquad \text{Let } x = 0. \\ &= 1000(2)^0 \\ &= 1000(1) \\ &= 1000 \end{aligned}$$

The pollution in 1985 was 1000 units.

(b) Assuming that growth continued at the same rate, estimate the pollution in 1995.

Here, $x = 10$ represents 1995.

$$\begin{aligned} y &= 1000(2)^{.3x} \\ &= 1000(2)^{(.3)(10)} \qquad \text{Let } x = 10. \\ &= 1000(2)^3 \\ &= 1000(8) \\ &= 8000 \end{aligned}$$

In 1995 the pollution was about 8000 units.

(c) Graph $y = 1000(2)^{.3x}$.

The scale on the y-axis must be quite large to allow for the very large y-values. A calculator can be used to find a few more ordered pairs. The y^x (or x^y) key on the calculator is used to find values of numbers to a variable power. For example, to find y if $x = 15$ the equation gives $y = 1000(2)^{4.5}$. To evaluate $2^{4.5}$ with a calculator, touch 2, then the $\boxed{y^x}$ key, then touch $\boxed{4.5}$, then the $\boxed{\text{Enter}}$ or $\boxed{=}$ key. You should get $2^{4.5} \approx 22.6$, so $y \approx 1000(22.6) = 22,600$. The graph is shown in Figure 9.

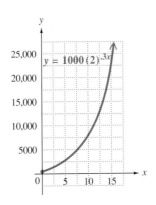

FIGURE 9

WORK PROBLEM 4 AT THE SIDE. ▶▶

4. The amount of a radioactive substance, in grams, present at time t is $A(t) = 100(3)^{-.5t}$, where t is in months. Find the amount present at

(a) $t = 0$;

(b) $t = 2$;

(c) $t = 10$.

(d) Graph the equation.

ANSWERS

4. (a) 100 grams **(b)** $33\frac{1}{3}$ grams

(c) .41 gram
(d)

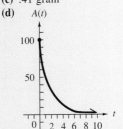

Numbers in the
Real World
a graphing calculator minicourse

Lesson 7: Solving a Nonlinear System of Equations

In Section 9.5 we saw how nonlinear systems of equations may be solved by elimination, substitution, or a combination of the methods. We can support our solutions graphically by graphing the equations that make up the system, and then finding the point(s) of intersection of the graphs. Suppose that we wish to solve the system found in Exercise 7 of Section 9.5 using graphical methods. The system is $y = x^2 + 6x + 9$, $x + y = 3$. The first equation has a parabola as its graph, and the second has a line as its graph. If we graph $y_1 = x^2 + 6x + 9$ and $y_2 = -x + 3$ on the same set of axes, we see that one point of intersection is $(-6, 9)$. The other is $(-1, 4)$. See the first figure. These are the same solutions that we obtain if we use an algebraic method.

A nonlinear system such as the one found in Example 1 of Section 9.5 is a bit more difficult to solve graphically, since the first equation of the system has a circle as its graph. A circle does not represent the graph of a function, so it will be necessary to solve $x^2 + y^2 = 9$ for y as follows: $x^2 + y^2 = 9$, $y^2 = 9 - x^2$, $y = \pm\sqrt{9 - x^2}$.

If we graph $y_1 = \sqrt{9 - x^2}$, $y_2 = -\sqrt{9 - x^2}$, and $y_3 = 2x - 3$, we can use the intersection-of-graphs capability of the calculator to find the solution set of the system $x^2 + y^2 = 9$, $2x - y = 3$. As seen in the second figure, one point of intersection is $(2.4, 1.8)$. The other is $(0, -3)$. This supports the algebraic results found earlier.

(In order to obtain an accurate depiction of a circle on a graphing calculator screen, we must use a "square" window. Consult your owner's manual for details.)

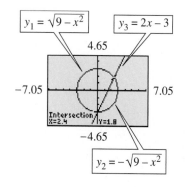

Graphing Calculator Explorations

1. The nonlinear system $y = x^2 + 6x$, $3y = 12x$ appears in Exercise 6 of Section 9.5.

   (a) Solve the system algebraically.

   (b) Graph the first equation as $y_1 = x^2 + 6x$ and the second as $y_2 = 4x$. Use the intersection-of-graphs capability of your calculator to support your algebraic solution from part (a).

2. The nonlinear system $x^2 + y^2 = 4$, $y = x^2 - 2$ appears in Exercise 18 of Section 9.5.

   (a) Solve the system algebraically. (*Hint:* There are three solutions, and two of them have irrational x-coordinates.)

   (b) Graph the circle $x^2 + y^2 = 4$ by graphing $y_1 = \sqrt{4 - x^2}$ and $y_2 = -\sqrt{4 - x^2}$. Use a square viewing window. Then graph the parabola $y_3 = x^2 - 2$. Support your results of part (a) using intersection of graphs.

   (c) Show that the irrational x-coordinates have decimal approximations equal to the values you find when you use the square root key of your calculator.

10.2 Exercises

Fill in the blanks with the correct responses.

1. For an exponential function $f(x) = a^x$, if $a > 1$, the graph _____ from left to right.
$\underset{\text{(rises/falls)}}{}$

If $0 < a < 1$, the graph _____ from left to right.
$\underset{\text{(rises/falls)}}{}$

2. The y-intercept of the graph of $y = a^x$ is _____ .

3. The graph of the exponential function $y = a^x$ _____ have an x-intercept.
$\underset{\text{(does/does not)}}{}$

4. The point $(2, \underline{\hspace{0.5cm}})$ is on the graph of $f(x) = 3^{4x-3}$.

Graph the following exponential functions. See Examples 1–3.

5. $f(x) = 3^x$

6. $f(x) = 5^x$

7. $g(x) = \left(\dfrac{1}{3}\right)^x$

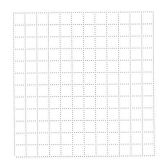
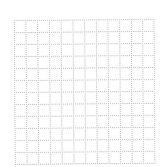
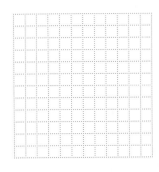

8. $g(x) = \left(\dfrac{1}{5}\right)^x$

9. $y = 2^{2x-2}$

10. $y = 2^{2x+1}$

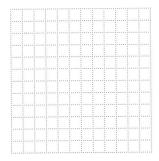
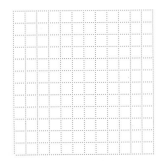
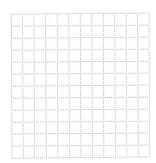

11. Based on your answer to Exercise 1, make a conjecture (an educated guess) about whether an exponential function $f(x) = a^x$ is one-to-one. Then decide whether it has an inverse based on the concepts of Section 10.1.

12. What is the domain of an exponential function $f(x) = a^x$? What is its range?

Solve the following equations. See Examples 4 and 5.

13. $6^x = 36$

14. $8^x = 64$

15. $100^x = 1000$

16. $8^x = 4$

17. $16^{2x+1} = 64^{x+3}$

18. $9^{2x-8} = 27^{x-4}$

19. $5^x = \dfrac{1}{125}$

20. $3^x = \dfrac{1}{81}$

21. $5^x = .2$

22. $10^x = .1$

23. $\left(\dfrac{3}{2}\right)^x = \dfrac{8}{27}$

24. $\left(\dfrac{4}{3}\right)^x = \dfrac{27}{64}$

Solve each of the following applications of exponential functions. Use a calculator with a y^x key. See Example 6.

25. The amount of radioactive material in an ore sample is given by the function

$$A(t) = 100(3.2)^{-.5t},$$

where $A(t)$ is the amount present, in grams, in the sample t months after the initial measurement.

(a) How much was present at the initial measurement? *(Hint: $t = 0$.)*

(b) How much was present 2 months later?

(c) How much was present 10 months later?

26. The population of Brazil, in millions, is approximated by the function

$$f(x) = 155.3(2)^{.025x},$$

where $x = 0$ corresponds to 1994, $x = 1$ corresponds to 1995, and so on.

(a) What will be the population of Brazil in the year 2000 according to this model?

(b) What will be the population in 2034?

(c) How will the population in 2034 compare to the population in 1994?

The bar graph shows the average annual major league baseball player's salary for each year since free agency began. Using a technique from statistics, it was determined that the function

$$S(x) = 74,741(1.17)^x$$

approximates the salary, where $x = 0$ corresponds to 1976, and so on, up to $x = 18$ representing 1994. (Salary is in dollars.)

27. Based on this model, what was the average salary in 1986?

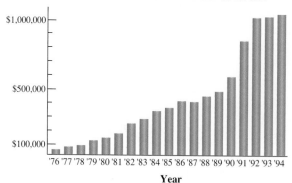

— **AVERAGE ANNUAL BASEBALL SALARY** —

Source: MLBPA

28. Based on the graph, in what year did the average salary first exceed $1,000,000?

29. The accompanying graphing calculator screen shows the function $S(x)$ graphed from $x = 0$ to $x = 20$. Interpret the display at the bottom of the screen.

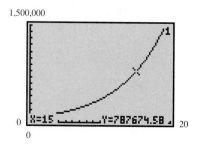

30. In the bar graph, we see that the tops of the bars rise from left to right, and in the calculator-generated graph, the curve rises from left to right. What part of the equation $S(x) = 74,741(1.17)^x$ indicates that during this time period, baseball salaries were *rising*?

10.3 Logarithmic Functions

The graph of $y = 2^x$ is the curve shown in **blue** in Figure 10. Because $y = 2^x$ is a one-to-one function, it has an inverse. By interchanging x and y, we get $x = 2^y$, the inverse of $y = 2^x$. As we saw in Section 10.1, the graph of the inverse is found by reflecting the graph of $y = 2^x$ about the line $y = x$. The graph of $x = 2^y$ is shown as a **red** curve in Figure 10.

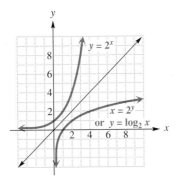

FIGURE 10

OBJECTIVES

1. Define a logarithm.
2. Write exponential statements in logarithmic form and logarithmic statements in exponential form.
3. Solve logarithmic equations of the form $\log_a b = k$ for a, b, or k.
4. Graph logarithmic functions.
5. Use logarithmic functions in applications.

FOR EXTRA HELP

 Tutorial   Tape 15   SSM, Sec. 10.3

OBJECTIVE 1 We cannot solve the equation $x = 2^y$ for the dependent variable y with the methods presented up to now. The following definition is used to solve $x = 2^y$ for y.

> ### Definition of Logarithm
>
> For all positive numbers a, $a \neq 1$, and all positive numbers x,
>
> $$y = \log_a x \text{ means the same as } x = a^y.$$

The abbreviation **log** is used for **logarithm** (LOG-uh-rith-im). Read $\log_a x$ as "the logarithm of x to the base a." **This key statement should be memorized.** To remember the location of the base and exponent in each form, refer to the diagram below.

$$\text{Logarithmic form: } y = \overset{\text{Exponent}}{\underset{\underset{\text{Base}}{\uparrow}}{\log_a}} x$$

$$\text{Exponential form: } x = a^{\overset{\text{Exponent}}{\downarrow} y}$$
$$\underset{\underset{\text{Base}}{\uparrow}}{}$$

When working with logarithmic (log-uh-RITH-mik) form and exponential form, remember the following.

> A logarithm is an exponent; $\log_a x$ is the exponent on the base a that yields the number x.

OBJECTIVE 2 We can use the definition of logarithm to write exponential statements in logarithmic form and logarithmic statements in exponential form.

E X A M P L E I Converting between Exponential and Logarithmic Form

The following list shows several pairs of equivalent statements. The same statement is written in both exponential and logarithmic form.

── **CONTINUED ON NEXT PAGE**

1. Complete the chart.

Exponential Form	Logarithmic Form
$2^5 = 32$	_____
$100^{1/2} = 10$	_____
_____	$\log_8 4 = \dfrac{2}{3}$
_____	$\log_6 \dfrac{1}{1296} = -4$

Exponential Form	Logarithmic Form
$3^2 = 9$	$\log_3 9 = 2$
$\left(\dfrac{1}{5}\right)^{-2} = 25$	$\log_{1/5} 25 = -2$
$10^5 = 100{,}000$	$\log_{10} 100{,}000 = 5$
$4^{-3} = \dfrac{1}{64}$	$\log_4 \dfrac{1}{64} = -3$

◀◀ **WORK PROBLEM 1 AT THE SIDE.**

OBJECTIVE ▶ A **logarithmic equation** is an equation with a logarithm in at least one term. We can solve logarithmic equations of the form $\log_a b = k$ for any of the three variables by first writing the equation in exponential form.

2. Solve each equation.

(a) $\log_3 27 = x$

(b) $\log_5 p = 2$

(c) $\log_m \dfrac{1}{16} = -4$

E X A M P L E   2   **Solving Logarithmic Equations**

Solve the following equations.

(a) $\log_4 x = -2$

$$\log_4 x = -2$$
$$x = 4^{-2} \qquad \text{Convert to exponential form.}$$
$$x = \dfrac{1}{16}$$

The solution set is $\left\{\dfrac{1}{16}\right\}$.

(b) $\log_{1/2} 16 = y$

$$\log_{1/2} 16 = y$$
$$\left(\dfrac{1}{2}\right)^y = 16 \qquad \text{Convert to exponential form.}$$
$$(2^{-1})^y = 2^4 \qquad \text{Write with same base.}$$
$$2^{-y} = 2^4 \qquad \text{Multiply exponents.}$$
$$-y = 4 \qquad \text{Set exponents equal.}$$
$$y = -4 \qquad \text{Solve.}$$

The solution set is $\{-4\}$.

◀◀ **WORK PROBLEM 2 AT THE SIDE.**

For any positive real number b, we know that $b^1 = b$ and $b^0 = 1$. Writing these two statements in logarithmic form gives the following two properties of logarithms.

> For any positive real number b, $b \neq 1$,
>
> $$\log_b b = 1 \quad \text{and} \quad \log_b 1 = 0.$$

E X A M P L E   3   **Using Properties of Logarithms**

(a) $\log_7 7 = 1$ **(b)** $\log_{\sqrt{2}} \sqrt{2} = 1$

(c) $\log_9 1 = 0$ **(d)** $\log_{.2} 1 = 0$

WORK PROBLEM 3 AT THE SIDE. ▶▶

Now we can define the logarithmic function with base a as follows.

Logarithmic Function

If a and x are positive numbers, with $a \neq 1$, then

$$f(x) = \log_a x$$

defines the **logarithmic function with base a.**

OBJECTIVE ▶4▶ To graph a logarithmic function, it is helpful to write it in exponential form first. Then we can plot selected ordered pairs to determine the graph.

E X A M P L E   4   **Graphing a Logarithmic Function**

Graph $y = \log_{1/2} x$.

Writing $y = \log_{1/2} x$ in its exponential form as $x = \left(\frac{1}{2}\right)^y$ helps us identify ordered pairs that satisfy the equation. Here it is easier to choose values for y and find the corresponding values of x. Doing this gives the following pairs.

x	$\frac{1}{4}$	$\frac{1}{2}$	1	2	4	8
y	2	1	0	-1	-2	-3

Be careful to get these in the right order.

Plotting these points and connecting them with a smooth curve, we get the graph in Figure 11. This graph is typical of logarithmic functions with base $0 < a < 1$. The graph of $x = 2^y$ in Figure 10, which is equivalent to $y = \log_2 x$, is typical of graphs of logarithmic functions with base $a > 1$.

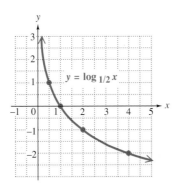

FIGURE 11

WORK PROBLEM 4 AT THE SIDE. ▶▶

Based on the graphs of the functions $y = \log_2 x$ in Figure 10 and $y = \log_{1/2} x$ in Figure 11, we can make the following generalizations about the graphs of logarithmic functions of the form $g(x) = \log_a x$.

Characteristics of the Graph of $g(x) = \log_a x$

1. The graph contains the point $(1, 0)$.
2. When $a > 1$, the graph will *rise* from left to right, from the fourth quadrant to the first. When $0 < a < 1$, the graph will *fall* from left to right, from the first quadrant to the fourth.
3. The graph will approach the y-axis, but never touch it. (It is an asymptote.)
4. The domain is $(0, \infty)$, and the range is $(-\infty, \infty)$.

---

3. Find the value of each of the following.

(a) $\log_{2/5} \dfrac{2}{5}$

(b) $\log_{.4} 1$

4. Graph.

(a) $y = \log_3 x$

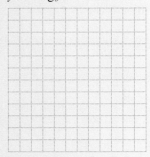

(b) $y = \log_{1/10} x$

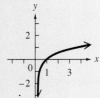

(b)

5. A population of mites in a laboratory is growing so that the population is

$$P(t) = 80 \log_{10}(t + 10),$$

where t is the number of days after a study began. Find $P(t)$ for the following values of t.

(a) $t = 0$

(b) $t = 90$

(c) $t = 990$

(d) Graph the function.

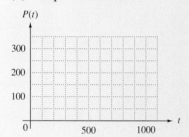

Compare these generalizations to the similar ones for exponential functions in Section 10.2.

OBJECTIVE 5 Logarithmic functions, like exponential functions, are used in applications to describe growth and decay.

E X A M P L E 5 Solving a Logarithmic Growth Problem

Sales (in thousands of units) of a new product are approximated by

$$S(t) = 100 + 30 \log_3(2t + 1),$$

where t is the number of years after the product is introduced.

(a) What were the sales after 1 year?

Let $t = 1$, and find $S(1)$.

$$S(t) = 100 + 30 \log_3(2t + 1)$$
$$S(1) = 100 + 30 \log_3(2 \cdot 1 + 1) \qquad \text{Let } t = 1.$$
$$= 100 + 30 \log_3 3$$
$$= 100 + 30(1) \qquad \log_3 3 = 1$$
$$= 130$$

Sales were 130 thousand units after 1 year.

(b) Find the sales after 13 years.

Find $S(13)$.

$$S(t) = 100 + 30 \log_3(2t + 1)$$
$$S(13) = 100 + 30 \log_3(2 \cdot 13 + 1) \qquad \text{Let } t = 13.$$
$$= 100 + 30 \log_3 27$$
$$= 100 + 30(3) \qquad \log_3 27 = 3$$
$$= 190$$

After 13 years, sales had increased to 190 thousand units.

(c) Graph $y = S(t)$.

Use the two ordered pairs $(1, 130)$ and $(13, 190)$ found above. Check that $(0, 100)$ and $(40, 220)$ also satisfy the equation. Use these ordered pairs and a knowledge of the general shape of the graph of a logarithmic function to get the graph in Figure 12.

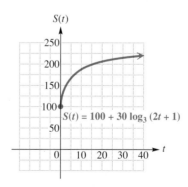

FIGURE 12

◀◀ **WORK PROBLEM 5 AT THE SIDE.**

10.3 Exercises

1. By definition $\log_a x$ is the exponent to which the base a must be raised in order to obtain x. Use this to simplify each of the following, without doing any written work. (Example: $\log_3 9$ is 2, because 2 is the exponent to which 3 must be raised in order to obtain 9.)
   (a) $\log_4 16$ (b) $\log_3 81$ (c) $\log_3(\frac{1}{3})$
   (d) $\log_{10} .01$ (e) $\log_5 \sqrt{5}$ (f) $\log_{12} 1$

2. Compare the summary of facts about the graph of $f(x) = a^x$ in Section 10.2 with the similar summary of facts about the graph of $g(x) = \log_a x$ in this section. Make a list of the facts that reinforce the concept that f and g are inverse functions.

Write in logarithmic form. See Example 1.

3. $4^5 = 1024$

4. $3^6 = 729$

5. $\left(\dfrac{1}{2}\right)^{-3} = 8$

6. $\left(\dfrac{1}{6}\right)^{-3} = 216$

7. $10^{-3} = .001$

8. $36^{1/2} = 6$

Write in exponential form. See Example 1.

9. $\log_4 64 = 3$

10. $\log_2 512 = 9$

11. $\log_{10} \dfrac{1}{10,000} = -4$

12. $\log_{100} 100 = 1$

13. $\log_6 1 = 0$

14. $\log_\pi 1 = 0$

15. When a student asked his teacher to explain to him how to evaluate $\log_9 3$ without showing any work, his teacher told him, "Think radically." Explain what the teacher meant by this hint.

16. A student told her teacher "I know that $\log_2 1$ is the exponent to which 2 must be raised in order to obtain 1, but I can't think of any such number." How would you explain to the student that the value of $\log_2 1$ is 0?

Solve each equation for x. See Examples 2 and 3.

17. $x = \log_{27} 3$

18. $x = \log_{125} 5$

19. $\log_x 9 = \dfrac{1}{2}$

20. $\log_x 5 = \dfrac{1}{2}$

21. $\log_x 125 = -3$

22. $\log_x 64 = -6$

23. $\log_{12} x = 0$

24. $\log_4 x = 0$

25. $\log_x x = 1$

26. $\log_x 1 = 0$

27. $\log_x \dfrac{1}{25} = -2$

28. $\log_x \dfrac{1}{10} = -1$

29. $\log_8 32 = x$

30. $\log_{81} 27 = x$

31. $\log_\pi \pi^4 = x$

32. $\log_{\sqrt{2}} \sqrt{2}^9 = x$

33. $\log_6 \sqrt{216} = x$

34. $\log_4 \sqrt{64} = x$

If the point (p, q) is on the graph of $f(x) = a^x$ (for $a > 0$ and $a \neq 1$), then the point (q, p) is on the graph of $f^{-1}(x) = \log_a x$. Use this fact, and refer to the graphs required in Exercises 5–8 in Section 10.2 to graph the following logarithmic functions. See Example 4.

35. $y = \log_3 x$

36. $y = \log_5 x$

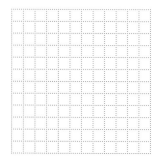

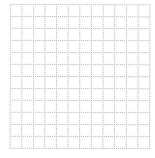

37. $y = \log_{1/3} x$

38. $y = \log_{1/5} x$

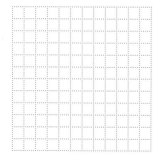

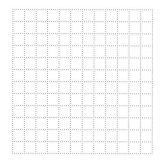

39. Graph the function $y = \log_{2.718} x$ using the exponential key of a calculator, and rewriting it as $2.718^y = x$. Choose -1, 0, and 1 as y values, and approximate x to the nearest tenth.

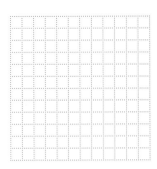

40. Use the exponential key of your calculator to find approximations for the expression $(1 + \frac{1}{x})^x$, using x values of 1, 10, 100, 1000, and 10,000. Explain what seems to be happening as x gets larger and larger. (*Hint*: Look at the base in Exercise 39.)

Solve each application of logarithmic functions. See Example 5.

41. A study showed that the number of mice in an old abandoned house was approximated by the function

$$M(t) = 6 \log_4 (2t + 4),$$

where t is measured in months and $t = 0$ corresponds to January, 1993. Find the number of mice in the house in
(a) January, 1993
(b) July, 1993
(c) July, 1995.
(d) Graph the function.

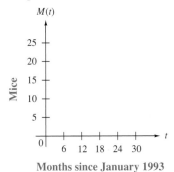

Months since January 1993

42. A supply of hybrid striped bass were introduced into a lake in January, 1980. Biologists researching the bass population over the next decade found that the number of bass in the lake was approximated by the function

$$B(t) = 500 \log_3 (2t + 3),$$

where $t = 0$ corresponds to January, 1980, $t = 1$ to January, 1981, $t = 2$ to January, 1982, and so on. Use this function to find the bass population in
(a) January, 1980
(b) January, 1983
(c) January, 1992.
(d) Graph the function for $0 \le t \le 12$.

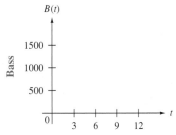

Years since January 1980

Use the graph below to predict the value of $f(t)$ for the given values of t.

43. $t = 0$

44. $t = 10$

45. $t = 60$

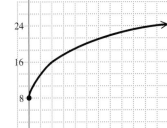

46. Show that the points determined in Exercises 43–45 lie on the graph of $f(t) = 8 \log_5(2t + 5)$.

47. Explain why 1 is not allowed as a base for a logarithmic function.

48. Explain why $\log_a 1$ is 0 for any value of a that is allowed as the base of a logarithm. Use a rule of exponents introduced earlier in your explanation.

49. The domain of $f(x) = a^x$ is $(-\infty, \infty)$, while the range is $(0, \infty)$. Therefore, because $g(x) = \log_a x$ is the inverse of f, the domain of g is _____, while the range of g is _____ .

50. The graphs of both $f(x) = 3^x$ and $g(x) = \log_3 x$ rise from left to right. Which one rises at a faster rate?

In the United States, the intensity of an earthquake is rated using the Richter *scale. The Richter scale rating of an earthquake of intensity x is given by*

$$R = \log_{10} \frac{x}{x_0},$$

where x_0 is the intensity of an earthquake of a certain (small) size. The figure shows Richter scale ratings for major Southern California earthquakes since 1920. As the figure indicates, earthquakes "come in bunches" and the 1990s have been an especially busy time.

51. The 1994 Northridge earthquake had a Richter scale rating of 6.7; the Landers earthquake had a rating of 7.3. How much more powerful was the Landers earthquake than the Northridge earthquake?

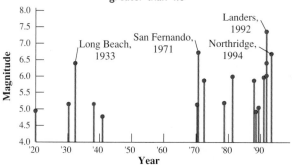

MAJOR SOUTHERN CALIFORNIA EARTHQUAKES

Earthquakes with magnitudes greater than 4.8

Sources: Caltech; U.S. Geological Survey

52. Compare the smallest rated earthquake in the figure (at 4.8) with the Landers quake. How much more powerful was the Landers quake?

10.4 Properties of Logarithms

Logarithms are very important in applications and in further work in mathematics. The properties that make logarithms so useful are given in this section.

OBJECTIVE 1 ▶ One way in which logarithms simplify problems is by changing a problem of multiplication into one of addition. This is done with the following property of logarithms.

Multiplication Property for Logarithms

If x, y, and b are positive real numbers, where $b \neq 1$, then

$$\log_b xy = \log_b x + \log_b y.$$

The logarithm of a product is the sum of the logarithms.

Note
The word statement of the product rule can be restated by replacing "logarithm" with "exponent." The rule then becomes the familiar rule for multiplying exponential expressions: The *exponent* of a product is equal to the sum of the *exponents* of the factors.

To prove this rule, let $m = \log_b x$ and $n = \log_b y$, and recall that

$$\log_b x = m \qquad \text{means} \qquad b^m = x,$$
$$\log_b y = n \qquad \text{means} \qquad b^n = y.$$

Now consider the product xy.

$$xy = b^m \cdot b^n \qquad \text{Substitution}$$
$$xy = b^{m+n} \qquad \text{Product rule for exponents}$$
$$\log_b xy = m + n \qquad \text{Convert to logarithmic form.}$$
$$\log_b xy = \log_b x + \log_b y \qquad \text{Substitution}$$

The last statement is the result we wish to prove.

EXAMPLE I   Using the Multiplication Property

Use the multiplication property to rewrite the following (assume $x > 0$).

(a) $\log_5(6 \cdot 9)$

By the multiplication property,

$$\log_5(6 \cdot 9) = \log_5 6 + \log_5 9.$$

(b) $\log_7 8 + \log_7 12$

$$\log_7 8 + \log_7 12 = \log_7(8 \cdot 12) = \log_7 96$$

(c) $\log_3(3x)$

$$\log_3(3x) = \log_3 3 + \log_3 x$$

Because $\log_3 3 = 1$,

$$\log_3(3x) = 1 + \log_3 x.$$

(d) $\log_4 x^3$

$$\log_4 x^3 = \log_4(x \cdot x \cdot x) \qquad x^3 = x \cdot x \cdot x$$
$$= \log_4 x + \log_4 x + \log_4 x$$
$$= 3 \log_4 x$$

OBJECTIVES

1 ▶ Use the multiplication property for logarithms.

2 ▶ Use the division property for logarithms.

3 ▶ Use the power property for logarithms.

4 ▶ Use the properties of logarithms to write logarithmic expressions in alternative forms.

FOR EXTRA HELP

Tutorial   Tape 15   SSM, Sec. 10.4

1. Use the multiplication property to rewrite the following.

(a) $\log_6(5 \cdot 8)$

(b) $\log_4 3 + \log_4 7$

(c) $\log_8 8k$ $(k > 0)$

(d) $\log_5 m^2$

2. Use the division property to rewrite the following.

(a) $\log_7 \dfrac{9}{4}$

(b) $\log_3 p - \log_3 q$
$(p > 0, q > 0)$

(c) $\log_4 \dfrac{3}{16}$

◀◀ **WORK PROBLEM 1 AT THE SIDE.**

OBJECTIVE ▶ The rule for division is similar to the rule for multiplication.

Division Property for Logarithms

If x, y, and b are positive real numbers, where $b \neq 1$, then

$$\log_b \frac{x}{y} = \log_b x - \log_b y.$$

The logarithm of the quotient is the difference of the logarithms.

The proof of this rule is very similar to the proof of the multiplication property.

E X A M P L E   2   Using the Division Property

Use the division property to rewrite the following.

(a) $\log_4 \dfrac{7}{9} = \log_4 7 - \log_4 9$

(b) If $x > 0$, then $\log_5 6 - \log_5 x = \log_5 \dfrac{6}{x}$.

(c) $\log_3 \dfrac{27}{5} = \log_3 27 - \log_3 5$

$$= 3 - \log_3 5 \qquad \log_3 27 = 3$$

◀◀ **WORK PROBLEM 2 AT THE SIDE.**

OBJECTIVE ▶ The next rule gives a method for evaluating powers and roots such as

$$2^{\sqrt{2}}, \quad (\sqrt{2})^{3/4}, \quad (.032)^{5/8}, \quad \text{and} \quad \sqrt[5]{12}.$$

This rule makes it possible to find approximations for numbers that could not be evaluated before. By the multiplication property for logarithms,

$$\log_5 2^3 = \log_5(2 \cdot 2 \cdot 2)$$
$$= \log_5 2 + \log_5 2 + \log_5 2$$
$$= 3 \log_5 2.$$

Also,

$$\log_2 7^4 = \log_2(7 \cdot 7 \cdot 7 \cdot 7)$$
$$= \log_2 7 + \log_2 7 + \log_2 7 + \log_2 7$$
$$= 4 \log_2 7.$$

Furthermore, we saw in Example 1(d) that $\log_4 x^3 = 3 \log_4 x$. These examples suggest the following generalization.

Power Property for Logarithms

If x and b are positive real numbers, where $b \neq 1$, and r is any real number, then

$$\log_b x^r = r(\log_b x).$$

The logarithm of a number to a power equals the exponent times the logarithm of the number.

As examples of this result,

$$\log_b m^5 = 5 \log_b m \quad \text{and} \quad \log_3 5^{3/4} = \frac{3}{4} \log_3 5.$$

To prove the power property, let

$$\log_b x = m.$$

$b^m = x$	Convert to exponential form.
$(b^m)^r = x^r$	Raise to the power r.
$b^{mr} = x^r$	Power rule for exponents
$\log_b x^r = mr$	Convert to logarithmic form.
$\log_b x^r = rm$	
$\log_b x^r = r \log_b x$	$m = \log_b x$

This is the statement to be proved.

As a special case of this rule, let $r = \frac{1}{p}$, so that $\log_b \sqrt[p]{x} = \log_b x^{1/p} = \frac{1}{p} \log_b x$. For example, using this result with $x > 0$,

$$\log_b \sqrt[5]{x} = \frac{1}{5} \log_b x \quad \text{and} \quad \log_b \sqrt[3]{x^4} = \frac{4}{3} \log_b x.$$

E X A M P L E 3 Using the Power Rule

Use the power rule to rewrite each of the following. Assume $b > 0$, $x > 0$, and $b \neq 1$.

(a) $\log_5 4^2 = 2 \log_5 4$

(b) $\log_b x^5 = 5 \log_b x$

(c) $\log_b \sqrt{7}$

When using the power rule with logarithms of expressions involving radicals, begin by rewriting the radical expression with a rational exponent, as shown earlier.

$$\log_b \sqrt{7} = \log_b 7^{1/2} \qquad \sqrt{x} = x^{1/2}$$

$$= \frac{1}{2} \log_b 7 \qquad \text{Power rule}$$

(d) $\log_2 \sqrt[5]{x^2} = \log_2 x^{2/5} = \frac{2}{5} \log_2 x$

> **WORK PROBLEM 3 AT THE SIDE.** ▶▶

Two special properties involving both exponential and logarithmic expressions come directly from the fact that logarithmic and exponential functions are inverses of each other.

If $b > 0$ and $b \neq 1$, then

$$b^{\log_b x} = x \quad (x > 0) \qquad \text{and} \qquad \log_b b^x = x.$$

To prove the first statement, let

$y = \log_b x.$	
$b^y = x$	Convert to exponential form.
$b^{\log_b x} = x$	Replace y with $\log_b x$.

The proof of the second statement is similar.

---

3. Use the power rule to rewrite the following. Assume $a > 0$, $b > 0$, $a \neq 1$, and $b \neq 1$.

(a) $\log_3 5^2$

(b) $\log_a x^4 \quad (x > 0)$

(c) $\log_b \sqrt{8}$

(d) $\log_2 \sqrt[3]{2}$

3. (a) $2 \log_3 5$ (b) $4 \log_a x$
(c) $\frac{1}{2} \log_b 8$ (d) $\frac{1}{3}$

4. Find the value of each expression.

(a) $\log_{10} 10^3$

(b) $\log_2 8$

(c) $5^{\log_5 3}$

EXAMPLE 4   Using the Special Properties

Find the value of the following logarithmic expressions.

(a) $\log_5 5^4 = 4$, by the second property.

(b) $\log_3 9$

$$\log_3 9 = \log_3 3^2 = 2$$

The second property was used in the last step.

(c) $4^{\log_4 10} = 10$

■

◀◀ WORK PROBLEM 4 AT THE SIDE.

OBJECTIVE ▶ The properties of logarithms are useful for writing expressions in an alternative form. This use of logarithms is important in calculus.

EXAMPLE 5   Writing Logarithms in Alternative Forms

Use the properties of logarithms to rewrite each expression. Assume all variables represent positive real numbers.

(a) $\log_4 4x^3 = \log_4 4 + \log_4 x^3$ Multiplication property

$\qquad\qquad = 1 + 3 \log_4 x$ $\log_4 4 = 1$; Power property

5. Write as a sum or difference of logarithms. Assume all variable expressions are positive.

(a) $\log_6 36m^5$

(b) $\log_7 \sqrt{\dfrac{m}{n}} = \log_7 \left(\dfrac{m}{n}\right)^{1/2}$

$\qquad\qquad = \dfrac{1}{2} \log_7 \dfrac{m}{n}$ Power property

$\qquad\qquad = \dfrac{1}{2}(\log_7 m - \log_7 n)$ Division property

(b) $\log_2 \sqrt{9z}$

(c) $\log_5 \dfrac{a}{bc} = \log_5 a - \log_5 bc$ Division property

$\qquad\qquad = \log_5 a - (\log_5 b + \log_5 c)$ Multiplication property

$\qquad\qquad = \log_5 a - \log_5 b - \log_5 c$

Notice the careful use of parentheses in the second step. Because we are subtracting the logarithm of a product and rewriting it as a sum of two terms, we must place parentheses around the sum.

(d) $\log_8(2p + 3r)$ cannot be rewritten by the properties of logarithms.

■

(c) $\log_q \dfrac{8r}{m-1}$ $(m \neq 1, q \neq 1)$

(d) $\log_4(3x + y)$

◀◀ WORK PROBLEM 5 AT THE SIDE.

Caution
Remember that there is no property of logarithms to rewrite the logarithm of a *sum*. That is,

$$\log_b(x + y) \neq \log_b x + \log_b y.$$

ANSWERS

4. (a) 3 **(b)** 3 **(c)** 3

5. (a) $2 + 5 \log_6 m$ **(b)** $\log_2 3 + \dfrac{1}{2} \log_2 z$

   (c) $\log_q 8 + \log_q r - \log_q (m - 1)$
   (d) cannot be rewritten

10.4 Exercises

Fill in the blanks with the correct responses in Exercises 1–4.

1. The logarithm of the product of two numbers is equal to the _____ of the logarithms of the numbers.

2. The logarithm of the quotient of two numbers is equal to the _____ of the logarithms of the numbers.

3. $\log_a b^k = \underline{\hspace{1cm}} \log_a b \quad (a > 0,\ a \neq 1,\ b > 0)$

4. The logarithm of the square root of a number is equal to _____ times the logarithm of the number.

Use the properties of logarithms introduced in this section to express each of the following as a sum or difference of logarithms, or as a single number if possible. Assume that all variables represent positive real numbers. See Examples 1–5.

5. $\log_7 \dfrac{4}{5}$

6. $\log_8 \dfrac{9}{11}$

7. $\log_2 8^{1/4}$

8. $\log_3 9^{3/4}$

9. $\log_4 \dfrac{3\sqrt{x}}{y}$

10. $\log_5 \dfrac{6\sqrt{z}}{w}$

11. $\log_3 \dfrac{\sqrt[3]{4}}{x^2 y}$

12. $\log_7 \dfrac{\sqrt[3]{13}}{pq^2}$

13. $\log_3 \sqrt{\dfrac{xy}{5}}$

14. $\log_6 \sqrt{\dfrac{pq}{7}}$

15. $\log_2 \dfrac{\sqrt[3]{x} \cdot \sqrt[5]{y}}{r^2}$

16. $\log_4 \dfrac{\sqrt[4]{z} \cdot \sqrt[5]{w}}{s^2}$

17. A student erroneously wrote $\log_a(x + y) = \log_a x + \log_a y$. When his teacher explained that this was indeed wrong, the student claimed he had used the distributive property. Write a few sentences explaining why the distributive property does not apply in this case.

18. Write a few sentences explaining how the rules for multiplying and dividing powers of the same base are similar to the rules for finding logarithms of products and quotients.

Use the properties of logarithms introduced in this section to express each of the following as a single logarithm. Assume all variables are defined in such a way that the variable expressions are positive, and bases are positive numbers not equal to 1. See Examples 1–5.

19. $\log_b x + \log_b y$

20. $\log_b 2 + \log_b z$

21. $3 \log_a m - \log_a n$

22. $5 \log_b x - \log_b y$

23. $(\log_a r - \log_a s) + 3 \log_a t$

24. $(\log_a p - \log_a q) + 2 \log_a r$

25. $3 \log_a 5 - 4 \log_a 3$

26. $3 \log_a 5 + \dfrac{1}{2} \log_a 9$

27. $\log_{10}(x + 3) + \log_{10}(x - 3)$

28. $\log_{10}(y + 4) + \log_{10}(y - 4)$

29. $3 \log_p x + \dfrac{1}{2} \log_p y - \dfrac{3}{2} \log_p z - 3 \log_p a$

30. $\dfrac{1}{3} \log_b x + \dfrac{2}{3} \log_b y - \dfrac{3}{4} \log_b s - \dfrac{2}{3} \log_b t$

31. Explain why the statement for the power rule for logarithms requires that x be a positive real number.

32. What is wrong with the following "proof" that $\log_2 16$ does not exist?

$$\log_2 16 = \log_2 (-4)(-4)$$
$$= \log_2(-4) + \log_2(-4)$$

Since the logarithm of a negative number is not defined, the final step cannot be evaluated, and so $\log_2 16$ does not exist.

MATHEMATICAL CONNECTIONS (EXERCISES 33–38)

Work Exercises 33–38 in order.

33. Evaluate $\log_3 81$.

34. Write the *meaning* of the expression $\log_3 81$.

35. Evaluate $3^{\log_3 81}$.

36. Write the *meaning* of the expression $\log_2 19$.

37. Evaluate $2^{\log_2 19}$.

38. Keeping in mind that a logarithm is an exponent, and using the results from Exercises 33–37, what is the simplest form of the expression $k^{\log_k m}$?

10.5 *Evaluating Logarithms*

As mentioned earlier, logarithms are important in many applications of mathematics to everyday problems, particularly in biology, engineering, economics, and social science. In this section we show how to find numerical approximations for logarithms. Traditionally base 10 logarithms have been used most extensively, because our number system is base 10. Logarithms to base 10 are called **common logarithms,** and $\log_{10} x$ is abbreviated as simply $\log x$, where the base is understood to be 10.

OBJECTIVE 1 We use calculators to evaluate common logarithms. In the next example we give the results of evaluating some common logarithms using a calculator with a log key. (This may be a second function key on some calculators.) For simple scientific calculators, just enter the number, then touch the log key. For graphing calculators, these steps are reversed. We will give all logarithms to four decimal places.

EXAMPLE 1   Evaluating Common Logarithms

Evaluate each logarithm.

(a) $\log 327.1 \approx 2.5147$

(b) $\log 437{,}000 \approx 5.6405$

(c) $\log .0615 \approx -1.2111$

In Example 1(c), $\log .0615 \approx -1.2111$, a negative result. The common logarithm of a number between 0 and 1 is always negative because the logarithm is the exponent on 10 that produces the number. For example,

$$10^{-1.2111} \approx .0615.$$

If the exponent (the logarithm) were positive, the result would be greater than 1 because $10^0 = 1$.

WORK PROBLEM 1 AT THE SIDE. ▶▶

OBJECTIVE 2 In chemistry, the **pH** (pronounced "p" "h") of a solution is defined as follows.

$$\mathbf{pH} = -\mathbf{log[H_3O^+]},$$

where $[H_3O^+]$ is the hydronium ion concentration in moles per liter.

The pH is a measure of the acidity (uh-SID-uh-tee) or alkalinity (al-kuh-LIN-uh-tee) of a solution; water, for example, has a pH of 7. In general, acids have pH numbers less than 7, and alkaline solutions have pH values greater than 7.

1. Find the following.

(a) log 41,600

(b) log 43.5

(c) log .442

2. Find the pH of solutions with the following hydronium ion concentrations.

(a) 3.7×10^{-8}

(b) 1.2×10^{-3}

E X A M P L E 2 **Finding pH**

Find the pH of grapefruit with a hydronium (hy-DROH-nee-um) ion (EYE-on) concentration of 6.3×10^{-4}.

Use the definition of pH.

$$pH = -\log(6.3 \times 10^{-4})$$
$$= -(\log 6.3 + \log 10^{-4}) \qquad \text{Multiplication property}$$
$$\approx -[.7993 - 4]$$
$$= -.7993 + 4 \approx 3.2$$

It is customary to round pH values to the nearest tenth.

■

◀◀ WORK PROBLEM 2 AT THE SIDE.

E X A M P L E 3 **Finding Hydronium Ion Concentration**

Find the hydronium ion concentration of drinking water with a pH of 6.5.

$$pH = 6.5 = -\log[H_3O^+]$$
$$\log[H_3O^+] = -6.5$$

This last line indicates that $10^{-6.5} = [H_3O^+]$. Using a calculator, we find

$$[H_3O^+] = 3.2 \times 10^{-7}.$$

■

3. Find the hydronium ion concentrations of solutions with the following pHs.

(a) 4.6

(b) 7.5

◀◀ WORK PROBLEM 3 AT THE SIDE.

OBJECTIVE 3▶ The most important logarithms used in applications are **natural logarithms,** which have as base the number e. The number e is irrational, like π: $e \approx 2.7182818$. Logarithms to base e are called natural logarithms because they occur in biology and the social sciences in natural situations that involve growth or decay. The base e logarithm of x is written ln x (read "el en x"). A graph of $y = \ln x$, the natural logarithm function, is given in Figure 13.

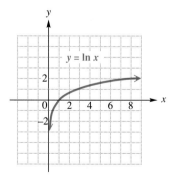

FIGURE 13

A calculator key labeled $\boxed{\ln x}$ is used to evaluate natural logarithms. If your calculator has an $\boxed{e^x}$ key, but not a key labeled $\boxed{\ln x}$, find natural logarithms by entering the number, touching the $\boxed{\text{INV}}$ key, and then touching the $\boxed{e^x}$ key. This works because $y = e^x$ is the inverse function of $y = \ln x$ (or $y = \log_e x$).

┌─ **E X A M P L E 4   Finding Natural Logarithms**

Find each of the following logarithms to the nearest ten-thousandth.

(a) ln .5841

$$\ln .5841 \approx -.5377$$

As with common logarithms, a number between 0 and 1 has a negative natural logarithm.

(b) $\ln 192.7 \approx 5.2611$

(c) $\ln 10.84 \approx 2.3832$

───────────

4. Find each logarithm to the nearest ten-thousandth.

(a) ln .01

(b) ln 27

(c) ln 529

> **WORK PROBLEM 4 AT THE SIDE.** ▶▶

OBJECTIVE ▶ One of the most common applications of exponential functions depends on the fact that in many situations involving growth or decay of a population, the amount or number of some quantity present at time t can be closely approximated by

$$y = y_0 e^{kt}$$

where y_0 is the amount or number present at time $t = 0$, k is a constant, and e is the base of natural logarithms mentioned earlier.

┌─ **E X A M P L E 5   Applying Natural Logarithms**

The population of Canada, in millions, is approximated by the exponential function with

$$f(t) = 29.5e^{.052t},$$

where t represents time in years from 1995. Thus, 1995 is represented by $t = 0$, 1996 is represented by $t = 1$, and so on. To find the population in 1995, we substitute 0 for t to get

$$f(0) = 29.5e^{(.052)(0)} = 29.5(1) = 29.5$$

million. Similarly, the population in 5 years (the year 2000) will be about

$$f(5) = 29.5e^{(.052)(5)} = 38.3$$

million.

5. In Example 5, find the population in 1998.

> **WORK PROBLEM 5 AT THE SIDE.** ▶▶

┌─ **E X A M P L E 6   Applying Natural Logarithms**

The number of years, $N(r)$, since two independently evolving languages split off from a common ancestral language is approximated by

$$N(r) = -5000 \ln r,$$

where r is the percent of words from the ancestral language common to both languages now. Find N if $r = 70\%$.

Write 70% as .7, and find $N(.7)$.

$$N(.7) = -5000 \ln .7 \approx 1783$$

Approximately 1800 years have passed since the two languages separated.

6. Find $N(.2)$ in Example 6.

ANSWERS

4. (a) -4.6052 **(b)** 3.2958 **(c)** 6.2710

5. about 34.5 million

6. about 8000 years

> **WORK PROBLEM 6 AT THE SIDE.** ▶▶

7. Find each logarithm.

(a) $\log_3 17$
(Use common logarithms.)

(b) $\log_9 121$
(Use natural logarithms.)

Note

In Examples 5 and 6, the final answers were obtained *without* rounding off the intermediate values. In general, it is best to wait until the final step to round off the answer; otherwise, a build-up of round-off error may cause the final answer to have an incorrect final decimal place digit.

OBJECTIVE 5 A calculator can be used to approximate the values of common logarithms (base 10) or natural logarithms (base e). However, sometimes we need to use logarithms to other bases. The following rule is used to convert logarithms from one base to another.

Change-of-Base Rule

If $a > 0$, $a \neq 1$, $b > 0$, $b \neq 1$, and $x > 0$, then

$$\log_a x = \frac{\log_b x}{\log_b a}.$$

Note

As an aid in remembering the change-of-base rule, notice that x is "above" a on both sides of the equation.

Any positive number other than 1 can be used for base b in the change-of-base rule, but usually the only practical bases are e and 10 because calculators give logarithms only for these two bases.

To prove the formula for change of base, let $\log_a x = m$. We use the fact that if $r = s$, $\log_b r = \log_b s$.

$$\log_a x = m$$
$$a^m = x \qquad \text{Change to exponential form.}$$
$$\log_b(a^m) = \log_b x \qquad \text{Take logs on both sides.}$$
$$m \log_b a = \log_b x \qquad \text{Use the power property.}$$
$$(\log_a x)(\log_b a) = \log_b x \qquad \text{Substitute for } m.$$
$$\log_a x = \frac{\log_b x}{\log_b a} \qquad \text{Divide both sides by } \log_b a.$$

E X A M P L E 7   Using the Change-of-Base Rule

Find each logarithm.

(a) $\log_5 12$

Use common logarithms and the rule for change of base.

$$\log_5 12 = \frac{\log 12}{\log 5}$$

Now evaluate this quotient.

$$\log_5 12 \approx 1.5440$$

(b) $\log_2 134$

Use natural logarithms and the change-of-base rule.

$$\log_2 134 = \frac{\ln 134}{\ln 2} \approx 7.0661$$

◀◀ **WORK PROBLEM 7 AT THE SIDE.**

ANSWERS

7. (a) 2.5789 **(b)** 2.1827

10.5 *Exercises*

Fill in the blanks.

1. The base in the expression log x is understood to be _____ .

2. The base in ln x is understood to be _____ .

3. We know that $10^{\underline{\quad}} = 1$ and $10^{\underline{\quad}} = 10$. Therefore, log 5 is between _____ and _____ .

4. We know that $e^{\underline{\quad}} = 2.718$ and $e^{\underline{\quad}} = 7.390$. Therefore, ln 4 is between _____ and _____ .

5. log $10^{19.2} =$ _____ . (Do not use a calculator.)

6. ln $e^{\sqrt{2}} =$ _____ . (Do not use a calculator.)

You will need a calculator for the remaining exercises in this set.

Find each logarithm. Give an approximation to the nearest ten-thousandth. See Examples 1 and 4.

7. log 43

8. log 98

9. log 328.4

10. log 457.2

11. log .0326

12. log .1741

13. $\log(4.76 \times 10^9)$

14. $\log(2.13 \times 10^4)$

15. ln 7.84

16. ln 8.32

17. ln .0556

18. ln .0217

19. ln 388.1

20. ln 942.6

21. $\ln(8.59 \times e^2)$

22. $\ln(7.46 \times e^3)$ **23.** $\ln 10$ **24.** $\log e$

Use the change-of-base rule (either with common or natural logarithms) to find the following logarithms. Give approximations to the nearest ten-thousandth. See Example 7.

25. $\log_6 13$ **26.** $\log_7 19$ **27.** $\log_{\sqrt{2}} \pi$

28. $\log_\pi \sqrt{2}$ **29.** $\log_{21} .7496$ **30.** $\log_{19} .8325$

31. Let m be the number of letters in your first name, and let n be the number of letters in your last name.
   (a) In your own words, explain what $\log_m n$ means.
   (b) Use your calculator to find $\log_m n$.
   (c) Raise m to the power indicated by the number you found in part (b). What is your result?

32. The equation $5^x = 7$ cannot be solved using the methods described in Section 10.2. However, in solving this equation, we must find the exponent to which 5 must be raised in order to obtain 7: this is $\log_5 7$.
   (a) Use the change-of-base rule and your calculator to find $\log_5 7$.
   (b) Raise 5 to the number you found in part (a). What is your result?
   (c) Using as many decimal places as your calculator gives, write the solution set of $5^x = 7$. (Equations of this type will be studied in more detail in Section 10.6.)

Use the formula $\mathrm{pH} = -\log[H_3O^+]$ *to find the pH of the substance with the given hydronium ion concentration. See Example 2.*

33. Ammonia, 2.5×10^{-12} **34.** Sodium bicarbonate, 4.0×10^{-9}

35. Grapes, 5.0×10^{-5} **36.** Tuna, 1.3×10^{-6}

Use the formula for pH to find the hydronium ion concentration of the substance with the given pH. See Example 3.

37. Human blood plasma, 7.4

38. Human gastric contents, 2.0

39. Spinach, 5.4

40. Bananas, 4.6

Solve the following problems. See Examples 5 and 6.

41. Suppose that the amount, in grams, of pluto-nium-241 present in a given sample is deter-mined by the function

$$A(t) = 2.00e^{-.053t},$$

where t is measured in years. Find the amount present in the sample after the given number of years.
(a) 4   **(b)** 10   **(c)** 20
(d) What was the initial amount present?

42. Suppose that the amount, in grams, of radium-226 present in a given sample is determined by the function

$$A(t) = 3.25e^{-.00043t},$$

where t is measured in years. Find the amount present in the sample after the given number of years.
(a) 20   **(b)** 100   **(c)** 500
(d) What was the initial amount present?

43. The number of books, in millions, sold per year in the United States between 1985 and 1990 can be approximated by the function

$$N(t) = 1757e^{.0264t},$$

where $t = 0$ corresponds to the year 1985. Based on this model, how many books were sold in 1994? (*Source:* Book Industry Study Group)

44. Personal consumption expenditures for recre-ation in billions of dollars in the United States during the years 1984 through 1990 can be approximated by the function

$$C(t) = 185.4e^{.0587t},$$

where $t = 0$ corresponds to the year 1984. Based on this model, how much were personal consumption expenditures in 1994? (*Source:* U.S. Bureau of Economic Analysis)

45. The number of Cesarean section deliveries in the United States has increased over the years. According to statistics provided by the U.S. National Center for Health Statistics, between the years 1980 and 1989, the number of such births, in thousands, can be approximated by the function

$$B(t) = 624.6e^{.0516t},$$

where $t = 1$ corresponds to the year 1980. Based on this model, how many such births were there in 1994?

46. According to an article in *The AMATYC Review* (Spring, 1993), the number of students enrolled, in thousands, in intermediate algebra in two-year colleges since 1966 can be approximated by the function

$$E(t) = 39.8e^{.073t},$$

where $t = 1$ corresponds to 1966. Based on this equation, how many students were enrolled in intermediate algebra at the two-year college level in 1995?

INTERPRETING TECHNOLOGY (EXERCISES 47–48)
---

47. The function $B(x) = 624.6e^{.0516t}$, described in Exercise 45 with $x = t$, is graphed in a graphing calculator-generated window in the accompanying figure. Interpret the meanings of x and y in the display at the bottom in the context of Exercise 45.

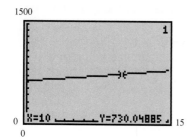

48. The function $E(x) = 39.8e^{.073x}$, described in Exercise 46 with $x = t$, is graphed in a graphing calculator-generated window in the accompanying figure. Interpret the meanings of x and y in the display at the bottom in the context of Exercise 46.

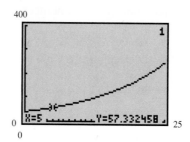

For Exercises 49–52, refer to Example 6 and use the function $N(r) = -5000 \ln r$. Round answers to the nearest hundred.

49. Find $N(.85)$ **50.** Find $N(.35)$ **51.** Find $N(.10)$

52. How many years have elapsed since the split if 75% of the words of the ancestral language are common to both languages today?

10.6 Exponential and Logarithmic Equations and Their Applications

As mentioned at the beginning of this chapter, exponential and logarithmic functions are important in many applications of mathematics. Using these functions in applications requires solving exponential and logarithmic equations. Some simple equations were solved in earlier sections of this chapter. More general methods for solving these equations depend on the following properties.

Properties of Exponential and Logarithmic Equations

For all real numbers $b > 0$, $b \neq 1$, and any real numbers x and y:

1. If $x = y$, then $b^x = b^y$.
2. If $b^x = b^y$, then $x = y$.
3. If $x = y$, and $x > 0$, $y > 0$, then $\log_b x = \log_b y$.
4. If $x > 0$, $y > 0$, and $\log_b x = \log_b y$, then $x = y$.

Property 2 was used to solve exponential equations earlier and property 3 was used in Section 10.5.

OBJECTIVE ▶ The first examples illustrate a general method for solving exponential equations using property 3.

E X A M P L E   1   Solving an Exponential Equation

Solve the equation $3^m = 12$.

$$3^m = 12$$

$$\log 3^m = \log 12 \qquad \text{Property 3}$$

$$m \log 3 = \log 12 \qquad \text{Power property}$$

$$m = \frac{\log 12}{\log 3} \qquad \text{Divide by } \log 3.$$

This quotient is the exact solution. To get a decimal approximation for the solution, use a calculator. A calculator gives

$$m \approx 2.262,$$

and the solution set is $\{2.262\}$. Check that $3^{2.262} \approx 12$.

Caution

Be careful: $\dfrac{\log 12}{\log 3}$ is *not* equal to $\log 4$ because $\log 4 \approx .6021$, but $\dfrac{\log 12}{\log 3} \approx 2.262$.

WORK PROBLEM 1 AT THE SIDE. ▶▶

When an exponential equation has e as the base, it is easiest to use base e logarithms.

---

1. Give decimal approximations to the nearest thousandth for the solutions.

(a) $2^p = 9$

(b) $10^k = 4$

699

2. Solve $e^{-.01t} = .38$.

EXAMPLE 2 Solving an Exponential Equation with Base e

Solve $e^{.003x} = 40$.

Take base e logarithms on both sides.

$$\ln e^{.003x} = \ln 40$$

$$.003x \ln e = \ln 40 \qquad \text{Power property}$$

$$.003x = \ln 40 \qquad \ln e = \ln e^1 = 1$$

$$x = \frac{\ln 40}{.003} \qquad \text{Divide by } .003.$$

$$x \approx 1230 \qquad \text{Use a calculator.}$$

The solution set is $\{1230\}$. Check that $e^{.003(1230)} \approx 40$.

◀◀ **WORK PROBLEM 2 AT THE SIDE.**

In summary, we can solve exponential equations by one of the following methods. (The method used depends upon the form of the equation.) Examples 1 and 2 illustrate Method 1. We gave examples of Method 2 in Section 10.2.

3. Solve $\log_5 \sqrt{x - 7} = 1$.

Methods for Solving Exponential Equations

1. Using property 3, take logarithms to the same base on each side; then use the power property of logarithms on one side or both sides.
2. Using property 2, write both sides as exponentials with the same base; then set the exponents equal.

OBJECTIVE ▶2 The next three examples illustrate some ways to solve logarithmic equations. The properties of logarithms from Section 10.4 are useful here, as is using the definition of a logarithm to change to exponential form.

EXAMPLE 3 Solving a Logarithmic Equation

Solve $\log_2(x + 5)^3 = 4$.

$$(x + 5)^3 = 2^4 \qquad \text{Convert to exponential form.}$$

$$(x + 5)^3 = 16$$

$$x + 5 = \sqrt[3]{16} \qquad \text{Take the cube root on both sides.}$$

$$x = -5 + \sqrt[3]{16}$$

Verify that the solution satisfies the equation, so the solution set is $\{-5 + \sqrt[3]{16}\}$.

Caution
Recall that the domain of $y = \log_b x$ is $(0, \infty)$. For this reason, *it is always necessary to check that the solution of a logarithmic equation is in the domain of the logarithmic expression.*

◀◀ **WORK PROBLEM 3 AT THE SIDE.**

EXAMPLE 4 Solving a Logarithmic Equation

Solve $\log_2(x + 1) - \log_2 x = \log_2 8$.

$$\log_2(x + 1) - \log_2 x = \log_2 8$$

$$\log_2 \frac{x + 1}{x} = \log_2 8 \qquad \text{Division property}$$

$$\frac{x + 1}{x} = 8 \qquad \text{Property 4}$$

$$8x = x + 1 \qquad \text{Multiply by } x.$$

$$x = \frac{1}{7} \qquad \text{Subtract } x; \text{ divide by 7.}$$

Check this solution by substituting in the given equation. Here, both $x + 1$ and x must be positive. If $x = \frac{1}{7}$, this condition is satisfied, and the solution set is $\{\frac{1}{7}\}$.

WORK PROBLEM 4 AT THE SIDE. ▶▶

EXAMPLE 5 Solving a Logarithmic Equation

Solve $\log x + \log(x - 21) = 2$.

For this equation, write the left side as a single logarithm. Then write in exponential form and solve the equation.

$$\log x + \log(x - 21) = 2$$

$$\log x(x - 21) = 2 \qquad \text{Product rule}$$

$$x(x - 21) = 10^2 \qquad \text{Log } x = \log_{10} x; \text{ write in exponential form.}$$

$$x^2 - 21x = 100$$

$$x^2 - 21x - 100 = 0 \qquad \text{Standard form}$$

$$(x - 25)(x + 4) = 0 \qquad \text{Factor.}$$

$$x - 25 = 0 \quad \text{or} \quad x + 4 = 0 \qquad \text{Set each factor equal to 0.}$$

$$x = 25 \quad \text{or} \qquad x = -4$$

The value -4 must be rejected as a solution, since it leads to the logarithm of a negative number in the original equation:

$$\log(-4) + \log(-4 - 21) = 2. \qquad \text{The left side is not defined.}$$

The only solution, therefore, is 25, and the solution set is $\{25\}$.

Caution
Do not reject a potential solution just because it is nonpositive. Reject any value which *leads to* the logarithm of a nonpositive number.

WORK PROBLEM 5 AT THE SIDE. ▶▶

4. Solve.
$$\log_8(2x + 5) + \log_8 3 = \log_8 33$$

5. Solve.
$$\log_3 2x - \log_3(3x + 15) = -2$$

6. Find the value of $2000 deposited at 5% compounded annually for 10 years.

In summary, we use the following steps to solve a logarithmic equation.

Solving a Logarithmic Equation

Step 1 Use the multiplication or division properties of logarithms to get a single logarithm on one side.

Step 2 **(a)** Use property 4: If $\log_b x = \log_b y$, then $x = y$. (See Example 4.)

       (b) Write the equation in exponential form: if $\log_b x = k$, then $x = b^k$. (See Example 3 and Example 5.)

OBJECTIVE 3 So far in this book, problems involving applications of interest have been limited to simple interest. In most cases, the interest paid or charged is compound (KAHM-pound) interest (interest paid on both principal and interest). The formula for compound interest is an important application of exponential functions.

Compound Interest

If P dollars is deposited in an account paying an annual rate of interest r compounded (paid) n times per year, the account will contain

$$A = P\left(1 + \frac{r}{n}\right)^{nt}$$

dollars after t years.

In the formula, r is usually expressed as a decimal.

E X A M P L E 6   **Solving a Compound Interest Problem**

How much money will there be in an account at the end of 5 years if $1000 is deposited at 6% compounded quarterly? (Assume no withdrawals are made.)

    Because interest is compounded quarterly, $n = 4$. The other values given in the problem are $P = 1000$, $r = .06$ (because $6\% = .06$), and $t = 5$. Substitute into the compound interest formula to get the value of A.

$$A = 1000\left(1 + \frac{.06}{4}\right)^{4 \cdot 5}$$

$$A = 1000(1.015)^{20}$$

Now use the y^x key on a calculator, and round the answer to the nearest cent.

$$A = 1346.86$$

The account will contain $1346.86. (The actual amount of interest earned is $1346.86 - $1000 = $346.86. Do you see why?)

◀◀ WORK PROBLEM 6 AT THE SIDE.

─E X A M P L E 7   **Solving an Exponential Decay Problem**

Nuclear energy derived from radioactive isotopes can be used to supply power to space vehicles. The output of the radioactive power supply for a certain satellite is given by the function

$$y = 40e^{-.004t},$$

where y is in watts and t is the time in days.

(a) How much power will be available at the end of 180 days?

Let $t = 180$ in the formula.

$$y = 40e^{-.004(180)}$$

$$y \approx 19.5 \qquad \text{Use a calculator.}$$

About 19.5 watts will be left.

(b) How long will it take for the amount of power to be half of its original strength?

The original amount of power is 40 watts. (Why?) Because half of 40 is 20, replace y with 20 in the formula, and solve for t.

$$20 = 40e^{-.004t}$$

$$.5 = e^{-.004t} \qquad \text{Divide by 40.}$$

$$\ln .5 = \ln e^{-.004t}$$

$$\ln .5 = -.004t \qquad \ln e^k = k$$

$$t = \frac{\ln .5}{-.004}$$

$$t \approx 173 \qquad \text{Use a calculator.}$$

After about 173 days, the amount of available power will be half of its original amount.

WORK PROBLEM 7 AT THE SIDE. ▶▶

7. (a) In Example 7, suppose the output is given by $y = 50e^{-.002t}$. How much power will be available in 180 days?

(b) How long will it take for the amount of power to be half its original strength?

7. (a) 34.9 watts **(b)** about 347 days

NUMBERS IN THE *Real World* *a graphing calculator minicourse*

Lesson 8: Inverse Functions

In Section 10.1 we described how inverses of one-to-one functions may be determined algebraically. We also explained how the graph of a one-to-one function f compares to the graph of its inverse f^{-1}: it is a reflection of the graph of f^{-1} across the line $y = x$. In Example 3(a) of that section we showed that the inverse of the one-to-one function $f(x) = 2x + 5$ is given by $f^{-1}(x) = \dfrac{x-5}{2}$. If we use a square viewing window and graph $y_1 = f(x) = 2x + 5$, $y_2 = f^{-1}(x) = \dfrac{x-5}{2}$, and $y_3 = x$, we can see how this reflection appears on the screen. See the first figure at the left.

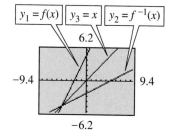

Some graphing calculators have the capability to draw the inverse of a function. The second figure shows the graphs of $f(x) = x^3 + 2$ and its inverse.

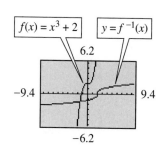

GRAPHING CALCULATOR EXPLORATIONS

1. Refer to the graph of $f(x) = x^3 + 2$ in the second figure.

   (a) Find f^{-1} algebraically.

   (b) Graph $y_1 = f(x) = x^3 + 2$, $y_2 = f^{-1}(x)$, and $y_3 = x$ on the same square viewing window. Describe what you see.

2. Graph $y_1 = 10^x$ and $y_2 = \log x$ on the same square viewing window.

   (a) Trace to the point where $x = .5$ on y_1, and find the approximation for y.

   (b) Find $\sqrt{10}$ using the square root key of your calculator. Compare it to the value you found in part (a) and explain why the approximations are the same.

   (c) Trace to the point where $x = 6$ on y_2, and find the approximation for y.

   (d) Find $\log 6$ using the common logarithm key of your calculator. Compare it to the value you found in part (c) and explain why the approximations are the same.

3. (a) Graph $y_1 = 4^x$ on your calculator.

   (b) The function y_1 is one-to-one and has an inverse: it is $y_2 = \log_4 x$. To graph a logarithmic function that is neither base ten nor base e on a graphing calculator, we must use the change-of-base rule. Graph y_2 as $\log x/\log 4$.

   (c) Trace to the point where $x = .5$ on y_1. What is the value of y?

   (d) Trace to the point on y_2 where x is equal to the value of y in part (c). What is the new value of y?

   (e) How do the results of parts (c) and (d) support the fact that y_1 and y_2 are inverses?

10.6 Exercises

The following equations were solved in Section 10.2 by writing both sides with the same base and then setting the exponents equal. Solve each equation using the method of Example 1 of this section.

1. $5^x = 125$

2. $9^x = 27$

3. By inspection, determine the solution set of $2^x = 64$. Now, without using a calculator, give the exact value of $\dfrac{\log 64}{\log 2}$.

4. Which one of the following is *not* a solution of $7^x = 23$?

   (a) $\dfrac{\log 23}{\log 7}$ **(b)** $\dfrac{\ln 23}{\ln 7}$

   (c) $\log_7 23$ **(d)** $\log_{23} 7$

Solve each of the following equations. Give solutions to the nearest thousandth. See Example 1.

5. $7^x = 5$

6. $4^x = 3$

7. $9^{-x+2} = 13$

8. $6^{-t+1} = 22$

9. $2^{y+3} = 5^y$

10. $6^{m+3} = 4^m$

Use natural logarithms to solve each of the following equations. Give solutions to the nearest thousandth. See Example 2.

11. $e^{.006x} = 30$

12. $e^{.012x} = 23$

13. $e^{-.103x} = 7$

14. $e^{-.205x} = 9$

15. $\ln e^{.04x} = \sqrt{3}$

16. $\ln e^{.45x} = \sqrt{7}$

17. Try solving one of the equations in Exercises 11–16 using common logarithms rather than natural logarithms. (You should get the same solution.) Explain why using natural logarithms is a better choice.

18. If you were asked to solve $10^{.0025x} = 75$, would natural or common logarithms be a better choice? Explain your answer.

Solve each of the following equations. Give the exact solution. See Example 3.

19. $\log_3(6x + 5) = 2$

20. $\log_5(12x - 8) = 3$

21. $\log_7(x + 1)^3 = 2$

22. $\log_4(y - 3)^3 = 4$

23. Suppose that in solving a logarithmic equation having the term $\log_4(x - 3)$ you obtain an apparent solution of 2. All algebraic work is correct. Explain why you must reject 2 as a solution of the equation.

24. Suppose that in solving a logarithmic equation having the term $\log_7(3 - x)$ you obtain an apparent solution of -4. All algebraic work is correct. Should you reject -4 as a solution of the equation? Explain why or why not.

Solve each of the following equations. Give exact solutions. See Examples 4 and 5.

25. $\log(6x + 1) = \log 3$

26. $\log(7 - x) = \log 12$

27. $\log_5(3t + 2) - \log_5 t = \log_5 4$

28. $\log_2(x + 5) - \log_2(x - 1) = \log_2 3$

29. $\log 4x - \log(x - 3) = \log 2$

30. $\log(-x) + \log 3 = \log(2x - 15)$

31. $\log_2 x + \log_2(x - 7) = 3$

32. $\log(2x - 1) + \log 10x = \log 10$

33. $\log 5x - \log(2x - 1) = \log 4$

34. $\log_3 x + \log_3(2x + 5) = 1$

35. $\log_2 x + \log_2(x - 6) = 4$

36. $\log_2 x + \log_2(x + 4) = 5$

Solve each problem. See Examples 6 and 7.

37. How much money will there be in an account at the end of 6 years if $2000.00 is deposited at 4% compounded quarterly? (Assume no withdrawals are made.)

38. How much money will there be in an account at the end of 7 years if $3000.00 is deposited at 3.5% compounded quarterly? (Assume no withdrawals are made.)

39. A sample of 400 grams of lead-210 decays to polonium-210 according to the function

$$A(t) = 400e^{-.032t},$$

where t is time in years. How much lead will be left in the sample after 25 years?

40. How long will it take the initial sample of lead in Exercise 39 to decay to half of its original amount? (This is called the *half-life* of the substance.)

*Banks sometimes compute interest based on what is known as **continuous compounding**. Rather than paying interest a finite number of times per year (as explained in the text just before Example 6), interest is earned at all times. As a result, the formula in this section cannot be applied because n is approaching infinity. The formula used to determine the amount A in an account having initial principal P compounded continuously at an annual rate r for t years is*

$$A = Pe^{rt}.$$

41. What will be the amount A in an account with initial principal $4000.00 if interest is compounded continuously at an annual rate of 3.5% for 6 years?

42. Refer to Exercise 38. Does the money grow to a larger value under those conditions, or when invested for 7 years at 3% compounded continuously?

43. How long would it take an initial principal P to double if it is invested at 4.5% compounded continuously? (This is called the *doubling time* of the money.)

44. How long would it take $4000 to grow to $6000 at 3.25% compounded continuously?

INTERPRETING TECHNOLOGY (EXERCISES 45–46)

---

45. The screens show the graphs of $f(x) = \log x^2$ and $g(x) = 2 \log x$. Why does the power rule from Section 10.4 not apply here?

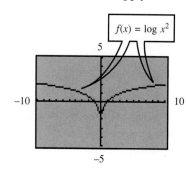
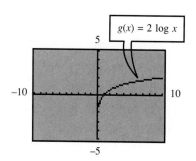

46. One of the graphs shown below is the graph of $f(t) = 2e^{-.125t}$; the other is the graph of $g(t) = 300e^{.4t}$. Which is the graph of f?

A.

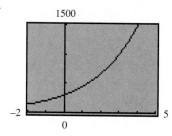

B.

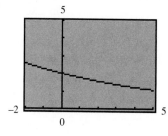

CHAPTER 10 SUMMARY

10.1 **one-to-one function**
A one-to-one function is a function in which each x-value corresponds to just one y-value and each y-value corresponds to just one x-value.

inverse of a function f
If f is a one-to-one function, the inverse of f is the set of all ordered pairs of the form (y, x), where (x, y) belongs to f.

10.2 **exponential equation**
An equation involving an exponential, where the variable is in the exponent, is an exponential equation.

10.3 **logarithm**
A logarithm is an exponent; $\log_a x$ is the exponent on the base a that gives the number x.

logarithmic equation
A logarithmic equation is an equation with a logarithm in at least one term.

10.5 **common logarithm**
A common logarithm is a logarithm to the base 10.

natural logarithm
A natural logarithm is a logarithm to the base e.

NEW SYMBOLS

f^{-1} the inverse of f

$\log_a x$ the logarithm of x to the base a

$\log x$ common (base 10) logarithm of x

$\ln x$ natural (base e) logarithm of x

e a constant, approximately 2.7182818

QUICK REVIEW

Concepts	Examples
10.1 Inverse Functions	
Horizontal Line Test If a horizontal line intersects the graph of a function in no more than one point, then the function is one-to-one.	Find f^{-1} if $f(x) = 2x - 3$. The graph of f is a straight line, so f is one-to-one by the horizontal line test.
Inverse Functions For a one-to-one function f defined by an equation $y = f(x)$, the defining equation of the inverse function f^{-1} is found by exchanging x and y, solving for y, and replacing y with $f^{-1}(x)$.	Exchange x and y in the equation $y = 2x - 3$. $$x = 2y - 3$$ Solve for y to get $\quad y = \frac{1}{2}x + \frac{3}{2}.$ Therefore, $f^{-1}(x) = \frac{1}{2}x + \frac{3}{2}.$
The graph of f^{-1} is a mirror image of the graph of f with respect to the line $y = x$.	The graphs of a function f and its inverse f^{-1} are given below. 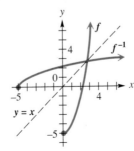

Concepts	*Examples*
10.2 Exponential Functions For $a > 0$, $a \neq 1$, $f(x) = a^x$ is an exponential function with base a. **Graph of $f(x) = a^x$** The graph contains the point $(0, 1)$. When $a > 1$, the graph rises from left to right. When $0 < a < 1$, the graph falls from left to right. The x-axis is an asymptote. The domain is $(-\infty, \infty)$; the range is $(0, \infty)$.	$f(x) = 10^x$ is an exponential function with base 10. 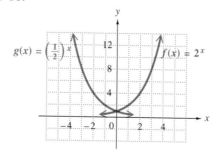
10.3 Logarithmic Functions $y = \log_a x$ means $x = a^y$. For $b > 0$, $b \neq 1$, $\log_b b = 1$ and $\log_b 1 = 0$. For $a > 0$, $a \neq 1$, $x > 0$, $f(x) = \log_a x$ is the logarithmic function with base a. **Graph of $g(x) = \log_a x$** The graph contains the point $(1, 0)$. When $a > 1$, the graph rises from left to right. When $0 < a < 1$, the graph falls from left to right. The y-axis is an asymptote. The domain is $(0, \infty)$; the range is $(-\infty, \infty)$.	$y = \log_2 x$ means $x = 2^y$. $\log_3 3 = 1$ $\log_5 1 = 0$ $f(x) = \log_6 x$ is the logarithmic function with base 6. 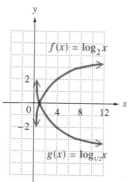
10.4 Properties of Logarithms **Multiplication Property** $$\log_a xy = \log_a x + \log_a y$$ **Division Property** $$\log_a \frac{x}{y} = \log_a x - \log_a y$$ **Power Property** $$\log_a x^r = r \log_a x$$ **Special Properties** $$b^{\log_b x} = x \quad \text{and} \quad \log_b b^x = x$$	$\log_2 3m = \log_2 3 + \log_2 m$ $\log_5 \frac{9}{4} = \log_5 9 - \log_5 4$ $\log_{10} 2^3 = 3 \log_{10} 2$ $6^{\log_6 10} = 10 \qquad \log_3 3^4 = 4$

Concepts	Examples
10.5 Evaluating Logarithms **Change-of-Base Rule** If $a > 0$, $a \neq 1$, $b > 0$, $b \neq 1$, $x > 0$, then $$\log_a x = \frac{\log_b x}{\log_b a}.$$	$$\log_3 17 = \frac{\ln 17}{\ln 3} = \frac{\log 17}{\log 3}$$
10.6 Exponential and Logarithmic Equations and Their Applications To solve exponential equations, use these properties $(b > 0, b \neq 1)$. **1.** If $b^x = b^y$, then $x = y$.	Solve $2^{3x} = 2^5$. $3x = 5$ $x = \dfrac{5}{3}$
2. If $x = y$, $(x > 0, y > 0)$, then $\log_b x = \log_b y$.	Solve $5^m = 8$. $\log 5^m = \log 8$ $m \log 5 = \log 8$ $m = \dfrac{\log 8}{\log 5}$
To solve logarithmic equations, use these properties, where $b > 0$, $b \neq 1$, $x > 0$, $y > 0$. First use the properties of 10.4, if necessary, to get the equation in the proper form. **1.** If $\log_b x = \log_b y$, then $x = y$.	Solve $\log_3 2x = \log_3(x + 1)$. $2x = x + 1$ $x = 1$
2. If $\log_b x = y$, then $b^y = x$.	Solve $\log_2(3a - 1) = 4$. $3a - 1 = 2^4 = 16$ $3a = 17$ $a = \dfrac{17}{3}$

NUMBERS IN THE

Real World   *a graphing calculator minicourse*

Lesson 9: Solution of Exponential and Logarithmic Equations

$y_1 = 4^x - 8$

In Lessons 1 and 6 we saw how the x-intercepts of the graph of a function f correspond to the real solutions of the equation $f(x) = 0$. The ideas presented there dealt with linear and quadratic functions. We can extend those ideas to exponential and logarithmic functions. For example, consider the equation $4^x = 8$. Using traditional methods, we can determine that the solution set of this equation is $\{1.5\}$. Now if we write the equation with 0 on one side, we get $4^x - 8 = 0$. Graphing $y_1 = 4^x - 8$ and finding the x-intercept supports this solution, as seen in the first figure at the left. The x-intercept is 1.5, as expected.

Exercise 32 of Section 10.6 requires the solution of the logarithmic equation $\log (2x - 1) + \log 10x = \log 10$. If we graph $y_1 = \log (2x - 1) + \log 10x - \log 10$, we see that the x-intercept is 1, supporting the result obtained when the equation is solved algebraically. See the second figure at the left.

$y_1 = \log (2x - 1) + \log 10x - \log 10$

GRAPHING CALCULATOR EXPLORATIONS

1. The equation solved in Example 1 of Section 10.6 is equivalent to $3^x = 12$.

   (a) Because x is the exponent to which 3 must be raised in order to obtain 12, x is the _____ to the base _____ of _____.

   (b) Graph $y_1 = 3^x - 12$, and find an approximation for the x-intercept.

   (c) Use the common logarithm key of your calculator to find an approximation for $\log 12/\log 3$. How does it compare to the x-intercept in part (b)?

   (d) Clear your screen, and graph $y_1 = \log_3 x = \log x/\log 3$. Then trace to where $x = 12$, and find the corresponding value of y. How does it compare to your values found in parts (b) and (c)?

2. Use a graph to support the fact that the solution set of $\log_3 x + \log_3(2x + 5) = 1$ is $\{\frac{1}{2}\}$. (*Hint:* Graph $y_1 = \log_3 x + \log_3(2x + 5) - 1$, and find the x-intercept.)

3. The formula for continuous compounding of interest is given just prior to Exercises 41–44 in Section 10.6. If we graph the function $y_1 = 1000e^{.04x}$, using a minimum x-value of 0, a maximum x-value of 20, a minimum y-value of 0, and a maximum y-value of 2500, we can observe how \$1000 will grow if compounded continuously at a rate of 4%. Use the graph to find the amount of money that will be in the account after (a) 10 years and (b) 15 years.

CHAPTER 10 REVIEW EXERCISES

[10.1] *Determine whether the graph is the graph of a one-to-one function.*

1.

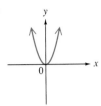

2.

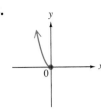

Determine whether the function is one-to-one. If it is, find its inverse.

3. $f(x) = -3x + 7$

4. $f(x) = \sqrt[3]{6x - 4}$

5. $f(x) = -x^2 + 3$

6. $f(x) = x$

Each function graphed below is one-to-one. Graph its inverse on the same set of axes.

7.

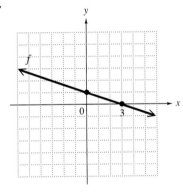

8.

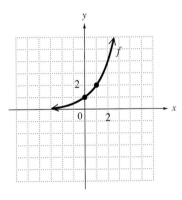

[10.2] *Graph each of the following functions.*

9. $f(x) = 3^x$

10. $f(x) = \left(\dfrac{1}{3}\right)^x$

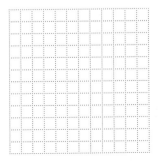

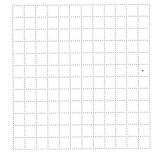

11. $y = 3^{x+1}$

12. $y = 2^{2x+3}$

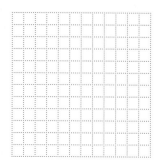

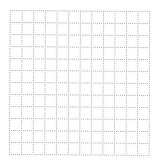

Solve each of the following equations.

13. $4^{3x} = 8^{x+4}$

14. $\left(\dfrac{1}{27}\right)^{x-1} = 9^{2x}$

15. $5^x = 1$

16. What is the y-intercept of the graph of $y = a^x$ $(a > 0, a \neq 1)$?

17. How does the answer to Exercise 16 reinforce the definition of 0 as an exponent?

[10.3] *Graph each of the following functions.*

18. $g(x) = \log_3 x$ (*Hint:* See Exercise 9.)

19. $g(x) = \log_{1/3} x$ (*Hint:* See Exercise 10.)

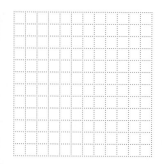

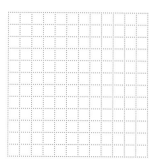

20. **(a)** Write in exponential form: $\log_5 625 = 4$.
   (b) Write in logarithmic form: $5^{-2} = .04$.

21. **(a)** In your own words, explain the meaning of $\log_b a$.
   (b) Based on your explanation above, simplify the expression $b^{\log_b a}$.

[10.4] *Apply the properties of logarithms introduced in Section 10.4 to express each of the following as a sum or difference of logarithms. Assume that all variables represent positive real numbers.*

22. $\log_4 3x^2$

23. $\log_2 \dfrac{p^2 r}{\sqrt{z}}$

Use the properties of logarithms introduced in Section 10.4 to express each of the following as a single logarithm. Assume that all variables represent positive real numbers, $b \neq 1$.

24. $\log_b 3 + \log_b x - 2 \log_b y$

25. $\log_3(x + 7) - \log_3(4x + 6)$

[10.5] *Find each logarithm. Give approximations to four decimal places.*

26. $\log 28.9$

27. $\log .257$

28. $\log 10^{4.8613}$

29. $\ln 28.9$

30. $\ln .257$

31. $\ln e^{4.8613}$

32. Use your calculator to find approximations of the following logarithms.
   (a) $\log 356.8$ **(b)** $\log 35.68$ **(c)** $\log 3.568$
   (d) Observe your answers and make a conjecture concerning the decimal values of the common logarithms of numbers greater than 1 that have the same digits.

Use the change-of-base rule (either with common or natural logarithms) to find each of the following logarithms. Give approximations to four decimal places.

33. $\log_{16} 13$

34. $\log_4 12$

35. $\log_{\sqrt{6}} \sqrt{13}$

36. A population of hares in a specific area is growing according to the function

$$H(t) = 500 \log_3(2t + 3),$$

where t is time in years after the population was introduced into the area. Find the number of hares for the following times.
   (a) $t = 0$ **(b)** $t = 3$ **(c)** $t = 12$
   (d) Is the change-of-base rule needed to work this problem?

Use the formula $\text{pH} = -\log[H_3 O^+]$ *to find the pH of the substances with the given hydronium ion concentrations.*

37. Milk, 4.0×10^{-7}

38. Crackers, 3.8×10^{-9}

39. The population of Cairo, in millions, is approximated by the function $C(t) = 9e^{.026t}$, where $t = 0$ represents 1990. Find the population in the following years.
   (a) 2000   **(b)** 1970

40. Suppose the quantity, measured in grams, of a radioactive substance present at time t is given by

$$Q(t) = 500e^{-.05t},$$

where t is measured in days. Find the quantity present at the following times.
   (a) $t = 0$ **(b)** $t = 4$

[10.6] *Solve each equation. Give solutions to the nearest thousandth.*

41. $3^x = 9.42$ **42.** $2^{x-1} = 15$ **43.** $e^{.06x} = 3$

Solve each equation. Give exact solutions.

44. $\log_3(9x + 8) = 2$ **45.** $\log_5(y + 6)^3 = 2$

46. $\log_3(p + 2) - \log_3 p = \log_3 2$ **47.** $\log(2x + 3) = \log x + 1$

48. $\log_4 x + \log_4(8 - x) = 2$ **49.** $\log_2 x + \log_2(x + 15) = 4$

Solve each problem.

50. Refer to Exercise 39. In what year did the population reach 10 million?

51. Refer to Exercise 40. What is the half-life of the substance? (That is, how long would it take for the initial amount to become half of what it was?)

52. How much would be in an account after 3 years if $6500.00 was invested at 3% annual interest, compounded daily (use $n = 365$)?

53. Which is a better plan?
Plan A: Invest $1000.00 at 4% compounded quarterly for 3 years
Plan B: Invest $1000.00 at 3.9% compounded monthly for 3 years

---

MIXED REVIEW EXERCISES

Solve.

54. $\log_3(x + 9) = 4$ **55.** $\log_2 32 = x$ **56.** $\log_x \dfrac{1}{81} = 2$ **57.** $27^x = 81$

58. $2^{2x-3} = 8$ **59.** $\log_3(x + 1) - \log_3 x = 2$ **60.** $\log(3x - 1) = \log 10$

1. Decide whether the function is or is not one-to-one.
   (a) $f(x) = x^2 + 9$ (b)

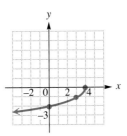

1. (a) _____

   (b) _____

2. Find $f^{-1}(x)$ for the one-to-one function $f(x) = \sqrt[3]{x + 7}$.

3. Graph the inverse of f, given the graph of f below.

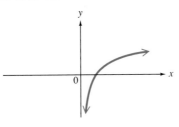

2. _____

3.

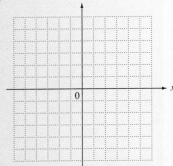

Graph the following functions.

4. $y = 6^x$

4.

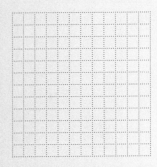

5. $y = \log_6 x$

5.

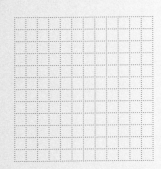

6. How can the graph of the function in Exercise 5 be obtained from the graph of the function in Exercise 4?

6. _____

Solve each equation. Give the exact solution.

7. _____

7. $5^x = \dfrac{1}{625}$

8. _____

8. $2^{3x-7} = 8^{2x+2}$

Solve the following problem.

9. (a) _____

(b) _____

(c) _____

(d)

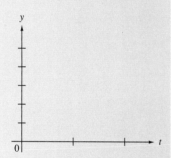

9. A small business estimates that the value y of a copy machine is decreasing according to the function

$$y = 5000(2)^{-.15t},$$

where t is the number of years that have elapsed since the machine was purchased, and y is in dollars.

(a) What was the original value of the machine? (*Hint:* Let $t = 0$.)

(b) What is the value of the machine 5 years after purchase? Give your answer to the nearest dollar.

(c) What is the value of the machine 10 years after purchase? Give your answer to the nearest dollar.

(d) Graph the function.

10. _____

10. Write in logarithmic form: $4^{-2} = .0625$

11. _____

11. Write in exponential form: $\log_7 49 = 2$

Solve each equation.

12. $\log_{1/2} x = -5$

12. _____

13. $x = \log_9 3$

13. _____

14. $\log_x 16 = 4$

14. _____

15. Fill in the blanks with the correct responses: The value of $\log_2 32$ is _____. This means that if we raise _____ to the _____ power, the result is _____.

15. _____

Use properties of logarithms to write the expressions in Exercises 16 and 17 as sums or differences of logarithms. Assume variables represent positive numbers.

16. $\log_3 x^2 y$

16. _____

17. $\log_5\left(\dfrac{\sqrt{x}}{yz}\right)$

17. _____

Use properties of logarithms to write the expressions in Exercises 18 and 19 as single logarithms. Assume variables represent positive real numbers, $b \neq 1$.

18. $3 \log_b s - \log_b t$

18. _____

19. $\dfrac{1}{4} \log_b r + 2 \log_b s - \dfrac{2}{3} \log_b t$

19. _____

Use a calculator to find an approximation to the nearest ten-thousandth for each of the following logarithms.

20. log 21.3

20. _____

21. ln .43

21. _____

22. (a) _____

(b) _____

22. (a) Between what two consecutive integers must the value of $\log_6 45$ be?
(b) Use a calculator to find an approximation of $\log_6 45$ to the nearest ten-thousandth.

23. _____

23. Solve for x, and give the solution correct to the nearest ten-thousandth.

$$3^x = 78$$

24. _____

24. Solve: $\log_8(x + 5) + \log_8(x - 2) = \log_8 8$.

25. (a) _____

(b) _____

25. Another way of writing the function of item 9 (describing the value of the office copy machine) is

$$y = 5000e^{-.104t}.$$

(a) What will be the value of the copy machine 15 years after purchase? Give your answer to the nearest dollar.
(b) To the nearest whole number, after how many years will the value of the machine be half of its original value?

Note: This cumulative review exercise set may be considered as a final examination for the course.

Let $S = \{-\frac{9}{4}, -2, -\sqrt{2}, 0, .6, \sqrt{11}, \sqrt{-8}, 6, \frac{30}{3}\}$. *List the elements of S that are members of the following sets.*

1. Integers

2. Rational numbers

3. Irrational numbers

4. Real numbers

Simplify the following.

5. $|-8| + 6 - |-2| - (-6 + 2)$

6. $-12 - |-3| - 7 - |-5|$

7. $2(-5) + (-8)(4) - (-3)$

Solve the following.

8. $7 - (3 + 4a) + 2a = -5(a - 1) - 3$

9. $2m + 2 \le 5m - 1$

10. $|2x - 5| = 9$

11. $|3p| - 4 = 12$

12. $|3k - 8| \leq 1$

13. $|4m + 2| > 10$

Perform the indicated operations.

14. $(2p + 3)(3p - 1)$

15. $(4k - 3)^2$

16. $(3m^3 + 2m^2 - 5m) - (8m^3 + 2m - 4)$

17. Divide $6t^4 + 17t^3 - 4t^2 + 9t + 4$ by $3t + 1$.

Factor as completely as possible.

18. $8x + x^3$

19. $24y^2 - 7y - 6$

20. $5z^3 - 19z^2 - 4z$

21. $16a^2 - 25b^4$

22. $8c^3 + d^3$

23. $16r^2 + 56rq + 49q^2$

Simplify as much as possible in Exercises 24–27.

24. $\dfrac{(5p^3)^4(-3p^7)}{2p^2(4p^4)}$

25. $\dfrac{x^2 - 9}{x^2 + 7x + 12} \div \dfrac{x - 3}{x + 5}$

26. $\dfrac{2}{k + 3} - \dfrac{5}{k - 2}$

27. $\dfrac{3}{p^2 - 4p} - \dfrac{4}{p^2 + 2p}$

28. Candy worth \$1.00 per pound is to be mixed with 10 pounds of candy worth \$1.96 per pound to get a mixture that will be sold for \$1.60 per pound. How many pounds of the \$1.00 candy should be used?

Simplify in Exercises 29–31.

29. $\left(\dfrac{5}{4}\right)^{-2}$

30. $\dfrac{6^{-3}}{6^{2}}$

31. $2\sqrt{32} - 5\sqrt{98}$

32. Multiply: $(5 + 4i)(5 - 4i)$.

Solve the equations or inequalities in Exercises 33 and 34.

33. $10p^2 + p - 2 = 0$

34. $k^2 + 2k - 8 > 0$

35. Recently the U.S. population has been growing according to the equation

$$y = 1.7x + 230$$

where y gives the population (in millions) in year x, measured from year 1980. For example, in 1980 $x = 0$ and $y = 1.7(0) + 230 = 230$. This means that the population was about 230 million in 1980. To find the population in 1985, let $x = 5$, and so on. Find the population in each of the following years.
(a) 1982   **(b)** 1985   **(c)** 1990
(d) In what year will the population reach 315 million, based on this equation?

36. The long-term debt of the Port of New Orleans dropped from $70,000,000 in 1986 to $25,300,000 in 1994. What was the rate of change in debt? (*Source:* Division of Finance and Accounting, Port of New Orleans)

37. Find the standard form of the equation of the line through $(5, -1)$ and parallel to the line with equation $3x - 4y = 12$.

Graph the following.

38. $y = -2.5x + 5$

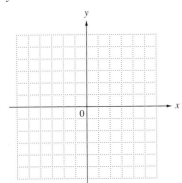

39. $-4x + y \leq 5$

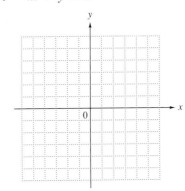

40. $y = \dfrac{1}{3}(x - 1)^2 + 2$

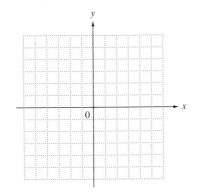

41. $\dfrac{x^2}{9} + \dfrac{y^2}{16} = 1$

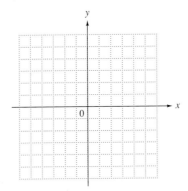

42. $25x^2 - 16y^2 = 400$

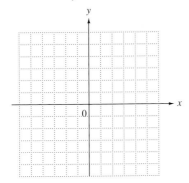

43. (a) Which of the *equations* in Exercises 38–42 define functions?
  (b) Of the functions described in part (a), which have inverses?

Solve each system of equations.

44. $5x - 3y = 14$
$2x + 5y = 18$

45. $x + 2y + 3z = 11$
$3x - y + z = 8$
$2x + 2y - 3z = -12$

46. Evaluate $\begin{vmatrix} 2 & 4 & 5 \\ 1 & 3 & 0 \\ 0 & -1 & -2 \end{vmatrix}$.

47. Graph $f(x) = 2^x$.

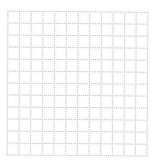

48. Solve $5^{x+3} = \left(\dfrac{1}{25}\right)^{3x+2}$.

49. Graph $f(x) = \log_3 x$.

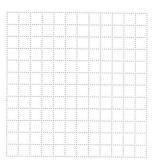

50. Solve $\log_5 x + \log_5(x + 4) = 1$.

Answers to Selected Exercises

The solutions to selected odd-numbered exercises are given in the section beginning on page A-37.

In this section we provide the answers that we think most students will obtain when they work the exercises using the methods explained in the text. If your answer does not look exactly like the one given here, it is not necessarily wrong. In many cases there are equivalent forms of the answer that are correct. For example, if the answer section shows $\frac{3}{4}$ and your answer is .75, you have obtained the right answer but written it in a different (yet equivalent) form. Unless the directions specify otherwise, .75 is just as valid an answer as $\frac{3}{4}$.

In general, if your answer does not agree with the one given in the text, see whether it can be transformed into the other form. If it can, then it is the correct answer. If you still have doubts, talk with your instructor.

CHAPTER 1

SECTION 1.1 (page 7)

1. true **3.** true **5.** False; absolute value is never negative.

7. **9.**

11. The graph of a number is a point on the number line. The coordinate of a point on the number line is the number that corresponds to the point.
13. $\{1, 2, 3, 4, 5\}$ **15.** $\{5, 6, 7, 8, \ldots\}$
17. $\{10, 12, 14, 16, \ldots\}$ **19.** $\emptyset$ **21.** $\{-4, 4\}$
23. $\{0, 3, 6, 9, \ldots\}$ **25.** yes
27. $\{x \mid x$ is a multiple of 4 greater than 0$\}$
29. $\{x \mid x$ is an even natural number less than or equal to 8$\}$
31. **(a)** $4, 5, 17, \frac{40}{2}$ (or 20) **(b)** $0, 4, 5, 17, \frac{40}{2}$

(c) $-8, 0, 4, 5, 17, \frac{40}{2}$ **(d)** $-8, -.6, 0, \frac{3}{4}, 4, 5, \frac{13}{2}, 17, \frac{40}{2}$

(e) $-\sqrt{5}, \sqrt{3}$ **(f)** All are real numbers except $\frac{1}{0}$. **(g)** $\frac{1}{0}$

33. **(a)** -6 **(b)** 6 **35.** **(a)** 12 **(b)** 12
37. **(a)** 0 **(b)** 0
39. **(a)** $-\frac{6}{5}$ **(b)** $\frac{6}{5}$ **41.** 8 **43.** -5 **45.** -2
47. -4.5 **49.** 5 **51.** 6 **53.** 22 **55.** 0
57. False; some are integers, but some, like $\frac{3}{4}$, are not.
59. False; no irrational number is an integer.
61. true **63.** False; no rational numbers are irrational.
65. true **67.** true
69. Pacific Ocean, Indian Ocean, Caribbean Sea, South China Sea, Gulf of California **71.** true
73. all real numbers **75.** all real numbers greater than or equal to 0

SECTION 1.2 (page 15)

1. true **3.** false **5.** true **7.** true
9. $2 > -3$. Both are true because -3 is less than 2 and 2 is greater than -3.
11. In $x < 5$, x represents any real number less than (but not equal to) 5. In $x \leq 5$, x represents any real number less than or equal to 5.
13. $7 > y$ **15.** $3t - 4 \leq 10$
17. $5 \geq 5$ **19.** $-3 < t < 5$ **21.** $-3 \leq 3x < 4$
23. $5x + 3 \neq 0$ **25.** true **27.** true
29. true **31.** $3 \geq 2$ **33.** $-3 \leq -3$ **35.** $5 \not< 3$
37. $(-2, \infty)$
39. $(-\infty, 6]$
41. $(0, 3.5)$
43. $[2, 7]$
45. $(-4, 3]$
47. 1989, 1990, 1993, 1994, 1995
49. greater than

SECTION 1.3 (page 23)

1. the numbers are additive inverses; $4 + (-4) = 0$
3. negative; $-7 + (-21) = -28$
5. the positive number has a larger absolute value; $15 + (-2) = 13$
7. the one with smaller absolute value is subtracted from the one with larger absolute value; $-15 - (-3) = -12$
9. negative; $(-5)(15) = -75$ **11.** 9 **13.** -19
15. $-\frac{19}{12}$ **17.** -1.85 **19.** -11 **21.** 21
23. -13 **25.** -10.18 **27.** $\frac{67}{30}$ **29.** -6
31. One example is $-4 - (-9) = -4 + 9 = 5$.
33. It is true for multiplication (and division). It is false for addition and for subtraction when the number to be subtracted has the smaller absolute value. A more precise statement is, "The product or quotient of two negatives is positive." **35.** -35 **37.** 40 **39.** 2 **41.** -12
43. $\frac{6}{5}$ **45.** 1 **47.** 5.88 **49.** -10.676 **51.** $\frac{1}{6}$
53. $-\frac{1}{7}$ **55.** $-\frac{3}{2}$ **57.** 50 **59.** -1000
61. 12 **63.** -7 **65.** 6 **67.** -4
69. undefined **71.** $\frac{25}{102}$ **73.** $-\frac{9}{13}$ **75.** -2.1
77. $10,000$ **79.** -11 **81.** 16 **83.** -4
85. -19 **87.** $112°$ Fahrenheit **89.** $-.5$ **91.** 2.9

SECTION 1.4 (page 33)

1. False; $-4^6 = -(4^6)$. **3.** true **5.** true **7.** true
9. False; the base is 3.

11. 16   **13.** .021952   **15.** $\dfrac{1}{125}$ **17.** $\dfrac{343}{1000}$

19. -125 **21.** 256   **23.** -729 **25.** -4096

27. $(-a)^n = -a^n$ if a is a real number and n is an odd natural number.

29. exponent: 7; base: -4.1 **31.** exponent: 7; base: 4.1

33. 9   **35.** 13   **37.** -20 **39.** $\dfrac{10}{11}$ **41.** $-.7$

43. not a real number

45. $\sqrt{16}$ is the positive (or principal) square root of 16, namely 4. -4 is the negative square root of 16, written $-\sqrt{16}$.

47. not a real number   **49.** -3 **51.** 5   **53.** 3

55. -3 **57.** 4   **59.** 14   **61.** -91 **63.** -48

65. 8   **67.** -2 **69.** undefined   **71.** $\dfrac{7}{5}$

73. 7   **75.** -1 **77.** 13   **79.** -96 **81.** $-\dfrac{15}{238}$

83. 42   **85.** $\dfrac{20}{7}$

SECTION 1.5 (page 43)

1. 1900   **3.** 75

5. The identity element for addition is 0.

7. Like terms are terms with the same variables to the same powers.

9. The commutative property is used to change the order of two terms or factors, while the associative property is used to change the grouping.

11. $8k$ **13.** $-2r$ **15.** $-8z + 4w$ (cannot be simplified)   **17.** $6a$ **19.** $2m + 2p$

21. $-10d + 5f$ **23.** $-6y + 3$ **25.** $p + 11$

27. $-2k + 15$ **29.** $m - 14$ **31.** -1

33. $2p + 7$ **35.** $-6z - 39$ **37.** $(5 + 8)x = 13x$

39. $(5 \cdot 9)r = 45r$ **41.** $9y + 5x$ **43.** 7

45. $8(-4) + 8x = -32 + 8x$

47. Answers will vary. One example is washing your face and brushing your teeth.

49. Yes. Any different nonzero numbers a and b that have the same absolute value satisfy $\frac{a}{b} = \frac{b}{a}$. For example, $a = -5$ and $b = 5$.

51. associative property of addition

52. associative property of addition

53. commutative property of addition

54. associative property of addition

55. distributive property

56. arithmetic facts

CHAPTER 1 REVIEW EXERCISES (page 49)

1.

2.

3. 16   **4.** 23   **5.** -4 **6.** -5 **7.** $0, \dfrac{12}{3}$ (or 4)

8. $-9, -\sqrt{4}$ (or -2), $0, \dfrac{12}{3}$ (or 4)

9. $-9, -\dfrac{4}{3}, -\sqrt{4}$ (or -2), $-.25, 0, .\overline{35}, \dfrac{5}{3}, \dfrac{12}{3}$ (or 4)

10. All are real numbers except $\sqrt{-9}$.

11. $\{4, 5, 6, 7, 8\}$ **12.** $\{0, 1, 2, 3\}$

13. true   **14.** false   **15.** true

16. $(-\infty, -5)$

17. $(-2, 3]$

18. $\dfrac{41}{24}$ **19.** $-\dfrac{1}{2}$ **20.** -3 **21.** -17.09

22. -39 **23.** -1 **24.** $\dfrac{23}{20}$ **25.** $-\dfrac{5}{18}$ **26.** -35

27. If the signs of the numbers being added are the same, the sign of the sum matches them. If the signs are different, the sign of the sum matches the one with the larger absolute value.

28. To subtract $a - b$, write as an addition problem, $a + (-b)$, and add.   **29.** -90

30. $\dfrac{2}{3}$ **31.** -11.408 **32.** 2

33. -15 **34.** -4 **35.** 3.21

36. $\dfrac{5}{7-7}$ is undefined.   **37.** 10,000   **38.** $\dfrac{27}{343}$

39. -125 **40.** -125 **41.** 2.89   **42.** 20

43. 3   **44.** -7 **45.** 3   **46.** not a real number

47. -4 **48.** 44   **49.** $\dfrac{7}{3}$ **50.** -2 **51.** -30

52. -30 **53.** -116 **54.** $-\dfrac{8}{51}$

55. Work within the parentheses first.

56. $(4 + 6)^2 = 10^2 = 100$; $4^2 + 6^2 = 52$; $100 \neq 52$

57. $21q$ **58.** $-4z$ **59.** $5m$ **60.** $4p$

61. $-2k - 6$ **62.** $6r + 18$ **63.** $18m + 27n$

64. $-3k + 4h$ **65.** $-p - 3q$ **66.** $-6x + 6$

67. $y + 1$ **68.** 0   **69.** 2   **70.** $-18m$

71. $(2 + 3)x = 5x$ **72.** -4 **73.** $(2 \cdot 4)x = 8x$

74. $13 + (-3) = 10$ **75.** 0   **76.** $5x + 5z$

77. 7   **78.** 1   **79.** $(3 + 5 + 6)a = 14a$ **80.** 0

81. $\dfrac{256}{625}$ **82.** 25   **83.** 31   **84.** 9   **85.** 0

86. -2 **87.** $\dfrac{4}{3}$ **88.** -6.16 **89.** -9

CHAPTER 1 TEST (page 53)

1.

2. $0, 3, \sqrt{25}$ (or 5), $\dfrac{24}{2}$ (or 12)

3. $-1, 0, 3, \sqrt{25}$ (or 5), $\dfrac{24}{2}$ (or 12)

4. $-1, -.5, 0, 3, \sqrt{25}$ (or 5), $7.5, \dfrac{24}{2}$ (or 12)

5. All are real numbers except $\sqrt{-4}$.

6. $(-\infty, -3)$ ←————————→
-3

7. $(-4, 2]$ ←————————→
$-4 \qquad 2$

8. 0   **9.** -26 **10.** 19   **11.** 1

12. $\dfrac{16}{7}$ **13.** $\dfrac{11}{23}$ **14.** 14   **15.** -15 **16.** -3

17. not a real number   **18.** if n is an odd number

19. 2   **20.** $-\dfrac{6}{23}$ **21.** $10k - 10$

22. Both terms change sign and are added to $3r + 8$; $7r + 2$.   **23.** B   **24.** E   **25.** D   **26.** A

27. F   **28.** C   **29.** C   **30.** E

CHAPTER 2

SECTION 2.1 (page 61)

1. Both sides are evaluated as 30.
3. For any number, we have a solution. For example, if the last name is *Lincoln,* we have $x = 7$. Both sides are evaluated as -48.   **5.** $\{-1\}$ **7.** $\{3\}$ **9.** $\{-7\}$

11. $\{0\}$ **13.** $\left\{-\dfrac{5}{3}\right\}$ **15.** $\left\{-\dfrac{1}{2}\right\}$ **17.** $\{2\}$

19. $\{-2\}$ **21.** $\{7\}$ **23.** $\{-5\}$ **25.** 12
27. Yes, you will get the correct solution. The coefficients will be larger, but in the end the solution will be the same.
29. $\{4\}$ **31.** $\{0\}$ **33.** $\{0\}$ **35.** $\{2000\}$
37. $\{25\}$ **39.** $\{40\}$ **41.** A conditional equation has one solution, an identity has infinitely many solutions, and a contradiction has no solution.   **43.** contradiction; $\emptyset$
45. conditional; $\{0\}$ **47.** identity; $\{$all real numbers$\}$
49. $\{-8\}$ **50.** $\{-8\}$ **51.** 33   **52.** 33   **53.** $\{-8\}$
54. -7 **55.** solution set   **57.** The solution set of the first equation is $\{-8\}$, while the solution set of the second equation is $\emptyset$. They are not the same.
59. The solution sets, $\{4\}$ and $\{-4, 4\}$, are not the same.
61. 2820 million dollars; 1996 (when $x = 7$)
63. (a) 230 billion dollars   **(b)** 275 billion dollars
(c) 300 billion dollars   **(d)** 335 billion dollars

SECTION 2.2 (page 71)

1. (a) $3x = 5x + 8$ **(b)** $tc = bt + k$
2. (a) $3x - 5x = 8$ **(b)** $tc - bt = k$
3. (a) $-2x = 8$ **(b)** $t(c - b) = k$; distributive

property   **4. (a)** $x = -4$ **(b)** $t = \dfrac{k}{c - b}$

5. $c \neq b$; If c is equal to b, the denominator is 0.
6. To solve an equation for a particular variable, such as solving the second equation for t, go through the same steps as you would in solving for x in the first equation. Treat all other variables as if they were constants.
7. yes; the distributive property

9. $r = \dfrac{I}{pt}$ **11.** $L = \dfrac{P - 2W}{2}$ **13.** $W = \dfrac{V}{LH}$

15. $r = \dfrac{C}{2\pi}$ **17.** $B = \dfrac{2A}{h} - b$ or $B = \dfrac{2A - bh}{h}$

19. $C = \dfrac{5}{9}(F - 32)$ **21.** $r = \dfrac{-2k - 3y}{a - 1}$ or $r = \dfrac{2k + 3y}{1 - a}$

23. $y = \dfrac{-x}{w - 3}$ or $y = \dfrac{x}{3 - w}$

25. Yes, because multiplying both numerator and denominator by -1 gives this equivalent answer.
27. about 3.108 hours   **29.** -40 degrees Fahrenheit
31. 230 meters   **33.** radius: 240 inches; diameter: 480 inches   **35.** 8 feet   **37. (a)** 174   **(b)** 106
(c) 123   **(d)** 73   **(e)** 84   **39. (a)** \$527,850,000
(b) \$596,700,000   **(c)** \$160,650,000
41. 75% water, 25% alcohol   **43.** 63,120
45. 3%   **47.** 5.1%

SECTION 2.3 (page 83)

1. *Step 1* Read the problem.   *Step 2* Choose a variable and identify what it represents.   *Step 3* Write an equation. *Step 4* Solve the equation.   *Step 5* Answer the question. *Step 6* Check your answer.

3. $x - 13$ **5.** $7 + x$ **7.** $8(x + 12)$ **9.** $\dfrac{x}{6}$

Let x represent the unknown in Exercises 11, 13, and 15.

11. $x + 6 = -31$; -37 **13.** $x - (-4x) = x + 9$; $\dfrac{9}{4}$

15. $12 - \dfrac{2}{3}x = 10$; 3   **17.** (d)   **19.** expression

21. equation   **23.** expression   **25.** width: 165 feet; length: 265 feet   **27.** 850 miles, 925 miles, 1300 miles
29. Boeing: 403; McDonnell Douglas: 293
31. General Motors: 121.1 billion dollars; Ford Motor: 92.4 billion dollars   **33.** 7.0%   **35.** \$2008
37. \$122.28   **39.** \$4000 at 3%; \$8000 at 4%
41. \$10,000 at 4.5%; \$19,000 at 3%   **43.** \$58,000
45. 5 liters   **47.** 4 liters   **49.** 1 gallon
51. 150 pounds   **53.** One cannot expect the final mixture to be worth more than each of the ingredients.
55. (a) $800 - x$ **(b)** $800 - y$
56. (a) $.05x$; $.10(800 - x)$ **(b)** $.05y$; $.10(800 - y)$
57. (a) $.05x + .10(800 - x) = 800(.0875)$
(b) $.05y + .10(800 - y) = 800(.0875)$
58. (a) \$200 at 5%; \$600 at 10%
(b) 200 liters of 5% acid; 600 liters of 10% acid
59. The processes are the same. The amounts of money in Problem A correspond to the amounts of pure acid in Problem B.

SECTION 2.4 (page 93)

1. \$4.10   **3.** 52 miles per hour   **5.** The problem asks for the *distance* to the workplace. To find this distance, we must multiply the rate, 10 miles per hour, by the time, $\frac{3}{4}$ hour.   **7.** 13 quarters, 23 half-dollars
9. 17 pennies, 17 dimes, 10 quarters   **11.** 11 two-cent pieces, 33 three-cent pieces   **13.** 305 students, 105 nonstudents   **15.** 10.35 meters per second

17. 8.29 meters per second   **19.** $1\dfrac{3}{4}$ hours

21. 10:00 A.M.   **23.** 18 miles   **25.** 55 miles per hour
27. 20°, 30°, 130°
29. 40°, 80°   **30.** 120°   **31.** The sum is equal to the measure of the angle found in Exercise 30.   **32.** The sum of the measures of angles ① and ② is equal to the measure of angle ③.   **33.** both measure 49°
35. 65°, 115°   **37.** 27, 28, 29, 30   **39.** 76, 77

Section 2.5 (page 105)

1. D   **3.** B   **5.** F   **7.** Use a parenthesis when an endpoint is not included; use a bracket when it is included.
9. $(-\infty, -3]$

-3

11. $[5, \infty)$

5

13. $(7, \infty)$

7

15. $(-4, \infty)$

-4

17. $(-\infty, -40]$

-40

19. $[3, \infty)$

3

21. $(-\infty, 4]$

4

23. $\left(-\infty, -\dfrac{15}{2}\right)$

$-\dfrac{15}{2}$

25. $\left[\dfrac{1}{2}, \infty\right)$

$\dfrac{1}{2}$

27. $(3, \infty)$

3

29. $(-\infty, 4)$

4

31. $\{-9\}$

-9

32. $(-9, \infty)$

-9

33. $(-\infty, -9)$

-9

34. We obtain the set of all real numbers.

-9

35. $(-\infty, -3)$
36. On the number line, the point that corresponds to a number separates those points on the line that correspond to numbers less than that number from those that are greater than that number.
37. $(1, 11)$

$1 \qquad 11$

39. $[-14, 10]$

$-14 \qquad 10$

41. $[-5, 6]$

$-5 \qquad 6$

43. $\left[-\dfrac{14}{3}, 2\right]$

$-\dfrac{14}{3} \qquad 2$

45. $\left[-\dfrac{1}{2}, \dfrac{35}{2}\right]$

$-\dfrac{1}{2} \qquad \dfrac{35}{2}$

47. $\left(-\dfrac{1}{3}, \dfrac{1}{9}\right]$

$-\dfrac{1}{3} \qquad \dfrac{1}{9}$

49. April, May, June, July   **51.** January, February, March, August, September, October, November, December
53. at least 80   **55.** 50 miles   **57.** 26 tapes

Section 2.6 (page 113)

1. true   **3.** true   **5.** False; 6 is not included in the union.   **7.** $\{4\}$ or D **9.** $\emptyset$ **11.** $\{1, 2, 3, 4, 5, 6\}$ or A **13.** $\{1, 3, 5, 6\}$ **15.** $\{1, 4, 6\}$ **17.** Each is equal to $\{1\}$. This illustrates the associative property of set intersection.   **19.** The intersection of two streets is the region common to *both* streets.
21. $(-3, 2)$

$-3 \qquad 2$

23. $(-\infty, 2]$

2

25. $\emptyset$

27. $[4, \infty)$

4

29. $[-1, 3]$

$-1 \qquad 3$

31. $[5, 9]$

$5 \qquad 9$

33. $(-\infty, 4]$

4

35. $(-\infty, 2] \cup [4, \infty)$

$2 \qquad 4$

37. $(-\infty, 8]$

8

39. $(-\infty, 1] \cup [10, \infty)$

$1 \qquad 10$

41. $[-2, \infty)$

-2

43. $(-\infty, \infty)$

0

45. $(-\infty, -5) \cup (5, \infty)$

$-5 \qquad 5$

47. $(-\infty, 2) \cup (2, \infty)$

2

49. intersection; $(-5, -1)$

$-5 \qquad -1$

51. union; $(-\infty, 4)$

4

53. intersection; $[4, 12]$

$4 \qquad 12$

55. union; $(-\infty, 0] \cup [2, \infty)$

$0 \qquad 2$

57. Mario, Joe   **58.** none of them
59. none of them   **60.** Luigi, Than

Section 2.7 (page 123)

1. E; C; D; B; A   **3.** Use *or* for the equality statement and the $>$ statement. Use *and* for the $<$ statement.
5. $\{-12, 12\}$ **7.** $\{-5, 5\}$ **9.** $\{-6, 12\}$

11. $\{-4, 3\}$ **13.** $\left\{-3, \dfrac{11}{2}\right\}$ **15.** $\left\{-\dfrac{19}{2}, \dfrac{9}{2}\right\}$

17. $\{-10, -2\}$ **19.** $\left\{-8, \dfrac{32}{3}\right\}$

21. $(-\infty, -3) \cup (3, \infty)$

23. $(-\infty, -4] \cup [4, \infty)$

25. $(-\infty, -12) \cup (8, \infty)$

27. $\left(-\infty, -\dfrac{7}{3}\right] \cup [3, \infty)$

29. $(-\infty, -2) \cup (8, \infty)$

31. (a)

  (b)

33. $[-3, 3]$

35. $(-4, 4)$

37. $[-12, 8]$

39. $\left(-\dfrac{7}{3}, 3\right)$

41. $[-2, 8]$

43. $(-\infty, -5) \cup (13, \infty)$

45. $\{-6, -1\}$

47. $\left[-\dfrac{10}{3}, 4\right]$

49. $\left[-\dfrac{7}{6}, -\dfrac{5}{6}\right]$

51. $\{-5, 5\}$ **53.** $\{-5, -3\}$ **55.** $(-\infty, -3) \cup (2, \infty)$

57. $[-10, 0]$ **59.** $\{-1, 3\}$ **61.** $\left\{-3, \dfrac{5}{3}\right\}$

63. $\left\{-\dfrac{1}{3}, -\dfrac{1}{15}\right\}$ **65.** $\left\{-\dfrac{5}{4}\right\}$ **67.** $\emptyset$

69. $\left\{-\dfrac{1}{4}\right\}$ **71.** $\emptyset$ **73.** $(-\infty, \infty)$ **75.** $\left\{-\dfrac{3}{7}\right\}$

77. $(-\infty, \infty)$ **79.** $\left(-\infty, -\dfrac{7}{10}\right) \cup \left(-\dfrac{7}{10}, \infty\right)$

81. 460.2 feet   **82.** Federal Office Building, City Hall, Kansas City Power and Light, Hyatt Regency

83. Southwest Bell Telephone, City Center Square, Commerce Tower, Federal Office Building, City Hall, Kansas City Power and Light, Hyatt Regency
84. (a) $\left| x - 460.2 \right| \geq 75$ **(b)** $x \geq 535.2$ or $x \leq 385.2$
(c) Pershing Road Associates, AT&T Town Pavilion, One Kansas City Place **(d)** It makes sense because it includes all buildings *not* listed earlier.

SUMMARY ON SOLVING LINEAR AND ABSOLUTE VALUE EQUATIONS AND INEQUALITIES   (page 129)

1. $\{12\}$ **3.** $\{7\}$ **5.** $\emptyset$ **7.** $\left[-\dfrac{2}{3}, \infty\right)$

9. $\{-3\}$ **11.** $(-\infty, 5]$ **13.** $\{2\}$ **15.** $\emptyset$

17. $(-5.5, 5.5)$ **19.** $\left\{-\dfrac{96}{5}\right\}$ **21.** $(-\infty, -24)$

23. $\left\{\dfrac{7}{2}\right\}$ **25.** $(-\infty, \infty)$ **27.** $(-\infty, -4) \cup (7, \infty)$

29. $\left\{-\dfrac{1}{5}\right\}$ **31.** $\left[-\dfrac{1}{3}, 3\right]$ **33.** $\left\{-\dfrac{1}{6}, 2\right\}$

35. $(-\infty, -1] \cup \left[\dfrac{5}{3}, \infty\right)$ **37.** $\left\{-\dfrac{5}{2}\right\}$

39. $\left[-\dfrac{9}{2}, \dfrac{15}{2}\right]$ **41.** $(-\infty, \infty)$ **43.** $(-\infty, \infty)$

45. $\{-2\}$ **47.** $(-\infty, -1) \cup (2, \infty)$

CHAPTER 2 REVIEW EXERCISES   (page 135)

1. $\left\{-\dfrac{9}{5}\right\}$ **2.** $\left\{\dfrac{1}{3}\right\}$ **3.** $\{10\}$ **4.** $\left\{-\dfrac{7}{5}\right\}$

5. $\emptyset$ **6.** $\{0\}$ **7.** $\{16\}$ **8.** $\{300\}$ **9.** (b)
10. Begin by subtracting 5 from both sides. Then divide both sides by -2.
11. identity; $(-\infty, \infty)$ **12.** contradiction; $\emptyset$
13. conditional; $\{0\}$

14. $H = \dfrac{V}{LW}$ **15.** $h = \dfrac{2A}{B + b}$ **16.** $d = \dfrac{C}{\pi}$

17. 6 feet   **18.** 9.8%   **19.** 6.5%
20. 20°   **21.** 37.4%   **22.** 100 millimeters

23. $9 - \dfrac{1}{3}x$ **24.** $\dfrac{4x}{x + 9}$

25. length: 13 meters; width: 8 meters
26. 17 inches, 17 inches, 19 inches
27. 12 kilograms   **28.** 30 liters
29. $10,000 at 6%; $6000 at 4%   **30.** 66.6%
31. (a)   **32.** (a) 530 miles **(b)** 328 miles
33. 2.2 hours   **34.** 50 kilometers per hour; 65 kilometers per hour   **35.** 1 hour
36. 46 miles per hour
37. 850 reserved; 246 general admission
38. 249 student; 62 nonstudent
39. 40°, 45°, 95°   **40.** (c)   **41.** 105° for each angle
42. 46, 47, 48

43. $(-9, \infty)$
-9

44. $(-\infty, -3]$
-3

45. $\left(\dfrac{3}{2}, \infty\right)$
$\dfrac{3}{2}$

46. $\left(-\infty, -\dfrac{14}{9}\right)$
$-\dfrac{14}{9}$

47. $[-3, \infty)$
-3

48. $[-3, 12]$
-3 12

49. $[3, 5)$
3 5

50. $\left(-3, \dfrac{7}{2}\right)$
-3 $\dfrac{7}{2}$

51. any grade greater than or equal to 61%
52. Because the statement $-8 < -13$ is *false*, the inequality has no solution.
53. $\{3, 9\}$ **54.** $\{1, 3, 5, 6, 7, 9, 12\}$
55. $(6, 9)$
6 9

56. $(8, 14)$
8 14

57. $(-\infty, -3] \cup (5, \infty)$
-3 5

58. $(-\infty, \infty)$
0

59. $\emptyset$

60. $(-\infty, -2] \cup [7, \infty)$
-2 7

61. 1988, 1989, 1990, 1991, 1992
62. 1981, 1982, 1983, 1984, 1985, 1988, 1989, 1990, 1991, 1992
63. $\{-7, 7\}$ **64.** $\{-11, 7\}$
65. $\left\{-\dfrac{1}{3}, 5\right\}$ **66.** $\emptyset$ **67.** $\{0, 7\}$ **68.** $\left\{-\dfrac{3}{2}, \dfrac{1}{2}\right\}$
69. $\left\{-\dfrac{3}{4}, \dfrac{1}{2}\right\}$ **70.** $\left\{-\dfrac{1}{2}\right\}$
71. $(-14, 14)$
-14 14

72. $[-1, 13]$
-1 13

73. $[-3, -2]$
-3 -2

74. $(-\infty, \infty)$
0

75. $\left(-\infty, -\dfrac{8}{5}\right) \cup (2, \infty)$
$-\dfrac{8}{5}$ 2

76. $(-\infty, \infty)$
0

77. $\left(-\infty, \dfrac{7}{6}\right]$ **78.** $[-4, 5)$ **79.** $\emptyset$
80. 6 inches, 12 inches, 16 inches   **81.** $(-\infty, 2]$
82. $(-\infty, -1) \cup \left(\dfrac{11}{7}, \infty\right)$
83. $\{-5, 15\}$ **84.** $[-16, 10]$
85. 6 inches   **86.** 683 votes and 532 votes

CHAPTER 2 TEST   (page 143)

1. $\{-19\}$ **2.** $\{5\}$ **3.** $(-\infty, \infty)$
4. $L = \dfrac{P - 2W}{2}$ or $L = \dfrac{P}{2} - W$
5. 3.2 hours   **6.** 3.75%   **7.** 9696 residents
8. $8000 at 3%; $20,000 at 5%   **9.** faster car: 60 miles per hour; slower car: 45 miles per hour
10. 40°, 40°, 100°
11. We must reverse the direction of the inequality symbol.
12. $[1, \infty)$
1

13. $(-\infty, 28)$
28

14. $[-3, 3]$
-3 3

15. 94 or greater
16. $[2, 9)$
2 9

17. $\left[-\dfrac{5}{2}, 1\right]$
$-\dfrac{5}{2}$ 1

18. $\left(-\infty, -\dfrac{7}{6}\right) \cup \left(\dfrac{17}{6}, \infty\right)$
$-\dfrac{7}{6}$ $\dfrac{17}{6}$

19. $\left\{-\dfrac{5}{3}, 3\right\}$ **20.** $\left\{-\dfrac{5}{7}, \dfrac{11}{3}\right\}$

CUMULATIVE REVIEW 1–2   (page 145)

1. $9, \sqrt{36}$ (or 6)   **2.** $0, 9, \sqrt{36}$ (or 6)
3. $-8, 0, 9, \sqrt{36}$ (or 6)
4. $-8, -\dfrac{2}{3}, 0, \dfrac{4}{5}, 9, \sqrt{36}$ (or 6)   **5.** $-\sqrt{6}$
6. All are real numbers.
7. $-\dfrac{22}{21}$ **8.** 8   **9.** 8   **10.** 0
11. -243 **12.** $\dfrac{216}{343}$ **13.** 1   **14.** -4096
15. $\sqrt{-36}$ is not a real number.
16. $\dfrac{4 + 4}{4 - 4}$ is undefined.
17. -16 **18.** -34 **19.** 184

20. $\dfrac{27}{16}$ **21.** $-20r + 17$ **22.** $13k + 42$

23. commutative property   **24.** distributive property

25. inverse property   **26.** $-\dfrac{3}{2}$ **27.** $\{5\}$ **28.** $\{30\}$

29. $\{15\}$ **30.** $b = P - a - c$

31. $[-14, \infty)$ ———————▶
$\qquad\qquad\qquad -14$

32. $\left[\dfrac{5}{3}, 3\right)$ ———————▶
$\qquad\qquad \dfrac{5}{3} \qquad 3$

33. $(-\infty, 0) \cup (2, \infty)$ ◀———————▶
$\qquad\qquad\qquad 0 \quad 2$

34. $\left(-\infty, -\dfrac{1}{7}\right] \cup [1, \infty)$ ◀———————▶
$\qquad\qquad\qquad -\dfrac{1}{7} \quad 1$

35. managerial and professional specialty

36. mathematical and computer scientists

37. 2 liters   **38.** 9 cents, 12 nickels, 8 quarters

CHAPTER 3

SECTION 3.1 (page 153)

1. **(a)** between 1989 and 1990   **(b)** between 1991 and 1992   **(c)** 1991   **3.** Another name is the Cartesian system, named after René Descartes.   **5.** origin
7. y; x **9.** two   **11.** **(a)** I   **(b)** III   **(c)** II   **(d)** IV
(e) none   **13.** **(a)** I or III   **(b)** II or IV   **(c)** II or IV
(d) I or III
15–24.

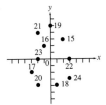

25. -3; 3; 2; -1

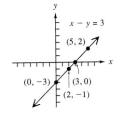

27. $\dfrac{5}{2}$; 5; $\dfrac{3}{2}$; 1

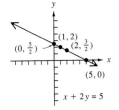

29. -4; 5; $-\dfrac{12}{5}$; $\dfrac{5}{4}$

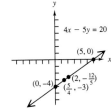

31. $x = 0$
33. $(6, 0)$; $(0, 4)$

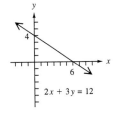

35. $(6, 0)$; $(0, -2)$

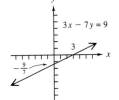

37. $(3, 0)$; $\left(0, -\dfrac{9}{7}\right)$

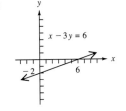

39. none; $(0, 5)$

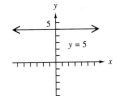

41. $(2, 0)$; none

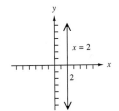

43. $(0, 0)$; $(0, 0)$

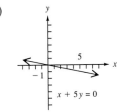

45. 218.753 miles per hour   **47.** (c)   **49.** (d)

SECTION 3.2 (page 163)

1. 2 **3.** undefined   **5.** (a), (b), (c), (d), (f)

7. 8 **9.** $\frac{5}{6}$ **11.** 0

13. undefined

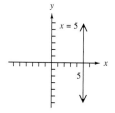

15. 0

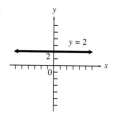

17. B   **19.** A

21. $-\frac{1}{2}$

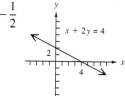

23. 1

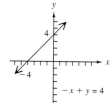

25. $-\frac{6}{5}$

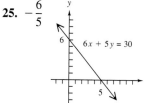

27. $\frac{5}{2}$

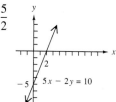

29. 4

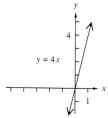

31. 0

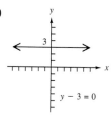

33.

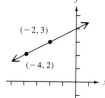

35.

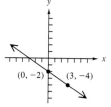

37.

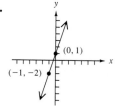

39. perpendicular   **41.** parallel   **43.** neither

45. $\frac{7}{10}$ **47.** (a) \$.92   (b) It means an *increase* in price.

49. 6.25 billion minutes per year

51. $\frac{1}{3}$ **52.** $\frac{1}{3}$ **53.** $\frac{1}{3}$ **54.** $\frac{1}{3} = \frac{1}{3} = \frac{1}{3}$ is true.

55. They are collinear.   **56.** They are not collinear.

SECTION 3.3 (page 177)

1. A   **3.** C   **5.** H   **7.** B   **9.** $3x + 4y = 10$

11. $2x + y = 18$ **13.** $x - 2y = -13$ **15.** $y = 12$

17. $x = 9$ **19.** $x = .5$ **21.** $2x - y = 2$

23. $x + 2y = 8$ **25.** $2x - 13y = -6$ **27.** $y = 5$

29. $x = 7$ **31.** $y = -\frac{5}{2}x + 10; -\frac{5}{2}; (0, 10)$

33. $y = \frac{2}{3}x - \frac{10}{3}; \frac{2}{3}; \left(0, -\frac{10}{3}\right)$

35. $y = 5x + 15$ **37.** $y = -\frac{2}{3}x + \frac{4}{5}$

39. $y = \frac{2}{5}x + 5$ **41.** $3x - y = 19$

43. $x - 2y = 2$ **45.** $x + 2y = 18$ **47.** $y = 7$

49. $y = 45x$; (0, 0), (5, 225), (10, 450)

51. $y = 1.30x$; (0, 0), (5, 6.50), (10, 13.00)

53. $y = 3x + 15$; (0, 15), (5, 30), (10, 45)

55. $y = .10x + 25.00$; (0, 25.00), (5, 25.50), (10, 26.00)

57. $69 = 3x + 15$; 18 days

59. $42.30 = .10x + 25.00$; 173 miles

61. $y = -\dfrac{829}{8}x + 29{,}557$

63. A is y_1 and B is y_2.   **65.** 8

67. 32; 212   **68.** (0, 32) and (100, 212)   **69.** $\dfrac{9}{5}$

70. $F = \dfrac{9}{5}C + 32$ **71.** $C = \dfrac{5}{9}(F - 32)$

72. When the Celsius temperature is 50°, the Fahrenheit temperature is 122°.

SECTION 3.4 (page 189)

1. solid; below   **3.** dashed; above

5. 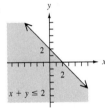   **7.**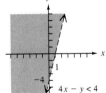

9.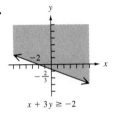

11.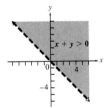

13.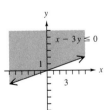

15.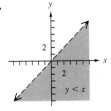

17. If the connecting word is "and," use intersection. If the connecting word is "or," use union.

19.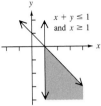

21.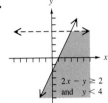

23.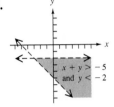

25.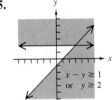

27.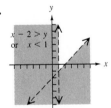

29.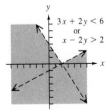

31.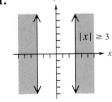

33.

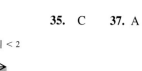

35. C   **37.** A

SECTION 3.5 (page 201)

1. (1990, 746,220), (1992, 661,391), (1993, 590,324)
3. (1990, 285.7), (1992, 318.8), (1993, 339.9)
5. The vertical line test is used to determine whether a graph is that of a function. Any vertical line will intersect the graph of a function in at most one point.
7. function; domain: $\{5, 3, 4, 7\}$; range: $\{1, 2, 9, 3\}$
9. not a function; domain: $\{2, 0\}$; range: $\{4, 2, 6\}$
11. not a function; domain: $\{1, 2, 3, 5\}$; range: $\{10, 15, 19, -27\}$
13. function; domain: $(-\infty, \infty)$; range: $(-\infty, 4]$
15. not a function; domain: $[-4, 4]$; range: $[-3, 3]$
17. function; domain: $(-\infty, \infty)$ **19.** not a function; domain: $[0, \infty)$ **21.** not a function; domain: $(-\infty, \infty)$
23. function; domain: $[0, \infty)$ **25.** function; domain: $(-\infty, 0) \cup (0, \infty)$ **27.** function (also a linear function); domain: $(-\infty, \infty)$ **29.** function; domain: $\left[-\frac{1}{2}, \infty\right)$ **31.** function; domain: $(-\infty, 9) \cup (9, \infty)$
33. 4   **35.** -11 **37.** $-3p + 4$ **39.** $3x + 4$
41. $-3x - 2$ **43.** -8 **45.** No. In general, $f(g(x)) \neq g(f(x))$.   **47.** line; -2; linear; $-2x + 4$; -2; $3; -2$
49. domain: $(-\infty, \infty)$; range: $(-\infty, \infty)$

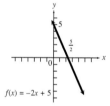

51. domain: $(-\infty, \infty)$; range: $(-\infty, \infty)$

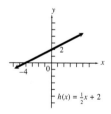

53. domain: $(-\infty, \infty)$; range: $\{2\}$

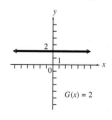

55. (a) 29,562   **(b)** 29,316   **(c)** 28,947
57. In 1986 (when $x = 2$), the number of post offices was approximately 29,439.   **59.** $f(3) = 7$
61. $f(x) = -3x + 5$

SECTION 3.6 (page 213)

1. 36   **3.** .625   **5.** $222\frac{2}{9}$ **7.** increases; decreases
9. $\$1.09\frac{9}{10}$ **11.** 8 pounds   **13.** 100 cycles per second
15. 3 footcandles   **17.** $420   **19.** 800 gallons
21. 25   **23.** 9   **25.** $(0, 0)$, $(1, 1.25)$ **26.** 1.25
27. $y = 1.25x + 0$ or $y = 1.25x$ **28.** $a = 1.25, b = 0$
29. It is the price per gallon and the slope of the line.
30. It can be written in the form $y = kx$ (where $k = a$). The value of a is called the constant of variation.

CHAPTER 3 REVIEW EXERCISES (page 221)

1. $3; 2; 0; \dfrac{10}{3}$

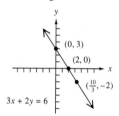

2. $-4; 3; -5; 4$

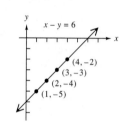

3. $(3, 0)$; $(0, 4)$

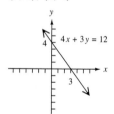

4. $(3, 0)$; $\left(0, \dfrac{15}{7}\right)$

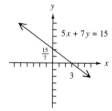

5. $-\dfrac{8}{5}$ **6.** 2   **7.** $\dfrac{3}{4}$ **8.** 0
9. $-\dfrac{2}{3}$ **10.** $-\dfrac{1}{3}$ **11.** positive slope
12. negative slope   **13.** 0 slope   **14.** undefined slope
15. $y = \dfrac{3}{5}x - 8$ **16.** $y = -\dfrac{1}{3}x + 5$ **17.** $y = 12$
18. $x = 2$ **19.** $y = 4$ **20.** $x = .3$ **21.** $9x + y = 13$
22. $7x - 5y = -16$ **23.** $4x - y = 26$

24. $5x + 2y = 2$ **25. (a)** $9014.25   **(b)** 1984
(c)

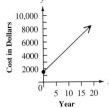

26. (a) 216.25 miles per hour   **(b)** .860 mile per hour; there is a discrepancy because the equation is only an *approximate* linear model.

27.

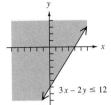

$3x - 2y \le 12$

28.

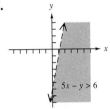

$5x - y > 6$

29.

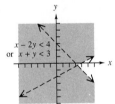

$x \ge 2$

30.

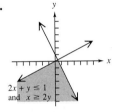

$2x + y \le 1$ and $x \ge 2y$

31.

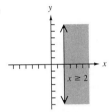

$x - 2y < 4$ or $x + y < 3$

32.

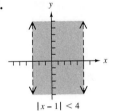

$|x - 1| < 4$

33. domain: $\{-4, 1\}$; range: $\{2, -2, 5, -5\}$; not a function
34. domain: $\{-14, 91, 17, 75, -23\}$;
range: $\{9, 12, 18, 70, 56, 5\}$; not a function
35. domain: $[-4, 4]$; range: $[0, 2]$; function   **36.** -6
37. -15 **38.** -96 **39.** $-8p^2 + 6p - 6$
40. function; linear function; domain: $(-\infty, \infty)$
41. not a function; domain: $(-\infty, \infty)$ **42.** function;
domain: $(-\infty, \infty)$ **43.** function; domain: $\left[-\dfrac{7}{4}, \infty\right)$
44. not a function; domain: $[0, \infty)$
45. function; domain: $(-\infty, 36) \cup (36, \infty)$
46.

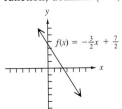

$f(x) = -\frac{3}{2}x + \frac{7}{2}$

47. 15   **48.** 5   **49.** .850 ohm
50. 430 millimeters   **51.** 5.625 kilometers per second
52. $.71\pi$ seconds

CHAPTER 3 TEST (page 225)

1. $\dfrac{1}{2}$ **2.** $\dfrac{3}{2}$; $\left(\dfrac{13}{3}, 0\right)$; $\left(0, -\dfrac{13}{2}\right)$ **3.** 0; none; (0, 5)
4. The graph is a vertical line.   **5.** $y = 14$
6. $5x + y = 19$ **7. (a)** $3x + 5y = -11$
(b) $x + 2y = -3$
8.

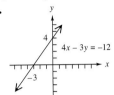

$4x - 3y = -12$

9.

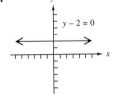

$y - 2 = 0$

10.

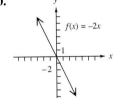

$f(x) = -2x$

11.

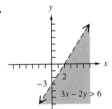

$3x - 2y > 6$

12. D   **13. (a)** 2   **(b)** $[3, \infty)$ **14.** 11,989
15. 256 feet   **16.** 800 pounds

CUMULATIVE REVIEW 1–3 (page 227)

1. true   **2.** true   **3.** false   **4.** 4   **5.** .64
6. -4 **7.** $\dfrac{8}{5}$ **8.** $-2m + 6$ **9.** $4m - 3$
10. $2x^2 + 5x + 4$ **11.** $(2, \infty)$ **12.** $(-\infty, 1]$
13. $(-3, 5]$ **14.** no   **15.** -24 **16.** 204
17. 56   **18.** undefined   **19.** 10   **20.** $\left\{\dfrac{7}{6}\right\}$
21. $\{-1\}$ **22.** $\left(-\infty, \dfrac{15}{4}\right]$ **23.** $\left(-\dfrac{1}{2}, \infty\right)$
24. (2, 3)   **25.** $(-\infty, 2) \cup (3, \infty)$ **26.** $\left\{-\dfrac{16}{5}, 2\right\}$
27. $(-11, 7)$ **28.** $(-\infty, -2] \cup [7, \infty)$ **29.** 2 hours
30. 4 white pills   **31.** $(0, -3), (4, 0), \left(2, -\dfrac{3}{2}\right)$

32. x-intercept: $(-2, 0)$; y-intercept: $(0, 4)$

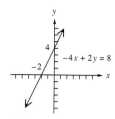

33. $-\dfrac{3}{2}$ **34.** $-\dfrac{3}{4}$ **35.** $3x + 4y = -4$
36. $y = -2$ **37.** $4x + 3y = 7$
38. (a) $(-\infty, \infty)$ (b) 22 **39.** $105{,}666\dfrac{2}{3}$
40. the segment for 1988 through 1991

CHAPTER 4
SECTION 4.1 (page 237)

1. $3; -6$ **3.** $\emptyset$ **5.** 0 **7.** yes   **9.** no
11. $\{(2, 2)\}$

13. $\{(3, -1)\}$ **15.** $\{(2, -3)\}$
17. $\left\{\left(\dfrac{3}{2}, -\dfrac{3}{2}\right)\right\}$ **19.** $\{(x, y)\mid 7x + 2y = 6\}$; dependent equations   **21.** $\{(2, -4)\}$ **23.** $\emptyset$; inconsistent system
25. $y = -\dfrac{3}{7}x + \dfrac{4}{7}; y = -\dfrac{3}{7}x + \dfrac{3}{14}; 0$
27. Both are $y = -\dfrac{2}{3}x + \dfrac{1}{3}$; infinitely many
29. $\{(1, 2)\}$ **31.** $\left\{\left(\dfrac{22}{9}, \dfrac{22}{3}\right)\right\}$ **33.** $\{(5, 4)\}$
35. $\left\{\left(-5, -\dfrac{10}{3}\right)\right\}$ **37.** $\{(2, 6)\}$
39. $\{(1, 3)\}$ **40.** $f(x) = -3x + 6$; linear
41. $g(x) = \dfrac{2}{3}x + \dfrac{7}{3}$; linear   **42.** one; 1; 3; 1; 3; 1; 3
43. (a) years 0 to 10 (b) year 10; about $690
45. (a) 1978 and 1982 (b) just less than 500,000
47. (b)   **49.** (a) $y_1 = 4x + 4$
(b) $y_2 = -2x + 7$ (c) $\left\{\left(\dfrac{1}{2}, 6\right)\right\}$

SECTION 4.2 (page 249)

1. The statement means that when -1 is substituted for x, 2 is substituted for y, and 3 is substituted for z in the three equations, the resulting three statements are true.

3. $\{(1, 4, -3)\}$ **5.** $\{(0, 2, -5)\}$ **7.** $\left\{\left(-\dfrac{7}{3}, \dfrac{22}{3}, 7\right)\right\}$
9. $\{(4, 5, 3)\}$ **11.** $\{(2, 2, 2)\}$ **13.** $\left\{\left(\dfrac{8}{3}, \dfrac{2}{3}, 3\right)\right\}$
15. Answers will vary. Some possible answers are (a) two perpendicular walls and the ceiling in a normal room (b) the floors of three different levels of an office building (c) three pages of this book (since they intersect in the spine).   **17.** $\emptyset$
The solution sets in Exercises 19 and 21 may be given in other equivalent forms.
19. $\{(x, y, z)\mid x - y + 4z = 8\}$
21. $\{(x, y, z)\mid 2x + y - z = 6\}$ **23.** $\{(0, 0, 0)\}$
25. $128 = a + b + c$ **26.** $140 = 2.25a + 1.5b + c$
27. $80 = 9a + 3b + c$
28. $a + b + c = 128$
   $2.25a + 1.5b + c = 140$
   $9a + 3b + c = 80$; $\{(-32, 104, 56)\}$
29. $f(x) = -32x^2 + 104x + 56$ **30.** 56 feet
31. $\{(2, 1, 5, 3)\}$

SECTION 4.3 (page 259)

1. Texas: 93; Florida: 41   **3.** $x = 40, y = 50$, so the angles measure 40° and 50°.   **5.** length: 78 feet; width: 36 feet   **7.** CGA monitor: $400; VGA monitor: $500   **9.** 6 units of yarn; 2 units of thread
11. dark clay: $5 per kilogram; light clay: $4 per kilogram
13. (a) 3.2 ounces (b) 8 ounces (c) 12.8 ounces
(d) 16 ounces   **15.** $.89x$
17. 6 gallons of 25%; 14 gallons of 35%
19. 6 liters of pure acid; 48 liters of 10% acid
21. 14 kilograms of nuts; 16 kilograms of cereal
23. 76 general admission; 108 with student ID
25. 28 dimes; 66 quarters   **27.** $1000 at 2%; $2000 at 4%   **29.** $25y$ miles   **31.** freight train: 50 kilometers per hour; express train: 80 kilometers per hour
33. top speed: 2100 miles per hour; wind speed: 300 miles per hour   **35.** $x + y + z = 180$; angle measures: 70°, 30°, 80°   **37.** first: 20°; second: 70°; third: 90°   **39.** shortest: 12 centimeters; middle: 25 centimeters; longest: 33 centimeters   **41.** A: 180 cases; B: 60 cases; C: 80 cases   **43.** $2a + b + c = -5$
44. $-a + c = -1$ **45.** $3a + 3b + c = -18$
46. $a = 1, b = -7, c = 0$; $x^2 + y^2 + x - 7y = 0$
47. The relation is not a function because a vertical line intersects its graph more than once.

SECTION 4.4 (page 269)

1. true   **3.** true   **5.** false   **7.** -3 **9.** 14
11. 0   **13.** 59   **15.** 14   **17.** 1   **19.** -22
21. 0   **23.** 0   **25.** 20   **27.** 0   **29.** -22
31. 0   **33.** $\dfrac{y_2 - y_1}{x_2 - x_1}$ **34.** $y - y_1 = \dfrac{y_2 - y_1}{x_2 - x_1}(x - x_1)$
35. $x_2y - x_1y - x_2y_1 - xy_2 + x_1y_2 + xy_1 = 0$
36. The result is the same as in Exercise 35.
37. 52   **39.** 9

SECTION 4.5 (page 277)

1. $\{(4, 6)\}$ **3.** $\{(-5, 2)\}$ **5.** $\{(1, 0)\}$

7. $\left\{\left(\frac{11}{58}, -\frac{5}{29}\right)\right\}$ **9.** $\{(-2, 3, 5)\}$ **11.** $\{(5, 0, 0)\}$

13. Cramer's rule does not apply.

15. $\{(20, -13, -12)\}$ **17.** $\left\{\left(\frac{62}{5}, -\frac{1}{5}, \frac{27}{5}\right)\right\}$

19. $\{(-1, 2, 5, 1)\}$
21. The systems will vary. One such system is
$5x + 6y = 7$
$8x + 9y = 10$.
In all cases, the solution set is $\{(-1, 2)\}$.
22.

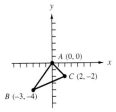

23. $\frac{1}{2}\begin{vmatrix} 0 & 0 & 1 \\ -3 & -4 & 1 \\ 2 & -2 & 1 \end{vmatrix}$ **24.** 7

CHAPTER 4 REVIEW EXERCISES (page 283)

1. $\{(2, 2)\}$

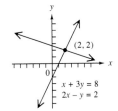

2. $\{(0, 1)\}$

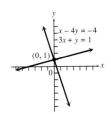

3. $\{(-1, 2)\}$ **4.** $\{(-6, 3)\}$
5. $\left\{\left(\frac{68}{13}, -\frac{31}{13}\right)\right\}$ **6.** $\{(0, 4)\}$
7. $\{(x, y) \mid -3x + y = 6\}$ **8.** $\emptyset$
9. $\left\{\left(-\frac{8}{9}, -\frac{4}{3}\right)\right\}$ **10.** $\{(2, 4)\}$
11. Answers may vary.

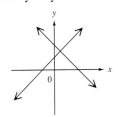

12. Answers may vary.

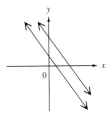

13. Answers may vary.

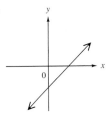

14. Because the lines have the same slope (3) but different
y-intercepts $((0, 2)$ and $(0, -4))$, the lines do not intersect.
Thus the system has no solution. **15.** $\{(1, -5, 3)\}$
16. $\emptyset$ **17.** $\{(1, 2, 3)\}$ **18.** $\{(0, 0, 0)\}$ **19.** length:
6 feet; width: 4 feet **20.** 3 weekend days; 5 weekdays
21. plane: 300 miles per hour; wind: 20 miles per hour
22. 30 pounds of $2-a-pound candy; 70 pounds of
$1-a-pound candy **23.** 85°; 60°; 35° **24.** $40,000 at
10%; $100,000 at 6%; $140,000 at 5% **25.** 10 pounds
of jelly beans; 2 pounds of chocolate eggs; 3 pounds of
marshmallow chicks **26.** 8 fish at $20; 15 fish at $40;
6 fish at $65 **27.** 80 **28.** -21
29. -38 **30.** 0 **31.** 2 **32.** 0 **33.** Cramer's
rule does not apply if $D = 0$.
34. $\left\{\left(\frac{37}{11}, \frac{14}{11}\right)\right\}$ **35.** $\left\{\left(\frac{39}{10}, \frac{6}{5}\right)\right\}$ **36.** $\{(3, -2, 1)\}$
37. $\left\{\left(\frac{172}{67}, -\frac{14}{67}, -\frac{87}{67}\right)\right\}$ **38.** $\{(12, 9)\}$ **39.** $\emptyset$
40. $\{(3, -1)\}$ **41.** $\{(5, 3)\}$ **42.** $\{(0, 4)\}$
43. $\left\{\left(\frac{82}{23}, -\frac{4}{23}\right)\right\}$ **44.** 20 liters **45.** 38°, 66°, 76°

CHAPTER 4 TEST (page 287)

1. $\{(6, 1)\}$

2. $\{(3, 3)\}$ **3.** $\{(0, -2)\}$ **4.** $\emptyset$
5. $\left\{\left(-\frac{2}{3}, \frac{4}{5}, 0\right)\right\}$ **6.** $\{(6, -4)\}$
7. $\{(x, y) \mid 12x - 5y = 8\}$ **8.** 180 votes; 225 votes
9. 4 liters of 20%; 8 liters of 50% **10.** AC adaptor: $8;
rechargeable flashlight: $15 **11.** 60 ounces of Orange
Pekoe; 30 ounces of Irish Breakfast; 10 ounces of Earl Grey
12. 3 **13.** 0 **14.** $\left\{\left(-\frac{9}{4}, \frac{5}{4}\right)\right\}$ **15.** $\{(-1, 2, 3)\}$

CUMULATIVE REVIEW 1–4 (page 289)

1. 81   **2.** -81 **3.** -81 **4.** $.7$ **5.** $-.7$
6. not a real number   **7.** 4   **8.** -4 **9.** -199
10. 455   **11.** 14   **12.** $\left\{-\dfrac{15}{4}\right\}$ **13.** $\left\{\dfrac{2}{3}, 2\right\}$
14. $x = \dfrac{d - by}{a - c}$ or $x = \dfrac{by - d}{c - a}$ **15.** $\{11\}$
16. $\left(-\infty, \dfrac{240}{13}\right]$ **17.** $\left[-2, \dfrac{2}{3}\right]$ **18.** $(-\infty, \infty)$
19. 45 miles per hour, 75 miles per hour   **20.** 6 meters
21. 35 pennies, 29 nickels, 30 dimes   **22.** $46°, 46°, 88°$
23. $y = 6$ **24.** $x = 4$ **25.** $-\dfrac{4}{3}$ **26.** $\dfrac{3}{4}$
27. $4x + 3y = 10$ **28.** $f(x) = -\dfrac{4}{3}x + \dfrac{10}{3}$
29. $5x - 3y = -15$
30.

$(-1, -3)$ $(2, -1)$

31.

$-3x - 2y \le 6$

32. $\{(3, -3)\}$ **33.** $\{(5, 3, 2)\}$ **34.** 33   **35.** 5 pounds of oranges; 1 pound of apples   **36.** video rental firm: $60,000; bond and money market fund: $10,000 each
37. $x = 8$ or 800 parts; $3000   **38.** about $500

CHAPTER 5

SECTION 5.1 (page 297)

1. The bases must be the same.   **3.** exp: 4; base: 5
5. exp: 4; base: -5 **7.** exp: -1; base: p **9.** 1
11. -1 **13.** 1   **15.** -2 **17.** $\dfrac{1}{y^2}$
19. $\dfrac{1}{(-7)^2}$ or $\dfrac{1}{7^2}$ **21.** x^2 **23.** 5^{-2} **25.** x^{-3}
27. $(-4)^{-3}$ or -4^{-3} **29.** $-a^n = (-a)^n$ when n is an odd number. When n is even, $-a^n \ne (-a)^n$.   **31.** $\dfrac{4}{9}$
33. $\dfrac{1}{64}$ **35.** $-\dfrac{1}{64}$ **37.** $-\dfrac{1}{64}$ **39.** 9   **41.** $-\dfrac{16}{3}$
43. $\dfrac{27}{8}$ **45.** $\dfrac{5}{6}$ **47.** 2^{16} **49.** 3^3 **51.** $\dfrac{1}{3^3}$ **53.** $\dfrac{1}{9^2}$
55. $\dfrac{1}{t^7}$ **57.** r^5 **59.** $\dfrac{1}{a^5}$ **61.** x^{11} **63.** r^6
65. $-\dfrac{56}{k^2}$ **67.** $\dfrac{1}{5}; \dfrac{5}{6}; \dfrac{1}{5} \ne \dfrac{5}{6}$ **69.** 25; 13; 25 $\ne$ 13

SECTION 5.2 (page 305)

1. false   **3.** true   **5.** (d)   **7.** incorrect; a^2b^2
9. correct   **11.** incorrect; $\dfrac{4^5}{a^5}$ **13.** incorrect; z^{20}

15. $\left(\dfrac{4}{3}\right)^2$ **17.** $\dfrac{5}{6}$ **19.** $\dfrac{1}{2^9 \cdot 5^3}$ **21.** $5^{12} \cdot 6^6$
23. $\dfrac{1}{k^2}$ **25.** $-4r^6$ **27.** $\dfrac{5^4}{a^{10}}$ **29.** $\dfrac{z^4}{x^3}$ **31.** $\dfrac{1}{5p^{10}}$
33. $\dfrac{4}{a^2}$ **35.** $\dfrac{1}{6y^{13}}$ **37.** $\dfrac{2^2 k^5}{m^2}$ **39.** Write the fraction as its reciprocal raised to the negative of the negative power.   **41.** 5.3×10^2 **43.** 8.3×10^{-1}
45. 6.92×10^{-6} **47.** -3.85×10^4 **49.** 72,000
51. .00254   **53.** $-60,000$ **55.** .000012
57. .0000025   **59.** 200,000   **61.** 2×10^4 hours
63. $3838.38   **65.** 1987

SECTION 5.3 (page 315)

1. (c)   **3.** (a)   **5.** 7; 1   **7.** -15; 2   **9.** 1; 4
11. -1; 6   **13.** Because $4^5 = 1024$, it is a constant, and so it has degree 0.   **15.** neither   **17.** ascending
19. descending   **21.** monomial; 0   **23.** binomial; 1
25. trinomial; 3   **27.** none of these; 5   **29.** $8z^4$
31. $7m^3$ **33.** $5x$ **35.** already simplified
37. $7y - 3y^2$ **39.** $8k^2 + 2k - 7$
41. $-2n^4 - n^3 + n^2$ **43.** $-9p^2 + 11p - 9$
45. $5a + 18$ **47.** $14m^2 - 13m + 6$
49. $13z^2 + 10z - 3$ **51.** $10y^3 - 7y^2 + 5y + 8$
53. $-5a^4 - 6a^3 + 9a^2 - 11$ **55.** $r + 13$
57. $8x^2 + x - 2$ **59.** $-2a^2 - 2a - 7$
61. $-3z^5 + z^2 + 7z$ **63.** $6y^4 - 10y^2 + 12$
65. $-9m^3 - 6m^2 + 6m$ **67.** (a) -10 (b) 8
69. (a) 8 (b) 2   **71.** (a) 8 (b) 74
73. $8z^2 + 4z + 2$ **75.** 31.6   **77.** 29.8
79. 253.2; No, it is more than 100%. The mathematical model only applied to the years 1990 through 1994.

SECTION 5.4 (page 325)

1. commutative; associative   **3.** term; term; like terms
5. difference; squares   **7.** $-24m^5$
9. $-6x^2 + 15x$ **11.** $-2q^3 - 3q^4$
13. $18k^4 + 12k^3 + 6k^2$ **15.** $6m^3 + m^2 - 14m - 3$
17. $-d^4 + 6d^3 + 2d^2 - 13d + 6$ **19.** $6y^2 + y - 12$
21. $-2b^3 + 2b^2 + 18b + 12$ **23.** $25m^2 - 9n^2$
25. $8z^4 - 14z^3 + 17z^2 + 20z - 3$
27. $6p^4 + p^3 + 4p^2 - 27p - 6$ **29.** $m^2 - 3m - 40$
31. $12k^2 + k - 6$ **33.** $3z^2 + zw - 4w^2$
35. $12c^2 + 16cd - 3d^2$ **37.** $.1x^2 + .63x - .13$
39. $3r^2 - \dfrac{23}{4}ry - \dfrac{1}{2}y^2$
41. The product of two binomials is the sum of the product of the first terms, the product of the outer terms, the product of the inner terms, and the product of the last terms.
43. $4p^2 - 9$ **45.** $25m^2 - 1$ **47.** $9a^2 - 4c^2$
49. $16x^2 - \dfrac{4}{9}$ **51.** $16m^2 - 49n^4$ **53.** $25y^6 - 4$
55. $y^2 - 10y + 25$ **57.** $4p^2 + 28p + 49$
59. $16n^2 - 24nm + 9m^2$ **61.** $k^2 - \dfrac{10}{7}kp + \dfrac{25}{49}p^2$
63. $(x + y)^2 = x^2 + 2xy + y^2$, because it is a perfect square trinomial. Thus, it differs from $x^2 + y^2$ by $2xy$.

65. $25x^2 + 10x + 1 + 60xy + 12y + 36y^2$
67. $4a^2 + 4ab + b^2 - 9$ **69.** $4h^2 - 4hk + k^2 - j^2$
71. $125r^3 - 75r^2s + 15rs^2 - s^3$
73. $m^4 - 4m^3p + 6m^2p^2 - 4mp^3 + p^4$ **75.** $\frac{1}{2}x^2 - 2y^2$
77. $15x^2 + 13x - 6$ **79.** $(2 + 3)^3 \neq 2^3 + 3^3$ because
$125 \neq 35$; $(x + y)^3 = x^3 + 3x^2y + 3xy^2 + y^3$

SECTION 5.5 (page 333)

1. quotient; exponents **3.** descending powers
5. $3y + 4 - \dfrac{5}{y}$ **7.** $3m + 5 + \dfrac{6}{m}$ **9.** $n - \dfrac{3n^2}{2m} + 2$
11. $r^2 - 7r + 6$ **13.** $y - 3$ **15.** $t + 5$
17. $z^2 + 3$ **19.** $x^2 + 2x - 3 + \dfrac{6}{4x + 1}$
21. $2x - 5 + \dfrac{-4x + 5}{3x^2 - 2x + 4}$ **23.** $2k^2 + 3k - 1$
25. $9z^2 - 4z + 1 + \dfrac{-z + 6}{z^2 - z + 2}$ **27.** $\dfrac{2}{3}x - 1$
29. $\dfrac{3}{4}a - 2 + \dfrac{1}{4a + 3}$ **31.** $2x + 7$
33. $t^3 + 6t^2 + 5t + 4$ **34.** $t + 4$
35. $t^2 + 2t - 3 + \dfrac{16}{t + 4}$ **36.** $118\dfrac{1}{7}$ **37.** $118\dfrac{1}{7}$
38. They are the same.

SECTION 5.6 (page 339)

1. Synthetic division provides a quick, easy way to divide a polynomial by a binomial. **3.** Since the variables are not present, a missing term will not be noticed in synthetic division, so the quotient will be wrong if place holders are not inserted.
5. $x - 5$ **7.** $4m - 1$ **9.** $2a + 4 + \dfrac{5}{a + 2}$
11. $p - 4 + \dfrac{9}{p + 1}$ **13.** $4a^2 + a + 3$
15. $x^4 + 2x^3 + 2x^2 + 7x + 10 + \dfrac{18}{x - 2}$
17. $-4r^5 - 7r^4 - 10r^3 - 5r^2 - 11r - 8 + \dfrac{-8}{r - 1}$
19. $-3y^4 + 8y^3 - 21y^2 + 36y - 72 + \dfrac{143}{y + 2}$
21. $y^2 + y + 1 + \dfrac{2}{y - 1}$ **23.** 7
25. -2 **27.** 0 **29.** yes **31.** no **33.** no

SECTION 5.7 (page 345)

1. To factor a polynomial means to write it as the product of two or more polynomials. **3.** $z^2(m + n)^4$ **5.** $3m$
7. $3(r + t)^2$ **9.** (a) **11.** $8k(k^2 + 3)$
13. $xy(3 - 5y)$ **15.** $-2p^2q^4(2p + q)$
17. $7x^3(3x^2 + 5x - 2)$ **19.** $5ac(3ac^2 - 5c + a)$
21. cannot be factored **23.** $(m - 4)(2m + 5)$
25. $11(2z - 1)$ **27.** $-y^5(r + w + yz + yk)$
29. $(2 - x)(10 - x - x^2)$
31. $14z^3(3 - 4z)$; $-14z^3(-3 + 4z)$
33. $5a^3(-1 + 2a - 3a^2)$; $-5a^3(1 - 2a + 3a^2)$

35. $(k + h)(2 + j)$ **37.** $(a + b)(3m + 2b)$
39. $(z - 6)(z + 9)$ **41.** $(r + 3w)(r - 3t)$
43. $(2k + 7)(k - 3)$ **45.** $(-3r + 4s)(2r - 3s)$
47. $(4m^2 + p)(-4m + p^2)$ **49.** $(a - 4)(2b^2 + 1)$
51. $(x^3 - 3)(y^2 + 1)$ **53.** $(-2b^2 - 1)(4 - a)$;
$-(2b^2 + 1)(-a + 4)$ (Other answers are possible.)
55. $k^{-4}(k^2 + 2)$

SECTION 5.8 (page 353)

1. true **3.** false **5.** $(y - 3)(y + 10)$
7. $(p - 8)(p + 7)$ **9.** $(m - 10)(-m + 6)$
11. $(a + 5b)(a - 7b)$ **13.** prime
15. $(xy + 9)(xy + 2)$ **17.** $(-6m + 5)(m + 3)$
19. $(5x - 6)(2x + 3)$ **21.** $(4k + 3)(5k + 8)$
23. $(3a - 2b)(5a - 4b)$ **25.** $(6m - 5)^2$ **27.** prime
29. $(2xz - 1)(3xz + 4)$ **31.** $3(4x + 5)(2x + 1)$
33. $5(a + 6)(3a - 4)$ **35.** $11x(x - 6)(x - 4)$
37. $2xy^3(x - 12y)(x - 12y)$ **39.** She lost some credit because the factor $4x + 10$ can be factored further as $2(2x + 5)$. **41.** $(2x^2 + 3)(x^2 - 6)$
43. $(4x^2 + 3)(4x^2 + 1)$ **45.** $(6p^3 - r)(2p^3 - 5r)$
47. $(5k + 4)(2k + 1)$ **49.** $(3m + 3p + 5)(m + p - 4)$
51. $(a + b)^2(a - 3b)(a + 2b)$ **53.** no
54. 1, 3, 5, 9, 15, 45; no **55.** no
56. $(5x + 2)(2x + 5)$; no **57.** Since k is odd, 2 is not a factor of $2x^2 + kx + 8$, and because 2 is a factor of $2x + 4$, the binomial $2x + 4$ cannot be a factor.
58. $3y + 15$ cannot be a factor of $12y^2 - 11y - 15$ because 3 is a factor of $3y + 15$, but 3 is not a factor of $12y^2 - 11y - 15$.

SECTION 5.9 (page 359)

1. (a), (d) **3.** (b), (c) **5.** The sum of two squares can be factored only if the binomial has a common factor.
7. $(p + 4)(p - 4)$ **9.** $(5x + 2)(5x - 2)$
11. $(3a + 7b)(3a - 7b)$
13. $4(4m^2 + y^2)(2m + y)(2m - y)$
15. $(y + z + 9)(y + z - 9)$
17. $(4 + x + 3y)(4 - x - 3y)$
19. $4pq$ **21.** $(k - 3)^2$ **23.** $(2z + w)^2$
25. $(4m - 1 + n)(4m - 1 - n)$
27. $(2r - 3 + s)(2r - 3 - s)$
29. $(x + y - 1)(x - y + 1)$ **31.** $2(7m + 3n)^2$
33. $(p + q + 1)^2$ **35.** $(a - b + 4)^2$
37. $(2x - y)(4x^2 + 2xy + y^2)$
39. $(4g + 3h)(16g^2 - 12gh + 9h^2)$
41. $3(2n + 3p)(4n^2 - 6np + 9p^2)$
43. $(y + z - 4)(y^2 + 2yz + z^2 + 4y + 4z + 16)$
45. $(m^2 - 5)(m^4 + 5m^2 + 25)$ **47.** $2b(3a^2 + b^2)$
48. $(x - y)(x^2 + xy + y^2)(x + y)(x^2 - xy + y^2)$
49. $(x - y)(x + y)(x^4 + x^2y^2 + y^4)$
50. The product of the trinomials in Exercise 48 must equal the trinomial in Exercise 49.
$(x^2 + xy + y^2)(x^2 - xy + y^2) = x^4 + x^2y^2 + y^4$
51. Start by factoring as the difference of squares.

SUMMARY ON FACTORING METHODS (page 361)

1. $(10a + 3b)(10a - 3b)$ **3.** $6p^3(3p^2 - 4 + 2p^3)$
5. $(x + 7)(x - 5)$ **7.** $(7z + 4)(7z - 4)$
9. $(x - 10)(x^2 + 10x + 100)$ **11.** prime
13. $(3t - 7u)(2t + 11u)$ **15.** $8p(5p - 4)$

17. $(2k + 7r)^2$ **19.** $(m - 2)(n + 5)$
21. $(3m - 5n + p)(3m - 5n - p)$
23. $7(2k - 5)(4k^2 + 10k + 25)$
25. $16z^2x(zx - 2)$ **27.** $(m + 2)(m - 2)^2$
29. $3(3m + 8n)^2$ **31.** $(5m^2 + 6)(25m^4 - 30m^2 + 36)$
33. $(2m + 5n)(m - 3n)$ **35.** $4y(y - 2)$
37. $(7z + 2k)(2z - k)$ **39.** $16(4b + 5c)(4b - 5c)$
41. $8(5z + 4)(25z^2 - 20z + 16)$
43. $(5r - s)(2r + 5s)$ **45.** $8x^2(4 + 2x - 3x^3)$
47. $(7x + 5q)(2x - 5q)$ **49.** $(y + 5)(y - 2)$
51. $2a(a^2 + 3a - 2)$ **53.** $(9p - 5r)(2p + 7r)$
55. $(x - 2y + 2)(x - 2y - 2)$ **57.** $(5r + 2s - 3)^2$
59. $(z + 2)(z - 2)(z^2 - 5)$

SECTION 5.10 (page 369)

1. First rewrite the equation so one side is zero. Factor the other side and set each factor equal to 0. Each solution of one of these linear equations is a solution of the quadratic equation.   **3.** The two linear equations lead to $x = -3$ and $x = 2$, but -3 and 2 are not solutions. The correct solution can be found by multiplying the factors on the left side of the equation, and subtracting 1 from both sides to get zero on one side.   **5.** $\{5, -10\}$

7. $\left\{-\dfrac{8}{3}, \dfrac{5}{2}\right\}$ **9.** $\{-2, 5\}$ **11.** $\{-6, -3\}$

13. $\left\{-\dfrac{1}{2}, 4\right\}$ **15.** $\left\{-\dfrac{1}{3}, \dfrac{4}{5}\right\}$ **17.** $\{-3, 4\}$

19. $\left\{-5, -\dfrac{1}{5}\right\}$ **21.** $\{0, 6\}$ **23.** $\{-3, 3\}$ **25.** $\{-2, 2\}$

27. $4x^2 - 13x + 3 = 0$ **29.** width: 10 meters; length: 14 meters   **31.** 8 seconds   **33.** 50 feet by 100 feet
35. length: 15 inches; width: 9 inches

CHAPTER 5 REVIEW EXERCISES (page 375)

1. 64   **2.** $\dfrac{1}{81}$ **3.** .04   **4.** 18   **5.** $\dfrac{81}{16}$

6. In the expression $(-6)^0$, the base is -6, so $(-6)^0 = 1$. In -6^0, however, the base is 6 and $6^0 = 1$, so $-6^0 = -1$.

7. 7^8 **8.** 3^8 **9.** m^{11} **10.** $\dfrac{90}{x^3}$ **11.** 4^3 **12.** y^4

13. (a) true   **(b)** $a^4 = (-a)^4$ **(c)** $5^{-2} = \dfrac{1}{5^2}$ **(d)** true

14. 2.790×10^3 **15.** 8.5×10^{-6} **16.** 2.96×10^{-1}
17. 36,000   **18.** .00571   **19.** .0904   **20.** By definition, multiplying by 10^{-a} equals multiplying by $1/10^a$ which is equivalent to dividing by 10^a.

21. 5.449×10^3 **22.** 63 square miles   **23.** $\dfrac{5^2}{5^2} = 5^0$;

Since $\dfrac{25}{25} = 1$ and $\dfrac{25}{25} = 5^0$, it follows that $5^0 = 1$.

24. $\dfrac{1}{4^6}$ **25.** $\dfrac{1}{k^{10}}$ **26.** 5^3z^{10} **27.** $\dfrac{1}{35p^8}$ **28.** $\dfrac{3^5}{2^6a^6}$
29. $\dfrac{3 \cdot 5^2 r^5 q^5}{2}$ **30. (a)** $11k^3 - 3k^2 + 9k$
(b) trinomial   **(c)** 3   **31. (a)** $-9m^7 + 14m^6$
(b) binomial   **(c)** 7   **32. (a)** $-5y^4 + 3y^3 + 7y^2 - 2y$
(b) none of these   **(c)** 4   **33. (a)** 27   **(b)** -3

34. (a) -11 **(b)** 4   **35.** $-x^2 - 3x + 1$
36. $18m - 10$ **37.** $-5y^3 - 4y^2 + 6y - 12$
38. $6a^3 - 4a^2 - 16a + 15$ **39.** $8y^2 - 9y + 5$
40. $-15b^3 - 50b$ **41.** $-12k^3 - 42k$
42. $15m^2 - 7m - 2$ **43.** $14y^2 + 5y - 24$
44. $6w^2 - 13wt + 6t^2$ **45.** $10p^4 + 30p^3 - 8p^2 - 24p$
46. $3q^3 - 13q^2 - 14q + 20$
47. $9z^4 - 12z^3 + 16z^2 - 11z + 2$ **48.** $36r^4 - 1$
49. $z^2 - \dfrac{9}{25}$ **50.** $16m^2 + 24m + 9$
51. $4n^2 - 40n + 100$
52. $\dfrac{3x}{2} - \dfrac{8}{3} + \dfrac{1}{x}$ **53.** $y^2 - 3y + \dfrac{5}{4}$
54. $3x - 4 + \dfrac{-10}{2x - 3}$ **55.** $5k + 7 + \dfrac{-3}{3k - 2}$
56. $p^2 + 6p + 9 + \dfrac{54}{2p - 3}$ **57.** $4y^2 + 1 + \dfrac{-2y}{3y^2 + 1}$
58. $x^2 - 7x + 6$ **59.** $p^2 + 3p - 6$ **60.** $3p + 2$
61. $10k - 23 + \dfrac{31}{k + 2}$ **62.** $2k^2 + k + 3 + \dfrac{21}{k - 3}$
63. $-a^3 + 4a^2 + 3a + 6 + \dfrac{-9}{a + 4}$ **64.** yes
65. no   **66.** -13 **67.** -5 **68.** $6p(2p - 1)$
69. $7y(3y + 5)$ **70.** $4qb(3q + 2b - 5q^2b)$
71. $6rt(r^2 - 5rt + 3t^2)$ **72.** $(x + 3)(x - 3)$
73. $(z + 1)(3z - 1)$ **74.** $(m + q)(4 + n)$
75. $(x + y)(x + 5)$ **76.** $(3p - 4)(p + 1)$
77. $(3k - 2)(2k + 5)$ **78.** $(3r + 1)(4r - 3)$
79. $(2m + 5)(5m + 6)$ **80.** $(2k - h)(5k - 3h)$
81. prime   **82.** $2x(4 + x)(3 - x)$
83. $3b(2b - 5)(b + 1)$ **84.** $(y^2 + 4)(y^2 - 2)$
85. $(2k^2 + 1)(k^2 - 3)$ **86.** The answer given is a sum, not a product. The correct answer requires another step to get $(x^2 + 5)(y^2 - 6)$.   **87. (a)** true
(b) $x^2 + y^2$ is prime or $x^2 + 2xy + y^2 = (x + y)(x + y)$
(c) $x^3 + y^3 = (x + y)(x^2 - xy + y^2)$
(d) $x^3 - y^3 = (x - y)(x^2 + xy + y^2)$
88. $(4p + 3)(4p - 3)$ **89.** $(3z + 10)(3z - 10)$
90. $4(3t + 4z + 7)(3t - 4z - 1)$ **91.** prime
92. $(4x - 5)^2$ **93.** $(2z + 3m)^2$
94. $(1 - y)(1 + y + y^2)$
95. $(5x - 2)(25x^2 + 10x + 4)$
96. $4(r + 3)(r^2 - 3r + 9)$ **97.** To use the zero-factor property, one side of the equation must be zero. The correct solutions are $x = -1$ and $x = 3$.
98. $\left\{-4, \dfrac{2}{3}\right\}$ **99.** $\left\{-\dfrac{1}{4}, \dfrac{3}{4}\right\}$ **100.** $\left\{-2, \dfrac{5}{3}\right\}$
101. 3 feet   **102.** length: 60 feet; width: 40 feet
103. 10 meters   **104.** 6 inches
105. $8x^2 - 2x - 3$
106. $\dfrac{2^2}{3^2}$ **107.** $\dfrac{1}{2^4y^{18}}$ **108.** $19w^2 - 5w + 21$
109. $m - 3$ **110.** -9 **111.** $2500z^6$
112. $-3k^2 + 4k - 7$ **113.** $21p^9 + 7p^8 + 14p^7$
114. $2y^2x + \dfrac{3y^3}{2x} + \dfrac{5x^2}{2}$ **115.** 13
116. $(5m - 3)(2m + 1)$ **117.** $12z(1 - 6z)$
118. $(4 - p)(16 + 4p + p^2)$ **119.** $(z - 4x)(2z + x)$
120. $5d^2(3d^2 - 2)$ **121.** $(5c + 3)^2$

CHAPTER 5 TEST (page 381)

1. For example, if $a = 4$, $(2 \cdot 4)^{-3} = 8^{-3} = \dfrac{1}{8^3}$ and

$\dfrac{2}{a^3} = \dfrac{2}{4^3}$; $\dfrac{1}{8^3} \neq \dfrac{2}{4^3}$ **2.** $\dfrac{2^3}{3^3}$ **3.** $-\dfrac{m^2}{4n^4}$ **4.** $\dfrac{a^7}{5^3 \cdot 2}$

5. $x^3 - 2x^2 - 10x - 13$ **6.** $10x^2 - x - 3$

7. $6m^3 - 7m^2 - 30m + 25$ **8.** $36x^2 - y^2$

9. $9k^2 + 6kq + q^2$ **10.** $4p - 8 + \dfrac{6}{p}$

11. $3q^3 - 4q^2 + q + 4 + \dfrac{-2}{3q - 2}$ **12. (a)** -24

(b) yes **13.** $11z(z - 4)$ **14.** $2(h - 1)(h + 1)$

15. $(x + y)(3 + b)$ **16.** $(4p - q)(p + q)$

17. prime **18.** $(y - 6)(y^2 + 6y + 36)$

19. $(3k + 11j)(3k - 11j)$

20. $(2k^2 - 5)(3k^2 + 7)$ **21. (d)**

22. $\left\{-2, -\dfrac{2}{3}\right\}$ **23.** $\left\{\dfrac{1}{5}, \dfrac{3}{2}\right\}$ **24.** $\left\{-\dfrac{2}{5}, 1\right\}$

25. 8 feet

CUMULATIVE REVIEW 1–5 (page 383)

1. $\sqrt{7}$ **2.** $-5, -\dfrac{3}{4}, 0, 1.6, 9$ **3.** $-5, 0, 9$

4. $(-1, 7]$ **5.** -1 **6.** $\dfrac{4}{9}$ **7.** $\varnothing$ **8.** $\{1, 6\}$

9. $k = \dfrac{4p + v^2}{3}$ **10.** $\left\{-\dfrac{16}{3}\right\}$ **11.** $\left(-\dfrac{3}{2}, \infty\right)$

12. $\left[-5, -\dfrac{2}{3}\right]$ **13.** II **14.** IV **15.** III

16. no quadrant

17.

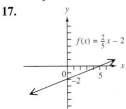

18.

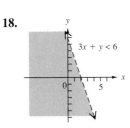

19. 6 **20.** -3 **21.** $\{(1, -1, 3)\}$ **22.** If x is the length of a shorter side, and y is the length of a longer side, $y = x + 2, 2x + 2y = 68$. **23.** $\{(1, -1, 3)\}$

24. $\dfrac{1}{343}$ **25.** -1 **26.** 16 **27.** $\dfrac{y}{18x}$

28. $\dfrac{5my^4}{3}$ **29.** $10p^3 + 7p^2 - 28p - 24$

30. $x^3 + 12x^2 - 3x - 7$

31. $(2w + 7z)(8w - 3z)$

32. $(2x - 1 + y)(2x - 1 - y)$ **33.** $(2y - 9)^2$

34. $(10x^2 + 9)(10x^2 - 9)$

35. $(2p + 3)(4p^2 - 6p + 9)$ **36.** $\left\{-4, -\dfrac{3}{2}, 1\right\}$

37. $\left\{\dfrac{1}{3}\right\}$ **38.** $\{-3, 3\}$ **39.** 4 years **40.** 4 meters

CHAPTER 6

SECTION 6.1 (page 391)

1. C **3.** F **5.** D **7.** 7 **9.** $-\dfrac{1}{7}$ **11.** 0

13. $-2, \dfrac{3}{2}$ **15.** none **17.** none **19. (a)** numerator: $x^2, 4x$; denominator: $x, 4$ **(b)** Factor the numerator as $x(x + 4)$. Then divide the numerator and denominator both by $x + 4$ to get x. **21.** $\dfrac{4x}{3y}$ **23.** $\dfrac{x - 3}{x + 5}$

25. $\dfrac{x + 3}{2x(x - 3)}$ **27.** already in lowest terms **29.** $\dfrac{7}{8}$

31. $\dfrac{6}{7}$ **33.** $\dfrac{z}{6}$ **35.** $\dfrac{2}{t - 3}$ **37.** $\dfrac{b + 1}{2}$

39. $\dfrac{x - 3}{x + 1}$ **41.** $\dfrac{4x + 1}{4x + 3}$ **43.** $a^2 - ab + b^2$

45. $\dfrac{c + 6d}{c - d}$ **47.** $\dfrac{a + b}{a - b}$ **49.** -1

In Exercises 51, 53, and 55, there are other acceptable ways to express the answer.

51. $-(x + y)$ **53.** $-\dfrac{x + y}{x - y}$ **55.** $-\dfrac{1}{2}$

57. already in lowest terms **59. (b) and (d)**

61. $\dfrac{3y}{x^2}$ **63.** $\dfrac{3a^3b^2}{4}$ **65.** $\dfrac{7x}{6}$ **67.** $-\dfrac{35}{8}$

69. $-\dfrac{a + 1}{2a}$ **71.** $\dfrac{-m(m + 7)}{m + 1}$ **73.** -2

75. $\dfrac{x + 4}{x - 4}$ **77.** $\dfrac{3x - y}{3x + 2y}$ **79.** $\dfrac{k + 5p}{2k + 5p}$

81. $(k - 1)(k - 2)$ **83.** $\dfrac{(a + 5)(a - 1)}{3a^2 - 2a + 1}$

85. $\dfrac{16.5}{73.2}$; .23 **87.** $\dfrac{26}{705}$; .04

SECTION 6.2 (page 403)

1. To add or subtract rational expressions that have a common denominator, first add or subtract the numerators, then place the result over the common denominator. Write the answer in lowest terms.

3. $\dfrac{9}{t}$ **5.** $\dfrac{2}{x}$ **7.** 1 **9.** $x - 5$ **11.** $\dfrac{1}{p + 3}$

13. $a - b$ **15.** $72x^4y^5$ **17.** $z(z - 2)$

19. $2(y + 4)$ **21.** $(x + 9)^2(x - 9)$

23. $(m + n)(m - n)$ **25.** $x(x - 4)(x + 1)$

27. $(t + 5)(t - 2)(2t - 3)$ **29.** $2y(y + 3)(y - 3)$

31. Yes, they are both correct, because the expressions are equivalent. Multiplying $\dfrac{3}{5-y}$ by 1 in the form $\dfrac{-1}{-1}$ gives $\dfrac{-3}{y-5}$. **33.** $\dfrac{31}{3t}$ **35.** $\dfrac{5-22x}{12x^2y}$ **37.** $\dfrac{1}{x(x-1)}$

39. $\dfrac{5a^2-7a}{(a+1)(a-3)}$ **41.** 3 **43.** $\dfrac{3}{x-4}$ or $\dfrac{-3}{4-x}$

45. $\dfrac{w+z}{w-z}$ or $\dfrac{-w-z}{z-w}$ **47.** $\dfrac{-13}{12(3+x)}$

49. $\dfrac{2(2x-1)}{x-1}$ **51.** $\dfrac{7}{y}$ **53.** $\dfrac{6}{x-2}$

55. $\dfrac{3x-2}{x-1}$ **57.** $\dfrac{4x-7}{x^2-x+1}$ **59.** $\dfrac{2x+1}{x}$

61. $\dfrac{2x(x+12y)}{(x+2y)(x-y)(x+6y)}$ **63.** $\dfrac{3r-2s}{(2r-s)(3r-s)}$

65. $\dfrac{8}{9}$ **66.** $\dfrac{3}{7}+\dfrac{5}{9}-\dfrac{6}{63}$; They are the same.

67. $\dfrac{8}{9};\dfrac{8}{9}$; yes **68.** Answers will vary. Suppose the name is Hart, so that $x=4$. The problem is $\dfrac{3}{2}+\dfrac{5}{4}-\dfrac{6}{8}$. The predicted answer is $\dfrac{8}{4}=2$, which is correct. **69.** It causes $\dfrac{3}{x-2}$ and $\dfrac{6}{x^2-2x}$ to be undefined. **70.** 0

Section 6.3 (page 411)

1. Begin by simplifying the numerator. Then simplify the denominator. Write as a division problem, and proceed.

3. $\dfrac{2x}{x-1}$ **5.** $\dfrac{2(k+1)}{3k-1}$ **7.** $\dfrac{5x^2}{9z^3}$ **9.** $\dfrac{1+x}{-1+x}$

11. $\dfrac{y+x}{y-x}$ **13.** $4x$ **15.** $x+4y$ **17.** $\dfrac{3y}{2}$

19. $\dfrac{x^2+5x+4}{x^2+5x+10}$ **21.** $\dfrac{m^2+6m-4}{m(m-1)}$

22. $\dfrac{m^2-m-2}{m(m-1)}$ **23.** $\dfrac{m^2+6m-4}{m^2-m-2}$ **24.** $m(m-1)$

25. $\dfrac{m^2+6m-4}{m^2-m-2}$ **26.** Method 1 involves simplifying the numerator and the denominator separately and then performing a division. Method 2 involves multiplying the fraction by a form of 1, the identity element for multiplication. (Preferences will vary.)

27. $\dfrac{x^2y^2}{y^2+x^2}$ **29.** $\dfrac{y^2+x^2}{xy^2+x^2y}$ or $\dfrac{y^2+x^2}{xy(y+x)}$ **31.** $\dfrac{rs}{s+r}$

Section 6.4 (page 415)

1. $-1,2$ **3.** $-\dfrac{5}{3},0,-\dfrac{3}{2}$ **5.** 0 **7.** $4,\dfrac{7}{2}$

9. There are no numbers that would have to be rejected.

11. $\{8\}$ **13.** $\{-3\}$ **15.** $\left\{-\dfrac{7}{12}\right\}$

17. $\varnothing$ **19.** $\{1\}$ **21.** $\{-6,4\}$

23. $\{-3\}$ **25.** $\{5\}$ **27.** $\{5\}$ **29.** $\varnothing$

31. $\left\{\dfrac{27}{56}\right\}$ **33.** $\varnothing$ **35.** $\{-10\}$ **37.** $\varnothing$ **39.** $\{0\}$

41. $\left\{x\,\middle|\,x\neq-\dfrac{3}{2},x\neq\dfrac{3}{2}\right\}$

44. $C=-4$; $\{-2\}$; -1 is rejected. **45.** $C=24$; $\{-4\}$; 3 is rejected. **46.** Answers will vary. However, in every case, $-B$ will be the rejected solution, and $\{-A\}$ will be the solution set. **47.** Answers will vary. For example, if the answer (solution) chosen is 2, then the equation is $\dfrac{x}{4}-\dfrac{x}{6}=\dfrac{1}{6}$, and the solution set is $\{2\}$.

Summary on Operations and Equations Involving Rational Expressions (page 419)

1. equation; $\{20\}$ **3.** expression; $\dfrac{2(x+5)}{5}$

5. expression; $\dfrac{y+x}{y-x}$ **7.** equation; $\{7\}$

9. equation; $\{1\}$ **11.** expression; $\dfrac{25}{4(r+2)}$

13. expression; $\dfrac{24p}{p+2}$ **15.** equation; $\{0\}$

17. expression; $\dfrac{5}{3z}$ **19.** equation; $\{2\}$

21. expression; $\dfrac{-x}{3x+5y}$ **23.** expression; $\dfrac{3}{2s-5r}$

25. equation; $\left\{\dfrac{5}{4}\right\}$ **27.** expression; $\dfrac{2z-3}{2z+3}$

29. expression; $\dfrac{t-2}{8}$ **31.** expression; $\dfrac{13x+28}{2x(x+4)(x-4)}$ **33.** expression; $\dfrac{k(2k^2-2k+5)}{(k-1)(3k^2-2)}$

Section 6.5 (page 423)

1. (a) **3.** (a) **5.** 270 **7.** 1.349 **9.** 24
11. (a) $5x=4+2x+4$
(b) $5x-2x=4+2x+4-2x$ (c) $(5-2)x=8$
(d) $3x=8$ (e) $x=\dfrac{8}{3}$ **12.** (a) $bx=c+dx+2d$
(b) $bx-dx=c+dx+2d-dx$
(c) $x(b-d)=c+2d$ (d) $x=\dfrac{c+2d}{b-d}$ **13.** They were not similar terms, so they could not be combined.
14. There is no difference in the procedure used. In solving an equation, we find a numerical solution, and in solving a formula, we get a variable expression.

15. $G=\dfrac{Fd^2}{Mm}$ **17.** $a=\dfrac{bc}{c+b}$ **19.** $v=\dfrac{PVt}{pT}$

21. $P=\dfrac{A}{1+rt}$ **23.** $b=\dfrac{2A}{h}-B$ or $b=\dfrac{2A-Bh}{h}$

Section 6.6 (page 431)

1. 4 to 5 or $\dfrac{4}{5}$ **3.** 5 to 13 or $\dfrac{5}{13}$ **5.** 15 girls, 5 boys

7. $\dfrac{1}{2}$ job per hour **9.** 24,400,000 **11.** $300

13. 25,000 **15.** 3 miles per hour **17.** $4\dfrac{1}{2}$ miles

19. 190 miles   **21.** To solve problems about distance, rate, and time, we use the formula $d = rt$. To solve problems about work, we use the similar formula $A = rt$. **23.** $\frac{40}{13}$ or $3\frac{1}{13}$ hours   **25.** $17\frac{1}{2}$ hours   **27.** 36 hours **29.** 240 minutes   **31.** $x = 3.5$; $AC = 8$; $DF = 12$

Chapter 6 Review Exercises (page 439)

1. -3 **2.** $-6, 6$ **3.** $2, 5$ **4.** $\frac{11n^2}{2m}$

5. $\frac{x}{2}$ **6.** $\frac{y+5}{y-3}$ **7.** $\frac{5m+n}{5m-n}$

8. $\frac{-1}{2+r}$ (There are other ways.)   **9.** $\frac{3y}{2}$ **10.** $\frac{10p^3}{3q}$

11. $-\frac{3}{10}$ **12.** $\frac{3y^2(2y+3)}{2y-3}$ **13.** $\frac{-3(w+4)}{w}$

14. $\frac{y(y+5)}{y-5}$ **15.** $\frac{z(z+2)}{z+5}$ **16.** $\frac{2p+3}{6}$ **17.** 1

18. The terms x and 5 in the denominator cannot be separated into denominators of two fractions. The correct simplified form is x.

19. $60x$ **20.** $96b^5$ **21.** $9r^2(3r+1)$ **22.** $45(2k+1)$ **23.** $(3x-1)(2x+5)(3x+4)$

24. $\frac{19}{8y}$ **25.** $\frac{16z-3}{2z^2}$ **26.** $\frac{8}{t-2}$ or $\frac{-8}{2-t}$

27. 12   **28.** $\frac{71}{30(a+2)}$ **29.** $\frac{3x+2}{x-5}$

30. $\frac{13r^2+5rs}{(5r+s)(2r-s)(r+s)}$ **31.** $\frac{3+2t}{4-7t}$ **32.** $\frac{mn^4}{2}$

33. $\frac{1}{32}$ **34.** $\frac{3-5x}{6x+1}$ **35.** $\frac{1}{3q+2p}$ **36.** $\frac{x+1}{x-1}$

37. $\{1\}$ **38.** $\{-3\}$ **39.** $\{-2\}$ **40.** $\{0\}$

41. $\left\{\frac{1}{3}\right\}$ **42.** $\emptyset$

43. Although her algebra was correct, 3 is not a solution because it causes a denominator to be 0 in the original equation. It must be rejected. $\emptyset$ is correct.   **44.** In simplifying the expression we are combining to get a single term in x, while in solving the equation we are finding a value for x that makes the equation true.

45. $\frac{15}{2}$ **46.** 30   **47.** $m = \frac{Fd^2}{GM}$

48. $l = \frac{2S}{n} - a$ or $l = \frac{2S - na}{n}$

49. $m = \frac{Mv - \mu M}{\mu}$ or $m = \frac{Mv}{\mu} - M$

50. $R = \frac{nE - Inr}{I}$ or $R = \frac{nE}{I} - nr$

51. \$21.06   **52.** 10,725   **53.** $\frac{24}{5}$ or $4\frac{4}{5}$ minutes

54. $\frac{18}{5}$ or $3\frac{3}{5}$ hours   **55.** bus: 50 miles per hour;

train: 60 miles per hour   **56.** 16 kilometers per hour

57. $\frac{6m+5}{3m^2}$ **58.** $\frac{k-3}{36k^2+6k+1}$ **59.** $\frac{x^2-6}{2(2x+1)}$

60. $\frac{x(9x+1)}{3x+1}$ **61.** $\frac{11}{3-x}$ or $\frac{-11}{x-3}$ **62.** $\frac{1}{3}$

63. $\frac{s^2+t^2}{st(s-t)}$ **64.** $\frac{2(2a^2+3ab+6b^2)}{(a+3b)(a-2b)(a+b)}$

65. $r = \frac{AR}{R-A}$ or $r = \frac{-AR}{A-R}$ **66.** $\{1, 4\}$

67. $\left\{-\frac{14}{3}\right\}$ **68.** $\frac{60}{7}$ or $8\frac{4}{7}$ minutes

Chapter 6 Test (page 443)

1. $-2, \frac{4}{3}$ **2.** $\frac{2x-5}{x(3x-1)}$ **3.** $\frac{3x}{2y^8}$ **4.** $\frac{y+4}{y-5}$

5. $\frac{x+5}{x}$ **6.** $t^2(t+3)(t-2)$ **7.** $\frac{7-2t}{6t^2}$

8. $\frac{13x+35}{(x-7)(x+7)}$ **9.** $\frac{4}{x+2}$ **10.** $\frac{72}{11}$ **11.** $\frac{-1}{a+b}$

12. (a) $\frac{11(x-6)}{12}$ **(b)** $\{6\}$ **13.** $\{6\}$ **14.** $\left\{\frac{1}{2}\right\}$

15. $\{5\}$ **16.** $r = \frac{En - IRn}{I}$ or $r = \frac{En}{I} - Rn$

17. $\frac{48}{5}$ **18.** $\frac{45}{14}$ or $3\frac{3}{14}$ hours   **19.** 15 miles per hour

20. 48,000 fish

Cumulative Review 1–6 (page 445)

1. $(-\infty, \infty)$
2. $[-3, 3]$
3. \$4000 at 4%; \$8000 at 3%
4. 78 or greater   **5.** $\emptyset$ **6.** 0
7. $(-\infty, 1] \cup [2, \infty)$ **8.** $\{11\}$
9. $(-6, \infty)$
10. $(-5, 4)$

11. x-intercept: $(-5, 0)$; y-intercept: $(0, 4)$
12. no x-intercept; y-intercept: $(0, -4)$
13. x-intercept and y-intercept: $(0, 0)$
14. $\{(4, -2)\}$ **15.** 3   **16.** 7.6×10^{-5}
17. 5,600,000,000   **18.** $\frac{21y^7}{x^9}$ **19.** $\frac{a^{10}}{b^{10}}$ **20.** $\frac{m}{n}$

21. $-25x^3 - 2x^2 - 36x + 114$ **22.** $12f^2 + 5f - 3$

23. $x^3 + y^3$ **24.** $49t^6 - 64$ **25.** $\frac{1}{16}x^2 + \frac{5}{2}x + 25$

26. 20   **27.** $2x^3 + 5x^2 - 3x - 2$
28. $(2x+5)(x-9)$ **29.** $25(2t^2+1)(2t^2-1)$

30. $\left\{-\frac{7}{3}, 1\right\}$ **31.** $-1, 4$ **32.** $\frac{2x-3}{2(x-1)}$ **33.** 3

34. $\frac{-a-5b}{(a+b)(a-b)}$ **35.** $\frac{2(x+2)}{2x-1}$

36. $\frac{5+3x-3y}{(x-y)(x^2+xy+y^2)}$ **37.** $\{-4\}$

38. $q = \frac{fp}{p-f}$ or $q = \frac{-fp}{f-p}$ **39.** 150 miles per hour

40. $\frac{6}{5}$ or $1\frac{1}{5}$ hours

CHAPTER 7

SECTION 7.1 (page 453)

1. 6 **3.** 3 **5.** 5 **7.** 2, −2 **9.** 11, −11
11. 42, −42 **13.** no real number square roots
15. (c) **17.** (c) **19.** 6 **21.** 47
23. not a real number **25.** −38
27. not a real number
29. (a) not a real number (b) negative (c) zero
31. −9 **33.** 6 **35.** −4 **37.** −8 **39.** 6
41. −3 **43.** not a real number **45.** 2
47. −9 **49.** $\dfrac{8}{9}$ **51.** $|x|$ **53.** x **55.** x^5
57. 97.381 **59.** 16.863 **61.** −4 and 4 **62.** 4
63. 4, −4; 4 **64.** {−4, 4} **65.** 3 and −3 **66.** 3
67. 3, −3; 3 **68.** {−3, 3} **69.** $\pm\sqrt{25}$ represents
the two numbers: $\sqrt{25} = 5$ and $-\sqrt{25} = -5$.
70. $\sqrt[4]{x^4}$ is always nonnegative, so it must be simplified as
$|x|$. For example if $x = -2$, $\sqrt[4]{x^4} = \sqrt[4]{(-2)^4} = \sqrt[4]{16} =$
$2 \neq x$. However, $|x| = |-2| = 2$. This problem does not
occur with $\sqrt[3]{x^3}$, because if x is negative, so is $\sqrt[3]{x^3}$.

SECTION 7.2 (page 459)

1. true **3.** False; $\sqrt{3^2 + 4^2} = \sqrt{9 + 16} = \sqrt{25} = 5$,
but $3 + 4 = 7$. **5.** (a) **7.** 13 **9.** 9 **11.** 2
13. $\dfrac{8}{9}$ **15.** −3 **17.** 1000 **19.** −12
21. not a real number **23.** $\dfrac{1}{512}$ **25.** $\dfrac{2}{5}$ **27.** $\dfrac{9}{4}$
29. not a real number **31.** 13 **33.** 1000
35. 25 **37.** 9 **39.** 4 **41.** y **43.** $k^{2/3}$
45. $9x^8y^{10}$ **47.** $\dfrac{1}{x^{10/3}}$ **49.** $\dfrac{1}{m^{1/4}n^{3/4}}$ **51.** $\dfrac{c^{11/3}}{b^{11/4}}$
53. 5; 12; 3; 4; 625 **55.** 64 **57.** 64 **59.** x^{10}
61. $\sqrt[6]{x^5}$ **63.** $y\sqrt{7y}$ **65.** $\sqrt[15]{t^8}$ **67.** $x^{-1/2}$
68. $m^{5/2}$ **69.** $k^{-3/4}$ **70.** $x^{-1/2}(3 - 4x)$
71. $m^{5/2}(m^{1/2} - 3)$ **72.** $k^{-3/4}(9 + 2k^{1/2})$
73. $t^{-1/2}(4 + 7t^2)$ **74.** $x^{-1/3}(8x - 5)$ **75.** 4.5 hours
77. No. Both quantities are irrational numbers with
unending decimal representations. They might disagree if
more decimal places are shown. (In fact, they disagree in
the next decimal place!)

SECTION 7.3 (page 469)

1. $\sqrt{mn}$ **3.** radical; denominator **5.** Because there
are only two factors of $\sqrt[3]{x}$, $\sqrt[3]{x} \cdot \sqrt[3]{x} = (\sqrt[3]{x})^2$ or $\sqrt[3]{x^2}$.
7. $\sqrt{30}$ **9.** $\sqrt[3]{14xy}$ **11.** $\sqrt[4]{36}$ **13.** $\dfrac{8}{11}$
15. $\dfrac{\sqrt{3}}{5}$ **17.** $\dfrac{\sqrt{x}}{5}$ **19.** $\dfrac{p^3}{9}$ **21.** $\dfrac{3}{4}$ **23.** $-\dfrac{\sqrt[3]{r^2}}{2}$
25. $2\sqrt{3}$ **27.** $12\sqrt{2}$ **29.** $-4\sqrt{2}$ **31.** $-2\sqrt{7}$
33. not a real number **35.** $4\sqrt[4]{2}$ **37.** $-2\sqrt[3]{2}$
39. $2\sqrt[3]{5}$ **41.** $-4\sqrt[4]{2}$ **43.** $2\sqrt[3]{2}$ **45.** His
reasoning was incorrect. Here 8 is a term, not a factor.
47. $6k\sqrt{2}$ **49.** $\dfrac{3\sqrt[3]{3}}{4}$ **51.** $11x^3$ **53.** $-3t^4$

55. $-10m^4z^2$ **57.** $5a^2b^3c^4$ **59.** $\dfrac{1}{2}r^2t^5$
61. $5x\sqrt{2x}$ **63.** $-10r^5\sqrt{5r}$ **65.** $x^3y^4\sqrt{13x}$
67. $2z^2w^3$ **69.** $-2zt^2\sqrt[3]{2z^2t}$ **71.** $3x^3y^4$
73. $-3r^3s^2\sqrt[4]{2r^3s^2}$ **75.** $\dfrac{y^5\sqrt{y}}{6}$ **77.** $\dfrac{x^5\sqrt[3]{x}}{3}$
79. $4\sqrt{3}$ **81.** $x^2\sqrt{x}$ **83.** $\sqrt[6]{432}$ **85.** $\sqrt[12]{6912}$
87. 5 **89.** $8\sqrt{2}$ **91.** $\sqrt{37}$ **93.** $2\sqrt{10}$
95. $6\sqrt{2}$ **97.** $\sqrt{5y^2 - 2xy + x^2}$
99. $8\sqrt{5}$ feet; 17.9 feet

SECTION 7.4 (page 475)

1. (b) **3.** 15; each radicand is a whole number power
corresponding to the index of the radical. **5.** −4
7. $7\sqrt{3}$ **9.** $24\sqrt{2}$ **11.** 0 **13.** $20\sqrt{5}$
15. $12\sqrt{2x}$ **17.** $-11m\sqrt{2}$ **19.** $\sqrt[3]{2}$ **21.** $2\sqrt[3]{x}$
23. $19\sqrt[4]{2}$ **25.** $x\sqrt[4]{xy}$ **27.** $(4 + 3xy)\sqrt[3]{xy^2}$
29. $\dfrac{7\sqrt{2}}{6}$ **31.** $\dfrac{5\sqrt{2}}{3}$ **33.** Both are approximately
11.3137085. **35.** Both are approximately 31.6227766.
37. (a) **39.** $12\sqrt{5} + 5\sqrt{3}$ inches
41. $58\sqrt{2} + 10\sqrt{3}$ centimeters

SECTION 7.5 (page 483)

1. $\sqrt{ab}$ **2.** $x^2 - y^2$ **3.** $x^2 - y$ **4.** $x - y$
5. $x^2 + 2xy + y^2$ **6.** $x + 2\sqrt{xy} + y$
7. $6 - 4\sqrt{3}$ **9.** $6 - \sqrt{6}$ **11.** 2 **13.** 9
15. $3\sqrt{2} - 5\sqrt{3} + 2\sqrt{6} - 10$ **17.** $3x - 4$
19. $4x - y$ **21.** $16x + 24\sqrt{x} + 9$ **23.** $81 - \sqrt[3]{4}$
25. $6 - 4\sqrt{3}$ is not equal to $2\sqrt{3}$ because 6 and $4\sqrt{3}$ are
not like terms, so they cannot be combined. **27.** $\sqrt{7}$
29. $5\sqrt{3}$ **31.** $\dfrac{\sqrt{6}}{2}$ **33.** $\dfrac{9\sqrt{15}}{5}$ **35.** $-\sqrt{2}$
37. $\dfrac{-8\sqrt{3k}}{k}$ **39.** $\dfrac{6\sqrt{3}}{y}$ **41.** Both methods lead to
the same result, $\dfrac{6\sqrt{3}}{y}$, but multiplying the numerator and
denominator by $\sqrt{y}$ produces this result more directly,
with less simplification required. **43.** $\dfrac{\sqrt{14}}{2}$
45. $-\dfrac{\sqrt{14}}{10}$ **47.** $\dfrac{2\sqrt{6x}}{x}$ **49.** $-\dfrac{7r\sqrt{2rs}}{s}$
51. $\dfrac{12x^3\sqrt{2xy}}{y^5}$ **53.** $\dfrac{\sqrt[3]{18}}{3}$ **55.** $\dfrac{\sqrt[3]{12}}{3}$
57. $-\dfrac{\sqrt[3]{2pr}}{r}$ **59.** $\dfrac{2\sqrt[4]{x^3}}{x}$ **61.** We must multiply the
numerator and denominator by $4 - \sqrt{3}$, so the
denominator becomes $(4 + \sqrt{3})(4 - \sqrt{3}) = 16 - 3 =$
13, a rational number. **63.** $\dfrac{2(4 - \sqrt{3})}{13}$
65. $3(\sqrt{5} - \sqrt{3})$ **67.** $\sqrt{3} + \sqrt{7}$
69. $\sqrt{7} - \sqrt{6} - \sqrt{14} + 2\sqrt{3}$
71. $\dfrac{4\sqrt{x}(\sqrt{x} + 2\sqrt{y})}{x - 4y}$ **73.** $\dfrac{x - 2\sqrt{xy} + y}{x - y}$

75. Square both sides to show that each is equal to $\dfrac{2 - \sqrt{3}}{4}$.   **77.** $\dfrac{5 + 2\sqrt{6}}{4}$ **79.** $\dfrac{4 + 2\sqrt{2}}{3}$

81. $\dfrac{6 + 2\sqrt{6x}}{3}$ **83.** Both expressions are approximately equal to .2588190451.

85. $\dfrac{319}{6(8\sqrt{5} + 1)}$ **86.** $\dfrac{9a - b}{(\sqrt{b} - \sqrt{a})(3\sqrt{a} - \sqrt{b})}$

87. $\dfrac{(3\sqrt{a} + \sqrt{b})(\sqrt{b} + \sqrt{a})}{b - a}$

88. In Exercise 86, we multiply the numerator and denominator by the conjugate of the numerator, while in Exercise 87 we multiply by the conjugate of the denominator.

SECTION 7.6 (page 491)

1. No. There is no solution.   **3.** $\{19\}$ **5.** $\left\{\dfrac{38}{3}\right\}$

7. $\emptyset$ **9.** $\{5\}$ **11.** $\{1\}$ **13.** $\{9\}$ **15.** You cannot just square each term. The right-hand side should be $(8 - x)^2 = 64 - 16x + x^2$.   **17.** $\{4\}$ **19.** $\{-3, -1\}$
21. $\emptyset$ **23.** $\{5\}$ **25.** $\{7\}$ **27.** $\emptyset$ **29.** 3
31. $\{-13\}$ **33.** $\{14\}$ **35.** $\emptyset$ **37.** $\{7\}$ **39.** $\{2, 14\}$
41. $\left\{\dfrac{1}{4}, 1\right\}$ **43.** $\{4\}$ **45.** $\{5\}$ **47.** 5.1 feet

SECTION 7.7 (page 501)

1. i **3.** $\sqrt{b}$ **5.** $a + bi$ is a complex number if a and b are real numbers and i is the imaginary unit. Therefore, for every real number a, if $b = 0$, $a = a + 0i$ is a complex number.   **7.** $13i$ **9.** $-12i$ **11.** $i\sqrt{5}$
13. $4i\sqrt{3}$ **15.** -15 **17.** -10 **19.** $\sqrt{3}$
21. $5i$ **23.** $-1 + 7i$ **25.** $0 + 0i$ **27.** $7 + 3i$
29. $-2 + 0i$ **31.** $1 + 13i$ **33.** $6 + 6i$
35. $4 + 2i$ **37.** -81 **39.** -16 **41.** $-10 - 30i$
43. $10 - 5i$ **45.** $-9 + 40i$ **47.** 153
49. (a) $a - bi$ (b) a^2; b^2 **51.** $1 + i$
53. $-1 + 2i$ **55.** $2 + 2i$ **57.** $-\dfrac{5}{13} - \dfrac{12}{13}i$
59. (a) $4x + 1$ (b) $4 + i$ **60.** (a) $-2x + 3$
(b) $-2 + 3i$ **61.** (a) $3x^2 + 5x - 2$ (b) $5 + 5i$
62. (a) $-\sqrt{3} + \sqrt{6} + 1 - \sqrt{2}$ (b) $\dfrac{1}{5} - \dfrac{7}{5}i$

63. Because $i^2 = -1$, two pairs of like terms can be combined in Exercise 61(b).   **64.** Because $i^2 = -1$, terms can be combined in the numerator, and the denominator is changed.   **65.** $\dfrac{5}{41} + \dfrac{4}{41}i$ **67.** -1 **69.** i

71. 1   **73.** $-i$ **75.** Since $i^{20} = (i^4)^5 = 1^5 = 1$, the student multiplied by 1, which is justified by the identity property of multiplication.   **77.** $\dfrac{1}{2} + \dfrac{1}{2}i$

79. $(1 + 5i)^2 - 2(1 + 5i) + 26$ will simplify to 0 when the operations are applied.

CHAPTER 7 REVIEW EXERCISES (page 509)

1. 42   **2.** -17 **3.** not a real number   **4.** 6
5. -5 **6.** $-3z^4$ **7.** -2 **8.** n must be even, and a must be negative.   **9.** 6.325   **10.** 8.775

11. 17.607   **12.** 32   **13.** -4 **14.** $-\dfrac{216}{125}$

15. -32 **16.** $\dfrac{1000}{27}$ **17.** 49   **18.** 96

19. $\dfrac{k^{17/12}}{2}$ **20.** By a power rule for exponents and the definition of $x^{1/n}$, $a^{m/n} = (a^m)^{1/n} = \sqrt[n]{a^m}$.   **21.** 3^9

22. $7^4\sqrt{7}$ **23.** $m^4\sqrt[3]{m}$ **24.** $k^2\sqrt[4]{k}$ **25.** $\sqrt{66}$
26. $\sqrt{5r}$ **27.** $\sqrt[3]{30}$ **28.** $\sqrt[4]{21}$ **29.** $2\sqrt{5}$
30. $5\sqrt{3}$ **31.** $-5\sqrt{5}$ **32.** $-3\sqrt[3]{4}$ **33.** $10y^3\sqrt{y}$
34. $4pq^2\sqrt[3]{p}$ **35.** $\dfrac{7}{9}$ **36.** $\dfrac{y\sqrt{y}}{12}$ **37.** $\dfrac{m^5}{3}$
38. $\dfrac{\sqrt[3]{r^2}}{2}$ **39.** $\sqrt{130}$ **40.** $\sqrt{53}$ **41.** $-11\sqrt{2}$

42. $23\sqrt{5}$ **43.** $7\sqrt{3y}$ **44.** $26m\sqrt{6m}$ **45.** $19\sqrt[3]{2}$
46. $-8\sqrt[4]{2}$ **47.** $1 - \sqrt{3}$ **48.** 2   **49.** $9 - 7\sqrt{2}$
50. $86 + 8\sqrt{55}$ **51.** $15 - 2\sqrt{26}$ **52.** $12 - 2\sqrt{35}$
53. $.1678404338 \neq 59.16079783$ **54.** $\dfrac{\sqrt{30}}{5}$

55. $-3\sqrt{6}$ **56.** $\dfrac{3\sqrt{7py}}{y}$ **57.** $\dfrac{\sqrt{22}}{4}$ **58.** $-\dfrac{\sqrt[3]{45}}{5}$

59. $\dfrac{3m\sqrt[3]{4n}}{n^2}$ **60.** $\dfrac{\sqrt{2} - \sqrt{7}}{-5}$ **61.** $\dfrac{-5(\sqrt{6} + \sqrt{3})}{3}$

62. $\{2\}$ **63.** $\{6\}$ **64.** $\emptyset$ **65.** $\{0, 5\}$ **66.** $\{9\}$

67. $\{3\}$ **68.** $\{7\}$ **69.** $\left\{-\dfrac{1}{2}\right\}$ **70.** $\{6\}$ **71.** $5i$

72. $10i\sqrt{2}$ **73.** no   **74.** $-10 - 2i$ **75.** $14 + 7i$
76. $-\sqrt{35}$ **77.** -45 **78.** 3   **79.** $5 + i$
80. $32 - 24i$ **81.** $1 - i$ **82.** $4 + i$ **83.** $-i$
84. 1   **85.** $-i$ **86.** -4 **87.** $-13ab^2$

88. $\dfrac{1}{100}$ **89.** $\dfrac{1}{y^{1/2}}$ **90.** $\dfrac{x^{3/4}}{z^{3/4}}$ **91.** k^6

92. $3z^3t^2\sqrt[3]{2t^2}$ **93.** $57\sqrt{2}$ **94.** $6x\sqrt[3]{y^2}$

95. $\sqrt{35} + \sqrt{15} - \sqrt{21} - 3$ **96.** $-\dfrac{\sqrt{3}}{6}$

97. $\dfrac{\sqrt[3]{60}}{5}$ **98.** $\dfrac{2\sqrt{z}(\sqrt{z} + 2)}{z - 4}$ **99.** $7i$

100. $3 - 7i$ **101.** $-5i$ **102.** 1   **103.** $\{5\}$

104. $\left\{\dfrac{3}{2}\right\}$ **105.** $\sqrt{101}$ **106.** 7.9 feet   **107.** 8.2 miles

CHAPTER 7 TEST (page 513)

1. -29 **2.** 15   **3.** (c)   **4.** 12.09   **5.** $\dfrac{1}{256}$

6. $\dfrac{9y^{3/10}}{x^2}$ **7.** $3x^2y^3\sqrt{6x}$ **8.** $2ab^3\sqrt[4]{2a^3b}$
9. $\sqrt[6]{200}$ **10.** $26\sqrt{5}$ **11.** $66 + \sqrt{5}$
12. $-2(\sqrt{7} - \sqrt{5})$ **13.** $\dfrac{-\sqrt{10}}{4}$ **14.** $\dfrac{2\sqrt[3]{25}}{5}$
15. $\sqrt{26}$ **16.** 59.8   **17.** $\{6\}$ **18.** $-5 - 8i$
19. $3 + 4i$ **20.** $-i$

CUMULATIVE REVIEW 1–7 (page 515)

1. $\dfrac{4}{5}$ **2.** $\left\{\dfrac{11}{10}, \dfrac{7}{2}\right\}$ **3.** $(-6, \infty)$ **4.** $(1, 3)$

5. $(-2, 1)$ **6.** $12x + 11y = 18$ **7.** (c)

8. (a) $(0, 6)$ **(b)** $(2, 0)$ **9.** $120

10.

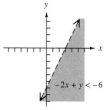

11. Both angles measure $80°$. **12.** $\emptyset$ **13.** infinite number of solutions **14.** 0 **15.** -27

16. $-k^3 - 3k^2 - 8k - 9$ **17.** $8x^2 + 17x - 21$

18. $z - 2 + \dfrac{3}{z}$ **19.** $3y^3 - 3y^2 + 4y + 1 + \dfrac{-10}{2y + 1}$

20. $(2p - 3q)(p - q)$ **21.** $(3k^2 + 4)(6k^2 - 5)$

22. $(x + 8)(x^2 - 8x + 64)$ **23.** $\dfrac{y}{y + 5}$

24. $\dfrac{4x + 2y}{(x + y)(x - y)}$ **25.** $-\dfrac{9}{4}$ **26.** $\dfrac{-1}{a + b}$

27. $\left\{-3, -\dfrac{5}{2}\right\}$ **28.** $\left\{-\dfrac{2}{5}, 1\right\}$ **29.** $\dfrac{1}{243}$ **30.** $x^{1/12}$

31. $8\sqrt{5}$ **32.** $\dfrac{-9\sqrt{5}}{20}$ **33.** $4(\sqrt{6} + \sqrt{5})$ **34.** $6\sqrt[3]{4}$

35. $\sqrt{29}$ **36.** $\{6\}$ **37.** 15 miles per hour

38. Brenda: 8 miles per hour; Chuck: 4 miles per hour

39. 17 dimes and 12 quarters **40.** $\dfrac{80}{39}$ or $2\dfrac{2}{39}$ liters

CHAPTER 8

SECTION 8.1 (page 523)

1. Divide both sides by 2.
3. Square root property for $(2x + 1)^2 = 5$; completing the square for $x^2 + 4x = 12$
5. $\{-9, 9\}$ **7.** $\{-\sqrt{17}, \sqrt{17}\}$ **9.** $\{-7, 3\}$

11. $\left\{\dfrac{1 + \sqrt{7}}{3}, \dfrac{1 - \sqrt{7}}{3}\right\}$

13. $\left\{\dfrac{-1 + 2\sqrt{6}}{4}, \dfrac{-1 - 2\sqrt{6}}{4}\right\}$ **15.** $\{-2i\sqrt{3}, 2i\sqrt{3}\}$

17. $\{5 + i\sqrt{3}, 5 - i\sqrt{3}\}$

19. $\left\{\dfrac{1 + 2i\sqrt{2}}{6}, \dfrac{1 - 2i\sqrt{2}}{6}\right\}$ **21.** $\{-4, 6\}$

23. $\left\{-3, \dfrac{8}{3}\right\}$ **25.** $\left\{\dfrac{-5 + \sqrt{41}}{4}, \dfrac{-5 - \sqrt{41}}{4}\right\}$

27. $\{-2 + 3i, -2 - 3i\}$ **29.** $\left\{\dfrac{4 + \sqrt{3}}{3}, \dfrac{4 - \sqrt{3}}{3}\right\}$

31. $\left\{\dfrac{2 + \sqrt{3}}{3}, \dfrac{2 - \sqrt{3}}{3}\right\}$

33. $\left\{\dfrac{-2 + 2i\sqrt{2}}{3}, \dfrac{-2 - 2i\sqrt{2}}{3}\right\}$

35. $\{1 + \sqrt{2}, 1 - \sqrt{2}\}$ **37.** $\{-3 + i\sqrt{3}, -3 - i\sqrt{3}\}$

39. x^2 **40.** x **41.** $6x$ **42.** 1 **43.** 9

44. $(x + 3)^2$ or $x^2 + 6x + 9$

SECTION 8.2 (page 531)

1. true **3.** true **5.** true **7.** $\{3, 5\}$

9. $\left\{\dfrac{-2 + \sqrt{2}}{2}, \dfrac{-2 - \sqrt{2}}{2}\right\}$

11. $\left\{\dfrac{1 + \sqrt{3}}{2}, \dfrac{1 - \sqrt{3}}{2}\right\}$ **13.** $\{5 + \sqrt{7}, 5 - \sqrt{7}\}$

15. $\left\{\dfrac{-2 + \sqrt{10}}{2}, \dfrac{-2 - \sqrt{10}}{2}\right\}$

17. $\{-1 + 3\sqrt{2}, -1 - 3\sqrt{2}\}$ **19.** $\{-i\sqrt{47}, i\sqrt{47}\}$

21. $\{3 + i\sqrt{5}, 3 - i\sqrt{5}\}$ **23.** $\left\{\dfrac{1 + i\sqrt{6}}{2}, \dfrac{1 - i\sqrt{6}}{2}\right\}$

25. $\left\{\dfrac{-2 + i\sqrt{2}}{3}, \dfrac{-2 - i\sqrt{2}}{3}\right\}$

27. $\{4 + 3i\sqrt{2}, 4 - 3i\sqrt{2}\}$

29. $\left\{\dfrac{-1 + i\sqrt{3}}{2}, \dfrac{-1 - i\sqrt{3}}{2}\right\}$

31. at 3 seconds and 5 seconds **33.** It reaches its *maximum* height at 5 seconds because this is the only time it reaches 400 feet.

35. 2117 **36.** 2243 **37.** 2356 **38.** Find $f(3)$.
39. 1992 ($x = 4$ corresponds to 1992) **40.** 1999
41. (b) **43. (c)** **45. (a)** **47. (d)**
49. $(6x + 5)(4x - 9)$ **51.** cannot be factored
53. $(12x + 1)(x - 7)$ **55.** -4 and 4

SECTION 8.3 (page 541)

1. Multiply by the LCD, x.
3. Make a substitution for $r^2 + r$. **5.** $\{-4, 7\}$

7. $\left\{-\dfrac{2}{3}, 1\right\}$ **9.** $\left\{-\dfrac{14}{17}, 5\right\}$ **11.** $\left\{-\dfrac{11}{7}, 0\right\}$

13. $\left\{\dfrac{-1 + \sqrt{13}}{2}, \dfrac{-1 - \sqrt{13}}{2}\right\}$ **15.** $\dfrac{1}{m}$ job per hour

17. 25 miles per hour **19.** 80 miles per hour
21. 9 minutes **23.** $\{3\}$

25. $\left\{\dfrac{8}{9}\right\}$ **27.** $\{16\}$ **29.** $\left\{\dfrac{2}{5}\right\}$ **31.** $\{-3, 3\}$

33. $\left\{-\dfrac{3}{2}, -1, 1, \dfrac{3}{2}\right\}$ **35.** $\{-6, -5\}$ **37.** $\{-4, 1\}$

39. $\left\{-\dfrac{1}{3}, \dfrac{1}{6}\right\}$ **41.** $\left\{-\dfrac{1}{2}, 3\right\}$ **43.** $\{-8, 1\}$ **45.** $\{25\}$

47. It would cause both denominators to be 0, and division by 0 is undefined. **48.** The solution is $\dfrac{12}{5}$.

49. $\left(\dfrac{x}{x - 3}\right)^2 + 3\left(\dfrac{x}{x - 3}\right) - 4 = 0$ **50.** The numerator can never equal the denominator, since the denominator is 3 less than the numerator.

51. $\left\{\dfrac{12}{5}\right\}$; The values for t are -4 and 1. The value 1 is impossible because it leads to a contradiction $\left(\text{since } \dfrac{x}{x - 3} \text{ is never equal to } 1\right)$. **52.** $\left\{\dfrac{12}{5}\right\}$; The values for s are $\dfrac{1}{x}$ and $\dfrac{-4}{x}$. The value $\dfrac{1}{x}$ is impossible, since $\dfrac{1}{x} \neq \dfrac{1}{x - 3}$ for all x.

SECTION 8.4 (page 549)

1. $m = \sqrt{p^2 - n^2}$ **3.** $t = \dfrac{\pm\sqrt{dk}}{k}$ **5.** $d = \dfrac{\pm\sqrt{skI}}{I}$

7. $v = \dfrac{\pm\sqrt{kAF}}{F}$ **9.** $r = \dfrac{\pm\sqrt{3\pi Vh}}{\pi h}$

11. $t = \dfrac{-B \pm \sqrt{B^2 - 4AC}}{2A}$ **13.** $h = \dfrac{D^2}{k}$

15. $\ell = \dfrac{p^2 g}{k}$ **17.** If g is positive, the only way to have a real value for p is to have kl positive, since the quotient of two positive numbers is positive. If k and l have different signs, their product is negative, leading to a negative radicand.   **19.** eastbound ship: 80 miles; southbound ship: 150 miles   **21.** 20 meters, 21 meters, 29 meters   **23.** 12 centimeters by 20 centimeters   **25.** 1 foot   **27.** 4.3 hours   **29.** 7.8 hours and 8.3 hours   **31.** $4814.10   **33.** 1987   **35.** 9.2 seconds   **37.** 1 second and 8 seconds   **39.** .035 or 3.5%   **41.** 5.5 meters per second   **43.** 5 or 14

SECTION 8.5 (page 559)

1. Include the endpoints if the symbol is $\geq$ or $\leq$. Exclude the endpoints if the symbol is $>$ or $<$.
3. $(-\infty, -4] \cup [3, \infty)$
5. $(-\infty, -1) \cup (5, \infty)$

7. $(-4, 6)$

9. $(-\infty, 1] \cup [3, \infty)$

11. $\left(-\infty, -\dfrac{3}{2}\right] \cup \left[\dfrac{3}{5}, \infty\right)$

13. $\left(-\dfrac{2}{3}, \dfrac{1}{3}\right)$

15. $\left(-\infty, -\dfrac{1}{2}\right] \cup \left[\dfrac{1}{3}, \infty\right)$

17. $(-\infty, 3 - \sqrt{3}] \cup [3 + \sqrt{3}, \infty)$

19. $(-\infty, 1) \cup (2, 4)$

21. $\left[-\dfrac{3}{2}, \dfrac{1}{3}\right] \cup [4, \infty)$

23. $(-\infty, 1) \cup (4, \infty)$

25. $\left[-\dfrac{3}{2}, 5\right)$

27. $(2, 6]$

29. $\left(-\infty, \dfrac{1}{2}\right) \cup \left(\dfrac{5}{4}, \infty\right)$

31. $[-4, -2)$

33. $\left(0, \dfrac{1}{2}\right) \cup \left(\dfrac{5}{2}, \infty\right)$

35. $(-\infty, \infty)$ **37.** $\emptyset$
39. 3 seconds and 13 seconds   **40.** between 3 seconds and 13 seconds   **41.** at 0 seconds (the time when it is initially projected) and at 16 seconds (the time when it hits the ground)   **42.** between 0 and 3 seconds and also between 13 and 16 seconds

CHAPTER 8 REVIEW EXERCISES (page 567)

1. $\{-11, 11\}$ **2.** $\{-\sqrt{3}, \sqrt{3}\}$
3. $\left\{-\dfrac{15}{2}, \dfrac{5}{2}\right\}$ **4.** $\left\{\dfrac{2 + 5i}{3}, \dfrac{2 - 5i}{3}\right\}$
5. $\{-2 + \sqrt{19}, -2 - \sqrt{19}\}$ **6.** $\left\{\dfrac{1}{2}, 1\right\}$
7. $\left\{-\dfrac{7}{2}, 3\right\}$ **8.** $\left\{\dfrac{-5 + \sqrt{53}}{2}, \dfrac{-5 - \sqrt{53}}{2}\right\}$
9. $\left\{\dfrac{1 + \sqrt{41}}{2}, \dfrac{1 - \sqrt{41}}{2}\right\}$
10. $\left\{\dfrac{7}{3}\right\}$ **11.** $\left\{\dfrac{2 + i\sqrt{2}}{3}, \dfrac{2 - i\sqrt{2}}{3}\right\}$
12. $\left\{\dfrac{-7 + \sqrt{37}}{2}, \dfrac{-7 - \sqrt{37}}{2}\right\}$
13. $\left\{\dfrac{-3 + i\sqrt{23}}{4}, \dfrac{-3 - i\sqrt{23}}{4}\right\}$
14. The student was incorrect since the fraction bar should extend under the term $-b$.
15. 4 seconds and 12 seconds
16. The rock reaches the height 768 feet twice: once on its way up and again on its way down.
17. $k = 1$ **18.** all k such that $k < 1$
19. (c)   **20.** (a)   **21.** (d)   **22.** (b)
23. It factors as $(6x - 5)(4x - 9)$.
24. It cannot be factored.
25. $\left\{-\dfrac{5}{2}, 3\right\}$ **26.** $\left\{-\dfrac{1}{2}, 1\right\}$
27. $\left\{-\dfrac{11}{6}, -\dfrac{19}{12}\right\}$ **28.** $\{-4\}$ **29.** $\left\{-\dfrac{343}{8}, 64\right\}$
30. $\{-3, -2, 1, 2\}$
31. 40 miles per hour   **32.** Linda: 7.2 hours; Ed: 5.2 hours   **33.** Because x appears on the left side alone, and because it is equal to the nonnegative square root of $2x + 4$, it cannot be negative.   **34.** Take the positive and negative square roots of each value obtained from the formula.
35. $d = \dfrac{\pm\sqrt{SkI}}{I}$ **36.** $v = \dfrac{\pm\sqrt{rFkw}}{kw}$

ANSWERS

37. $r = \dfrac{-\pi h \pm \sqrt{\pi^2 h^2 + 2\pi S}}{2\pi}$

38. $t = \dfrac{3m \pm \sqrt{9m^2 + 24m}}{2m}$

39. 5.2 seconds   **40.** .7 second and 4.0 seconds
41. 10 meters by 12 meters   **42.** 9 feet, 12 feet, 15 feet
43. 68 and 69   **44.** 1 inch   **45.** 3 minutes
46. $.50   **47.** 4.5%   **48.** Jim: 7.5 hours;
Jake: 8.5 hours

49. $\left(-\infty, -\dfrac{3}{2}\right) \cup (4, \infty)$

50. $[-4, 3]$

51. $\left(-\infty, -\dfrac{1}{2}\right) \cup (3, \infty)$

52. $[-3, 2)$

53. $(-\infty, -5] \cup [-2, 3]$

54. $\emptyset$

55. $R = \dfrac{\pm\sqrt{Vh - r^2 h}}{h}$ **56.** $\left\{\dfrac{3 + i\sqrt{3}}{3}, \dfrac{3 - i\sqrt{3}}{3}\right\}$

57. $\{-2, -1, 3, 4\}$ **58.** $(-\infty, -6) \cup \left(-\dfrac{3}{2}, 1\right)$

59. $\left\{\dfrac{-11 + \sqrt{7}}{3}, \dfrac{-11 - \sqrt{7}}{3}\right\}$ **60.** $y = \dfrac{6p^2}{z}$

61. $\left\{\dfrac{3}{5}, 1\right\}$ **62.** $\left\{-\dfrac{5}{3}, -\dfrac{3}{2}\right\}$ **63.** $\left(-5, -\dfrac{23}{5}\right]$

64. 3 hours and 6 hours   **65.** 7 miles per hour
66. $\left\{\dfrac{4}{11}, 5\right\}$ **67.** $\{-11i\sqrt{2}, 11i\sqrt{2}\}$ **68.** $(-\infty, \infty)$

CHAPTER 8 TEST (page 573)

1. $\{-3\sqrt{6}, 3\sqrt{6}\}$ **2.** $\left\{-\dfrac{8}{7}, \dfrac{2}{7}\right\}$

3. $\{-1 + \sqrt{2}, -1 - \sqrt{2}\}$

4. $\left\{\dfrac{3 + \sqrt{17}}{4}, \dfrac{3 - \sqrt{17}}{4}\right\}$

5. $\left\{\dfrac{2 + i\sqrt{11}}{3}, \dfrac{2 - i\sqrt{11}}{3}\right\}$ **6.** $\left\{\dfrac{2}{3}\right\}$

7. Maretha: 11.1 hours; Lillaana: 9.1 hours   **8.** (a)
9. discriminant: 88; There are two irrational solutions.

10. $\left\{-\dfrac{2}{3}, 6\right\}$ **11.** $\left\{\dfrac{-7 + \sqrt{97}}{8}, \dfrac{-7 - \sqrt{97}}{8}\right\}$

12. $\left\{-2, -\dfrac{1}{3}, \dfrac{1}{3}, 2\right\}$ **13.** $\left\{-\dfrac{5}{2}, 1\right\}$

14. $r = \dfrac{\pm\sqrt{\pi S}}{2\pi}$ **15.** 1985

16. 7 miles per hour   **17.** 2 feet   **18.** 16 meters

19. $(-\infty, -5) \cup \left(\dfrac{3}{2}, \infty\right)$

20. $(-\infty, 4) \cup [9, \infty)$

CUMULATIVE REVIEW 1–8 (page 575)

1. $-2, 0, 7$ **2.** $-\dfrac{7}{3}, -2, 0, .7, 7, \dfrac{32}{3}$

3. all except $\sqrt{-8}$ **4.** All are complex numbers.

5. 6   **6.** 41   **7.** $\{1\}$ **8.** $\left[2, \dfrac{8}{3}\right]$

9. $\left(-\infty, -\dfrac{9}{4}\right) \cup \left(\dfrac{5}{4}, \infty\right)$

10. slope: $\dfrac{1}{2}$; y-intercept: $\left(0, -\dfrac{7}{4}\right)$ **11.** $x + 3y = -1$

12. $\dfrac{x^8}{y^4}$ **13.** $\dfrac{4}{xy^2}$ **14.** $\sqrt{15}$; $\left[\dfrac{5}{2}, \infty\right)$

15. $\{(1, -4)\}$ **16.** $\left\{\left(\dfrac{1}{3}, \dfrac{1}{4}, -\dfrac{1}{2}\right)\right\}$

17. $\dfrac{4}{9}t^2 + 12t + 81$ **18.** $-3t^3 + 5t^2 - 12t + 15$

19. $4x^2 - 6x + 11 + \dfrac{4}{x + 2}$ **20.** $x(4 + x)(4 - x)$

21. $(4m - 3)(6m + 5)$ **22.** $(3x - 5y)^2$

23. $(2x + 3y)(4x^2 - 6xy + 9y^2)$ **24.** $\dfrac{x - 5}{x + 5}$

25. $-\dfrac{5}{18}$ **26.** $-\dfrac{8}{k}$ **27.** $\dfrac{r - s}{r}$ **28.** $\dfrac{3\sqrt[3]{4}}{4}$

29. $\sqrt{7} + \sqrt{5}$ **30.** $\left\{\dfrac{2}{3}\right\}$ **31.** $\emptyset$

32. $\left\{\dfrac{7 + \sqrt{177}}{4}, \dfrac{7 - \sqrt{177}}{4}\right\}$ **33.** $\{-2, -1, 1, 2\}$

34. 7 inches by 3 inches   **35.** southbound car: 57 miles;
eastbound car: 76 miles
36. suppliers I and III: 22 units; supplier II: 56 units
37. .3 second and 4.7 seconds   **38.** $\{-3, 5\}$
39. $\left\{\dfrac{2 + \sqrt{10}}{2}, \dfrac{2 - \sqrt{10}}{2}\right\}$ **40.** $(5, 13)$

CHAPTER 9

SECTION 9.1 (page 585)

1. (a) B   (b) C   (c) A   (d) D   **3.** $(0, 0)$
5. $(0, 4)$ **7.** $(1, 0)$ **9.** $(-3, -4)$
11. downward; wider   **13.** upward; narrower
15. In Exercise 9, the parabola is shifted 3 units to the left
and 4 units down. The parabola in Exercise 10 is shifted
5 units to the right and 8 units down.   **17.** (a) I
(b) IV   (c) II   (d) III

19.

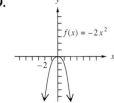

21.

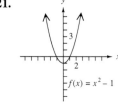

23.

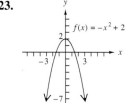

25.

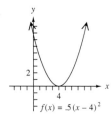

27.

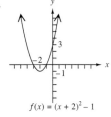

29.

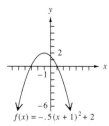

31. It is shifted 6 units upward.

32.

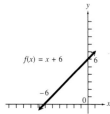

33. It is shifted 6 units upward.

34. It is shifted 6 units to the right.

35.

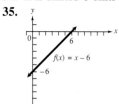

36. It is shifted 6 units to the right.

SECTION 9.2 (page 595)

1. If x is squared, it has a vertical axis; if y is squared, it has a horizontal axis.   **3.** Use the discriminant of the corresponding quadratic equation. If it is positive there are two x-intercepts. If it is zero, there is just one intercept (the vertex), and if it is negative, there are no x-intercepts.
5. $(-1, 3)$; upward; narrower; no x-intercepts
7. $\left(\dfrac{5}{2}, \dfrac{37}{4}\right)$; downward; same; two x-intercepts
9. $(-3, -9)$; to the right; wider
11. domain: $(-\infty, \infty)$; range: $[-1, \infty)$

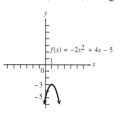

13. domain: $(-\infty, \infty)$; range: $(-\infty, -3]$

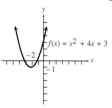

15.   **17.**

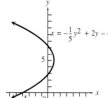   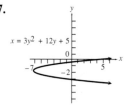

19. B   **21.** A   **23.** 160 feet by 320 feet
25. 16 feet; 2 seconds   **27.** 30 and 30   **29.** The parabola opens downward, indicating $a < 0$.   **31.** Since $x = 2.1 \approx 2$, the year corresponding to 2, which is 1990, is the year of maximum number of jobs.
33. (a) $\{1, 3\}$ **(b)** $(-\infty, 1) \cup (3, \infty)$ **(c)** $(1, 3)$
34. (a) $\left\{-4, \dfrac{2}{3}\right\}$ **(b)** $(-\infty, -4] \cup \left[\dfrac{2}{3}, \infty\right)$ **(c)** $\left(-4, \dfrac{2}{3}\right)$
35. (a) $\{-2, 5\}$ **(b)** $[-2, 5]$ **(c)** $(-\infty, -2] \cup [5, \infty)$
36. (a) $\left\{-3, \dfrac{5}{2}\right\}$ **(b)** $\left[-3, \dfrac{5}{2}\right]$ **(c)** $(-\infty, -3] \cup \left[\dfrac{5}{2}, \infty\right)$

Section 9.3 (page 607)

1. 0   **3.** 0, 0   **5.** B   **7.** A

9.

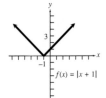

$f(x) = |x + 1|$

11.

$f(x) = \frac{1}{x} + 1$

13.

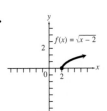

$f(x) = \sqrt{x - 2}$

15.

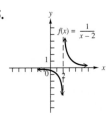

$f(x) = \frac{1}{x - 2}$

17.

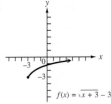

$f(x) = \sqrt{x + 3} - 3$

19. B   **21.** D   **23.** There are some x-values that yield two y-values. In a function, every x yields one and only one y. (A circle also fails the vertical line test.)
25. $(4, 6)$, $r = 7$ **27.** $(1, -2)$, $r = 3$
29. $(-5, -4)$, $r = 6$
31.

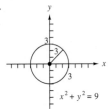

$x^2 + y^2 = 9$

33.

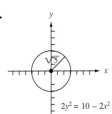

$2y^2 = 10 - 2x^2$

35.

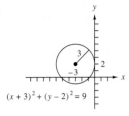

$(x + 3)^2 + (y - 2)^2 = 9$

37.

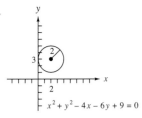

$x^2 + y^2 - 4x - 6y + 9 = 0$

39.

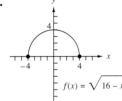

$f(x) = \sqrt{16 - x^2}$

41.

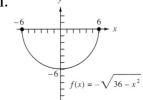

$f(x) = -\sqrt{36 - x^2}$

43. (a) $|x + p|$ **(b)** The distance from the point to the origin should equal the distance from the line to the origin. **(c)** $\sqrt{(x - p)^2 + y^2}$ **(d)** $y^2 = 4px$

Section 9.4 (page 619)

1. C   **3.** D   **5.** When written in one of the forms given in the box "Equations of Hyperbolas" in this section, it will open up and down if the $-$ sign precedes the x^2 term; it will open left and right if the $-$ sign precedes the y^2 term.
7.

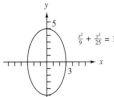

$\frac{x^2}{9} + \frac{y^2}{25} = 1$

9.

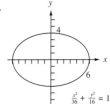

$\frac{x^2}{36} + \frac{y^2}{16} = 1$

11.

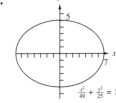

$\frac{x^2}{49} + \frac{y^2}{25} = 1$

13. The points (a, b), $(a, -b)$, $(-a, -b)$, $(-a, b)$ or the points (b, a), $(-b, a)$, $(-b, -a)$, $(b, -a)$ are used as corners of a rectangle. The diagonals of the rectangle are drawn, and they are used as asymptotes for the hyperbola.

15.

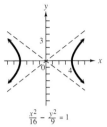

$$\frac{x^2}{16} - \frac{y^2}{9} = 1$$

17.

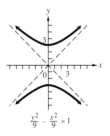

$$\frac{y^2}{9} - \frac{x^2}{9} = 1$$

19.

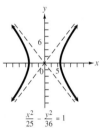

$$\frac{x^2}{25} - \frac{y^2}{36} = 1$$

21. hyperbola

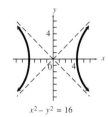

$x^2 - y^2 = 16$

23. ellipse

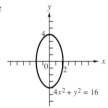

$4x^2 + y^2 = 16$

25. circle

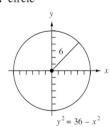

$y^2 = 36 - x^2$

27. hyperbola

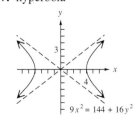

$9x^2 = 144 + 16y^2$

29. hyperbola

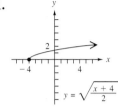

$y^2 = 4 + x^2$

31.

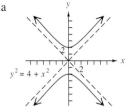

$y = \sqrt{\dfrac{x + 4}{2}}$

33. The two thumbtacks act as foci, and the length of the string is constant, satisfying the requirements of the definition of an ellipse.

35. **(a)** $\sqrt{43} + \sqrt{5013} \approx 77.4$ million miles
(b) $\sqrt{5013} - \sqrt{43} \approx 64.2$ million miles

SECTION 9.5 (page 627)

1. Substitute $x - 1$ for y in the first equation. Then solve for x. Find the corresponding y-values by substituting back into $y = x - 1$.   **3.** No, it is not possible. The maximum number of points of intersection is two.

5. $\left\{(0, 0), \left(\dfrac{1}{2}, \dfrac{1}{2}\right)\right\}$ **7.** $\{(-6, 9), (-1, 4)\}$

9. $\left\{\left(-\dfrac{1}{5}, \dfrac{7}{5}\right), (1, -1)\right\}$ **11.** $\left\{(-2, -2), \left(-\dfrac{4}{3}, -3\right)\right\}$

13. $\{(-3, 1), (1, -3)\}$ **15.** $\left\{\left(-\dfrac{3}{2}, -\dfrac{9}{4}\right), (-2, 0)\right\}$

17. $\{(-\sqrt{3}, 0), (\sqrt{3}, 0), (-\sqrt{5}, 2), (\sqrt{5}, 2)\}$

19. $\{(-2, 0), (2, 0)\}$ **21.** $\{(i\sqrt{2}, -3i\sqrt{2}),$
$(-i\sqrt{2}, 3i\sqrt{2}), (-\sqrt{6}, -\sqrt{6}), (\sqrt{6}, \sqrt{6})\}$

23. $\{(-2i\sqrt{2}, -2\sqrt{3}), (-2i\sqrt{2}, 2\sqrt{3}), (2i\sqrt{2}, -2\sqrt{3}),$
$(2i\sqrt{2}, 2\sqrt{3})\}$ **25.** $\{(-\sqrt{5}, -\sqrt{5}), (\sqrt{5}, \sqrt{5})\}$

27. $\{(i, 2i), (-i, -2i), (2, -1), (-2, 1)\}$

29. $(-1, 2)$ **31.** $(-4, 8)$ **33.** length: 12 feet; width: 7 feet   **35.** \$20; 800 calculators

SECTION 9.6 (page 635)

1. **(a)** B **(b)** D **(c)** A **(d)** C
3.

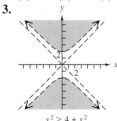

$y^2 > 4 + x^2$

5.

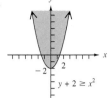

$y + 2 \geq x^2$

7.

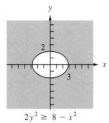

$2y^2 \geq 8 - x^2$

9.

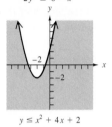

$y \leq x^2 + 4x + 2$

11.

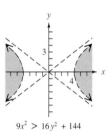

$9x^2 > 16y^2 + 144$

13.

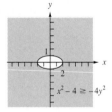

$x^2 - 4 \geq -4y^2$

15.

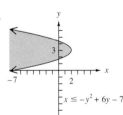

$x \leq -y^2 + 6y - 7$

17.

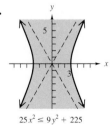

$25x^2 \leq 9y^2 + 225$

19. (c)

21.

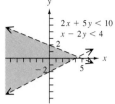

$2x + 5y < 10$
$x - 2y < 4$

23.

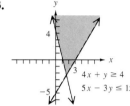

$4x + y \geq 4$
$5x - 3y \leq 15$

25.

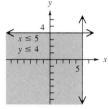

$x \leq 5$
$y \leq 4$

27.

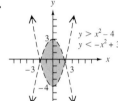

$y > x^2 - 4$
$y < -x^2 + 3$

29.

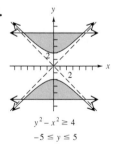

$y^2 - x^2 \geq 4$
$-5 \leq y \leq 5$

31.

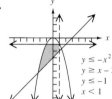

$y \leq -x^2$
$y \geq x - 3$
$y \leq -1$
$x < 1$

33. A   **35.** B

Chapter 9 Review Exercises (page 645)

1. $(0, -2)$ **2.** $(0, 6)$ **3.** $(-2, 0)$ **4.** $(1, 0)$
5. $(3, 7)$ **6.** $(2, 1)$ **7.** $(3, -2)$ **8.** $(-4, 3)$

9. domain: $(-\infty, \infty)$; range: $[-3, \infty)$

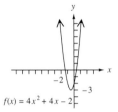

$f(x) = 4x^2 + 4x - 2$

10. domain: $(-\infty, \infty)$; range: $(-\infty, 3]$

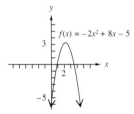

$f(x) = -2x^2 + 8x - 5$

11.

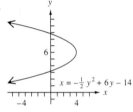

$x = -\frac{1}{2}y^2 + 6y - 14$

12.

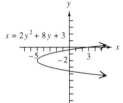

$x = 2y^2 + 8y + 3$

13. If the discriminant $b^2 - 4ac$ is positive, there are two x-intercepts. If it is zero, there is only one x-intercept (the vertex). If it is negative, there are no x-intercepts.
14. (a)   **15.** 1986; $.1 billion   **16.** 5 seconds; 400 feet   **17.** length: 50 meters; width: 50 meters
18. 5 and 5
19.

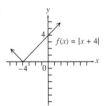

$f(x) = |x + 4|$

20.

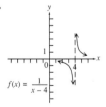

$f(x) = \dfrac{1}{x - 4}$

21.

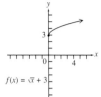

$f(x) = \sqrt{x} + 3$

22. $(x + 2)^2 + (y - 4)^2 = 9$
23. $(x + 1)^2 + (y + 3)^2 = 25$
24. $(x - 4)^2 + (y - 2)^2 = 36$ **25.** $(-3, 2)$, $r = 4$
26. $(4, 1)$, $r = 2$ **27.** $(-1, -5)$, $r = 3$
28. $(3, -2)$, $r = 5$
29.

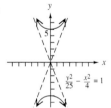

$\dfrac{x^2}{16} + \dfrac{y^2}{9} = 1$

30.

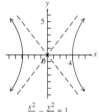

$\dfrac{x^2}{49} + \dfrac{y^2}{25} = 1$

31.

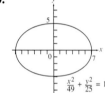

$\dfrac{x^2}{16} - \dfrac{y^2}{25} = 1$

32.

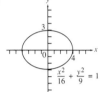

$\dfrac{y^2}{25} - \dfrac{x^2}{4} = 1$

33.

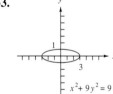

$x^2 + 9y^2 = 9$

34.

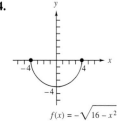

$f(x) = -\sqrt{16 - x^2}$

ANSWERS

35. circle   **36.** parabola
37. hyperbola   **38.** ellipse   **39.** parabola
40. hyperbola   **41.** $\{(6, -9), (-2, -5)\}$
42. $\{(1, 2), (-5, 14)\}$ **43.** $\{(4, 2), (-1, -3)\}$
44. $\{(-2, -4), (8, 1)\}$ **45.** $\{(-\sqrt{2}, 2), (-\sqrt{2}, -2),$
$(\sqrt{2}, -2), (\sqrt{2}, 2)\}$ **46.** $\{(-\sqrt{6}, -\sqrt{3}), (-\sqrt{6}, \sqrt{3}),$
$(\sqrt{6}, -\sqrt{3}), (\sqrt{6}, \sqrt{3})\}$ **47.** 0, 1, or 2   **48.** 0, 1, 2,
3, or 4

49.

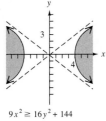

$9x^2 \geq 16y^2 + 144$

50.

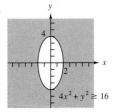

$4x^2 + y^2 \geq 16$

51.

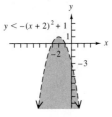

$y < -(x + 2)^2 + 1$

52.

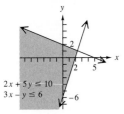

$2x + 5y \leq 10$
$3x - y \leq 6$

53.

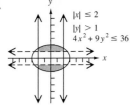

$|x| \leq 2$
$|y| > 1$
$4x^2 + 9y^2 \leq 36$

54.

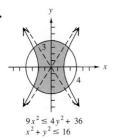

$9x^2 \leq 4y^2 + 36$
$x^2 + y^2 \leq 16$

55.

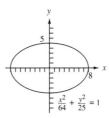

$\frac{x^2}{64} + \frac{y^2}{25} = 1$

56.

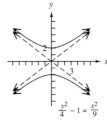

$\frac{y^2}{4} - 1 = \frac{x^2}{9}$

57.

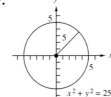

$x^2 + y^2 = 25$

58.

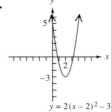

$y = 2(x - 2)^2 - 3$

59.

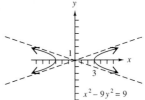

$x^2 - 9y^2 = 9$

60.

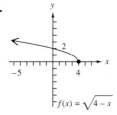

$f(x) = \sqrt{4 - x}$

61.

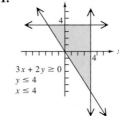

$3x + 2y \geq 0$
$y \leq 4$
$x \leq 4$

62.

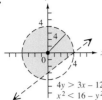

$$4y > 3x - 12$$
$$x^2 < 16 - y^2$$

CHAPTER 9 TEST (page 651)

1. $(0, -2)$

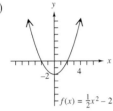

$$f(x) = \tfrac{1}{2}x^2 - 2$$

2. vertex: $(2, 3)$; domain: $(-\infty, \infty)$; range: $(-\infty, 3]$

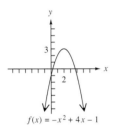

$$f(x) = -x^2 + 4x - 1$$

3. **(a)** 31,600 **(b)** 1984; 36,000   **4.** B

5.

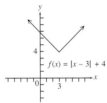

$$f(x) = |x - 3| + 4$$

6. center: $(2, -3)$; radius: 4

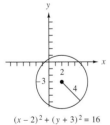

$$(x - 2)^2 + (y + 3)^2 = 16$$

7. center: $(-4, 1)$; radius: 5   **8.** One variable is squared; the other is to the first power.

9.

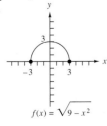

$$f(x) = \sqrt{9 - x^2}$$

10.

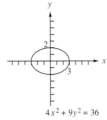

$$4x^2 + 9y^2 = 36$$

11.

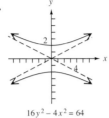

$$16y^2 - 4x^2 = 64$$

12.

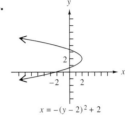

$$x = -(y - 2)^2 + 2$$

13. ellipse   **14.** hyperbola

15. parabola   **16.** $\left\{ \left(-\tfrac{1}{2}, -10\right), (5, 1) \right\}$

17. $\left\{ (-2, -2), \left(\tfrac{14}{5}, -\tfrac{2}{5}\right) \right\}$ **18.** $\{(-\sqrt{22}, -\sqrt{3}),$
$(-\sqrt{22}, \sqrt{3}), (\sqrt{22}, -\sqrt{3}), (\sqrt{22}, \sqrt{3})\}$

19.

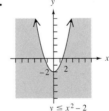

$$y \le x^2 - 2$$

20.

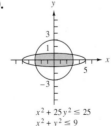

$$x^2 + 25y^2 \le 25$$
$$x^2 + y^2 \le 9$$

CUMULATIVE REVIEW 1–9 (page 655)

1. -4 **2.** $\left\{\tfrac{2}{3}\right\}$ **3.** $\left(-\infty, \tfrac{3}{5}\right]$ **4.** $\{-4, 4\}$

5. $(-\infty, -5) \cup (10, \infty)$ **6.** $\tfrac{2}{3}$ **7.** $3x + 2y = -13$

8. $25y^2 - 30y + 9$ **9.** $12r^2 + 40r - 7$

10. $4x^3 - 4x^2 + 3x + 5 + \dfrac{3}{2x + 1}$

11. $(3x + 2)(4x - 5)$ **12.** $(2y^2 - 1)(y^2 + 3)$
13. $(z^2 + 1)(z + 1)(z - 1)$

14. $(a - 3b)(a^2 + 3ab + 9b^2)$ **15.** $\dfrac{40}{9}$

16. $\dfrac{y - 1}{y(y - 3)}$ **17.** $\dfrac{3c + 5}{(c + 5)(c + 3)}$ **18.** $\dfrac{1}{p}$

19. $\dfrac{6}{5}$ or $1\dfrac{1}{5}$ hours   **20.** $\dfrac{3}{4}$ **21.** $\dfrac{a^5}{4}$ **22.** $2\sqrt[3]{2}$

23. $\dfrac{3\sqrt{10}}{2}$ **24.** $\dfrac{7}{5} + \dfrac{11}{5}i$ **25.** $\emptyset$ **26.** $\left\{\dfrac{1}{5}, -\dfrac{3}{2}\right\}$

27. $\left\{\dfrac{1 + 2\sqrt{2}}{4}, \dfrac{1 - 2\sqrt{2}}{4}\right\}$

28. $\left\{\dfrac{3 + \sqrt{33}}{6}, \dfrac{3 - \sqrt{33}}{6}\right\}$

29. $\left\{-\dfrac{\sqrt{6}}{2}, \dfrac{\sqrt{6}}{2}, -\sqrt{7}, \sqrt{7}\right\}$ **30.** $v = \dfrac{\pm\sqrt{rFkw}}{kw}$

31.

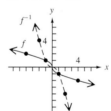

$f(x) = -3x + 5$

32.

$f(x) = -2(x - 1)^2 + 3$

33.

$f(x) = \sqrt{x - 2}$

34.

$\dfrac{x^2}{4} - \dfrac{y^2}{16} = 1$

35.

$\dfrac{x^2}{25} + \dfrac{y^2}{16} \leq 1$

36. $\{(3, -3)\}$ **37.** $\{(4, 1, -2)\}$

38. $\left\{(-1, 5), \left(\dfrac{5}{2}, -2\right)\right\}$

39. 40 miles per hour   **40.** 15 fives and 7 twenties

CHAPTER 10

SECTION 10.1 (page 665)

1. no   **3.** (a)   **5.** Two or more siblings might be in the class.   **7.** $\{(6, 3), (10, 2), (12, 5)\}$ **9.** not a one-to-one function   **11.** $f^{-1}(x) = \dfrac{x - 4}{2}$

13. $g^{-1}(x) = x^2 + 3, x \geq 0$ **15.** not one-to-one
17. $f^{-1}(x) = \sqrt[3]{x + 4}$ **19.** Yes, by the horizontal line test, $f(x) = mx + b$ is one-to-one and so it has an inverse.
21. (a) 8   (b) 3   **23.** (a) 1   (b) 0
25. (a) one-to-one
(b)

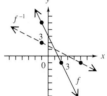

27. (a) not one-to-one
29. (a) one-to-one
(b)

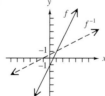

31.

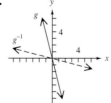

33.

35.

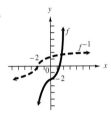

37.

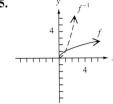

39. inverses; $y_2 = \sqrt[3]{x - 5}$ **41.** It is not a function.

Section 10.2 (page 675)

1. rises; falls   **3.** does not

5.

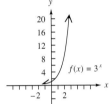

7.

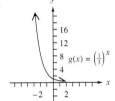

9.

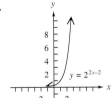

11. It is one-to-one and thus has an inverse.

13. {2}   **15.** $\left\{\dfrac{3}{2}\right\}$ **17.** {7}   **19.** {−3}

21. {−1}   **23.** {−3}   **25.** **(a)** 100 grams
(b) 31.25 grams   **(c)** .30 gram (to the nearest hundredth)
27. about $360,000   **29.** In $1976 + 15 = 1991$, the
average salary was about $800,000.

Section 10.3 (page 681)

1. **(a)** 2   **(b)** 4   **(c)** −1   **(d)** −2   **(e)** $\frac{1}{2}$ **(f)** 0
3. $\log_4 1024 = 5$ **5.** $\log_{1/2} 8 = -3$
7. $\log_{10} .001 = -3$ **9.** $4^3 = 64$

11. $10^{-4} = \dfrac{1}{10,000}$

13. $6^0 = 1$ **15.** Since the radical $\sqrt{9} = 9^{1/2} = 3$, the
exponent to which 9 must be raised is 1/2.   **17.** $\left\{\dfrac{1}{3}\right\}$

19. {81}   **21.** $\left\{\dfrac{1}{5}\right\}$ **23.** {1}

25. $\{x \mid x > 0, x \ne 1\}$ **27.** {5}   **29.** $\left\{\dfrac{5}{3}\right\}$ **31.** {4}

33. $\left\{\dfrac{3}{2}\right\}$

35.

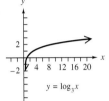

37.

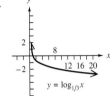

39.

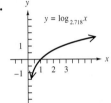

41. **(a)** 6   **(b)** 12   **(c)** 18
(d)

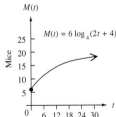

Months since January 1993

43. 8   **45.** 24
47. Since every real number power of 1 equals 1,
$f(x) = \log_1 x$ is not a logarithmic function.   **49.** $(0, \infty)$;
$(-\infty, \infty)$ **51.** about 4 times more powerful

Section 10.4 (page 689)

1. sum   **3.** k **5.** $\log_7 4 - \log_7 5$
7. $\dfrac{1}{4} \log_2 8$ or $\dfrac{3}{4}$ **9.** $\log_4 3 + \dfrac{1}{2} \log_4 x - \log_4 y$

11. $\dfrac{1}{3} \log_3 4 - 2 \log_3 x - \log_3 y$

13. $\frac{1}{2} \log_3 x + \frac{1}{2} \log_3 y - \frac{1}{2} \log_3 5$

15. $\frac{1}{3} \log_2 x + \frac{1}{5} \log_2 y - 2 \log_2 r$

17. The distributive property tells us that the *product* $a(x + y)$ equals the sum $ax + ay$. In the notation $\log_a (x + y)$, the parentheses do not indicate multiplication. They indicate that $x + y$ is the result of raising a to some power. **19.** $\log_b xy$ **21.** $\log_a \frac{m^3}{n}$ **23.** $\log_a \frac{rt^3}{s}$

25. $\log_a \frac{125}{81}$ **27.** $\log_{10}(x^2 - 9)$ **29.** $\log_p \frac{x^3 y^{1/2}}{z^{3/2} a^3}$

31. For the power rule $\log_b x^r = r \log_b x$ to be true, x must be in the domain of $g(x) = \log_b x$.

33. 4 **34.** It is the exponent to which 3 must be raised in order to obtain 81. **35.** 81 **36.** It is the exponent to which 2 must be raised in order to obtain 19. **37.** 19 **38.** m

SECTION 10.5 (page 695)

1. 10 **3.** 0; 1; 0; 1 **5.** 19.2 **7.** 1.6335
9. 2.5164 **11.** -1.4868 **13.** 9.6776
15. 2.0592 **17.** -2.8896 **19.** 5.9613
21. 4.1506 **23.** 2.3026 **25.** 1.4315 **27.** 3.3030
29. $-.0947$ **31.** Answers will vary. Suppose the name is Paul Bunyan, with $m = 4$ and $n = 6$. **(a)** $\log_4 6$ is the exponent to which 4 must be raised in order to obtain 6.
(b) 1.29248125 **(c)** 6 (the value of n) **33.** 11.6
35. 4.3 **37.** 4.0×10^{-8} **39.** 4.0×10^{-6}
41. **(a)** 1.62 grams **(b)** 1.18 grams **(c)** .69 gram
(d) 2.00 grams **43.** 2228 million (or approximately 2.2 billion) **45.** 1354.4 thousand (or 1,354,400)
47. In 1989 (when $x = 10$), the number of births was about 730,000. **49.** about 800 years **51.** about 11,500 years

SECTION 10.6 (page 705)

1. $\{3\}$ **3.** $\{6\}$; $\frac{\log 64}{\log 2} = 6$ **5.** $\{.827\}$ **7.** $\{.833\}$
9. $\{2.269\}$ **11.** $\{566.866\}$ **13.** $\{-18.892\}$
15. $\{43.301\}$ **17.** Natural logarithms are a better choice because $\ln e^x = x$. **19.** $\left\{\frac{2}{3}\right\}$

21. $\{-1 + \sqrt[3]{49}\}$ **23.** 2 cannot be a solution, because $\log(2 - 3) = \log(-1)$, which is not acceptable.
25. $\left\{\frac{1}{3}\right\}$ **27.** $\{2\}$ **29.** $\emptyset$ **31.** $\{8\}$ **33.** $\left\{\frac{4}{3}\right\}$
35. $\{8\}$ **37.** \$2539.47 **39.** about 180 grams
41. \$4934.71 **43.** 15.4 years
45. The domain of f is $(-\infty, 0) \cup (0, \infty)$, while the domain of g is $(0, \infty)$.

CHAPTER 10 REVIEW EXERCISES (page 713)

1. not one-to-one **2.** one-to-one
3. $f^{-1}(x) = \frac{x - 7}{-3}$ or $\frac{7 - x}{3}$ **4.** $f^{-1}(x) = \frac{x^3 + 4}{6}$
5. not one-to-one **6.** $f^{-1}(x) = x$

7.

8.

9.

10.

11.

12.

13. $\{4\}$ **14.** $\left\{\frac{3}{7}\right\}$ **15.** $\{0\}$ **16.** $(0, 1)$
17. It says that $a^0 = 1$, which agrees with the definition.
18.

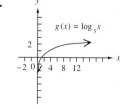

19.

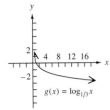

$g(x) = \log_{1/3} x$

20. (a) $5^4 = 625$ **(b)** $\log_5 .04 = -2$
21. (a) $\log_b a$ represents the exponent on b that equals a.
(b) a **22.** $\log_4 3 + 2 \log_4 x$
23. $2 \log_2 p + \log_2 r - \dfrac{1}{2} \log_2 z$ **24.** $\log_b \dfrac{3x}{y^2}$
25. $\log_3 \dfrac{x + 7}{4x + 6}$ **26.** 1.4609 **27.** $-.5901$
28. 4.8613 **29.** 3.3638 **30.** -1.3587
31. 4.8613 **32. (a)** 2.552424846 **(b)** 1.552424846
(c) 0.552424846 **(d)** The whole number parts will vary,
but the decimal parts are the same. **33.** $.9251$
34. 1.7925 **35.** 1.4315 **36. (a)** 500 **(b)** 1000
(c) 1500 **(d)** no **37.** 6.4 **38.** 8.4
39. (a) 11.7 million **(b)** 5.4 million
40. (a) 500 grams **(b)** about 409 grams
41. $\{2.042\}$ **42.** $\{4.907\}$ **43.** $\{18.310\}$
44. $\left\{\dfrac{1}{9}\right\}$ **45.** $\{-6 + \sqrt[3]{25}\}$ **46.** $\{2\}$
47. $\left\{\dfrac{3}{8}\right\}$ **48.** $\{4\}$ **49.** $\{1\}$ **50.** about 1994
51. almost 14 years **52.** $\$7112.11$ **53.** Plan A; it
would pay $\$2.92$ more. **54.** $\{72\}$ **55.** $\{5\}$
56. $\left\{\dfrac{1}{9}\right\}$ **57.** $\left\{\dfrac{4}{3}\right\}$ **58.** $\{3\}$ **59.** $\left\{\dfrac{1}{8}\right\}$ **60.** $\left\{\dfrac{11}{3}\right\}$

CHAPTER 10 TEST (page 717)

1. (a) not one-to-one **(b)** one-to-one
2. $f^{-1}(x) = x^3 - 7$
3.

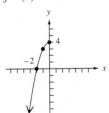

4.

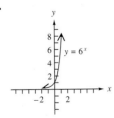

$y = 6^x$

5.

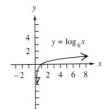

$y = \log_6 x$

6. Interchange the x- and y-values of the ordered pairs,
because they are inverses of each other. **7.** $\{-4\}$
8. $\left\{-\dfrac{13}{3}\right\}$ **9. (a)** $\$5000$ **(b)** $\$2973$ **(c)** $\$1768$
(d)

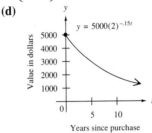

$y = 5000(2)^{-.15t}$

Value in dollars / Years since purchase

10. $\log_4 .0625 = -2$ **11.** $7^2 = 49$
12. $\{32\}$ **13.** $\left\{\dfrac{1}{2}\right\}$ **14.** $\{2\}$ **15.** $5; 2; 5\text{th}; 32$
16. $2 \log_3 x + \log_3 y$
17. $\dfrac{1}{2} \log_5 x - \log_5 y - \log_5 z$ **18.** $\log_b \dfrac{s^3}{t}$
19. $\log_b \dfrac{r^{1/4} s^2}{t^{2/3}}$ **20.** 1.3284 **21.** $-.8440$
22. (a) 2 and 3 **(b)** 2.1245 **23.** $\{3.9656\}$ **24.** $\{3\}$
25. (a) $\$1051$ **(b)** after 7 years (rounded up)

CUMULATIVE REVIEW 1–10 (page 721)

1. $-2, 0, 6, \dfrac{30}{3}$ (or 10) **2.** $-\dfrac{9}{4}, -2, 0, .6, 6, \dfrac{30}{3}$ (or 10)
3. $-\sqrt{2}, \sqrt{11}$ **4.** all except $\sqrt{-8}$ **5.** 16
6. -27 **7.** -39 **8.** $\left\{-\dfrac{2}{3}\right\}$ **9.** $[1, \infty)$
10. $\{-2, 7\}$ **11.** $\left\{-\dfrac{16}{3}, \dfrac{16}{3}\right\}$ **12.** $\left[\dfrac{7}{3}, 3\right]$
13. $(-\infty, -3) \cup (2, \infty)$ **14.** $6p^2 + 7p - 3$
15. $16k^2 - 24k + 9$ **16.** $-5m^3 + 2m^2 - 7m + 4$
17. $2t^3 + 5t^2 - 3t + 4$ **18.** $x(8 + x^2)$
19. $(3y - 2)(8y + 3)$ **20.** $z(5z + 1)(z - 4)$
21. $(4a + 5b^2)(4a - 5b^2)$
22. $(2c + d)(4c^2 - 2cd + d^2)$ **23.** $(4r + 7q)^2$
24. $-\dfrac{1875p^{13}}{8}$ **25.** $\dfrac{x + 5}{x + 4}$ **26.** $\dfrac{-3k - 19}{(k + 3)(k - 2)}$
27. $\dfrac{22 - p}{p(p - 4)(p + 2)}$ **28.** 6 pounds **29.** $\dfrac{16}{25}$
30. $\dfrac{1}{6^5}$ **31.** $-27\sqrt{2}$ **32.** 41 **33.** $\left\{-\dfrac{1}{2}, \dfrac{2}{5}\right\}$
34. $(-\infty, -4) \cup (2, \infty)$ **35. (a)** 233.4 million
(b) 238.5 million **(c)** 247 million **(d)** 2030
36. $-5,587,500$ dollars per year **37.** $3x - 4y = 19$

38.

$y = -2.5x + 5$

39.

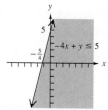

$-4x + y \leq 5$

$-\frac{5}{4}$

40.

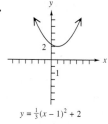

$y = \frac{1}{3}(x - 1)^2 + 2$

41.

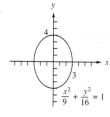

$\frac{x^2}{9} + \frac{y^2}{16} = 1$

42.

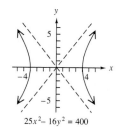

$25x^2 - 16y^2 = 400$

43. **(a)** those in Exercises 38 and 40   **(b)** only the one in Exercise 38   **44.** $\{(4, 2)\}$ **45.** $\{(1, -1, 4)\}$ **46.** -9

47.

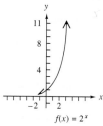

$f(x) = 2^x$

48. $\{-1\}$

49.

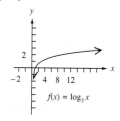

$f(x) = \log_3 x$

50. $\{1\}$

CHAPTER 1

SECTION 1.1 (page 7)

1. True: for example, $10 \div 0$ is undefined whereas $0 \div 10 = 0$.

5. False: the absolute value of every number is positive or zero.

9. Place dots as needed to indicate the points $-\frac{2}{3}$, 0, $\frac{4}{5}$, $\frac{12}{5} = 2.4$, $\frac{9}{2} = 4.5$, and 4.8. See the graph in the answer section.

13. The set of natural numbers is $\{1, 2, 3, \ldots\}$, so the set of natural numbers less than 6 is $\{1, 2, 3, 4, 5\}$.

17. The set of even integers is $\{\ldots, -2, 0, 2, 4, \ldots\}$, so the set of even integers greater than 8 is $\{10, 12, 14, 16, \ldots\}$.

21. $|4| = 4$ and $|-4| = 4$, so the set of numbers whose absolute value is 4 is $\{-4, 4\}$.

25. Yes, the choice of variables is not important. The sets have the same description so they represent the same set.

29. Answers may vary. One possible answer is: $\{x \mid x$ is an even natural number less than or equal to 8$\}$

33. (a) The additive inverse of 6 is -6 (change the sign).
(b) The absolute value of 6 is $|6| = 6$.

37. $5 - 5$ is 0. The additive inverse of 0 is $-0 = 0$. The absolute value of 0 is $|0| = 0$.

41. Use the definition of absolute value: $|-8| = -(-8) = 8$.

45. Since $|-2| = 2$, $-|-2| = -(2) = -2$.

49. $|-2| + |3| = 2 + 3 = 5$

53. $|-9| + |-13| = 9 + 13 = 22$

57. Since $\frac{1}{2}$ is a rational number but not an integer, the statement is false.

61. True, because every whole number corresponds to a point on the number line, which corresponds to a real number.

65. True: for example, $5 = \frac{5}{1}$ is a rational number that is also an integer.

69. Bodies of water from deepest to shallowest:

Pacific Ocean	$-12,925$ feet
Indian Ocean	$-12,598$ feet
Caribbean Sea	-8448 feet
South China Sea	-4802 feet
Gulf of California	-2375 feet

73. The absolute value of any real number equals the absolute value of its negative. Therefore, the statement is true for all real numbers.

SECTION 1.2 (page 15)

1. True: -6 is to the left of -2 so -6 is less than -2.

5. True: 3 is to the right of -2 so 3 is greater than -2.

9. $-3 < 2$ may be written as $2 > -3$. Both of these inequalities are true because -3 is less than 2 and 2 is greater than -3.

13. $7 > y$: the arrow points to the smaller number y.

17. $5 \geq 5$: the symbol $\geq$ represents "is greater than or equal to."

21. $-3 \leq 3x$ (-3 is included) and also $3x < 4$ (4 is not included). This can be written as $-3 \leq 3x < 4$.

25. True, because $7 + 3 = 10$ and -6 is less than 10.

29. True, because $|-3| = 3$, so $-|-3| = -3$, and "$\geq$" includes "$=$."

33. "$-3 \not> -3$" is read "-3 is not greater than -3," which is equivalent to "-3 is less than or equal to -3," written "$-3 \leq -3$."

For Exercises 37–45, see the graphs in the answer section.

37. $\{x \mid x > -2\}$ includes all numbers greater than -2, written $(-2, \infty)$. Place a parenthesis at -2 since -2 is not an element of the set.

41. $\{x \mid 0 < x < 3.5\}$ includes all numbers between 0 and 3.5, written $(0, 3.5)$. Parentheses are placed at 0 and 3.5 since these numbers are not included.

45. $\{x \mid -4 < x \leq 3\}$ includes all numbers between -4 and 3, written $(-4, 3]$. A bracket is placed at 3 to show that 3 is included.

49. The number of Oregon permits in 1995 was 26,201, and the number of Utah permits in 1995 was 20,898. So the number of Oregon permits was greater than the number of Utah permits in that year.

SECTION 1.3 (page 23)

1. The sum of a positive number and a negative number is zero if the numbers are additive inverses.

For example, $4 + (-4) = 0$.

5. The sum of a positive number and a negative number is positive if the positive number has a larger absolute value.

For example, $15 + (-2) = 13$.

9. The product of two numbers with unlike signs is negative.

For example, $(-5)(15) = -75$.

13. $-6 + (-13) = -(6 + 13) = -19$

17. The difference between 2.3 and .45 is 1.85. The larger number in absolute value, -2.3, is negative, so the answer is negative. Thus, $-2.3 + .45 = -1.85$.

21. $8 - (-13) = 8 + 13 = 21$

25. $-12.31 - (-2.13) = -12.31 + 2.13 = -10.18$

29. $-2 - |-4| = -2 - 4 = -2 + (-4) = -6$

33. The statement, "Two negatives give a positive," is true for multiplication and division. It is false for addition and for subtraction when the number to be subtracted has the smaller absolute value. A more precise statement is "The product or quotient of two negatives is positive."

37. The numbers have the same sign, so $-8(-5) = (8 \cdot 5) = 40$.

41. The numbers have different signs, so $\frac{3}{4}(-16) =$
$-\left(\frac{3}{4} \cdot \frac{16}{1}\right) = -\frac{48}{4} = -12.$

45. $-\frac{3}{8}\left(-\frac{24}{9}\right) = \frac{3}{8} \cdot \frac{24}{9} = \frac{72}{72} = 1$

49. $3.4(-3.14) = -(3.4 \cdot 3.14) = -10.676$

53. The reciprocal of $-7 = -\frac{7}{1}$ is $-\frac{1}{7}$.

57. The reciprocal of $.02 = \frac{.02}{1}$ is $\frac{1}{.02}$. However,
$1 \div .02 = 50.$

61. Since $.08\overline{3} = \frac{1}{12}$, the reciprocal is $\frac{1}{.08\overline{3}} = \frac{1}{\frac{1}{12}} = 12.$

65. $\frac{-24}{-4} = -24\left(-\frac{1}{4}\right) = 6$

69. Division by zero is undefined, so $\frac{5}{0}$ is undefined.

73. $\dfrac{\frac{12}{13}}{-\frac{4}{3}} = \frac{12}{13} \div \left(-\frac{4}{3}\right) = \frac{12}{13}\left(-\frac{3}{4}\right)$
$= -\frac{36}{52} = -\frac{9}{13}$

77. $\frac{-100}{-.01} = \frac{100}{.01} = 100 \div \frac{1}{100} = 100(100) = 10,000$

81. $6 - (-2) + 8 = 6 + 2 + 8 = 16$

85. $-4 - [[(-4 - 6) + 12] - 13$
$= -4 - [-10 + 12] - 13$
$= -4 - 2 - 13$
$= -4 + (-2) + (-13)$
$= -19$

89. In 1986, the percent change in the U.S. was 2.0 and the percent change in California was 2.5. Subtract these values to find the difference.

$$2.0 - 2.5 = -.5$$

In 1986, the difference between the percent change for the U.S. and for California was $-.5$.

SECTION 1.4 (page 33)

1. False: $-4^6 = -4 \cdot 4 \cdot 4 \cdot 4 \cdot 4 \cdot 4 = -4096$ whereas $(-4)^6 = (-4)(-4)(-4)(-4)(-4)(-4) = 4096.$

5. True: $(-2)^7$ gives an odd number of negative factors, so the expression is negative.

9. False: 3, not -3, is the base. If the problem was written $(-3)^5$, then -3 would be the base.

13. $.28^3 = (.28)(.28)(.28) = .021952$

17. $\left(\frac{7}{10}\right)^3 = \left(\frac{7}{10}\right)\left(\frac{7}{10}\right)\left(\frac{7}{10}\right) = \frac{343}{1000}$

21. $(-2)^8 = (-2)(-2)(-2)(-2)(-2)(-2)(-2)(-2)$
$= 256$

25. $-8^4 = -(8 \cdot 8 \cdot 8 \cdot 8) = -4096$

29. The exponent is 7; the base is -4.1.

33. $\sqrt{81} = 9$ because $9^2 = 81.$

37. $-\sqrt{400} = -20$ because $20^2 = 400.$

41. $-\sqrt{.49} = -.7$ because $(.7)^2 = (.7)(.7) = .49.$

45. It is incorrect to say that $\sqrt{16}$ is equal to 4 or -4 because $\sqrt{16}$ is the positive (or principal) square root of 16, namely 4. -4 is the negative square root of 16, written $-\sqrt{16}$.

49. $\sqrt[3]{-27} = -3$ because $(-3)^3 = -27.$

53. $\sqrt[5]{243} = 3$ because $3^5 = 243.$

57. $2[-5 - (-7)] = 2[-5 + 7] = 2 \cdot 2 = 4$

61. $(-3)(5)^2 - (-2)(-8) = (-3)(25) - (-2)(-8)$
$= -75 - 16$
$= -91$

65. $\frac{-8 + (-16)}{-3} = \frac{-24}{-3} = 8$

69. $\frac{2(-5) + (-3)(-2)}{-6 + 5 + 1} = \frac{-10 + 6}{0}$
Answer: undefined

73. $-7\left[6 - \frac{5}{8}(24) + 3\left(\frac{8}{3}\right)\right] = -7[6 - 15 + 8]$
$= -7[-1] = 7$

77. $\sqrt[3]{b} + c - a = \sqrt[3]{64} + 6 - (-3) = 4 + 6 + 3$
$= 13$

81. $\frac{2c + a^3}{4b + 6a} = \frac{2(6) + (-3)^3}{4(64) + 6(-3)} = \frac{12 + (-27)}{256 + (-18)} = -\frac{15}{238}$

85. $(4(-3) - 2^3)/(5 - 4(3)) = (4(-3) - 8)/(5 - 4(3))$
$= (-12 - 8)/(5 - 12)$
$= -20/-7$
$= 20/7$

SECTION 1.5 (page 43)

1. Using the distributive property, we can easily calculate the problem mentally.

$$96 \cdot 19 + 4 \cdot 19 = 19(96 + 4)$$
$$= 19(100)$$
$$= 1900$$

5. The identity element for addition is 0.

9. The commutative property is used to change the order of two terms or factors, while the associative property is used to change the grouping.

13. $-9r + 7r = (-9 + 7)r$ Distributive property
$= -2r$

17. $-a + 7a = -1a + 7a$ Identity property
$= (-1 + 7)a$ Distributive property
$= 6a$

21. $-5(2d - f) = -5(2d) + (-5)(-f)$ Distributive property
$= -10d + 5f$

25. $-6p + 11p - 4p + 6 + 5$
$= (-6 + 11 - 4)p + (6 + 5)$
$= p + 11$

29. $-2(m + 1) + 3(m - 4)$
$= -2(m) + (-2)(1) + 3(m) + (3)(-4)$ Distributive property
$= -2m + 3m - 2 - 12$ Commutative property
$= (-2 + 3)m + [-2 + (-12)]$ Distributive property
$= m - 14$

33. $-(2p + 5) + 3(2p + 4) - 2p$
$= -2p - 5 + 6p + 12 - 2p$
$= (-2 + 6 - 2)p + (-5 + 12)$
$= 2p + 7$

37. $5x + 8x = (5 + 8)x = 13x$

41. $5x + 9y = 9y + 5x$: Reverse the order of addition.

45. $8(-4 + x) = 8(-4) + 8(x) = -32 + 8x$

49. Yes. Let $a = 2$, $b = -2$. Then $\frac{2}{-2} = \frac{-2}{2} = -1$. Any different nonzero numbers a and b that have the same absolute value satisfy $\frac{a}{b} = \frac{b}{a}$.

53. Commutative property of addition:
$$(4 + 2x) = (2x + 4)$$

CHAPTER 2

SECTION 2.1 (page 61)

1. $3(x + 4) = 5x$
$3(6 + 4) = 5(6)$ Let $x = 6$.
$3(10) = 30$?
$30 = 30$ True

5. $\quad 7k + 8 = 1$
$7k + 8 - \mathbf{8} = 1 - \mathbf{8}$ Subtract 8.
$\quad\quad 7k = -7$
$\quad\quad \dfrac{7k}{7} = \dfrac{-7}{7}$ Divide by 7.
$\quad\quad k = -1$
The solution set is $\{-1\}$.

9. $7y - 5y + 15 = y + 8$
$2y + 15 = y + 8$ Combine terms.
$2y + 15 - \mathbf{15} = y + 8 - \mathbf{15}$ Subtract 15.
$2y = y - 7$
$2y - \mathbf{y} = y - \mathbf{y} - 7$ Subtract y.
$y = -7$
The solution set is $\{-7\}$.

13. $\quad\quad 2(x + 3) = -4(x + 1)$
$\quad\quad \mathbf{2x + 6} = \mathbf{-4x - 4}$ Clear parentheses.
$2x + \mathbf{4x} + 6 - \mathbf{6} = -4x + \mathbf{4x} - 4 - \mathbf{6}$ Add $4x$; subtract 6.
$\quad\quad 6x = -10$
$\quad\quad \dfrac{6x}{6} = \dfrac{-10}{6}$ Divide by 6.
$\quad\quad x = -\dfrac{5}{3}$
The solution set is $\{-\frac{5}{3}\}$.

17. $\quad 2x + 3(x - 4) = 2(x - 3)$
$\quad 2x + \mathbf{3x - 12} = \mathbf{2x - 6}$ Clear parentheses.
$\quad 5x - 12 = 2x - 6$
$5x - \mathbf{2x} - 12 + \mathbf{12} = 2x - \mathbf{2x} - 6 + \mathbf{12}$ Subtract $2x$; add 12.
$\quad\quad 3x = 6$
$\quad\quad \dfrac{3x}{3} = \dfrac{6}{3}$ Divide by 3.
$\quad\quad x = 2$
The solution set is $\{2\}$.

21. $-[2z - (5z + 2)] = 2 + (2z + 7)$
$-[2z - 5z - 2] = 2 + \mathbf{2z + 7}$ Clear parentheses.
$\mathbf{-2z + 5z + 2} = 2 + 2z + 7$ Clear brackets.
$\mathbf{3z + 2} = 2z + \mathbf{9}$ Combine terms.
$3z - \mathbf{2z} + 2 - \mathbf{2} = 2z - \mathbf{2z} + 9 - \mathbf{2}$ Subtract $2z$; subtract 2.
$z = 7$
The solution set is $\{7\}$.

25. The denominators of the fractions are 3, 4, and 1. The least common denominator is $(3)(4)(1) = 12$, since it is the smallest number into which each denominator can divide without a remainder.

29. $\quad\quad \dfrac{3x}{4} + \dfrac{5x}{2} = 13$
$\quad 4\left(\dfrac{3x}{4} + \dfrac{5x}{2}\right) = 4(13)$ Multiply by 4.
$4\left(\dfrac{3x}{4}\right) + 4\left(\dfrac{5x}{2}\right) = 4(13)$
$\quad\quad 3x + 10x = 52$ Clear parentheses.
$\quad\quad 13x = 52$ Combine terms.
$\quad\quad \dfrac{13x}{13} = \dfrac{52}{13}$ Divide by 13.
$\quad\quad x = 4$
The solution set is $\{4\}$.

33. $\quad \dfrac{4t + 1}{3} = \dfrac{t + 5}{6} + \dfrac{t - 3}{6}$
$6\left(\dfrac{4t + 1}{3}\right) = 6\left(\dfrac{t + 5}{6}\right) + 6\left(\dfrac{t - 3}{6}\right)$ Multiply by 6.
$2(4t + 1) = t + 5 + t - 3$
$8t + 2 = 2t + 2$ Clear parentheses; combine terms.
$6t = 0$ Subtract $2t$; subtract 2.
$t = 0$ Divide by 6.
The solution set is $\{0\}$. (Zero can be in the numerator but not in the denominator.)

37. $.02(50) + .08r = .04(50 + r)$
$2(50) + 8r = 4(50 + r)$ Multiply by 100.
$100 + 8r = 200 + 4r$ Clear parentheses.
$4r = 100$ Subtract $4r$; subtract 100.
$\dfrac{4r}{4} = \dfrac{100}{4}$ Divide by 4.
$r = 25$
The solution set is $\{25\}$.

41. A conditional equation has one solution, an identity has infinitely many solutions, and a contradiction has no solution.

45. $-11m + 4(m - 3) + 6m = 4m - 12$
$-11m + 4m - 12 + 6m = 4m - 12$ Clear parentheses.
$-m - 12 = 4m - 12$ Combine terms.
$0 = 5m$ Add m; add 12.
$0 = m$ Divide by 5.
This is a conditional equation with solution set $\{0\}$.

49. $2[3x + (x - 2)] = 9x + 4$
$2[3x + x - 2] = 9x + 4$ Clear parentheses.
$2(4x - 2) = 9x + 4$ Combine terms.
$8x - 4 = 9x + 4$ Distributive property
$-8 = x$ Subtract $8x$; subtract 4.
The solution set is $\{-8\}$.

53. From Exercise 52, k must equal 33, so solve the equation
$$-4(x + 2) - 3(x + 5) = 33.$$
$-4x - 8 - 3x - 15 = 33$ Clear parentheses.
$-7x - 23 = 33$ Combine terms.
$-7x = 56$ Add 23.
$\dfrac{-7x}{-7} = \dfrac{56}{-7}$ Divide by -7.
$x = -8$

The solution set is $\{-8\}$.

SOLUTIONS

57. Replacing −8 for y in the second equation will cause the denominator to equal 0. Division by 0 is undefined. The solution set of the first equation is $\{-8\}$ and of the second is $\emptyset$, so the equations are not equivalent.

61. Since $x = 5$ corresponds to 1994, let $x = 5$ in the linear equation $y = 420x + 720$ and solve for y.

$$y = 420(5) + 720$$
$$= 2100 + 720$$
$$= 2820$$

So, in 1994, 2820 million dollars were lost due to credit card fraud.
Now let $y = 3660$ and solve for x.

$$3660 = 420x + 720$$
$$2940 = 420x$$
$$7 = x$$

So, in 1996 (when $x = 7$), credit card fraud losses reached 3660 million dollars.

SECTION 2.2 (page 71)

1. (a) $x = \dfrac{5x + 8}{3}$ **(b)** $t = \dfrac{bt + k}{c}$

$3 \cdot x = 3\left[\dfrac{5x + 8}{3}\right]$ $c \cdot t = c\left[\dfrac{bt + k}{c}\right]$

$3x = 5x + 8$ $ct = bt + k$

5. The answer for the second equation is

$$t = \frac{k}{c - b}.$$

$c \neq b$ because if c is equal to b, the denominator is 0. Zero is not allowed in the denominator of a fraction.

9. $I = prt$

$\dfrac{I}{pt} = \dfrac{prt}{pt}$ Divide by pt.

$\dfrac{I}{pt} = r$ or $r = \dfrac{I}{pt}$

13. $V = LWH$

$\dfrac{V}{LH} = \dfrac{LWH}{LH}$ Divide by LH.

$\dfrac{V}{LH} = W$ or $W = \dfrac{V}{LH}$

17. $A = \dfrac{1}{2}h(B + b)$

$2A = 2\left[\dfrac{1}{2}h(B + b)\right]$ Multiply by 2.

$2A = Bh + bh$ Clear parentheses.

$2A - bh = Bh$ Subtract bh.

$\dfrac{2A - bh}{h} = \dfrac{Bh}{h}$ Divide by h.

$\dfrac{2A - bh}{h} = B$ or $B = \dfrac{2A - bh}{h}$

Note that $\dfrac{2A - bh}{h} = \dfrac{2A}{h} - \dfrac{bh}{h} = \dfrac{2A}{h} - b$, an equivalent answer.

21. $2k + ar = r - 3y$

$ar = r - 3y - 2k$ Subtract $2k$.

$ar - r = -3y - 2k$ Subtract r.

$r(a - 1) = -3y - 2k$ Use distributive property.

$r = \dfrac{-3y - 2k}{a - 1}$ or $r = \dfrac{-2k - 3y}{a - 1}$

Using the distributive property in the numerator and denominator gives an alternative answer:

$$\frac{-(2k + 3y)}{-(1 - a)} = \frac{2k + 3y}{1 - a}$$

25. Multiply her answer by $\dfrac{-1}{-1}$ or 1.

$\dfrac{-1}{-1} \cdot \dfrac{z}{2 - m} = \dfrac{-1(z)}{-1(2 - m)} = \dfrac{-z}{m - 2}$. Yes, her answer was correct.

29. Use the formula $F = \dfrac{9}{5}C + 32$.

$$F = \frac{9}{5}(-40) + 32 \quad \text{Substitute } -40 \text{ for C.}$$
$$= -72 + 32 \quad \text{Multiply.}$$
$$= -40$$

The temperature was −40 degrees Fahrenheit.

33. Use $C = 2\pi r$ and solve for r to obtain $r = \dfrac{C}{2\pi}$. With $C = 480\pi$, then $r = \dfrac{480\pi}{2\pi}$. The radius is 240 inches. Since $d = 2r$, then $d = 2(240)$. The diameter is 480 inches.

37. (a) According to the graph, 31% of HMOs have less than 15,000 members. Since there are about 560 HMOs, we must find 31% of 560.

$$.31(560) = 173.6$$

Therefore, about 174 of the HMOs had less than 15,000 members.
(b) Similarly, find 19% of 560.

$$.19(560) = 106.4$$

About 106 of the HMOs had 15,000–24,999 members.
(c) Find 22% of 560.

$$.22(560) = 123.2$$

About 123 of the HMOs had 25,000–49,999 members.
(d) Find 13% of 560.

$$.13(560) = 72.8$$

About 73 of the HMOs had 50,000–99,999 members.
(e) Find 15% of 560.

$$.15(560) = 84$$

So, 84 of the HMOs had 100,000 members or more.
41. Out of 36 ounces of liquid, 9 ounces are pure alcohol. Therefore, 27 ounces must be water. Let x represent the percent of water. Then the percent of water in the mixture is

$$x = \frac{27}{36}$$

$$x = .75 = \frac{75}{100} = 75\%.$$

Let y represent the percent of alcohol. Then the percent of alcohol in the mixture is

$$y = \frac{9}{36}$$

$$= .25 = \frac{25}{100} = 25\%.$$

45. Use $\frac{a}{b} = \frac{P}{100}$ with $a = 6300$, $b = 210{,}000$ to obtain:

$$\frac{6300}{210{,}000} = \frac{P}{100}$$

$$\frac{630{,}000}{210{,}000} = P \qquad \text{Multiply by 100.}$$

$$3 = P \qquad \text{Reduce the fraction.}$$

The agent received a 3% rate of commission.

Section 2.3 (page 83)

Answers may vary. A sample answer follows:

1. *Step 1*   Read the problem.
Step 2   Choose a variable and write what it represents.
Step 3   Write an equation.
Step 4   Solve the equation.
Step 5   Answer the question.
Step 6   Check your answer.

5. "Increased by" indicates addition. The phrase "7 increased by a number" translates as $7 + x$.

9. "Quotient" indicates division. Rather than $x \div 6$, the phrase "the quotient of a number and 6" is usually written as a fraction $\frac{x}{6}$.

13. Restate the problem as: "from the number x subtract the product of the number x and -4, and set this equal to 9 added to the number x." This translates into the equation

$$x - (-4x) = x + 9.$$

Solve this equation as follows:

$$x + 4x = x + 9 \qquad \text{Multiply.}$$

$$5x = x + 9 \qquad \text{Combine terms.}$$

$$4x = 9 \qquad \text{Subtract } x.$$

$$x = \frac{9}{4} \qquad \text{Divide by 4.}$$

The number is $\frac{9}{4}$.

17. 20% can be written as $.20 = .2 = \frac{20}{100} = \frac{1}{5}$, so "20% of a number" can be written as $.20x$, $.2x$, and $\frac{x}{5}$. We see that "20% of a number" cannot be written as $20x$, selection (d).

21. $5(x + 3) - 8(2x - 6) = 12$ has an equal sign, so this represents an equation.

25. *Step 1*   What must be found? The dimensions of the base. What is given? The length of the base is 65 feet less than twice the width; the perimeter is 860 feet.
Make a sketch.

$2W - 65$

Step 2   Choose a variable: let $W =$ the width; then $2W - 65 =$ the length.
Step 3   Write an equation. The perimeter of a rectangle is given by the formula $P = 2L + 2W$, so

$$860 = 2(2W - 65) + 2W. \quad \text{Substitute.}$$

Step 4   Solve the equation.

$$860 = 2(2W - 65) + 2W$$

$$860 = 4W - 130 + 2W$$

$$860 = 6W - 130$$

$$990 = 6W$$

$$165 = W$$

Step 5   Answer the question: the width is 165 feet; the length is $2(165) - 65 = 265$ feet.
Step 6   Check the answer by substituting these dimensions into the words of the original problem.

29. *Step 1*   What must be found? The seating capacity of each model. What is given? The Boeing seats 110 more passengers than the McDonnell Douglas; together they seat 696 passengers.
Step 2   Choose a variable.

$m =$ the number of passengers
the McDonnell Douglas seats
$m + 110 =$ the number of passengers
the Boeing seats

Step 3   Write an equation. Together the two models seat 696 passengers, so we can write

$$m + (m + 110) = 696.$$

Step 4   Solve the equation.

$$m + (m + 110) = 696$$

$$2m + 110 = 696$$

$$2m = 586$$

$$m = 293$$

Step 5   Answer the question. Since m represents the number of passengers the McDonnell Douglas seats, the McDonnell Douglas seats 293 passengers. Boeing seats $m + 110 = 293 + 110 = 403$ passengers.
Step 6   Check the answer. 403 is 110 more than 293, and the sum of 403 and 293 is 696.

33. Find: the percent decrease in scores from 1970 to 1980
Given: the scores dropped from 19.9 to 18.5
Let $P =$ the percent decrease.
Use the formula: $\frac{\text{amount decrease}}{\text{base score}} = \frac{P}{100}$ with amount decrease $= 19.9 - 18.5$ and base score $= 19.9$.
Solve the equation $\frac{19.9 - 18.5}{19.9} = \frac{P}{100}$

$$\frac{1.4}{19.9} = \frac{P}{100}$$

$$\frac{140}{19.9} = P \qquad \text{Multiply by 100.}$$

$$7.0 = P$$

The percent decrease was 7.0% .

37. Let $x =$ the motel receipts before tax is added in. In words, the equation is: the motel receipts before tax + 8% tax = $1650.78. The equation is written and solved as follows.

$$x + .08x = 1650.78$$

$$\begin{array}{ll} 1.08x = 1650.78 & \text{Add } 1x \text{ and } .08x. \\ x = 1528.50 & \text{Divide by } 1.08. \end{array}$$

This gives the motel receipts before the taxes. The sales tax is 8% of 1528.50 or $.08(1528.50) = \$122.28$.

41. Let $x =$ the amount invested at 4.5%; then $2x - 1000 =$ the amount invested at 3%. The formula for interest is $I = prt$ with $t = 1$ year. The information is organized in the following table:

% as a Decimal	Amount Invested	Interest in 1 Year
.045	x	$.045x$
.03	$2x - 1000$	$.03(2x - 1000)$
		Total interest: 1020

The last column gives the equation:

$.045x + .03(2x - 1000) = 1020.$

$$\begin{array}{ll} .045x + .06x - 30 = 1020 & \text{Clear parentheses.} \\ .105x - 30 = 1020 & \text{Combine terms.} \\ .105x = 1050 & \text{Add 30.} \\ x = 10,000 & \text{Divide by } .105. \end{array}$$

He invested $10,000 at 4.5% and $2(10,000) - 1000 = \$19,000$ at 3%.

45. Let $x =$ the number of liters of 10% solution used. Make a table:

Strength	Liters of Solution	Liters of Pure Acid
4%	10	$.04(10) = .4$
10%	x	$.10x$
6%	$x + 10$	$.06(x + 10)$

The amount of pure acid from the 4% solution plus the amount of pure acid from the 10% solution must equal the amount of pure acid in the final (6%) solution. From the last column, the equation is:

$.4 + .10x = .06(x + 10).$

$$\begin{array}{ll} .4 + .10x = .06x + .6 & \text{Clear parentheses.} \\ .04x = .2 & \text{Subtract } .06x; \text{ subtract } .4. \\ x = 5 & \text{Divide by } .04. \end{array}$$

Five liters of the 10% solution are needed.

49. Let $x =$ the amount of pure dye used (pure dye is 100%). Make a table:

Strength	Gallons of Solution	Gallons of Pure Dye
100%	x	$1.00x = x$
25%	4	$.25(4) = 1$
40%	$x + 4$	$.40(x + 4)$

Write the equation from the last column in the table.

$$\begin{array}{ll} x + 1 = .4(x + 4) & .4 = .40 \\ x + 1 = .4x + 1.6 & \text{Clear parentheses.} \\ .6x = .6 & \text{Subtract } .4x; \text{ subtract } 1. \\ x = 1 & \end{array}$$

One gallon of pure (100%) dye is needed.

53. Answers will vary. A sample answer follows: One cannot expect the final mixture to be worth more than each of the ingredients.

57. (a) The sum of the two expressions found in part (a) of Exercise 56 is

$$.05x + .10(800 - x).$$

This expression must be equal to the total amount of interest earned in one year. This is found by multiplying the amount invested, $800, by the interest rate, 8.75%. Therefore, the equation is

$$.05x + .10(800 - x) = 800(.0875).$$

(b) Similarly, the sum of the two expressions found in part (b) of Exercise 56 is

$$.05y + .10(800 - y).$$

This expression must be equal to the amount of pure acid in the final mixture. This is found by multiplying the number of liters in the mixture, 800, by the percent of acid in the mixture, 8.75%. Therefore, the equation is

$$.05y + .10(800 - y) = 800(.0875).$$

Section 2.4 (page 93)

1. The total amount is given by:

$$46(.05) + 18(.10) = 2.30 + 1.80 = \$4.10$$

5. The problem asks for the *distance* to the workplace. To find this distance, we must multiply the rate, 10 miles per hour, by the time, $\frac{3}{4}$ hour.

9. Let $x =$ the number of pennies and $x =$ the number of dimes. Then, $44 - 2x =$ the number of quarters. Make a table.

Denomination	Number of Coins	Value
.01	x	$.01x$
.10	x	$.10x$
.25	$44 - 2x$	$.25(44 - 2x)$
	Total:	4.37

Write the equation from the last column in the table.

$.01x + .10x + .25(44 - 2x) = 4.37$

$$\begin{array}{ll} x + 10x + 25(44 - 2x) = 437 & \text{Multiply by 100.} \\ x + 10x + 1100 - 50x = 437 & \text{Clear parentheses.} \\ -39x + 1100 = 437 & \text{Combine terms.} \\ -39x = -663 & \text{Subtract 1100.} \\ x = 17 & \text{Divide by } -39. \end{array}$$

There are 17 pennies, 17 dimes, and $44 - 2(17) = 10$ quarters.

13. Let x = number of student tickets sold. Then, $410 - x$ = number of non-student tickets sold. Make a table.

Cost of Ticket	Number Sold	Amount Collected
1.50	x	$1.50x$
3.50	$410 - x$	$3.50(410 - x)$

Total: 825

Write the equation from the last column in the table.

$$1.50x + 3.50(410 - x) = 825$$

$1.5x + 1435 - 3.5x = 825$ Clear parentheses.

$-2x = -610$ Combine terms; subtract 1435.

$x = 305$ Divide by -2.

305 student tickets were sold; $410 - 305 = 105$ non-student tickets were sold.

17. Since distance = rate · time, rate = distance ÷ time.

$$r = \frac{d}{t}$$

$$r = \frac{400}{48.25} \quad \text{Let } d = 400; t = 48.25.$$

$r \approx 8.29$ meters per second

21. Let x = time for Lois to commute to work. Since Clark leaves 15 minutes after Lois, or $\frac{15}{60} = \frac{1}{4}$ hour, then

$$x - \frac{1}{4} = \text{time for Clark.}$$

Complete the table using the formula $rt = d$.

	r	t	d
Lois	35	x	$35x$
Clark	40	$x - \dfrac{1}{4}$	$40\left(x - \dfrac{1}{4}\right)$

Since Lois and Clark are going in opposite directions, then we add their distances to get 140 miles, or:

$$35x + 40\left(x - \frac{1}{4}\right) = 140$$

$35x + 40x - 10 = 140$ Clear parentheses.

$75x = 150$ Combine terms; add 10.

$x = 2.$ Divide by 75.

They will be 140 miles apart at 8 A.M. + 2 hours = 10 A.M.

25. Let x = her rate (speed) during the first part of the trip. Then, $x - 25$ = her rate during rush hour traffic. Complete the table using the formula $rt = d$.

	r	t	d
First part	x	2	$2x$
Rush hour	$x - 25$	$\dfrac{1}{2}$	$\dfrac{1}{2}(x - 25)$

Since the entire trip took $2\frac{1}{2}$ hours, add the distances together and set equal to 125 miles.

$$2x + \frac{1}{2}(x - 25) = 125$$

$4x + x - 25 = 250$ Multiply by 2 to clear the fraction.

$5x = 275$ Combine terms; add 25.

$x = 55$

The speed during the first part of the trip was 55 mph.

29. Since the three marked angles are angles of a triangle, their sum must be $180°$.

$$x + 2x + 60 = 180$$

$3x + 60 = 180$ Combine like terms.

$3x = 120$ Subtract 60.

$x = 40$ Divide by 3.

One angle measures $40°$ and the other measures $2x = 2(40) = 80°$. Since $40° + 80° + 60° = 180°$, the answers are correct.

33. Since vertical angles have equal measures, set the two expressions representing the angles equal.

$$9 - 5x = 25 - 3x$$

$-2x = 16$ Add $3x$; subtract 9.

$x = -8$ Divide by -2.

Since both angles are equal, substitute $x = -8$ into either expression to find the measure of each angle: $9 - 5(-8) = 9 + 40 = 49$ degrees.

37. Let x represent the first of the unknown integers. Then $x + 1$ will be the second, $x + 2$ will be the third, and $x + 3$ will be the fourth. Now write an equation representing the problem.

Sum of the first three	is	54 more than the fourth
↓	↓	↓

$$x + (x + 1) + (x + 2) = (x + 3) + 54$$

$3x + 3 = x + 57$ Combine terms.

$2x = 54$ Subtract x; subtract 3.

$x = 27$ Divide by 2.

The first integer is $x = 27$, the second is $27 + 1 = 28$, the third is $27 + 2 = 29$, and the fourth is $27 + 3 = 30$. The four integers are 27, 28, 29, and 30. Check by substituting these numbers back into the words of the original problem.

$$27 + 28 + 29 = 30 + 54 \quad ?$$

$84 = 84$ True

SECTION 2.5 (page 105)

1. $x \le 3$ is written as $(-\infty, 3]$ in interval notation. The correct choice is D.

5. $-3 \le x \le 3$ is written as $[-3, 3]$ in interval notation. The correct choice is F.

See the graphs in the answer section for Exercises 9–45.

9. $5r \le -15$

$r \le -3$ Divide by 5.

Solution set: $(-\infty, -3]$

13. $\dfrac{3k - 1}{4} > 5$

$3k - 1 > 20$ Multiply by 4.

$3k > 21$ Add 1.

$k > 7$ Divide by 3.

Solution set: $(7, \infty)$

17. $-\dfrac{3}{4}r \ge 30$

$-3r \ge 120$ Multiply by 4.

$r \le -40$ Divide by -3; reverse the inequality sign.

Solution set: $(-\infty, -40]$

21. $-1.3m \ge -5.2$

$m \le 4$ Divide by -1.3; reverse the inequality sign.

Solution set: $(-\infty, 4]$

25. $y + 4(2y - 1) \ge y$

$y + 8y - 4 \ge y$ Clear parentheses.

$9y - 4 \ge y$ Combine terms.

$8y \ge 4$ Subtract y; add 4.

$y \ge \dfrac{1}{2}$ Divide by 8; reduce the fraction.

Solution set: $\left[\dfrac{1}{2}, \infty\right)$

29. $-3(z - 6) > 2z - 2$

$-3z + 18 > 2z - 2$ Clear parentheses.

$-5z > -20$ Subtract $2z$; subtract 18.

$z < 4$ Divide by -5; reverse the inequality sign.

Solution set: $(-\infty, 4)$

33. $5(x + 3) - 2(x - 4) < 2(x + 7)$

$5x + 15 - 2x + 8 < 2x + 14$ Clear parentheses.

$3x + 23 < 2x + 14$ Combine terms.

$x < -9$ Subtract $2x$; subtract 23.

Solution set: $(-\infty, -9)$

37. $-4 < x - 5 < 6$

$1 < x < 11$ Add 5 to each of the three parts of the inequality.

Solution set: $(1, 11)$

41. $-6 \le 2z + 4 \le 16$

$-10 \le 2z \le 12$ Subtract 4 from all three parts.

$-5 \le z \le 6$ Divide each part by 2.

Solution set: $[-5, 6]$

45. $-1 \le \dfrac{2x - 5}{6} \le 5$

$-6 \le 2x - 5 \le 30$ Multiply each part by 6 to clear the fraction.

$-1 \le 2x \le 35$ Add 5 to each part.

$-\dfrac{1}{2} \le x \le \dfrac{35}{2}$ Divide each part by 2.

Solution set: $\left[-\frac{1}{2}, \frac{35}{2}\right]$

49. In the months of April, May, June, and July, the percent of tornadoes exceeded 7.7%. Notice that in August, the percent of tornadoes was *exactly* 7.7%. This does not *exceed* 7.7%, so the month of August is not included.

53. Let $x =$ her score on the third test. Her average must be 84 or greater. To find the average of three numbers, add them and divide by 3.

$$\dfrac{90 + 82 + x}{3} \ge 84$$

$$\dfrac{172 + x}{3} \ge 84 \quad \text{Combine terms.}$$

$$172 + x \ge 252 \quad \text{Multiply by 3.}$$

$$x \ge 80 \quad \text{Subtract 172.}$$

She must score at least 80 on her third test.

57. The business will show a profit only when $R > C$ or

$$24x > 20x + 100$$

$$4x > 100 \quad \text{Subtract } 20x.$$

$$x > 25. \quad \text{Divide by 4.}$$

The company will show a profit upon selling 26 tapes.

SECTION 2.6 (page 113)

1. This statement is true. The set of real numbers is the set of rational and irrational numbers combined.

5. This statement is false. In the union of the sets $(-\infty, 6)$ and $(6, \infty)$, 6 is not included.

9. $B \cap \emptyset$, the intersection of set B and the set of no elements (empty set), is the set of no elements or $\emptyset$.

13. $B \cup C$, the union of B and C, contains all elements in either B or C or both B and C: $\{1, 3, 5, 6\}$.

17. $B \cap C =$ the set of elements common to both B and $C = \{1\}$. $A \cap (B \cap C) = A \cap \{1\} = \{1\}$. $A \cap B = \{1, 3, 5\}$. $(A \cap B) \cap C = \{1, 3, 5\} \cap C = \{1\}$. Therefore, $A \cap (B \cap C) = (A \cap B) \cap C$. This illustrates the associative property of set intersection.

See the graphs for Exercises 21–53 in the answer section.

21. $x < 2$ and $x > -3$ is true whenever $-3 < x < 2$. The solution set is $(-3, 2)$.

25. There is no number x such that $x \le 3$ and at the same time $x \ge 6$. Therefore, the solution set is $\emptyset$.

29. $x \ge -1$ and $x \le 3$ is true whenever $-1 \le x \le 3$. The solution set is $[-1, 3]$.

33. Solve each inequality separately.

$$3x - 4 \le 8 \quad \text{and} \quad 4x - 1 \le 15$$

$$3x \le 12 \quad \text{and} \quad 4x \le 16$$

$$x \le 4 \quad \text{and} \quad x \le 4$$

The solution set is $(-\infty, 4]$.

37. $x \le 1$ or $x \le 8$ is true when $x \le 8$ (since $x \le 1$ is already included in the set $x \le 8$). The solution set is $(-\infty, 8]$.

41. $x \ge -2$ or $x \ge 5$ is true when $x \ge -2$ (since $x \ge 5$ is already included in the set $x \ge -2$). The solution set is $[-2, \infty)$.

45. Solve each inequality separately.

$$x + 2 > 7 \quad \text{or} \quad x - 1 < -6$$

$$x > 5 \quad \text{or} \quad x < -5$$

The solution set is $(-\infty, -5) \cup (5, \infty)$.

(Note that the symbol $\cup$ replaces the word "or.")

49. The word "and" means the same as "intersection."
$x < -1$ and $x > -5$ is true only when
$-5 < x < -1$.
The solution set is $(-5, -1)$.

53. The word "and" means the same as "intersection."
Solve each inequality.

$$x + 1 \geq 5 \quad \text{and} \quad x - 2 \leq 10$$
$$x \geq 4 \quad \text{and} \quad x \leq 12$$

This last statement is true only when $4 \leq x \leq 12$.
The solution set is $[4, 12]$.

57. First find the area and perimeter of each of the residents' yards.
Luigi's yard:

$$A = l \cdot w \qquad\qquad P = 2l + 2w$$
$$= 50 \cdot 30 \qquad\qquad = 2(50) + 2(30)$$
$$= 1500 \text{ square feet} \qquad = 100 + 60$$
$$= 160 \text{ feet}$$

Mario's yard:

$$A = 40 \cdot 35 \qquad\qquad P = 2(40) + 2(35)$$
$$= 1400 \text{ square feet} \qquad = 80 + 70$$
$$= 150 \text{ feet}$$

Than's yard:

$$A = 60 \cdot 50 \qquad\qquad P = 2(60) + 2(50)$$
$$= 3000 \text{ square feet} \qquad = 120 + 100$$
$$= 220 \text{ feet}$$

Joe's yard:

$$A = \frac{1}{2}bh \qquad\qquad P = 30 + 40 + 50$$
$$= \frac{1}{2}(40)(30) \qquad\qquad = 120 \text{ feet}$$
$$= 600 \text{ square feet}$$

Since each resident has 150 feet of fencing and enough sod to cover 1400 square feet of lawn, Mario's and Joe's yards can be fenced *and* sodded.

SECTION 2.7 (page 123)

1. $|x| = 5$ has two solutions: $x = 5$ or $x = -5$.
Choice E
$|x| < 5$ is written $-5 < x < 5$.
Choice C uses parentheses.
$|x| > 5$ is written $x < -5$ or $x > 5$.   Choice D uses parentheses.
$|x| \leq 5$ is written $-5 \leq x \leq 5$.
Choice B
$|x| \geq 5$ is written $x \leq -5$ or $x \geq 5$.
Choice A

5. $|x| = 12$ gives two equations: $x = 12$ or $x = -12$.
The solution set is $\{-12, 12\}$.

9. $|y - 3| = 9$ gives two equations:

$$y - 3 = 9 \quad \text{or} \quad y - 3 = -9$$
$$y = 12 \quad \text{or} \quad y = -6.$$

Solution set: $\{-6, 12\}$

13. $|4r - 5| = 17$ gives two equations:

$$4r - 5 = 17 \qquad \text{or} \quad 4r - 5 = -17.$$
$$4r = 22 \qquad \text{or} \qquad 4r = -12$$
$$r = \frac{22}{4} = \frac{11}{2} \quad \text{or} \qquad r = -3.$$

Solution set: $\{-3, \frac{11}{2}\}$

17. $\left|\frac{1}{2}x + 3\right| = 2$ gives two equations:

$$\frac{1}{2}x + 3 = 2 \quad \text{or} \quad \frac{1}{2}x + 3 = -2.$$
$$\frac{1}{2}x = -1 \quad \text{or} \qquad \frac{1}{2}x = -5$$
$$x = -2 \quad \text{or} \qquad x = -10$$

Solution set: $\{-10, -2\}$

See the graphs in the answer section for Exercises 21–49.

21. $|x| > 3$ is written as $x < -3$ or $x > 3$.
Solution set: $(-\infty, -3) \cup (3, \infty)$

25. $|t + 2| > 10$ is written as

$$t + 2 < -10 \quad \text{or} \quad t + 2 > 10.$$
$$t < -12 \quad \text{or} \qquad t > 8$$

Solution set: $(-\infty, -12) \cup (8, \infty)$

29. $|3 - x| > 5$ is written as

$$3 - x < -5 \quad \text{or} \quad 3 - x > 5.$$
$$-x < -8 \quad \text{or} \qquad -x > 2$$
$$x > 8 \qquad \text{or} \qquad x < -2 \quad \text{Multiply by } -1 \text{ and reverse the signs.}$$

Solution set: $(-\infty, -2) \cup (8, \infty)$

33. $|x| \leq 3$ is written as $-3 \leq x \leq 3$.
Solution set: $[-3, 3]$

37. $|t + 2| \leq 10$ is written as

$$-10 \leq t + 2 \leq 10.$$
$$-12 \leq t \leq 8 \qquad \text{Subtract 2 from all parts.}$$

Solution set: $[-12, 8]$

41. $|3 - x| \leq 5$ is written as

$$-5 \leq 3 - x \leq 5.$$
$$-8 \leq -x \leq 2$$
$$8 \geq x \geq -2 \qquad \text{Multiply by } -1 \text{ and reverse the inequality signs.}$$

The solution set is $[-2, 8]$.

45. $|7 + 2z| = 5$ gives two equations

$$7 + 2z = 5 \quad \text{or} \quad 7 + 2z = -5.$$
$$2z = -2 \quad \text{or} \qquad 2z = -12 \quad \text{Subtract 7.}$$
$$z = -1 \quad \text{or} \qquad z = -6 \quad \text{Divide by 2.}$$

Solution set: $\{-6, -1\}$

SOLUTIONS

49. $|-6x - 6| \le 1$ is written as

$$-1 \le -6x - 6 \le 1.$$

$$5 \le -6x \le 7 \qquad \text{Add 6 to all parts.}$$

$$-\frac{5}{6} \ge x \ge -\frac{7}{6} \qquad \text{Divide by } -6; \text{ reverse signs.}$$

Solution set: $\left[-\frac{7}{6}, -\frac{5}{6}\right]$

53. $|x + 4| + 1 = 2$ becomes $|x + 4| = 1$.

$$x + 4 = 1 \quad \text{or} \quad x + 4 = -1$$

$$x = -3 \quad \text{or} \quad x = -5$$

Solution set: $\{-5, -3\}$

57. $|x + 5| - 6 \le -1$ becomes $|x + 5| \le 5$, written as

$$-5 \le x + 5 \le 5$$

$$-10 \le x \le 0. \qquad \text{Add } -5 \text{ to all parts.}$$

Solution set: $[-10, 0]$

61. $\left|m - \frac{1}{2}\right| = \left|\frac{1}{2}m - 2\right|$ gives two equations

$$m - \frac{1}{2} = \frac{1}{2}m - 2 \quad \text{or} \quad m - \frac{1}{2} = -\left(\frac{1}{2}m - 2\right).$$

$$2m - 1 = m - 4 \quad \text{or} \quad 2m - 1 = -m + 4$$

$$\text{Multiply by 2.}$$

$$m = -3 \quad \text{or} \quad 3m = 5$$

$$m = -3 \quad \text{or} \quad m = \frac{5}{3}$$

Solution set: $\{-3, \frac{5}{3}\}$

65. $|2p - 6| = |2p + 11|$ gives two equations

$$2p - 6 = 2p + 11 \quad \text{or} \quad 2p - 6 = -(2p + 11).$$

$$-6 = 11 \qquad \text{or} \quad 2p - 6 = -2p - 11$$

$$\text{no solution} \qquad \qquad 4p = -5$$

$$p = -\frac{5}{4}$$

There is one solution from the second equation: $\{-\frac{5}{4}\}$.

69. $|4x + 1| = 0$ gives only one equation.

$$4x + 1 = 0$$

$$4x = -1$$

$$x = -\frac{1}{4}$$

Solution set: $\{-\frac{1}{4}\}$

73. Since the absolute value of an expression is always nonnegative (positive or zero), then the inequality $|x + 5| > -9$ is true for any real number x. The solution set is $(-\infty, \infty)$.

77. $|5x - 2| \ge 0$ includes two statements.
(A) $|5x - 2| = 0$, which becomes $5x - 2 = 0$ or $x = \frac{2}{5}$.
(B) $|5x - 2| > 0$, which is true for any real number x except 2/5, since the absolute value is always greater than 0. Since the solution set includes the elements in (A) or (B) above, then the solution set is $(-\infty, \infty)$.

81. Add the heights of the buildings and then divide by 10 (the number of buildings).

$$\begin{aligned} \text{average} &= (626 + 590 + 504 + 476 + 443 + 413 \\ &\quad + 402 + 402 + 394 + 352) \div 10 \\ &= 4602 \div 10 \\ &= 460.2 \end{aligned}$$

For the ten tallest buildings in Kansas City, Missouri, the average height is 460.2 feet.

SUMMARY ON SOLVING LINEAR AND ABSOLUTE VALUE EQUATIONS AND INEQUALITIES (page 129)

1. $4z + 1 = 49$
$$4z = 48 \qquad \text{Subtract 1.}$$
$$z = 12 \qquad \text{Divide by 4.}$$
Solution set: $\{12\}$

5. Since the absolute value of an expression is always nonnegative, there is no number a such that $|a + 3| = -4$. Therefore, the solution set is $\emptyset$.

9. $2q - 1 = -7$
$$2q = -6 \qquad \text{Add 1.}$$
$$q = -3 \qquad \text{Divide by 2.}$$
Solution set: $\{-3\}$

13. $9y - 3(y + 1) = 8y - 7$
$$9y - 3y - 3 = 8y - 7 \qquad \text{Clear parentheses.}$$
$$6y - 3 = 8y - 7 \qquad \text{Combine terms.}$$
$$4 = 2y \qquad \text{Subtract } 6y; \text{ add 7.}$$
$$2 = y \qquad \text{Divide by 2.}$$
Solution set: $\{2\}$

17. $|q| < 5.5$ is written as $-5.5 < q < 5.5$. Solution set: $(-5.5, 5.5)$

21. $$\frac{1}{4}p < -6$$

$$4\left(\frac{1}{4}p\right) < 4(-6) \qquad \text{Multiply by 4.}$$

$$p < -24 \qquad \text{Clear the fraction; simplify.}$$

Solution set: $(-\infty, -24)$

25. $r + 9 + 7r = 4(3 + 2r) - 3$
$$8r + 9 = 12 + 8r - 3 \qquad \text{Combine terms; clear parentheses.}$$
$$8r + 9 = 8r + 9$$
$$0 = 0 \qquad \text{Subtract } 8r \text{ and 9.}$$
This last statement is true for any real number r.
Solution set: $(-\infty, \infty)$

29. The expression $|5a + 1|$ is never less than 0 since an absolute value expression must be nonnegative. However, $|5a + 1| = 0$ is the same as $5a + 1 = 0$.

$$5a + 1 = 0$$

$$5a = -1 \qquad \text{Subtract 1.}$$

$$a = \frac{-1}{5} \qquad \text{Divide by 5.}$$

Solution set: $\{-\frac{1}{5}\}$

33. $|7z - 1| = |5z + 3|$ gives two equations.

$$7z - 1 = 5z + 3 \quad \text{or} \quad 7z - 1 = -(5z + 3)$$
$$2z = 4 \qquad\qquad 7z - 1 = -5z - 3$$
$$z = 2 \qquad\qquad 12z = -2$$
$$\qquad\qquad z = \frac{-2}{12} = \frac{-1}{6}$$

Solution set: $\{-\frac{1}{6}, 2\}$

37. $-(m + 4) + 2 = 3m + 8$

$$
\begin{array}{ll}
-m - 4 + 2 = 3m + 8 & \text{Clear parentheses.} \\
-m - 2 = 3m + 8 & \text{Combine terms.} \\
-10 = 4m & \text{Add } m\text{; subtract 8.} \\
\dfrac{-5}{2} = m & \text{Divide by 4 and reduce.}
\end{array}
$$

Solution set: $\{-\frac{5}{2}\}$

41. $|y - 1| \geq -6$ is true for all real numbers y since the absolute value is always greater than or equal to 0. Solution set: $(-\infty, \infty)$.

45. $|r - 5| = |r + 9|$ gives two equations.

$$r - 5 = r + 9 \quad \text{or} \quad r - 5 = -(r + 9)$$
$$-5 = 9 \qquad\qquad r - 5 = -r - 9$$
$$\text{no solution} \qquad\qquad 2r = -4$$
$$\qquad\qquad r = -2$$

The solution set from the second equation is $\{-2\}$.

CHAPTER 3

SECTION 3.1 (page 153)

1. **(a)** The percentage was approximately the same between 1989 and 1990.
(b) The increase was greatest between 1991 and 1992.
(c) In 1991 the percent was about 16.5%.

5. The point with coordinates $(0, 0)$ is called the origin of a rectangular coordinate system.

9. To graph a straight line, we must find a minimum of two points.

13. **(a)** If $xy > 0$, then both x and y have the same sign.
(x, y) is in quadrant I if x and y are positive.
(x, y) is in quadrant III if x and y are negative.
(b) If $xy < 0$, then x and y have different signs.
(x, y) is in quadrant II if $x < 0$ and $y > 0$.
(x, y) is in quadrant IV if $x > 0$ and $y < 0$.

(c) If $\dfrac{x}{y} < 0$, then x and y have different signs.
(x, y) is in either quadrant II or IV (see part b).

(d) If $\dfrac{x}{y} > 0$, then x and y have the same sign.
(x, y) is in either quadrant I or III (see part a).

17. Plot $(-3, -2)$ by going three units to the left of the origin along the x-axis, then going directly two units down parallel to the y-axis.

21. Plot $(-2, 4)$ by going two units to the left of the origin along the x-axis and four units up parallel to the y-axis.

See the graphs in the answer section of your text for Exercises 25–41.

25. If $x = 0$, $x - y = 3$ becomes $0 - y = 3$ or $y = -3$.
If $y = 0$, $x - y = 3$ becomes $x - 0 = 3$ or $x = 3$.
If $x = 5$, $x - y = 3$ becomes $5 - y = 3$.
So $-y = -2$, or $y = 2$.
If $x = 2$, $x - y = 3$ becomes $2 - y = 3$.
So $-y = 1$, or $y = -1$.
The ordered pairs are $(0, -3)$, $(3, 0)$, $(5, 2)$, and $(2, -1)$.

29. If $x = 0$, $4x - 5y = 20$ becomes $4(0) - 5y = 20$.
So $-5y = 20$, or $y = -4$.
If $y = 0$, $4x - 5y = 20$ becomes $4x - 5(0) = 20$.
So $4x = 20$, or $x = 5$.
If $x = 2$, $4x - 5y = 20$ becomes $4(2) - 5y = 20$.
So $8 - 5y = 20$, $-5y = 12$, and $y = -\frac{12}{5}$.
If $y = -3$, $4x - 5y = 20$ becomes $4x - 5(-3) = 20$.
So $4x + 15 = 20$, $4x = 5$, and $x = \frac{5}{4}$.
The ordered pairs are $(0, -4)$, $(5, 0)$, $(2, -\frac{12}{5})$, and $(\frac{5}{4}, -3)$.

33. For the x-intercept, let $y = 0$ in $2x + 3y = 12$ to get $2x = 12$ or $x = 6$. The ordered pair is $(6, 0)$. For the y-intercept, let $x = 0$ in $2x + 3y = 12$ to get $3y = 12$ or $y = 4$. The ordered pair is $(0, 4)$.

37. For the x-intercept, let $y = 0$ in $3x - 7y = 9$ to get $3x = 9$ or $x = 3$. The ordered pair is $(3, 0)$. For the y-intercept, let $x = 0$ in $3x - 7y = 9$ to get $-7y = 9$ or $y = -\frac{9}{7}$. The ordered pair is $(0, -\frac{9}{7})$.

41. The graph of $x = 2$ is a vertical line through the x-intercept $(2, 0)$. There is no y-intercept since this vertical line is parallel to the y-axis, and never intersects it.

45. Let $x = 8$ (corresponding to 1988) in the linear model $y = 2.503x + 198.729$ and solve for y.

$$
\begin{aligned}
y &= 2.503(8) + 198.729 \\
&= 20.024 + 198.729 \\
&= 218.753
\end{aligned}
$$

In 1988, the winning speed for the Indianapolis 500 was approximately 218.753 miles per hour.

49. **(a)** $y = 2x - 3$ does not correspond to the table of points because when $x = 0$, $y = 2(0) - 3 = -3$, not 3.
(b) $y = -2x - 3$ does not correspond to the table of points because when $x = 0$, $y = -2(0) - 3 = -3$, not 3.
(c) $y = 2x + 3$ does not correspond to the table of points because when $x = 1$, $y = 2(1) + 3 = 5$, not 1.
(d) $y = -2x + 3$ *does* correspond to the table of points because when each x-value is substituted into the equation, the result is the corresponding y-value shown in the table.

SECTION 3.2 (page 163)

1. To move from point A to point B in line segment AB, we must "rise" 2 units and "run" 1 unit. Therefore, the

$$\text{Slope of } AB = \frac{\text{rise}}{\text{run}} = \frac{2}{1} = 2.$$

5. Slope $= \dfrac{\text{vertical rise}}{\text{horizontal run}} = \dfrac{2}{10}$.

$$\frac{2}{10} = .2; \quad \frac{2}{10} = \frac{1}{5}; \quad \frac{2}{10} = 20\%; \quad \frac{2}{10} = \frac{20}{100} = 20\%$$

The correct choices are (a), (b), (c), (d), and (f).

9. Let $(x_1, y_1) = (-4, 1)$ and $(x_2, y_2) = (2, 6)$. Then

$$m = \frac{y_2 - y_1}{x_2 - x_1} = \frac{6 - 1}{2 - (-4)} = \frac{5}{6}.$$

See the graphs in the answer section for Exercises 13 and 21–37.

13. The graph of $x = 5$ is a vertical line with x-intercept $(5, 0)$. The slope of a vertical line is undefined because the denominator equals zero in the slope formula.

17. The line with positive slope is choice B since when the change in y is positive the change in x is also positive. (As the graph moves from left to right, the line goes up.)

21. Two points on the graph of $x + 2y = 4$ are the x- and y-intercepts $(4, 0)$ and $(0, 2)$.

$$\text{The slope } m = \frac{y_2 - y_1}{x_2 - x_1} = \frac{2 - 0}{0 - 4} = -\frac{2}{4} = -\frac{1}{2}.$$

25. Two points on the graph of $6x + 5y = 30$ are the intercepts $(5, 0)$ and $(0, 6)$.

$$\text{The slope } m = \frac{y_2 - y_1}{x_2 - x_1} = \frac{6 - 0}{0 - 5} = -\frac{6}{5}.$$

29. In $y = 4x$, replace x with 0 and then x with 1 to get the ordered pairs $(0, 0)$ and $(1, 4)$, respectively. (There are other possibilities for ordered pairs.)

$$\text{The slope } m = \frac{y_2 - y_1}{x_2 - x_1} = \frac{4 - 0}{1 - 0} = 4.$$

33. Locate $(-4, 2)$ on your graph. The slope is given as $\frac{1}{2}$. From $(-4, 2)$, go up 1 unit. Then go 2 units to the right to get to $(-2, 3)$. Draw the line through $(-4, 2)$ and $(-2, 3)$.

37. Locate $(-1, -2)$ on your graph. The slope is given as $m = 3 = \frac{3}{1}$. From $(-1, -2)$, go up 3 units. Then go 1 unit to the right to get to $(0, 1)$. Draw the line through $(-1, -2)$ and $(0, 1)$.

41. The slope of the line through $(4, 6)$ and $(-8, 7)$ is

$$m_1 = \frac{y_2 - y_1}{x_2 - x_1} = \frac{7 - 6}{-8 - 4} = \frac{1}{-12} = -\frac{1}{12}.$$

The slope of the line through $(-5, 5)$ and $(7, 4)$ is

$$m_2 = \frac{y_2 - y_1}{x_2 - x_1} = \frac{4 - 5}{7 - (-5)} = \frac{-1}{12} = -\frac{1}{12}.$$

Since the slopes m_1 and m_2 are equal, the two lines are parallel.

45. Since the front of the upper deck is 160 feet from home plate and the back of the upper deck is 250 feet from home plate, the deck is $250 - 160 = 90$ feet long. We are also given that the deck is 63 feet high. To find the slope of the deck, find the

$$\frac{\text{vertical change}}{\text{horizontal change}} = \frac{63}{90} = \frac{7}{10}.$$

The slope of the upper deck at the new Comiskey Park is $\frac{7}{10}$.

49. Let $x = 0$ correspond to 1986, $x = 1$ correspond to 1987, and so on. Then 1993 corresponds to $x = 7$ and 1995 corresponds to $x = 9$. Since in 1993 there were

47.5 billion minutes, this corresponds to the ordered pair, $(7, 47.5)$. Similarly, since in 1995 there were 60 billion minutes, this corresponds to the ordered pair, $(9, 60)$. To find the average rate of change, let $(x_1, y_1) = (7, 47.5)$ and $(x_2, y_2) = (9, 60)$ and find the slope.

$$m = \frac{y_2 - y_1}{x_2 - x_1} = \frac{60 - 47.5}{9 - 7} = \frac{12.5}{2} = 6.25$$

The average rate of change is 6.25 billion minutes per year.

53. To find the slope of segment AC, let $(x_1, y_1) = (3, 1)$ and $(x_2, y_2) = (9, 3)$. Then

$$m = \frac{y_2 - y_1}{x_2 - x_1} = \frac{3 - 1}{9 - 3} = \frac{2}{6} = \frac{1}{3}.$$

SECTION 3.3 (page 177)

1. In $y = 2x + 3$, the slope is $m = 2$ and the y-intercept is $(0, 3)$. Choice A is the only graph with a positive slope that has a positive y-intercept.

5. Write $y = 2x$ as $y = 2x + 0$. The slope $m = 2$ and $b = 0$, so the y-intercept is $(0, 0)$. Choice H is the only graph with a positive slope that intersects the y-axis at $(0, 0)$.

9.

$$
\begin{array}{ll}
y - y_1 = m(x - x_1) & \text{Point-slope form} \\[4pt]
y - 4 = -\frac{3}{4}[x - (-2)] & \text{Let } x_1 = -2, y_1 = 4, \\
& m = -\frac{3}{4}. \\[4pt]
y - 4 = -\frac{3}{4}(x + 2) & \\[4pt]
4(y - 4) = -3(x + 2) & \text{Multiply by 4.} \\
4y - 16 = -3x - 6 & \text{Clear parentheses.} \\
3x + 4y = 10 & \text{Standard form}
\end{array}
$$

13.

$$
\begin{array}{ll}
y - y_1 = m(x - x_1) & \text{Point-slope form} \\[4pt]
y - 4 = \frac{1}{2}[x - (-5)] & \text{Let } x_1 = -5, y_1 = 4, m = \frac{1}{2}. \\[4pt]
y - 4 = \frac{1}{2}(x + 5) & \\[4pt]
2(y - 4) = x + 5 & \text{Multiply by 2.} \\
2y - 8 = x + 5 & \\
-13 = x - 2y & \text{Subtract 5 and } 2y. \\
x - 2y = -13 & \text{Standard form}
\end{array}
$$

17. A vertical line has undefined slope, so the equation is of the form $x = k$, where k is the x-coordinate of a point on the line. Since $(9, 10)$ is a point on the line, the equation is $x = 9$.

21. The slope of the line through $(3, 4)$ and $(5, 8)$ is

$$m = \frac{y_2 - y_1}{x_2 - x_1} = \frac{8 - 4}{5 - 3} = \frac{4}{2} = 2.$$

Use $m = 2$, $x_1 = 3$, $y_1 = 4$ in $y - y_1 = m(x - x_1)$ to get

$$
\begin{array}{ll}
y - 4 = 2(x - 3) & \\
y - 4 = 2x - 6 & \text{Clear parentheses.} \\
2 = 2x - y & \text{Add } 6 - y \text{ to both sides.} \\
2x - y = 2. & \text{Standard form}
\end{array}
$$

25. The slope of the line through $\left(-\frac{2}{5}, \frac{2}{5}\right)$ and $\left(\frac{4}{3}, \frac{2}{3}\right)$ is

$$m = \frac{y_2 - y_1}{x_2 - x_1} = \frac{\frac{2}{3} - \frac{2}{5}}{\frac{4}{3} - \left(-\frac{2}{5}\right)} = \frac{\frac{4}{15}}{\frac{26}{15}} = \frac{4}{26} = \frac{2}{13}.$$

Use $m = \frac{2}{13}, x_1 = \frac{4}{3}, y_1 = \frac{2}{3}$ to get

$$y - \frac{2}{3} = \frac{2}{13}\left(x - \frac{4}{3}\right)$$

$$39\left(y - \frac{2}{3}\right) = 39 \cdot \frac{2}{13}\left(x - \frac{4}{3}\right) \qquad \text{The LCD is 39.}$$

$$39y - 26 = 6\left(x - \frac{4}{3}\right)$$

$$39y - 26 = 6x - 8 \qquad \text{Clear parentheses.}$$

$$6x - 39y = -18$$

$$2x - 13y = -6. \qquad \text{Divide both sides by 3.}$$

29. The line through $(7, 6)$ and $(7, -8)$ is a vertical line, because both points have the same x-value. The equation of a vertical line is $x = k$, where k is the common x-value. So the equation is $x = 7$.

33. Put $2x - 3y = 10$ into $y = mx + b$ form.

$$2x - 3y = 10$$

$$-3y = -2x + 10 \qquad \text{Subtract } 2x.$$

$$y = \frac{-2}{-3}x + \frac{10}{-3} \qquad \text{Divide by } -3.$$

$$y = \frac{2}{3}x - \frac{10}{3}$$

From the last equation, $m = \frac{2}{3}$ and $b = -\frac{10}{3}$. The slope is $\frac{2}{3}$ and the y-intercept is $\left(0, -\frac{10}{3}\right)$.

37. $y = mx + b$ $\qquad$ Slope-intercept form

$$y = -\frac{2}{3}x + \frac{4}{5} \qquad \text{Let } m = -\frac{2}{3} \text{ and } b = \frac{4}{5}.$$

41. Since parallel lines have equal slopes, this line will have the same slope as $3x - y = 8$.
Write $3x - y = 8$ in $y = mx + b$ form to get $y = 3x - 8$. The slope is 3.
Use $x_1 = 7, y_1 = 2,$ and $m = 3$ in $y - y_1 = m(x - x_1)$ to get

$$y - 2 = 3(x - 7)$$

$$y - 2 = 3x - 21 \qquad \text{Clear parentheses.}$$

$$3x - y = 19. \qquad \text{Put into standard form.}$$

45. Put $2x - y = 7$ into $y = mx + b$ form to get $y = 2x - 7$. The slope is 2.
The slope of the line perpendicular to $2x - y = 7$ is the negative reciprocal of 2 or $-\frac{1}{2}$.
Use $x_1 = 8, y_1 = 5,$ and $m = -\frac{1}{2}$ in $y - y_1 = m(x - x_1)$ to get

$$y - 5 = -\frac{1}{2}(x - 8)$$

$$2(y - 5) = 2\left[-\frac{1}{2}(x - 8)\right] \qquad \text{Multiply by 2.}$$

$$2y - 10 = -x + 8 \qquad \text{Clear parentheses.}$$

$$x + 2y = 18. \qquad \text{Put into standard form.}$$

49. If x represents the number of hours traveled at 45 miles per hour, and y represents the distance traveled, then

$$y = 45x.$$

When $x = 0, y = 45(0) = 0$.

$$\text{Ordered pair: } (0, 0)$$

When $x = 5, y = 45(5) = 225$.

$$\text{Ordered pair: } (5, 225)$$

When $x = 10, y = 45(10) = 450$.

$$\text{Ordered pair: } (10, 450)$$

53. Since it costs \$15 plus \$3 per day to rent a chain saw, it costs $3x + 15$ dollars for x days. Thus, if y represents the charge to the user (in dollars), the equation is $y = 3x + 15$.
When $x = 0, y = 3(0) + 15 = 0 + 15 = 15$.

$$\text{Ordered pair: } (0, 15)$$

When $x = 5, y = 3(5) + 15 = 15 + 15 = 30$.

$$\text{Ordered pair: } (5, 30)$$

When $x = 10, y = 3(10) + 15 = 30 + 15 = 45$.

$$\text{Ordered pair: } (10, 45)$$

57. In Exercise 53, $y = 3x + 15$, where y represents the total charge for renting a saw for x days. If the total charge is \$69.00, then

$$69 = 3x + 15$$

$$54 = 3x$$

$$18 = x.$$

The saw was rented for 18 days.

61. If we let $x = 0$ represent the year 1985, then $x = 8$ represents the year 1993. Use the graph to find two ordered pairs representing the data. Let $(x_1, y_1) = (0, 29{,}557)$ and $(x_2, y_2) = (8, 28{,}728)$ and find the slope.

$$m = \frac{y_2 - y_1}{x_2 - x_1} = \frac{28{,}728 - 29{,}557}{8 - 0} = \frac{-829}{8}$$

To find the equation of the line in the form $y = mx + b, m = -\frac{829}{8}$ and $b = 29{,}557$ (when $x = 0$), so

$$y = -\frac{829}{8}x + 29{,}557.$$

65. We are given the slope of the line and a point on the line. Find the equation of the line using the point slope form.

$$y - y_1 = m(x - x_1)$$

$$y - (-1) = 3(x - 1) \qquad m = 3; (x_1, y_1) = (1, -1)$$

$$y + 1 = 3x - 3 \qquad \text{Clear parentheses.}$$

$$y = 3x - 4$$

The equation of the line is $y = 3x - 4$. To find the y-coordinate of the point on the line whose x-coordinate is 4, let $x = 4$ and solve for y.

SOLUTIONS

$$y = 3(4) - 4$$
$$y = 12 - 4$$
$$y = 8$$

When $x = 4$, $y = 8$.

69. In Exercise 68, the two points were $(0, 32)$ and $(100, 212)$. To find the slope, let $(x_1, y_1) = (0, 32)$ and $(x_2, y_2) = (100, 212)$.

$$m = \frac{y_2 - y_1}{x_2 - x_1} = \frac{212 - 32}{100 - 0} = \frac{180}{100} = \frac{9}{5}$$

SECTION 3.4 (page 189)

1. The boundary of the graph of $y \le -x + 2$ will be a solid line, and the shading will be below the line.

See the graphs for Exercises 5–33 in the answer section of your text.

5. To graph $x + y \le 2$,
(a) graph the line $x + y = 2$ by drawing a solid line through the intercepts $(2, 0)$ and $(0, 2)$;
(b) test a point not on this line, such as $(0, 0)$, in $x + y \le 2$ to get $0 + 0 \le 2$, a true statement;
(c) shade that side of the line containing the test point $(0, 0)$.

9. To graph $x + 3y \ge -2$,
(a) graph the solid line $x + 3y = -2$ through the intercepts $(-2, 0)$ and $(0, -\frac{2}{3})$;
(b) test a point not on this line, such as $(0, 0)$, in $x + 3y \ge -2$ to get $0 + 0 \ge -2$, a true statement;
(c) shade that side of the line containing the test point $(0, 0)$.

13. To graph $x - 3y \le 0$,
(a) graph the solid line $x - 3y = 0$ through the points $(0, 0)$ and $(3, 1)$;
(b) test a point not on the line, such as $(1, 1)$, in $x - 3y \le 0$ to get $1 - 3(1) \le 0$, a true statement;
(c) shade that side of the line containing the test point $(1, 1)$.

17. When graphing compound inequalities, if the connecting word is "and," use intersection. If the connecting word is "or," use union.

21. To graph $2x - y \ge 2$,
(a) graph the solid line $2x - y = 2$ through the intercepts $(1, 0)$ and $(0, -2)$;
(b) test a point not on the line, such as $(0, 0)$, in $2x - y \ge 2$ to get $0 - 0 \ge 2$, a false statement;
(c) shade that side of the graph *not* containing the test point $(0, 0)$.
To graph $y < 4$ on the same axes,
(a) graph the dashed horizontal line through $(0, 4)$;
(b) test a point, such as $(0, 0)$, in $y < 4$ to get $0 < 4$, a true statement;
(c) shade that side of the dashed line containing the test point $(0, 0)$.
The word "and" indicates the intersection of the two graphs given above. The final solution set consists of the region where the two shaded regions overlap.

25. To graph $x - y \ge 1$,
(a) graph the solid line $x - y = 1$ through the intercepts $(1, 0)$ and $(0, -1)$;
(b) test a point not on this line, such as $(0, 0)$, in $x - y \ge 1$ to get $0 - 0 \ge 1$, a false statement;

(c) shade that side of the line *not* containing $(0, 0)$.
To graph $y \ge 2$ on the same axes,
(a) graph the solid horizontal line through $(0, 2)$;
(b) test a point, such as $(0, 0)$, in $y \ge 2$ to get $0 \ge 2$, a false statement;
(c) shade that side of the line *not* containing $(0, 0)$.
The word "or" indicates the union of the graphs given above. All shaded regions are included in the union (rather than considering only where the graphs overlap).

29. To graph $3x + 2y < 6$,
(a) draw a dashed line $3x + 2y = 6$ through the intercepts $(2, 0)$ and $(0, 3)$;
(b) test a point not on this line, such as $(0, 0)$, in $3x + 2y < 6$ to get $0 + 0 < 6$, a true statement;
(c) shade that side of the line containing $(0, 0)$.
To graph $x - 2y > 2$ on the same axes,
(a) draw a dashed line $x - 2y = 2$ through the intercepts $(2, 0)$ and $(0, -1)$;
(b) test a point not this line, such as $(0, 0)$, in $x - 2y > 2$ to get $0 - 0 > 2$, a false statement;
(c) shade that side of the line *not* containing $(0, 0)$.
The word "or" indicates the union of the two graphs given above. All shaded regions are included in the union.

33. Rewrite $|y + 1| < 2$ as $-2 < y + 1 < 2$. Subtract 1 from all parts to get $-3 < y < 1$.
Draw two horizontal dashed lines, $y = -3$ and $y = 1$, one through $(0, -3)$ and one through $(0, 1)$.
Since $-3 < y < 1$, the shaded portion will fall between the two dashed lines.

37. The slope of the line $y = -3x - 6$ is -3 and the y-intercept is $(0, -6)$. This boundary line is given in A and D. Since the inequality is $\le$, the graph includes the points below the line. Therefore, the graph is given in A.

SECTION 3.5 (page 201)

1. In the table, the three ordered pairs with the first entries 1990, 1992 and 1993 are $(1990, 746{,}220)$, $(1992, 661{,}391)$, $(1993, 590{,}324)$.

5. The vertical line test is used to determine whether a graph is that of a function. Any vertical line will intersect the graph of a function in at most one point.

9. In the relation $\{(2, 4), (0, 2), (2, 6)\}$, the x-value 2 has two y-values of 4 and 6. Therefore, this relation is not a function. The domain is the set of x-values $\{0, 2\}$. The range is the set of y-values $\{2, 4, 6\}$.

13. Using the vertical line test, we find any vertical line will cross the graph at most once. This indicates that the graph represents a function.
This graph extends indefinitely to the left $(-\infty)$ and indefinitely to the right (∞). Therefore, the domain is $(-\infty, \infty)$. This graph extends indefinitely downward $(-\infty)$, and reaches a high point at $y = 4$. Therefore, the range is $(-\infty, 4]$.

17. In $y = x^2$, each value of x corresponds to one y-value. For example, if $x = 3$, then $y = (3)^2 = 9$. Therefore, $y = x^2$ defines y as a function of x.
Since any x-value, positive, negative, or zero, can be squared, the domain is $(-\infty, \infty)$.

21. In $x + y < 4$, for a particular x-value, more than one y-value can be selected to satisfy $x + y < 4$. For example, if $x = 2$, $y = 0$, then

$$2 + 0 < 4, \text{ a true statement.}$$

Now, if $x = 2$, and $y = 1$,

$$2 + 1 < 4, \text{ also a true statement.}$$

Therefore, $x + y < 4$ does not define y as a function of x. The graph of $x + y < 4$ consists of the shaded region below the dashed line $x + y = 4$ which extends indefinitely from left to right. Therefore, the domain is $(-\infty, \infty)$.

25. Rewrite $xy = 1$ as $y = \frac{1}{x}$.
Note that x can never equal zero, otherwise the denominator would equal zero.
The domain is $(-\infty, 0) \cup (0, \infty)$.
Each nonzero x-value gives exactly one y-value.
Therefore, $xy = 1$ defines y as a function of x.

29. To determine the domain of $y = \sqrt{4x + 2}$, recall that

$$4x + 2 \geq 0$$

$$x \geq -\frac{1}{2}.$$

Therefore, the domain is $[-\frac{1}{2}, \infty)$.
Each x-value from the domain $[-\frac{1}{2}, \infty)$ produces exactly one y-value. Therefore, $y = \sqrt{4x + 2}$ defines a function.

33. If $f(x) = -3x + 4$, $f(0) = -3(0) + 4 = 4$.

37. If $f(x) = -3x + 4$,
$f(p) = -3p + 4$.

41. If $f(x) = -3x + 4$,
$f(x + 2) = -3(x + 2) + 4$
$= -3x - 6 + 4$
$= -3x - 2$.

45. No, in general $f(g(x)) \neq g(f(x))$. (However, there are cases where $f(g(x)) = g(f(x))$ such as when $f(x) = g(x)$. Another example is the functions $f(x) = x^3$ and $g(x) = \sqrt[3]{x}$.)

See the graphs in the answer section of your text for Exercises 49 and 53.

49. To graph the linear function $f(x) = -2x + 5$, replace $f(x)$ with y and then graph the equation $y = -2x + 5$. The graph of $y = -2x + 5$ is a line with slope -2 and y-intercept $(0, 5)$. The domain and range are both $(-\infty, \infty)$.

53. For every value of x, the corresponding y-value is 2 since $G(x) = 2$. The graph is a horizontal line with y-intercept $(0, 2)$. The domain is $(-\infty, \infty)$ and the range is $\{2\}$.

57. The graphing calculator screen shows that in 1986 (when $x = 2$), the number of post offices was approximately 29,439.

61. First find the slope of the line. Let $(x_1, y_1) = (-1, 8)$ and $(x_2, y_2) = (4, -7)$.

$$m = \frac{y_2 - y_1}{x_2 - x_1} = \frac{-7 - 8}{4 - (-1)} = \frac{-15}{5} = -3$$

Now use the point-slope form to find the equation of the line.

$y - y_1 = m(x - x_1)$	Point-slope form
$y - 8 = -3(x - (-1))$	Let $x_1 = -1, y_1 = 8, m = -3$
$y - 8 = -3(x + 1)$	
$y - 8 = -3x - 3$	Clear parentheses.
$y = -3x + 5$	Add 8.

Since $y = f(x)$ the equation describing this function is given by $f(x) = -3x + 5$.

SECTION 3.6 (page 213)

1. Since x varies directly as y,

$$x = ky, \text{ for some constant } k.$$

Substitute $x = 9$ and $y = 3$ into $x = ky$ to get

$$9 = k(3) \quad \text{or} \quad k = 3.$$

Substitute $k = 3$ and $y = 12$ into $x = ky$ to get

$$x = (3)(12) = 36.$$

5. Since p varies jointly as q and r^2, $p = kqr^2$, for some constant k. Given that $p = 200$ when $q = 2$ and $r = 3$, solve for k.

$$200 = k(2)(3)^2$$

$$200 = 18k$$

$$k = \frac{200}{18} = \frac{100}{9}$$

Using $k = \frac{100}{9}$, $q = 5$, and $r = 2$, find p.

$$p = \left(\frac{100}{9}\right)(5)(2)^2 = \frac{100}{9}(20) = \frac{2000}{9} = 222\frac{2}{9}.$$

9. Let $g =$ the number of gallons he bought and $P =$ the amount paid for the gasoline. Then P varies directly as g, and

$$P = kg.$$

Here k represents the price per gallon of gasoline. Since $P = 8.79$ when $g = 8$,

$$8.79 = 8k$$

$$k \approx 1.099$$

Therefore, the price per gallon of gasoline is about $\$1.09\frac{9}{10}$.

13. Let f represent the frequency and L the length of the string. Then

$$f = \frac{k}{L}$$

where k is the constant of variation. Find k by substituting 2 for L and 250 for f.

$$250 = \frac{k}{2}$$

$$k = 500 \quad \text{Multiply by 2.}$$

Now, substitute $k = 500$ and $L = 5$ in $f = \frac{k}{L}$ to get

$$f = \frac{500}{5} = 100 \text{ cycles per second.}$$

17. Let I represent the simple interest, P the principal, and t the time. Then

$$I = kPt \text{ for a given interest rate, } k.$$

Find k by substituting \$2000 for P, 4 for t, and 280 for I to get

$$280 = k(2000)(4)$$
$$280 = 8000k$$
$$k = \frac{280}{8000} = \frac{35}{1000}.$$

Now, substitute $k = \frac{35}{1000}$, $P = 2000$, and $t = 6$ to get

$$I = \frac{35}{1000}(2000)(6)$$
$$= \frac{35(12,000)}{1000} = 35(12) = 420.$$

\$420 interest would be earned in 6 years.

21. Let B represent the body-mass index (BMI), w the weight, and h the height. Then

$$B = \frac{kw}{h^2}.$$

In Example 7, k was determined to be 694. So, the equation we use is

$$B = \frac{694w}{h^2}.$$

Let $w = 205$ and $h = 75$ (6 feet, 3 inches = 75 in.).

$$B = \frac{694(205)}{75^2}$$
$$= \frac{142,270}{5625}$$
$$\approx 25$$

Ken's body-mass index is 25.

25. The two ordered pairs of the form (gallons, price) are

$$(0, 0), (1, 1.25).$$

29. The value of a from Exercise 28 is 1.25. This number is the price per gallon of gasoline and is the slope of the line.

CHAPTER 4

SECTION 4.1 (page 237)

1. If $(3, -6)$ is a solution of a linear system in two variables, then substituting 3 for x and substituting -6 for y leads to a true statement in *both* equations.

5. If the two lines forming a system have the same slope and different y-intercepts, the system has zero solutions. (The lines are parallel.)

9. Replace x with 5 and y with 2 in each equation of the system.

$$
\begin{array}{ll}
2x - y = 8 & 3x + 2y = 20 \\
2(5) - 2 = 8 \ \ ? & 3(5) + 2(2) = 20 \ \ ? \\
10 - 2 = 8 \ \ ? & 15 + 4 = 20 \ \ ? \\
\text{True} & \text{False}
\end{array}
$$

The ordered pair $(5, 2)$ is not a solution of the system since it does not make both equations true.

13.
$$2x - 5y = 11$$
$$3x + \ y = 8$$

$2x - 5y = 11$	Leave the first equation unchanged.
$15x + 5y = 40$	Multiply the second equation by 5.
$\overline{17x \qquad = 51}$	Add.
$x = 3$	Divide by 17.

Substitute $x = 3$ into $2x - 5y = 11$.

$6 - 5y = 11$	
$-5y = 5$	Subtract 6.
$y = -1$	Divide by -5.

Solution set: $\{(3, -1)\}$

17.
$$3x + 3y = 0$$
$$4x + 2y = 3$$

$6x + 6y = 0$	Multiply the first equation by 2.
$-12x - 6y = -9$	Multiply the second equation by -3.
$\overline{-6x \qquad = -9}$	Add.
$x = \dfrac{-9}{-6} = \dfrac{3}{2}$	Divide by -6.

Substitute $x = \frac{3}{2}$ into $3x + 3y = 0$.

$$3\left(\frac{3}{2}\right) + 3y = 0$$
$$\frac{9}{2} + 3y = 0$$

$3y = -\dfrac{9}{2}$	Subtract $\frac{9}{2}$.
$\dfrac{1}{3}(3y) = \dfrac{1}{3}\left(-\dfrac{9}{2}\right)$	Multiply by $\frac{1}{3}$.
$y = -\dfrac{3}{2}$	

Solution set: $\{(\frac{3}{2}, -\frac{3}{2})\}$

21.
$$\frac{x}{2} + \frac{y}{3} = -\frac{1}{3}$$
$$\frac{x}{2} + 2y = -7$$

Use the LCD 6 for both equations.

$-3x - \ 2y = \ \ 2$	Multiply the first equation by -6.
$3x + 12y = -42$	Multiply the second equation by 6.
$\overline{10y = -40}$	Add.
$y = -4$	Divide by 10.

Substitute $y = -4$ into $-3x - 2y = 2$.

$$-3x - 2(-4) = 2$$
$$-3x + 8 = 2$$
$$-3x = -6 \quad \text{Subtract 8.}$$
$$x = 2 \quad \text{Divide by } -3.$$

Solution set: $\{(2, -4)\}$

25. Write $3x + 7y = 4$ in $y = mx + b$ form.

$$7y = -3x + 4 \quad \text{Subtract } 3x.$$
$$y = -\frac{3}{7}x + \frac{4}{7} \quad \text{Divide by 7.}$$

Write $6x + 14y = 3$ in $y = mx + b$ form.

$$14y = -6x + 3 \quad \text{Subtract } 6x.$$
$$y = -\frac{6}{14}x + \frac{3}{14} \quad \text{Divide by 14.}$$
$$y = -\frac{3}{7}x + \frac{3}{14} \quad \begin{array}{l}\text{Write the fractions in}\\ \text{lowest terms.}\end{array}$$

Since the equations both have a slope of $-\frac{3}{7}$, but different y-intercepts, the graphs are parallel lines. The system has zero solutions.

29. $4x + y = 6$
$\quad\ \ y = 2x$

Substitute $2x$ for y in the first equation to get

$$4x + 2x = 6$$
$$6x = 6$$
$$x = 1.$$

Substitute 1 for x in $y = 2x$ to get

$$y = 2(1) = 2.$$

Solution set: $\{(1, 2)\}$

33. $5x - 4y = 9$
$\quad\ \ 3 - 2y = -x$
Solve the second equation for x.

$$3 - 2y = -x$$
$$2y - 3 = x \quad \text{Multiply by } -1.$$

Substitute $2y - 3$ for x in the first equation.

$$5(2y - 3) - 4y = 9$$
$$10y - 15 - 4y = 9 \quad \text{Clear parentheses.}$$
$$6y - 15 = 9 \quad \text{Combine terms.}$$
$$6y = 24 \quad \text{Add 15.}$$
$$y = 4 \quad \text{Divide by 6.}$$

Substitute 4 for y in $x = 2y - 3$.

$$x = 2(4) - 3 = 5$$

Solution set: $\{(5, 4)\}$

37. $\dfrac{1}{2}x + \dfrac{1}{3}y = 3$
$\qquad\qquad y = 3x$

Substitute $3x$ for y in the first equation.

$$\frac{1}{2}x + \frac{1}{3}(3x) = 3$$
$$\frac{1}{2}x + x = 3$$

Multiply by 2 to clear the fraction.

$$x + 2x = 6$$
$$3x = 6$$
$$x = 2 \quad \text{Divide by 3.}$$

Substitute 2 for x in the second equation.

$$y = 3(2) = 6$$

Solution set: $\{(2, 6)\}$

41. Solve for y:

$$-2x + 3y = 7$$
$$3y = 2x + 7 \quad \text{Add } 2x.$$
$$y = \frac{2}{3}x + \frac{7}{3}. \quad \text{Divide by 3.}$$

Rename it $g(x)$:

$$g(x) = \frac{2}{3}x + \frac{7}{3}.$$

g is a linear function since it is in the form $g(x) = mx + b$.

45. (a) The number of degrees in all fields for men and women reach equal numbers between the years 1978 and 1982 (since this is where the lines intersect).
(b) On the vertical axis, the lines intersect at just less than 500,000 bachelor's degrees.

49. (a) To find the equation of y_1, begin by finding the slope of the line. Choose any two points on the line. Let $(x_1, y_1) = (0, 4)$ and $(x_2, y_2) = (1, 8)$. Then

$$m = \frac{y_2 - y_1}{x_2 - x_1} = \frac{8 - 4}{1 - 0} = \frac{4}{1} = 4.$$

Since the line passes through the point $(0, 4)$, the y-intercept, we can use the slope-intercept form of an equation, $y_1 = mx + b$, where $m = 4$ and $b = 4$. So,

$$y_1 = 4x + 4.$$

(b) To find the equation of y_2, first find the slope. Let $(x_1, y_1) = (0, 7)$ and $(x_2, y_2) = (1, 5)$. Then

$$m = \frac{y_2 - y_1}{x_2 - x_1} = \frac{5 - 7}{1 - 0} = \frac{-2}{1} = -2.$$

Again, we can use the slope-intercept form of an equation, $y_2 = mx + b$, where $m = -2$ and $b = 7$. So,

$$y_2 = -2x + 7.$$

(c) Solve the system by substitution.

$$y = 4x + 4 \quad (1)$$
$$y = -2x + 7 \quad (2)$$

SOLUTIONS

Substitute $-2x + 7$ for y in (1).

$$-2x + 7 = 4x + 4$$

$$3 = 6x \qquad \text{Add } 2x; \text{ subtract 4.}$$

$$\frac{1}{2} = x \qquad \text{Divide by 6 and reduce.}$$

Let $x = \frac{1}{2}$ in (2) and solve for y.

$$y = -2\left(\frac{1}{2}\right) + 7$$

$$y = -1 + 7$$

$$y = 6$$

The solution set is $\{(\frac{1}{2}, 6)\}$.

SECTION 4.2 (page 249)

1. The statement means that when -1 is substituted for x, 2 is substituted for y, and 3 is substituted for z in the three equations, the resulting three statements are true.

5.
$$2x + 5y + 2z = 0 \quad (1)$$
$$4x - 7y - 3z = 1 \quad (2)$$
$$3x - 8y - 2z = -6 \quad (3)$$

Add equations (1) and (3) to eliminate z.

$$
\begin{array}{l}
2x + 5y + 2z = 0 \quad (1) \\
\underline{3x - 8y - 2z = -6} \quad (3) \\
5x - 3y \phantom{{}+ 2z} = -6 \quad (4)
\end{array}
$$

Multiply equation (1) by 3 and multiply equation (2) by 2. Then add the results to eliminate z again.

$$
\begin{array}{l}
6x + 15y + 6z = 0 \\
\underline{8x - 14y - 6z = 2} \\
14x + y \phantom{{}+ 6z} = 2 \quad (5)
\end{array}
$$

Solve the system
$$5x - 3y = -6 \quad (4)$$
$$14x + y = 2. \quad (5)$$

Multiply equation (5) by 3 then add this result to (4).

$$
\begin{array}{l}
5x - 3y = -6 \\
\underline{42x + 3y = 6} \\
47x \phantom{{}+ 3y} = 0
\end{array}
$$

$$x = 0$$

Substitute $x = 0$ into $5x - 3y = -6$ to get $y = 2$. Substitute $x = 0$ and $y = 2$ into $2x + 5y + 2z = 0$ to get $2(0) + 5(2) + 2z = 0$. Solve this last equation to get $z = -5$.
Solution set: $\{(0, 2, -5)\}$

9.
$$2x - 3y + 2z = -1 \quad (1)$$
$$x + 2y + z = 17 \quad (2)$$
$$2y - z = 7 \quad (3)$$

Multiply equation (2) by -2, then add to equation (1).

$$
\begin{array}{l}
2x - 3y + 2z = -1 \\
\underline{-2x - 4y - 2z = -34} \\
-7y \phantom{{}+ 2z} = -35
\end{array}
$$

$$y = 5 \qquad \text{Divide by } -7.$$

Substitute 5 for y in equation (3).

$$2(5) - z = 7$$

$$-z = -3 \qquad \text{Subtract 10.}$$

$$z = 3 \qquad \text{Multiply by } -1.$$

Substitute $y = 5$ and $z = 3$ into equation (1).

$$2x - 3(5) + 2(3) = -1$$

$$2x - 9 = -1 \qquad \text{Combine terms.}$$

$$2x = 8 \qquad \text{Add 9.}$$

$$x = 4 \qquad \text{Divide by 2.}$$

Solution set: $\{(4, 5, 3)\}$

13.
$$2x + y \phantom{{}- 2z} = 6 \quad (1)$$
$$3y - 2z = -4 \quad (2)$$
$$3x \phantom{{}+ y} - 5z = -7 \quad (3)$$

To eliminate y, multiply equation (1) by -3, then add to equation (2).

$$
\begin{array}{l}
-6x - 3y \phantom{{}- 2z} = -18 \\
\underline{ 3y - 2z = -4} \\
-6x \phantom{{}- 3y} - 2z = -22 \quad (4)
\end{array}
$$

Multiply equation (3) by 2 then add to equation (4).

$$
\begin{array}{l}
6x - 10z = -14 \\
\underline{-6x - 2z = -22} \\
-12z = -36
\end{array}
$$

$$z = 3 \qquad \text{Divide by } -12.$$

Substitute $z = 3$ into equation (3).

$$3x - 5(3) = -7$$

$$3x = 8 \qquad \text{Add 15.}$$

$$x = \frac{8}{3} \qquad \text{Divide by 3.}$$

Substitute $z = 3$ into equation (2).

$$3y - 2(3) = -4$$

$$3y = 2 \qquad \text{Add 6.}$$

$$y = \frac{2}{3} \qquad \text{Divide by 3.}$$

Solution set: $\{(\frac{8}{3}, \frac{2}{3}, 3)\}$

17.
$$2x + 2y - 6z = 5 \quad (1)$$
$$-3x + y - z = -2 \quad (2)$$
$$-x - y + 3z = 4 \quad (3)$$

Multiply equation (3) by 2 then add to equation (1).

$$
\begin{array}{l}
2x + 2y - 6z = 5 \\
\underline{-2x - 2y + 6z = 8} \\
0 = 13, \text{ a false statement}
\end{array}
$$

Solution set: $\emptyset$

21.
$$2x + y - z = 6 \quad (1)$$
$$4x + 2y - 2z = 12 \quad (2)$$
$$-x - \frac{1}{2}y + \frac{1}{2}z = -3 \quad (3)$$

Multiplying equation (1) by 2 gives equation (2). Multiplying equation (3) by -4 also results in equation (2). Because of this, the three equations are dependent. Solution set: $\{(x, y, z) \mid 2x + y - z = 6\}$

25. Let $x = 1$ and $f(x) = 128$ and simplify.

$$f(x) = ax^2 + bx + c$$
$$128 = a(1)^2 + b(1) + c \quad \text{Substitute.}$$
$$128 = a + b + c$$

So, one equation is $128 = a + b + c$.

29. In Exercise 28, the solution set of the system was $\{(-32, 104, 56)\}$. The function f for this particular projectile is of the form $f(x) = ax^2 + bx + c$. Now substitute $a = -32$, $b = 104$, and $c = 56$ to get

$$f(x) = -32x^2 + 104x + 56.$$

SECTION 4.3 (page 259)

1. Let x = the number of daily newspapers in Texas and y = the number of daily newspapers in Florida. Since the two states had a total number of 134 dailies,

$$x + y = 134. \quad (1)$$

Since Texas had 52 more daily newspapers than Florida,

$$x = y + 52. \quad (2)$$

Substitute $y + 52$ for x in equation (1).

$$(y + 52) + y = 134$$
$$2y + 52 = 134$$
$$2y = 82$$
$$y = 41$$

Substitute $y = 41$ into equation (2).

$$x = 41 + 52$$
$$x = 93$$

Texas had 93 daily newspapers and Florida had 41.

5. Let W = the width and L = the length of the rectangle. Since the length is 42 feet more than the width,

$$L = W + 42. \quad (1)$$

The perimeter of a rectangle is given by

$$2W + 2L = P.$$

With perimeter $P = 228$ feet,

$$2W + 2L = 228. \quad (2)$$

Substitute $W + 42$ for L in equation (2).

$$2W + 2(W + 42) = 228$$
$$2W + 2W + 84 = 228$$
$$4W + 84 = 228$$
$$4W = 144$$
$$W = 36$$

Substitute $W = 36$ into equation (1).

$$L = 36 + 42 = 78$$

The length is 78 feet; the width is 36 feet.

9. Let x = the number of units of yarn, and
  y = the number of units of thread.
Make a chart to organize the information in the problem.

	Yarn	Thread	Hours
Hours on Machine A	1	1	8
Hours on Machine B	2	1	14

From the chart, write a system of equations.

$$x + y = 8 \quad (1)$$
$$2x + y = 14 \quad (2)$$

Solve the system.

$$
\begin{array}{ll}
-x - y = -8 & \text{Multiply equation (1) by } -1. \\
\underline{2x + y = 14} & \text{Equation (2) unchanged} \\
x \quad\quad = 6 & \text{Add.}
\end{array}
$$

Substitute $x = 6$ into equation (1) to get $y = 2$. The factory should make 6 units of yarn and 2 units of thread to keep its machines running at capacity.

13. Use the formula (rate or percent) · (base amount) = amount (percentage) of pure acid to compute parts (a)–(d).
  (a) $(.10)(32) = 3.2$ ounces
  (b) $(.25)(32) = 8$ ounces
  (c) $(.40)(32) = 12.8$ ounces
  (d) $(.50)(32) = 16$ ounces

17. Let x = the amount of 25% alcohol solution, and
  y = the amount of 35% alcohol solution.
Complete the table in your text. The percent times the amount of solution gives the amount of pure alcohol in the third column.

Kind	Amount	Pure Alcohol
25%	x	$.25x$
35%	y	$.35y$
32%	20	$(.32)(20) = 6.4$

The third row gives the total amounts of solution and pure alcohol. From the table, write a system of equations.

$$x + y = 20 \quad (1)$$
$$.25x + .35y = 6.4 \quad (2)$$

Solve the system.

$$
\begin{array}{ll}
-25x - 25y = -500 & \text{Multiply (1) by } -25. \\
\underline{25x + 35y = \quad 640} & \text{Multiply (2) by 100.} \\
10y = \quad 140 & \text{Add.}
\end{array}
$$

$$y = 14$$

Substitute $y = 14$ into equation (1).

$$x + 14 = 20 \quad \text{so} \quad x = 6$$

Six gallons of 25% solution and fourteen gallons of 35% solution should be used.

21. Let x = the amount of nuts and y = the amount of cereal. From the table in your text, write a system of equations.

Total amount of ingredients: $x + y = 30$ (1)

Total value of ingredients: $2.50x + 1.00y$
$$= 1.70(30) = 51 \ (2)$$

Solve the system.

$$
\begin{array}{ll}
-25x - 25y = -750 & \text{Multiply (1) by } -25. \\
\underline{25x + 10y = 510} & \text{Multiply (2) by 10.} \\
{-15y} = -240 & \text{Add.} \\
y = 16 & \text{Divide by } -15.
\end{array}
$$

Substitute $y = 16$ into equation (1).

$$x + 16 = 30$$
$$x = 14$$

The party mix requires 14 kilograms of nuts and 16 kilograms of cereal.

25. Let x = the number of dimes and
y = the number of quarters.
Make a table using the information in the problem.

Coin	Amount	Value
Dimes ($.10)	x	$.10x$
Quarters ($.25)	y	$.25y$
Totals	94	19.30

From the table, write a system of equations.
Total amount of coins: $x + y = 94$ (1)
Total value of coins: $.10x + .25y = 19.30$ (2)
Solve the system.

$$
\begin{array}{ll}
-10x - 10y = -940 & \text{Multiply (1) by } -10. \\
\underline{10x + 25y = 1930} & \text{Multiply (2) by 100.} \\
15y = 990 & \text{Add.} \\
y = 66 & \text{Divide by 15.}
\end{array}
$$

Substitute $y = 66$ into equation (1).

$$x + 66 = 94$$
$$x = 28$$

He has 28 dimes and 66 quarters.

29. In the formula $d = rt$, substitute 25 for r (rate or speed) and y for t (time in hours) to get $d = 25y$ miles.

33. Let x = the top speed of the snow speeder, and
y = the speed of the wind.
Furthermore,

$$\text{rate into headwind} = x - y$$

and $\quad$ rate with tailwind $= x + y$.

Use these rates and the information in the problem to complete a table.

	d	r	t
Into headwind	3600	$x - y$	2
With tailwind	3600	$x + y$	1.5

From the table and using the formula $d = rt$, write a system of equations.

$$3600 = 2(x - y)$$
$$3600 = 1.5(x + y)$$

Remove parentheses and move the variables to the left side.

$$2x - 2y = 3600 \quad (1)$$
$$1.5x + 1.5y = 3600 \quad (2)$$

Solve the system.

$$
\begin{array}{ll}
6x - 6y = 10{,}800 & \text{Multiply (1) by 3.} \\
\underline{6x + 6y = 14{,}400} & \text{Multiply (2) by 4.} \\
12x = 25{,}200 & \text{Add.} \\
x = 2100 & \text{Divide by 12.}
\end{array}
$$

Substitute $x = 2100$ into equation (1).

$$
\begin{array}{ll}
2(2100) - 2y = 3600 & \\
4200 - 2y = 3600 & \text{Multiply.} \\
-2y = -600 & \text{Subtract 4200.} \\
y = 300 & \text{Divide by } -2.
\end{array}
$$

The top speed of the snow speeder is 2100 miles per hour and the speed of the wind is 300 miles per hour.

37. Let x = the measure of the first angle,
y = the measure of the second angle, and
z = the measure of the third angle.
Write a system of equations from the information in the problem.
The sum of the angles in a triangle equals 180°, so

$$x + y + z = 180. \quad (1)$$

The measure of the second angle is 10° more than 3 times that of the first angle, so

$$y = 3x + 10. \quad (2)$$

The third angle is equal to the sum of the other two, so

$$z = x + y. \quad (3)$$

Solve the system. Substitute z for $x + y$ in equation (1).

$$z + z = 180$$
$$2z = 180$$
$$z = 90$$

Substitute $z = 90$ and $3x + 10$ for y in equation (3).

$$90 = x + 3x + 10$$
$$90 = 4x + 10$$
$$80 = 4x$$
$$20 = x$$

Substitute $x = 20$ and $z = 90$ into equation (3).

$$90 = 20 + y$$
$$70 = y$$

The three angles have measures of 20°, 70°, and 90°.

41. Let x = the number of cases sent to wholesaler A,
y = the number of cases sent to wholesaler B, and
z = the number of cases sent to wholesaler C.

The total output is 320 cases per day, so

$$x + y + z = 320. \quad (1)$$

The number of cases to A is three times that sent to B, so

$$x = 3y. \quad (2)$$

Wholesaler C gets 160 cases less than the sum sent to A and B, so

$$z = x + y - 160. \quad (3)$$

Solve equation (3) for $x + y$ to get

$$x + y = z + 160. \quad (4)$$

From (4), substitute $z + 160$ for $x + y$ in equation (1).

$$z + 160 + z = 320$$

$$2z = 160 \quad \text{Combine terms; subtract 160.}$$

$$z = 80 \quad \text{Divide by 2.}$$

With $z = 80$, from (2) substitute $3y$ for x in (3).

$$80 = 3y + y - 160$$

$$240 = 4y \quad \text{Combine terms; add 160.}$$

$$60 = y \quad \text{Divide by 4.}$$

Substitute $y = 60$ into (2).

$$x = 3(60) = 180$$

She should send 180 cases to Wholesaler A, 60 cases to Wholesaler B, and 80 cases to Wholesaler C.

45. In the equation $x^2 + y^2 + ax + by + c = 0$, let $x = 3$ and $y = 3$.

$$3^2 + 3^2 + a(3) + b(3) + c = 0$$

$$9 + 9 + 3a + 3b + c = 0$$

$$18 + 3a + 3b + c = 0$$

$$3a + 3b + c = -18$$

SECTION 4.4 (page 269)

1. It is true that a matrix is an array of numbers, while a determinant is just a number.

5. This statement is false. The value of $\begin{vmatrix} a & a \\ a & a \end{vmatrix}$ for *any* replacement of a is 0. For example, if $a = 2$,

$$\begin{vmatrix} 2 & 2 \\ 2 & 2 \end{vmatrix} = (2)(2) - (2)(2) = 0.$$

9. $\begin{vmatrix} 1 & -2 \\ 7 & 0 \end{vmatrix} = 1 \cdot 0 - (-2)(7) = 0 + 14 = 14$

13. $\begin{vmatrix} -1 & 2 & 4 \\ -3 & -2 & -3 \\ 2 & -1 & 5 \end{vmatrix}$

Expand by minors about the first column.

$$= -1\begin{vmatrix} -2 & -3 \\ -1 & 5 \end{vmatrix} - (-3)\begin{vmatrix} 2 & 4 \\ -1 & 5 \end{vmatrix} + 2\begin{vmatrix} 2 & 4 \\ -2 & -3 \end{vmatrix}$$

$$= -1[-2(5) - (-1)(-3)]$$
$$+ 3[2(5) - (-1)4] + 2[2(-3) - (-2)4]$$

$$= -1(-13) + 3(14) + 2(2)$$

$$= 13 + 42 + 4 = 59$$

17. $\begin{vmatrix} 1 & 0 & 0 \\ 0 & 1 & 0 \\ 0 & 0 & 1 \end{vmatrix} = 1\begin{vmatrix} 1 & 0 \\ 0 & 1 \end{vmatrix} - 0 + 0$ Expand by minors about the first column.

$$= 1[(1)(1) - (0)(0)] = 1$$

21. $\begin{vmatrix} 2 & 0 & 1 \\ -1 & 0 & 2 \\ 5 & 0 & 4 \end{vmatrix}$ Expand about column 2.

$$= -0\begin{vmatrix} -1 & 2 \\ 5 & 4 \end{vmatrix} + 0\begin{vmatrix} 2 & 1 \\ 5 & 4 \end{vmatrix} - 0\begin{vmatrix} 2 & 1 \\ -1 & 2 \end{vmatrix} = 0$$

25. $\begin{vmatrix} 3 & 5 & -2 \\ 1 & -4 & 1 \\ 3 & 1 & -2 \end{vmatrix}$ Expand about column 1.

$$= 3\begin{vmatrix} -4 & 1 \\ 1 & -2 \end{vmatrix} - 1\begin{vmatrix} 5 & -2 \\ 1 & -2 \end{vmatrix} + 3\begin{vmatrix} 5 & -2 \\ -4 & 1 \end{vmatrix}$$

$$= 3[(-4)(-2) - 1(1)] - 1[5(-2) - (-2)(1)]$$
$$+ 3[5(1) - (-2)(-4)]$$

$$= 3(7) - 1(-8) + 3(-3)$$

$$= 21 + 8 - 9 = 20$$

29. $\begin{vmatrix} 4 & 1 & 0 & 2 \\ 1 & 0 & 0 & -2 \\ 3 & 4 & 1 & -3 \\ -2 & 1 & 1 & -1 \end{vmatrix}$ Expand about column 3.

$$= 0 - 0 + 1\begin{vmatrix} 4 & 1 & 2 \\ 1 & 0 & -2 \\ -2 & 1 & -1 \end{vmatrix} - 1\begin{vmatrix} 4 & 1 & 2 \\ 1 & 0 & -2 \\ 3 & 4 & -3 \end{vmatrix}$$

$$\begin{vmatrix} 4 & 1 & 2 \\ 1 & 0 & -2 \\ -2 & 1 & -1 \end{vmatrix}$$ Expand about column 2.

$$= -1\begin{vmatrix} 1 & -2 \\ -2 & -1 \end{vmatrix} + 0 - 1\begin{vmatrix} 4 & 2 \\ 1 & -2 \end{vmatrix}$$

$$= -1(-1 - 4) + 0 - 1(-8 - 2)$$

$$= 5 + 10 = 15$$

$$\begin{vmatrix} 4 & 1 & 2 \\ 1 & 0 & -2 \\ 3 & 4 & -3 \end{vmatrix}$$ Expand about row 2.

$$= -1\begin{vmatrix} 1 & 2 \\ 4 & -3 \end{vmatrix} + 0 - (-2)\begin{vmatrix} 4 & 1 \\ 3 & 4 \end{vmatrix}$$

$$= -1(-3 - 8) + 0 + 2(16 - 3)$$

$$= 11 + 26 = 37$$

Thus, the given determinant is $1(15) - 1(37) = -22$.

33. The expression for the slope of a line passing through the points (x_1, y_1) and (x_2, y_2) is

$$\frac{y_2 - y_1}{x_2 - x_1}.$$

37. The calculator screen is displaying the

matrix $\begin{bmatrix} 3 & -5 \\ 8 & 4 \end{bmatrix}$.

$$\begin{vmatrix} 3 & -5 \\ 8 & 4 \end{vmatrix} = 12 - (-40) = 52$$

SECTION 4.5 (page 277)

1. $x = \dfrac{D_x}{D} = \dfrac{112}{28} = 4$ and $y = \dfrac{D_y}{D} = \dfrac{168}{28} = 6$

Solution set: $\{(4, 6)\}$

5. $3x + 2y = 3$
$\quad 2x - 4y = 2$

$D = \begin{vmatrix} 3 & 2 \\ 2 & -4 \end{vmatrix} = -12 - 4 = -16$

$D_x = \begin{vmatrix} 3 & 2 \\ 2 & -4 \end{vmatrix} = -12 - 4 = -16$

$D_y = \begin{vmatrix} 3 & 3 \\ 2 & 2 \end{vmatrix} = 6 - 6 = 0$

$x = \dfrac{D_x}{D} = \dfrac{-16}{-16} = 1$ and $y = \dfrac{D_y}{D} = \dfrac{0}{-16} = 0$

Solution set: $\{(1, 0)\}$

9. $2x + 3y + 2z = 15$
$\quad x - y + 2z = 5$
$\quad x + 2y - 6z = -26$

$D = \begin{vmatrix} 2 & 3 & 2 \\ 1 & -1 & 2 \\ 1 & 2 & -6 \end{vmatrix}$ Expand about row 1.

$= 2\begin{vmatrix} -1 & 2 \\ 2 & -6 \end{vmatrix} - 3\begin{vmatrix} 1 & 2 \\ 1 & -6 \end{vmatrix} + 2\begin{vmatrix} 1 & -1 \\ 1 & 2 \end{vmatrix}$

$D = 2(2) - 3(-8) + 2(3) = 34$

$D_x = \begin{vmatrix} 15 & 3 & 2 \\ 5 & -1 & 2 \\ -26 & 2 & -6 \end{vmatrix}$ Expand about row 2.

$= -5\begin{vmatrix} 3 & 2 \\ 2 & -6 \end{vmatrix} - 1\begin{vmatrix} 15 & 2 \\ -26 & -6 \end{vmatrix} - 2\begin{vmatrix} 15 & 3 \\ -26 & 2 \end{vmatrix}$

$= -5(-22) - (-38) - 2(108)$

$D_x = 110 + 38 - 216 = -68$

$x = \dfrac{D_x}{D} = \dfrac{-68}{34} = -2$

$D_y = \begin{vmatrix} 2 & 15 & 2 \\ 1 & 5 & 2 \\ 1 & -26 & -6 \end{vmatrix}$ Expand about column 1.

$= 2\begin{vmatrix} 5 & 2 \\ -26 & -6 \end{vmatrix} - \begin{vmatrix} 15 & 2 \\ -26 & -6 \end{vmatrix} + \begin{vmatrix} 15 & 2 \\ 5 & 2 \end{vmatrix}$

$= 2(22) - (-38) + (20)$

$D_y = 44 + 38 + 20 = 102$

$y = \dfrac{D_y}{D} = \dfrac{102}{34} = 3$

Let $x = -2$ and $y = 3$ in the second equation.

$$x - y + 2z = 5$$
$$-2 - 3 + 2z = 5$$
$$2z = 10 \quad \text{Add 5.}$$
$$z = 5 \quad \text{Divide by 2.}$$

Solution set: $\{(-2, 3, 5)\}$

13. $2x - 3y + 4z = 8$
$\quad 6x - 9y + 12z = 24$
$\quad -4x + 6y - 8z = -16$

$D = \begin{vmatrix} 2 & -3 & 4 \\ 6 & -9 & 12 \\ -4 & 6 & -8 \end{vmatrix}$ Expand about row 1.

$= 2\begin{vmatrix} -9 & 12 \\ 6 & -8 \end{vmatrix} + 3\begin{vmatrix} 6 & 12 \\ -4 & -8 \end{vmatrix} + 4\begin{vmatrix} 6 & -9 \\ -4 & 6 \end{vmatrix}$

$D = 2(0) + 3(0) + 4(0) = 0$

Because $D = 0$, Cramer's rule does not apply.

17. $\quad x - 3y \quad\quad = 13$
$\quad\quad\quad 2y + z = 5$
$\quad -x \quad\quad + z = -7$

$D = \begin{vmatrix} 1 & -3 & 0 \\ 0 & 2 & 1 \\ -1 & 0 & 1 \end{vmatrix}$ Expand about column 1.

$= 1\begin{vmatrix} 2 & 1 \\ 0 & 1 \end{vmatrix} - 0 - 1\begin{vmatrix} -3 & 0 \\ 2 & 1 \end{vmatrix}$

$D = 2 - 0 - (-3) = 5$

$D_x = \begin{vmatrix} 13 & -3 & 0 \\ 5 & 2 & 1 \\ -7 & 0 & 1 \end{vmatrix}$ Expand about column 3.

$= 0 - 1\begin{vmatrix} 13 & -3 \\ -7 & 0 \end{vmatrix} + 1\begin{vmatrix} 13 & -3 \\ 5 & 2 \end{vmatrix}$

$= 0 - (-21) + (26 + 15)$

$D_x = 21 + 41 = 62$

$x = \dfrac{D_x}{D} = \dfrac{62}{5}$

$D_y = \begin{vmatrix} 1 & 13 & 0 \\ 0 & 5 & 1 \\ -1 & -7 & 1 \end{vmatrix}$ Expand about column 1.

$= 1\begin{vmatrix} 5 & 1 \\ -7 & 1 \end{vmatrix} - 0 - 1\begin{vmatrix} 13 & 0 \\ 5 & 1 \end{vmatrix}$

$D_y = 12 - 0 - 13 = -1$

$y = \dfrac{D_y}{D} = -\dfrac{1}{5}$

Let $y = -\frac{1}{5}$ in the second equation.

$$2\left(-\frac{1}{5}\right) + z = 5$$

$$-\frac{2}{5} + z = 5$$

$$-2 + 5z = 25 \quad \text{Multiply by 5.}$$

$$5z = 27 \quad \text{Add 2.}$$

$$z = \frac{27}{5} \quad \text{Divide by 5.}$$

Solution set: $\{(\frac{62}{5}, -\frac{1}{5}, \frac{27}{5})\}$

21. Answers will vary for the systems. One example follows.

$\quad x + 2y = 3$
$\quad 4x + 5y = 6$

$D = \begin{vmatrix} 1 & 2 \\ 4 & 5 \end{vmatrix} = 5 - 8 = -3$

$$D_x = \begin{vmatrix} 3 & 2 \\ 6 & 5 \end{vmatrix} = 15 - 12 = 3$$

$$D_y = \begin{vmatrix} 1 & 3 \\ 4 & 6 \end{vmatrix} = 6 - 12 = -6$$

$$x = \frac{D_x}{D} = \frac{3}{-3} = -1 \quad \text{and} \quad y = \frac{D_y}{D} = \frac{-6}{-3} = 2$$

Solution set: $\{(-1, 2)\}$
In all cases, the solution set is $\{(-1, 2)\}$.

CHAPTER 5

SECTION 5.1 (page 297)

1. To use the product rule or the quotient rule with two exponential expressions, the bases must be the same.

5. The exponent is 4 and the base is -5, since -5 is in parentheses.

9. $25^0 = 1$ since $x^0 = 1$ for any nonzero base x.

13. $(-15)^0 = 1$ since -15 is in parentheses.

17. $y^{-2} = \dfrac{1}{y^2}$ A negative exponent leads to a reciprocal.

21. $\dfrac{1}{x^{-2}} = \dfrac{1}{\frac{1}{x^2}} = x^2$

25. $\dfrac{1}{x^3} = x^{-3}$ A reciprocal can be rewritten using a negative exponent.

29. Let $a = 2$ and $n = 2$.

$$-2^2 = -4 \qquad (-2)^2 = 4 \quad \text{Opposites}$$

Let $a = 2$ and $n = 3$.

$$-2^3 = -8 \qquad (-2)^3 = -8 \quad \text{Equal}$$

Let $a = 2$ and $n = 4$.

$$-2^4 = -16 \qquad (-2)^4 = 16 \quad \text{Opposites}$$

Let $a = 2$ and $n = 5$.

$$-2^5 = -32 \qquad (-2)^5 = -32 \quad \text{Equal}$$

So, when n is odd, $-a^n = (-a)^n$ and when n is even, $-a^n$ is the opposite of $(-a)^n$.

33. $4^{-3} = \dfrac{1}{4^3} = \dfrac{1}{64}$ $\quad 4^3 = 4 \cdot 4 \cdot 4 = 64$

37. $(-4)^{-3} = \dfrac{1}{(-4)^3} = \dfrac{1}{-64} = -\dfrac{1}{64}$

41. $\dfrac{-3^{-1}}{4^{-2}} = \dfrac{-4^2}{3^1} = \dfrac{-16}{3} = -\dfrac{16}{3}$

45. $3^{-1} + 2^{-1} = \dfrac{1}{3} + \dfrac{1}{2} = \dfrac{2}{6} + \dfrac{3}{6} = \dfrac{5}{6}$

49. $\dfrac{3^5}{3^2} = 3^{5-2} = 3^3$ Quotient rule

53. $\dfrac{9^{-1}}{9} = \dfrac{9^{-1}}{9^1} = 9^{-1-1} = 9^{-2} = \dfrac{1}{9^2}$ Quotient rule

57. $r^4 r = r^4 r^1 = r^{4+1} = r^5$ Product rule

61. $\dfrac{x^7}{x^{-4}} = x^{7-(-4)} = x^{7+4} = x^{11}$

65. $7k^2(-2k)(4k^{-5}) = (7)(-2)(4)(k^2 \cdot k^1 \cdot k^{-5})$ Regroup factors.

$$= -56k^{2+1+(-5)}$$

Multiply; use Product rule.

$$= -56k^{-2} \quad \text{Simplify.}$$

$$= -\frac{56}{k^2} \quad \text{Write with a positive exponent.}$$

69. $(2 + 3)^2 = 5^2 = 25;$ $2^2 + 3^2 = 4 + 9 = 13;$
$25 \neq 13$.

SECTION 5.2 (page 305)

1. This statement is false.

$$(5^2)^3 = 5^{(2)(3)} = 5^6 \quad \text{and}$$
$$5^2 \cdot 5^3 = 5^{2+3} = 5^5$$

5. (a) Incorrect: $\left(\dfrac{3}{4}\right)^{-1} = \left(\dfrac{4}{3}\right)^1$ not $-\dfrac{3}{4}$

(b) Incorrect: $\dfrac{3^{-1}}{4^{-1}} = \dfrac{4^1}{3^1} = \left(\dfrac{4}{3}\right)^1$ not $\left(\dfrac{4}{3}\right)^{-1}$

(c) Incorrect: $\dfrac{3^{-1}}{4} = \dfrac{1}{3 \cdot 4} = \dfrac{1}{12}$ and $\dfrac{3}{4^{-1}} = 3 \cdot 4 = 12$

(d) Correct: $\dfrac{3^{-1}}{4^{-1}} = \dfrac{4}{3}$ and $\left(\dfrac{3}{4}\right)^{-1} = \dfrac{4}{3}$

9. Correct: $(5x)^3 = 5^3 x^3$

13. Incorrect: $(z^4)^5 = z^{4 \cdot 5} = z^{20}$ not z^9

17. $\left(\dfrac{6}{5}\right)^{-1} = \left(\dfrac{5}{6}\right)^1 = \dfrac{5}{6}$

21. $(5^{-4} \cdot 6^{-2})^{-3} = 5^{-4(-3)} \cdot 6^{-2(-3)}$ Power rule
$$= 5^{12} \cdot 6^6$$

25. $-4r^{-2}(r^4)^2 = -4 \cdot r^{-2} \cdot r^8$ Power rule
$$= -4r^{-2+8} \quad \text{Product rule}$$
$$= -4r^6$$

29. $(z^{-4}x^3)^{-1} = z^{-4(-1)}x^{3(-1)}$ Power rule
$$= z^4 x^{-3}$$
$$= \dfrac{z^4}{x^3} \quad \text{Write answer with a positive exponent.}$$

33. $\dfrac{4a^5(a^{-1})^3}{(a^{-2})^{-2}} = \dfrac{4 \cdot a^5 \cdot a^{-3}}{a^4}$ Power rule
$$= \dfrac{4a^{5+(-3)}}{a^4} \quad \text{Product rule}$$
$$= \dfrac{4a^2}{a^4}$$
$$= 4a^{2-4} \quad \text{Quotient rule}$$
$$= 4a^{-2}$$
$$= \dfrac{4}{a^2} \quad \text{Write answer with a positive exponent.}$$

37. $\dfrac{(2k)^2 m^{-5}}{(km)^{-3}} = \dfrac{2^2 k^2 m^{-5}}{k^{-3} m^{-3}}$ Power rule
$$= 2^2 k^{2-(-3)} m^{-5-(-3)} \quad \text{Quotient rule}$$
$$= 2^2 k^5 m^{-2}$$
$$= \dfrac{2^2 k^5}{m^2} \quad \text{Write answer with a positive exponent.}$$

41. $530 = 5_\wedge 30.$ Place the caret to the right of 5, count two places to the decimal point. Since 5.30 is less than 530, the exponent on base ten is 2, rather than -2. So

$$530 = 5.3 \times 10^2$$

SOLUTIONS

45. .000006$_\wedge$92   Place the caret to the right of 6, count six places to the decimal point. Since 6.92 is larger than .00000692, the exponent on base ten is -6, rather than 6. So .00000692 $= 6.92 \times 10^{-6}$.

49. $7.2 \times 10^4 = 72{,}000$ Move the decimal point 4 places to the right because of the positive exponent.

53. $-6 \times 10^4 = -60{,}000$ Move the decimal point 4 places to the right because of the positive exponent.

57. $\dfrac{3 \times 10^{-2}}{12 \times 10^3} = \dfrac{3}{12} \times \dfrac{10^{-2}}{10^3} = .25 \times 10^{-2-3}$

$\qquad\qquad\qquad\qquad = .25 \times 10^{-5}$

$\qquad\qquad\qquad\qquad = .0000025$

61. Use the formula: time $= \dfrac{\text{distance}}{\text{rate}}$

with distance $= (6.7 \times 10^7) - (3.6 \times 10^7)$

$\qquad\qquad\quad = (6.7 - 3.6) \times 10^7$ Distributive property

$\qquad\qquad\quad = 3.1 \times 10^7$ miles

and rate $= 1.55 \times 10^3$ hours.

So, time $= \dfrac{3.1 \times 10^7}{1.55 \times 10^3} = 2 \times 10^4$ (20,000) hours.

65. $7 \times 10^9 = 7{,}000{,}000{,}000$ or 7 billion

According to the graph, in 1987 approximately 7×10^9 calls were made on the AT&T System.

SECTION 5.3 (page 315)

1. (c) $5m^{-2} + m^2$ is not a polynomial because one of the exponents on the variable m is negative.

5. $7z = 7z^1$ so the coefficient is 7 and the degree is 1.

9. $x^4 = 1x^4$ so the coefficient is 1 and the degree is 4.

13. The degree of 4^5 is not 5 because $4^5 = 1024$. 1024 is a constant, so it has a degree of 0.

17. The polynomial is in ascending powers, since the exponents are increasing from left to right.

21. Write 24 as $24x^0$. This is a monomial of degree 0.

25. $2r^3 + 3r^2 + 5r$ has three terms and the largest exponent is 3. This is a trinomial of degree 3.

29. $5z^4 + 3z^4 = (5 + 3)z^4 = 8z^4$

33. $x + x + x + x + x = (1 + 1 + 1 + 1 + 1)x = 5x$

37. $y^2 + 7y - 4y^2 = (1 - 4)y^2 + 7y$ Combine like terms.

$\qquad\qquad\qquad = -3y^2 + 7y$

$\qquad\qquad\qquad = 7y - 3y^2$

41. $n^4 - 2n^3 + n^2 - 3n^4 + n^3$

$\qquad = (1 - 3)n^4 + (-2 + 1)n^3 + n^2$ Combine like terms.

$\qquad = -2n^4 - n^3 + n^2$

45. Change the signs in the original second line, then add column by column to obtain the result on the bottom line.

$$\begin{array}{r} 12a + 15 \\ -7a + 3 \leftarrow \text{Change signs on this line and add.} \\ \hline 5a + 18 \leftarrow \text{Answer} \end{array}$$

49. Add column by column to obtain the result on the bottom line.

$$\begin{array}{r} 12z^2 - 11z + 8 \\ 5z^2 + 16z - 2 \\ -4z^2 + 5z - 9 \\ \hline 13z^2 + 10z - 3 \leftarrow \text{Answer} \end{array}$$

53. $-5a^4 \qquad\quad + 8a^2 - 9$

$\qquad\quad - 6a^3 + a^2 - 2 \leftarrow$ Change the signs and add.

$\overline{-5a^4 - 6a^3 + 9a^2 - 11} \leftarrow$ Answer

57. $(5x^2 + 7x - 4) + (3x^2 - 6x + 2)$

$\quad = 5x^2 + 3x^2 + 7x - 6x - 4 + 2$ Remove parentheses, then group like terms.

$\quad = 8x^2 + x - 2$ Combine like terms.

61. $(z^5 + 3z^2 + 2z) - (4z^5 + 2z^2 - 5z)$

$\quad = z^5 + 3z^2 + 2z - 4z^5 - 2z^2 + 5z$ Change the signs of all terms in the second polynomial and add.

$\quad = z^5 - 4z^5 + 3z^2 - 2z^2 + 2z + 5z$ Rearrange terms.

$\quad = -3z^5 + z^2 + 7z$ Combine like terms.

65. $[-(4m^2 - 8m + 4m^3) - (3m^2 + 2m + 5m^3)] + m^2$

$\quad = [-4m^2 + 8m - 4m^3 - 3m^2 - 2m - 5m^3] + m^2$

      Change signs of terms inside brackets.

$\quad = -7m^2 + 6m - 9m^3 + m^2$ Combine terms inside brackets. Remove brackets.

$\quad = -9m^3 - 6m^2 + 6m$ Combine terms, put in descending powers.

69. $P(x) = x^2 - 3x + 4$

(a) $P(-1) = (-1)^2 - 3(-1) + 4 = 1 + 3 + 4 = 8$

(b) $P(2) = (2)^2 - 3(2) + 4 = 4 - 6 + 4 = 2$

73. $Q(z) + R(z) - P(z)$

$\quad = (3z^2 + 2) + (6z + 2z^2) - (2z - 3z^2)$

$\quad = 3z^2 + 2 + 6z + 2z^2 - 2z + 3z^2$ Clear parentheses.

$\quad = 3z^2 + 2z^2 + 3z^2 + 6z - 2z + 2$ Rearrange terms.

$\quad = 8z^2 + 4z + 2$ Combine like terms.

77. $x = 3$ corresponds to 1993.

$P(x) = 1.06x^3 - 6.00x^2 + 7.86x + 31.6$

$P(3) = 1.06(3)^3 - 6.00(3)^2 + 7.86(3) + 31.6$

$\qquad = 1.06(27) - 6.00(9) + 7.86(3) + 31.6$

$\qquad = 28.62 - 54.00 + 23.58 + 31.6$

$\qquad = 29.8$

In 1993, 29.8% of all recorded music sales were represented by rock.

SECTION 5.4 (page 324)

1. The commutative and associative properties of multiplication are used to rewrite the product $(4m^3)(6m^5)$ as $(4 \cdot 6)(m^3 \cdot m^5)$.

5. The product of the sum and difference of two terms equals the difference of the squares of the terms. $(x + y)(x - y) = x^2 - y^2$

9. $3x(-2x + 5) = 3x(-2x) + 3x(5)$ Distributive property

$= -6x^2 + 15x$

13. $6k^2(3k^2 + 2k + 1) = 6k^2(3k^2) + 6k^2(2k) + 6k^2(1)$

Distributive property

$= 18k^4 + 12k^3 + 6k^2$

17. Multiply $(-d + 6)(d^3 - 2d + 1)$ vertically as follows:

$$
\begin{array}{l}
\quad d^3 \qquad\quad -2d + 1 \\
\quad\qquad\qquad\;\; -d + 6 \\
\hline
6d^3 \qquad\quad -12d + 6 \qquad \text{Multiply the top row by 6.} \\
-d^4 \qquad +2d^2 - \quad d \qquad \text{Multiply the top row by } -d. \\
\hline
-d^4 + 6d^3 + 2d^2 - 13d + 6 \qquad \text{Combine like terms.}
\end{array}
$$

21.
$$
\begin{array}{l}
\qquad\qquad -b^2 + 3b + 3 \\
\qquad\qquad\qquad\;\; 2b + 4 \\
\hline
\qquad -4b^2 + 12b + 12 \qquad \text{Multiply the top row by 4.} \\
-2b^3 + 6b^2 + 6b \qquad\qquad \text{Multiply the top row by } 2b. \\
\hline
-2b^3 + 2b^2 + 18b + 12 \qquad \text{Combine like terms.}
\end{array}
$$

25.
$$
\begin{array}{l}
\qquad\quad 2z^3 - 5z^2 + 8z - 1 \\
\qquad\qquad\qquad\qquad\; 4z + 3 \\
\hline
\qquad 6z^3 - 15z^2 + 24z - 3 \qquad \text{Multiply top row by 3.} \\
8z^4 - 20z^3 + 32z^2 - 4z \qquad\qquad \text{Multiply top row by } 4z. \\
\hline
8z^4 - 14z^3 + 17z^2 + 20z - 3 \qquad \text{Combine like terms.}
\end{array}
$$

29. $(m + 5)(m - 8)$

$= m^2 - 8m + 5m - 40$ Use FOIL method.

$\qquad\uparrow \qquad\;\; \uparrow \qquad \uparrow \qquad \uparrow$

$\qquad\text{F} \qquad \text{O} \qquad \text{I} \qquad \text{L}$

$= m^2 - 3m - 40$ Combine terms.

33. $(z - w)(3z + 4w) = 3z^2 + 4zw - 3zw - 4w^2$

$= 3z^2 + zw - 4w^2$

37. $(.2x + 1.3)(.5x - .1) = .1x^2 - .02x + .65x - .13$

$= .1x^2 + .63x - .13$

41. Answers will vary. A sample answer follows:
The product of two binomials is the sum of the product of the first terms, the product of the outer terms, the product of the inner terms, and the product of the last terms.

45. $(5m - 1)(5m + 1) = (5m)^2 - (1)^2$ Product of sum

$= 25m^2 - 1$ and difference of terms

49. $\left(4x - \dfrac{2}{3}\right)\left(4x + \dfrac{2}{3}\right) = (4x)^2 - \left(\dfrac{2}{3}\right)^2$

$= 16x^2 - \dfrac{4}{9}$

53. $(5y^3 + 2)(5y^3 - 2) = (5y^3)^2 - (2)^2$

$= 25y^6 - 4$

57. $(2p + 7)^2 = (2p)^2 + 2(2p)(7) + (7)^2$ Square of a

$= 4p^2 + 28p + 49$ binomial

61. $\left(k - \dfrac{5}{7}p\right)^2 = (k)^2 - 2 \cdot k \cdot \dfrac{5}{7}p + \left(\dfrac{5}{7}p\right)^2$

$= k^2 - \dfrac{10}{7}kp + \dfrac{25}{49}p^2$

65. $[(5x + 1) + 6y]^2$

$= (5x + 1)^2 + 2(5x + 1)(6y) + (6y)^2$

Square of a binomial

$= 25x^2 + 10x + 1 + 60xy + 12y + 36y^2$

Simplify each term.

69. $[(2h - k) + j][(2h - k) - j]$

$= (2h - k)^2 - j^2$ Product of the sum and difference of terms

$= 4h^2 - 4hk + k^2 - j^2$ Square of a binomial

73. $(m - p)^4 = (m - p)^2(m - p)^2$

$= (m^2 - 2mp + p^2)(m^2 - 2mp + p^2)$

$= m^4 - 2m^3p + m^2p^2 - 2m^3p + 4m^2p^2$

$\qquad - 2mp^3 + m^2p^2 - 2mp^3 + p^4$

$= m^4 - 4m^3p + 6m^2p^2 - 4mp^3 + p^4$

77. The formula for the area of a parallelogram is $A = bh$. So, with $b = 5x + 6$ and $h = 3x - 1$, it follows that

$A = (5x + 6)(3x - 1)$

$= 15x^2 - 5x + 18x - 6$ Multiply the binomials.

$= 15x^2 + 13x - 6.$ Combine terms.

SECTION 5.5 (page 333)

1. The quotient of two monomials is found using the quotient rule of exponents.

5. $\dfrac{9y^2 + 12y - 15}{3y} = \dfrac{9y^2}{3y} + \dfrac{12y}{3y} - \dfrac{15}{3y}$ Divide each term by $3y$.

$= 3y + 4 - \dfrac{5}{y}$ Write in lowest terms.

9. $\dfrac{14m^2n^2 - 21mn^3 + 28m^2n}{14m^2n}$

$= \dfrac{14m^2n^2}{14m^2n} - \dfrac{21mn^3}{14m^2n} + \dfrac{28m^2n}{14m^2n}$

$= n - \dfrac{3n^2}{2m} + 2$

13.
$$
\begin{array}{r}
y - 3 \\
y + 6 \overline{) y^2 + 3y - 18} \\
\underline{y^2 + 6y } \\
-3y - 18 \\
\underline{-3y - 18} \\
0
\end{array}
$$

Answer: $y - 3$

17.
$$
\begin{array}{r}
z^2 + 3 \\
2z - 5 \overline{) 2z^3 - 5z^2 + 6z - 15} \\
\underline{2z^3 - 5z^2 } \\
6z - 15 \\
\underline{6z - 15} \\
0
\end{array}
$$

Answer: $z^2 + 3$

21.
$$
\begin{array}{r}
2x - 5 \\
3x^2 - 2x + 4 \overline{) 6x^3 - 19x^2 + 14x - 15} \\
\underline{6x^3 - 4x^2 + 8x } \\
-15x^2 + 6x - 15 \\
\underline{-15x^2 + 10x - 20} \\
-4x + 5
\end{array}
$$

Answer: $2x - 5 + \dfrac{-4x + 5}{3x^2 - 2x + 4}$

25.
$$
\begin{array}{r}
9z^2 - 4z + 1 \\
z^2 - z + 2 \overline{) 9z^4 - 13z^3 + 23z^2 - 10z + 8} \\
\underline{9z^4 - 9z^3 + 18z^2 } \\
-4z^3 + 5z^2 - 10z \\
\underline{-4z^3 + 4z^2 - 8z} \\
z^2 - 2z + 8 \\
\underline{z^2 - z + 2} \\
-z + 6
\end{array}
$$

Answer: $9z^2 - 4z + 1 + \dfrac{-z + 6}{z^2 - z + 2}$

29.
$$4a + 3 \overline{)3a^2 - (23/4)a - 5} \quad\quad \frac{(3/4)a - 2}{}$$

$$\underline{3a^2 + (9/4)a}$$
$$-8a - 5$$
$$\underline{-8a - 6}$$
$$1$$

Answer: $\dfrac{3}{4} a - 2 + \dfrac{1}{4a + 3}$

33. $1654 = 1000 + 600 + 50 + 4$
$$= t^3 + 6t^2 + 5t + 4 \quad\quad \text{Let } t = 10.$$

37. The answer to Exercise 35 was $t^2 + 2t - 3 + \dfrac{16}{t + 4}$.

Substitute $t^2 = 100$ and $t = 10$ to get

$$100 + 2(10) - 3 + \frac{16}{10 + 4}$$

$$= 100 + 20 - 3 + \frac{16}{14}$$

$$= 117 + 1\frac{1}{7}$$

$$= 118\frac{1}{7}$$

SECTION 5.6 (page 339)

1. Synthetic division provides a quick, easy way to divide a polynomial by a binomial.

5. $\dfrac{x^2 - 6x + 5}{x - 1}$ is divided synthetically as follows.

$$1\overline{)1 \quad -6 \quad 5} \leftarrow \text{Coefficients of numerator}$$
$$\underline{1 \quad -5}$$
$$1 \quad -5 \quad 0 \quad \text{Write the answer from the bottom row.}$$
$$x \quad - \quad 5$$

9. $\dfrac{2a^2 + 8a + 13}{a + 2}$ is rewritten and divided as follows.

$$-2\overline{)2 \quad\quad 8 \quad\quad 13}$$
$$\underline{-4 \quad -8}$$
$$2 \quad\quad 4 \quad\quad 5$$

Answer from the bottom row: $2a + 4 + \dfrac{5}{a + 2}$

13. $\dfrac{4a^3 - 3a^2 + 2a - 3}{a - 1}$ is rewritten and divided as follows.

$$1\overline{)4 \quad -3 \quad 2 \quad -3}$$
$$\underline{4 \quad 1 \quad 3}$$
$$4 \quad 1 \quad 3 \quad 0$$

Answer from the bottom row: $4a^2 + a + 3$

17. $(-4r^6 - 3r^5 - 3r^4 + 5r^3 - 6r^2 + 3r) \div (r - 1)$ becomes

$$1\overline{)-4 \quad -3 \quad -3 \quad 5 \quad -6 \quad 3 \quad 0}$$
$$\underline{-4 \quad -7 \quad -10 \quad -5 \quad -11 \quad -8}$$
$$-4 \quad -7 \quad -10 \quad -5 \quad -11 \quad -8 \quad -8.$$

Answer:

$$-4r^5 - 7r^4 - 10r^3 - 5r^2 - 11r - 8 + \frac{-8}{r - 1}$$

21. $\dfrac{y^3 + 1}{y - 1} = \dfrac{y^3 + 0y^2 + 0y + 1}{y - 1}$ becomes

$$1\overline{)1 \quad 0 \quad 0 \quad 1}$$
$$\underline{1 \quad 1 \quad 1}$$
$$1 \quad 1 \quad 1 \quad 2.$$

Answer: $y^2 + y + 1 + \dfrac{2}{y - 1}$

25. From the remainder theorem if $P(r) = -r^3 - 5r^2 - 4r - 2$ is divided by $r - (-4)$, the remainder is $P(-4)$.

$$-4\overline{)-1 \quad -5 \quad -4 \quad -2}$$
$$\underline{4 \quad 4 \quad 0}$$
$$-1 \quad -1 \quad 0 \quad -2$$

The remainder is -2, so $P(-4) = -2$.

29.
$$-2\overline{)1 \quad -2 \quad -3 \quad 10}$$
$$\underline{-2 \quad 8 \quad -10}$$
$$1 \quad -4 \quad 5 \quad 0 \leftarrow \text{Remainder is zero.}$$
Yes, -2 is a solution since the remainder is zero.

33.
$$-2\overline{)3 \quad 2 \quad -2 \quad 11}$$
$$\underline{-6 \quad 8 \quad -12}$$
$$3 \quad -4 \quad 6 \quad -1 \leftarrow \text{Remainder is not zero.}$$
No, -2 is not a solution since the remainder is not zero.

SECTION 5.7 (page 345)

1. Answers will vary. A sample answer follows:
To factor a polynomial means to write it as the product of two or more polynomials.

5. Rewrite the list of terms $9m^3$, $3m^2$, $15m$ as $3m(3m^2), 3m(m)$, and $3m(5)$. The greatest common factor for these terms is $3m$.

9. $6x^3y^4 - 12x^5y^2 + 24x^4y^8$
The numerical part of the greatest common factor is 6. Find the variable part by writing each variable with its least degree. Here, 3 is the least exponent that appears on x, while 2 is the least exponent on y. So, $6x^3y^2$ is the greatest common factor. Therefore,

$$6x^3y^4 - 12x^5y^2 + 24x^4y^8 = 6x^3y^2(y^2 - 2x^2 + 4xy^6)$$

Choice (a) is the correct response.

13. $3xy - 5xy^2 = xy \cdot 3 - xy \cdot 5y$
$$= xy(3 - 5y)$$

17. $21x^5 + 35x^4 - 14x^3$
$$= 7x^3 \cdot 3x^2 + 7x^3 \cdot 5x - 7x^3 \cdot 2$$
$$= 7x^3(3x^2 + 5x - 2)$$

21. $-27m^3p^5 + 5r^4s^3 - 8x^5z^4$ cannot be factored because there is no common factor (except ± 1).

25. $(2z - 1)(z + 6) - (2z - 1)(z - 5)$
$$= (2z - 1)[(z + 6) - (z - 5)] \quad \text{Factor out } 2z - 1.$$
$$= (2z - 1)[z + 6 - z + 5] \quad \text{Clear parentheses inside brackets.}$$
$$= (2z - 1)(11) \quad \text{Combine terms inside brackets.}$$
$$= 11(2z - 1)$$

29. $5(2 - x)^2 - (2 - x)^3 + 4(2 - x)$
$\quad = (2 - x)[5(2 - x) - (2 - x)^2 + 4]$ Factor out
$\quad\quad\quad\quad\quad\quad\quad\quad\quad\quad\quad\quad\quad\quad\quad 2 - x.$
$\quad = (2 - x)[10 - 5x - (4 - 4x + x^2) + 4]$
$\quad = (2 - x)[10 - 5x - 4 + 4x - x^2 + 4]$
$\quad\quad\quad\quad\quad\quad\quad\quad$ Clear parentheses inside brackets.
$\quad = (2 - x)(10 - x - x^2)$ Combine terms.

33. Use a common factor of $5a^3$:

$$-5a^3 + 10a^4 - 15a^5 = 5a^3(-1 + 2a - 3a^2)$$

Use a common factor of $-5a^3$:

$$-5a^3 + 10a^4 - 15a^5 = -5a^3(1 - 2a + 3a^2)$$

37. $3ma + 3mb + 2ab + 2b^2$
$\quad = (3ma + 3mb) + (2ab + 2b^2)$ Group terms.
$\quad = 3m(a + b) + 2b(a + b)$ Factor out the
$\quad\quad\quad\quad\quad\quad\quad\quad\quad\quad\quad\quad\quad$ common
$\quad\quad\quad\quad\quad\quad\quad\quad\quad\quad\quad\quad\quad$ factors.
$\quad = (a + b)(3m + 2b)$ Factor out $a + b$.

41. $r^2 - 9tw + 3rw - 3rt$
$\quad = r^2 + 3rw - 3rt - 9tw$ Rearrange terms.
$\quad = (r^2 + 3rw) + (-3rt - 9tw)$ Group terms.
$\quad = r(r + 3w) - 3t(r + 3w)$ Factor out
$\quad\quad\quad\quad\quad\quad\quad\quad\quad\quad\quad\quad\quad$ common
$\quad\quad\quad\quad\quad\quad\quad\quad\quad\quad\quad\quad\quad$ factors.
$\quad = (r + 3w)(r - 3t)$ Factor out
$\quad\quad\quad\quad\quad\quad\quad\quad\quad\quad\quad\quad\quad r + 3w.$

45. $-6r^2 + 9rs + 8rs - 12s^2$
$\quad = (-6r^2 + 9rs) + (8rs - 12s^2)$
$\quad = -3r(2r - 3s) + 4s(2r - 3s)$
$\quad = (-3r + 4s)(2r - 3s)$

49. $2ab^2 - 4 - 8b^2 + a$
$\quad = 2ab^2 + a - 8b^2 - 4$ Rearrange terms.
$\quad = (2ab^2 + a) + (-8b^2 - 4)$
$\quad = a(2b^2 + 1) - 4(2b^2 + 1)$
$\quad = (a - 4)(2b^2 + 1)$

53. Rewrite $a - 4$ as $-(4 - a)$. Then

$(2b^2 + 1)(a - 4) = -(2b^2 + 1)(4 - a)$ First form
$\quad\quad\quad\quad\quad\quad\quad = (-2b^2 - 1)(4 - a)$ Second form

(Other answers are possible.)

SECTION 5.8 (page 353)

1. This statement is true. A prime polynomial is a polynomial that cannot be factored.

5. To factor $y^2 + 7y - 30$, we need two integer factors whose sum is 7 (coefficient of the middle term) and whose product is -30 (the last term). Since $-3 + 10 = 7$ and $-3 \cdot 10 = -30$, we have

$$y^2 + 7y - 30 = (y - 3)(y + 10).$$

9. $-m^2 + 16m - 60 = -(m^2 - 16m + 60)$
To factor $m^2 - 16m + 60$, we need two numbers whose sum is -16 and product is 60. Since $-10 + (-6) = -16$ and $-10 \cdot (-6) = 60$, then

$$m^2 - 16m + 60 = (m - 10)(m - 6).$$
$$-m^2 + 16m - 60 = -(m - 10)(m - 6)$$
$$= (m - 10)(-m + 6)$$

13. Although $-5 \cdot 3 = -15$, the sum $-5 + 3 = -2$, which is not the coefficient of the middle term, -3. Therefore, $y^2 - 3yq - 15q^2$ is prime (not factorable).

17. To factor $-6m^2 - 13m + 15$, find the product $-6(15) = -90$. Two integers whose product is -90 and sum is -13 are 5 and -18.

$$-6m^2 - 13m + 15 = -6m^2 + 5m - 18m + 15$$
$$= m(-6m + 5) + 3(-6m + 5)$$
$$\text{Factor by grouping.}$$
$$= (-6m + 5)(m + 3)$$

21. To factor $20k^2 + 47k + 24$, find $20(24) = 480$. Two integers whose product is 480 and sum is 47 are 15 and 32.

$$20k^2 + 47k + 24 = 20k^2 + 15k + 32k + 24$$
$$= 5k(4k + 3) + 8(4k + 3)$$
$$= (4k + 3)(5k + 8)$$

25. Use the alternative method. To factor $36m^2 - 60m + 25$, write $36m^2$ as $6m \cdot 6m$ and 25 as $5 \cdot 5$. Use these factors in the binomial factors to obtain

$$36m^2 - 60m + 25 = (6m - 5)(6m - 5)$$
$$= (6m - 5)^2.$$

Check that the middle term is $-60m$, as required.

29. To factor $6x^2z^2 + 5xz - 4$, write $6x^2z^2$ as $2xz \cdot 3xz$ and -4 as $-1 \cdot 4$. Use these factors in the binomial factors to obtain

$$6x^2z^2 + 5xz - 4 = (2xz - 1)(3xz + 4).$$

Check that the middle term is $8xz - 3xz = 5xz$, as required.

33. $15a^2 + 70a - 120$
$\quad = 5(3a^2 + 14a - 24)$ Factor out 5.
$\quad = 5(a + 6)(3a - 4)$ Factor $3a^2 + 14a - 24.$

37. $2x^3y^3 - 48x^2y^4 + 288xy^5$
$\quad = 2xy^3(x^2 - 24xy + 144y^2)$ Factor out $2xy^3$.
$\quad = 2xy^3(x - 12y)(x - 12y)$ Factor the trinomial.
$\quad = 2xy^3(x - 12y)^2$

41. In $2x^4 - 9x^2 - 18$, let $y = x^2$ to obtain

$$2y^2 - 9y - 18 = (2y + 3)(y - 6).$$

Substitute x^2 for y in this last result to obtain

$$2x^4 - 9x^2 - 18 = (2x^2 + 3)(x^2 - 6).$$

45. In $12p^6 - 32p^3r + 5r^2$, let $x = p^3$ to obtain

$$12x^2 - 32xr + 5r^2 = (6x - r)(2x - 5r).$$

Substitute p^3 for x in this last result to obtain

$$12p^6 - 32p^3r + 5r^2 = (6p^3 - r)(2p^3 - 5r).$$

49. In $3(m + p)^2 - 7(m + p) - 20$, let $x = m + p$ to obtain

$$3x^2 - 7x - 20 = (3x + 5)(x - 4).$$

Substitute $m + p$ for x in this last result to obtain

$[3(m + p) + 5][(m + p) - 4]$
$\quad = (3m + 3p + 5)(m + p - 4).$ Clear parentheses inside brackets.

53. Since $45 = 3^2 \cdot 5$ and neither 3 nor 5 has 2 as a factor, 2 is not a factor of 45.

57. 2 is a factor of $2x + 4$. So $2x + 4$ cannot be a factor of $2x^2 + kx + 8$ if k is an odd integer, because 2 would not be a factor of $2x^2 + kx + 8$ in that case. If a number is not a factor of a polynomial, it cannot be a factor of any factor of the polynomial.

SECTION 5.9 (page 359)

1. (a) Yes, 64 and m^2 are squares.
    (b) $2x^2$ is not a square.
    (c) $k^2 + 9$ is a *sum* of squares.
    (d) Yes, $4z^2$ and 49 are squares.

5. The sum of two squares can be factored only if the binomial has a common factor. For example,
$$9x^2 + 81 = 9(x^2 + 9).$$

9. $25x^2 - 4 = (5x)^2 - (2)^2$
$$= (5x + 2)(5x - 2)$$

13. $64m^4 - 4y^4 = 4(16m^4 - y^4)$ Factor out 4.
$$= 4[(4m^2)^2 - (y^2)^2]$$
$$= 4(4m^2 + y^2)(4m^2 - y^2)$$
         Factor the difference of squares.
$$= 4(4m^2 + y^2)(2m + y)(2m - y)$$
         Factor the difference of squares.

17. Let $z = (x + 3y)$. So
$$16 - (x + 3y)^2 = 16 - z^2$$
$$= (4 + z)(4 - z) \quad \text{Difference of squares}$$

Substitute $(x + 3y)$ for z into this last result to obtain
$$[4 + (x + 3y)][4 - (x + 3y)]$$
$$= (4 + x + 3y)(4 - x - 3y) \quad \text{Clear parentheses inside brackets.}$$

21. $k^2 - 6k + 9 = (k)^2 - 2(k)(3) + (3)^2$
$$= (k - 3)^2 \quad \text{Factor the square trinomial.}$$

25. $16m^2 - 8m + 1 - n^2$
$$= (16m^2 - 8m + 1) - n^2 \quad \text{Group the first three terms.}$$
$$= [(4m)^2 - 2(4m)(1) + 1^2] - n^2$$
$$= (4m - 1)^2 - n^2 \quad \text{Factor the square trinomial.}$$
$$= (4m - 1 + n)(4m - 1 - n) \quad \text{Factor the difference of squares.}$$

29. $x^2 - y^2 + 2y - 1$
$$= x^2 - (y^2 - 2y + 1) \quad \text{Group the last three terms.}$$
$$= x^2 - (y - 1)^2 \quad \text{Factor the square trinomial.}$$
$$= [x + (y - 1)][x - (y - 1)] \quad \text{Factor the difference of squares.}$$
$$= (x + y - 1)(x - y + 1) \quad \text{Clear parentheses inside brackets.}$$

33. In $(p + q)^2 + 2(p + q) + 1$, let $x = p + q$ to obtain $x^2 + 2x + 1 = (x + 1)^2$ by factoring the square trinomial. Substitute $p + q$ for x into this last result to get $(p + q + 1)^2$.

37. $8x^3 - y^3 = (2x)^3 - (y)^3$
$$= [2x - y][(2x)^2 + (2x)(y) + (y)^2]$$
         Factor the difference of cubes.
$$= (2x - y)(4x^2 + 2xy + y^2)$$

41. $24n^3 + 81p^3$
$$= 3(8n^3 + 27p^3)$$
$$= 3[(2n)^3 + (3p)^3]$$

$$= 3[2n + 3p][(2n)^2 - (2n)(3p) + (3p)^2]$$
         Factor the sum of cubes.
$$= 3(2n + 3p)(4n^2 - 6np + 9p^2)$$

45. $m^6 - 125 = (m^2)^3 - (5)^3$
$$= [m^2 - 5][(m^2)^2 + (m^2)(5) + (5)^2]$$
         Factor the difference of cubes.
$$= (m^2 - 5)(m^4 + 5m^2 + 25)$$

49. $x^6 - y^6 = (x^2 - y^2)(x^4 + x^2y^2 + y^4)$
$$= (x - y)(x + y)(x^4 + x^2y^2 + y^4)$$
         Factor the difference of squares.

SUMMARY ON FACTORING METHODS (page 361)

1. $100a^2 - 9b^2 = (10a)^2 - (3b)^2$ Difference of squares
$$= (10a + 3b)(10a - 3b)$$

5. To factor $x^2 + 2x - 35$, we need two integer factors whose sum is 2 and product is -35. So
$$x^2 + 2x - 35 = (x + 7)(x - 5).$$

9. $x^3 - 1000 = (x)^3 - (10)^3$ Difference of cubes
$$= (x - 10)(x^2 + 10x + 100)$$

13. To factor $6t^2 + 19tu - 77u^2$, write $6t^2$ as $3t \cdot 2t$ and $-77u^2$ as $-7u \cdot 11u$. Use these factors in the binomial factors to obtain
$$6t^2 + 19tu - 77u^2 = (3t - 7u)(2t + 11u).$$

17. $4k^2 + 28kr + 49r^2$
$$= (2k)^2 + 2(2k)(7r) + (7r)^2 \quad \text{Square trinomial}$$
$$= (2k + 7r)^2$$

21. $9m^2 - 30mn + 25n^2 - p^2$
$$= (9m^2 - 30mn + 25n^2) - p^2 \quad \text{Group the first 3 terms.}$$
$$= [(3m)^2 - 2(3m)(5n) + (5n)^2] - p^2$$
$$= (3m - 5n)^2 - p^2 \quad \text{Factor the square trinomial.}$$
$$= [(3m - 5n) + p][(3m - 5n) - p]$$
         Factor the difference of squares.
$$= (3m - 5n + p)(3m - 5n - p)$$
         Clear parentheses inside brackets.

25. $16z^3x^2 - 32z^2x$
$$= 16z^2x \cdot zx - 16z^2x \cdot 2$$
$$= 16z^2x(zx - 2) \quad \text{Factor out } 16z^2x.$$

29. $27m^2 + 144mn + 192n^2$
$$= 3(9m^2 + 48mn + 64n^2) \quad \text{Factor out 3.}$$
$$= 3[(3m)^2 + 2(3m)(8n) + (8n)^2]$$
$$= 3(3m + 8n)^2 \quad \text{Factor the square trinomial.}$$

33. To factor $2m^2 - mn - 15n^2$, write $2m^2$ as $2m \cdot m$ and $-15n^2$ as $5n(-3n)$. Use these factors in the binomial factors to obtain
$$2m^2 - mn - 15n^2 = (2m + 5n)(m - 3n).$$

37. In $14z^2 - 3zk - 2k^2$, write $14z^2$ as $7z \cdot 2z$ and $-2k^2$ as $2k(-k)$. Use these factors in the binomial factors to obtain
$$14z^2 - 3zk - 2k^2 = (7z + 2k)(2z - k).$$

41. $1000z^3 + 512$
$$= 8(125z^3 + 64) \quad \text{Factor out 8.}$$
$$= 8[(5z)^3 + (4)^3]$$
$$= 8[5z + 4][(5z)^2 - (5z)(4) + (4)^2]$$
$$= 8(5z + 4)(25z^2 - 20z + 16) \quad \text{Factor the sum of cubes.}$$

45. $32x^2 + 16x^3 - 24x^5$
$\quad = 8x^2(4 + 2x - 3x^3)$ Factor out $8x^2$.
The trinomial $4 + 2x - 3x^3$ cannot be factored further.

49. To factor $y^2 + 3y - 10$, we need two integer factors whose sum is 3 and product is -10. Since $5 + (-2) = 3$ and $5(-2) = -10$, then

$$y^2 + 3y - 10 = (y + 5)(y - 2).$$

53. To factor $18p^2 + 53pr - 35r^2$, write $18p^2$ as $9p \cdot 2p$ and $-35r^2$ as $-5r \cdot 7r$. Use these factors in the binomial factors to obtain

$$18p^2 + 53pr - 35r^2 = (9p - 5r)(2p + 7r).$$

57. In $(5r + 2s)^2 - 6(5r + 2s) + 9$, let $x = 5r + 2s$ to get $x^2 - 6x + 9 = (x - 3)^2$. Factor the square trinomial. Substitute $5r + 2s$ for x in this last result to get $(5r + 2s - 3)^2$.

Section 5.10 (page 369)

1. Answers will vary. A sample answer follows:
To solve a quadratic equation using the zero-factor property, first rewrite the equation so one side is zero. Factor the other side and set each factor equal to zero. Each solution of these linear equations is a solution of the quadratic equation.

5. $(x - 5)(x + 10) = 0$
$\quad x - 5 = 0$ or $x + 10 = 0$
$\quad\quad x = 5$ or $\quad\quad x = -10$
Solution set: $\{5, -10\}$

9. $m^2 - 3m - 10 = 0$
$\quad (m + 2)(m - 5) = 0$ Factor.
$\quad m + 2 = 0$ or $m - 5 = 0$
$\quad\quad m = -2$ or $\quad\quad m = 5$
Solution set: $\{-2, 5\}$

13. $\quad\quad 2x^2 = 7x + 4$
$\quad 2x^2 - 7x - 4 = 0$ Get zero on one side.
$\quad (2x + 1)(x - 4) = 0$ Factor.
$\quad 2x + 1 = 0$ or $x - 4 = 0$

$$x = -\frac{1}{2} \quad \text{or} \quad x = 4$$

Solution set: $\{-\frac{1}{2}, 4\}$

17. $2y^2 - 12 - 4y = y^2 - 3y$
$\quad y^2 - y - 12 = 0$ Get zero on one side and collect terms.
$\quad (y + 3)(y - 4) = 0$ Factor.
$\quad y + 3 = 0$ or $y - 4 = 0$
$\quad\quad y = -3$ or $\quad\quad y = 4$
Solution set: $\{-3, 4\}$

21. $6m^2 - 36m = 0$
$\quad 6m(m - 6) = 0$ Factor.
$\quad 6m = 0$ or $m - 6 = 0$
$\quad\quad m = 0$ or $\quad\quad m = 6$
Solution set: $\{0, 6\}$

25. $\quad\quad 4p^2 - 16 = 0$
$\quad (2p + 4)(2p - 4) = 0$
$\quad 2p + 4 = 0$ or $2p - 4 = 0$
$\quad\quad p = -2$ or $\quad\quad p = 2$
Solution set: $\{-2, 2\}$

29. Let W = width of the floor. Then $W + 4$ = length of floor. Use $A = LW$.

$(W + 4)W = 140$
$\quad W^2 + 4W = 140$ Clear parentheses.

$\quad W^2 + 4W - 140 = 0$ Get zero on one side.

$\quad (W + 14)(W - 10) = 0$ Factor.
$W + 14 = 0$ or $W - 10 = 0$
$\quad W = -14$ or $\quad\quad W = 10$

Reject -14 as a possible width of the rectangle. Therefore, the width is 10 meters and the length is $10 + 4 = 14$ meters.

33. Let W = width of rectangle and L = length of rectangle. Now, $2W + 2L$ = perimeter of a rectangle, so

$$2W + 2L = 300.$$

Solve for L as follows:

$2L = 300 - 2W$ Subtract $2W$.
$\quad L = 150 - W$ Divide each term by 2.

LW = area of a rectangle, so, with $L = 150 - W$,

$(150 - W)W = 5000$
$\quad 150W - W^2 = 5000$ Clear parentheses.
$-W^2 + 150W - 5000 = 0$ Get zero on one side.
$\quad W^2 - 150W + 5000 = 0$ Multiply by -1.
$\quad (W - 50)(W - 100) = 0$ Factor.
$W - 50 = 0$ or $W - 100 = 0$
$\quad W = 50$ or $\quad\quad W = 100$
If $W = 50$, then $L = 150 - 50 = 100$. If $W = 100$, then $L = 150 - 100 = 50$. (However, length should be greater than width.) The dimensions are 50 feet by 100 feet.

CHAPTER 6

Section 6.1 (page 391)

1. $\dfrac{x - 3}{x + 4} = \dfrac{-1(x - 3)}{-1(x + 4)} = \dfrac{-x + 3}{-x - 4} = \dfrac{3 - x}{-x - 4}$

So, $\dfrac{x - 3}{x + 4} = \dfrac{3 - x}{-x - 4}$, choice C.

5. $\dfrac{-x - 3}{x + 4} = \dfrac{-1(-x - 3)}{-1(x + 4)} = \dfrac{x + 3}{-x - 4}$

So, $\dfrac{-x - 3}{x + 4} = \dfrac{x + 3}{-x - 4}$, choice D.

9. $f(x) = \dfrac{3x - 6}{7x + 1}$ is not defined when the denominator is zero.

$$7x + 1 = 0$$
$$7x = -1 \quad \text{Subtract 1.}$$
$$x = -\frac{1}{7} \quad \text{Divide by 7.}$$

The rational expression is undefined when $x = -\dfrac{1}{7}$.

13. $f(x) = \dfrac{3x + 1}{2x^2 + x - 6}$ is not defined when the denominator is zero.

$$2x^2 + x - 6 = 0$$
$$(2x - 3)(x + 2) = 0 \quad \text{Factor.}$$
$$2x - 3 = 0 \quad \text{or} \quad x + 2 = 0$$
$$2x = 3 \quad \text{or} \quad x = -2$$
$$x = \frac{3}{2} \quad \text{or} \quad x = -2$$

The rational expression is undefined when $x = \frac{3}{2}$ or $x = -2$.

17. $f(x) = \dfrac{2x^2 - 3x + 4}{3x^2 + 8}$ is not defined when the denominator is zero.

$$3x^2 + 8 = 0$$
$$3x^2 = -8$$
$$x^2 = -\frac{8}{3}$$

The square of any real number x is positive or zero. There are no real numbers which make this rational expression undefined.

21. $\dfrac{24x^2y^4}{18xy^5} = \dfrac{4x \cdot 6xy^4}{3y \cdot 6xy^4} = \dfrac{4x}{3y} \cdot 1 = \dfrac{4x}{3y}$

25. $\dfrac{4x(x + 3)}{8x^2(x - 3)} = \dfrac{(x + 3) \cdot 4x}{2x(x - 3) \cdot 4x} = \dfrac{x + 3}{2x(x - 3)}$

29. $\dfrac{7x - 7}{8x - 8} = \dfrac{7(x - 1)}{8(x - 1)} = \dfrac{7}{8}$

33. $\dfrac{3z^2 + z}{18z + 6} = \dfrac{z(3z + 1)}{6(3z + 1)} = \dfrac{z}{6}$

37. $\dfrac{4b^2 - 4}{8b - 8} = \dfrac{4(b + 1)(b - 1)}{8(b - 1)} = \dfrac{b + 1}{2}$

41. $\dfrac{8x^2 - 10x - 3}{8x^2 - 6x - 9} = \dfrac{(4x + 1)(2x - 3)}{(4x + 3)(2x - 3)} = \dfrac{4x + 1}{4x + 3}$

45. $\dfrac{2c^2 + 2cd - 60d^2}{2c^2 - 12cd + 10d^2} = \dfrac{2(c^2 + cd - 30d^2)}{2(c^2 - 6cd + 5d^2)}$
$$= \dfrac{2(c + 6d)(c - 5d)}{2(c - d)(c - 5d)}$$
$$= \dfrac{c + 6d}{c - d}$$

49. $\dfrac{7 - b}{b - 7} = \dfrac{-1(b - 7)}{b - 7} = -1$

53. $\dfrac{(a - 3)(x + y)}{(3 - a)(x - y)} = \dfrac{(a - 3)(x + y)}{-1(a - 3)(x - y)} = -\dfrac{x + y}{x - y}$

57. $\dfrac{a^2 - b^2}{a^2 + b^2} = \dfrac{(a + b)(a - b)}{a^2 + b^2}$

The numerator and denominator have no common factors except 1. Therefore, the original expression is already in lowest terms.

61. $\dfrac{x^3}{3y} \cdot \dfrac{9y^2}{x^5} = \dfrac{3y \cdot 3x^3y}{x^2 \cdot 3x^3y} = \dfrac{3y}{x^2} \cdot 1 = \dfrac{3y}{x^2}$

65. $\dfrac{4x}{8x + 4} \cdot \dfrac{14x + 7}{6} = \dfrac{4x \cdot 7(2x + 1)}{4(2x + 1) \cdot 6}$
$$= \dfrac{7x}{6}$$

69. $\dfrac{a^2 - 1}{4a} \cdot \dfrac{2}{1 - a} = \dfrac{(a - 1)(a + 1) \cdot 2}{4a \cdot (1 - a)}$
$$= \dfrac{2(-1)(1 - a)(a + 1)}{4a(1 - a)}$$
$$= \dfrac{-2(a + 1)}{4a}$$
$$= -\dfrac{a + 1}{2a}$$

73. $\dfrac{12x - 10y}{3x + 2y} \cdot \dfrac{6x + 4y}{10y - 12x} = \dfrac{2(6x - 5y) \cdot 2(3x + 2y)}{(3x + 2y) \cdot 2(5y - 6x)}$
$$= \dfrac{4(-1)(5y - 6x)(3x + 2y)}{2(5y - 6x)(3x + 2y)}$$
$$= \dfrac{-4}{2} = -2$$

77. $\dfrac{15x^2 - xy - 2y^2}{15x^2 + 11xy + 2y^2} \cdot \dfrac{15x^2 + xy - 2y^2}{15x^2 + 4xy - 4y^2}$
$$= \dfrac{(5x - 2y)(3x + y)}{(5x + 2y)(3x + y)} \cdot \dfrac{(5x + 2y)(3x - y)}{(5x - 2y)(3x + 2y)}$$
$$= \dfrac{3x - y}{3x + 2y}$$

81. $\left(\dfrac{6k^2 - 13k - 5}{k^2 + 7k} \div \dfrac{2k - 5}{k^3 + 6k^2 - 7k}\right) \cdot \dfrac{k^2 - 5k + 6}{3k^2 - 8k - 3}$
$$= \left[\dfrac{(3k + 1)(2k - 5)}{k(k + 7)} \div \dfrac{2k - 5}{k(k + 7)(k - 1)}\right]$$
$$\cdot \dfrac{(k - 2)(k - 3)}{(3k + 1)(k - 3)}$$
$$= \dfrac{(3k + 1)(2k - 5)}{k(k + 7)} \cdot \dfrac{k(k + 7)(k - 1)}{2k - 5}$$
$$\cdot \dfrac{(k - 2)(k - 3)}{(3k + 1)(k - 3)}$$
$$= (k - 1)(k - 2)$$

85. The greatest amount of money invested in the U.S. by the Japanese was in 1988 at \$16.5 billion. Add the amounts invested from 1985 to 1993 to find the total investment.

$$\text{Total} = 1.9 + 7.5 + 12.8 + 16.5 + 14.8 + 13.1$$
$$+ 5.1 + 0.8 + 0.7$$
$$= 73.2$$

The ratio of the greatest number of dollars (in billions) invested to the total investment from 1985 to 1993 is

$$\dfrac{16.5}{73.2} \approx .23.$$

SECTION 6.2 (page 403)

1. To add or subtract rational expressions that have a common denominator, first add or subtract the numerators and then place the result over the common denominator. Write the answer in lowest terms.

5. $\dfrac{11}{5x} - \dfrac{1}{5x} = \dfrac{11 - 1}{5x} = \dfrac{10}{5x} = \dfrac{2}{x}$

9. $\dfrac{x^2}{x + 5} - \dfrac{25}{x + 5} = \dfrac{x^2 - 25}{x + 5} = \dfrac{(x + 5)(x - 5)}{x + 5}$
$$= x - 5$$

13. $\dfrac{a^3}{a^2 + ab + b^2} - \dfrac{b^3}{a^2 + ab + b^2}$

$\qquad = \dfrac{a^3 - b^3}{a^2 + ab + b^2} = \dfrac{(a - b)(a^2 + ab + b^2)}{a^2 + ab + b^2}$

$\qquad = a - b$

17. The LCD for $\dfrac{7}{z - 2}$ and $\dfrac{5}{z}$ is $z(z - 2)$.

21. To find the LCD for $\dfrac{x}{x^2 - 81}$ and $\dfrac{3x}{x^2 + 18x + 81}$,

factor each denominator:

$$x^2 - 81 = (x - 9)(x + 9)$$
$$x^2 + 18x + 81 = (x + 9)(x + 9) \quad \text{or} \quad (x + 9)^2.$$

The LCD is $(x + 9)^2(x - 9)$.

25. To find the LCD of $\dfrac{x + 8}{x^2 - 3x - 4}$ and $\dfrac{-9}{x + x^2}$, factor

each denominator:

$$x^2 - 3x - 4 = (x - 4)(x + 1)$$
$$x + x^2 = x(1 + x).$$

The LCD is $x(x - 4)(x + 1)$.

29. To find the LCD of $\dfrac{y}{2y + 6}, \dfrac{3}{y^2 - 9}$, and $\dfrac{6}{y}$, factor the

denominators:

$$2y + 6 = 2(y + 3)$$
$$y^2 - 9 = (y + 3)(y - 3).$$

The LCD is $2y(y + 3)(y - 3)$.

33. The LCD is $3t$, so

$$\frac{8}{t} + \frac{7}{3t} = \frac{8 \cdot 3}{t \cdot 3} + \frac{7}{3t} = \frac{24 + 7}{3t} = \frac{31}{3t}.$$

37. The LCD is $x(x - 1)$.

$$\frac{1}{x - 1} - \frac{1}{x} = \frac{1 \cdot x}{(x - 1)x} - \frac{1 \cdot (x - 1)}{x(x - 1)}$$

$$= \frac{x - (x - 1)}{x(x - 1)}$$

$$= \frac{x - x + 1}{x(x - 1)}$$

$$= \frac{1}{x(x - 1)}$$

41. $\dfrac{17y + 3}{9y + 7} - \dfrac{-10y - 18}{9y + 7} = \dfrac{17y + 3 - (-10y - 18)}{9y + 7}$

$$= \frac{17y + 3 + 10y + 18}{9y + 7}$$

$$= \frac{27y + 21}{9y + 7}$$

$$= \frac{3(9y + 7)}{9y + 7}$$

$$= 3$$

45. $\dfrac{w}{w - z} - \dfrac{z}{z - w} = \dfrac{w}{w - z} - \dfrac{z}{-(w - z)}$

$$= \frac{w}{w - z} + \frac{z}{w - z} = \frac{w + z}{w - z}$$

49. The LCD is $(x - 1)(x + 1)$.

$$\frac{4x}{x - 1} - \frac{2}{x + 1} - \frac{4}{x^2 - 1}$$

$$= \frac{4x(x + 1)}{(x - 1)(x + 1)} - \frac{2(x - 1)}{(x + 1)(x - 1)}$$

$$\quad - \frac{4}{(x + 1)(x - 1)}$$

$$= \frac{4x^2 + 4x - (2x - 2) - 4}{(x + 1)(x - 1)}$$

$$= \frac{4x^2 + 4x - 2x + 2 - 4}{(x - 1)(x + 1)}$$

$$= \frac{4x^2 + 2x - 2}{(x - 1)(x + 1)} = \frac{2(2x^2 + x - 1)}{(x - 1)(x + 1)}$$

$$= \frac{2(2x - 1)(x + 1)}{(x - 1)(x + 1)} = \frac{2(2x - 1)}{x - 1}$$

53. The LCD is $x(x - 2)$.

$$\frac{5}{x - 2} + \frac{1}{x} + \frac{2}{x^2 - 2x}$$

$$= \frac{5x}{(x - 2)x} + \frac{1(x - 2)}{x(x - 2)} + \frac{2}{x(x - 2)}$$

$$= \frac{5x + x - 2 + 2}{x(x - 2)}$$

$$= \frac{6x}{x(x - 2)} = \frac{6}{x - 2}$$

57. The LCD is $(x + 1)(x^2 - x + 1)$.

$$\frac{4}{x + 1} + \frac{1}{x^2 - x + 1} - \frac{12}{x^3 + 1}$$

$$= \frac{4(x^2 - x + 1)}{(x + 1)(x^2 - x + 1)} + \frac{1 \cdot (x + 1)}{(x^2 - x + 1)(x + 1)}$$

$$\quad - \frac{12}{(x + 1)(x^2 - x + 1)}$$

$$= \frac{4x^2 - 4x + 4 + x + 1 - 12}{(x + 1)(x^2 - x + 1)}$$

$$= \frac{4x^2 - 3x - 7}{(x + 1)(x^2 - x + 1)}$$

$$= \frac{(4x - 7)(x + 1)}{(x + 1)(x^2 - x + 1)} = \frac{4x - 7}{x^2 - x + 1}$$

61. Factor each denominator:

$$x^2 + xy - 2y^2 = (x + 2y)(x - y) \quad \text{and}$$
$$x^2 + 5xy - 6y^2 = (x + 6y)(x - y).$$

The LCD is $(x + 2y)(x - y)(x + 6y)$, so

$$\frac{5x}{(x + 2y)(x - y)} - \frac{3x}{(x + 6y)(x - y)}$$

$$= \frac{5x(x + 6y)}{(x + 2y)(x - y)(x + 6y)}$$

$$\quad - \frac{3x(x + 2y)}{(x + 6y)(x - y)(x + 2y)}$$

(continued)

SOLUTIONS

$$= \frac{5x^2 + 30xy - (3x^2 + 6xy)}{(x + 2y)(x - y)(x + 6y)}$$

$$= \frac{5x^2 + 30xy - 3x^2 - 6xy}{(x + 2y)(x - y)(x + 6y)}$$

$$= \frac{2x^2 + 24xy}{(x + 2y)(x - y)(x + 6y)}$$

$$= \frac{2x(x + 12y)}{(x + 2y)(x - y)(x + 6y)}.$$

65. The LCD is 63.

$$\frac{3}{7} + \frac{5}{9} - \frac{6}{63} = \frac{27}{63} + \frac{35}{63} - \frac{6}{63}$$

$$= \frac{56}{63}$$

$$= \frac{8}{9}$$

69. The problem in Example 6 is

$$\frac{3}{x - 2} + \frac{5}{x} - \frac{6}{x^2 - 2x}.$$

$x = 2$ will not work for the problem because if $x = 2$, $\dfrac{3}{x - 2}$ and $\dfrac{6}{x^2 - 2x}$ will be undefined, since their denominators will become zero.

SECTION 6.3 (page 411)

1. Answers will vary. A sample answer follows:
To simplify a complex fraction using Method 1, begin by simplifying the numerator. Then simplify the denominator, write as a division problem, and proceed.

5. $\dfrac{\dfrac{k + 1}{2k}}{\dfrac{3k - 1}{4k}} = \dfrac{k + 1}{2k} \cdot \dfrac{4k}{3k - 1} = \dfrac{4k(k + 1)}{2k(3k - 1)}$

$$= \frac{2(k + 1)}{3k - 1}.$$

9. $\dfrac{\dfrac{1}{x} + 1}{-\dfrac{1}{x} + 1} = \dfrac{x\left(\dfrac{1}{x} + 1\right)}{x\left(-\dfrac{1}{x} + 1\right)}$

$$= \frac{1 + x}{-1 + x}$$

13. $\dfrac{\dfrac{8x - 24y}{10}}{\dfrac{x - 3y}{5x}} = \dfrac{8x - 24y}{10} \cdot \dfrac{5x}{x - 3y}$

$$= \frac{8(x - 3y)5x}{10(x - 3y)}$$

$$= \frac{5x \cdot 8(x - 3y)}{10(x - 3y)}$$

$$= \frac{40x(x - 3y)}{10(x - 3y)} = 4x$$

17. The LCD is $9y$.

$$\frac{y - \dfrac{y - 3}{3}}{\dfrac{4}{9} + \dfrac{2}{3y}} = \frac{9y\left(y - \dfrac{y - 3}{3}\right)}{9y\left(\dfrac{4}{9} + \dfrac{2}{3y}\right)}$$

$$= \frac{9y^2 - 3y(y - 3)}{4y + 6}$$

$$= \frac{9y^2 - 3y^2 + 9y}{4y + 6}$$

$$= \frac{6y^2 + 9y}{4y + 6}$$

$$= \frac{3y(2y + 3)}{2(2y + 3)} = \frac{3y}{2}$$

21. The LCD is $m(m - 1)$.

$$\frac{4}{m} + \frac{m + 2}{m - 1} = \frac{4(m - 1)}{m(m - 1)} + \frac{(m + 2)m}{(m - 1)m}$$

$$= \frac{4m - 4 + m^2 + 2m}{m(m - 1)}$$

$$= \frac{m^2 + 6m - 4}{m(m - 1)}$$

25. The answer to Exercise 24 is $m(m - 1)$.

$$\frac{\dfrac{4}{m} + \dfrac{m + 2}{m - 1}}{\dfrac{m + 2}{m} - \dfrac{2}{m - 1}} = \frac{m(m - 1)\left[\dfrac{4}{m} + \dfrac{m + 2}{m - 1}\right]}{m(m - 1)\left[\dfrac{m + 2}{m} - \dfrac{2}{m - 1}\right]}$$

$$= \frac{4(m - 1) + m(m + 2)}{(m + 2)(m - 1) - 2m}$$

$$= \frac{4m - 4 + m^2 + 2m}{m^2 + m - 2 - 2m}$$

$$= \frac{m^2 + 6m - 4}{m^2 - m - 2}$$

29. $\dfrac{x^{-2} + y^{-2}}{x^{-1} + y^{-1}} = \dfrac{\dfrac{1}{x^2} + \dfrac{1}{y^2}}{\dfrac{1}{x} + \dfrac{1}{y}}$

$$= \frac{x^2 y^2 \left(\dfrac{1}{x^2} + \dfrac{1}{y^2}\right)}{x^2 y^2 \left(\dfrac{1}{x} + \dfrac{1}{y}\right)}$$

$$= \frac{y^2 + x^2}{xy^2 + x^2 y} \quad \text{or}$$

$$= \frac{y^2 + x^2}{xy(y + x)}$$

SECTION 6.4 (page 415)

1. In $\dfrac{1}{x+1} - \dfrac{1}{x-2} = 0$, set each denominator equal to zero and solve.

$$x + 1 = 0 \quad \text{or} \quad x - 2 = 0$$
$$x = -1 \quad \text{or} \quad x = 2$$

Solutions of -1 and 2 would be rejected since these values would make a denominator of the original equation equal to zero.

5. In $\dfrac{1}{3x} + \dfrac{1}{2x} = \dfrac{x}{3}$, set the denominators $3x$ and $2x$ equal to zero and solve.

$$3x = 0 \quad \text{or} \quad 2x = 0, \text{ so } x = 0.$$

A solution of zero would be rejected.

9. The denominators in $\dfrac{x+5}{10} - \dfrac{2x+3}{5} = \dfrac{x}{20}$ contain no variable factors, so the denominators can never equal zero. There are no solutions that would have to be rejected.

13. $\dfrac{x+8}{5} = \dfrac{6+x}{3}$

Multiply each side of the equation by the least common denominator 15.

$$15\left(\frac{x+8}{5}\right) = 15\left(\frac{6+x}{3}\right)$$
$$3(x+8) = 5(6+x)$$
$$3x + 24 = 30 + 5x$$
$$-2x = 6$$
$$x = -3$$

Solution set: $\{-3\}$

All apparent solutions in Exercises 17–41 should be checked in the original equation to identify any extraneous solutions.

17. $\dfrac{3x+1}{x-4} = \dfrac{6x+5}{2x-7}$

Multiply each term by the LCD $(x-4)(2x-7)$.

$$(x-4)(2x-7)\left(\frac{3x+1}{x-4}\right)$$
$$= (x-4)(2x-7)\left(\frac{6x+5}{2x-7}\right)$$
$$(2x-7)(3x+1) = (x-4)(6x+5)$$
$$6x^2 - 19x - 7 = 6x^2 - 19x - 20$$
$$-7 = -20$$

The last statement is impossible so this equation has no solution.
Solution set: $\varnothing$

21. $x - \dfrac{24}{x} = -2$

Multiply each side by the LCD x.

$$x\left(x - \frac{24}{x}\right) = -2 \cdot x$$
$$x^2 - 24 = -2x$$
$$x^2 + 2x - 24 = 0$$
$$(x+6)(x-4) = 0$$
$$x + 6 = 0 \quad \text{or} \quad x - 4 = 0$$
$$x = -6 \quad \text{or} \quad x = 4$$

Solution set: $\{-6, 4\}$

25. $\dfrac{-2}{3t-6} - \dfrac{1}{36} = \dfrac{-3}{4t-8}$

$$\frac{-2}{3(t-2)} - \frac{1}{36} = \frac{-3}{4(t-2)}$$

Multiply each term by the LCD $36(t-2)$.

$$36(t-2)\left(\frac{-2}{3(t-2)} - \frac{1}{36}\right) = 36(t-2)\left(\frac{-3}{4(t-2)}\right)$$
$$12(-2) - 1(t-2) = 9(-3)$$
$$-24 - t + 2 = -27$$
$$-t - 22 = -27$$
$$-t = -5$$
$$t = 5$$

Solution set: $\{5\}$

29. $\dfrac{1}{y+2} + \dfrac{3}{y+7} = \dfrac{5}{y^2+9y+14}$

$$\frac{1}{y+2} + \frac{3}{y+7} = \frac{5}{(y+2)(y+7)}$$

Multiply each term by the LCD $(y+2)(y+7)$.

$$(y+2)(y+7)\left(\frac{1}{y+2} + \frac{3}{y+7}\right)$$
$$= (y+2)(y+7)$$
$$\cdot \left(\frac{5}{(y+2)(y+7)}\right)$$
$$y + 7 + 3(y+2) = 5$$
$$y + 7 + 3y + 6 = 5$$
$$4y + 13 = 5$$
$$4y = -8$$
$$y = -2$$

Replace y with -2 in the original equation to get

$$\frac{1}{0} + \frac{3}{5} = \frac{5}{4 - 18 + 14}$$

Since 0 cannot appear in a denominator, this equation has no solution.
Solution set: $\varnothing$

SOLUTIONS

33. $\dfrac{6}{w+3} + \dfrac{-7}{w-5} = \dfrac{-48}{w^2 - 2w - 15}$

$\dfrac{6}{w+3} + \dfrac{-7}{w-5} = \dfrac{-48}{(w+3)(w-5)}$

Multiply each term by the LCD $(w+3)(w-5)$.

$(w+3)(w-5)\left(\dfrac{6}{w+3} + \dfrac{-7}{w-5}\right)$

$\qquad = (w+3)(w-5)\left(\dfrac{-48}{(w+3)(w-5)}\right)$

$6(w-5) - 7(w+3) = -48$

$6w - 30 - 7w - 21 = -48$

$-w - 51 = -48$

$-w = 3$

$w = -3$

Replace w with -3 in the original equation to get

$\dfrac{6}{0} + \dfrac{-7}{-8} = \dfrac{-48}{9 + 6 - 15}$

Since 0 cannot appear in the denominator, this equation has no solution.
Solution set: $\emptyset$

37. $\dfrac{6}{x-4} + \dfrac{5}{x} = \dfrac{-20}{x^2 - 4x}$

$\dfrac{6}{x-4} + \dfrac{5}{x} = \dfrac{-20}{x(x-4)}$

Multiply each term by the LCD $x(x-4)$.

$x(x-4)\left(\dfrac{6}{x-4} + \dfrac{5}{x}\right) = x(x-4)\left(\dfrac{-20}{x(x-4)}\right)$

$6x + 5(x-4) = -20$

$6x + 5x - 20 = -20$

$11x = 0$

$x = 0$

Replace x with 0 in the original equation to get

$\dfrac{6}{-4} + \dfrac{5}{0} = \dfrac{-20}{0}$

Since 0 cannot appear in the denominator, this equation has no solution.
Solution set: $\emptyset$

41. $\dfrac{4x-7}{4x^2-9} = \dfrac{-2x^2+5x-4}{4x^2-9} + \dfrac{x+1}{2x+3}$

$\dfrac{4x-7}{(2x+3)(2x-3)} = \dfrac{-2x^2+5x-4}{(2x+3)(2x-3)} + \dfrac{x+1}{2x+3}$

Multiply each term by the LCD $(2x+3)(2x-3)$.

$4x - 7 = -2x^2 + 5x - 4 + (2x-3)(x+1)$

$4x - 7 = -2x^2 + 5x - 4 + 2x^2 - x - 3$

$4x - 7 = 4x - 7$

$0 = 0$

This last statement is true for all real numbers x except those that make the denominator 0 in the original equation. Set each factor in the original denominators equal to zero and solve.

$2x + 3 = 0 \quad$ or $\quad 2x - 3 = 0$

$2x = -3 \quad$ or $\quad 2x = 3$

$x = -\dfrac{3}{2} \quad$ or $\quad x = \dfrac{3}{2}$

Use set-builder notation to write the solution set.
Solution set: $\{x \mid x \neq -\frac{3}{2}, x \neq \frac{3}{2}\}$

45. If $A = 4$ and $B = -3$, then
$C = -2AB = -2(4)(-3) = 24$.
The equation becomes

$\dfrac{4}{x-3} + \dfrac{x}{x+3} = \dfrac{24}{x^2-9}$.

Multiply each term by the LCD $(x-3)(x+3)$.

$(x-3)(x+3)\left[\dfrac{4}{x-3} + \dfrac{x}{x+3}\right]$

$\qquad = (x-3)(x+3)\left[\dfrac{24}{x^2-9}\right]$

$4(x+3) + x(x-3) = 24$

$4x + 12 + x^2 - 3x = 24$

$x^2 + x + 12 = 24$

$x^2 + x - 12 = 0$

$(x+4)(x-3) = 0$

$x + 4 = 0 \quad$ or $\quad x - 3 = 0$

$x = -4 \quad$ or $\quad x = 3$

3 is rejected, so the solution set is $\{-4\}$.

SUMMARY ON OPERATIONS AND EQUATIONS INVOLVING RATIONAL EXPRESSIONS (page 419)

The solutions in Exercises 9 and 25 should be checked.

1. equation

$\dfrac{x}{2} - \dfrac{x}{4} = 5$

$4\left(\dfrac{x}{2} - \dfrac{x}{4}\right) = 4 \cdot 5$

$2x - x = 20$

$x = 20$

Solution set: $\{20\}$

5. expression

$\dfrac{\dfrac{1}{x} + \dfrac{1}{y}}{\dfrac{1}{x} - \dfrac{1}{y}} = \dfrac{xy\left(\dfrac{1}{x} + \dfrac{1}{y}\right)}{xy\left(\dfrac{1}{x} - \dfrac{1}{y}\right)}$

$= \dfrac{y + x}{y - x}$

9. equation

$\dfrac{4}{x} - \dfrac{8}{x+1} = 0$

$x(x+1)\left(\dfrac{4}{x} - \dfrac{8}{x+1}\right) = x(x+1) \cdot 0$

$$4(x + 1) - 8x = 0$$
$$4x + 4 - 8x = 0$$
$$-4x = -4$$
$$x = 1$$

Solution set: $\{1\}$

13. expression

$$\frac{3p^2 - 6p}{p + 5} \div \frac{p^2 - 4}{8p + 40}$$
$$= \frac{3p(p - 2)}{p + 5} \cdot \frac{8(p + 5)}{(p + 2)(p - 2)}$$
$$= \frac{24p(p - 2)(p + 5)}{(p + 2)(p - 2)(p + 5)}$$
$$= \frac{24p}{p + 2}$$

17. expression

$$\frac{10z^2 - 5z}{3z^3 - 6z^2} \div \frac{2z^2 + 5z - 3}{z^2 + z - 6}$$
$$= \frac{5z(2z - 1)}{3z^2(z - 2)} \cdot \frac{(z + 3)(z - 2)}{(2z - 1)(z + 3)}$$
$$= \frac{5z(2z - 1)(z + 3)(z - 2)}{3z^2(2z - 1)(z + 3)(z - 2)}$$
$$= \frac{5}{3z}$$

21. expression

$$\frac{\frac{5}{x} - \frac{3}{y}}{\frac{9x^2 - 25y^2}{x^2y}} = \frac{x^2y\left(\frac{5}{x} - \frac{3}{y}\right)}{x^2y\left(\frac{9x^2 - 25y^2}{x^2y}\right)}$$
$$= \frac{5xy - 3x^2}{9x^2 - 25y^2}$$
$$= \frac{-x(3x - 5y)}{(3x + 5y)(3x - 5y)}$$
$$= \frac{-x}{3x + 5y}$$

25. equation

$$\frac{8}{3k + 9} - \frac{8}{15} = \frac{2}{5k + 15}$$
$$\frac{8}{3(k + 3)} - \frac{8}{15} = \frac{2}{5(k + 3)}$$

Multiply each term by the LCD $15(k + 3)$.

$$5(8) - 8(k + 3) = 3(2)$$
$$40 - 8k - 24 = 6$$
$$-8k + 16 = 6$$
$$-8k = -10$$
$$k = \frac{-10}{-8} = \frac{5}{4}$$

Solution set: $\{\frac{5}{4}\}$

29. expression

$$\frac{\frac{t}{4} - \frac{1}{t}}{1 + \frac{t + 4}{t}} = \frac{4t\left(\frac{t}{4} - \frac{1}{t}\right)}{4t\left(1 + \frac{t + 4}{t}\right)}$$
$$= \frac{t^2 - 4}{4t + 4t + 16}$$
$$= \frac{t^2 - 4}{8t + 16}$$
$$= \frac{(t + 2)(t - 2)}{8(t + 2)}$$
$$= \frac{t - 2}{8}$$

33. expression

$$\frac{2k + \frac{5}{k - 1}}{3k - \frac{2}{k}} = \frac{k(k - 1)\left(2k + \frac{5}{k - 1}\right)}{k(k - 1)\left(3k - \frac{2}{k}\right)}$$
$$= \frac{2k^2(k - 1) + 5k}{3k^2(k - 1) - 2(k - 1)}$$
$$= \frac{2k^3 - 2k^2 + 5k}{3k^3 - 3k^2 - 2k + 2}$$
$$= \frac{k(2k^2 - 2k + 5)}{(3k^3 - 3k^2) + (-2k + 2)}$$
$$= \frac{k(2k^2 - 2k + 5)}{3k^2(k - 1) - 2(k - 1)}$$
$$= \frac{k(2k^2 - 2k + 5)}{(k - 1)(3k^2 - 2)}$$

SECTION 6.5 (page 423)

1. **(a)** Correct: $r = \frac{d}{t}$ is equivalent to $rt = d$.

(b) Incorrect: $t = \frac{d}{r}$, not $\frac{r}{d}$.

(c) Incorrect: $r = \frac{d}{t}$, not $\frac{t}{d}$.

(d) Incorrect: $d = rt$, not $\frac{t}{r}$.

5. $m = \frac{F}{a}$

$30 = \frac{F}{9}$ Let $m = 30$, $a = 9$.

$270 = F$ or $F = 270$

9. $\frac{1}{a} = \frac{1}{b} + \frac{1}{c}$

$\frac{1}{8} = \frac{1}{b} + \frac{1}{12}$ Let $a = 8$, $c = 12$.

$24b\left(\frac{1}{8}\right) = 24b\left(\frac{1}{b} + \frac{1}{12}\right)$
$3b = 24 + 2b$
$b = 24$

13. In Exercise 12, there were no similar terms, so they could not be combined.

17.

$$\frac{1}{a} = \frac{1}{b} + \frac{1}{c}$$

$$abc\left(\frac{1}{a}\right) = abc\left(\frac{1}{b} + \frac{1}{c}\right)$$

$$bc = ac + ab$$

$$bc = a(c + b)$$

$$\frac{bc}{c + b} = a \quad \text{or} \quad a = \frac{bc}{c + b}$$

21.

$$A = P + Prt$$

$$A = P(1 + rt)$$

$$\frac{A}{1 + rt} = P \quad \text{or} \quad P = \frac{A}{1 + rt}$$

SECTION 6.6 (page 431)

1. In 1986, the box office revenue was 4 billion and the home video revenue was 5 billion. The ratio of the box office revenue to the home video revenue was 4 to 5 or $\frac{4}{5}$.

5. Let x = the number of girls in the class. Write and solve a proportion.

$$\frac{3}{4} = \frac{x}{20}$$

Multiply each term by 20.

$$20\left(\frac{3}{4}\right) = 20\left(\frac{x}{20}\right)$$

$$15 = x \quad \text{or} \quad x = 15$$

There are 15 girls and $20 - 15 = 5$ boys in the class.

9. Let x = the number of households composed of a married couple with children under 18. Write and solve a proportion.

$$\frac{25.5}{100} = \frac{x}{95,700,000}$$

Multiply each side by 95,700,000.

$$95,700,000\left(\frac{25.5}{100}\right) = 95,700,000\left(\frac{x}{95,700,000}\right)$$

$$957,000(25.5) = x$$

$$x \approx 24,400,000$$

There are about 24,400,000 married couples with children under 18 (to the nearest hundred thousand).

13. Let x = the number of fish in the lake. Write and solve a proportion.

$$\text{Tagged in sample} \rightarrow \frac{8}{400} = \frac{500}{x} \leftarrow \text{Tagged in lake}$$
$$\text{Total in sample} \rightarrow \qquad\qquad \leftarrow \text{Total in lake}$$

Multiply each term by the LCD $400x$.

$$400x\left(\frac{8}{400}\right) = 400x\left(\frac{500}{x}\right)$$

$$8x = 200,000$$

$$x = 25,000$$

There are approximately 25,000 fish in the lake.

17. Let x = the distance between his office and home. Make a table using the information in the problem and the formula $t = \frac{d}{r}$.

	d	r	t
Bike	x	12	$\frac{x}{12}$
Car	x	36	$\frac{x}{36}$

From the problem, the equation is stated in words: Car driving time = Bike riding time less $\frac{1}{4}$ hour. Use the car time and the bike time given in the table to write the equation.

$$\frac{x}{36} = \frac{x}{12} - \frac{1}{4}$$

$$36\left(\frac{x}{36}\right) = 36\left(\frac{x}{12}\right) - 36\left(\frac{1}{4}\right)$$

$$x = 3x - 9$$

$$-2x = -9$$

$$x = 4\frac{1}{2}$$

The distance between his office and home is $4\frac{1}{2}$ miles.

21. To solve problems about distance, rate and time, we use the formula $d = rt$. To solve problems about work, we use the similar formula $A = rt$.

25. Let x = the number of hours it will take Sandi to refinish the table alone. So, in one hour, Sandi does $1/x$ of the job. Since Cary can do the job in 7 hours, in one hour, he does $1/7$ of the job. Together they can do it in 5 hours, so in one hour, they can do $1/5$ of the job. Use this information to write an equation.

Amount Sandi does in 1 hour		Amount Cary does in 1 hour		Amount done together in 1 hour
↓		↓		↓
$\frac{1}{x}$	$+$	$\frac{1}{7}$	$=$	$\frac{1}{5}$

Solve this equation. The least common denominator is $35x$.

$$35x\left(\frac{1}{x} + \frac{1}{7}\right) = 35x \cdot \frac{1}{5}$$

$$35 + 5x = 7x$$

$$35 = 2x$$

$$\frac{35}{2} = x$$

Working alone, Sandi can refinish the table in $\frac{35}{2}$ or $17\frac{1}{2}$ hours.

29. Let x = the number of hours it will take to fill the pond if both pipes are left on the same time. So, $1/x$ is the amount filled in one hour when both pipes are turned on together. Since the inlet pipe can fill the barrel in 60 minutes, in one minute, $1/60$ would be filled. Since the outlet pipe can empty the pond in 80 minutes, in

one minute 1/80 would be emptied. In this case, emptying translates to subtraction. Use this information to write an equation.

Amount inlet pipe fills in 1 hour	Amount outlet pipe empties in 1 hour	Amount filled in 1 hour with both pipes on
$\dfrac{1}{60}$	$-\quad\dfrac{1}{80}$	$=\quad\dfrac{1}{x}$

Solve this equation. The least common denominator is $240x$.

$$240x\left(\frac{1}{60} - \frac{1}{80}\right) = 240x \cdot \frac{1}{x}$$
$$4x - 3x = 240$$
$$x = 240$$

If both pipes are left on the same time, it will take 240 minutes to fill the pond.

CHAPTER 7

SECTION 7.1 (page 453)

1. The principal square root of 36 is 6.
5. If a whole number is a four-digit perfect square and its units digit is 5, the units digit of its real square roots must be 5. For example, the real square roots of
$$\underbrace{2025}_{\text{Units digits are 5.}} \text{ are } 45 \text{ and } -45.$$

9. The square roots of 121 are 11 and -11 since $11^2 = 121$ and $(-11)^2 = 121$. (Note that the symbol $\sqrt{121}$ represents only the principal root, 11.)
13. There are no real number square roots of -225 since no real number can be squared to give -225 ($15^2 = 225$ and $(-15)^2 = 225$).
17. Area of a rectangle $=$ Length $\cdot$ Width. Approximate the length $\sqrt{98}$ with $\sqrt{100} = 10$, and the width $\sqrt{26}$ with $\sqrt{25} = 5$. The best estimate for the area is $10 \cdot 5 = 50$: choice (c).
21. Use a calculator to find $\sqrt{2209} = 47$.
25. $-\sqrt{1444} = -(38) = -38$ since $38^2 = 1444$.
29. (a) For $a > 0$, let $a = 4$. So $-\sqrt{-4}$ is not a real number since it is not possible to take the square root of -4.
    (b) For $a < 0$, let $a = -4$. Then $-\sqrt{-(-4)} = -\sqrt{4} = -2$. So, for $a < 0$, $-\sqrt{-a}$ is a negative number.
    (c) For $a = 0$, then $-\sqrt{-0} = -\sqrt{0} = -0 = 0$.
33. $\sqrt[3]{216} = 6$ since $6^3 = 216$.
37. $-\sqrt[3]{512} = -(8) = -8$ since $8^3 = 512$.
41. $-\sqrt[4]{81} = -(3) = -3$ since $3^4 = 81$.
45. $\sqrt[6]{(-2)^6} = |-2| = 2$.
49. $\sqrt{\dfrac{64}{81}} = \dfrac{8}{9}$, since $\left(\dfrac{8}{9}\right)^2 = \dfrac{64}{81}$.
53. $\sqrt[3]{x^3} = x$ ($|x|$ is not required for an odd index.)
57. $\sqrt{9483} \approx 97.380696$ Use a calculator.
    $\qquad\qquad \approx 97.381$ Round to three decimal places.

61. The square roots of 16 are 4 and -4 since $4^2 = 16$ and $(-4)^2 = 16$.
65. The fourth roots of 81 are 3 and -3 since $3^4 = 81$ and $(-3)^4 = 81$.
69. $\pm\sqrt{25}$ represents the two numbers $\sqrt{25} = 5$ and $-\sqrt{25} = -5$.

SECTION 7.2 (page 459)

1. The statement is true.
$$8^{2/3} = (\sqrt[3]{8})^2 = 2^2 = 4$$

5. (a) $(-27)^{2/3} = (\sqrt[3]{-27})^2 = (-3)^2 = 9$
   (b) $(-64)^{5/3} = (\sqrt[3]{-64})^5 = (-4)^5 = -1024$
   (c) $(-100)^{1/2} = \sqrt{-100}$, not a real number.
   (d) $(-32)^{1/5} = \sqrt[5]{-32} = -2$
   $(-27)^{2/3}$ is a positive number, so (a) is the correct choice.
9. $729^{1/3} = \sqrt[3]{729} = 9$
13. $\left(\dfrac{64}{81}\right)^{1/2} = \sqrt{\dfrac{64}{81}} = \dfrac{8}{9}$
17. $100^{3/2} = (100^{1/2})^3 = (\sqrt{100})^3 = 10^3 = 1000$
21. $(-144)^{1/2} = \sqrt{-144}$, not a real number.
25. $\left(\dfrac{625}{16}\right)^{-1/4} = \left(\dfrac{16}{625}\right)^{1/4} = \sqrt[4]{\dfrac{16}{625}} = \dfrac{2}{5}$
29. $\left(-\dfrac{4}{9}\right)^{-1/2} = \left(-\dfrac{9}{4}\right)^{1/2}$, not a real number.
33. Use the following keystrokes on your calculator:

    $\boxed{1}\,\boxed{0}\,\boxed{0}\,\boxed{y^x}\,\boxed{(}\,\boxed{3}\,\boxed{\div}\,\boxed{2}\,\boxed{)}\,\boxed{=}$

    The calculator display is 1000.
37. $3^{1/2} \cdot 3^{3/2} = 3^{1/2+3/2}$ Product rule
    $\qquad\qquad = 3^{4/2} = 3^2 = 9$
41. $y^{7/3} \cdot y^{-4/3} = y^{(7/3)+(-4/3)} = y^{3/3} = y$
45. $(27^1 x^{12} y^{15})^{2/3} = 27^{2/3} x^{12(2/3)} y^{15(2/3)}$ Power rule
    $\qquad\qquad = (27^{1/3})^2 x^8 y^{10}$
    $\qquad\qquad = 3^2 x^8 y^{10} = 9x^8 y^{10}$
49. $\dfrac{m^{3/4} n^{-1/4}}{(m^2 n^1)^{1/2}} = \dfrac{m^{3/4} n^{-1/4}}{m^1 n^{1/2}}$ Power rule
    $\qquad\qquad = m^{(3/4)-1} n^{(-1/4)-(1/2)}$ Quotient rule
    $\qquad\qquad = m^{(3/4)-(4/4)} n^{(-1/4)-(2/4)}$
    $\qquad\qquad = m^{-1/4} n^{-3/4}$
    $\qquad\qquad = \dfrac{1}{m^{1/4} n^{3/4}}$ Write with positive exponents.
53. One way to evaluate $\sqrt[3]{5^{12}}$ is to raise 5 to the power obtained when 12 is divided by 3. The resulting exponent is 4, which gives a final answer of 625.
57. $\sqrt[3]{4^9} = 4^{9/3} = 4^3 = 64$
61. $\sqrt[3]{x} \cdot \sqrt{x} = x^{1/3} \cdot x^{1/2} = x^{2/6} \cdot x^{3/6} = x^{5/6} = \sqrt[6]{x^5}$
65. $\dfrac{\sqrt[3]{t^4}}{\sqrt[5]{t^4}} = \dfrac{t^{4/3}}{t^{4/5}} = \dfrac{t^{20/15}}{t^{12/15}} = t^{(20/15)-(12/15)} = t^{8/15} = \sqrt[15]{t^8}$
69. In the expression $9k^{-3/4} + 2k^{-1/4}$, the common factor is $k^{-3/4}$ since $-\frac{3}{4}$ is smaller than $-\frac{1}{4}$.
73. $\dfrac{4}{\sqrt{t}} + 7\sqrt{t^3} = \dfrac{4}{t^{1/2}} + 7t^{3/2}$
    $\qquad\qquad = 4t^{-1/2} + 7t^{3/2}$
    $\qquad\qquad = t^{-1/2}(4 + 7t^2)$

77. The conclusion that $\sqrt[4]{\dfrac{2143}{22}} = \pi$ is incorrect. Both quantities are irrational numbers with unending decimal representations. They might disagree if more decimal places are shown. In fact, they disagree in the next decimal place. It would, however, be correct to conclude that $\sqrt[4]{\dfrac{2143}{22}} \approx \pi$.

SECTION 7.3 (page 469)

1. By the product rule, $\sqrt{m} \cdot \sqrt{n} = \sqrt{mn}$.

5. $\sqrt[3]{x} \cdot \sqrt[3]{x}$ is not equal to x because there are only two factors of $\sqrt[3]{x}$.
$$\sqrt[3]{x} \cdot \sqrt[3]{x} = (\sqrt[3]{x})^2 \quad \text{or} \quad \sqrt[3]{x^2}$$

9. $\sqrt[3]{7x} \cdot \sqrt[3]{2y} = \sqrt[3]{7x \cdot 2y} = \sqrt[3]{14xy}$

13. $\sqrt{\dfrac{64}{121}} = \dfrac{\sqrt{64}}{\sqrt{121}} = \dfrac{8}{11}$

17. $\sqrt{\dfrac{x}{25}} = \dfrac{\sqrt{x}}{\sqrt{25}} = \dfrac{\sqrt{x}}{5}$

21. $\sqrt[3]{\dfrac{27}{64}} = \dfrac{\sqrt[3]{27}}{\sqrt[3]{64}} = \dfrac{3}{4}$

25. $\sqrt{12} = \sqrt{4 \cdot 3} = \sqrt{4} \cdot \sqrt{3} = 2\sqrt{3}$

29. $-\sqrt{32} = -\sqrt{16 \cdot 2} = -\sqrt{16} \cdot \sqrt{2} = -4\sqrt{2}$

33. $\sqrt{-300}$ is not a real number, because the negative sign is under the radical.

37. $\sqrt[3]{-16} = \sqrt[3]{-8 \cdot 2} = \sqrt[3]{-8} \cdot \sqrt[3]{2} = -2\sqrt[3]{2}$

41. $-\sqrt[4]{512} = -\sqrt[4]{256 \cdot 2} = -\sqrt[4]{256} \cdot \sqrt[4]{2} = -4\sqrt[4]{2}$

45. His reasoning was incorrect. The radicand 14 must be written as a *product* of two factors (not a sum of two terms) where one of the two factors is a perfect cube.

49. $\sqrt[3]{\dfrac{81}{64}} = \dfrac{\sqrt[3]{81}}{\sqrt[3]{64}} = \dfrac{\sqrt[3]{27 \cdot 3}}{4} = \dfrac{3\sqrt[3]{3}}{4}$

53. $-\sqrt[3]{27t^{12}} = -\sqrt[3]{27} \cdot \sqrt[3]{t^{12}} = -3t^4$

57. $-\sqrt[3]{-125a^6b^9c^{12}} = -\sqrt[3]{-125} \cdot \sqrt[3]{a^6} \cdot \sqrt[3]{b^9} \cdot \sqrt[3]{c^{12}}$
$$= -(-5) \cdot a^2 \cdot b^3 \cdot c^4$$
$$= 5a^2b^3c^4$$

61. $\sqrt{50x^3} = \sqrt{25 \cdot x^2 \cdot 2x} = \sqrt{25} \cdot \sqrt{x^2} \cdot \sqrt{2x}$
$$= 5x\sqrt{2x}$$

65. $\sqrt{13x^7y^8} = \sqrt{x^6y^8 \cdot 13x} = \sqrt{x^6 \cdot y^8} \cdot \sqrt{13x}$
$$= x^3y^4\sqrt{13x}$$

69. $\sqrt[3]{-16z^5t^7} = \sqrt[3]{-8z^3t^6 \cdot 2z^2t} = -2zt^2\sqrt[3]{2z^2t}$

73. $-\sqrt[4]{162r^{15}s^{10}} = -\sqrt[4]{81r^{12}s^8 \cdot 2r^3s^2}$
$$= -\sqrt[4]{81r^{12}s^8} \cdot \sqrt[4]{2r^3s^2}$$
$$= -3r^3s^2\sqrt[4]{2r^3s^2}$$

77. $\sqrt[3]{\dfrac{x^{16}}{27}} = \dfrac{\sqrt[3]{x^{15}x}}{\sqrt[3]{27}} = \dfrac{x^5\sqrt[3]{x}}{3}$

81. $\sqrt[10]{x^{25}} = x^{25/10} = x^{5/2} = \sqrt{x^5} = \sqrt{x^4} \cdot \sqrt{x} = x^2\sqrt{x}$

85. $\sqrt[4]{3} \cdot \sqrt[3]{4}$ has a common index of $4 \cdot 3 = 12$. Rewrite each radical with an index of 12.
$$\sqrt[4]{3} = 3^{1/4} = 3^{3/12} = \sqrt[12]{3^3}$$
$$\sqrt[3]{4} = 4^{1/3} = 4^{4/12} = \sqrt[12]{4^4}$$

$$\sqrt[4]{3} \cdot \sqrt[3]{4} = \sqrt[12]{3^3} \cdot \sqrt[12]{4^4}$$
$$= \sqrt[12]{3^3 \cdot 4^4}$$
$$= \sqrt[12]{27 \cdot 256} = \sqrt[12]{6912}$$

89. By the formula, the length of the missing leg is
$$a = \sqrt{c^2 - b^2}$$
$$= \sqrt{12^2 - 4^2} \quad \text{Let } c = 12 \text{ and } b = 4.$$
$$= \sqrt{144 - 16}$$
$$= \sqrt{128}$$
$$= \sqrt{64 \cdot 2} \quad \text{Factor.}$$
$$= \sqrt{64} \cdot \sqrt{2} \quad \text{Product rule.}$$
$$= 8\sqrt{2}.$$

93. Let $(x_1, y_1) = (-1, 5)$ and $(x_2, y_2) = (5, 3)$.
$$d = \sqrt{(x_2 - x_1)^2 + (y_2 - y_1)^2}$$
$$= \sqrt{[5 - (-1)]^2 + (3 - 5)^2}$$
$$= \sqrt{6^2 + (-2)^2}$$
$$= \sqrt{36 + 4}$$
$$= \sqrt{40}$$
$$= 2\sqrt{10}$$

97. Let $(x_1, y_1) = (x + y, y)$ and $(x_2, y_2) = (x - y, x)$.
$$d = \sqrt{(x_2 - x_1)^2 + (y_2 - y_1)^2}$$
$$= \sqrt{[(x - y) - (x + y)]^2 + (x - y)^2}$$
$$= \sqrt{(-2y)^2 + x^2 - 2xy + y^2}$$
$$= \sqrt{4y^2 + x^2 - 2xy + y^2}$$
$$= \sqrt{5y^2 - 2xy + x^2}$$

SECTION 7.4 (page 475)

1. Only (b) has like radical factors, so it can be simplified without first simplifying the individual radical expressions: $3\sqrt{6} + 9\sqrt{6} = 12\sqrt{6}$.

5. $\sqrt{36} - \sqrt{100} = 6 - 10 = -4$

9. $6\sqrt{18} - \sqrt{32} + 2\sqrt{50}$
$$= 6\sqrt{9 \cdot 2} - \sqrt{16 \cdot 2} + 2\sqrt{25 \cdot 2}$$
$$= 6 \cdot 3\sqrt{2} - 4\sqrt{2} + 2 \cdot 5\sqrt{2}$$
$$= 18\sqrt{2} - 4\sqrt{2} + 10\sqrt{2}$$
$$= 24\sqrt{2}$$

13. $2\sqrt{5} + 3\sqrt{20} + 4\sqrt{45}$
$$= 2\sqrt{5} + 3\sqrt{4 \cdot 5} + 4\sqrt{9 \cdot 5}$$
$$= 2\sqrt{5} + 3 \cdot 2\sqrt{5} + 4 \cdot 3\sqrt{5}$$
$$= 2\sqrt{5} + 6\sqrt{5} + 12\sqrt{5}$$
$$= 20\sqrt{5}$$

17. $3\sqrt{72m^2} - 5\sqrt{32m^2} - 3\sqrt{18m^2}$
$$= 3\sqrt{36m^2 \cdot 2} - 5\sqrt{16m^2 \cdot 2} - 3\sqrt{9m^2 \cdot 2}$$
$$= 3 \cdot 6m\sqrt{2} - 5 \cdot 4m\sqrt{2} - 3 \cdot 3m\sqrt{2}$$
$$= 18m\sqrt{2} - 20m\sqrt{2} - 9m\sqrt{2}$$
$$= -11m\sqrt{2}$$

21. $2\sqrt[3]{27x} - 2\sqrt[3]{8x}$
$$= 2\sqrt[3]{27} \cdot \sqrt[3]{x} - 2\sqrt[3]{8} \cdot \sqrt[3]{x}$$
$$= 2 \cdot 3 \cdot \sqrt[3]{x} - 2 \cdot 2 \cdot \sqrt[3]{x}$$
$$= 6\sqrt[3]{x} - 4\sqrt[3]{x}$$
$$= 2\sqrt[3]{x}$$

25. $3\sqrt[4]{x^5 y} - 2x\sqrt[4]{xy}$
$$= 3\sqrt[4]{x^4 \cdot xy} - 2x\sqrt[4]{xy}$$
$$= 3\sqrt[4]{x^4} \cdot \sqrt[4]{xy} - 2x\sqrt[4]{xy}$$
$$= 3x\sqrt[4]{xy} - 2x\sqrt[4]{xy}$$
$$= x\sqrt[4]{xy}$$

29. $\sqrt{\dfrac{8}{9}} + \sqrt{\dfrac{18}{36}}$
$$\approx \frac{\sqrt{8}}{\sqrt{9}} + \frac{\sqrt{18}}{\sqrt{36}}$$
$$\approx \frac{2\sqrt{2}}{3} + \frac{3\sqrt{2}}{6}$$
$$\approx \frac{4\sqrt{2}}{6} + \frac{3\sqrt{2}}{6}$$
$$\approx \frac{4\sqrt{2} + 3\sqrt{2}}{6}$$
$$\approx \frac{7\sqrt{2}}{6}$$

33. $3\sqrt{32} - 2\sqrt{8}$
$$\approx 3(5.656854249) - 2(2.828427125)$$
$$\approx 16.97056275 - 5.65685425$$
$$\approx 11.3137085$$

$$8\sqrt{2} \approx 8(1.414213562)$$
$$\approx 11.3137085$$

The fact that both $3\sqrt{32} - 2\sqrt{8}$ and $8\sqrt{2}$ are each approximately equal to 11.3137085 supports the statement $3\sqrt{32} - 2\sqrt{8} = 8\sqrt{2}$.

37. Approximate the length $\sqrt{192}$ with $\sqrt{196} = 14$, and the width $\sqrt{48}$ with $\sqrt{49} = 7$. The best estimate for the yard's dimensions is 14 by 7: choice (a).

41. To find the perimeter, add the lengths of the sides.
$$4\sqrt{18} + 4\sqrt{32} + 6\sqrt{50} + 5\sqrt{12}$$
$$= 4\sqrt{9 \cdot 2} + 4\sqrt{16 \cdot 2} + 6\sqrt{25 \cdot 2} + 5\sqrt{4 \cdot 3}$$
$$= 4 \cdot 3\sqrt{2} + 4 \cdot 4\sqrt{2} + 6 \cdot 5\sqrt{2} + 5 \cdot 2\sqrt{3}$$
$$= 12\sqrt{2} + 16\sqrt{2} + 30\sqrt{2} + 10\sqrt{3}$$
$$= 58\sqrt{2} + 10\sqrt{3} \text{ centimeters}$$

SECTION 7.5 (page 483)

1. $\sqrt{a} \cdot \sqrt{b} = \sqrt{ab}$

5. $(x + y)^2 = x^2 + 2xy + y^2$

9. $\sqrt{2}(\sqrt{18} - \sqrt{3}) = \sqrt{2} \cdot \sqrt{18} - \sqrt{2} \cdot \sqrt{3}$
$$= \sqrt{36} - \sqrt{6}$$
$$= 6 - \sqrt{6}$$

13. $(\sqrt{12} - \sqrt{3})(\sqrt{12} + \sqrt{3}) = (\sqrt{12})^2 - (\sqrt{3})^2$
$$= 12 - 3 = 9$$

17. $(\sqrt{3x} + 2)(\sqrt{3x} - 2) = (\sqrt{3x})^2 - 2^2$
$$= 3x - 4$$

21. $(4\sqrt{x} + 3)^2 = (4\sqrt{x})^2 + 2(4\sqrt{x})(3) + 3^2$
$$= 16x + 24\sqrt{x} + 9$$

25. $6 - 4\sqrt{3}$ is not equal to $2\sqrt{3}$ because 6 and $4\sqrt{3}$ are not like terms, so they cannot be combined.

29. $\dfrac{15}{\sqrt{3}} = \dfrac{15}{\sqrt{3}} \cdot \dfrac{\sqrt{3}}{\sqrt{3}} = \dfrac{15 \cdot \sqrt{3}}{\sqrt{3} \cdot \sqrt{3}} = \dfrac{15\sqrt{3}}{3} = 5\sqrt{3}$

33. $\dfrac{9\sqrt{3}}{\sqrt{5}} = \dfrac{9\sqrt{3} \cdot \sqrt{5}}{\sqrt{5} \cdot \sqrt{5}} = \dfrac{9\sqrt{15}}{5}$

37. $\dfrac{-8\sqrt{3}}{\sqrt{k}} = \dfrac{-8\sqrt{3} \cdot \sqrt{k}}{\sqrt{k} \cdot \sqrt{k}} = \dfrac{-8\sqrt{3k}}{k}$

41. Multiply the numerator and denominator by $\sqrt{y}$:
$$\frac{6\sqrt{3y}}{\sqrt{y^3}} = \frac{6\sqrt{3y} \cdot \sqrt{y}}{\sqrt{y^3} \cdot \sqrt{y}} = \frac{6\sqrt{3y^2}}{\sqrt{y^4}} = \frac{6y\sqrt{3}}{y^2} = \frac{6\sqrt{3}}{y}.$$

Multiply the numerator and denominator by $\sqrt{y^3}$:
$$\frac{6\sqrt{3y}}{\sqrt{y^3}} = \frac{6\sqrt{3y} \cdot \sqrt{y^3}}{\sqrt{y^3} \cdot \sqrt{y^3}} = \frac{6\sqrt{3y^4}}{\sqrt{y^6}} = \frac{6y^2\sqrt{3}}{y^3} = \frac{6\sqrt{3}}{y}.$$

Answers will vary. A sample answer follows: Both methods lead to the same result, $\dfrac{6\sqrt{3}}{y}$, but multiplying the numerator and denominator by $\sqrt{y}$ produces this result more directly, with less simplification required.

45. $-\sqrt{\dfrac{7}{50}} = -\dfrac{\sqrt{7}}{\sqrt{25 \cdot 2}} = -\dfrac{\sqrt{7}}{5\sqrt{2}}$
$$= -\frac{\sqrt{7} \cdot \sqrt{2}}{5\sqrt{2} \cdot \sqrt{2}} = -\frac{\sqrt{14}}{10}$$

49. $-\sqrt{\dfrac{98r^3}{s}} = -\dfrac{\sqrt{49r^2 \cdot 2r}}{\sqrt{s}} = -\dfrac{7r\sqrt{2r} \cdot \sqrt{s}}{\sqrt{s} \cdot \sqrt{s}}$
$$= -\frac{7r\sqrt{2rs}}{s}$$

53. $\sqrt[3]{\dfrac{2}{3}} = \dfrac{\sqrt[3]{2}}{\sqrt[3]{3}} \cdot \dfrac{\sqrt[3]{9}}{\sqrt[3]{9}} = \dfrac{\sqrt[3]{18}}{\sqrt[3]{27}} = \dfrac{\sqrt[3]{18}}{3}$

57. $-\sqrt[3]{\dfrac{2p}{r^2}} = -\dfrac{\sqrt[3]{2p} \cdot \sqrt[3]{r}}{\sqrt[3]{r^2} \cdot \sqrt[3]{r}} = -\dfrac{\sqrt[3]{2pr}}{\sqrt[3]{r^3}} = -\dfrac{\sqrt[3]{2pr}}{r}$

61. To rationalize the denominator of the expression $\dfrac{2}{4 + \sqrt{3}}$, we must multiply the numerator and denominator by $4 - \sqrt{3}$. The denominator becomes $(4 + \sqrt{3})(4 - \sqrt{3}) = 16 - 3 = 13$, which is a rational number.

65. $\dfrac{6}{\sqrt{5} + \sqrt{3}} = \dfrac{6(\sqrt{5} - \sqrt{3})}{(\sqrt{5} + \sqrt{3})(\sqrt{5} - \sqrt{3})}$
$$= \frac{6(\sqrt{5} - \sqrt{3})}{(\sqrt{5})^2 - (\sqrt{3})^2}$$
$$= \frac{6(\sqrt{5} - \sqrt{3})}{5 - 3} = \frac{6(\sqrt{5} - \sqrt{3})}{2}$$
$$= 3(\sqrt{5} - \sqrt{3})$$

69. $\dfrac{1 - \sqrt{2}}{\sqrt{7} + \sqrt{6}} = \dfrac{(1 - \sqrt{2})(\sqrt{7} - \sqrt{6})}{(\sqrt{7} + \sqrt{6})(\sqrt{7} - \sqrt{6})}$
$$= \frac{\sqrt{7} - \sqrt{6} - \sqrt{14} + \sqrt{12}}{(\sqrt{7})^2 - (\sqrt{6})^2}$$
$$= \frac{\sqrt{7} - \sqrt{6} - \sqrt{14} + \sqrt{4 \cdot 3}}{7 - 6}$$
$$= \sqrt{7} - \sqrt{6} - \sqrt{14} + 2\sqrt{3}$$

SOLUTIONS

73. $\dfrac{\sqrt{x} - \sqrt{y}}{\sqrt{x} + \sqrt{y}} = \dfrac{(\sqrt{x} - \sqrt{y})(\sqrt{x} - \sqrt{y})}{(\sqrt{x} + \sqrt{y})(\sqrt{x} - \sqrt{y})}$

$= \dfrac{(\sqrt{x})^2 - 2\sqrt{x}\sqrt{y} + (\sqrt{y})^2}{(\sqrt{x})^2 - (\sqrt{y})^2}$

$= \dfrac{x - 2\sqrt{xy} + y}{x - y}$

77. $\dfrac{25 + 10\sqrt{6}}{20} = \dfrac{5(5 + 2\sqrt{6})}{20} = \dfrac{5 + 2\sqrt{6}}{4}$

81. $\dfrac{6x + \sqrt{24x^3}}{3x} = \dfrac{6x + \sqrt{4x^2 \cdot 6x}}{3x}$

$= \dfrac{6x + 2x\sqrt{6x}}{3x}$

$= \dfrac{x(6 + 2\sqrt{6x})}{3x}$

$= \dfrac{6 + 2\sqrt{6x}}{3}$

85. $\dfrac{8\sqrt{5} - 1}{6} = \dfrac{(8\sqrt{5} - 1)(8\sqrt{5} + 1)}{6(8\sqrt{5} + 1)}$

$= \dfrac{64 \cdot 5 - 1}{6(8\sqrt{5} + 1)}$

$= \dfrac{320 - 1}{6(8\sqrt{5} + 1)}$

$= \dfrac{319}{6(8\sqrt{5} + 1)}$

Section 7.6 (page 491)

1. $\sqrt{9} = 3$, not -3. There is no solution of $\sqrt{x} = -3$.

5.
$$\sqrt{3k - 2} = 6$$
$$(\sqrt{3k - 2})^2 = 36 \quad \text{Square both sides.}$$
$$3k - 2 = 36$$
$$3k = 38$$
$$k = \frac{38}{3}$$

This answer checks in the original equation.
Solution set: $\left\{\frac{38}{3}\right\}$

9.
$$\sqrt{3x - 6} - 3 = 0$$
$$\sqrt{3x - 6} = 3 \quad \text{Get the radical term alone.}$$
$$(\sqrt{3x - 6})^2 = (3)^2 \quad \text{Square both sides.}$$
$$3x - 6 = 9$$
$$3x = 15$$
$$x = 5$$

This answer checks in the original equation.
Solution set: $\{5\}$

13.
$$3\sqrt{x} = \sqrt{8x + 9}$$
$$(3\sqrt{x})^2 = (\sqrt{8x + 9})^2 \quad \text{Square both sides.}$$
$$9x = 8x + 9$$
$$x = 9$$

This answer checks in the original equation.
Solution set: $\{9\}$

17.
$$\sqrt{3x + 4} = 8 - x$$
$$(\sqrt{3x + 4})^2 = (8 - x)^2 \quad \text{Square both sides.}$$
$$3x + 4 = (8)^2 - 2 \cdot 8 \cdot x + x^2 \quad \text{Square the binomial.}$$

$$3x + 4 = 64 - 16x + x^2$$
$$0 = x^2 - 19x + 60 \quad \text{Standard form}$$
$$0 = (x - 4)(x - 15) \quad \text{Factor.}$$
$$x - 4 = 0 \quad \text{or} \quad x - 15 = 0$$
$$x = 4 \quad \text{or} \qquad x = 15$$

Check $x = 4$: $\sqrt{3(4) + 4} = 8 - 4$?
$$\sqrt{16} = 4, \text{ a true statement.}$$
Check $x = 15$: $\sqrt{3(15) + 4} = 8 - 15$?
$$\sqrt{49} = -7, \text{ a false statement.}$$

Solution set: $\{4\}$

21.
$$\sqrt{r^2 - 15r + 15} + 5 = r$$
$$\sqrt{r^2 - 15r + 15} = r - 5 \quad \begin{array}{l}\text{Get the} \\ \text{radical} \\ \text{alone.}\end{array}$$
$$(\sqrt{r^2 - 15r + 15})^2 = (r - 5)^2 \quad \begin{array}{l}\text{Square both} \\ \text{sides.}\end{array}$$
$$r^2 - 15r + 15 = r^2 - 10r + 25 \quad \begin{array}{l}\text{Square the} \\ \text{binomial.}\end{array}$$
$$-5r = 10$$
$$r = -2$$

Check $r = -2$:

$$\sqrt{(-2)^2 - 15(-2) + 15} + 5 = -2 \quad ?$$
$$\sqrt{4 + 30 + 15} + 5 = -2 \quad ?$$
$$\sqrt{49} + 5 = -2 \quad ?$$
$$12 = -2 \quad ?$$

The last statement is false so $r = -2$ is not a solution.
Solution set: $\emptyset$

25.
$$\sqrt{11 + 2q} + 1 = \sqrt{5q + 1}$$
$$(\sqrt{11 + 2q} + 1)^2 = (\sqrt{5q + 1})^2$$
$$\text{Square both sides.}$$
$$(\sqrt{11 + 2q})^2 + 2 \cdot \sqrt{11 + 2q} + 1 = 5q + 1$$
$$\text{Square the binomial.}$$
$$11 + 2q + 2\sqrt{11 + 2q} + 1 = 5q + 1$$
$$2\sqrt{11 + 2q} = 3q - 11$$
$$\text{Get radical alone.}$$
$$(2\sqrt{11 + 2q})^2 = (3q - 11)^2$$
$$\text{Square both sides.}$$
$$4(11 + 2q) = 9q^2 - 66q + 121$$
$$44 + 8q = 9q^2 - 66q + 121$$
$$0 = 9q^2 - 74q + 77$$
$$\text{Standard form}$$
$$0 = (9q - 11)(q - 7)$$
$$\text{Factor.}$$
$$9q - 11 = 0 \quad \text{or} \quad q - 7 = 0$$
$$q = \frac{11}{9} \quad \text{or} \qquad q = 7$$

Check $q = \frac{11}{9}$:

$$\sqrt{11 + 2\left(\frac{11}{9}\right)} + 1 = \sqrt{5\left(\frac{11}{9}\right) + 1} \quad ?$$
$$\sqrt{\frac{99}{9} + \frac{22}{9}} + 1 = \sqrt{\frac{55}{9} + \frac{9}{9}} \quad ?$$
$$\sqrt{\frac{121}{9}} + 1 = \sqrt{\frac{64}{9}} \quad ?$$
$$\frac{11}{3} + 1 = \frac{8}{3} \quad ?$$

The last statement is false so $q = \frac{11}{9}$ is not a solution.

Check $q = 7$: $\sqrt{11 + 2(7)} + 1 = \sqrt{5(7) + 1}$?

$$\sqrt{25} + 1 = \sqrt{36} ?$$

$$5 + 1 = 6 ?$$

The last statement is true so 7 is a solution.
Solution set: $\{7\}$

29. Rewrite $\sqrt[3]{x + 3} = \sqrt[3]{5 + 4x}$ as
$(x + 3)^{1/3} = (5 + 4x)^{1/3}$. To eliminate the $1/3$ power, raise both sides to a power of 3 to get

$$[(x + 3)^{1/3}]^3 = [(5 + 4x)^{1/3}]^3,$$

so $x + 3 = 5 + 4x$ by the power rule. Thus, 3 is the smallest power to which you can raise both sides to eliminate the radicals.

33. $\sqrt[3]{1 - 2k} - \sqrt[3]{-k - 13} = 0$

$$\sqrt[3]{1 - 2k} = \sqrt[3]{-k - 13}$$

$$(\sqrt[3]{1 - 2k})^3 = (\sqrt[3]{-k - 13})^3$$

$$1 - 2k = -k - 13$$

$$-k = -14$$

$$k = 14$$

This answer checks in the original equation.
Solution set: $\{14\}$

37. $\sqrt[4]{x + 7} = \sqrt[4]{2x}$

$$(\sqrt[4]{x + 7})^4 = (\sqrt[4]{2x})^4$$

$$x + 7 = 2x$$

$$7 = x$$

This answer checks in the original equation.
Solution set: $\{7\}$

41. $(2w - 1)^{2/3} - w^{1/3} = 0$

$$(2w - 1)^{2/3} = w^{1/3}$$

$$[(2w - 1)^{2/3}]^3 = (w^{1/3})^3$$

$$(2w - 1)^2 = w$$

$$4w^2 - 4w + 1 = w$$

$$4w^2 - 5w + 1 = 0$$

$$(4w - 1)(w - 1) = 0$$

$$4w - 1 = 0 \text{or} w - 1 = 0$$

$$w = \frac{1}{4} \text{or} w = 1$$

Both answers check. Solution set: $\{\frac{1}{4}, 1\}$

45. The solution set of the equation
$\sqrt{x + 4} = 2 + \sqrt{x - 4}$ is $\{5\}$ since the x-coordinate of the point of intersection of the graphs of
$f(x) = \sqrt{x + 4}$ and $g(x) = 2 + \sqrt{x - 4}$ is 5.

Section 7.7 (page 501)

1. i is the imaginary unit.

5. $a + bi$ is a complex number if a and b are real numbers and i is the imaginary unit. Therefore, for every real number a, if $b = 0$, $a = a + 0i$ is a complex number.

9. $-\sqrt{-144} = -\sqrt{144(-1)} = -\sqrt{144} \cdot \sqrt{-1} = -12i$

13. $\sqrt{-48} = \sqrt{48(-1)} = \sqrt{16 \cdot 3} \cdot \sqrt{-1} = 4\sqrt{3}i$
$$= 4i\sqrt{3}$$

17. $\sqrt{-4} \cdot \sqrt{-25} = 2i \cdot 5i = 10i^2 = 10(-1) = -10$

21. $\dfrac{\sqrt{-75}}{\sqrt{3}} = \dfrac{\sqrt{75} \cdot \sqrt{-1}}{\sqrt{3}} = \sqrt{\dfrac{75}{3}} \cdot i = \sqrt{25}i = 5i$

25. $(5 - i) + (-5 + i) = (5 - 5) + (-1 + 1)i$
$$= 0 + 0i$$

29. $(-3 - 4i) - (-1 - 4i) = (-3 + 1) + (-4 + 4)i$
$$= -2 + 0i$$

33. $[(7 + 3i) - (4 - 2i)] + (3 + i)$
$$= [(7 - 4) + (3 + 2)i] + (3 + i)$$
$$= (3 + 5i) + (3 + i)$$
$$= (3 + 3) + (5 + 1)i$$
$$= 6 + 6i$$

37. $(3i)(27i) = 81i^2 = 81(-1) = -81$

41. $5i(-6 + 2i) = (5i)(-6) + (5i)(2i)$
$$= -30i + 10i^2$$
$$= -30i + 10(-1)$$
$$= -10 - 30i$$

45. $(4 + 5i)^2 = 4^2 + 2(4)(5i) + (5i)^2$
$$= 16 + 40i + 25i^2$$
$$= 16 + 40i + 25(-1)$$
$$= -9 + 40i$$

49. (a) The conjugate of $a + bi$ is $a - bi$.
(b) If we multiply $a + bi$ by its conjugate, we get $a^2 + b^2$, which is always a real number.

53. $\dfrac{-7 + 4i}{3 + 2i} = \dfrac{(-7 + 4i)(3 - 2i)}{(3 + 2i)(3 - 2i)}$

$$= \dfrac{-21 + 14i + 12i - 8i^2}{3^2 - 4i^2}$$

$$= \dfrac{-21 + 26i - 8(-1)}{9 - 4(-1)}$$

$$= \dfrac{-21 + 26i + 8}{9 + 4}$$

$$= \dfrac{-13 + 26i}{13}$$

$$= \dfrac{13(-1 + 2i)}{13}$$

$$= -1 + 2i$$

57. $\dfrac{2 - 3i}{2 + 3i} = \dfrac{(2 - 3i)(2 - 3i)}{(2 + 3i)(2 - 3i)}$

$$= \dfrac{4 - 6i - 6i + 9i^2}{2^2 - 9i^2}$$

$$= \dfrac{4 - 12i + 9(-1)}{4 - 9(-1)}$$

$$= \dfrac{4 - 12i - 9}{4 + 9}$$

$$= \dfrac{-5 - 12i}{13}$$

$$= -\dfrac{5}{13} - \dfrac{12}{13}i$$

61. (a) $(x + 2)(3x - 1) = 3x^2 - x + 6x - 2$
$$= 3x^2 + 5x - 2$$
(b) $(1 + 2i)(3 - i) = 3 - i + 6i - 2i^2$
$$= 3 + 5i + 2$$
$$= 5 + 5i$$

65. The reciprocal of $5 - 4i$ is

$$\dfrac{1}{5 - 4i} = \dfrac{1 \cdot (5 + 4i)}{(5 - 4i)(5 + 4i)}$$

$$= \dfrac{5 + 4i}{25 - 16i^2}$$

$$= \dfrac{5 + 4i}{25 + 16} = \dfrac{5 + 4i}{41} = \dfrac{5}{41} + \dfrac{4}{41}i.$$

69. $i^{89} = i^{88} \cdot i$

$\quad = (i^4)^{22} \cdot i$

$\quad = 1^{22} \cdot i = 1 \cdot i = i$

73. $i^{-5} = i^{-4} \cdot i^{-1} = (i^4)^{-1} \cdot i^{-1} = 1 \cdot i^{-1}$

$\quad = \dfrac{1}{i} = \dfrac{1 \cdot i}{i \cdot i} = \dfrac{i}{i^2} = \dfrac{i}{-1} = -i$

77. $I = \dfrac{E}{R + (X_L - X_c)i}$

$\quad = \dfrac{2 + 3i}{5 + (4 - 3)i}$

$\quad = \dfrac{(2 + 3i)(5 - i)}{(5 + i)(5 - i)}$

$\quad = \dfrac{10 - 2i + 15i - 3i^2}{25 - i^2}$

$\quad = \dfrac{10 + 13i - 3(-1)}{25 - (-1)}$

$\quad = \dfrac{13 + 13i}{26}$

$\quad = \dfrac{13}{26} + \dfrac{13}{26}i = \dfrac{1}{2} + \dfrac{1}{2}i$

CHAPTER 8

SECTION 8.1 (page 523)

1. The first step in solving $2x^2 + 8x = 9$ by completing the square is to divide both sides by 2.

5. $x^2 = 81$

$\quad x = 9 \quad \text{or} \quad x = -9$

Solution set: $\{-9, 9\}$

9. $(x + 2)^2 = 25$

$\quad x + 2 = 5 \quad \text{or} \quad x + 2 = -5$

$\quad\quad x = 3 \quad \text{or} \quad\quad x = -7$

Solution set: $\{-7, 3\}$

13. $(4p + 1)^2 = 24$

$\quad 4p + 1 = \sqrt{24} = 2\sqrt{6} \quad \text{or} \quad 4p + 1 = -\sqrt{24}$

$\quad\quad\quad\quad\quad\quad\quad\quad\quad\quad\quad\quad\quad\quad\quad = -2\sqrt{6}$

$\quad 4p = -1 + 2\sqrt{6} \quad \text{or} \quad\quad 4p = -1 - 2\sqrt{6}$

$\quad p = \dfrac{-1 + 2\sqrt{6}}{4} \quad \text{or} \quad\quad p = \dfrac{-1 - 2\sqrt{6}}{4}$

Solution set: $\left\{ \dfrac{-1 + 2\sqrt{6}}{4}, \dfrac{-1 - 2\sqrt{6}}{4} \right\}$

17. $(r - 5)^2 = -3$

$\quad r - 5 = \sqrt{-3} \quad\quad \text{or} \quad r - 5 = -\sqrt{-3}$

$\quad r - 5 = i\sqrt{3} \quad\quad \text{or} \quad r - 5 = -i\sqrt{3}$

$\quad\quad r = 5 + i\sqrt{3} \quad \text{or} \quad\quad r = 5 - i\sqrt{3}$

Solution set: $\{5 + i\sqrt{3}, 5 - i\sqrt{3}\}$

21. $x^2 - 2x - 24 = 0$

$\quad x^2 - 2x = 24 \quad$ Get the variable terms alone.

Half of -2 is -1; $(-1)^2 = 1$; add 1 to each side.

$x^2 - 2x + 1 = 24 + 1$

$\quad (x - 1)^2 = 25 \quad$ Factor the left side.

$\quad x - 1 = 5 \quad \text{or} \quad x - 1 = -5 \quad$ Square root property

$\quad\quad x = 6 \quad \text{or} \quad\quad x = -4$

Solution set: $\{-4, 6\}$

25. $2k^2 + 5k - 2 = 0$

$\quad 2k^2 + 5k = 2$

$\quad k^2 + \dfrac{5}{2}k = 1$

$\quad \left(\dfrac{1}{2} \cdot \dfrac{5}{2} \right)^2 = \left(\dfrac{5}{4} \right)^2 = \dfrac{25}{16}; \text{add } \dfrac{25}{16} \text{ to each side.}$

$\quad k^2 + \dfrac{5}{2}k + \dfrac{25}{16} = 1 + \dfrac{25}{16} = \dfrac{16}{16} + \dfrac{25}{16}$

$\quad \left(k + \dfrac{5}{4} \right)^2 = \dfrac{41}{16}$

$\quad k + \dfrac{5}{4} = \dfrac{\sqrt{41}}{\sqrt{16}} = \dfrac{\sqrt{41}}{4} \quad \text{or} \quad k + \dfrac{5}{4} = -\dfrac{\sqrt{41}}{4}$

$\quad k = \dfrac{-5}{4} + \dfrac{\sqrt{41}}{4} = \dfrac{-5 + \sqrt{41}}{4} \quad \text{or}$

$\quad k = \dfrac{-5 - \sqrt{41}}{4}$

Solution set: $\left\{ \dfrac{-5 + \sqrt{41}}{4}, \dfrac{-5 - \sqrt{41}}{4} \right\}$

29. $9x^2 - 24x = -13$

$\quad x^2 - \dfrac{24}{9}x = \dfrac{-13}{9} \quad$ Divide by 9.

$\quad x^2 - \dfrac{8}{3}x = \dfrac{-13}{9} \quad -\dfrac{24}{9} = -\dfrac{8}{3}.$

$\quad \left(\dfrac{1}{2} \cdot \dfrac{8}{3} \right)^2 = \left(\dfrac{8}{6} \right)^2 = \left(\dfrac{4}{3} \right)^2 = \dfrac{16}{9} \quad \text{Add } \dfrac{16}{9} \text{ to each side.}$

$\quad x^2 - \dfrac{8}{3}x + \dfrac{16}{9} = \dfrac{-13}{9} + \dfrac{16}{9}$

$\quad \left(x - \dfrac{4}{3} \right)^2 = \dfrac{3}{9} \quad$ Factor.

$\quad x - \dfrac{4}{3} = \sqrt{\dfrac{3}{9}} \quad \text{or} \quad x - \dfrac{4}{3} = -\sqrt{\dfrac{3}{9}}$

$\quad x = \dfrac{4}{3} + \dfrac{\sqrt{3}}{3} \quad \text{or} \quad x = \dfrac{4}{3} - \dfrac{\sqrt{3}}{3}$

$\quad x = \dfrac{4 + \sqrt{3}}{3} \quad \text{or} \quad x = \dfrac{4 - \sqrt{3}}{3}$

Solution set: $\left\{ \dfrac{4 + \sqrt{3}}{3}, \dfrac{4 - \sqrt{3}}{3} \right\}$

33. $3r^2 + 4r + 4 = 0$

$\quad 3r^2 + 4r = -4 \quad$ Get the variable terms alone.

$\quad r^2 + \dfrac{4}{3}r = \dfrac{-4}{3} \quad$ Divide by 3.

$\quad \left(\dfrac{1}{2} \cdot \dfrac{4}{3} \right)^2 = \left(\dfrac{2}{3} \right)^2 = \dfrac{4}{9} \quad \text{Add } \dfrac{4}{9} \text{ to each side.}$

$\quad r^2 + \dfrac{4}{3}r + \dfrac{4}{9} = \dfrac{-4}{3} + \dfrac{4}{9} = \dfrac{-12}{9} + \dfrac{4}{9}$

$\quad \left(r + \dfrac{2}{3} \right)^2 = \dfrac{-8}{9} \quad$ Factor.

$\quad r + \dfrac{2}{3} = \dfrac{\sqrt{-8}}{\sqrt{9}} \quad \text{or} \quad r + \dfrac{2}{3} = -\dfrac{\sqrt{-8}}{\sqrt{9}}$

$\quad\quad \sqrt{-8} = \sqrt{-1} \cdot \sqrt{4} \cdot \sqrt{2} = 2i\sqrt{2}$

$\quad r = -\dfrac{2}{3} + \dfrac{2i\sqrt{2}}{3} \quad \text{or} \quad r = -\dfrac{2}{3} - \dfrac{2i\sqrt{2}}{3}$

$$r = \frac{-2 + 2i\sqrt{2}}{3} \quad \text{or} \quad r = \frac{-2 - 2i\sqrt{2}}{3}$$

Solution set: $\left\{ \dfrac{-2 + 2i\sqrt{2}}{3}, \dfrac{-2 - 2i\sqrt{2}}{3} \right\}$

37. $-m^2 - 6m - 12 = 0$

$m^2 + 6m + 12 = 0 \qquad$ Multiply each side by -1.

$m^2 + 6m = -12 \qquad$ Get the variable terms alone.

$$\left(\frac{1}{2} \cdot 6 \right)^2 = 3^2 = 9 \qquad \text{Add 9 to each side.}$$

$m^2 + 6m + 9 = -12 + 9$

$(m + 3)^2 = -3 \quad$ Factor.

$m + 3 = \sqrt{-3} \qquad$ or $\quad m + 3 = -\sqrt{-3}$

$m = -3 + i\sqrt{3} \quad$ or $\quad m = -3 - i\sqrt{3}$

Solution set: $\{ -3 + i\sqrt{3}, -3 - i\sqrt{3} \}$

41. From Exercise 40, we know that the area of each strip is x. Therefore, the total area of the six strips is 6 times x, or $6x$.

SECTION 8.2 (page 531)

1. This statement is true. The equation $(x + 4)^2 = -8$ has no real solutions because when applying the square root property, we get

$x + 4 = \sqrt{-8} \qquad$ or $\quad x + 4 = -\sqrt{-8}$

$x + 4 = 2i\sqrt{2} \qquad$ or $\quad x + 4 = -2i\sqrt{2}$

$x = -4 + 2i\sqrt{2} \quad$ or $\qquad x = -4 - 2i\sqrt{2}$

The solution set is $\{ -4 + 2i\sqrt{2}, -4 - 2i\sqrt{2} \}$, two imaginary solutions.

5. This statement is true. For example, if p is 3, a prime number, we have $x^2 = 3$. Apply the square root property and $x = \sqrt{3}$ or $x = -\sqrt{3}$, which are two irrational solutions. A prime number is a whole number greater than 1 whose only factors are 1 and itself. The square root of *any* prime number is irrational.

9. For $2k^2 + 4k + 1 = 0$, use $a = 2$, $b = 4$, and $c = 1$ in the quadratic formula.

$$k = \frac{-(4) \pm \sqrt{(4)^2 - 4(2)(1)}}{2(2)}$$

$$= \frac{-4 \pm \sqrt{16 - 8}}{4}$$

$$= \frac{-4 \pm \sqrt{8}}{4}$$

$$= \frac{-4 \pm 2\sqrt{2}}{4} = \frac{2(-2 \pm \sqrt{2})}{4} = \frac{-2 \pm \sqrt{2}}{2}$$

Solution set: $\left\{ \dfrac{-2 + \sqrt{2}}{2}, \dfrac{-2 - \sqrt{2}}{2} \right\}$

13. Rewrite $x^2 + 18 = 10x$ as $x^2 - 10x + 18 = 0$. Use $a = 1$, $b = -10$, $c = 18$ in the quadratic formula.

$$x = \frac{-(-10) \pm \sqrt{(-10)^2 - 4(1)(18)}}{2(1)}$$

$$= \frac{10 \pm \sqrt{100 - 72}}{2}$$

$$= \frac{10 \pm \sqrt{28}}{2}$$

$$= \frac{10 \pm 2\sqrt{7}}{2}$$

$$= \frac{2(5 \pm \sqrt{7})}{2} = 5 \pm \sqrt{7}$$

Solution set: $\{ 5 + \sqrt{7}, 5 - \sqrt{7} \}$

17. Put $(r - 3)(r + 5) = 2$ into $ax^2 + bx + c = 0$ form.

$$(r - 3)(r + 5) = 2$$

$$r^2 + 2r - 15 = 2$$

$$r^2 + 2r - 17 = 0$$

Use $a = 1$, $b = 2$, $c = -17$ in the quadratic formula.

$$r = \frac{-(2) \pm \sqrt{(2)^2 - 4(1)(-17)}}{2(1)}$$

$$= \frac{-2 \pm \sqrt{4 + 68}}{2}$$

$$= \frac{-2 \pm \sqrt{72}}{2}$$

$$= \frac{-2 \pm 6\sqrt{2}}{2} = \frac{-2(1 \pm 3\sqrt{2})}{2} = -1 \pm 3\sqrt{2}$$

Solution set: $\{ -1 + 3\sqrt{2}, -1 - 3\sqrt{2} \}$

21. For $r^2 - 6r + 14 = 0$, use $a = 1$, $b = -6$, $c = 14$ in the quadratic formula.

$$r = \frac{-(-6) \pm \sqrt{(-6)^2 - 4(1)(14)}}{2(1)}$$

$$= \frac{6 \pm \sqrt{36 - 56}}{2}$$

$$= \frac{6 \pm \sqrt{-20}}{2}$$

$$= \frac{6 \pm 2i\sqrt{5}}{2}$$

$$= \frac{2(3 \pm i\sqrt{5})}{2} = 3 \pm i\sqrt{5}$$

Solution set: $\{ 3 + i\sqrt{5}, 3 - i\sqrt{5} \}$

25. $x(3x + 4) = -2$

$3x^2 + 4x = -2$

$3x^2 + 4x + 2 = 0 \qquad$ Standard form

Use $a = 3$, $b = 4$, $c = 2$ in the quadratic formula.

$$x = \frac{-4 \pm \sqrt{4^2 - 4(3)(2)}}{2(3)}$$

$$= \frac{-4 \pm \sqrt{16 - 24}}{6}$$

$$= \frac{-4 \pm \sqrt{-8}}{6}$$

$$= \frac{-4 \pm 2i\sqrt{2}}{6}$$

$$= \frac{2(-2 \pm i\sqrt{2})}{6} = \frac{-2 \pm i\sqrt{2}}{3}$$

Solution set: $\left\{ \dfrac{-2 + i\sqrt{2}}{3}, \dfrac{-2 - i\sqrt{2}}{3} \right\}$

29. $\dfrac{1}{x^2} + 1 = -\dfrac{1}{x}$

$x^2\left(\dfrac{1}{x^2} + 1\right) = x^2\left(-\dfrac{1}{x}\right)$ Multiply by x^2.

$1 + x^2 = -x$

$x^2 + x + 1 = 0$ Standard form

Use $a = 1$, $b = 1$, $c = 1$ in the quadratic formula.

$x = \dfrac{-1 \pm \sqrt{1^2 - 4(1)(1)}}{2(1)}$

$= \dfrac{-1 \pm \sqrt{1 - 4}}{2} = \dfrac{-1 \pm \sqrt{-3}}{2}$

$= \dfrac{-1 \pm i\sqrt{3}}{2}$

Solution set: $\left\{\dfrac{-1 + i\sqrt{3}}{2}, \dfrac{-1 - i\sqrt{3}}{2}\right\}$

33. In the function $s(t) = -16t^2 + 160t$, $s(t)$ represents the distance from the ground after a time of t seconds. Replace $s(t)$ with 400 in the equation above to get

$400 = -16t^2 + 160t$

or $16t^2 - 160t + 400 = 0$ in standard form. Use $a = 16$, $b = -160$, $c = 400$ in the quadratic formula.

$t = \dfrac{-(-160) \pm \sqrt{(-160)^2 - 4(16)(400)}}{2(16)}$

$= \dfrac{160 \pm \sqrt{25{,}600 - 25{,}600}}{32}$

$= \dfrac{160 \pm \sqrt{0}}{32}$

$= \dfrac{160 \pm 0}{32} = \dfrac{160}{32} = 5$

After 5 seconds, the rock reaches a height of 400 feet. Since this is the only time the rock reaches 400 feet, this is the rock's maximum height.

37. To find the number of prisoners sentenced to death in 1990, use the function $f(x) = -6.5x^2 + 132.5x + 2117$ and let $x = 2$.

$f(x) = -6.5x^2 + 132.5x + 2117$

$f(2) = -6.5(2)^2 + 132.5(2) + 2117$

$= -6.5(4) + 132.5(2) + 2117$

$= -26 + 265 + 2117$

$= 2356$

There were 2356 prisoners sentenced to death in 1990.

41. In $25x^2 + 70x + 49 = 0$, $a = 25$, $b = 70$, $c = 49$.

$b^2 - 4ac = (70)^2 - 4(25)(49)$

$= 4900 - 4900 = 0$

Choice (b): exactly one rational number.

45. Rewrite $3x^2 = 5x + 2$ as $3x^2 - 5x - 2 = 0$. So $a = 3$, $b = -5$, $c = -2$.

$b^2 - 4ac = (-5)^2 - 4(3)(-2) = 25 + 24 = 49$.

Since 49 is a perfect square, the answer is choice (a): two distinct rational numbers.

49. In $24x^2 - 34x - 45$, $a = 24$, $b = -34$, $c = -45$.

$b^2 - 4ac = (-34)^2 - 4(24)(-45)$

$= 1156 + 4320$

$= 5476$

Since 5476 is a perfect square ($74 \cdot 74 = 5476$), the trinomial $24x^2 - 34x - 45$ can be factored.

$24x^2 - 34x - 45 = (6x + 5)(4x - 9)$

53. In $12x^2 - 83x - 7$, $a = 12$, $b = -83$, $c = -7$.

$b^2 - 4ac = (-83)^2 - 4(12)(-7)$

$= 6889 + 336$

$= 7225$

Since 7225 is a perfect square ($85 \cdot 85 = 7225$), the trinomial $12x^2 - 83x - 7$ can be factored.

$12x^2 - 83x - 7 = (12x + 1)(x - 7)$

Section 8.3 (page 541)

1. To solve the equation $\dfrac{14}{x} = x - 5$, we would first multiply by the LCD, x.

5. $1 - \dfrac{3}{x} - \dfrac{28}{x^2} = 0$

$x^2(1) - x^2\left(\dfrac{3}{x}\right) - x^2\left(\dfrac{28}{x^2}\right) = x^2 \cdot 0$ Multiply by x^2.

$x^2 - 3x - 28 = 0$

$(x - 7)(x + 4) = 0$ Factor.

$x - 7 = 0$ or $x + 4 = 0$

$x = 7$ or $x = -4$

Solution set: $\{-4, 7\}$

9. $\dfrac{1}{x} + \dfrac{2}{x + 2} = \dfrac{17}{35}$

Multiply each term by the LCD $35x(x + 2)$.

$35x(x + 2)\left(\dfrac{1}{x}\right) + 35x(x + 2)\left(\dfrac{2}{x + 2}\right)$

$= 35x(x + 2)\left(\dfrac{17}{35}\right)$

$35x + 70 + 70x = 17x^2 + 34x$

$70 + 105x = 17x^2 + 34x$

$0 = 17x^2 - 71x - 70$

$0 = (17x + 14)(x - 5)$

$17x + 14 = 0$ or $x - 5 = 0$

$x = -\dfrac{14}{17}$ $x = 5$

Solution set: $\{-\tfrac{14}{17}, 5\}$

13. $\dfrac{3}{2x} - \dfrac{1}{2(x+2)} = 1$

Multiply by $2x(x+2)$.

$$3(x+2) - x = 2x(x+2)$$
$$3x + 6 - x = 2x^2 + 4x$$
$$0 = 2x^2 + 2x - 6$$

Use $a = 2$, $b = 2$, $c = -6$ in the quadratic formula.

$$x = \frac{-2 \pm \sqrt{2^2 - 4(2)(-6)}}{2(2)}$$

$$= \frac{-2 \pm \sqrt{4 + 48}}{4}$$

$$= \frac{-2 \pm \sqrt{52}}{4}$$

$$= \frac{-2 \pm 2\sqrt{13}}{4} = \frac{2(-1 \pm \sqrt{13})}{4} = \frac{-1 \pm \sqrt{13}}{2}$$

Solution set: $\left\{ \dfrac{-1 + \sqrt{13}}{2}, \dfrac{-1 - \sqrt{13}}{2} \right\}$

17. Let $x =$ rate of the boat in still water.
With the speed of the current at 15 mph, then
$x - 15 =$ rate going upstream and
$x + 15 =$ rate going downstream.
Complete a table using the information in the problem, the rates given above, and the formula $t = \frac{d}{r}$.

	d	r	t
Upstream	4	$x - 15$	$\dfrac{4}{x-15}$
Downstream	16	$x + 15$	$\dfrac{16}{x+15}$

The time 48 minutes is written as $\frac{48}{60} = \frac{4}{5}$ hour. The time upstream plus the time downstream equals $\frac{4}{5}$. So, from the table, the equation is written

$$\frac{4}{x-15} + \frac{16}{x+15} = \frac{4}{5}.$$

Multiply each term by the LCD $5(x-15)(x+15)$.

$$20(x+15) + 80(x-15) = 4(x-15)(x+15)$$
$$20x + 300 + 80x - 1200 = 4(x^2 - 225)$$
$$100x - 900 = 4x^2 - 900$$
$$0 = 4x^2 - 100x$$
$$0 = 4x(x - 25)$$
$$4x = 0 \quad \text{or} \quad x - 25 = 0$$
$$x = 0 \quad \text{or} \quad x = 25$$

Reject $x = 0$ mph as a possible boat speed.
Yoshiaki's boat had a top speed of 25 miles per hour.

21. Let $x =$ the number of minutes it takes for the cold water tap alone to fill the washer,
$x + 9 =$ the number of minutes it takes for the hot water tap alone to fill the washer.

Working together, both taps can fill the washer in 6 minutes.
Complete a table using the above information.

Tap	Rate	Time	Fractional Part of Washer Filled
Cold	$\dfrac{1}{x}$	6	$\dfrac{1}{x}(6)$
Hot	$\dfrac{1}{x+9}$	6	$\dfrac{1}{x+9}(6)$

Since together the hot and cold taps fill one washer, the sum of their fractional parts is 1, or

$$\frac{6}{x} + \frac{6}{x+9} = 1.$$

Multiply each term by the LCD $x(x+9)$.

$$6(x+9) + 6x = x(x+9)$$
$$6x + 54 + 6x = x^2 + 9x$$
$$0 = x^2 - 3x - 54$$
$$0 = (x - 9)(x + 6)$$
$$x - 9 = 0 \quad \text{or} \quad x + 6 = 0$$
$$x = 9 \quad \text{or} \quad x = -6$$

Reject -6 as a possible time. The cold water tap can fill the washer in 9 minutes.

25.
$$3y = \sqrt{16 - 10y}$$
$$(3y)^2 = (\sqrt{16 - 10y})^2 \quad \text{Square both sides.}$$
$$9y^2 = 16 - 10y$$
$$9y^2 + 10y - 16 = 0 \quad \text{Standard form}$$
$$(9y - 8)(y + 2) = 0$$
$$9y - 8 = 0 \quad \text{or} \quad y + 2 = 0$$
$$9y = 8$$
$$y = \frac{8}{9} \quad \text{or} \quad y = -2$$

Check $y = -2$: $3(-2) = \sqrt{16 - 10(-2)}$?
$$-6 = \sqrt{36}, \text{ a false statement}$$

Check $y = \dfrac{8}{9}$: $3\left(\dfrac{8}{9}\right) = \sqrt{16 - 10\left(\dfrac{8}{9}\right)}$?

$$\frac{8}{3} = \sqrt{\frac{144}{9} - \frac{80}{9}} \text{ ?}$$

$$\frac{8}{3} = \sqrt{\frac{64}{9}}, \text{ a true statement}$$

Solution set: $\left\{ \dfrac{8}{9} \right\}$

29.
$$m = \sqrt{\frac{6 - 13m}{5}}$$
$$m^2 = \frac{6 - 13m}{5} \quad \text{Square both sides.}$$
$$5m^2 = 6 - 13m$$
$$5m^2 + 13m - 6 = 0 \quad \text{Standard form}$$
$$(5m - 2)(m + 3) = 0$$
$$5m - 2 = 0 \quad \text{or} \quad m + 3 = 0$$
$$5m = 2$$
$$m = \frac{2}{5} \quad \text{or} \quad m = -3$$

Check $m = \dfrac{2}{5}$: $\quad \dfrac{2}{5} = \sqrt{\dfrac{6 - 13(\frac{2}{5})}{5}}$?

$\qquad\qquad \dfrac{2}{5} = \sqrt{\dfrac{4}{25}}$, a true statement

Check $m = -3$: $-3 = \sqrt{\dfrac{6 - 13(-3)}{5}}$?

$\qquad\qquad -3 = \sqrt{\dfrac{45}{5}} = \sqrt{9}$

$\qquad\qquad -3 = 3$, a false statement

Solution set: $\{\frac{2}{5}\}$

33. In $4k^4 - 13k^2 + 9 = 0$, let $u = k^2$ and $u^2 = k^4$ to get
$4u^2 - 13u + 9 = 0$.
$(4u - 9)(u - 1) = 0$ Factor.
$4u - 9 = 0 \quad$ or $\quad u - 1 = 0$
$\quad 4u = 9$
$\qquad u = \dfrac{9}{4} \quad$ or $\qquad u = 1$

So $\quad k^2 = \dfrac{9}{4} \quad$ or $\qquad k^2 = 1 \quad$ Recall that $u = k^2$.

and $\quad k = \pm\dfrac{3}{2} \quad$ or $\qquad k = \pm 1$.   Square root property

Verify that all answers satisfy the original equation.
Solution set: $\{-\frac{3}{2}, -1, 1, \frac{3}{2}\}$

37. In $(t + 5)^2 + 6 = 7(t + 5)$, let $u = t + 5$ to get

$$u^2 + 6 = 7u.$$
$$u^2 - 7u + 6 = 0$$
$$(u - 6)(u - 1) = 0$$
$$u - 6 = 0 \quad \text{or} \quad u - 1 = 0$$
$$u = 6 \quad \text{or} \qquad u = 1.$$

So $\quad t + 5 = 6 \quad$ or $\quad t + 5 = 1 \qquad$ Recall that $u = t + 5$.

$\qquad\qquad t = 1 \quad$ or $\qquad t = -4$.

Verify that both answers check.
Solution set: $\{-4, 1\}$

41. In $2 - 6(m - 1)^{-2} = (m - 1)^{-1}$, let $u = m - 1$ to get

$$2 - 6u^{-2} = u^{-1}.$$

$$2 - \dfrac{6}{u^2} = \dfrac{1}{u} \qquad u^{-2} = \dfrac{1}{u^2} \text{ and } u^{-1} = \dfrac{1}{u}$$

Multiply each term by u^2.

$$2u^2 - 6 = u$$
$$2u^2 - u - 6 = 0$$
$$(2u + 3)(u - 2) = 0$$
$$2u + 3 = 0 \quad \text{or} \quad u - 2 = 0$$
$$2u = -3$$
$$u = -\dfrac{3}{2} \quad \text{or} \qquad u = 2$$

So $\quad m - 1 = -\dfrac{3}{2} \quad$ or $\quad m - 1 = 2$.
    Recall $u = m - 1$.

$$m = -\dfrac{1}{2} \quad \text{or} \qquad m = 3$$

Both answers check in the original equation.
Solution set: $\{-\frac{1}{2}, 3\}$

45. In $2(1 + \sqrt{y})^2 = 13(1 + \sqrt{y}) - 6$, let $u = 1 + \sqrt{y}$ to get

$$2u^2 = 13u - 6.$$
$$2u^2 - 13u + 6 = 0$$
$$(2u - 1)(u - 6) = 0$$
$$2u - 1 = 0 \quad \text{or} \quad u - 6 = 0$$
$$u = \dfrac{1}{2} \quad \text{or} \qquad u = 6$$

Replace u with $1 + \sqrt{y}$ to get

$$1 + \sqrt{y} = \dfrac{1}{2} \quad \text{or} \quad 1 + \sqrt{y} = 6.$$

$$\sqrt{y} = -\dfrac{1}{2} \quad \text{or} \qquad \sqrt{y} = 5$$

Not possible, since $\quad y = 25 \quad$ Square both
$\sqrt{y}$ must be positive. $\qquad\qquad$ sides.

Check $y = 25$: $2(1 + \sqrt{25})^2 = 13(1 + \sqrt{25}) - 6$?
$\qquad\qquad\qquad 2(1 + 5)^2 = 13(1 + 5) - 6$?
$\qquad\qquad\qquad\quad 2(36) = 78 - 6$?
$\qquad\qquad\qquad\qquad 72 = 72$, a true statement

Solution set: $\{25\}$

49. The equation $\dfrac{x^2}{(x - 3)^2} + \dfrac{3x}{x - 3} - 4 = 0$ can be rewritten as:

$$\left(\dfrac{x}{x - 3}\right)^2 + 3\left(\dfrac{x}{x - 3}\right) - 4 = 0.$$

SECTION 8.4 (page 549)

1. Since the triangle is a right triangle, use the Pythagorean theorem with legs m and n and hypotenuse p.

$$m^2 + n^2 = p^2$$
$$m^2 = p^2 - n^2 \qquad \text{Subtract } n^2.$$
$$m = \sqrt{p^2 - n^2} \qquad \text{Square root property}$$

Only the positive square root is given since m represents the side of a triangle.

5. $\qquad I = \dfrac{ks}{d^2}$

$$d^2(I) = d^2\left(\dfrac{ks}{d^2}\right) \qquad \text{Multiply by } d^2.$$
$$d^2 I = ks$$
$$d^2 = \dfrac{ks}{I} \qquad\qquad \text{Divide by } I.$$

$$d = \pm\sqrt{\frac{ks}{I}}$$ Square root property

$$= \frac{\pm\sqrt{ks} \cdot \sqrt{I}}{\sqrt{I} \cdot \sqrt{I}}$$ Rationalize the denominator.

$$d = \frac{\pm\sqrt{skI}}{I}$$

9. $V = \frac{1}{3}\pi r^2 h$

$3V = 3\left(\frac{1}{3}\pi r^2 h\right)$ Multiply by 3.

$3V = \pi r^2 h$

$\frac{3V}{\pi h} = r^2$ Divide by πh.

$r = \pm\sqrt{\frac{3V}{\pi h}}$ Square root property

$= \frac{\pm\sqrt{3V} \cdot \sqrt{\pi h}}{\sqrt{\pi h} \cdot \sqrt{\pi h}}$ Rationalize the denominator.

$r = \frac{\pm\sqrt{3\pi Vh}}{\pi h}$

13. $D = \sqrt{kh}$
$D^2 = kh$ Square both sides.
$\frac{D^2}{k} = h$ or $h = \frac{D^2}{k}$ Divide by k.

17.   In the formula $p = \sqrt{\frac{k\ell}{g}}$, if g is a positive number and k and ℓ have *different* signs, the value of $\frac{k\ell}{g}$ is negative. The square root of a negative number gives an imaginary value for p. For the equation to have a real value of p, k and ℓ must have the same signs (either both positive or both negative), thus creating a positive number under the radical and a real value for p.

21.   Let x = the length of the shorter leg,
$x + 1$ = the length of the longer leg, and
$x + 9$ = the length of the hypotenuse.
By the Pythagorean theorem,

$$x^2 + (x+1)^2 = (x+9)^2$$
$x^2 + x^2 + 2x + 1 = x^2 + 18x + 81$ Square the binomials.

$2x^2 + 2x + 1 = x^2 + 18x + 81$

$x^2 - 16x - 80 = 0$ Standard form

$(x - 20)(x + 4) = 0$ Factor.

$x - 20 = 0$ or $x + 4 = 0$
$x = 20$ or $x = -4$

Reject the negative solution since x represents the side of a triangle.
The length of the shorter leg is 20 meters.
The length of the longer leg is $20 + 1 = 21$ meters.
The length of the hypotenuse is $20 + 9 = 29$ meters.

25.   Let x = the width of the uncovered strip of flooring.
From the problem,

(length of the rug) · (width of the rug) = 234.

From the sketch below, the rug is centered in the room a distance x from the walls (width of the strip x), so the length of the rug

$= $ length of the room $- 2 \cdot$ (width of the strip)
$= 20 - 2x$

and the width of the rug

$= $ width of the room $- 2 \cdot$ (width of the strip)
$= 15 - 2x.$

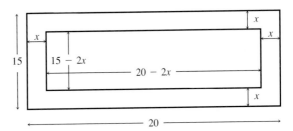

The equation (length of rug) · (width of rug) = 234 becomes

$$(20 - 2x)(15 - 2x) = 234$$
$300 - 70x + 4x^2 = 234$ Multiply the binomials.
$4x^2 - 70x + 66 = 0$ Standard form
$2(2x^2 - 35x + 33) = 0$ Factor.
$2(2x - 33)(x - 1) = 0$ Factor.
$2x - 33 = 0$ or $x - 1 = 0$
$x = \frac{33}{2}$ or $x = 1$

Reject $\frac{33}{2} = 16\frac{1}{2}$ since $16\frac{1}{2}$ is wider than the room itself. The width of the uncovered strip is 1 foot.

29.   Let x = the time in hours for pipe A to fill the tank,
$x + .5$ = the time in hours for pipe B to fill the tank,
and 4 = the time in hours for both pipes together to fill the tank.

Then $\frac{1}{x}$ = the rate for pipe A,

$\frac{1}{x + .5}$ = the rate for pipe B.

The part of the job done by A in 4 hours is $\frac{1}{x}(4) = \frac{4}{x}$.
The part of the job done by B in 4 hours is

$\frac{1}{x + .5}(4) = \frac{4}{x + .5}$. These parts must add to 1 whole job, so

$$\frac{4}{x} + \frac{4}{x + .5} = 1.$$
$4(x + .5) + 4x = x(x + .5)$ Multiply by $x(x + .5)$.
$4x + 2 + 4x = x^2 + .5x$
$x^2 - 7.5x - 2 = 0$ Standard form

Let $a = 1$, $b = -7.5$, and $c = -2$ in the quadratic formula.

$$x = \frac{-(-7.5) \pm \sqrt{(-7.5)^2 - 4(1)(-2)}}{2(1)}$$

$$= \frac{7.5 \pm \sqrt{64.25}}{2}$$

$$\approx \frac{7.5 \pm 8.0156}{2}$$

$$x \approx \frac{7.5 + 8.0156}{2} \quad \text{or} \quad x \approx \frac{7.5 - 8.0156}{2}$$

$$\approx 7.8 \qquad\qquad\qquad \approx -.3$$

Reject the solution $-.3$ since x represents time in hours. Pipe A takes 7.8 hours to fill the tank alone. Pipe B takes $7.8 + .5 = 8.3$ hours to fill the tank alone.

33. In the quadratic model, $f(x) = 18.7x^2 + 105.3x + 4814.1$, let $f(x) = 5300$ and solve for x.

$$5300 = 18.7x^2 + 105.3x + 4814.1$$
$$0 = 18.7x^2 + 105.3x - 485.9 \quad \text{Standard form}$$

Let $a = 18.7$, $b = 105.3$, and $c = -485.9$ in the quadratic formula.

$$x = \frac{-105.3 \pm \sqrt{(105.3)^2 - 4(18.7)(-485.9)}}{2(18.7)}$$

$$= \frac{-105.3 \pm \sqrt{11{,}088.09 + 36{,}345.32}}{37.4}$$

$$= \frac{-105.3 \pm \sqrt{47{,}433.41}}{37.4}$$

$$\approx \frac{-105.3 \pm 217.8}{37.4}$$

$$x = \frac{-105.3 + 217.8}{37.4} \quad \text{or} \quad x = \frac{-105.3 - 217.8}{37.4}$$

$$\approx 3 \qquad\qquad\qquad \approx -8.6$$

Reject the solution -8.6 since x represents time and $0 \le x \le 6$ for this model. $x = 3$ corresponds to 1987. So, in 1987, the poverty threshold was about $5300.

37. In $f(t) = 144t - 16t^2$, replace $f(t)$ with 128.

$$128 = 144t - 16t^2$$

Now solve for t.

$$16t^2 - 144t + 128 = 0 \quad \text{Standard form}$$
$$16(t^2 - 9t + 8) = 0 \quad \text{Factor.}$$
$$16(t - 8)(t - 1) = 0 \quad \text{Factor.}$$
$$t - 8 = 0 \quad \text{or} \quad t - 1 = 0$$
$$t = 8 \quad \text{or} \qquad t = 1$$

The object will be 128 feet above the ground at 1 second and at 8 seconds.

41. We use $\ell = 1.2$, $g = 9.8$, and the Froude number is 2.57.

$$2.57 = \frac{v^2}{9.8(1.2)}$$

$$2.57 = \frac{v^2}{11.76} \quad \text{Multiply.}$$

$$v^2 = 30.2232 \quad \text{Multiply by 11.76.}$$
$$v \approx 5.5 \qquad \text{Find the positive square root.}$$

The value of v is approximately 5.5 meters per second.

SECTION 8.5 (page 559)

1. When solving a quadratic or higher-degree inequality, include the endpoints if the symbol is $\ge$ or $\le$. Exclude the endpoints if the symbol is $>$ or $<$.

See the answer section for all final graphs in this section.

5. $(x + 1)(x - 5) > 0$
Solve $(x + 1)(x - 5) = 0$.

$$x + 1 = 0 \quad \text{or} \quad x - 5 = 0$$
$$x = -1 \quad \text{or} \qquad\qquad x = 5$$

The numbers -1 and 5 divide the number line as shown into three regions A, B, and C.

Test a point from each region in the inequality.

A: If $x = -2$,
$$(x + 1)(x - 5) = (-1)(-7) = 7 > 0. \quad \text{True}$$
B: If $x = 0$,
$$(x + 1)(x - 5) = (1)(-5) = -5 > 0. \quad \text{False}$$
C: If $x = 6$,
$$(x + 1)(x - 5) = (7)(1) = 7 > 0. \qquad \text{True}$$

Based on these results, the solution set includes regions A and C, excluding $x = -1$ and $x = 5$ (because we have an inequality sign $>$ rather than $\ge$).
The solution set is $(-\infty, -1) \cup (5, \infty)$.

9. $x^2 - 4x + 3 \ge 0$
Solve $x^2 - 4x + 3 = 0$.

$$(x - 1)(x - 3) = 0$$
$$x - 1 = 0 \quad \text{or} \quad x - 3 = 0$$
$$x = 1 \quad \text{or} \qquad\qquad x = 3$$

The numbers 1 and 3 divide the number line as shown into three regions, A, B, C.

Test a point from each region in the inequality.

A: If $x = 0$,
$$x^2 - 4x + 3 = 0^2 - 4(0) + 3$$
$$= 3 \ge 0. \qquad \text{True}$$
B: If $x = 2$,
$$x^2 - 4x + 3 = 2^2 - 4(2) + 3$$
$$= -1 \ge 0. \qquad \text{False}$$
C: If $x = 4$,
$$x^2 - 4x + 3 = 4^2 - 4(4) + 3$$
$$= 3 \ge 0. \qquad \text{True}$$

Based on these results, the solution set includes regions A and C, including $x = 1$ and $x = 3$ (because of the inequality sign $\geq$).
The solution set is $(-\infty, 1] \cup [3, \infty)$.

13. $9p^2 + 3p < 2$

Solve
$$9p^2 + 3p - 2 = 0.$$
$$(3p - 1)(3p + 2) = 0$$
$$3p - 1 = 0 \quad \text{or} \quad 3p + 2 = 0$$
$$p = \frac{1}{3} \quad \text{or} \quad p = -\frac{2}{3}$$

The numbers $\frac{1}{3}$ and $-\frac{2}{3}$ divide the number line as shown into three regions A, B, C.

Test a point from each region in the inequality.

A: If $p = -1$,
$$9p^2 + 3p = 9(-1)^2 + 3(-1) = 6 < 2. \quad \text{False}$$

B: If $p = 0$,
$$9p^2 + 3p = 9(0^2) + 3(0) = 0 < 2. \quad \text{True}$$

C: If $p = 1$,
$$9p^2 + 3p = 9(1^2) + 3(1) = 12 < 2. \quad \text{False}$$

Only region B will be included in the solution set, excluding $p = -\frac{2}{3}$ and $p = \frac{1}{3}$.
Solution set: $\left(-\frac{2}{3}, \frac{1}{3}\right)$

17. $y^2 - 6y + 6 \geq 0$

Solve $y^2 - 6y + 6 = 0$.
Since $y^2 - 6y + 6$ does not factor, let $a = 1$, $b = -6$, and $c = 6$ in the quadratic formula.
$$y = \frac{-(-6) \pm \sqrt{(-6)^2 - 4(1)(6)}}{2(1)}$$
$$= \frac{6 \pm \sqrt{12}}{2}$$
$$= \frac{6 \pm 2\sqrt{3}}{2} = \frac{2(3 \pm \sqrt{3})}{2} = 3 \pm \sqrt{3}$$

Approximate $\sqrt{3}$ as 1.7. Then
$$3 + \sqrt{3} \approx 3 + 1.7 = 4.7 \text{ and } 3 - \sqrt{3} \approx 3 - 1.7$$
$$= 1.3.$$

The numbers $3 - \sqrt{3}$ and $3 + \sqrt{3}$ divide the number line into regions A, B, and C as shown.

Test a point from each region.

A: If $y = 0$, $y^2 - 6y + 6 = 6 \geq 0$. True

B: If $y = 2$, $y^2 - 6y + 6 = -2 \geq 0$. False

C: If $y = 5$, $y^2 - 6y + 6 = 1 \geq 0$. True

The solution set includes regions A and C, including $y = 3 - \sqrt{3}$ and $y = 3 + \sqrt{3}$.
Solution set: $(-\infty, 3 - \sqrt{3}] \cup [3 + \sqrt{3}, \infty)$

21. $(a - 4)(2a + 3)(3a - 1) \geq 0$
Set each factor equal to 0 and then solve.
$$a - 4 = 0 \quad \text{or} \quad 2a + 3 = 0 \quad \text{or} \quad 3a - 1 = 0$$
$$a = 4 \quad \text{or} \quad a = -\frac{3}{2} \quad \text{or} \quad a = \frac{1}{3}$$

The numbers $-\frac{3}{2}, \frac{1}{3}$, and 4 divide the number line as shown into four regions A, B, C, and D.

Test a point from each region in the inequality.

A: If $a = -2$,
$$(a - 4)(2a + 3)(3a - 1) = -42 \geq 0. \quad \text{False}$$

B: If $a = 0$,
$$(a - 4)(2a + 3)(3a - 1) = 12 \geq 0. \quad \text{True}$$

C: If $a = 1$,
$$(a - 4)(2a + 3)(3a - 1) = -30 \geq 0. \quad \text{False}$$

D: If $a = 5$,
$$(a - 4)(2a + 3)(3a - 1) = 182 \geq 0. \quad \text{True}$$

The solution set includes regions B and D, including the points $a = -\frac{3}{2}$, $a = \frac{1}{3}$, and $a = 4$.
Solution set: $\left[-\frac{3}{2}, \frac{1}{3}\right] \cup [4, \infty)$

25. $\dfrac{2y + 3}{y - 5} \leq 0$

Write as an equation; then solve.
$$\frac{2y + 3}{y - 5} = 0$$
$$2y + 3 = 0 \qquad \text{Multiply by } y - 5.$$
$$y = -\frac{3}{2}$$

Set the denominator equal to 0 and solve.
$$y - 5 = 0$$
$$y = 5$$

The numbers $-\frac{3}{2}$ and 5 divide the number line as shown into three regions A, B, C.

Test a point from each region in the inequality.
A: Let $y = -2$.
$$\frac{2(-2) + 3}{(-2) - 5} \leq 0$$
$$\frac{1}{7} \leq 0 \quad \text{False}$$

B: Let $y = 0$.

$$\frac{2(0) + 3}{0 - 5} \leq 0$$

$$-\frac{3}{5} \leq 0 \quad \text{True}$$

C: Let $y = 6$.

$$\frac{2(6) + 3}{6 - 5} \leq 0$$

$$15 \leq 0 \quad \text{False}$$

Based on these results, the solution set includes the points in region B. The endpoint 5 is not included since it makes the denominator 0. The endpoint $-\frac{3}{2}$ is included because of the $\leq$ sign.
The solution set is $[-\frac{3}{2}, 5)$.

29. $\dfrac{3}{2t - 1} < 2$

Write as an equation; then solve.

$$\frac{3}{2t - 1} = 2$$

$$3 = 4t - 2 \quad \text{Multiply by } 2t - 1.$$

$$5 = 4t$$

$$\frac{5}{4} = t$$

Set the denominator equal to zero and solve.

$$2t - 1 = 0$$

$$t = \frac{1}{2}$$

The numbers $\frac{1}{2}$ and $\frac{5}{4}$ divide the number line as shown into three regions.

A   B   C

$\frac{1}{2}$ $\frac{5}{4}$

Test a point from each region in the inequality.
A: Let $t = 0$.

$$\frac{3}{2(0) - 1} < 2$$

$$-3 < 2 \quad \text{True}$$

B: Let $t = 1$.

$$\frac{3}{2(1) - 1} < 2$$

$$3 < 2 \quad \text{False}$$

C: Let $t = 2$.

$$\frac{3}{2(2) - 1} < 2$$

$$1 < 2 \quad \text{True}$$

Based on these results, the points in regions A and C are in the solution set. Neither endpoint $\frac{1}{2}$ nor $\frac{5}{4}$ is included.
The solution set is $(-\infty, \frac{1}{2}) \cup (\frac{5}{4}, \infty)$.

33. $\dfrac{4k}{2k - 1} < k$

Write as an equation; then solve.

$$\frac{4k}{2k - 1} = k$$

$$4k = 2k^2 - k \quad \text{Multiply by } 2k - 1.$$

$$0 = 2k^2 - 5k \quad \text{Standard form}$$

$$0 = k(2k - 5) \quad \text{Factor.}$$

$$k = 0 \quad \text{or} \quad 2k - 5 = 0$$

$$k = 0 \quad \text{or} \quad k = \frac{5}{2}$$

Set the denominator $2k - 1$ equal to zero and solve.

$$2k - 1 = 0$$

$$k = \frac{1}{2}$$

Based on these results, 0, $\frac{1}{2}$, and $\frac{5}{2}$ divide the number line as shown into four regions.

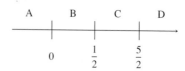

Test a point from each region.
A: Let $k = -1$.

$$\frac{4(-1)}{2(-1) - 1} < -1$$

$$\frac{4}{3} < -1 \quad \text{False}$$

B: Let $k = \frac{1}{4}$.

$$\frac{4\left(\frac{1}{4}\right)}{2\left(\frac{1}{4}\right) - 1} < \frac{1}{4}$$

$$-2 < \frac{1}{4} \quad \text{True}$$

C: Let $k = 1$.

$$\frac{4(1)}{2(1) - 1} < 1$$

$$4 < 1 \quad \text{False}$$

D: Let $k = 3$.

$$\frac{4(3)}{2(3) - 1} < 3$$

$$\frac{12}{5} < 3 \quad \text{True}$$

Based on these results, the points in regions B and D are in the solution set. None of the endpoints 0, $\frac{1}{2}$ and $\frac{5}{2}$ are included.
The solution set is $(0, \frac{1}{2}) \cup (\frac{5}{2}, \infty)$.

37. In $(3x + 5)^2 \le -4$, the squared expression, $(3x + 5)^2$, must be positive or zero. It can never be negative.
Solution set: $\emptyset$

41. To find the times the rock will be at ground level, use the quadratic function $s(t) = -16t^2 + 256t$ and set $s(t) = 0$.

$$0 = -16t^2 + 256t$$
$$16t^2 - 256t = 0$$
$$16t(t - 16) = 0$$
$$16t = 0 \quad \text{or} \quad t - 16 = 0$$
$$t = 0 \quad \text{or} \quad t = 16$$

The rock will be at ground level at 0 seconds (the time when it is initially projected) and at 16 seconds (the time when it hits the ground).

CHAPTER 9

SECTION 9.1 (page 585)

1. Since $a = 1$ in all four functions, all graphs open upward. Also, since $a = 1$, all four graphs are the same width as $f(x) = x^2$. As a result, we must consider the vertex of each.
 (a) $f(x) = (x + 2))^2 - 1$
  $= (x - (-2))^2 + (-1)$
 Thus, $h = -2$ and $k = -1$, so the vertex is $(h, k) = (-2, -1)$. The graph is given in B.
 (b) $f(x) = (x + 2)^2 + 1$
  $= (x - (-2))^2 + 1$
 Thus, $h = -2$ and $k = 1$, so the vertex is $(h, k) = (-2, 1)$. The graph is given in C.
 (c) In $f(x) = (x - 2)^2 - 1$, $h = 2$ and $k = -1$, so the vertex is $(h, k) = (2, -1)$. The graph is given in A.
 (d) In $f(x) = (x - 2)^2 + 1$, $h = 2$ and $k = 1$, so the vertex is $(h, k) = (2, 1)$. The graph is given in D.

5. Write $f(x) = x^2 + 4$ in the form $f(x) = a(x - h)^2 + k$ to get $f(x) = (x - 0)^2 + 4$. The vertex $(h, k) = (0, 4)$.

9. From the form $f(x) = a(x - h)^2 + k$, $f(x) = (x + 3)^2 - 4$ has vertex $(h, k) = (-3, -4)$.

13. The graph of $f(x) = 3x^2 + 1$ opens upward since $3 > 0$, and is narrower than $f(x) = x^2$ since $|3| > 1$.

17. The vertex of the graph of $f(x) = a(x - h)^2 + k$ is (h, k). Thus:
 (a) If $h > 0$ and $k > 0$, (h, k) is in quadrant I.
 (b) If $h > 0$ and $k < 0$, (h, k) is in quadrant IV.
 (c) If $h < 0$ and $k > 0$, (h, k) is in quadrant II.
 (d) If $h < 0$ and $k < 0$, (h, k) is in quadrant III.

See the answer section in your text for the graphs for Exercises 21–29.

21. The vertex of $f(x) = x^2 - 1$ is at $(0, -1)$; the graph opens upward and has the same shape as $y = x^2$ because $a = 1$. Two other points are $(-2, 3)$ and $(2, 3)$.

25. The vertex of $f(x) = .5(x - 4)^2$ is at $(4, 0)$; the graph opens upward and is wider than $y = x^2$ because $|a| = |.5| < 1$. Two other points are $(2, 2)$ and $(6, 2)$.

29. The vertex of $f(x) = -.5(x + 1)^2 + 2$ is at $(-1, 2)$; the graph opens downward and is wider than $y = x^2$ because $|a| = |-.5| < 1$. Two other points are $(3, -6)$ and $(-5, -6)$.

33. Since the graph of $f(x) = x + 6$ has a y-intercept of $(0, 6)$ and the graph of $g(x) = x$ has a y-intercept of $(0, 0)$, the graph of $f(x) = x + 6$ is shifted 6 units upward.

SECTION 9.2 (page 595)

1. In an equation of a parabola, if x is squared, the parabola has a vertical axis and if y is squared, the parabola has a horizontal axis.

5. $y = 2x^2 + 4x + 5$
$y = 2(x^2 + 2x) + 5$ Factor out 2.
$y = 2(x^2 + 2x + 1) + 5 - 2$ Complete the square; add $2(1) - 2$.
$y = 2(x + 1)^2 + 3$ Write in standard form.
This is a vertical parabola with vertex at $(-1, 3)$; the graph opens upward since $2 > 0$ and is narrower than $y = x^2$ since $|2| > 1$. To find the number of x-intercepts, use the discriminant $b^2 - 4ac$ with $a = 2$, $b = 4$, and $c = 5$. The value of the discriminant is

$$4^2 - 4(2)(5) = 16 - 40 = -24.$$

Since $-24 < 0$, there are no x-intercepts.

9. $x = \frac{1}{3}y^2 + 6y + 24$

$x = \frac{1}{3}(y^2 + 18y) + 24$ Factor out $\frac{1}{3}$.

$x = \frac{1}{3}(y^2 + 18y + 81) + 24 - 27$ Complete the square; add $\frac{1}{3}(81) - 27$.

$x = \frac{1}{3}(y + 9)^2 - 3$

$x = \frac{1}{3}(y - (-9))^2 - 3$ Write in standard form $x = a(y - k)^2 + h$.

This is a horizontal parabola with vertex at $(h, k) = (-3, -9)$; the graph opens to the right since $\frac{1}{3} > 0$ and is wider than $x = y^2$ since $\left|\frac{1}{3}\right| < 1$.

See the answer section in your text for the graphs for Exercises 13 and 17.

13. From $f(x) = -2x^2 + 4x - 5$, $a = -2$, and $b = 4$. Use the vertex formula to get

$$x = \frac{-b}{2a} = \frac{-4}{2(-2)} = \frac{-4}{-4} = 1.$$

So $y = f(1) = -2(1)^2 + 4(1) - 5 = -2 + 4 - 5 = -3$. The vertex is at $(1, -3)$. The graph of this vertical parabola opens downward, since $-2 < 0$.

Find the y-intercept by substituting $x = 0$ in $y = -2x^2 + 4x - 5$ to get $y = -5$. The y-intercept is $(0, -5)$. Find the x-intercept by substituting $y = 0$ in $y = -2x^2 + 4x - 5$ to get $0 = -2x^2 + 4x - 5$. Use the quadratic formula with $a = -2$, $b = 4$, and $c = -5$ to solve this equation.

$$x = \frac{-4 \pm \sqrt{4^2 - 4(-2)(-5)}}{2(-2)} = \frac{-4 \pm \sqrt{-24}}{-4}$$

The solutions are imaginary numbers, so there are no x-intercepts. Additional points are $(-1, -11)$, $(2, -5)$, and $(3, -11)$.

The domain is $(-\infty, \infty)$ since any real number can replace x. Because $(1, -3)$ is the highest point on the graph, -3 is the largest y-value and the range is $(-\infty, -3]$.

17. From $x = 3y^2 + 12y + 5$, completing the square on y gives

$$
\begin{aligned}
x &= 3(y^2 + 4y \qquad\quad) + 5 \qquad \text{Factor out 3.}\\
&= 3(y^2 + 4y + 4 - 4) + 5 \qquad \text{Add 0 } (4 - 4 = 0).\\
&= 3(y^2 + 4y + 4) - 12 + 5 \qquad \text{Distributive property}\\
&= 3(y + 2)^2 - 7 \qquad \text{Factor.}\\
&= 3[y - (-2)]^2 - 7.
\end{aligned}
$$

Because $3 > 0$, the graph opens to the right and since $|3| > 1$, it is narrower than the graph of $x = y^2$. The vertex is $(-7, -2)$. Additional points are $(-4, -1)$ and $(5, 0)$.

21. $f(x) = x^2 - 8x + 14$

$f(x) = (x^2 - 8x + 16) + 14 - 16$ Complete the square.

$f(x) = (x - 4)^2 - 2$ Write in standard form.

This is a vertical parabola with vertex at $(4, -2)$. The graph opens upward since $1 > 0$, so the vertex would have to be a minimum point. Choice A has a minimum point, with vertex at $(4, -2)$ and therefore represents the graph of $f(x) = x^2 - 8x + 14$.

25. The graph of $h(t) = -16t^2 + 32t$ is a parabola that opens down ($a = -16 < 0$). The y-coordinate of the vertex represents the maximum height of the object. To find the vertex, use $\left(-\dfrac{b}{2a}, h\left(-\dfrac{b}{2a}\right)\right)$, where $a = -16$ and $b = 32$.

So

$$\frac{-b}{2a} = \frac{-32}{2(-16)} = \frac{-32}{-32} = 1$$

$$h\left(\frac{-b}{2a}\right) = h(1) = -16(1)^2 + 32(1)$$

$$= -16 + 32 = 16$$

The vertex is $(1, 16)$, so the maximum height is 16 feet.

To find the number of seconds it takes for the object to hit the ground, let $h(t) = 0$ and solve for t.

$$0 = -16t^2 + 32t$$
$$0 = -16t(t - 2)$$
$$-16t = 0 \quad \text{or} \quad t - 2 = 0$$
$$t = 0 \quad \text{or} \quad t = 2$$

It takes the object 2 seconds to reach the ground. (The solution 0 represents the time that it was on the ground before it was thrown.)

29. In the function $f(x) = -.10x^2 + .42x + 11.90$, the coefficient of x^2 is negative because the parabola opens downward, indicating that $a < 0$.

33. **(a)** The x-intercepts determine the solutions of the quadratic equation $x^2 - 4x + 3 = 0$. The solution set is $\{1, 3\}$.

(b) The x-values of the points on the graph that are *above* the x-axis form the solution set of $x^2 - 4x + 3 > 0$. As seen in the graph, this solution set is $(-\infty, 1) \cup (3, \infty)$.

(c) The x-values of the points on the graph that are *below* the x-axis form the solution set of $x^2 - 4x + 3 < 0$. Those x-values belong to the open interval $(1, 3)$.

SECTION 9.3 (page 607)

1. For the reciprocal function $f(x) = \dfrac{1}{x}$, 0 is the only real number not in the domain. Zero causes the function to be undefined.

5. The graph of $f(x) = |x - 2| + 2$ is found by shifting the graph of $f(x) = |x|$ two units to the right and two units up. Thus, the function $f(x) = |x - 2| + 2$ matches graph B.

See the answer section in your text for the graphs for Exercises 9–17.

9. The graph of $f(x) = |x + 1|$ is found by shifting the graph of $y = |x|$ one unit to the left. The following table of ordered pairs gives some specific points the graph passes through.

x	y
-2	1
-1	0
0	1
1	2
2	3

13. The graph of $f(x) = \sqrt{x - 2}$ is found by shifting the graph of $y = \sqrt{x}$ two units to the right. The following table of ordered pairs gives some specific points the graph passes through.

x	y
2	0
3	1
6	2

17. The graph of $f(x) = \sqrt{x + 3} - 3$ is found by shifting the graph of $y = \sqrt{x}$ three units to the left and three units down. The following table of ordered pairs gives some specific points the graph passes through.

x	y
-3	-3
-2	-2
1	-1

21. This circle, with equation $(x + 3)^2 + (y - 2)^2 = 25$, has a center at $(-3, 2)$ and a radius of 5. Graph D is its match.

25. Rearrange the terms in $x^2 + y^2 - 8x - 12y + 3 = 0$ to get

$$x^2 - 8x + y^2 - 12y = -3.$$

Complete the square for each variable.

$$(x^2 - 8x + 16 - 16) + (y^2 - 12y + 36 - 36) = -3$$
$$(x^2 - 8x + 16) - 16 + (y^2 - 12y + 36) - 36 = -3$$
$$(x^2 - 8x + 16) + (y^2 - 12y + 36) = -3 + 16 + 36$$

Add $16 + 36$ to both sides. Factor the left side; simplify the right side.

$$(x - 4)^2 + (y - 6)^2 = 49$$

From the last equation above, the center is at $(4, 6)$ and the radius is 7.

29. Divide each term of $2x^2 + 2y^2 + 20x + 16y + 10 = 0$ by 2 to get

$$x^2 + y^2 + 10x + 8y + 5 = 0.$$

Rearrange the terms in this last equation to get

$$x^2 + 10x + y^2 + 8y = -5.$$

Complete the square for each variable.

$$(x^2 + 10x + 25) + (y^2 + 8y + 16)$$
$$= -5 + 25 + 16$$

Factor the left side; simplify the right side.

$$(x + 5)^2 + (y + 4)^2 = 36$$

The center is at $(-5, -4)$ and the radius is 6.

For Exercises 33–41, see the graphs in the answer section of your textbook.

33. Rearrange the terms for $2y^2 = 10 - 2x^2$, then divide by 2 to get $x^2 + y^2 = 5$.
Write $x^2 + y^2 = 5$ in standard form to get

$$(x - 0)^2 + (y - 0)^2 = 5.$$

Graph the circle with center at $(0, 0)$ and radius $\sqrt{5} \approx 2.2$.

37. Rearrange the terms of $x^2 + y^2 - 4x - 6y + 9 = 0$ to get

$$x^2 - 4x + y^2 - 6y = -9.$$

Complete the square for each variable.

$$(x^2 - 4x + 4) + (y^2 - 6y + 9) = -9 + 4 + 9$$

Factor the left side; simplify the right side.

$$(x - 2)^2 + (y - 3)^2 = 4$$

Graph the circle with center at $(2, 3)$ and radius 2.

41. Let $f(x) = y$, then square both sides of $y = -\sqrt{36 - x^2}$.

$$y^2 = (-\sqrt{36 - x^2})^2$$
$$y^2 = 36 - x^2 \quad \text{The square of a}$$
$$\qquad\qquad\qquad \text{negative is positive.}$$
$$x^2 + y^2 = 36$$

$x^2 + y^2 = 36$ is a circle with center at $(0, 0)$ and radius 6. However, $y = -\sqrt{36 - x^2}$ is satisfied only if $y \le 0$. Therefore, the graph is a semicircle below the x-axis with x-intercepts $(-6, 0)$ and $(6, 0)$.

Section 9.4 (page 619)

1. $\dfrac{x^2}{25} + \dfrac{y^2}{9} = 1$ is the equation of an ellipse. In general, $\dfrac{x^2}{a^2} + \dfrac{y^2}{b^2} = 1$ is an ellipse with x-intercepts $(a, 0)$ and $(-a, 0)$ and y-intercepts $(0, b)$ and $(0, -b)$. So, in the equation $\dfrac{x^2}{25} + \dfrac{y^2}{9} = 1$, the x-intercepts are $(5, 0)$ and $(-5, 0)$ and the y-intercepts are $(0, 3)$ and $(0, -3)$. Thus, graph C matches the equation $\dfrac{x^2}{25} + \dfrac{y^2}{9} = 1$.

5. When an equation of a hyperbola is written in the form $\dfrac{x^2}{a^2} - \dfrac{y^2}{b^2} = 1$ (here the negative sign precedes the y^2 term), the hyperbola will open left and right.
When an equation of a hyperbola is written in the form $\dfrac{y^2}{b^2} - \dfrac{x^2}{a^2} = 1$ (here the negative sign precedes the x^2 term), the hyperbola will open up and down.

See the answer section in your text for the graphs for Exercises 9, 17–29.

9. $\dfrac{x^2}{36} + \dfrac{y^2}{16} = 1$

The x-intercepts are found from $\pm\sqrt{36}$: $(6, 0)$ and $(-6, 0)$.
The y-intercepts are found from $\pm\sqrt{16}$: $(0, 4)$ and $(0, -4)$.
Sketch the ellipse through these four points.

13. To sketch a hyperbola, the fundamental rectangle is used by first plotting the points (a, b), $(a, -b)$, $(-a, -b)$, $(-a, b)$ or the points (b, a), $(-b, a)$, $(-b, -a)$, $(b, -a)$. These points are used as corners of a rectangle. The diagonals of the rectangle are then drawn, and they are used as asymptotes for the hyperbola.

17. $\dfrac{y^2}{9} - \dfrac{x^2}{9} = 1$

y-intercepts: $(0, 3)$ and $(0, -3)$. There are no x-intercepts.
One asymptote passes through $(3, 3)$ and $(-3, -3)$.
The other asymptote passes through $(-3, 3)$ and $(3, -3)$.
Draw the asymptotes and sketch the hyperbola through the intercepts and approaching the asymptotes.

21. Divide each term in $x^2 - y^2 = 16$ by 16 to get

$$\frac{x^2}{16} - \frac{y^2}{16} = 1.$$

The graph is a hyperbola with x-intercepts $(4, 0)$ and $(-4, 0)$ and no y-intercepts. One asymptote passes

through (4, 4) and (−4, −4). The other asymptote passes through (−4, 4) and (4, −4).
Sketch the graph through the intercepts and approaching the asymptotes.

25. Rewrite $y^2 = 36 - x^2$ as $x^2 + y^2 = 36$.
The graph is a circle with center at (0, 0) and radius 6. Sketch the graph.

29. Rewrite $y^2 = 4 + x^2$ as $y^2 - x^2 = 4$. Divide each term by 4 to get

$$\frac{y^2}{4} - \frac{x^2}{4} = 1.$$

The graph is a hyperbola with y-intercepts at (0, 2) and (0, −2). One asymptote passes through (2, 2) and (−2, −2). The other asymptote passes through (−2, 2) and (2, −2). Sketch the graph.

33. This method of sketching an ellipse works because the two thumbtacks act as foci, and the length of the string is constant, satisfying the requirements of the definition of an ellipse.

Section 9.5 (page 627)

1. To solve the system

$$x^2 + y^2 = 25$$
$$y = x - 1$$

by the substitution method, substitute $x - 1$ for y in the first equation. Then solve for x. Find the corresponding y-values by substituting back into $y = x - 1$.

5. $y = 4x^2 - x$ (1)
$y = x$ (2)
Substitute x for y in equation (1) to get

$$x = 4x^2 - x$$
$$0 = 4x^2 - 2x \qquad \text{Get 0 alone on one side.}$$
$$0 = 2x(2x - 1) \qquad \text{Factor.}$$
$$2x = 0 \quad \text{or} \quad 2x - 1 = 0$$
$$x = 0 \quad \text{or} \qquad x = \frac{1}{2}.$$

Use equation (2) to find y for each x-value.
If $x = 0$, then $y = 0$.
If $x = \frac{1}{2}$, then $y = \frac{1}{2}$.
Solution set: $\{(0, 0), (\frac{1}{2}, \frac{1}{2})\}$

9. $x^2 + y^2 = 2$ (1)
$2x + y = 1$ (2)
Solve equation (2) for y to get

$$y = 1 - 2x. \quad (3)$$

Substitute $1 - 2x$ for y in equation (1).

$$x^2 + (1 - 2x)^2 = 2$$
$$x^2 + 1 - 4x + 4x^2 = 2 \qquad \text{Square the binomial.}$$
$$5x^2 - 4x - 1 = 0 \qquad \text{Combine terms; get 0 alone.}$$
$$(5x + 1)(x - 1) = 0 \qquad \text{Factor.}$$
$$5x + 1 = 0 \quad \text{or} \quad x - 1 = 0$$
$$x = -\frac{1}{5} \quad \text{or} \qquad x = 1$$

Use equation (3) to find y for each x-value.

If $x = -\frac{1}{5}, y = 1 - 2\left(-\frac{1}{5}\right) = \frac{5}{5} + \frac{2}{5} = \frac{7}{5}$.

If $x = 1, y = 1 - 2(1) = -1$.

Solution set: $\{(-\frac{1}{5}, \frac{7}{5}), (1, -1)\}$

13. $xy = -3$ (1)
$x + y = -2$ (2)
Solve equation (2) for y to get

$$y = -x - 2. \quad (3)$$

Substitute $-x - 2$ for y in equation (1).

$$x(-x - 2) = -3$$
$$-x^2 - 2x + 3 = 0 \qquad \text{Clear parentheses; get 0 alone.}$$
$$x^2 + 2x - 3 = 0 \qquad \text{Multiply by } -1.$$
$$(x + 3)(x - 1) = 0 \qquad \text{Factor.}$$
$$x + 3 = 0 \quad \text{or} \quad x - 1 = 0$$
$$x = -3 \quad \text{or} \qquad x = 1$$

Use equation (3) to find y for each x-value.

If $x = -3, y = -(-3) - 2 = 1$.
If $x = 1, y = -(1) - 2 = -3$.

Solution set: $\{(-3, 1), (1, -3)\}$

17. $2x^2 - y^2 = 6$ (1)
$y = x^2 - 3$ (2)
Substitute $x^2 - 3$ for y in equation (1).

$$2x^2 - (x^2 - 3)^2 = 6$$
$$2x^2 - (x^4 - 6x^2 + 9) = 6 \qquad \text{Square the binomial.}$$
$$-x^4 + 8x^2 - 15 = 0 \qquad \text{Clear parentheses; get 0 alone on one side.}$$
$$x^4 - 8x^2 + 15 = 0 \qquad \text{Multiply by } -1.$$

Let $z = x^2$ in the last equation above. Then

$$z^2 - 8z + 15 = 0$$
$$(z - 3)(z - 5) = 0 \qquad \text{Factor.}$$
$$z - 3 = 0 \quad \text{or} \quad z - 5 = 0$$
$$z = 3 \quad \text{or} \qquad z = 5.$$

Since $z = x^2$,

$$x^2 = 3 \quad \text{or} \quad x^2 = 5.$$

Use the square root property.

$$x = \pm\sqrt{3} \quad \text{or} \quad x = \pm\sqrt{5}$$

Use equation (2) to find y for each x-value.

If $x = \sqrt{3}$ or $-\sqrt{3}$, then $y = (\pm\sqrt{3})^2 - 3 = 0$.
If $x = \sqrt{5}$ or $-\sqrt{5}$, then $y = (\pm\sqrt{5})^2 - 3 = 2$.

Solution set: $\{(-\sqrt{3}, 0), (\sqrt{3}, 0), (-\sqrt{5}, 2), (\sqrt{5}, 2)\}$

21. $xy = 6$ (1)
$3x^2 - y^2 = 12$ (2)
Solve equation (1) for y to get

$$y = \frac{6}{x}. \quad (3)$$

Substitute $\frac{6}{x}$ for y in equation (2).

$$3x^2 - \left(\frac{6}{x}\right)^2 = 12$$

$$3x^2 - \frac{36}{x^2} = 12$$

$$3x^4 - 36 = 12x^2 \quad \text{Multiply by } x^2.$$

$$3x^4 - 12x^2 - 36 = 0 \quad \text{Get 0 alone.}$$

Let $z = x^2$ to get

$$3z^2 - 12z - 36 = 0$$

$$3(z^2 - 4z - 12) = 0$$

$$3(z + 2)(z - 6) = 0. \quad \text{Factor.}$$

$$z + 2 = 0 \quad \text{or} \quad z - 6 = 0$$

$$z = -2 \quad \text{or} \quad z = 6$$

Since $z = x^2$, $x^2 = -2$ or $x^2 = 6$.

Use the square root property.

$$x = \pm\sqrt{-2} = \pm i\sqrt{2} \quad \text{or} \quad x = \pm\sqrt{6}$$

Use equation (3) to find y.

If $x = i\sqrt{2}$, $y = \dfrac{6}{i\sqrt{2}} = \dfrac{6 \cdot i\sqrt{2}}{i\sqrt{2} \cdot i\sqrt{2}} = \dfrac{6i\sqrt{2}}{2i^2}$

$$= -3i\sqrt{2}.$$

If $x = -i\sqrt{2}$, $y = 3i\sqrt{2}$.

If $x = \sqrt{6}$, $y = \dfrac{6}{\sqrt{6}} = \dfrac{6 \cdot \sqrt{6}}{\sqrt{6} \cdot \sqrt{6}} = \dfrac{6\sqrt{6}}{6} = \sqrt{6}$.

If $x = -\sqrt{6}$, $y = -\sqrt{6}$.

Solution set: $\{(i\sqrt{2}, -3i\sqrt{2}), (-i\sqrt{2}, 3i\sqrt{2}),$
$(-\sqrt{6}, -\sqrt{6}), (\sqrt{6}, \sqrt{6})\}$

25. $x^2 + xy + y^2 = 15$
$x^2 + y^2 = 10$

Multiply the second equation by -1 then add to the first equation.

$$
\begin{array}{rcl}
x^2 + xy + y^2 &=& 15 \\
-x^2 \qquad\;\; - y^2 &=& -10 \\
\hline
xy &=& 5 \quad \text{Add.}
\end{array}
$$

$$y = \frac{5}{x} \quad \text{Solve for } y.$$

Substitute $\frac{5}{x}$ for y in $x^2 + y^2 = 10$.

$$x^2 + \left(\frac{5}{x}\right)^2 = 10$$

$$x^2 + \frac{25}{x^2} = 10$$

$$x^4 + 25 = 10x^2 \quad \text{Multiply by } x^2.$$

$$x^4 - 10x^2 + 25 = 0 \quad \text{Get 0 alone.}$$

Let $z = x^2$ to get

$$z^2 - 10z + 25 = 0$$

$$(z - 5)^2 = 0 \quad \text{Factor.}$$

$$z - 5 = 0$$

$$z = 5.$$

Since $z = x^2$,

$$x^2 = 5, \quad \text{and} \quad x = \pm\sqrt{5}.$$

Using the equation $y = \dfrac{5}{x}$, we get the following.

If $x = -\sqrt{5}$, $y = \dfrac{5}{-\sqrt{5}} = \dfrac{5 \cdot \sqrt{5}}{-\sqrt{5} \cdot \sqrt{5}} = \dfrac{5\sqrt{5}}{-5}$

$$= -\sqrt{5}.$$

If $x = \sqrt{5}$, $y = \sqrt{5}$.

Solution set: $\{(-\sqrt{5}, -\sqrt{5}), (\sqrt{5}, \sqrt{5})\}$

29. $y = x^2 + 1$ (1)
$x + y = 1$ (2)

Substitute $x^2 + 1$ for y in equation (2) to get

$$x + (x^2 + 1) = 1$$

$$x^2 + x = 0 \quad \text{Subtract 1.}$$

$$x(x + 1) = 0 \quad \text{Factor.}$$

$$x = 0 \quad \text{or} \quad x + 1 = 0$$

$$x = -1$$

From the graphing calculator screen, we already know one solution is $(0, 1)$.
Use equation (1) to find y when $x = -1$.

$$y = (-1)^2 + 1$$

$$y = 1 + 1$$

$$y = 2$$

The other solution is $(-1, 2)$.

33. Let $W =$ the width, and $L =$ the length.
The formula for the area of a rectangle is $W \cdot L = A$, so

$$W \cdot L = 84. \quad (1)$$

The perimeter of a rectangle is given by $2W + 2L = P$, so

$$2W + 2L = 38. \quad (2)$$

Solve equation (2) for L to get

$$L = 19 - W. \quad (3)$$

Substitute $19 - W$ for L in equation (1).

$$W(19 - W) = 84$$

$$-W^2 + 19W - 84 = 0 \quad \text{Clear parentheses; get 0 alone.}$$

$$W^2 - 19W + 84 = 0 \quad \text{Multiply by } -1.$$

$$(W - 7)(W - 12) = 0 \quad \text{Factor.}$$

$$W - 7 = 0 \quad \text{or} \quad W - 12 = 0$$

$$W = 7 \quad \text{or} \quad W = 12$$

Using equation (3), with $W = 7$, we get

$$L = 19 - 7 = 12.$$

(For $W = 12$, $L = 7$, the same two numbers, but L should be greater than W.)
The length is 12 feet, and the width is 7 feet.

SOLUTIONS

SECTION 9.6 (page 635)

1. In all cases, the boundary is the graph of the equation $y = x^2 + 4$. This is a parabola with the same shape as $y = x^2$ but shifted 4 units up.
   (a) Because of the $\geq$ symbol, the graph of $y \geq x^2 + 4$ has a solid boundary and is shaded above the parabola. This describes graph B.
   (b) Because of the $\leq$ symbol, the graph of $y \leq x^2 + 4$ has a solid boundary and is shaded below the parabola. This describes graph D.
   (c) Because of the $<$ symbol, the graph of $y < x^2 + 4$ has a dashed boundary and is shaded below the parabola. This describes graph A.
   (d) Because of the $>$ symbol, the graph of $y > x^2 + 4$ has a dashed boundary and is shaded above the parabola. This describes graph C.

See the answer section in your textbook for the graphs for Exercises 5–29.

5. Rewrite $y + 2 \geq x^2$ as $y \geq x^2 - 2$.
   (a) Graph the solid vertical parabola $y = x^2 - 2$ with the vertex at $(0, -2)$. Two other points are $(1, -1)$ and $(-2, 2)$.
   (b) Test a point not on the parabola, say $(0, 0)$, in $y \geq x^2 - 2$ to get $0 \geq -2$, a true statement.
   (c) Shade that portion of the graph that contains the point $(0, 0)$. This is the region above the parabola.

9. (a) Graph the solid vertical parabola $y = x^2 + 4x + 2$. Use the vertex formula $x = \frac{-b}{2a}$ to obtain the vertex at $(-2, -2)$. Two other points are $(0, 2)$ and $(1, 7)$.
   (b) Test a point not on the parabola, say $(0, 0)$, in $y \leq x^2 + 4x + 2$ to get $0 \leq 2$, a true statement.
   (c) Shade below the parabola, since this region contains $(0, 0)$.

13. Rewrite $x^2 - 4 \geq -4y^2$ as $\frac{x^2}{4} + \frac{y^2}{1} \geq 1$.
    (a) Graph the solid ellipse $\frac{x^2}{4} + \frac{y^2}{1} = 1$ through the x-intercepts $(2, 0)$, $(-2, 0)$ and the y-intercepts $(0, 1)$, $(0, -1)$.
    (b) Test a point not on the ellipse, say $(0, 0)$, in $x^2 - 4 \geq -4y^2$ to get $0 - 4 \geq 0$, a false statement.
    (c) Shade outside the ellipse, since this region does not contain $(0, 0)$.

17. Rewrite $25x^2 \leq 9y^2 + 225$ as $\frac{x^2}{9} - \frac{y^2}{25} \leq 1$.
    (a) Graph the solid hyperbola $\frac{x^2}{9} - \frac{y^2}{25} = 1$ through the x-intercepts $(3, 0)$ and $(-3, 0)$.
    One asymptote passes through $(3, 5)$ and $(-3, -5)$. The other asymptote passes through $(-3, 5)$ and $(3, -5)$.
    (b) Test a point not on the hyperbola, say $(0, 0)$, in $25x^2 \leq 9y^2 + 225$ to get $0 \leq 0 + 225$, a true statement.
    (c) Shade that part of the graph that contains $(0, 0)$. This is the region between the two branches of the hyperbola.

21. Graph $2x + 5y = 10$ as a dashed line through $(5, 0)$ and $(0, 2)$. Test $(0, 0)$ in $2x + 5y < 10$ to get $0 < 10$, a true statement. Shade the region containing $(0, 0)$.

Graph $x - 2y = 4$ as a dashed line through $(4, 0)$ and $(0, -2)$. Test $(0, 0)$ in $x - 2y < 4$ to get $0 < 4$, a true statement. Shade the region containing $(0, 0)$. The solution is the intersection of the two shaded regions.

25. Graph $x = 5$ as a solid vertical line through $(5, 0)$. Since $x \leq 5$, shade the left side of the line.
    Graph $y = 4$ as a solid horizontal line through $(0, 4)$. Since $y \leq 4$, shade below the line.
    The solution is the intersection of the two shaded regions.

29. Rewrite $y^2 - x^2 \geq 4$ as $\frac{y^2}{4} - \frac{x^2}{4} \geq 1$.
    Graph the solid hyperbola through the y-intercepts $(0, 2)$ and $(0, -2)$. The asymptotes pass through $(2, 2)$, $(-2, -2)$ and $(-2, 2)$, $(2, -2)$.
    Test $(0, 0)$ in $y^2 - x^2 \geq 4$ to get $0 \geq 4$, a false statement. Shade above and below the two branches of the hyperbola.
    Graph solid horizontal lines through $(0, 5)$ and $(0, -5)$. Since $-5 \leq y \leq 5$, shade the region between the two horizontal lines.
    The solution is the intersection of the shaded regions.

33. The graph of $y > x^2 + 2$ consists of all points that are above the graph of $y = x^2 + 2$. The graph is shown in A.

CHAPTER 10

SECTION 10.1 (page 665)

1. The function defined by $f(x) = x^2$ is not a one-to-one function because most y-values correspond to *two* x-values. For example, when $y = 9$, $x = 3$ or -3.

5. This function might not be a one-to-one function because two or more siblings might be in the class.

9. This is not a one-to-one function. The ordered pairs $(-1, 3)$ and $(4, 3)$ have the same y-values for two different x-values.

13. Write $g(x) = \sqrt{x - 3}$ as $y = \sqrt{x - 3}$. Since $x \geq 3$, $y \geq 0$. For each x-value, there is only one y-value, so the function is one-to-one. Exchange x and y to get $x = \sqrt{y - 3}$. Note that $y \geq 3$ so $x \geq 0$. Solve $x = \sqrt{y - 3}$ for y.

$$x = \sqrt{y - 3}$$
$$x^2 = y - 3 \qquad \text{Square both sides.}$$
$$x^2 + 3 = y \qquad \text{Add 3.}$$
$$g^{-1}(x) = x^2 + 3, x \geq 0 \qquad \text{Replace } y \text{ with } g^{-1}(x).$$

17. Write $f(x) = x^3 - 4$ as $y = x^3 - 4$. Exchange x and y to get $x = y^3 - 4$. Solve for y.

$$x = y^3 - 4$$
$$x + 4 = y^3 \qquad \text{Add 4.}$$
$$\sqrt[3]{x + 4} = y \qquad \text{Take the cube root of each side.}$$

The inverse is $f^{-1}(x) = \sqrt[3]{x + 4}$.

21. (a) $f(3) = 2^3 = (2)(2)(2) = 8$
    (b) Since f is one-to-one and $f(3) = 8$, $f^{-1}(8) = 3$.

For Exercises 25(b), 29(b), 33, and 37, refer to the graphs in the answer section of your textbook.

25. (a) The function is one-to-one since a horizontal line placed anywhere will intersect the graph at only one point.
(b) In the graph, the two points marked on the line are $(-1, 5)$ and $(2, -1)$.
Exchange x and y in each ordered pair to get $(5, -1)$ and $(-1, 2)$.
Plot these points, then draw a dashed line through them to obtain the graph of the inverse function.

29. (a) See the solution for Exercise 25(a).
(b) In the graph, the four points marked on the curve are $(-4, 2)$, $(-1, 1)$, $(1, -1)$, and $(4, -2)$.
Exchange x and y in each ordered pair to get $(2, -4)$, $(1, -1)$, $(-1, 1)$, and $(-2, 4)$.
Plot these points, then draw a dashed curve (symmetric to the original graph about the line $y = x$) through them to obtain the graph of the inverse.

33. To graph $g(x) = -4x$, plot the points $(0, 0)$ and $(1, -4)$, then draw a solid line through these points. Exchange x and y to get the points $(0, 0)$ and $(-4, 1)$. Draw a dashed line through these points to obtain the graph of the inverse function.

37. To graph $y = x^3 - 2$, complete the ordered pairs in the table: $(-1, -3)$, $(0, -2)$, $(1, -1)$, and $(2, 6)$. Draw a solid curve through these points.
Exchange x and y to get the following points: $(-3, -1)$, $(-2, 0)$, $(-1, 1)$, and $(6, 2)$. Draw a dashed curve (symmetric to the original graph about $y = x$) through these points to obtain the graph of the inverse function.

41. Although y_2 is the "inverse relation" of y_1, since y_1 is not one-to-one, y_2 is not a function.

SECTION 10.2 (page 675)

1. rises (see Example 1 in your text, $f(x) = 2^x$); falls (see Example 2, $g(x) = (\frac{1}{2})^x = 2^{-x}$).

For Exercises 5 and 9, refer to the graphs in the answer section of your text.

5. Complete some ordered pairs.

x	-3	-2	-1	0	1	2	3
$f(x)$	3^{-3}	3^{-2}	3^{-1}	3^0	3^1	3^2	3^3
	$= \frac{1}{27}$	$= \frac{1}{9}$	$= \frac{1}{3}$	$= 1$	$= 3$	$= 9$	$= 27$

Draw a graph through these ordered pairs.

9. Complete some ordered pairs.

x	-1	0	1	2
y	2^{-4}	2^{-2}	2^0	2^2
	$= \frac{1}{16}$	$= \frac{1}{4}$	$= 1$	$= 4$

Draw a graph through these ordered pairs.

13. $6^x = 36$
$6^x = 6^2$ Write both sides with the same base.
$x = 2$ Set the exponents equal.
Solution set: $\{2\}$

17. $16^{2x+1} = 64^{x+3}$
$(4^2)^{2x+1} = (4^3)^{x+3}$ Write both sides with the same base.
$4^{4x+2} = 4^{3x+9}$ Power rule for exponents
$4x + 2 = 3x + 9$ Set the exponents equal. If $a^x = a^y$, then $x = y$.
$x = 7$ Subtract $3x$; subtract 2.
Solution set: $\{7\}$

21. $5^x = .2$
$5^x = \dfrac{1}{5}$ $.2 = \frac{2}{10} = \frac{1}{5}$
$5^x = 5^{-1}$ Write using the same base.
$x = -1$ Set the exponents equal.
Solution set: $\{-1\}$

25. (a) $A(0) = 100(3.2)^{-.5(0)} = 100(3.2)^0 = 100(1)$
$= 100$ grams
(b) $A(2) = 100(3.2)^{-.5(2)}$
$= 100(3.2)^{-1}$
$= 100(.3125)$ Use a calculator.
$= 31.25$ grams
(c) $A(10) = 100(3.2)^{-.5(10)}$
$= 100(3.2)^{-5}$
$\approx 100(.002980)$ Use a calculator.
$\approx .298 \approx .30$ gram

29. According to the graphing calculator screen, in 1991 (when $x = 15$), the average salary was about $800,000.

SECTION 10.3 (page 681)

1. (a) $\log_4 16$ is 2 because $4^2 = 16$.
(b) $\log_3 81$ is 4 because $3^4 = 81$.
(c) $\log_3 \frac{1}{3} = \log_3 3^{-1}$ is -1.
(d) $\log_{10} .01 = \log_{10} \frac{1}{100} = \log_{10} 10^{-2}$ is -2.
(e) $\log_5 \sqrt{5} = \log_5 5^{1/2}$ is $\frac{1}{2}$.
(f) $\log_{12} 1$ is 0 because $12^0 = 1$.

5. In $(\frac{1}{2})^{-3}$, $\frac{1}{2}$ is the base and -3 is the exponent. So $(\frac{1}{2})^{-3} = 8$ becomes $\log_{1/2} 8 = -3$.

9. In $\log_4 64 = 3$, 4 is the base and 3 is the logarithm (exponent). So $\log_4 64 = 3$ becomes $4^3 = 64$.

13. In $\log_6 1 = 0$, 6 is the base and 0 is the logarithm (exponent). So $\log_6 1 = 0$ becomes $6^0 = 1$.

17. Write $x = \log_{27} 3$ in exponential form.

$27^x = 3$

$(3^3)^x = 3$ Write using the same base on each side.

$3^{3x} = 3^1$ Power rule; $3 = 3^1$

$3x = 1$ Set the exponents equal.

$x = \dfrac{1}{3}$

Solution set: $\{\frac{1}{3}\}$

21. Write $\log_x 125 = -3$ in exponential form.

$x^{-3} = 125$

$x^{-3} = 5^3 = \left(\dfrac{1}{5}\right)^{-3}$ Write both sides using the same exponent.

$x = \dfrac{1}{5}$ Set the bases equal.

Solution set: $\{\frac{1}{5}\}$

25. Write $\log_x x = 1$ in exponential form as $x^1 = x$. This is true for any $x > 0$ and $x \neq 1$. (The domain of $\log_a x$ is $x > 0$ and $x \neq 1$.)
Solution set: $\{x \mid x > 0, x \neq 1\}$

29. Write $\log_8 32 = x$ in exponential form.

$$8^x = 32$$
$$(2^3)^x = 2^5 \quad \text{Write with the same base.}$$
$$2^{3x} = 2^5 \quad \text{Power rule}$$
$$3x = 5 \quad \text{Set the exponents equal.}$$
$$x = \frac{5}{3}$$

Solution set: $\{\frac{5}{3}\}$

33. Write $\log_6 \sqrt{216} = x$ as $\log_6 216^{1/2} = x$.
Now write $\log_6 216^{1/2} = x$ in exponential form.

$$6^x = 216^{1/2}$$
$$6^x = (6^3)^{1/2} = 6^{3/2} \quad \text{Write with the same base.}$$
$$x = \frac{3}{2} \quad \text{Set the exponents equal.}$$

Solution set: $\{\frac{3}{2}\}$

37. Refer to 10.2 Exercise 7 for the graph of $y = \frac{1}{3}^x$. Since $y = \log_{1/3} x$ (or $\frac{1}{3}^y = x$) is the inverse of $y = \frac{1}{3}^x$, its graph is symmetric about the line $y = x$ to the graph of $y = \frac{1}{3}^x$ and can be plotted by reversing some ordered pairs that belong to $y = \frac{1}{3}^x$.
See the graph in the answer section of your textbook.

41. (a) $t = 0$ corresponds to January, 1993.

$$M(0) = 6 \log_4[2(0) + 4]$$
$$= 6 \log_4 4$$
$$= 6(1) \qquad \log_4 4 = 1$$
$$= 6 \text{ mice}$$

(b) $t = 6$ corresponds to July, 1993.

$$M(6) = 6 \log_4[2(6) + 4]$$
$$= 6 \log_4 16$$
$$= 6(2) \qquad \log_4 16 = 2$$
$$= 12 \text{ mice}$$

(c) $t = 30$ corresponds to July, 1995.

$$M(30) = 6 \log_4[2(30) + 4]$$
$$= 6 \log_4 64$$
$$= 6(3) \qquad \log_4 64 = 3$$
$$= 18 \text{ mice}$$

(d) Complete some ordered pairs from parts (a), (b), and (c).

t	0	6	30
$M(t)$	6	12	18

See the graph in the answer section of your textbook.

45. To find $f(60)$, find 60 on the t-axis, then go up to the graph and across to the $f(t)$ axis to read the value of $f(60)$: $f(60) = 24$.

49. The range of $f(x) = a^x$ is the domain of $g(x) = \log_a x$: $(0, \infty)$.
The domain of $f(x) = a^x$ is the range of $g(x) = \log_a x$: $(-\infty, \infty)$.

Section 10.4 (page 689)

1. sum $(\log_b xy = \log_b x + \log_b y)$

5. $\log_7 \frac{4}{5} = \log_7 4 - \log_7 5$ Division property

9. $\log_4 \dfrac{3\sqrt{x}}{y}$

$$= \log_4 \frac{3 \cdot x^{1/2}}{y}$$
$$= \log_4(3 \cdot x^{1/2}) - \log_4 y \qquad \text{Division property}$$
$$= \log_4 3 + \frac{1}{2} \log_4 x - \log_4 y \qquad \text{Multiplication and power properties}$$

13. $\log_3 \sqrt{\dfrac{xy}{5}}$

$$= \log_3 \left(\frac{xy}{5}\right)^{1/2}$$
$$= \frac{1}{2} \log_3 \left(\frac{xy}{5}\right) \qquad \text{Power property}$$
$$= \frac{1}{2}[\log_3 xy - \log_3 5] \qquad \text{Division property}$$
$$= \frac{1}{2}[\log_3 x + \log_3 y - \log_3 5] \qquad \text{Multiplication property}$$
$$= \frac{1}{2} \log_3 x + \frac{1}{2} \log_3 y - \frac{1}{2} \log_3 5 \qquad \text{Distributive property}$$

17. The distributive property tells us that the product $a(x + y)$ equals the sum $ax + ay$. In the notation $\log_a (x + y)$, the parentheses do not indicate multiplication. They indicate that $x + y$ is the result of raising a to some power.

21. $3 \log_a m - \log_a n$
$$= \log_a m^3 - \log_a n \quad \text{Power property}$$
$$= \log_a \frac{m^3}{n} \qquad \text{Division property}$$

25. $3 \log_a 5 - 4 \log_a 3$
$$= \log_a 5^3 - \log_a 3^4 \quad \text{Power property}$$
$$= \log_a \frac{5^3}{3^4} \qquad \text{Division property}$$
$$= \log_a \frac{125}{81}$$

29. $3 \log_p x + \frac{1}{2} \log_p y - \frac{3}{2} \log_p z - 3 \log_p a$
$$= \log_p x^3 + \log_p y^{1/2} - \log_p z^{3/2} - \log_p a^3$$
$$\qquad\qquad \text{Power property}$$
Group terms into sums.
$$= (\log_p x^3 + \log_p y^{1/2}) - (\log_p z^{3/2} + \log_p a^3)$$
$$= \log_p x^3 y^{1/2} - \log_p z^{3/2} a^3 \quad \text{Multiplication property}$$
$$= \log_p \frac{x^3 y^{1/2}}{z^{3/2} a^3} \qquad \text{Division property}$$

33. $\log_3 81 = \log_3 3^4 = 4$
The property that states "if $b > 0$ and $b \neq 1$, then $\log_b b^x = x$" was used in the last step.

37. $2^{\log_2 19} = 19$
Use the property that states "if $b > 0$ and $b \neq 1$, then $b^{\log_b x} = x \; (x > 0)$."

SECTION 10.5 (page 695)

1. The base in the expression $\log x$ is understood to be 10.

5. $\log 10^{19.2} = 19.2$.
The base is understood to be 10. Use the property that states "if $b > 0$ and $b \neq 1$, then $\log_b b^x = x$."

The directions given below for the calculator exercises may not work with some calculators. See your owner's manual for details.

9. Enter 328.4 in your calculator. Press the $\boxed{\log x}$ key to obtain 2.5164 (to the nearest ten-thousandth).

13. Enter 4.76×10^9 in your calculator as follows:
(1) Enter 4.76.
(2) Press the $\boxed{\text{EE}}$ or $\boxed{\text{EXP}}$ key.
(3) Enter the exponent 9.
Now press the $\boxed{\log x}$ key to obtain the result 9.6776 (to the nearest ten-thousandth).

17. Enter .0556 in your calculator.
Press the $\boxed{\ln x}$ key to obtain -2.8896 (to the nearest ten-thousandth).

21. Enter $8.59 \times e^2$ in your calculator as follows:
(1) Enter 8.59.
(2) Press the multiplication key.
(3) Assuming that $\boxed{e^x}$ is not on your calculator, enter 2, then the $\boxed{\text{INV}}$ key followed by the $\boxed{\ln x}$ key.
(4) Press the equals key.
Now press the $\boxed{\ln x}$ key to obtain the result 4.1506.

25. $\log_6 13 = \dfrac{\log 13}{\log 6} \approx \dfrac{1.11394}{.77815} \approx 1.4315$

29. $\log_{21} .7496 = \dfrac{\log .7496}{\log 21} \approx \dfrac{-.12517}{1.32222} \approx -.0947$

33. $\text{pH} = -\log [\text{H}_3\text{O}^+]$
$= -\log [2.5 \times 10^{-12}]$ $[\text{H}_3\text{O}^+] = 2.5 \times 10^{-12}$
$= -[\log 2.5 + \log 10^{-12}]$ Multiplication property
$\approx -[.4 + (-12)]$ Evaluate each logarithm.
$= -[-11.6] = 11.6$ Simplify.

37. $\text{pH} = -\log [\text{H}_3\text{O}^+]$
$7.4 = -\log [\text{H}_3\text{O}^+]$ $\text{pH} = 7.4$
$-7.4 = \log [\text{H}_3\text{O}^+]$ Multiply by -1.
$10^{-7.4} = [\text{H}_3\text{O}^+]$ Write in exponential form.
$[\text{H}_3\text{O}^+] \approx 4.0 \times 10^{-8}$ Use the $\boxed{10^x}$ key; write in scientific notation.

41. **(a)** $A(4) = 2.00e^{-.053(4)}$ Let $t = 4$.
$= 2.00e^{-.212}$
$\approx 2.00(.809)$ Use the $\boxed{e^x}$ key.
≈ 1.62 grams
(b) $A(10) = 2.00e^{-.053(10)}$ Let $t = 10$.
$= 2.00e^{-.53}$
$\approx 2.00(.589)$
≈ 1.18 grams
(c) $A(20) = 2.00e^{-.053(20)}$ Let $t = 20$.
$= 2.00e^{-1.06}$
$\approx 2.00(.346)$
$\approx .69$ gram
(d) The initial amount is the amount $A(t)$ present at time $t = 0$.
$A(0) = 2.00e^{-.053(0)} = 2.00e^0 = 2.00(1)$
$= 2.00$ grams

45. In 1994, $t = 15$.
$$B(15) = 624.6e^{.0516(15)}$$
$$= 624.6e^{.774}$$
$$\approx 624.6(2.1684) \quad \text{Use the } \boxed{e^x} \text{ key.}$$
$$\approx 1354.4 \text{ thousand}$$

49. $N(.85) = -5000 \ln(.85)$
$\approx -5000(-.1625)$ Use the $\boxed{\ln x}$ key.
≈ 800 years

SECTION 10.6 (page 705)

1. $5^x = 125$
$\log 5^x = \log 125$ Take the log of each side.
$x \log 5 = \log 125$ Power property
$x = \dfrac{\log 125}{\log 5}$ Divide by $\log 5$.
$\approx \dfrac{2.09691}{.69897}$ Use the $\boxed{\log x}$ key.
$= 3$ Divide.
Solution set: $\{3\}$

5. $7^x = 5$
$\log 7^x = \log 5$ Take the log of each side.
$x \log 7 = \log 5$ Power property
$x = \dfrac{\log 5}{\log 7}$ Divide by $\log 7$.
$\approx \dfrac{.69897}{.84510}$ Use the $\boxed{\log x}$ key.
$\approx .827$ Divide.
Solution set: $\{.827\}$

9. $2^{y+3} = 5^y$
$\log 2^{y+3} = \log 5^y$ Take the log of each side.
$(y + 3) \log 2 = y \log 5$ Power property
$y \log 2 + 3 \log 2 = y \log 5$ Distributive property
$y \log 2 - y \log 5 = -3 \log 2$ Subtract $y \log 5$; subtract $3 \log 2$.
$y(\log 2 - \log 5) = -3 \log 2$ Factor.
$y = \dfrac{-3 \log 2}{\log 2 - \log 5}$ Divide.
$\approx \dfrac{-3(.30103)}{.30103 - .69897}$ Use the $\boxed{\log x}$ key.
≈ 2.269
Solution set: $\{2.269\}$

13. $e^{-.103x} = 7$
$\ln e^{-.103x} = \ln 7$ Take the ln of each side.
$-.103x = \ln 7$ $\ln e^k = k$
$x = \dfrac{\ln 7}{-.103}$ Divide.
$x \approx -18.892$ Use the $\boxed{\ln x}$ key; divide.
Solution set: $\{-18.892\}$

17. Solve Exercise 13 using common logarithms.

$$e^{-.103x} = 7$$

$$\log e^{-.103x} = \log 7 \qquad \text{Take the log of each side.}$$

$$-.103x \log e = \log 7 \qquad \log_b x^r = r(\log_b x)$$

$$-.103x = \frac{\log 7}{\log e}$$

$$-.103x \approx \frac{.84510}{.43429} \qquad \text{Use the log key.}$$

$$-.103x \approx 1.9459 \qquad \text{Divide.}$$

$$x \approx -18.892 \qquad \text{Divide.}$$

Solution set: $\{-18.892\}$
Natural logarithms are a better choice because
$\ln e^x = x$. This makes the problem much easier.

21. Write $\log_7(x + 1)^3 = 2$ in exponential form.

$$(x + 1)^3 = 7^2$$

$$(x + 1)^3 = 49$$

$$x + 1 = \sqrt[3]{49} \qquad \text{Take the cube root on each side.}$$

$$x = -1 + \sqrt[3]{49} \qquad \text{Subtract 1.}$$

Check the answer $x = -1 + \sqrt[3]{49}$.

$$\log_7(-1 + \sqrt[3]{49} + 1)^3 = \log_7(\sqrt[3]{49})^3$$

$$= \log_7 49 = 2 \quad \text{True}$$

Solution set: $\{-1 + \sqrt[3]{49}\}$

25. $\log(6x + 1) = \log 3$

$$6x + 1 = 3 \qquad \text{If } \log x = \log y, \text{ then } x = y.$$

$$6x = 2 \qquad \text{Subtract 1.}$$

$$x = \frac{2}{6} = \frac{1}{3}.$$

Check the answer $x = \frac{1}{3}$.

$$\log\left[6\left(\frac{1}{3}\right) + 1\right] = \log(2 + 1) = \log 3 \quad \text{True}$$

Solution set: $\{\frac{1}{3}\}$

29. $\log 4x - \log(x - 3) = \log 2$

$$\log \frac{4x}{x - 3} = \log 2 \qquad \text{Division property}$$

$$\frac{4x}{x - 3} = 2 \qquad \text{If } \log x = \log y, \text{ then } x = y.$$

$$4x = 2x - 6 \qquad \text{Multiply by } x - 3.$$

$$2x = -6 \qquad \text{Subtract } 2x.$$

$$x = -3 \qquad \text{Divide by 2.}$$

Check the answer $x = -3$.

$$\log[4(-3)] - \log(-3 - 3) = \log(-12) - \log(-6)$$

This is not possible, since x must be positive when
evaluating $\log x$.
Solution set: $\emptyset$

33. $\log 5x - \log(2x - 1) = \log 4$

$$\log \frac{5x}{2x - 1} = \log 4 \qquad \text{Division property}$$

$$\frac{5x}{2x - 1} = 4 \qquad \begin{array}{l}\text{If } \log x = \log y,\\ \text{then } x = y.\end{array}$$

$$5x = 8x - 4 \qquad \text{Multiply by } 2x - 1.$$

$$4 = 3x \qquad \text{Subtract } 5x; \text{ add 4.}$$

$$\frac{4}{3} = x \qquad \text{Divide by 3.}$$

Check the answer $x = \frac{4}{3}$.

$$\log 5\left(\frac{4}{3}\right) - \log\left[2\left(\frac{4}{3}\right) - 1\right]$$

$$= \log \frac{20}{3} - \log \frac{5}{3}$$

$$= \log \frac{\dfrac{20}{3}}{\dfrac{5}{3}}$$

$$= \log\left(\frac{20}{3} \cdot \frac{3}{5}\right) = \log 4, \text{ a true statement.}$$

Solution set: $\{\frac{4}{3}\}$

37. Use the formula $A = P(1 + \frac{r}{n})^{nt}$ with $P = 2000$,
$r = .04$, $n = 4$, and $t = 6$.

$$A = 2000\left(1 + \frac{.04}{4}\right)^{(4)(6)}$$

$$= 2000(1.01)^{24}$$

$$\approx 2000(1.269734649) \qquad \text{Use the } \boxed{y^x} \text{ key.}$$

$$\approx \$2539.47$$

41. Use the formula $A = Pe^{rt}$ with $P = 4000$, $r = .035$,
and $t = 6$.

$$A = 4000e^{(.035)(6)}$$

$$= 4000e^{.21}$$

$$\approx 4000(1.23367806) \qquad \text{Use the } \boxed{e^x} \text{ key.}$$

$$\approx \$4934.71$$

45. The power rule does not apply here because the
domain of f is $(-\infty, 0) \cup (0, \infty)$, while the domain of g
is $(0, \infty)$.

Index

A

Absolute value, 5, 451
  definition of, 5
Absolute value equations, 117
  solving, 118
Absolute value function, 601
Absolute value inequalities, 117, 186
  solving, 119
Addition
  of complex numbers, 497
  identity element for, 39
  of polynomials, 311
  of radical expressions, 473
  of rational expressions, 397
  of real numbers, 17
Addition method for systems, 231
Addition property
  of equality, 56
  of inequality, 99
Additive inverse of a number, 4
Algebraic expressions, 55, 309
Algebraic fraction, 385
Angles
  complementary, 98
  supplementary, 98
  vertical, 97
Annual interest rate, 76
Applied problems, 77
  steps in solving, 78
Approximately equal symbol, 452
Array of signs for a determinant, 267
Ascending powers, 315
Associative properties, 40
Asymptotes, 601
  of a hyperbola, 614
Average rate of change, 162
Axis
  of a parabola, 580, 584, 594
  x-, 148
  y-, 148

B

Bar graph, 147
Base of an exponent, 27, 291
Binomials, 310
  multiplication of, 321, 477
  square of, 323
Bode's law, 308
Boundary line, 183
Braces, 1
Brackets, 13

C

Cartesian coordinate system, 148
Celsius to Fahrenheit equation, 73, 182
Center of a circle, 604
Change-of-base rule, 694
Circle, 579, 604
  center of, 604
  equation of, 605
  graph of, 604
  radius of, 604
Coefficient, 40, 309
  numerical, 40

Collinear, 168
Combined variation, 211
Combining like terms, 40, 311
Common logarithms, 691
  evaluating, 691
Commutative properties, 40
Complementary angles, 98
Completing the square, 519, 589
Complex fractions, 407
  simplifying, 407
Complex numbers, 496
  addition of, 497
  conjugate of, 498
  division of, 498
  imaginary part of, 496
  multiplication of, 498
  real part of, 496
  subtraction of, 497
Components of an ordered pair, 148
Compound inequalities, 103, 109
Compound interest, 702
Conditional equation, 59
Conic sections, 579
  summary of, 616
Conjugates, 480
  of a complex number, 498
Consecutive integers, 98
Constant of variation, 207
Continuous compounding, 708
Contradiction, 59
Coordinate of a point
  on a line, 2
  in a plane, 148
Coordinate system
  Cartesian, 148
  rectangular, 148
Counting numbers, 1
Cramer's rule, 273
Cube roots of a number, 29, 450
Cubes
  difference of two, 356
  of numbers, 28
  sum of two, 357

D

Degree
  of a polynomial, 310
  of a term, 310
Denominator
  least common, 397
  rationalizing, 477
Dependent equations, 231
Dependent variable, 194
Descartes, René, 148
Descending powers, 309
Determinant method for systems, 273
Determinants, 265
  array of signs for, 267
  Cramer's rule for, 273
  evaluating, 265, 266
  expansion of, 266
  minor of, 266
Difference, 18
  of two cubes, 356
  of two squares, 355
Direct variation, 207
  as a power, 208
Discriminant, 528, 592

The purpose of this guide is to show those exercises from the text that are used in the Real to Reel videotape series that accompanies *Intermediate Algebra with Early Functions and Graphing,* Sixth Edition.

Section	Exercises	Section	Exercises
1.1	51	6.1	9, 13, 63, 75, 79
1.2	21	6.2	31, 41, 43, 47, 49
1.3	55	6.3	5, 11, 13, 29
1.4	61	6.4	3, 11, 29, 39
1.5	33	6.5	15
		6.6	9, 11, 19
2.1	17, 37		
2.2	21, 27	7.1	55
2.3	15, 33, 45	7.2	23
2.4	23	7.3	83
2.5	19, 45	7.4	39
2.6	7	7.5	19, 35
2.7	29, 55	7.6	19, 25
		7.7	13, 19, 43, 53, 77
3.1	35		
3.2	35	8.1	19
3.3	59	8.2	31
3.4	19, 33	8.3	13, 27, 43
3.5	1, 11, 19	8.4	21, 35
3.6	11, 15	8.5	31
4.1	13	9.1	29
4.2	13	9.2	11
4.3	19	9.3	9
4.4	19, 29	9.4	17
4.5	3, 13	9.5	27, 33
		9.6	27
5.1	65		
5.2	37	10.1	17, 21
5.3	59	10.2	25
5.4	65	10.3	27
5.5	7	10.4	9, 23
5.6	15, 25	10.5	33, 43
5.7	29, 35, 47	10.6	7, 19, 37
5.8	15, 31		
5.9	43		
5.10	13, 19, 29		